MW00845906

Antimicrobial Resistance

Vinay Kumar · Varsha Shriram · Atish Paul · Mansee Thakur

Editors

Antimicrobial Resistance

Underlying Mechanisms and
Therapeutic Approaches

Editors
Vinay Kumar
Department of Biotechnology
Savitribai Phule Pune University, Modern
College of Arts, Science and Commerce
Ganeshkhind, Maharashtra, India

Varsha Shriram
Department of Botany
Savitribai Phule Pune University, Ramkrishna
More Arts, Commerce and Science College
Pune, Maharashtra, India

Atish Paul
Department of Pharmacy
Birla Institute of Technology and Science
Pilani, Rajasthan, India

Mansee Thakur
School of Biomedical Sciences
Mahatma Gandhi Mission Institute of Heal
Navi Mumbai, Maharashtra, India

ISBN 978-981-16-3119-1   ISBN 978-981-16-3120-7   (eBook)
https://doi.org/10.1007/978-981-16-3120-7

This Springer imprint is published by the registered company Springer Nature Singapore Pte Ltd.
The registered company address is: 152 Beach Road, #21-01/04 Gateway East, Singapore 189721,
Singapore

Preface

Antimicrobial resistance (AMR), a condition where microbes become resistant or do not respond to the conventional antimicrobial drugs (or antibiotics), has emerged as a serious public health threat globally. Though it is a natural and evolutionary phenomenon, injudicious use of antibiotics in human health, hygiene, agriculture, and allied industries has caused a rapid emergence and persistence of AMR phenotypes. Initially considered a nosocomial problem, AMR has now spread widely and reached each corner of the community and environment. Thus there is an urgent need for the discovery and development of new and potent antibiotics. However, the drying pipeline of new and effective antibiotics has further aggravated the problem. This has left a daunting task for the medical and research community to control these drug-resistant *Superbugs*. Considering its severity, the AMR has been declared as a *Global Risk* by the World Economic Forum. Gram-negative bacteria in particular are becoming a serious cause of concern owing to their inherent membrane-mediated drug-resistance abilities. Several research groups worldwide are engaged in research activities aimed to find out newer updates and deeper understandings of drug resistance, its determinants, persistence, and spread and identifying novel and potent therapies to address this problem. This has resulted in the advent of emerging and cutting-edge approaches and technologies useful in combating AMR. In recent years, plant natural products are hailed as potent agents to combat AMR as they are shown to be effective against the major determinants of AMR.

This comprehensive book aims to present the current knowledge and updates on AMR and underlying mechanisms in microbial pathogens including the members of threatening ESKAPE group. The chapters included in this book feature a wide spectrum of topics ranging from bacterial pathogenicity, AMR traits and molecular mechanisms in bacterial pathogens, role of persisters, and the effective strategies mediated through plant and microbial natural products and bio-functionalized nanomaterials for combating AMR with bactericidal and/or drug-resistance reversal (or re-sensitization) approaches. Besides, precise sequence-specific antimicrobials based on CRISPR have also been presented for tackling AMR menace.

The chapters are written by highly acclaimed experts of international repute working on different aspects of bacterial pathogenesis and AMR and targeting it with alternative approaches. We believe this collection to be a useful platform for students, scholars, scientists, academicians, and policymakers interested in, or engaged in, research or policy-making on combating AMR. This is an attempt to promote the research themed on these and other related areas via providing an integrated and comprehensive mix of basic and advanced information.

We highly appreciate the first-rate and timely contributions by the eminent authors presenting the relevant information on different aspects of antimicrobial resistance including the mechanisms underlying its emergence, persistence, and spread, besides alternative and novel approaches to tackle this serious global problem. We gratefully acknowledge the reviewers for their valuable comments that helped in the improvement of the scientific content and quality of the chapters.

We thank the Springer team comprising Dr. Madhurima Kahali, Ms. Rhea Dadra, and the entire production team for their consistent hard work and efforts in the publication of this book.

Ganeshkhind, India   Vinay Kumar
Pune, India   Varsha Shriram
Pilani, India   Atish Paul
Navi Mumbai, India   Mansee Thakur

Contents

About the Editors

Vinay Kumar is working as an Associate Professor in Biotechnology Department, Modern College (Savitribai Phule Pune University), Ganeshkhind, Pune, India. He did his Ph.D. in Biotechnology from Savitribai Phule Pune University in 2009. He has published more than 50 peer-reviewed research/review articles, 20 book chapters besides editing 6 books for Springer, Wiley, and Elsevier. He is a recipient of Young Scientist Award from SERB, Government of India. His research interests include antimicrobial resistance (AMR) in community settings, elucidating mechanisms underlying AMR and combating it with phytochemicals and nanomaterials.

Varsha Shriram has completed her Ph.D. in natural products chemistry and biotechnology from CSIR-National Chemical Laboratory, Pune and Savitribai Phule Pune University, Pune, India. She is currently working as an Associate Professor, Botany Department, Prof. Ramkrishna More College, Akurdi (Savitribai Phule Pune University), Pune, India. She has published more than 30 peer-reviewed research/review articles and is on the reviewer board for reputed journals. Her areas of research interest include medicinal plants, phytochemicals, bioactivities including anticancer and antimicrobial resistance reversal. She has completed extramural research projects in these and allied areas.

Atish Paul pursued Ph.D. in Natural Products from the National Institute of Pharmaceutical Education and Research, India. For his post-doctoral research, he joined Prof. Ikhlas Khan at the National Center for Natural Product Research (University of Mississippi, USA). He is an Associate Professor and former head at the Department of Pharmacy, the Birla Institute of Technology and Science, Pilani, India. His area of expertise is Natural Product Chemistry. He has published several research articles in reputed journals like Journal of Medicinal Chemistry, Bioorganic Medicinal Chemistry, Medicinal Chemistry Research, Journal of Chromatography-A, and contributed 24 Monographs. He is a reviewer for various journals of repute.

Mansee Thakur is a Gold Medalist during her MSc at Pt. R. S. S. University, Raipur and obtained her Ph.D. in Biotechnology from the University of Raipur, India in 2007. She is working as a Professor and Director, Mahatma Gandhi Missions Institute of Health Sciences, MGM School of Biomedical Sciences, Navi Mumbai. She has published more than 30 peer-reviewed research articles and has 4 patents to her credit. Her areas of research include natural products and their antimicrobial potencies, molecular biology and toxicity studies. She has completed many extramural research projects and guided students for their Masters and Doctoral studies.

Antimicrobial Resistance Traits and Resistance Mechanisms in Bacterial Pathogens

1

Deepjyoti Paul, Jyoti Verma, Anindita Banerjee, Dipasri Konar, and Bhabatosh Das

Contents

D. Paul
Department of Microbiology, Assam University, Silchar, Assam, India

J. Verma · B. Das (✉)
Molecular Genetics Laboratory, Infection and Immunology Division, Translational Health Science and Technology Institute, Faridabad, India
e-mail: bhabatosh@thsti.res.in

A. Banerjee
St. Xavier's College, 30, Mother Teresa Sarani, Kolkata, India

D. Konar
Department of Microbiology, Jan Swasthya Sahyog, Bilaspur, Chattisgarh, India

© The Author(s), under exclusive license to Springer Nature Singapore Pte Ltd. 2022
V. Kumar et al. (eds.), *Antimicrobial Resistance*,
https://doi.org/10.1007/978-981-16-3120-7_1

1

Abstract

Antimicrobial resistance is one of the most urgent health crises in the twenty-first century. Multiple factors including population explosion, industrial expansion, lifestyle changes, ease of global migration, and over-the-counter access to antimicrobials in the pharmacy contribute to the rapid emergence and spread of resistant pathogens. The resistant microbes are ubiquitously distributed both in the hospitals and in other environments. However, the prevalence of antibiotic resistance traits is very high in the six bacterial species, which are named as "ESKAPE" pathogens, an acronym for *Enterococcus faecium, Staphylococcus aureus, Klebsiella pneumoniae, Acinetobacter baumannii, Pseudomonas aeruginosa*, and *Enterobacter* species. Resistant bacteria neutralize the bactericidal or bacteriostatic activity of antibiotics either by altering amino acid sequences of the target molecule or by modifying the chemical structure of antibiotics. In addition, proteins located in the bacterial cell surface contribute significantly to antibiotic resistance by reducing the effective intracellular concentration of antibiotics by reducing membrane permeability or increasing expelling activity. In the ESKAPE pathogens, antibiotic resistance genes are most often genetically linked with mobile genetic elements and disseminate rapidly by horizontal and vertical gene transfer. In the present report, we have provided updated and comprehensive pictures of antibiotic resistance in the ESKAPE pathogens.

Keywords

ESKAPE · AMR · ARG · MGE · ICE

1.1 Introduction

Antimicrobial-resistant bacterial pathogens responsible for several infectious diseases are a predominant threat to human health and global economy. Several factors contributing to this threat include population explosion, industrial expansion, lifestyle changes, ease of global migration, lack of adequate medical facilities in the developing countries, and – the most alarming of all – emergence of antimicrobial resistance. Antimicrobial resistance (AMR) among community- and hospital-associated bacterial species is exponentially growing, and it has been estimated that, by 2050, it would be responsible for the death of one person every 3 s (Review on Antimicrobial Resistance 2016). The imprudent usage of antimicrobials (including antibiotics, antivirals, antifungals, etc.) has aggravated this lurking threat of AMR. At a global level, from 2000 to 2010, the usage of antimicrobials was observed to have increased by >30%, with the low- and middle-income countries (LMICs) being the major contributors, on account of high prevalence of infectious diseases, easy availability, and reduced cost of antimicrobials (Klein et al. 2021). Since the last 70 years, the compelling enrichment of the AMR genes in bacterial population has been ascribed to the activities of humans such as rampant antibiotic usage in hospitals, agriculture, poultry, and animal husbandry.

Antibiotics are known as the chemical compounds produced by the microorganisms, which either kill or inhibit the growth of other microbes in their vicinity and help the bacteria to outcompete other strains present in their niche and to gain fitness advantage (Fig. 1.1, Lew 2014). The development of antibiotic resistance against the pathogenic bacteria develops as soon as the concerned pathogen develops various resistance mechanisms rendering the action of the antibiotic to be ineffective (Fig. 1.1, Das et al. 2017). The microorganisms harboring the antimicrobial resistance genes (ARGs) are mostly referred to as the antibiotic-resistant strains or the superbugs. Hospitals are considered as the nodal points for the AMR dissemination as the antibiotic usage in hospitals exerts selective pressure and accelerates the progression to acquiring antibiotic resistance (Dijkshoorn et al. 2007). The AMR strains are ubiquitous in the hospitals as well as in other environments, such as water bodies, soil, human body, livestock, and industrial effluents. However, the problem of resistance is not uniformly distributed across all the bacterial pathogens. The Infectious Diseases Society of America (IDSA) has identified six different bacterial species to be threats due to their virulence and evolution of resistance mechanisms against commonly prescribed antibiotics. They are named as "ESKAPE" pathogens, an acronym for *Enterococcus faecium*, *Staphylococcus aureus*, *Klebsiella pneumoniae*, *Acinetobacter baumannii*, *Pseudomonas aeruginosa*, and *Enterobacter* species. This group of bacterial pathogens includes both Gram-negative and Gram-positive species that are proficient in "escaping" or evading the antimicrobial actions of conventional antibiotics. These pathogens are accountable for life-threatening infections, especially among the children, elderly, immunocompromised, as well as those who are critically ill. ESKAPE pathogens are associated with high mortality and morbidity resulting in higher healthcare costs (Founou et al. 2017). These pathogens have been recently listed in the list of 12 bacteria by WHO to accelerate the development of new antibiotics (Tacconelli et al. 2018). The pathogens have been divided into three categories, namely, critical, high, and medium priority, in accordance with the urgency of requirement for new antibiotics. Carbapenem-resistant *K. pneumoniae*, *A. baumannii*, and *P. aeruginosa* along with extended-spectrum β-lactamase (ESBL) *Enterobacter* spp. are listed in the critical priority. Vancomycin-resistant *E. faecium (VRE)* and methicillin- and vancomycin-resistant *S. aureus* (MRSA and VRSA) are listed high-priority group. Drawing

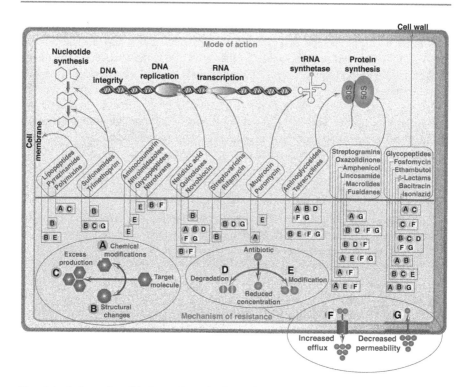

Fig. 1.1 Action of antibiotics and mechanism of antibiotic resistance. Target of different antibiotics is fairly distinct. Antibiotics can impede DNA replication, transcription, translation, cell wall biosynthesis, or integrity of nucleic acids or cell membrane by inhibiting activity of the key molecules involved in the metabolic or maintenance process. However, bacteria can neutralize the bactericidal or bacteriostatic activity of antibiotics by altering composition of the target molecules or by introducing chemical modification in the antibiotic molecules. Some enzymes involved in the antibiotic resistance process can hydrolyze the antibiotics. Several proteins located in the cell membrane confer antibiotic resistance by reducing the effective intracellular concentration of antibiotics by reducing permeability or increasing expelling activity. Few antibiotics are unable to kill bacteria because of the overproduction of their target molecules. Several mobile genetic elements disseminate antibiotic resistance genes rapidly through horizontal gene transfer

attention toward the ESKAPE pathogens would help to combat the mounting challenge of AMR.

The AMR crisis has been responsible for more than 35,000 deaths and around two million infections annually in the USA (https://www.cdc.gov/drugresistance/). The infection due to AMR bacteria has claimed the lives of 58,000 babies in India in a single year (Gandra et al. 2019). Indeed, the global epidemic of antibiotic resistance will unfold before all of us in no time. Hence, a comprehensive approach to unravel the origin of the resistance determinants in these ESKAPE pathogens, along with their arsenals of AMR including those associated with dissemination and maintenance of the resistance traits, will help to tackle this menace.

1.2 Antibiotics for the Treatment of Infections Caused by ESKAPE Pathogens

The commonly used treatment regimens for infections by ESKAPE pathogens include the antibiotics delivered singly or in combination (Table 1.1). Currently used treatment options are dwindling with the emergence of MDR strains, predisposing us toward the looming threat. Various alternative therapies such as bacteriophage therapy, nanoparticle-based antimicrobials and drug delivery systems, antimicrobial peptides, adjuvants, phytochemicals, and photodynamic therapy are looked as an option to combat AMR (Mandal et al. 2014).

Antibiotics used in combinations have often been found to be more efficient in combating infections by MDR organisms. When treating *E. faecium* and *S. aureus*,

Table 1.1 Antibiotic classes and commonly used antibiotics in the clinics

Class of antibiotic	Representative commonly used in clinics
Lipopeptide	Daptomycin
Pyrazinamide	Commonly used as first-line anti-tubercular
Polymyxins	Polymyxin B, colistin
Sulfonamide	Sulfadiazine, sulfasalazine, sulfamethoxazole
Trimethoprim	Used commonly in combination with sulfamethoxazole as cotrimoxazole
Aminocoumarin	Novobiocin, clorobiocin, coumermycin
Nitroimidazole	Metronidazole
Glycopeptide	Vancomycin, teicoplanin
Nitrofurans	Nitrofurantoin
Nalidixic acid	Itself is the representative; most used clinically
Quinolones	Ciprofloxacin, levofloxacin, ofloxacin
Novobiocin	Itself is the representative; most used clinically
Streptovaricins	Not commonly used as an antibacterial
Rifamycin	Rifampicin, rifapentine, rifabutin
Mupirocin	Itself is the representative; most used clinically
Puromycin	More often used in vitro
Aminoglycoside	Streptomycin, gentamicin, amikacin, tobramycin
Tetracycline	Minocycline, doxycycline, tigecycline
Streptogramin	Dalfopristin, quinupristin
Oxazolidinone	Linezolid
Amphenicol	Chloramphenicol
Lincosamide	Clindamycin
Macrolides	Azithromycin, erythromycin
Fusidanes	Fusidic acid
Fosfomycin	Itself is the representative; most used clinically
Ethambutol	Commonly used as first-line anti-tubercular
β-Lactams	Penicillin, cephalosporin, monobactams, carbapenems
Bacitracin	Itself is the representative; most used clinically
Isoniazid	Commonly used as first-line anti-tubercular

the Gram-positive members of the ESKAPE pathogens, with a combination of fosfomycin and daptomycin it has been observed to successfully clear the infection (Snyder et al. 2016; Coronado-Álvarez et al. 2019). On the other hand, antibiotic combinations checked against *S. aureus* include the antibiotic daptomycin or vancomycin and a newly added antibiotic ceftaroline. Colistin (polymyxin E) has been put into use as the last resort antibiotic against Gram-negative bacilli. A combination of colistin or tigecycline with other antibiotics has been tested to treat infections caused by *K. pneumoniae* and *A. baumannii* and has shown promising results.

1.3   Origin and Evolution of Antimicrobial Resistance Traits in ESKAPE Pathogens.

The origin of the antibiotics and the antibiotic resistance genes (ARGs) dates back to millions and billions of years (Martinez 2009, Baltz 2008). A study by D'Costa et al. (2011) showed the presence of resistance genes against β-lactams, glycopeptide, and tetracycline antibiotics in the metagenomic samples of 30,000-year-old permafrost, woolly mammoths, and animal and plant species of the Pleistocene. The antibiotic production and the selection of resistance genes are a result of Darwinian selection, which provides antibiotic-resistant bacteria with a selective advantage of niche exploitation (Waksman and Woodruff 1940). The producers of antibiotics are equipped with self-resistance gene determinants, and their expression is co-regulated with antibiotic biosynthesis gene. These resistance mechanisms can get disseminated to non-producer species in a niche where different strains co-exist together.

The golden era of antibiotics got shadowed due to the emergence of resistance to the frontline antimicrobial agents, such as sulfonamides (Davies and Davies 2010). The resistance against the "wonder drug" – penicillin – was observed even before its large-scale usage in the clinical settings, which was followed by the discovery of penicillinase in *S. aureus* and *S. pneumoniae* in 1940. This clearly depicted the presence of resistance genes conferring resistance against penicillin in the natural environment (Davies and Davies 2010; Ogawara 2016). Another analytical study of microbial DNA analysis from the dental plaque of ancient human remains showed the existence of ARG sequences sharing homology to the genes encoding resistance against the aminoglycosides, β-lactams, bacitracin, macrolides, and tetracycline in the genome of clinical strains (Warinner et al. 2014; Olaitan and Rolain 2016). The hypothesis of the acquisition of resistance genes from antibiotic producer organisms by the pathogens via horizontal gene transfer (HGT) was formerly proposed in the 1970s (Benveniste and Davies 1973). This was in accordance to the observations that the activity of aminoglycoside-modifying enzymes in actinomycetes shared similar biochemical activity to the different enzymes present in pathogenic strains. Another example includes the *vanHAX* genes – ARGs present in the antibiotic-producing bacteria found in soil were found in clinical isolates as well and shared considerable similarity between the protein sequences along with conserved organization and arrangement of genes (Barna and Williams 1984; Marshall et al. 1998).

The integration of the *van* operons on conjugative plasmids or transposons has led to an enhancement of their dissemination (Courvalin 2008). It has been speculated that while these enzymes have no direct linkage with resistance functions in producers, they may have a role in other metabolic functions (Benveniste and Davies 1973; Martínez 2018). This is supported by comparative genome sequence analysis studies depicting the diversity of ARGs, and these studies have shown that they are encoded by a diverse group of unrelated genes, which suggest that their origin is through multiple convergent paths that results in a similar function (Shaw et al. 1993).

Recent reports have provided an evidence of the transfer of intrinsic resistance gene from the environmental strains to the pathogenic strains. For example, the plasmid-encoded CTX-M, an extended-spectrum β-lactamase (ESBL) of class A gene of pathogenic bacteria, showed similarity with CTX-M gene found in the genome of a nonpathogenic environmental strain of *Kluyvera* species (Humeniuk et al. 2002; Cantón and Coque 2006). Similarly, the quinolone resistance determinant *qnr*, originally located in the genome of a nonpathogenic environmental *Shewanella* and *Vibrio* species, was lately found to be translocated to a conjugative plasmid in *Klebsiella* (Poirel et al. 2005). Another gene called *qepA* encoding a major facilitator superfamily (MFS) efflux pump was initially linked to a conjugative plasmid in clinical isolates of *E. coli* (Périchon et al. 2007). However, it was recently found in the metagenomic samples of river sediments polluted by untreated urban wastewaters (Cummings et al. 2011).

The introduction and usage of the antibiotics into the clinical settings was observed to have two opposite effects: first, the rapid and desired one is that of microbial growth inhibition by targeting the essential cellular processes and, second, the undesirable one is the emergence of AMR in pathogens by elimination of the sensitive variants (Das et al. 2020). The development of resistance is a natural evolutionary process with various factors such as the type of environment; the microbial community density in that specific habitat; the antibiotic regimen in health, food, agriculture, and animal husbandry sectors; etc. influencing its frequency of emergence (Holmes et al. 2016).

1.4   Genomic Insights into Antimicrobial-Resistant ESKAPE Pathogens

These ESKAPE pathogens encompassing both Gram-positive and Gram-negative bacteria are usually characterized as the carrier of drug resistance genes and are common causes for many life-threatening nosocomial infections. Both chromosome and plasmid carry several horizontally acquired resistance genes and act as potential reservoirs for other bacterial species. Prevalence of different classes of ARGs in the genome of ESKAPE pathogens varies widely (Fig. 1.2). Maximum prevalence of most of the ARGs is reported in the genome of *P. aeruginosa* followed by *A. baumannii*.

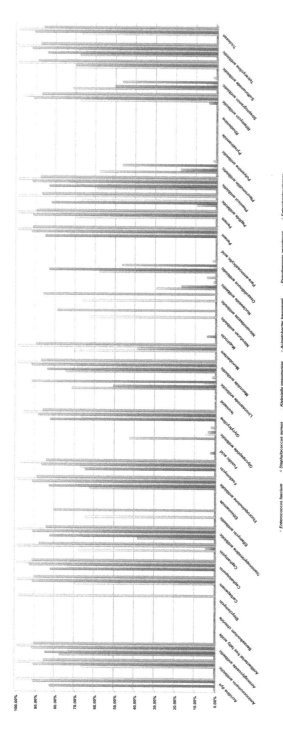

Fig. 1.2 Prevalence of antibiotic resistance genes in the genome of ESKAPE pathogens. A total of 28,222 whole genome sequences (W) and 7566 plasmid (P) sequences isolated from *Enterococcus faecium* [N = 1738 (W), 668 (P)], *Staphylococcus aureus* [N = 8724 (W), 1943 (P)], *Klebsiella pneumoniae* [N = 8385 (W), 4392(P)], *Acinetobacter baumannii* [N = 4336 (W), 424 (P)], *Pseudomonas aeruginosa* [N = 4527 (W), 52 (P)], and *Enterobacter cloacae* [N = 512(W), =87(P)] were analyzed to understand the prevalence of antibiotic resistance genes in the genome of ESKAPE pathogens. Data source: The Comprehensive Antibiotic Resistance Database (https://card.mcmaster.ca/genomes)

1.4.1 Gram-Positive ESKAPE (GP-ESKAPE) Pathogens

ESKAPE pathogens that belong to Gram-positive bacteria include *Enterococci* spp. and *Staphylococcus aureus*.

1.4.1.1 Vancomycin-Resistant *Enterococcus*
The increasing rate of antimicrobial resistance among *Enterococcus* is a serious concern, especially the incidence of vancomycin-resistant *Enterococcus* (VRE), which is mainly associated with *E. faecium* and *E. faecalis*. *E. faecium* has been classified into two major clades, viz., Clade A and Clade B, based on the CRISPR analysis, multi-locus sequence typing (MLST), and whole genome database to understand the clonal pattern. Clade A (Clades A1 and A2) set of organisms are mainly found in hospitals, whereas Clade B group of organisms are of nonclinical background and mainly associated with community origin. Clade A1 includes epidemic hospital strains and is also known to harbor virulence and antibiotic resistance genes, and Clade A2 include the strains responsible for causing animal and sporadic human infections (Santajit and Indrawattana 2016). Multi-locus sequence typing of *E. faecium* identified a global polyclonal cluster, which includes the sequence type ST17 followed by the descendant types ST16, ST78, ST63, ST64, and ST174. *E. faecium* 17 (CC17) is often characterized as the carrier of different multidrug-resistant gene markers and known to cause many serious infections. Apart from the CC17 clonal complex, many other significant sequence types such as ST203, ST910, ST78, and ST192 are also responsible for the evolution, emergence, and rising proportion of VRE infections (De Oliveira et al. 2020). But in the case of *E. faecalis*, MLST analysis revealed the most prevalent clones responsible for clinical infection and outbreak that include ST6, ST9, ST16, ST21, ST28, ST40, and ST87, and along with the hospital-derived infections, these STs are also frequently reported in the community, farm animals, and food products.

The glycopeptide resistance in *Enterococcus* is mainly mediated by the vancomycin resistance operon (*Van* operon) and is attributed by the inducible production of peptidoglycan precursors D-alanyl-D-lactate (D-Ala-D-Lac), and the substitution of D-Ala-D-Lac for the D-Ala-D-Ala dipeptide confers a higher level of vancomycin resistance to *Enterococcus* spp. The *Van* operon mainly consists of *vanS-vanR*, a response regulator; *vanX*, a D-Ala-D-Ala dipeptidase gene; *vanH*, a D-lactate dehydrogenase gene; and a variable ligase gene that is located either in chromosome or within plasmid. *VanA* and *VanB* are the most predominant type among these nine types of VREs (*vanA, vanB, vanC, vanD, vanE, vanG, vanL, vanM,* and *vanN*) and also showed enhanced resistance pattern toward all glycopeptide antibiotics (Ahmed and Baptiste 2018). The *vanA* gene cluster consists of seven open reading frames transcribed from two different promoters, and this *vanA* operon is typically linked with transposons (Tn) which was originally detected on the non-conjugative highly conserved transposon *Tn1546*-like element. The transposons carrying *vanA* gene cluster are generally carried on self-transferable conjugative plasmids, and it appears to be responsible for the spread of glycopeptide resistance in *Enterococci*. The *vanB* operon has a similar genetic backbone to *vanA* but is generally carried by large elements of 90–250 kb, and the transfer of *vanB* resistance alleles occurs through the acquisition or transfer of transposons such as Tn*1547*, Tn*1549*, and Tn*5382*. The conjugative *vanB* transposon, known as Tn*1549*, is widely prevalent among *VanB*-type *Enterococcus* and is mainly a chromosomal transposon. The major dissemination of *VanB*-type resistance is primarily due to the spread of *vanB2* cluster carried on Tn*916*-like conjugative transposons. The *vanC* operon is genetically different from *vanA* and *vanB* and is also less virulent than those of *Enterococci* spp. carrying inducible *vanA* and *vanB* operon. Similar to *vanC*, *vanD* and *vanE* gene clusters are also detected in various *Enterococcus* species and are located in the chromosome of the host organism, whereas *vanG* operon has been found to be transferable via a conjugative plasmid from *E. faecalis*. The family of *van* genes was expanded after the discovery of *vanL*, *vanM*, and *vanN* operons in *Enterococcus* spp. which were classified mainly based on their genetic structure, transferability, inducibility, and resistance profile against glycopeptides.

1.4.1.2 *Staphylococcus aureus*

Methicillin-resistant *Staphylococcus aureus* (MRSA) is a major clinical concern and responsible for the cause of many nosocomial and community-acquired infections. The acquisition of methicillin resistance *mecA* gene transforms a methicillin-susceptible *S. aureus* (MSSA) to a MRSA strain. This *mecA* gene encodes the additional penicillin-binding protein 2a (PBP2a), a peptidoglycan transpeptidase, and is located on a mobile genetic element *Staphylococcal* cassette chromosome *mec* (*SCCmec*) which is a site-specific transposon-like element exclusively found among *Staphylococcus* species. The SCCmec elements are integrated in the chromosomes of MRSA strains and are composed of *Mec*I (a repressor), *Mec*R1 (a transducer), and cassette chromosome recombinase (CCR) gene complex that mediates the integration and excision of the element from the chromosome.

The conserved structure of SCC*mec* and *mecA* has facilitated the molecular detection of methicillin resistance, and as a result, five major MRSA clones, *SCCmec* I–V, have been identified. Hospital-associated MRSA (HA-MRSA) isolates are characterized by the larger *SCCmec* types I–III and are mostly limited to healthcare settings, but community-associated MRSA (CA-MRSA) isolates are associated with novel, small variant of *SCCmec* types IV and V (Naorem et al. 2020). *mecC*, another *mec* allele which shows approximately 70% nucleotide sequence homology with the classical *mecA* gene, and are is mainly found in the livestock-associated MRSA (LA-MRSA). Borderline oxacillin-resistant *S. aureus* (BORSA) are resistant to oxacillin but do not carry *mecA* and *mecC* genes, and this may be due to the occurrence of mutations in both the promoter and coding sequence of *mecA*.

The nonsense or nonsynonymous mutation in *mecA* gene may be responsible for ceftaroline resistance in MRSA which is mainly observed in *S. aureus* ST239 strain, whereas the daptomycin resistance is mainly linked to the mutation in *mprF* gene which encodes for an enzyme called lysyl-phosphatidylglycerol synthetase and is one of the common mutations observed in MRSA strains. MRSA isolates have also developed resistance toward mupirocin due to the mutations in the chromosomal gene *ileS* or the plasmid-located genes *mupA* and *mupB*. Fluoroquinolone resistance in MRSA is mediated either by efflux pump genes, viz., *norA, norB, norC, mdeA, qacA*, and *qacB*, or by point mutations in the ParC subunit of topoisomerase IV (De Oliveria et al. 2020). Tetracycline resistance is primarily observed due to the acquisition of *tet* and *otr* genes with the most common tetracycline resistance mechanism being mediated by *tetA, tetM*, and *tetK* genes. MRSA strains acquire tigecycline resistance due to mutations in the transcriptional regulator *mepR* and in the efflux pump *mepA* resulting in the increased efflux of the drug. Vancomycin resistance in MRSA is successfully achieved by horizontal transfer of a plasmid-borne *vanA* gene transposon from vancomycin-resistant *Enterococcus* which results into the emergence of vancomycin-resistant *S. aureus* (VRSA). On the other hand, the chloramphenicol resistance in *S. aureus* can occur through the acquisition of *cfr* gene, while the *poxtA* gene is found to be responsible for oxazolidinone and phenicol resistance in MRSA (Antonelli et al. 2018).

Mobile genetic elements (MGEs) such as SCC*mec*, plasmids, and transposons are the critical factors for the acquisition and dissemination of the resistance determinants in *S. aureus*. In MRSA, the *mecA* or *mecC* genes, carried within SCC element, can be horizontally transferable. The association of *mecA* with SCC is not only important for *mecA* acquisition or transfer but is also a key factor for the co-existence of multiple resistance determinants in the same locus. This is enabled by the plasticity of the SCC*mec* element that can host several resistant genes that are associated with high risk of horizontal transmission.

In *S. aureus*, transposons (Tn) predominantly encode antibiotic resistance genes, and they are either inserted into the chromosome or in the mobile genetic element such as Tn*552*-related transposon (harbor penicillin-resistant gene *blaZ*), Tn*554* (encode resistance to spectinomycin and macrolide-lincosamide-streptogramin B antibiotics), Tn5801 (a conjugative transposon carrying *tetM* gene), Tn*4001* (associated with several multi-resistance plasmids and SCC*mec* elements), and the plasmid-borne transposon Tn*1546* (encodes the *vanA* operon).

The major *S. aureus* clones responsible for the emergence of multidrug resistance are identified as ST239, ST59, ST398, ST8, and ST9. ST239 is found to responsible for the global dissemination of HA-MRSA strains (Giulieri et al. 2020). The novel *S. aureus* clone ST772, also known as Bengal Bay clone ST772, is reported as a multidrug-resistant *S. aureus* lineage in community and healthcare-associated environments. While a variety of clonal complexes such as CC1, CC8, CC30, and CC45 have been identified among *S. aureus* strains, majority of the MRSA isolates are detected among CC398, CC9, and CC8 clonal complex (Li et al. 2019).

1.4.2 Gram-Negative ESKAPE (GN-ESKAPE) Pathogens

The ESKAPE pathogens that come under the umbrella of Gram-negative organisms include *Klebsiella pneumoniae, Acinetobacter baumannii, Pseudomonas aeruginosa*, and *Enterobacter* species.

1.4.2.1 *Klebsiella pneumoniae*

Among *Enterobacteriaceae*, *K. pneumoniae* is known to be an invasive and virulent pathogen and is broadly classified into two subtypes: classical *K. pneumoniae* (c*Kp*) and non-classical *K. pneumoniae* (nc*Kp*). The clones of nc*Kp* can cause severe infections, which are difficult to treat due to their continuous mutation and acquisition of antibiotic-resistant genes. Carbapenem-resistant *K. pneumoniae* (CRKP) strains are clinically prominent and have been found to be associated with many severe infections. The spread of CRKP that has been driven by many clones is largely attributed to ST307, ST11, ST15, ST101, and ST258 strains, along with the ST258 derivative ST512 (De Oliveira et al. 2020).

1.4.2.2 *Acinetobacter baumannii*

Multidrug-resistant *A. baumannii* typically cause several major infections in hospitalized patients, especially in the intensive care setting. The spread of carbapenem-resistant *A. baumannii* (CRAB) isolates is largely associated with three international clonal lineages: CC1, CC2, and CC3. CC1 is prevalent worldwide, while CC2 and CC3 are highly prevalent in Europe and North America (Dagher et al. 2019).

1.4.2.3 *Pseudomonas aeruginosa*

P. aeruginosa is an opportunistic pathogen causing complicated and life-threatening infections. The flexibility and adaptability of *P. aeruginosa* genome is a key feature in pathogens' high stability, ability to persist and evade antibiotic action. *P. aeruginosa* lineages ST235 and ST175 have emerged as high-risk dispersed clones and remain a major contributor of hospital-acquired infection due to their enhanced capacity to acquire and maintain foreign antibiotic resistance determinants (Treepong et al. 2018).

1.4.2.4 *Enterobacter* Species

Enterobacter is an extremely diverse group of bacteria, and multidrug-resistant *Enterobacter* species are responsible for an increasing cause of hospital-acquired infection. Currently, in the USA, *E. aerogenes* ST4 and ST93 and carbapenem-resistant *E. cloacae* ST178 and ST78 represent the prevalent lineages associated with nosocomial infection (De Oliveira et al. 2020).

Among these GN-ESKAPE pathogens, the most common drug resistance mechanism employed is the production of enzymes that can degrade or destroy the antibiotics. Beta-lactamase is one such enzyme, which has the ability to hydrolyze all the β-lactam antibiotics including the last-line drug carbapenem. They comprise of extended-spectrum β-lactamases (ESBLs), AmpC, and carbapenemases. Among the ESBLs, *TEM*, *SHV*, and *CTX-M* are most commonly found in GN-ESKAPE pathogens. The *TEM* beta-lactamase hydrolyzes cephalosporins and penicillins and is widely spread in *K. pneumoniae*, *Enterobacter* spp., and non-fermentative *P. aeruginosa*. *CTX-M*s have been identified among all the ESKAPE pathogens, while among SHV enzymes, *SHV-1* is the most clinically relevant and commonly found in *K. pneumoniae*. Other class A ESBLs, viz., *PER*, *VEB*, *GES*, and OXA families, have also been reported across all Gram-negative ESKAPE pathogens. *K. pneumoniae* carbapenemase (bla_{KPC}), a serine carbapenemase enzyme able to degrade all the beta-lactams including carbapenems, is predominantly detected in clinical isolates of *K. pneumoniae*. Furthermore, OXA-type carbapenemases are commonly found in *Acinetobacter* spp., followed by *K. pneumoniae* and *Enterobacter* spp. The most prominent metallo-β-lactamases (MBLs) encountered among GN-ESKAPE pathogens are imipenemase metallo-β-lactamases (IMP), Verona integron-encoded metallo-β-lactamases (VIM), and New Delhi metallo-beta-lactamase-1 (NDM-1). IMP- and VIM-type MBLs were first detected in clinical *P. aeruginosa* isolates and later have been identified in other ESKAPE pathogens. Similarly, NDM-type enzymes have also been detected across all the Gram-negative ESKAPE bacteria. AmpC beta-lactamases including penicillinase and cephalosporinase are usually identified in many *Enterobacter* spp., *P. aeruginosa*, and *Acinetobacter* spp.

Resistance to aminoglycoside antibiotic in GN-ESKAPE organism occurs mainly through the production of aminoglycoside-modifying enzymes (AMEs) which are further classified into three groups, i.e., aminoglycoside acetyltransferases (AACs), aminoglycoside phosphotransferases (APHs), and aminoglycoside nucleotidyl-transferases (ANTs). The 16S rRNA methylases (*NpmA* and *Rmt* gene family) encoded on the plasmids also confer resistance to all aminoglycosides in GN-ESKAPE pathogens (Ishizaki et al. 2018).

Efflux pump also plays an important role in the mechanism of antimicrobial resistance in GN-ESKAPE pathogens, and till date, six major families of efflux pumps have been characterized, which include resistance-nodulation-division (RND), multidrug and toxic compound extrusion (MATE), small multidrug resistance (SMR), proteobacterial antimicrobial compound efflux (PACE), major facilitator superfamily (MFS), and ATP-binding cassette (ABC) families. *AcrAB-TolC* and *MexAB-OprM*, belonging to the RND-type efflux pump-mediated resistance, are of major concern as they play an important role in the multidrug resistance phenotype among Gram-negative bacteria. *MexAB-OprM* efflux system is mainly observed in *P. aeruginosa* that exhibits resistance toward fluoroquinolone, aminoglycoside, and beta-lactams. Similarly, the overproduction of *AcrAB-TolC* is characteristic of multidrug-resistant *K. pneumoniae* and *Enterobacter* strains. The overexpression of *AdeABC*, *AdeFGH*, and *AdeIJK* RND-type efflux pumps leads to the development of MDR in *A. baumannii* (Santajit and Indrawattana 2016). The chromosomally encoded *OqxAB* efflux pump observed in *K. pneumoniae* and *Enterobacter* spp. contributes to reduced susceptibility to quinolone and chloramphenicol. The alterations in *AcrAB-TolC* and *KpnEF* efflux pump systems as well as the loss of putative porin, *KpnO*, may drive the aminoglycoside resistance in *K. pneumoniae* (Navon-Venezia et al. 2017).

The dissemination of resistance determinants among the GN-ESKAPE pathogens is mainly mediated by plasmids, insertion sequences, transposon, and other genetic elements. Insertion sequences (IS) are capable of self-transposition and mobilizing the neighboring genes. Significant among them are IS*Aba*125, which is mainly associated with NDM gene, and IS*Ecp*1, which appears to be responsible for mobilizing many antibiotic-resistant genes like bla_{CTX-M}, bla_{CMY-2}, bla_{OXA}, and bla_{ACC}. Other IS elements such as IS$Aba1$, IS1247, IS$Kpn23$, IS26, and IS$Enca1$ are also found to be associated with mobilization of the resistance determinants. The composite transposons that are found to be carrying AMR genes include Tn9 (IS1, chloramphenicol resistance), Tn10 (IS10, tetracycline resistance), Tn5 (IS50, aminoglycoside and bleomycin resistance), and, more recently, Tn6330 (IS$Apl1$), which is responsible for mobilizing the colistin resistance gene $mcr-1$. Among ESKAPE pathogens, AMR genes are frequently associated with the Tn3 family (Tn1, Tn2, and Tn3), Tn7-like unit transposons, and Tn552-like elements. The IS elements also play a significant role in the deactivation of the uptake system, i.e., $ompK36$ porin in $K.$ $pneumoniae$, which results in elevated carbapenem resistance. Similarly, the insertional inactivation of the $mgrB$ regulatory gene in $K.$ $pneumoniae$ leads to the overexpression of the $pmrHFIJKLM$ operon, which results in colistin resistance (Kumar et al. 2018). Integron, a gene-capture system, is known to be an important tool for horizontal dissemination of diverse resistance determinant, and recently, carbapenemase genes bla_{KPC}, bla_{VIM}, and bla_{NDM} are found to be carried within class 1 integron. Plasmids are known to be the most significant vehicle for the transfer of resistance determinants such as ESBLs (bla_{CTX-M}, bla_{TEM}, bla_{VEB}, bla_{PER}), AmpC (bla_{DHA}, bla_{CMY}), carbapenemases (bla_{KPC}, bla_{IMP}, bla_{NDM}, bla_{VIM}, bla_{OXA-23}), and colistin resistance gene (mcr) in GN-ESKAPE pathogens. Plasmids can harbor several IS and MGEs associated with AMR genes, and multidrug-resistant GN-ESKAPE pathogens carry the resistant determinants within a wide variety of plasmid incompatibility groups, viz., IncF, IncP, IncL, IncN, IncH, and IncX3, that facilitate the horizontal transfer within intra- and interspecies level (De Oliveira et al. 2020).

1.5   Ecology of Antimicrobial Resistance Genes in ESKAPE Pathogens

Comprehensive genomic studies of pathogenic and nonpathogenic bacterial species and the Comprehensive Antibiotic Resistance Database (CARD) catalogue 175,753 alleles with potential resistance functions against 249 clinically important antibiotics (https://card.mcmaster.ca). Most of the AMR genes reported in ESKAPE pathogens are acquired through HGT and are physically linked with replicative and/or integrative MGEs. HGT leads to evolution of bacteria by introducing multiple fitness factors even in case of single event of acquisition. Various studies have reported MGEs as the preexisting source of AMR genes, which are widely distributed in the genome of environmental, clinical, as well as human-associated microbial species (Partridge et al. 2018). Six major classes of MGEs are often found in the genome of ESKAPE pathogens, namely, transposons, gene cassettes and integrons, genomic islands, plasmids, bacteriophages, and ICEs (Partridge et al. 2018). These MGEs are involved in the exchange of AMR genes between various pathogens as well as between pathogenic and commensal bacterial population living commonly or transiently in similar environmental niches.

1.5.1   Plasmids

Plasmids are the self-replicating extrachromosomal modules that often endow the host with various fitness factors. However, they do not confer essentiality for the survival of bacteria under optimal growth conditions (Actis et al. 1999; Heuer and Smalla 2012; Nojiri 2013). They are ubiquitous in different domains of life and encode adaptive functions, such as heavy metal or antibiotic resistance, pathogenicity, or the ability to survive in a particular environmental niche and degradation of xenobiotics. The basic structure of the plasmid consists of genes encoding replicative functions and other accessory genes, involved in functions for segregation and selective advantage to the host. The accessory functions of AMR are encoded by either one or more resistance genes and may be linked with MGEs such as integron (In), insertion sequence (IS), and/or transposon (Tn) located in the plasmid backbone (Partridge et al. 2018).

The excessive usage of antibiotics such as ciprofloxacin and norfloxacin in the late 1980s led to the emergence of resistance mechanisms in Gram-negative bacteria. The emergence of plasmid-mediated quinolone resistance (PMQR) was observed since 1998. The first plasmid-mediated conferring quinolone resistance (presently known as *qnrA1*) was reported in *Klebsiella pneumoniae* (Martínez-Martínez et al. 1998). Thereafter, the other plasmid-encoded genes such as *qnrA*, *qnrB*, *qnrC*, *qnrD*, *qnrS*, and *qnrVC* were described in several Gram-negative pathogens (Poirel et al. 2017). These genes encode proteins that belong to the pentapeptide repeat family that provides protection to DNA gyrase and topoisomerase IV from the action of quinolone antibiotics. Smillie et al. (2010) reported 14% of the sequenced plasmids to be conjugative in nature that act as conveyors for AMR traits and virulence factors. The emergence of *mcr-1*, the plasmid-mediated colistin resistance gene in the genome of Gram-negative pathogens, lately is an evidence of continuous evolution of bacterial species through HGT (Liu et al. 2016). The dissemination of OXA β-lactamase genes illustrates their horizontal transfer through plasmids among different bacterial phyla dating back to millions of years (Barlow and Hall 2002).

1.5.2   Bacteriophages

Phages constitute the most abundant and rapidly replicating biological entities on Earth (Jackson et al. 2011) having an estimated number of $\sim 10^{30}$–10^{32} (Chibani-Chennoufi et al. 2004), outnumbering the bacterial population by a factor of 10 (Hendrix et al. 1999). They exploit host resources for their multiplication, and most of them are lytic in nature as the host is killed; hence, they are also known as antibacterial agents (Clokie et al. 2011). Lysogenic phages can integrate their genome into the host chromosome and replicate passively as a part of the host chromosome (Marti et al. 2014).

A recent study found 70% of phage DNA samples from 80 healthy individuals to be positive for various ARGs like *bla*-CTXM-1, *bla*TEM, and *qnr* genes (Quirós et al. 2014). Most of the enteric pathogens such as *Enterococcus* species (Mazaheri et al. 2010), *Salmonella enterica*, and *P. aeruginosa* (Blahová et al. 1992) harbor phages, which have been shown in vitro to act as disseminator of ARGs.

1.5.3   Transposons

Transposons are the jumping genetic elements which hop and integrate into the host chromosomal DNA. These elements show either intracellular or intercellular movement using the functions of their own integrase. Usually, for their integration, there is no requirement of sequence homology between the insertion site and the transposon element. However, in some cases, transposons may require specificity to a particular nucleotide sequence at the insertion site (Craig 1997). The insertion of transposon within a gene leads to disruption of the gene function and may modulate the gene expression if inserted into the regulatory sequences.

Members of ESKAPE contain various transposons such as Tn5 which encodes resistance to kanamycin and neomycin and Tn10 which encodes for tetracycline resistance. Transposons have been linked to the spread of bla_{NDM-1} gene in ESKAPE pathogens (Nordmann et al. 2012), which is a growing public threat.

1.5.4   Integrative and Conjugative Elements (ICEs)

ICEs are members of a class of self-transmissible bacterial MGEs also called as conjugative transposons or constins (Burrus and Waldor 2004). Their size ranges from ~20 kb to 500 kb, and they carry genes that confer phenotypes like resistance to antibiotics and heavy metals, bacteriocin synthesis, bacterial pathogenesis, carbon source utilization, biofilm formation, restriction modification, and resistance to bacteriophage infection (Johnson and Grossman 2015; Wozniak and Waldor 2010; Van Houdt et al. 2013). Their integration and excision are facilitated either by serine, tyrosine, or DDE recombinases (Johnson and Grossman 2015; Brochet et al. 2009). The integrases consist of highly conserved Arg-His-Arg-Tyr catalytic tetrad located in the carboxy-terminal end of the protein. The insertion takes place at a specific DNA sequence, attB in the bacterial chromosome, mostly located at the downstream of tRNA *gene*, and harbors multiple modules that are associated with distinct functions including (1) integration/excision module, (2) replication/DNA processing module, (3) regulation module, (4) DNA secretion module, and (5) auxiliary modules. The auxiliary modules are often linked with evolutionarily important traits, including antimicrobial resistance genes, heavy metal resistance genes, and virulence- or toxin-encoding genes, and genetic traits involved in the alternative catabolic pathways. The extrachromosomal ssDNA-ICE formation starts at the origin of transfer (oriT) and, after the recognition, is covalently linked with the 5′ end of the DNA relaxase. The linear ssDNA-ICEs with the DNA relaxase covalently linked get transferred to the recipient cell, and the circularization of ssDNA-ICEs is mediated by relaxase. Complementary strand synthesis occurs before its integration into the host chromosome.

The discovery and identification of ICEs started with emergence of multidrug-resistant bacterial pathogens. Identification of heavy metal and antibiotic resistance determinants transferred via conjugation and their location on the chromosomes instead of in a stable plasmid of various pathogens like *Enterococcus* (Franke and Clewell 1981), *Bacteroides* species (Rashtchian et al. 1982), and *Clostridium* species (Magot 1983; Smith et al. 1981) led to a speculation about the ICEs. Analysis of 1000 genomes identified putative 335 ICEs and 180 conjugative plasmids depicting both the presence of ICEs in most clades of bacteria and their higher frequency of occurrence compared to conjugative plasmids (Guglielmini et al. 2011). Conjugative transposons belonging to Tn*916*/Tn*1545* family in *Clostridia*, *Enterococcus*, and *Streptococci* carry various tetracycline, macrolide, and aminoglycoside resistance genes providing an insight into resistance gene transfer among these group of bacteria (Roberts and Mullany 2011).

1.5.5 Integrons

Integrons are the gene acquisition elements that capture exogenous ORFs via site-specific recombination and provide an active transcription platform to convert the integrated cassette into a functional gene (Hall and Collis 1995). They utilize their own integrase (IntI) for the integration of the cassette. In general, these integrons have three basic modules: (1) integrase encoding gene, (2) a specific integration site (*attI*), and (3) a functional promoter (Pc). The site-specific integration between the double-stranded host chromosome (*attC*) and the folded single-stranded exogenous cassette (*attI*) is catalyzed by integrase enzyme. The integrons identified can be categorized into two classes: the mobile integrons and superintegrons. The physical linkage of mobile integrons with the MGEs (insertion sequences, transposons, ICEs, conjugative plasmids) is involved in the dissemination of antibiotic resistance genes in bacteria. ARGs encoding resistance against aminoglycosides, β-lactams, chloramphenicol, trimethoprim, erythromycin, fosfomycin, rifampicin, quinolones, and several antiseptic compounds are physically linked with the mobile integrons as reported in several studies (Holmes et al. 2016). Superintegrons are a part of bacterial chromosome and represent a "core" gene-capture system in the Gram-negative bacteria that enables adaptation to diverse environments (Mazel 2006). The large number of gene cassettes linked with the integron having an identity of >80% between the attachment sites of these cassettes and no association with mobile genetic elements are the characteristic features of a superintegron subset.

1.6   Resistance Mechanisms of ESKAPE Pathogens

The principal antibiotic resistance mechanisms of ESKAPE pathogens include the following three major mechanisms: (1) the prevention of access to the target site by either change in membrane permeability or efflux of antibiotics from bacterial cell, (2) the inactivation of antibiotic molecule either by hydrolysis or chemical modification, and (3) the alteration of the target site (either protection/replacement/ absence/mutation or enzymatic modification of the target site) (Alekshun and Levy 2007; Blair et al. 2015) (Fig. 1.1). These resistance mechanisms could be either intrinsic mechanisms or the acquired resistance mechanisms. Another peculiar resistance mechanism includes the biofilm formation that helps the pathogen to evade the antimicrobial action as well as suppress the effect of immune response cells of the host. Additionally, biofilms also protect the dormant cells called persister cells that cause difficult-to-treat recalcitrant infections (Lewis 2008).

Intrinsic resistance mechanisms are the chromosomally encoded mechanisms that include the accumulation of point mutations in the target ORF or in the regulatory region, various nonspecific efflux pump proteins, antibiotic-inactivating enzymes, and various other mechanisms related to permeability barriers (Fajardo and Martínez 2008; Cox and Wright 2013). These mechanisms provide a low level of antibiotic resistance in the original host such as environmental bacteria or normal commensal bacteria that may behave as an opportunistic pathogen in immunocompromised patients (Wright 2007). On the other hand, the acquired resistance mechanisms lead to the acquisition of new functionalities by import of genetic systems to the bacterial cell, which may get integrated into the genome (genomic recombination) or maintained as the extrachromosomal elements (plasmid acquisition) (van Hoek et al. 2011).

1.6.1   Resistance Due to Decreased Permeability or Active Ejection of Antibiotics

An optimal intracellular concentration of an antibiotic is required for its antimicrobial action for the treatment of an infection. In a bacterium, there are two major factors that prevent the optimum accumulation of the antibiotic: (1) the state of the cell membrane permeability and (2) the action of selective efflux pump proteins (Fig. 1.1).

Reduced membrane permeability to the selective compounds is an important defense mechanism utilized by bacteria, which prevents the antibiotic entry into the cytoplasm (Munita and Arias 2016). The permeability threshold of the peptidoglycan layer of Gram-positive bacteria is in the range of 30–57 kDa (Scherrer and Gerhardt 1971), and its coarse meshwork allows the smaller molecules to penetrate easily, which makes Gram-positive organisms susceptible to various antibiotics (Randall et al. 2013). On the other hand, due to the presence of a much finer molecular sieve, the outer membrane of Gram-negative bacteria (Vaara 1992) comprising of lipid molecules (namely, lipid A) covalently linked to polysaccharide units and tightly packaged with saturated fatty acids has reduced membrane fluidity (Nikaido 2003) and increased permeability threshold. Hence, lipopolysaccharide units of Gram-negative bacteria, such as *P. aeruginosa*, *Salmonella enterica*, *V. cholerae*, etc., act as a major permeability barrier and are responsible for intrinsic resistance in these bacteria to several antibiotics such as azithromycin, erythromycin, polymyxin B, and rifamycin. The modifications of the membrane lipid barrier and alterations of the expression pattern of various outer membrane porins lead to the severalfold reduction in the permeability of antibiotics. Reduction in expression or loss or replacement of the channel proteins of outer membrane also reduces the permeability of selective antibiotics (Blair et al. 2015).

Various genes encoding efflux pump proteins, which may be part of chromosome or part of some acquired genetic element like plasmid, are ubiquitously present in bacteria and encode proteins that can actively transport various antimicrobial compounds by utilization of energy in the form of ATP or through transmembrane ion gradients (Li and Nikaido 2009). The bacterial efflux pump mediates the extrusion of the antibiotics out of the bacterial cell, thereby decreasing the effective concentration of the antibiotic inside the cell. The efflux systems have been confirmed to be present in archaea and prokaryotic and eukaryotic species (Van Bambeke et al. 2003). The principal role of efflux systems in a cell is the ejection of undesirable compounds such as heavy metals (Nies 2003), dyes (Kaatz et al. 1993), organic solvents (Ramos et al. 2002), amphiphilic detergents (Mahamoud et al. 2007), biocides (Costa et al. 2013), metabolites (Van Dyk et al. 2004), and quorum sensing molecules (Pearson et al. 1999) in addition to antibiotics. The presence of efflux pumps has also been attributed in biofilm formation and also in contributing toward AMR in many bacteria (Alav et al. 2018).

The efflux pumps can be divided into single-component transporters or multi-component transporters differing in terms of structural conformation, range of substrates that they may export, and their distribution in different types of bacterial organisms.

1.6.2 Resistance Due to Inactivation of the Antibiotics

Bacteria can modify or destroy the antibiotic scaffolds by either transfer of a chemical group to the scaffolds or hydrolysis of the core structure (Fig. 1.1). The modification of antibiotics by the enzymatic functions of the acquired genetic traits is the most common antibiotic resistance mechanisms in pathogenic bacteria.

1.6.2.1 Inactivation of Antibiotics by Hydrolysis.

The inactivation of β-lactam antibiotics is one of the most common antibiotic resistance mechanisms, which is mediated by the hydrolase activity of β-lactamases (De Pascale and Wright 2010). β-Lactams including penicillins, monobactams, cephalosporins, and carbapenem have a common β-lactam ring and are commonly prescribed antibiotics across the globe. The metallo-β-lactamase or serine β-lactamase hydrolyzes the β-lactam ring that confers the resistance against β-lactams in most of the bacterial species. In a β-lactam-sensitive bacterial cell, the presence of an active site serine residue of the penicillin-binding protein acts as a site for the nucleophilic attack and inactivation of the enzymes due to formation of a slowly hydrolyzing covalent β-lactam-PBP intermediate (Andersson and Hughes 2010). The hydrolysis of antibiotic scaffolds such as macrolide macrocycle lactone ring and fosfomycin epoxide ring and the amidohydrolysis of bacitracin undecaprenyl pyrophosphate have been implicated in the development of antibiotic resistance (Das et al. 2020).

1.6.2.2 Inactivation of Antibiotics by Chemical Modifications.

The modification of antibiotic scaffolds via enzymatic modification through the covalent transfer of different chemical groups is another very common resistance mechanism (Fig. 1.1). Various enzymes involved in drug resistance by modification of the antibiotic structure have been characterized (De Pascale and Wright 2010). These enzymes lead to inactivation of antibiotics using any of the seven modifications:

1. Phosphate group transfer to the antibiotic scaffold from either ATP or GTP, called O-phosphorylation. Resistance against chloramphenicol, fosfomycin, macrolide, and rifampicin antibiotics has been reported due to phosphorylation process.
2. Modification of the antibiotics by the transfer of adenosine monophosphate (AMP) moiety, called O-nucleotidylylation. This confers resistance against different members of aminoglycosides and lincosamide classes of antibiotics.
3. A glycosyl moiety addition to the antibiotics such as macrolides and rifampin called O-glycosylation.
4. ADP-ribose transfer from NAD leads to inactivation of rifampin known as O-ribosylation.
5. Transfer of an acetyl group from acetyl-CoA to the antibiotic, called O- and N-acetylation. Several antibiotics are inactivated due to acetylation such as aminoglycosides, chloramphenicol, fluoroquinolone, and streptothricin.
6. Addition of a hydroxyl group (-OH), called hydroxylation, leads to inactivation of an antibiotic.
7. There can be a chemical complex formation that prevents antibiotic access to the target (Das et al. 2020).

1.6.3 Resistance due to Alteration of the Target Site of the Antibiotic

1.6.3.1 Mutations of the Target Site

Most antibiotics target the essential cellular functions of a bacterial cell (Fig. 1.1). The antibiotics act on specific target molecules, and accumulations of point mutations in the target gene alter the amino acid composition of the translated protein, and therefore, the affinity of antibiotic to its target is reduced or lost leading to a reduction in antibiotic efficiency (Andersson and Hughes 2010). Such mechanisms of resistance are widely prevalent in enteric pathogens against commonly prescribed antibiotics that inhibit (1) the DNA replication machinery and repair in the quinolone resistance-determining region by affecting *parC-parE* and *gyrA-gyrB* gene functions (Hooper 2001), (2) transcription by altering *rpoB*-encoded protein (rifampicin), (3) protein synthesis by altering *rpsL* gene function (streptomycin), (4) alteration of cell wall biosynthesis by modification of *pbp* gene function (β-lactam), and (5) various metabolic enzymes by altering *embB*, *katG*, and *mshA* functions (SXT).

1.6.3.2 Enzymatic Modification of the Target Site

The RNA modification enzymes such as rRNA methyltransferase, which transfer a methyl group to some specific nucleotides of 16S or 23S rRNA gene, confer resistance against several protein synthesis inhibitors such as aminoglycoside and macrolide antibiotics routinely used in clinical settings to treat Gram-negative and Gram-positive bacterial infections (Holmes et al. 2016). Several genes (*aviR*, *cfr*, *emtA*, *ermA*, *ermB*, *ermC*) encode 23S rRNA methyltransferases that confer resistance against lincosamides, phenicols, oxazolidinones, pleuromutilins, and streptogramin-A antibiotics in Gram-negative bacteria (Das et al. 2020). However, the absence of Ksg methyltransferase in *E. coli* leads to kasugamycin resistance. In a similar manner, inactivation of pseudouridine synthase *rulC* gene that modifies the 23S rRNA confers significant resistance against clindamycin, linezolid, and tiamulin in various enteric pathogens (Das et al. 2020).

1.6.3.3 Target Protection

The genetic determinants that encode proteins involved in the protection of target have been characterized in the bacterial chromosome as well as MGEs. Fluoroquinolones (Qnr), tetracycline (Tet[M] and Tet[O]), and fusidic acid (FusB and FusC) are the antibiotics affected by this mechanism. Tet[M] and Tet [O] proteins are a member of the translation factor superfamily of GTPases (Fig. 1.1). They bear homology with the elongation factors (EF-G and EF-Tu) required in the synthesis of protein, and their interaction with the ribosome leads to dislodging of the tetracycline antibiotic from its binding site (Munita and Arias 2016). Another example of target protection includes the *qnr* family of quinolone resistance genes present on the plasmids in various pathogens encoding pentapeptide repeat proteins that bind and protect DNA gyrase and topoisomerase IV from the bactericidal action of quinolones (Blair et al. 2015). The resistance against cationic peptides such as polymyxins is most commonly linked to the alterations in the expression of regulators that affect LPS production and lead to alterations in the target or even absence of the LPS production that reduces binding of the drug (Poirel et al. 2017).

1.6.3.4 Substitution or Bypassing of the Target Site

Bacteria that easily evolve new targets with similar biochemical functions to the original target but differ from the target molecules of the antimicrobial molecule utilize this strategy (Fig. 1.1). The most common clinical examples are methicillin resistance as seen in *S. aureus* owing to the acquisition of PBP (PBP2a) and peptidoglycan modifications due to *van* gene clusters leading to vancomycin resistance in enterococci. Another route to evade the action of antibiotics is to "bypass" the metabolic pathway either by overproduction of the enzyme encoded by the antibiotic target or amino acid changes in the enzyme as seen in resistance to trimethoprim-sulfamethoxazole (TMP-SMX) (Munita and Arias 2016).

Deeper insights of the existing and new mechanisms of antibiotic resistance are required to design novel therapeutic agents. Advances in genomics, structural biology, and systems biology can be tapped to circumvent this problem. However, the mobilization of these various resistance genes via MGEs into the pathogens has made the treatment of several infections challenging due to the lack of appropriate antibiotics. Deciphering the mobile genetic elements and their propagation within the indigenous flora and the invading pathogens could pave the way for a better understanding of pathogen biology and emergence of drug resistance in the pathogens of public health importance.

1.7   Conclusion

Antimicrobial resistance is set to have devastating effects on global public health as the emergence and evolution of drug-resistant pathogens continue to threaten modern medicine and simultaneously the food security, agricultural sectors, and animal health. ESKAPE pathogens represent the peak for resistance and pathogenesis and a major cause of life-threatening infections and are characterized by their exacerbated ability for acquisition of multiple resistant genes. So, there is an urgent need for a rapid diagnostic assay, which will help to reduce the unnecessary consumption, blanket prescription, and release of antibiotics in the environment. This will ultimately minimize the selective pressure in the environment. The improved diagnostic solution will help to treat patients effectively and will reduce transmission dynamics of the resistance determinants. An interdisciplinary and one health approach is also the need of the hour to bring the development of the solutions to the global antimicrobial resistance challenge as well as the treatment and management of ESKAPE pathogens. Along with the quest and urgency for developing novel therapeutics to treat multidrug-resistant ESKAPE infections, it is also important to explore other novel alternative therapeutic strategies such as antimicrobial peptides, nanoparticles, phage therapy, and anti-virulence therapy to combat drug resistance pathogens as the currently available antibiotics have often been found lacking and there is a dearth of new drugs in the pipeline. It is necessary to maintain sustainable stewardship practices, AMR surveillance, and patient education in order to stay ahead of the emerging drug-resistant ESKAPE pathogens and to control the AMR crisis in both clinical and community settings.

References

Actis LA, Tolmasky ME, Crosa JH (1999) Bacterial plasmids: replication of extrachromosomal genetic elements encoding resistance to antimicrobial compounds. Front Biosci 4:D43–D62
Ahmed MO, Baptiste KE (2018) Vancomycin-resistant enterococci: a review of antimicrobial resistance mechanisms and perspectives of human and animal health. Microb Drug Resist (Larchmont, N.Y.) 24(5):590–606
Alav I, Sutton JM, Rahman KM (2018) Role of bacterial efflux pumps in biofilm formation. J Antimicrob Chemother 73:2003–2020

Alekshun MN, Levy SB (2007) Molecular mechanisms of antibacterial multidrug resistance. Cell 128:1037–1050

Andersson DI, Hughes D (2010) Antibiotic resistance and its cost: is it possible to reverse resistance? Nat Rev Microbiol 8:260–271

Antonelli A, D'Andrea MM, Brenciani A et al (2018) Characterization of poxtA, a novel phenicol-oxazolidinone-tetracycline resistance gene from an MRSA of clinical origin. J Antimicrob Chemother 73:1763–1769

Baltz RH (2008) Renaissance in antibacterial discovery from actinomycetes. Curr Opin Pharmacol 8:557–563

Barlow M, Hall BG (2002) Phylogenetic analysis shows that the OXA beta-lactamase genes have been on plasmids for millions of years. J Mol Evol 55:314–321

Barna JC, Williams DH (1984) The structure and mode of action of glycopeptide antibiotics of the vancomycin group. Annu Rev Microbiol 38:339–357

Benveniste R, Davies J (1973) Aminoglycoside antibiotic-inactivating enzymes in actinomycetes similar to those present in clinical isolates of antibiotic-resistant bacteria. Proc Natl Acad Sci U S A 70:2276–2280

Blahová J, Hupková M, Krčméry V, Schäfer V (1992) Imipenem and cefotaxime resistance: transduction by wild-type phages in hospital strains of Pseudomonas aeruginosa. J Chemother 4:335–337

Blair JMA, Webber MA, Baylay AJ et al (2015) Molecular mechanisms of antibiotic resistance. Nat Rev Microbiol 13:42–51

Brochet M, Da Cunha V, Couvé E et al (2009) Atypical association of DDE transposition with conjugation specifies a new family of mobile elements. Mol Microbiol 71:948–959

Burrus V, Waldor MK (2004) Shaping bacterial genomes with integrative and conjugative elements. Res Microbiol 155:376–386

Cantón R, Coque TM (2006) The CTX-M beta-lactamase pandemic. Curr Opin Microbiol 9:466–475

Centers for Disease Control and Prevention (2019) CDC's antibiotic resistance threats in the United States (AR Threats Report). https://www.cdc.gov/drugresistance/

Chibani-Chennoufi S, Bruttin A, Dillmann M-L, Brüssow H (2004) Phage-host interaction: an ecological perspective. J Bacteriol 186:3677–3686

Clokie MRJ, Millard AD, Letarov AV, Heaphy S (2011) Phages in nature. Bacteriophage 1:31–45

Comprehensive Antibiotic Resistance Database (2020) Nucleic acids research. https://card.mcmaster.ca/

Coronado-Álvarez NM, Parra D, Parra-Ruiz J (2019) Clinical efficacy of fosfomycin combinations against a variety of gram-positive cocci. Enferm Infecc Microbiol Clin 37:4–10

Costa SS, Viveiros M, Amaral L, Couto I (2013) Multidrug efflux pumps in Staphylococcus aureus: an update. Open Microbiol J 7:59–71

Courvalin P (2008) Predictable and unpredictable evolution of antibiotic resistance. J Intern Med 264:4–16

Cox G, Wright GD (2013) Intrinsic antibiotic resistance: mechanisms, origins, challenges and solutions. Int J Med Microbiol 303:287–292

Craig NL (1997) Target site selection in transposition. Annu Rev Biochem 66:437–474

Cummings DE, Archer KF, Arriola DJ et al (2011) Broad dissemination of plasmid-mediated quinolone resistance genes in sediments of two urban coastal wetlands. Environ Sci Technol 45:447–454

D'Costa VM, King CE, Kalan L et al (2011) Antibiotic resistance is ancient. Nature 477:457–461

Dagher TN, Al-Bayssari C, Chabou S et al (2019) Investigation of multidrug-resistant ST2 Acinetobacter baumannii isolated from Saint George hospital in Lebanon. BMC Microbiol 19 (1):29

Das B, Chaudhuri S, Srivastava R et al (2017) Fostering research into antimicrobial resistance in India. BMJ 358:j3535

Das B, Verma J, Kumar P et al (2020) Antibiotic resistance in Vibrio cholerae: understanding the ecology of resistance genes and mechanisms. Vaccine 38(Suppl 1):A83–A92

Davies J, Davies D (2010) Origins and evolution of antibiotic resistance. Microbiol Mol Biol Rev 74:417–433

De Oliveira DMP, Forde BM, Kidd TJ et al (2020) Antimicrobial resistance in ESKAPE pathogens. Clin Microbiol Rev 33(3):e00181. https://doi.org/10.1128/CMR.00181-19

De Pascale G, Wright GD (2010) Antibiotic resistance by enzyme inactivation: from mechanisms to solutions. Chembiochem 11:1325–1334

Dijkshoorn L, Nemec A, Seifert H (2007) An increasing threat in hospitals: multidrug-resistant Acinetobacter baumannii. Nat Rev Microbiol 5:939–951

Fajardo A, Martínez JL (2008) Antibiotics as signals that trigger specific bacterial responses. Curr Opin Microbiol 11:161–167

Founou RC, Founou LL, Essack SY (2017) Clinical and economic impact of antibiotic resistance in developing countries: a systematic review and meta-analysis. PLoS One 12:e0189621

Franke AE, Clewell DB (1981) Evidence for conjugal transfer of a Streptococcus faecalis transposon (Tn916) from a chromosomal site in the absence of plasmid DNA. Cold Spring Harb Symp Quant Biol 45(Pt 1):77–80

Gandra S, Tseng KK, Arora A et al (2019) The mortality burden of multidrug-resistant pathogens in India: a retrospective, observational study. Clin Infect Dis 69:563–570

Giulieri SG, Tong SYC, Williamson DA (2020) Using genomics to understand meticillin- and vancomycin-resistant Staphylococcus aureus infections. Microb Genom 6(1):e000324

Guglielmini J, Quintais L, Garcillán-Barcia MP et al (2011) The repertoire of ICE in prokaryotes underscores the unity, diversity, and ubiquity of conjugation. PLoS Genet 7:e1002222

Hall RM, Collis CM (1995) Mobile gene cassettes and integrons: capture and spread of genes by site-specific recombination. Mol Microbiol 15:593–600

Hendrix RW, Smith MCM, Burns RN et al (1999) Evolutionary relationships among diverse bacteriophages and prophages: all the world's a phage. Proc Natl Acad Sci 96:2192–2197

Heuer H, Smalla K (2012) Plasmids foster diversification and adaptation of bacterial populations in soil. FEMS Microbiol Rev 36:1083–1104

Holmes AH, Moore LSP, Sundsfjord A et al (2016) Understanding the mechanisms and drivers of antimicrobial resistance. Lancet 387:176–187

Hooper DC (2001) Mechanisms of action of antimicrobials: focus on fluoroquinolones. Clin Infect Dis 32(Suppl 1):S9–S15

Humeniuk C, Arlet G, Gautier V et al (2002) Beta-lactamases of Kluyvera ascorbata, probable progenitors of some plasmid-encoded CTX-M types. Antimicrob Agents Chemother 46:3045–3049

Ishizaki Y, Shibuya Y, Hayashi C et al (2018) Instability of the 16S rRNA methyltransferase-encoding npmA gene: why have bacterial cells possessing npmA not spread despite their high and broad resistance to aminoglycosides? J Antibiot 71:798–807

Jackson RW, Vinatzer B, Arnold DL et al (2011) The influence of the accessory genome on bacterial pathogen evolution. Mob Genet Elements 1:55–65

Johnson CM, Grossman AD (2015) Integrative and conjugative elements (ICEs): what they do and how they work. Annu Rev Genet 49:577–601

Kaatz GW, Seo SM, Ruble CA (1993) Efflux-mediated fluoroquinolone resistance in Staphylococcus aureus. Antimicrob Agents Chemother 37:1086–1094

Klein EY, Milkowska-Shibata M, Tseng KK et al (2021) Assessment of WHO antibiotic consumption and access targets in 76 countries, 2000-15: an analysis of pharmaceutical sales data. Lancet Infect Dis 21:107–115

Kumar A, Biswas L, Omgy N et al (2018) Colistin resistance due to insertional inactivation of the mgrB in Klebsiella pneumoniae of clinical origin: first report from India. Rev Esp Quimioter 31:406–410

Lew K (2014) Antibiotics, 1st edn. Marshall Cavendish Benchmark, New York

Lewis K (2008) Multidrug tolerance of biofilms and persister cells. Curr Top Microbiol Immunol 322:107–131

Li H, Andersen PS, Stegger M et al (2019) Antimicrobial resistance and virulence gene profiles of methicillin-resistant and -susceptible Staphylococcus aureus from food products in Denmark. Front Microbiol 10:2681

Li X-Z, Nikaido H (2009) Efflux-mediated drug resistance in Bacteria. Drugs 69:1555–1623

Liu Y-Y, Wang Y, Walsh TR et al (2016) Emergence of plasmid-mediated colistin resistance mechanism MCR-1 in animals and human beings in China: a microbiological and molecular biological study. Lancet Infect Dis 16:161–168

Magot M (1983) Transfer of antibiotic resistances from Clostridium innocuum to Clostridium perfringens in the absence of detectable plasmid DNA. FEMS Microbiol Lett 18:149–151

Mahamoud A, Chevalier J, Alibert-Franco S et al (2007) Antibiotic efflux pumps in gram-negative bacteria: the inhibitor response strategy. J Antimicrob Chemother 59:1223–1229

Mandal SM, Roy A, Ghosh AK et al (2014) Challenges and future prospects of antibiotic therapy: from peptides to phages utilization. Front Pharmacol 5:105

Marshall CG, Lessard IA, Park I, Wright GD (1998) Glycopeptide antibiotic resistance genes in glycopeptide-producing organisms. Antimicrob Agents Chemother 42:2215–2220

Marti E, Variatza E, Balcázar JL (2014) Bacteriophages as a reservoir of extended-spectrum β-lactamase and fluoroquinolone resistance genes in the environment. Clin Microbiol Infect 20:O456–O459

Martinez JL (2009) The role of natural environments in the evolution of resistance traits in pathogenic bacteria. Proc Biol Sci 276:2521–2530

Martínez JL (2018) Ecology and evolution of chromosomal gene transfer between environmental microorganisms and pathogens. Microbiol Spectr 6(1):141–160. https://doi.org/10.1128/microbiolspec.MTBP-0006-2016

Martínez-Martínez L, Pascual A, Jacoby GA (1998) Quinolone resistance from a transferable plasmid. Lancet 351:797–799

Mazaheri Nezhad Fard R, Barton MD, Heuzenroeder MW (2010) Novel bacteriophages in Enterococcus spp. Curr Microbiol 60:400–406

Mazel D (2006) Integrons: agents of bacterial evolution. Nat Rev Microbiol 4:608–620

Munita JM, Arias CA (2016) Mechanisms of antibiotic resistance. Microbiol Spectr 4:34

Naorem RS, Urban P, Goswami G, Fekete C (2020) Characterization of methicillin-resistant Staphylococcus aureus through genomics approach. 3 Biotech 10:401

Navon-Venezia S, Kondratyeva K, Carattoli A (2017) Klebsiella pneumoniae: a major worldwide source and shuttle for antibiotic resistance. FEMS Microbiol Rev 41:252–275

Nies DH (2003) Efflux-mediated heavy metal resistance in prokaryotes. FEMS Microbiol Rev 27:313–339

Nikaido H (2003) Molecular basis of bacterial outer membrane permeability revisited. Microbiol Mol Biol Rev 67:593–656

Nojiri H (2013) Impact of catabolic plasmids on host cell physiology. Curr Opin Biotechnol 24:423–430

Nordmann P, Gniadkowski M, Giske CG et al (2012) Identification and screening of carbapenemase-producing Enterobacteriaceae. Clin Microbiol Infect 18:432–438

Ogawara H (2016) Self-resistance in streptomyces, with special reference to β-lactam antibiotics. Molecules 21(5):605. https://doi.org/10.3390/molecules21050605

Olaitan AO, Rolain J-M (2016) Ancient resistome. Paleomicrobiol Hum 4:75–80

Partridge SR, Kwong SM, Firth N, Jensen SO (2018) Mobile genetic elements associated with antimicrobial resistance. Clin Microbiol Rev 31(4):e00088. https://doi.org/10.1128/CMR.00088-17

Pearson JP, Van Delden C, Iglewski BH (1999) Active efflux and diffusion are involved in transport of Pseudomonas aeruginosa cell-to-cell signals. J Bacteriol 181:1203–1210

Périchon B, Courvalin P, Galimand M (2007) Transferable resistance to aminoglycosides by methylation of G1405 in 16S rRNA and to hydrophilic fluoroquinolones by QepA-mediated efflux in Escherichia coli. Antimicrob Agents Chemother 51:2464–2469

Poirel L, Jayol A, Nordmann P (2017) Polymyxins: antibacterial activity, susceptibility testing, and resistance mechanisms encoded by plasmids or chromosomes. Clin Microbiol Rev 30:557–596

Poirel L, Rodriguez-Martinez J-M, Mammeri H et al (2005) Origin of plasmid-mediated quinolone resistance determinant QnrA. Antimicrob Agents Chemother 49:3523–3525

Quirós P, Colomer-Lluch M, Martínez-Castillo A et al (2014) Antibiotic resistance genes in the bacteriophage DNA fraction of human fecal samples. Antimicrob Agents Chemother 58:606–609

Ramos JL, Duque E, Gallegos M-T et al (2002) Mechanisms of solvent tolerance in gram-negative bacteria. Annu Rev Microbiol 56:743–768

Randall CP, Mariner KR, Chopra I, O'Neill AJ (2013) The target of Daptomycin is absent from Escherichia coli and other gram-negative pathogens. Antimicrob Agents Chemother 57:637–639

Rashtchian A, Dubes GR, Booth SJ (1982) Tetracycline-inducible transfer of tetracycline resistance in Bacteroides fragilis in the absence of detectable plasmid DNA. J Bacteriol 150:141–147

Review on Antimicrobial Resistance (2016) Tackling drug-resistant infections globally: final report and recommendations

Roberts AP, Mullany P (2011) Tn916-like genetic elements: a diverse group of modular mobile elements conferring antibiotic resistance. FEMS Microbiol Rev 35:856–871

Santajit S, Indrawattana N (2016) Mechanisms of antimicrobial resistance in ESKAPE pathogens. Biomed Res Int 2016:1–8

Scherrer R, Gerhardt P (1971) Molecular sieving by the Bacillus megaterium cell wall and protoplast. J Bacteriol 107:718–735

Shaw KJ, Rather PN, Hare RS, Miller GH (1993) Molecular genetics of aminoglycoside resistance genes and familial relationships of the aminoglycoside-modifying enzymes. Microbiol Rev 57:138–163

Smillie C, Garcillán-Barcia MP, Francia MV et al (2010) Mobility of plasmids. Microbiol Mol Biol Rev 74:434–452

Smith MD, Hazum S, Guild WR (1981) Homology among tet determinants in conjugative elements of streptococci. J Bacteriol 148:232–240

Snyder ADH, Hall Snyder AD, Werth BJ et al (2016) Fosfomycin enhances the activity of Daptomycin against vancomycin-resistant enterococci in an in vitro pharmacokinetic-Pharmacodynamic model. Antimicrob Agents Chemother 60:5716–5723

Tacconelli E, Carrara E, Savoldi A et al (2018) Discovery, research, and development of new antibiotics: the WHO priority list of antibiotic-resistant bacteria and tuberculosis. Lancet Infect Dis 18:318–327

Treepong P, Kos VN, Guyeux C et al (2018) Global emergence of the widespread Pseudomonas aeruginosa ST235 clone. Clin Microbiol Infect 24:258–266

Vaara M (1992) Agents that increase the permeability of the outer membrane. Microbiol Rev 56:395–411

Van Bambeke F, Glupczynski Y, Plésiat P et al (2003) Antibiotic efflux pumps in prokaryotic cells: occurrence, impact on resistance and strategies for the future of antimicrobial therapy. J Antimicrob Chemother 51:1055–1065

Van Dyk TK, Templeton LJ, Cantera KA et al (2004) Characterization of the Escherichia coli AaeAB efflux pump: a metabolic relief valve? J Bacteriol 186:7196–7204

van Hoek AHAM, Mevius D, Guerra B et al (2011) Acquired antibiotic resistance genes: an overview. Front Microbiol 2:203

Van Houdt R, Toussaint A, Ryan MP et al (2013) The Tn4371 ICE family of bacterial Mobile genetic elements. Landes Biosci 9:242

Waksman SA, Woodruff HB (1940) Bacteriostatic and bactericidal substances produced by a soil Actinomyces. Exp Biol Med 45:609–614

Warinner C, Rodrigues JFM, Vyas R et al (2014) Pathogens and host immunity in the ancient human oral cavity. Nat Genet 46:336–344

Wozniak RAF, Waldor MK (2010) Integrative and conjugative elements: mosaic mobile genetic elements enabling dynamic lateral gene flow. Nat Rev Microbiol 8:552–563

Wright GD (2007) The antibiotic resistome: the nexus of chemical and genetic diversity. Nat Rev Microbiol 5:175–186

Bacterial Multidrug Tolerance and Persisters: Understanding the Mechanisms, Clinical Implications, and Treatment Strategies

2

Mamta Singla, Vikas Chaudhary, and Anirban Ghosh

Contents

Mamta Singla · V. Chaudhary · A. Ghosh (✉)
Molecular Biophysics Unit, Indian Institute of Science, Bangalore, India
e-mail: ganirban@iisc.ac.in

© The Author(s), under exclusive license to Springer Nature Singapore Pte
Ltd. 2022
V. Kumar et al. (eds.), *Antimicrobial Resistance*,
https://doi.org/10.1007/978-981-16-3120-7_2

29

Abstract

Antibiotic resistance is a massive problem in today's world and a serious threat to human civilization. Apart from being genetically resistant to antibiotics, the other important mechanism by which pathogenic bacteria can evade antibiotics is multidrug tolerance through the formation of persisters. In contrast to antibiotic-resistant bacteria, persister cells require no heritable genetic changes for surviving antibiotic treatment but become drug tolerant due to transient growth arrest. Since most antibiotics only target active metabolic pathways in the growing cells, non-growing persisters can escape the bactericidal effects of antibiotics and resume growth once the antibiotic is withdrawn resulting in treatment failure. Here in this book chapter, we try to highlight different aspects of persisters starting from unique characteristics of persisters, mechanisms of persister formation, clinical relevance, different techniques to study persisters, and some promising treatment options. Finally, we emphasize how focused research in this particular field can have a big impact to eradicate antimicrobial resistance.

Keywords

Persisters · Dormancy · Antibiotic tolerance · Resistance · Biofilm · Stringent response · Quorum sensing · SOS response

2.1   Introduction: Bacterial Multidrug Tolerance and Persister Formation

Bacterial infections are one of the leading causes of mortality and morbidity worldwide. The first antibiotic, penicillin, was discovered almost a century ago to fight these disease-causing pathogens. However, over time, bacteria have evolved to develop resistance mechanisms against such antimicrobial drugs, posing a major public health threat. The other innate strategy that bacteria employ to evade antibiotic action is "persistence." This phenomenon is generally exhibited by a small subpopulation of dormant bacterial cells called "persisters." Joseph Bigger first described the existence of persister cells in 1944, where he observed that even a high dose of penicillin treatment could not sterilize the *S. aureus* infection (Bigger 1944). He hypothesized that the surviving bacteria are the WT non-growers rather than the heritable genetic mutants. However, only a decade ago, Nathalie Balaban and her colleagues confirmed that persister cells are truly non-proliferating. Persisters "get stuck" in the non-growing state before or during the course of antibiotic administration, survive the treatment, and later switch back to the growing cells. Persisters thus contribute to the multidrug tolerance resulting in prolonged therapy and treatment failure.

Persister cells are extensively studied and found among both Gram-positive (*Staphylococcus aureus*, *Bacillus subtilis*, etc.) and Gram-negative bacteria (*Acinetobacter baumannii*, *Pseudomonas aeruginosa*, etc.) and *mycobacteria* (reviewed later in the chapter). Furthermore, they are also found in eukaryotes like yeast, where they cause tolerance to antifungals (Wuyts et al. 2018). In addition to this, cancer cell lines also are described to form persister-like cells, which can then survive chemotherapy and spread resistance (Ramirez et al. 2016). Most of the bacterial and fungal infections caused by persisters often involve biofilm formation. The outer matrix of the biofilm provides a physical barrier against the host immune system and antibiotics, and the inner niche provides the optimum condition of dormancy, thus thriving for a prolonged time until the favorable growth condition becomes available (Lewis 2008). Consequently, persisters become responsible for recalcitrant chronic infections and treatment failure.

There have been numerous studies in recent years, focusing on the mechanism of persister formation and their importance in clinical settings (Wuyts et al. 2018). However, there is a pressing need to understand the multidrug tolerance phenotype and ambiguity in their phenotypic behavior. This might be promising and could open avenues of research toward developing new treatment options to eradicate persisters and ultimately reduce infection relapse rates. This chapter elucidates the current understanding of persister growth and survival mechanisms, the technology used to study the persisters, anti-persister strategies, and their clinical relevance.

2.2 Similarity and Difference Between Persisters and Resistant Mutants

Resistance, tolerance, and persistence are the three different but interrelated mechanisms that help bacterial populations to escape the antimicrobial activity of antibiotics. *Resistance* is the inherited ability of bacteria to survive and replicate in the presence of an antibacterial drug that would otherwise prevent growth. It is generally achieved by preventing antibiotic binding to its target and can be quantified as an increase in MIC irrespective of the treatment duration of antibiotics. In comparison, *tolerance* is the ability of the whole population to survive the transient exposure to raised concentrations of an antibiotic without any MIC modification.

The other form of phenotypic resistance (see Box 2.1) is *persistence*, which is merely an extension of tolerance and different in terms of the fraction of cells involved, i.e., tolerance and resistance are the phenomena of the whole population. In contrast, persistence is the phenomenon for a subpopulation. Persistence is shown by a small group of cells called persisters. Persister is derived from the Latin word *persister* (*per-* + *sister*), which means "to stand," so in biological terms, they can survive drug treatment but cannot grow. These cells are the phenotypic variant of a tolerant bacterial population from the same genetic makeup. They overcome the antibiotic attack by shutting down the targets, which essentially makes them non-targetable.

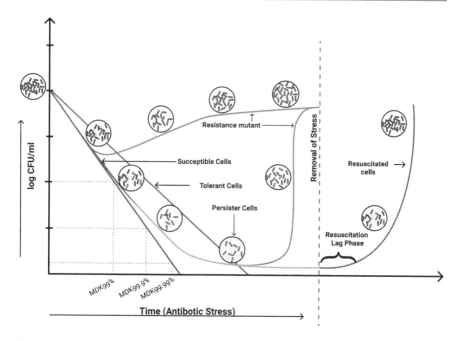

Fig. 2.1 The biphasic killing curve and the difference between resistance, tolerance, and persistence; with the advent of antibiotic treatment to a homogenous bacterial population, a large fraction of susceptible cells (purple line) gets killed rapidly, whereas tolerant cells (blue line) take longer time than susceptible cells with more MDK$_{99\%}$ value. Meanwhile, a tolerant subpopulation of persister cells (green descending curve) slowly emerges within the bacterial population that leads to a typical biphasic killing pattern. However, resistant mutants (orange ascending curve) can divide during the treatment, while persister can only survive (slowly descending curve). After removal of antibiotic stress, the persisters start dividing following an extended lag phase resulting in a homogeneous population of wild-type cells, as their tolerance phenotype is transient and non-heritable. Sometimes, persister cells can also give rise to resistant mutants

The major difference between persisters and resistant mutant is that the former one does not acquire any heritable genetic mutation, unlike the latter one.

Both persister cells and resistant mutants show distinct responses to antibiotic treatment. During the time-kill assays, the biphasic killing curve (Fig. 2.1) is the hallmark of antibiotic persistence, where a sensitive bulk population of cells gets killed in a logarithmic manner leaving behind the minor subpopulation of persisters, which are killed at a significantly slower rate. In the same assay, resistant mutants can divide, while persisters can only survive (dominant phenotype) during the antibiotic treatment. After the antibiotic is stopped, regrowth of surviving persisters results in an antibiotic-sensitive WT population with a similar minor fraction of tolerant cells, unlike uniformly antibiotic-insensitive, resistant population in the case of resistant mutants. The most widespread measure of resistance is the MIC (minimum inhibitory concentration) of the antibiotic, which cannot distinguish between resistant mutants and persisters. However, the other assay called MDK (minimum duration for killing) can measure the tolerance (see Box 2.1). For persister cells,

$MDK_{99.99\%}$ is higher than the susceptible strain. Higher antibiotic concentration can kill resistant mutants, whereas persisters often do not respond to an even higher concentration of antibiotics; rather more prolonged treatment can kill them.

Despite all the differences between persister and resistant mutants, one cannot deny that both are the results of an acquired defense mechanism by bacteria to survive antibiotic-mediated killing. They show enough overlapping, and hence, their cross talk can be explained by their shared mechanism/phenotype against antibiotic killing. For example, efflux pumps are an important player in antibiotic resistance (see Box 2.1); recently, their role has been demonstrated in persister formation in *mycobacteria* (Adams et al. 2011) and E. coli (Pu et al. 2016) as well. Similarly, the stringent and SOS response are responsible for persister formation, but in *E. coli*, inhibition of stringent response reduces resistance development, suggesting an overlap between both survival strategies. Also, resistant mutants can show reduced antibiotic uptake (aminoglycosides) just like persisters, rather than permanent arrest generally seen in most cases. Finally, the evolution of antibiotic-resistant mutations from the persistence shows a positive correlation even though they are mostly mechanistically different (reviewed in Sect. 2.5).

Box 2.1

Phenotypic resistance is the development of antibiotic resistance without any genetic alteration. This type of non-inherited resistance is generally associated with biofilm growth. Apart from persistence, the various other probable transient variants of phenotypic persistence include the change in bacterial permeability, biofilm, and drug indifference (Corona and Martinez 2013).

Resistant mutants are bacterial cells that can resist the antibiotic's effect and carry the inheritable mutation.

MDK (minimum duration for killing) is defined as the minimum time required to kill a large fraction or certain percentile of the population; for example, $MDK_{99\%}$ is the minimum duration to kill 99% of the population or $MDK_{99.9\%}$ the minimum duration for 99.9% of the population. It is a parameter to characterize the slower killing of a tolerant population, even at the high concentration of a drug. A high MDK value represents more time required to kill and hence more tolerance (Brauner et al. 2016).

MIC (minimum inhibitory concentration) is the lowest concentration of antibiotics required to clear the bacterial population.

The mechanism in antibiotic resistance: The main mechanism by which resistant mutants resist antibiotic killing is by altering (increase or decrease) antibiotic flux through the membrane, modifying the antibiotic targets, or modifying the antibiotic, in which multidrug efflux pumps are the primary role player to export antibiotics from the cell (Petchiappan and Chatterji 2017; Li and Nikaido 2009).

2.3   Different Mechanisms of Persister Formation

Persister formation is considered to be a stochastic event. In a laboratory setup of exponentially growing cultures, two types of persisters are often seen: (1) triggered or type I persisters, which are carried over from the stationary-phase preculture, and (2) spontaneous or type II persisters, which are formed during the exponential growth at a random manner (Balaban et al. 2004). Microscopic observation can differentiate the two types, and they exhibit different susceptibility patterns (Goneau et al. 2014). Knowledge about the stochasticity of the mechanism of persister formation is rather limited where it is still confusing how cells can go through a sudden "phenotypic switch" without any adverse events. Scientists speculate that many of such stochastic events happen accidentally due to some error in cell division or metabolism. Bruce Levin and colleagues famously termed these events as *persistence as stuff happens* (PaSH) (Levin et al. 2014). So, drug-tolerant persisters within a clonal bacterial population can be formed by a completely random way or by some deterministic mechanism or by both of these, and this balance between stochastic and deterministic events is known as "responsive diversification" (Kotte et al. 2014).

Bacterial persister formation has been studied to a large extent in different organisms. The more it was studied, the more it was learned that there is no single universal factor responsible for persister formation. In fact, in a single microorganism like *E. coli* K-12 MG1655, persister formation happens by different redundant mechanisms. In a way, we can say that bacterial cells do not rely on a single pathway/mechanism; it all depends on what kind of stress they encounter, and accordingly, they trigger the corresponding pathway. On some occasions, it is found that one particular stress can activate multiple persister formation pathways in an interconnected manner. All in all, the molecular mechanism of persister formation is diverse, debatable, and lacking comprehensive understanding at the moment.

To understand this section better, we can broadly categorize it into two different parts:

2.3.1   Nonspecific Determinants

There are certain cellular factors often linked to stress tolerance, which contribute to persister formation in a general, nonspecific manner. The molecular details to pinpoint one phenotype-one gene have not been studied in great detail in such cases. Stresses often found to be directly related to persister formation are hypoxia, nutrient starvation, and acidic pH (Nathan 2012). Under nutrient-limiting conditions, depending on the limiting nutrient like carbon, nitrogen, phosphorus, etc., several things can impact the formation of persisters, which are as follows:

1. Depleted ATP formation results because of impaired glycolysis, and TCA cycle results in dormancy phenotype, where cells become tolerant to lethal doses of

antibiotics which are meant to corrupt essential pathways in the metabolically active cell (Shan et al. 2015).

2. In *E. coli*, it is found that carbon starvation is linked to a significant change in metabolism patterns involving cyclic AMP (cAMP), phosphoenolpyruvate (PEP), and fructose 1,6-bisphosphate (FBP). Altered metabolism affects cellular physiology and antibiotic efficacy (Chubukov and Sauer 2014).

3. Accumulation of ppGpp is also observed under amino acid, nitrogen, and phosphorus starvation which plays a key role in persister formation (Potrykus and Cashel 2008).

4. Transcriptional regulation of genes governed by stationary-phase sigma factor RpoS is a major adaptation process often seen in nutrient-derived *E. coli* culture. RpoS makes sure of the downregulation of genes responsible for rapid growth and upregulation of genes involved in stress survival. Different factors, including ppGpp, control cellular RpoS levels and activity. Many genes in the RpoS regulon have been linked to persister formation in *E. coli* (Bougdour et al. 2008; Keseler et al. 2017).

5. *E. coli* cells are equipped to check the rate of translation to cope with nutrient limitation in the stationary phase. When *E. coli* cells enter the non-growing phase, the Ribosome modulation factor (RMF) and Hibernation promoting factor (HPF) deactivate the 70S ribosomal complex to block translation. Several fail-safe mechanisms like ribosome frameshift and stop codon read-through are well studied, which make sure cell survives with a basic set of proteins and block biosynthesis of unnecessary proteins at the same time (Wenthzel et al. 1998; Ueta et al. 2008; Barak et al. 1996).

2.3.2 Specific Determinants

1. Stringent Response.
   Secondary messenger guanosine pentaphosphate or tetraphosphate, collectively known as (p)ppGpp, has been studied in great detail and found that it contributes to growth retardation and stress tolerance under amino acid-limiting conditions, known as the stringent response. In *E. coli* and many other bacteria, (p)ppGpp is synthesized by Rel enzyme, which senses amino acid starvation by detecting uncharged tRNAs entering the ribosomal A site. Stress alarmone ppGpp interacts with RpoS and transcription factor DksA to regulate growth and finally gets hydrolyzed by SpoT hydrolase once the growth condition is favorable. (p)ppGpp is involved in dealing with different stresses and possibly forming non-growing persisters (Xiao et al. 1991; Rao et al. 1998; Haseltine and Block 1973; Boutte and Crosson 2013; Battesti and Bouveret 2006). High intracellular (p)ppGpp concentration has shown to play a positive correlation with persister formation in the case of HipA7 toxin and ObgE (Verstraeten et al. 2015; Korch et al. 2003), whereas deletion mutant of *relAspoT* was found to form fewer persisters (Fig. 2.2) (Fung et al. 2010; Amato et al. 2013). In the persister fraction of the cells, both

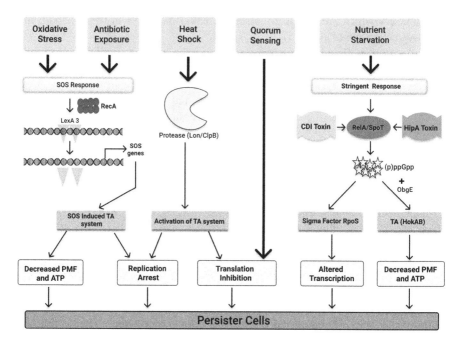

Fig. 2.2 An illustration of various mechanisms involved in persister formation. When the bacterial population gets exposed to oxidative and antibiotic stress, it triggers SOS response which leads to the activation of specific TA systems, resulting in a decrease of intracellular level of ATP and membrane potential. Other stressors like heat shock and quorum-sensing phenomenon can increase cellular dormancy and persistence through altering the cellular processes such as replication and translation. During nutrient starvation, secondary messengers like (p)ppGpp get activated via many intracellular molecules, which can also result in persister formation. Abbreviations: TA, toxin-antitoxin; PMF, proton motive force; CDI, contact-dependent inhibition

relA and *spoT* genes have found to be upregulated highlighting the direct role of (p)ppGpp in the formation and maintenance of persistence (Pu et al. 2016).

2. SOS Response.

   The sole function of the bacterial SOS response is to repair the DNA damage and temporarily halt the cell division until the damage is repaired. In *E. coli*, RecA and LexA are important proteins that take part in SOS response, where RecA binds to damaged ssDNA inducing autocleavage of LexA, resulting in derepression of SOS genes. Several studies have disclosed the possible link between SOS response and persistence. In general, it has been found that downregulation of RecA or expression of non-cleavable LexA3 protein causes a decreased number of persister (Fig. 2.2) (Dorr et al. 2009; Luidalepp et al. 2011; Wu et al. 2015). RecA protein is also found to be upregulated in isolated persister fraction of cells (Keren et al. 2004). Other SOS proteins like RuvA, RuvB, UvrD, and DinG have also been indirectly linked to persister formation as corresponding deletion mutants form fewer persisters, possibly because of their direct role in DNA repair and resulting growth stasis (Theodore et al. 2013).

Similarly, *umuC* and u*muD* genes were found to be positively induced in persister subpopulation, and a deletion mutant of *umuD* results in less persisters compared to WT. To learn more downstream events, researchers found that some of the SOS gene products are directly involved in persister formation like TisB (Dorr et al. 2010). TisB serves as a membrane toxin that causes a collapse in membrane potential and subsequent intracellular ATP depletion leading to persister formation (Berghoff et al. 2017).

3. Toxin-antitoxin (TA) Systems

As the name suggests, TA systems are composed of a stable toxin and a relatively unstable antitoxin and exist in a single operon. There are different subclasses, and often the nature of the toxin (DNase, mRNase, tRNAse, ionophore) dictates the physiological relevance of the toxin. TA systems have been originally discovered to play a critical role in plasmid segregation, but the role of chromosomal TA modules is often linked to persister formation or a survival strategy under stress per se. TA modules can be stochastically induced and/or stress-induced, and toxin concentration needs to reach a certain threshold to affect growth regulation. This can be achieved by induction of the whole TA module under particular stresses and subsequent degradation of the unstable antitoxin component. Overexpression of many toxins has been shown to increase the number of persisters (Kim and Wood 2010; Wilmaerts et al. 2018), but the deletion of the same TA modules hardly affects persister fraction in standard condition (Michiels et al. 2016) possibly because multiple TA modules get induced in that particular condition and/or possible cross-activation. However, there are some reports available where single deletion mutants of TA modules form fewer persisters under very specific conditions (Kim and Wood 2010; Norton and Mulvey 2012). Type I TA system toxins HokB and HokA, which are part of ObgE-directed persistence pathway, have shown to be contributing to persister formation in *E. coli* by membrane pore formation (Verstraeten et al. 2015). Among type II TA modules, *hipAB* has been studied quite extensively, and it has been found that the mutant version of the toxin (*hipA7*) contributes to ~10,000-fold increased persister formation (Korch et al. 2003; Vazquez-Laslop et al. 2006). HipA toxin targets glutamyl tRNA synthetase by phosphorylation and results in uncharged tRNA^{glu}, thus further inducing ppGpp synthesis via *RelA* (Germain et al. 2013). Apart from HipA, several mRNAse toxins have been shown to have an impact on persister formation, e.g., RelE, YafQ, and MazF (Pedersen et al. 2003; Prysak et al. 2009; Yamaguchi et al. 2009; Zhang et al. 2005). Overexpression of RelE toxin has been shown to modulate persister formation when cells are exposed to nutritional stress (Tashiro et al. 2012). Ectopic expression of YafQ toxin induces persister formation in both planktonic cells and biofilms (Harrison et al. 2009). Similarly, MazF toxin is also reported to contribute a higher fraction of persister formation (Tripathi et al. 2014). Since most of these TA modules work as a complex regulatory network, deletion of one module often does not have a large impact on persister formation, especially when tested in a non-optimized condition. In a recent publication, Ghosh et al. have shown for the first time that bacterial cells use deliverable toxins with a different range of activities to induce a high level of persister formation within the population in a very deterministic way. The population-level survival increases up to 10,000-fold when the toxin delivery happens in a density-dependent manner (Ghosh et al. 2018). They have shown that the toxic activity of contact-dependent inhibition (CDI) toxins increases both cellular (p)ppGpp levels and RNA polymerase sigmaS activity, which are the two key features of stringent response. CDI toxins have also been shown to alter gene expression levels and promote biofilm formation, resulting in more cooperative communication between kin cells.

4. Quorum Sensing

   It has been shown on multiple occasions that bacterial cell signaling molecules play a crucial role in behavioral change once it reaches a certain threshold concentration. Cell-density-dependent release of autoinducers into the outside media controls gene regulation and hence can contribute to persistence phenomenon. Quorum-sensing molecule pyocyanin and acyl-homoserine lactone elevate persister formation in *P. aeruginosa* (Moker et al. 2010). 2-Aminoacetophenone is another example of a quorum-sensing molecule that induces persister formation in *P. aeruginosa* by translation inhibition (Que et al. 2013). Indole acts as a quorum-sensing molecule and promotes persister formation in *E. coli* by upregulating OxyR (Vega et al. 2012) protein. Indole is shown to contribute to cross-species induction of persistence in *Salmonella typhimurium*. In *Streptococcus mutans*, competence-stimulating factor induces bacteriocin production which has an indirect effect on growth retardation and hence contributes to persister formation (Leung and Levesque 2012).

5. Heat Shock

   Several proteins involved in heat shock response in *E. coli* are often found to modulate the persister fraction, Lon, ClpB, DnaK, DnaJ, etc. (Pu et al. 2016; Luidalepp et al. 2011; Wu et al. 2015; Germain et al. 2013; Hansen et al. 2008; Liu et al. 2017), where deletion of such proteases decreases persister cell fraction (Fig. 2.2). These heat shock response proteases are thought to be involved in the degradation of antitoxins of TA modules and hence contribute to persister formation. Apart from being heat shock proteins, they generally have a defined cellular role which might indicate their dual function.

6. Oxidative Stress

   Oxidative stress has been linked to persister formation in *E. coli*. There are several target proteins that are known to be directly involved in oxidative stress response as well as modulating persister formation. Deletion mutant of oxidative stress regulator protein OxyR results in decreased persister formation (Wu et al. 2015). Similarly, SoxS regulator of the superoxide response regulon is shown to be upregulated in the persister fraction of cells (Pu et al. 2016). Apart from this, oxidative stress induces several multidrug osmolyte pump subunit genes of Acr, Emr, and Mac series that indirectly contributes to drug tolerance by reducing the intracellular drug concentration (Pu et al. 2016; Wu et al. 2012).

7. Other Mechanisms

   Over the years, few proteins have been unexpectedly identified to contribute to persister formation without any role in stress response. Some examples include structural proteins of flagella, both basal body and hook proteins (Shan et al. 2015), indicating the presence of flagellar structure and function is possibly related to persister formation. Deletion mutants of these proteins were found to form fewer persisters. Similarly, transposon mutagenesis has identified membrane sensor kinase gene *qseC*, known for type I pili activation, to be involved in persister formation (Kostakioti et al. 2009). Deletion mutants of genes having defined activity in chromosome segregation *xerCD* have shown to form fewer persisters (Dorr et al. 2009). Glycerol-3-phosphate dehydrogenase complex has been shown to contribute to persistence as overexpression of the subunits and deletion mutants has been shown to form more and fewer persisters, respectively (Spoering et al. 2006). In *S. aureus*, it has been shown that the stochastic variation in the synthesis of TCA cycle products promotes persistence (Zalis et al. 2019). In each of these cases, detailed studies to find the molecular mechanism linking to persister formation are still missing.

2.4   Regrowth of Persisters

Persistence is a transient phenomenon exhibited by a small subpopulation of cells by which they survive antibiotics and other stresses. During this non-growing phase, they maintain a basic level of energy metabolism, protein synthesis, and DNA winding to sustain viability. Theoretically, cells should resume growth and metabolic activity when the condition favors. Few things need to fall in place for successful resuscitation of persister cells like detection of nutrient availability, rearrangement of their gene expression pattern, resuming active metabolism, and, finally, cell division. If the cell fails in any of those steps, it cannot form a colony and is hence called viable but non-culturable (VBNC). Resuscitation of persisters has been mainly studied in laboratory conditions in both broth and solid media.

In the real environment, things may work differently, but in the in vitro setup, several factors play a key role in governing the duration of the lag phase during the heterogeneous transition of individual cells from non-growing to growing state and, hence, overall growth resumption kinetics.

1. Preculture conditions.
   In any given growing bacterial culture, the frequency of the persisters depends on many things, including the growth phase of preculture, which has been used to inoculate the same growing culture. If the preculture is from the stationary phase, persister regrowth happens with longer lag times (Luidalepp et al. 2011; Joers et al. 2010; Joers and Tenson 2016).
2. Culture medium in the growing culture.
   The frequency of persisters and further regrowth depends on the treatment culture medium. If the culture medium is rich, persister resuscitation happens faster with a shorter lag phase than the poor culture medium (Luidalepp et al. 2011).
3. Stress conditions.

In the case of induced persistence, it depends on what kind of stress cells have been exposed to before they formed persisters. It has been seen that resuscitation kinetics varies significantly from one stress to another even though the culture medium remains the same.

4. Posttreatment outgrowth media composition.

    Like the preculture media, the growth resumption depends heavily on the richness of outgrowth media. Bacteria usually wake up faster in rich media than poor media (Joers and Tenson 2016). The growth resumption timing varies from one individual cell to another, described by bet-hedging strategy, where cells with longer lag phases act as insurance policies for the whole clonal population (de Jong et al. 2011). It has been shown that a lower concentration of glucose, which is a preferential carbon source, supports faster resuscitation compared to a 100 times higher concentration of gluconate (Joers and Tenson 2016). Similar differences have been observed by plating ampicillin-treated *E. coli* persister cells in LB agar and R2A agar, where LB stimulates faster resuscitation and more colony formation (Kaldalu et al. 2016).

5. Genetic factors.

    Some signaling proteins play a key role in persister resuscitation. Actively growing cells of *Micrococcus luteus* secrete a signaling molecule called resuscitation-promoting factor (Rpf), which induces the growth resumption of the dormant cells (Mukamolova et al. 1998; Mukamolova et al. 2006). Another example includes Rpf protein in *M. tuberculosis* which might play a defining role in latent TB infection (Rosser et al. 2017). Distant Rpf subfamily proteins have been identified in many actinobacteria and firmicutes (Ravagnani et al. 2005; Nikitushkin et al. 2015). Peptidoglycan muropeptides also serve as signaling factors for growth stimulation of non-growing *Bacillus* sp. cells (Levin-Reisman et al. 2017).

2.5   Persisters as Predecessors of Resistant Mutants

As we have discussed earlier, persisters are characteristically different from resistant mutants in many ways. Recently, it was discovered that persisters could act as predecessors of resistant mutants both in laboratory conditions and in clinical settings. Persisters are formed when cells go to the transient non-growing phase, and if they sustain in such a hostile environment for a long time, they can acquire de novo mutations in the genome.

With the help of in vitro evolution experiments, Balaban group showed that resistant mutants emerge from the tolerant subpopulation of persisters with higher probability because by becoming tolerant first bacterial cells gain more time to acquire genetic mutations (Levin-Reisman et al. 2017). Increased tolerance to antibiotics acts as a stepping stone of evolution to become genetically resistant to the same drug. A similar study from Michiels' group found out that high-persister mutants are more likely to emerge as resistant mutants over time than low-persister mutants (Windels et al. 2019; Van den Bergh et al. 2016).

Some studies justify the link between antibiotic persistence and antibiotic resistance in the context of infection (Cohen et al. 2013). With the help of whole genome sequencing, it has been learned that *M. tuberculosis* can undergo several genetic mutations while it is thriving as a dormant cell (Ford et al. 2011). Mfd translocase enzyme in *M. tuberculosis* can potentially induce genetic mutations by nucleotide excision repair mechanism when persister cells face DNA damage inside host macrophage (Ragheb et al. 2019). Macrophage internalization of *Salmonella* cells forms non-growing persisters, where cells undergo massive stress and can potentially acquire beneficial mutations (Helaine et al. 2014). Gut inflammation in the mouse model of *Salmonella* infection causes SOS response and contributed to horizontal gene transfer (HGT), another mode of acquiring antibiotic resistance (Diard et al. 2017; Beaber et al. 2004). RpoS plays an essential role in maintaining retarded growth phase of persisters and simultaneously also can be involved in the selection of beta-lactam-resistant mutants (Gutierrez et al. 2013).

2.6   Different Techniques to Study Bacterial Persisters

The study and characterization of persister cells are challenging and debatable even after 76 years from their discovery, owing to their transient nature and reversibility of the physiological state. Moreover, they contribute to a small fraction of the whole population, making them difficult to isolate. Persister studies mainly involve characterizing different bacteria's growth behavior in the isogenic population and anticipating the molecular mechanism. It constitutes the in vitro and in vivo assays to visualize, isolate, and characterize them.

2.6.1   Time-Kill Curves

Assays to enumerate the persisters relied on their characteristic response to tolerate the antibiotic concentration up to the lethal dose for susceptible populations of the same genetic makeup. The bacterial cultures/cells are exposed to the bactericidal antibiotics for a short period (i.e., 2–6 h), and then growth dynamics are observed by colony-forming unit (CFU) (Keren et al. 2004; Hong et al. 2012). In such assays, the persister count may vary depending upon the type of antibiotics, age of cultures, phase at which antibiotic was administered, culture media, and antibiotic exposure time (Luidalepp et al. 2011). The number of cells is usually minimal, making it difficult to isolate and demanding to learn new techniques to isolate the persisters.

2.6.2   Single-Cell-Level Studies

1. Microfluidic with time-lapse microscopy.
   It is an advanced tool to study population heterogeneity in bacteria and has been used to demonstrate that the preexisting non-growers of the heterogeneous

subpopulation (Balaban et al. 2004). From this approach, we can isolate single cells and track their trajectory by time-lapse microscopy. The cells can be studied for a longer duration, even for months, at the same or different time points using this approach. It is generally composed of a microfluidic chamber containing culture media for the continuous flow of bacteria, and fluorescently labeled bacterial cells are tightly confined over the flow cell matrix. It has been extensively used to analyze growth dynamics and antibiotic killing and resuscitation of persister cells.

2. Scan lag

Scan lag, Start Growth Time (SGT) are a few alternatives of CFU estimation, developed to enumerate persister in a better way. Scan lag is an automated method that involves a repeated scan of Petri dishes over time and monitoring colony appearance through the lag phase's length. It has been used to detect the delayed growth and to measure the slow recovery of persisters. This can also be used to quantify the bacterial heterogeneity during the transition of the growth phase (Levin-Reisman et al. 2010; Fridman et al. 2014).

3. Fluorescent growth reporters.

They have been used to measure the persisters' growth dynamics and metabolic activity. GFP variant proteins are driven from the ribosomal promoter (rRNA promoter), which serves as an indicator of protein synthesis and is an alternative to persister isolation (Shah et al. 2006). Also, fluorescent indicators for cell division can distinguish between persister cells and other contaminated cell types, like VBNCs. More fluorescence-based techniques were developed, like the fluorescence dilution technique, to track bacterial divisions at the single-cell level based on the principle of carboxyfluorescein succinimidyl ester (CFSE) staining (Shah et al. 2006; Roostalu et al. 2008; Henry and Brynildsen 2016; Orman and Brynildsen 2013a).

4. FACS.

This is one of the most extensively used techniques to measure metabolic activity. It is an alternative approach that allows single-cell-level analysis of persisters in a high-throughput manner. There are many variations of FACS (Fluorescence-activated cell sorting); selection depends upon the type of fluorescent growth reporter used and the condition in which the experiment is conducted (Balaban et al. 2004; Rowe et al. 2016).

2.6.3 Population-Level Studies.

1. Genetic screen.

The classical genetic approach of mutational studies has identified multiple genes and pathways involved in persister generation and existence. However, different persister cells show overlapping mechanisms due to compensatory mutations and redundant pathways, limiting the use of this technique (Kim and Wood 2010). However, overexpression libraries are more reliable for identifying candidate persister genes (Scherrer and Moyed 1988; Zhang 2014).

2. Evolution.

An experimental evolutionary approach has been applied on multiple occasions to reveal resistant mutants from antibiotic-tolerant populations (Van den Bergh et al. 2016; Fridman et al. 2014).

3. Mathematical modeling.

Apart from the abovementioned techniques, in silico prediction using mathematical modeling can also allow predicting persister behavior and mechanism. For example, some models have been built in *B. subtilis* to explain the cryptic episodic selection for transiently non-growing population, which further maintains transformation and competency. It can also differentiate persistence from population-wide tolerance (Brauner et al. 2016; Balaban et al. 2004; Wilmaerts et al. 2019).

4. Transcriptomics.

Transcriptomics allows for tracking the global transcriptional pattern change, which helps to study the gene regulation in various persister formation and survival mechanisms. First, transcriptome profiling was done to identify several TA genes in *E. coli* through exposure to a high concentration of lytic antibiotic (ampicillin) and the selection of persisters followed by an expression profile, which was determined using microarray (Keren et al. 2004).

It has been used to show the decreased metabolism in persisters, interpreted from downregulation of metabolic genes in *M. tuberculosis* or *E. coli* (Keren et al. 2011; Prax and Bertram 2014).

To study the persisters in high number in vitro, it is possible to artificially generate persister cells by arresting protein synthesis (Manina et al. 2015), toxin production (Hong et al. 2012; Stapels et al. 2018), and pretreatment with stress such as hydrogen peroxide (Hong et al. 2012), but it has led to many of the erroneous interpretation in the field (Amato et al. 2013). While studying biphasic killing curves, experiments conducted with stressed cells but metabolically active cells often lead to false-positive results, so to avoid such errors, we should analyze the quiescent stage with more precision (Hong et al. 2012; Manina et al. 2015). Advanced microscopic technique such as TEM is more preferable to count the absolute viable cells than FACS because dead cells may still have an intact membrane without the cytosol (Dhar and McKinney 2010). While isolating persisters, both persisters and VBNC can get stained by propidium iodide (PI), exhibit metabolic activity, and are non-replicating during antibiotic stress during the growth. So, to differentiate between these two populations, FACS with rRNA reporter can be used, assuming that the persister will show a low level of protein synthesis, but that is not always the case. This issue was handled by persister-FACseq, which can quantify the physiology and heterogeneity of persisters and normal cells simultaneously without any need for persister isolation (Balaban et al. 2004; Henry and Brynildsen 2016).

2.6.4   In Vivo Models and their Relevance

The survival and response of persisters to antibiotics are dependent on and highly correlated with the host system's complexity. So, the mechanistic studies done for the persisters in in vitro conditions under the artificial environment fail to replicate the in vivo bacterial system and become clinically less relevant (Christensen et al. 2007). Moreover, persisters show retention of metabolic activity in in vivo models, unlike a completely dormant state in vitro (Allison et al. 2011a).

Besides that, the in vitro model fails to consider intercellular interactions, host environment, biomechanical cues, or the relevant physiological proteins in culture media. Because of these factors, there is a gap between in vitro and in vivo assay outcomes, limiting the therapeutic potential. Thus, to gain a deeper understanding of molecular mechanisms involved in persister formation in infection-relevant conditions, we have to develop and increase the use of in vivo models. So, in the table below, we try to summarize several in vivo models used in the persister study and the future approaches that might be more accurate to test the new theories regarding bacterial multidrug tolerance (Table 2.1).

2.6.5   Future Perspectives of in Vivo Techniques

Time-kill curves are labor-intensive and rarely used in clinical settings to quantify tolerance. So, the more formal system used is MDK calculation, which enables the classification of bacterial strains as tolerant, resistant, or persistent (Brauner et al. 2017) and can specify the treatment protocol. Tolerance disk test (TD test) can be used to measure the distinct levels of antibiotic tolerance in clinical isolates (Gefen et al. 2017). Single-cell sorting devices allow visual detection of surviving single cells using large-scale femtoliter droplet arrays (Iino et al. 2013). ^{13}C isotope labeling can provide insights into the intracellular carbon metabolism and its link to the expression of virulence genes (Eisenreich et al. 2010).

On the other hand, the development of high-throughput sequencing technologies with bioinformatic platforms has provided a great opportunity to study persisters more efficiently and precisely from the host perspective (Cabral et al. 2018). 16S rRNA profiling, community metagenomics, and RNA-Seq provide an excellent platform to study persister formation both in human patients and murine models. For example, the dual RNA-Seq approach is used to sequence simultaneously the transcriptomes of *Salmonella* persisters and the macrophage host in which it resides during infection (Westermann et al. 2017; Saliba et al. 2016). Further, to understand the critical biological mechanisms for persister formation, we can use the combined technology of in vivo fluorescent imaging and next-generation sequencing (Pu et al. 2016). Transposon sequencing (Tn-seq) can identify persister-associated gene targets under specific stress conditions in vivo. It can be achieved by animal infection with high-density transposon insertion libraries and Tn-seq, which further provides information on persister formation at the single organism level (van Opijnen et al. 2009). To study biofilm, we can use patient biopsy-derived biofilm

Table 2.1 Different techniques used for in vivo infection models to study persister

Bacterial species	The method used in the study	Remarks	References
Mouse			
M. Tuberculosis	• Microfluidic and time-lapse microscopy. • Mtb replication and ribosomal activity using fluorescent reporter strain rRNA-GFP (where a number of ribosomes will represent the growth of bacteria).	• Amplified bacterial phenotypic study. • Heterogeneity in the drug pressure and immunity (interferon-γ)-dependent persistence development. • Non-growing but metabolically active (NGMA) cells, a potential source of recurrent infection.	Manina et al. (2015); Stapels et al. (2018)
M. Tuberculosis	• Genetic screen (transposon mutagenesis) to identify persistence-impairing and persistence-enhancing mutations.	• Transposon mutations in *rv0096-rv0101* gene cluster. • Increase the lung persisters and tolerance to isoniazid (first-line drug).	Dhar and McKinney (2010)
P. aeruginosa	• Pre-colonized biomaterials (biofilm model). • Using mutant strains and quorum-sensing inhibitors.	• Importance of quorum sensing in biofilm persistence.	Christensen et al. (2007)
E. coli, *P. aeruginosa*, *L. monocytogenes*, *S. aureus*	• Non-persister cells killed with an antibiotic, resuspended in fresh media. • Checked for fluorescently labeled antibiotic uptake with carbon source by counting CFU. • Analyzed cells by FACS.	• Metabolite-induced uptake of aminoglycosides. • Improve the treatment of chronic urinary tract infection in the mouse model with increased aminoglycoside uptake.	Allison et al. (2011a)
Salmonella	• In cellular persister assays. • Growth rate reporter called TIMER. • Fast-growing bacteria were largely killed, and slow-growing or non-growing *Salmonella* were more likely to survive antibiotic treatment.	• A role for slow-growing or non-growing antibiotic persisters found within host immune cells.	Balaban et al. (2019)
S. aureus	• Deep-seated infection (biofilm formation). • Visualization – Gram staining and electron micrograph. • Antibiotic administration; CFU plating.	• Adep4 with rifampicin eradicates biofilm in a mouse model.	Conlon et al. (2013)

(continued)

Table 2.1 (continued)

Bacterial species	The method used in the study	Remarks	References
S. Typhimurium	• Single-cell growth reporter – TIMER. • Mouse typhoid fever model.	• Antibiotic killing correlated with single-cell division rates. • Selected extensive phenotypic variation of *Salmonella* in host tissues. • Slow growers are more sensitive to chemotherapy.	Claudi et al. (2014)
Guinea pig			
M. Tuberculosis	• Triple mutant strain of Mtb.	• MazF ribonucleases in Mtb stress adaptation. • Drug tolerance and virulence.	Tiwari et al. (2015)
Nematode – *C. elegans*			
E. coli, S. aureus, P. aeruginosa	• Antibiotic treatments enhanced the survival of the EHEC-infected worms. • MMC in an animal model and a wound model.	• Substantiating the clinical applicability of MMC against bacterial infections.	Kwan et al. (2015)

(as mentioned in the table) or use FISH (fluorescence in situ hybridization) labeling to visualize persisters' 3D context of biofilm (Brileya et al. 2014; Nistico et al. 2009). Also, LCMD (laser capture microdissection) sectioning followed by transcriptome profiling can help to study persisters harboring biofilm. The evolutionary approach of metagenomics and RNA-Seq in animal model infections over antibiotic therapy helps to study the selective pressure on persister genes (Henry and Brynildsen 2016; Orman and Brynildsen 2013b; MacGilvary and Tan 2018).

2.7 Persister Formation in Gram-Negative Bacteria

Gram-negative bacteria (GNB) are among the most significant public health concerns globally and pose significant challenges in the clinical control of many bacterial infections. They are known to have a wide range of mechanisms and strategies to form persisters and evade antibiotic treatment and persist in the host for a longer time, which results in treatment failure.

In this section, we will be discussing a few clinically important Gram-negative bacteria from the mechanism of persister formation as well as clinical importance point of view:

1. *Helicobacter pylori* have emerged as an important example of a persistent bacterial pathogen. Currently, its prevalence is around 40% of the US population and substantially higher in underdeveloped regions. It infects the stomach, and even after the chronic immune and inflammatory responses, *H. pylori* infection generally persists for the life of the host. It has been shown that *H. pylori* form intracellular persisters by inhibiting the NO production which is an essential part of innate immunity and thus subvert the adaptive immune response by blocking the antigen-dependent proliferation of T cells. Deletion mutant of *rocF* gene encodes for the arginase enzyme, responsible for NO production inhibition, and has been shown to have lesser drug tolerance in the ex vivo model (Gobert et al. 2001). *H. pylori* also have been shown to activate type I TA system *aapA1/isoA1* under stress and lead to growth inhibition and possibly persister formation during infection (Arnion et al. 2017).

2. *Salmonella enterica* ser. *Typhimurium* is responsible for many human diseases ranging from gastroenteritis to systemic infections. *Salmonella enterica* is a facultative intracellular pathogen that can replicate inside the macrophages and actively modulate the outer environment. Helaine and colleagues have previously shown that TA systems can contribute to macrophage-induced non-replicating persisters in *Salmonella*, tolerant against gentamicin and cefotaxime (Helaine et al. 2014). Another article has shown that bacteria deployed a specialized type III secretory system called SPI-2 to deliver virulence factors, including SteE, into host cells. SteE changed the cytokine profile of the infected macrophages to reprogram them into a noninflammatory and infection-permissive state where *Salmonella* persisters can survive for a long time and later reemerge and cause disease when antibiotic treatment stops (Stapels et al. 2018). Later, Cheverton and colleagues reported that the TacT toxin in *Salmonella* promotes non-growing persisters inside macrophages by acetylation of tRNA and subsequent translation inhibition (Cheverton et al. 2016). *Salmonella typhimurium*'s antibiotic tolerance was found to be increased in response to the bacterial signaling molecule indole (Vega et al. 2013). Even though *S. typhimurium* does not natively produce indole, these results suggest that this intestinal pathogen can benefit from indole signaling by *E. coli* and other commensal bacteria.

3. *Burkholderia pseudomallei* is the primary cause of melioidosis infection endemic in Southeast Asia and northern Australia. A major concern in the management of disease caused by these organisms is that an aggressive antibiotic regimen is required, and even after that, the recurrence rate is very common due to the persisters. In the stationary phase, cells overexpress HicA toxin, which is responsible for growth arrest and increases the persister population. Conversely, the deletion of mutant *hicAB* locus forms fewer persisters to ciprofloxacin (Butt et al. 2014). Similarly, in the case of *Burkholderia cenocepacia*, the role of TA modules was investigated by plasmid overexpression, and it was found that 7 out of 13 TA modules contribute to persister formation and the same toxins are also shown to have upregulated in the stationary phase (Van Acker et al. 2014).

4. *Vibrio cholerae* continues to be a major public health threat, particularly in countries where safe drinking water, sanitation, and hygiene are suboptimal. The mechanism underlying the ability to survive as a persister in nutrient-poor environments is largely unknown. Several morphological studies of these persisters are done which is demonstrated by their small size with a higher tendency of aggregation, which are often involved in excessive flagella production. Most importantly, the morphological reversion of these CAI-1 signaling-driven persister cells to WT happened when transferred to a nutrient-rich environment (Jubair et al. 2012). In a separate report, it has been shown that ciprofloxacin efficacy can be increased by glucose supplementation-driven sensitization of non-growing *V. cholerae* persister cells (Paranjape and Shashidhar 2020).

5. *Pseudomonas aeruginosa* is a part of the normal gut flora where the carriage rate is generally low, but in an immunocompromised state, it becomes higher and shows an intrinsic reduced susceptibility to several antibacterial agents, as well as a propensity to become persistent. *P. aeruginosa* persisters contribute heavily to the case of cystic fibrosis patients. Some genetic determinants have been linked to this drug tolerance phenomenon. One of the genes responsible is *dnpA* (de-*N*-acetylase) which is involved in non-inherited fluoroquinolone tolerance. *dnpA* deletion mutant is shown to form less persisters in both planktonic and biofilm growth models, whereas conditional overexpression of *dnpA* in the wild-type background forms more persisters (Liebens et al. 2014). A separate study indicated *hip* mutants, linked to MexB efflux pump, contributed to persister formation in cystic fibrosis clinical isolates of *P. aeruginosa* (Mulcahy et al. 2010). *P. aeruginosa* has also been shown to modulate persister fraction by *carB*-driven regulation of intracellular ATP concentration (Cameron et al. 2018), which used extensive transposon mutagenesis screening and revealed nine novel genes involved in *P. aeruginosa* persister formation against ofloxacin (De Groote et al. 2009).

6. *Acinetobacter baumannii* is a leading cause of hospital- and community-acquired infections that can rapidly acquire diverse mechanisms and undergo genetic modifications that confer persistence. In the recently published Tigecycline Evaluation and Surveillance Trial (TEST) data, 44% of *A. baumannii* shows multidrug-resistant (MDR) characteristics, which is the highest rate among Gram-negative pathogens. In polymicrobial infections, the production of quorum-sensing molecule 2-aminoacetophenone by *Pseudomonas aeruginosa* promotes intraspecies antibiotic tolerance (Que et al. 2013).

7. In *Sinorhizobium meliloti*, it has been shown that during starvation, they form two discrete cell types. The old-pole daughter cell retains most of the resource, polyhydroxybutyrate (PHB), and behaves similarly to persisters, which is characterized by metabolic dormancy and antibiotic tolerance. In contrast, the low-PHB, new-pole daughter cell is capable of quickly resuming growth when the starvation period ends (Ratcliff and Denison 2011).

2.8   Persister Formation in Gram-Positive Bacteria

Gram-positive bacteria are among the most common causes of clinical infection. This is due to their association with a diverse spectrum of pathology, ranging from mild skin and soft tissue infections (SSTIs) to life-threatening systemic sepsis and meningitis. Although a few antimicrobial agents already exist to treat such diseases, emerging issues such as persistence make them ineffective; hence, it is imperative to understand the mechanism behind the formation of such dormant variants.

1. *Bacillus subtilis*—In the stationary phase, they differentiate stochastically and become persisters which are associated with decreased growth rate and relatively less fitness, but when exposed to antibiotics inhibiting cell wall synthesis, transcription, and translation, they show a selective advantage of multidrug tolerance and recurrent infection (Liu et al. 2018a; Yuksel et al. 2016).
2. *Enterococcus* sp.—Enterococci have emerged as critical healthcare-associated pathogens during the last two decades. *Enterococcus faecium* and *Enterococcus faecalis* are the most clinically relevant pathogens. *E. faecalis* possess a mechanism to detoxify host-induced ROS and at the same time produce ROS for growth check; thus, they are able to survive as non-growing persisters inside the host (Portela et al. 2014). Proteomic studies have shown that *E. faecalis* produces approximately 200 different stress proteins involved in growth regulation when growing under different physical and chemical stresses (Giard et al. 2001). The clinical isolate of a relA mutant strain of *Enterococcus faecium* with constitutively high ppGpp concentration has contributed to an elevated level of persisters against linezolid and daptomycin (Honsa et al. 2017).
3. *Staphylococcus epidermidis*—*Staphylococcus epidermidis* is a frequent cause of healthcare-associated infections, which is attributed primarily due to its biofilm formation on the surfaces of implanted medical devices. Several studies have been done, and it's shown that in *S. epidermidis*, the highest percentages of persisters were obtained during the stationary phase in planktonic cultures and biofilm setup (Shapiro et al. 2011).
4. *Staphylococcus aureus*, a member of the ESKAPE pathogen group, is a major threat to public health due to the emergence of persistent and multiple drug-resistant strains. *S. aureus* strains have shown to be equipped with specific toxins in defense against ROS and thus exist as dormant variants inside the host (Foster 2005). *S. aureus* have shown to form persisters by reducing intracellular ATP concentration (Conlon et al. 2016). Another recent report described the role of *mazEF* TA module in the formation of antibiotic-tolerant cells and biofilms in *S. aureus* (Ma et al. 2019).
5. *Streptococcus mutans* are known to form an increased number of persisters by ectopic expression of type II toxin-antitoxin systems, MazEF and RelBE (Leung and Levesque 2012). Previously, the same group validated the role of the quorum-sensing peptide CSP pheromone, in the formation of stress-induced multidrug-tolerant persisters (Perry et al. 2009).

6. *Listeria monocytogenes* cause listeriosis, which has many successful treatment options, but their mortality is also very high; hence, understanding the response of *L. monocytogenes* to antibiotic exposure is critical to ensure better treatment. It has been recently discovered that *L. monocytogenes* can form persisters against nisin in in vitro stationary-phase culture (Wu et al. 2017).

7. *Clostridium difficile* caused *C. difficile* infection (CDI) that is a major healthcare-associated problem. The ability of these bacteria to form biofilm-harboring persister cells is often associated with recurring infection. It has been shown that the virulence-associated protein Cwp84 and a known quorum-sensing regulator LuxS are responsible for biofilm formation, whereas transcription factor *spoOA* mutant is known with impaired biofilm formation (Ethapa et al. 2013).

8. *Streptococcus pyogenes* is a causative agent for a wide range of diseases from mild infections, such as impetigo, to severe diseases like toxic shock syndrome. During the infection, they can undergo phenotypic switching which includes losing their cell wall and virulence factors and subsequently entering into a quiescent state under stress (Wood et al. 2005).

2.9 Persister Formation in *Mycobacteria*

Mycobacteria are among the first microbes to be discovered, and over the years, they have been intensely studied. Most individuals resolve infection soon after the onset of adaptive immunity, but in some cases, the infection is never cleared by the immune response which leads to the long-term infection; hence, the treatment duration becomes lengthy. For example, in *Mtb*, the therapy can require up to 12 months, whereas, in NTM, such as *Mycobacterium avium* complex (MAC), it can last up to 2 years of treatment. Understanding the molecular and cellular biology of persister formation in *mycobacteria* is critical to find new drugs to shorten the treatment.

In *Mycobacterium leprae*, there are some random reports regarding antibiotic persistence, but in the case of tuberculosis, it has been studied in real depth.

Mycobacterium tuberculosis causes infection in one-third of the human population resulting in tuberculosis. Around 90% of the cases are latent, with no symptoms, but they remain a point of concern due to their potential for reactivation. They can form cells of different sizes in different growth phases, and this variation is also involved in persister formation. LamA plays an important role in creating asymmetry during cell division by inhibiting the growth of nascent poles and is highly conserved among mycobacterial species. It's been shown that deletion of LamA in *M. tuberculosis* gives rise to both a more homogeneous population and a more uniform killing (Rego et al. 2017). In another study, HupB, a nucleoid-associated protein, is found to be important, and its deletion leads to the loss of both size variants and phenotypically drug-tolerant subpopulation (Sakatos et al. 2018). The survival of this pathogen in the host is generally associated with cell wall thickening, and these characteristics are also shared by persisters (Kieser and Rubin 2014). The accumulation of polyphosphate in the cell wall decreases the cell wall permeability for antibiotics and other hydrophilic compounds (Chuang et al. 2015). Similarly, mutations in genes that are involved in phospholipid biosynthesis (*plsB2* and *cdh*) are detected in the persister population (Torrey et al. 2016). Increasing the error rates in other housekeeping mechanisms can also contribute to the formation of persisters. A higher mistranslation rate of glutamine and asparagine in RNA polymerase is associated with increased tolerance to rifampicin (Javid et al. 2014). PhoY proteins in *M. tuberculosis* have been shown to promote persister formation by phosphate sensing (Namugenyi et al. 2017). Using *M. marinum* as a model system to understand mycobacterial pathogenesis, it's shown that persistent *mycobacteria* exist in a metabolically and replicative active state (Chan et al. 2002).

2.10 Anti-Persister Strategies

The first anti-persister strategy, which was explained by Bigger in 1944, involved pulse dosing of *S. aureus* cells with penicillin (Bigger 1944) and is still relevant. Furthermore, this has been validated by mathematical modeling and in vitro experiments in other bacterial strains (Lewis 2008; Sharma et al. 2015; De Leenheer and Cogan 2009; Meyer et al. 2020). However, drug tolerance and further resistance development stemming from persister subpopulation remain a major setback in clinical scenarios (Fauvart et al. 2011). The following section classifies anti-persister strategies broadly into ones that inhibit the formation of persister cells and others that kill existing persister cells, albeit strict distinction between these strategies is hard to make.

2.10.1 Inhibiting Formation of Persisters

The global cellular stress pathways, which are important for persister generation, can be a rational answer to the recalcitrant chronic infections. This can be achieved through targeting either specific determinant or nonspecific determinant (Sect. 2.3) factors responsible for persister cell formation. Hindering the formation by using inhibitors against (p)ppGpp enzymes (Fung et al. 2010), quorum sensing (He et al. 2012), efflux pumps (Kim and Wood 2010; Norton and Mulvey 2012), TA modules (Norton and Mulvey 2012), etc. is observed by various groups. An example of the former is the use of a combination of ampicillin, kanamycin, and Hipa7 inhibitors, which has shown to decrease persister levels in *E. coli* (Li et al. 2016), while impairing persister formation in *E. coli* by limiting protein aggregates with the addition of osmolytes is an example of the latter (Leszczynska et al. 2013). Apart from this, targeting stationary-phase respiration (Orman and Brynildsen 2015) and trans-translation (Shi et al. 2011) has also proven to be an effective way to block persister formation.

2.10.2 Direct Killing of Persister Cells

Due to the slightly varied mechanism of persisters, which involves slowing down cellular mechanisms to increase tolerance to drugs, one of the most effective anti-persister strategies is to target growth-independent regions yet indispensable to bacterial survival like cell membrane or DNA. Generally, these can be categorized as target dependent or target independent based on their modus operandi.

2.10.2.1 Target Independent

This involves using electrostatic methods such as membrane-penetrating cationic peptides to cause membrane disruption and cell wall death, thereby eliminating *E. coli* cells, among many others (Lei et al. 2019). HT61 is a quinolone-like compound that can interrupt the membrane integrity and clear the MRSA isolates without any toxicity. This compound and others, similar to targeting cells' membrane integrity, have reached clinical trials (EU Clinical Trials Register: 2009–017398-39) (Amison et al. 2020). Apart from this, physical methods like direct current (Niepa et al. 2012; Khan et al. 2020), UV, small molecules (Wang et al. 2013), antimicrobial peptides (Chen et al. 2011), and phages (Defraine et al. 2016) can alter the membrane potential and show bactericidal effects to eliminate persisters as well as in sensitization of cells by disrupting the physical integrity of the bacterial cell. AMPs like defensins and cathelicidins can kill bacteria by disrupting their lipid bilayer. Nevertheless, they have the disadvantage of being toxic to the host membrane and exhibit a high manufacturing cost (Marr et al. 2006).

2.10.2.2 Target Dependent

This approach involves targeting essential enzymes, genes regulating persister formation, and metabolic pathways essential for bacterial virulence and survival. For example, *phoU* gene, a regulator of *relA/spoT* metabolism upon deletion, showed increased susceptibility to antibiotics like ampicillin, gentamicin, and norfloxacin (Li and Zhang 2007). Antimicrobial peptides like peptide 1018 (Reffuveille et al. 2014), peptide SAAP-148 (de Breij et al. 2018), P5, and P9 (Li et al. 2019) have shown the potential to block (p)ppGpp synthesis and cause biofilm inhibition and thus treat MRSA successfully. Acyldepsipeptide (ADEP4) binds the ClpP protease and promotes nonspecific protein degradation, which further can eradicate chronic biofilms of both the osteomyelitis-associated strain UAMS-1 and *Staphylococcus aureus* (Li et al. 2019).

There have been several successful studies for an effective therapeutic strategy to eliminate persisters. Some of them are discussed in the following section.

1. Potentiation of already existing antibiotics happens when conventional antibiotics are combined with adjuvants or drug carriers like an antibody, bioactive compounds, polymers, lipids, nanoparticles, etc., to increase antibiotics' killing capacity (Kalan and Wright 2011; Grassi et al. 2017; Kim et al. 2017; Maiden et al. 2018). These provide a controllable and stable delivery system for the drugs and hence enhance its killing capacity (Koo et al. 2017).

Recently, an antibody-antibiotic conjugate platform has been established to potentiate drugs against the intracellular pathogenic bacteria (Lehar et al. 2015) where a specific antibody provides the drug an opportunity to attack the niche of a bacteria. Antimicrobial peptides (AMPs) can be also used as anti-persister adjuvants. For example, conjugation of tobramycin to the AMP upgrades the aminoglycoside antibiotic to a self-transporting system, which further increases persister killing in *E. coli* and *S. aureus* (Schmidt et al. 2014). Colistin with another antimicrobial peptide, amikacin, and gentamycin can kill *A. baumannii* and *E. coli* persister cells, respectively, by altering the cell membrane potential (Tang et al. 2019). Pentobra (Schmidt et al. 2015) and P14KanS (Mohamed et al. 2017) are a few successful adjuvants which open avenues to develop more and more combination of strategies with the drug carriers. Moreover, based on the synergistic property of antibiotics, the combined use of two or more antibiotics has an added advantage while dealing with polymicrobial infections. Combining cefoperazone or cefuroxime and doxycycline drug against growing culture with potent drug daptomycin or daunomycin for non-growing culture proves to be the most successful cocktail against *B. burgdorferi* persister biofilms (Khan et al. 2020; Feng et al. 2016). For *M. tuberculosis*, augmentin (containing amoxicillin) when given in combination with clavulanate, a β-lactamase inhibitor, increases the activity of β-lactam antibiotic clear persister (Li et al. 2018). Similarly, triclosan, when combined with tobramycin, can inhibit persister cell growth by suppressing RNA polymerase synthesis in *P. aeruginosa* (Chen et al. 2011).

2. Awakening of persister cells.

The transition of persister cells to an antibiotic-susceptible state can be triggered by specific types of sugars (glucose, fructose, mannitol, etc.), amino acids (L-Arg and L-Lys) (Deng et al. 2020), and fatty acids (cis-2-decenoic acids) (Marques et al. 2014) (see Table 2.2). Sugars like mannitol and fructose depend on increasing the proton motive force in cells like *E. coli* and *P. aeruginosa* to induce uptake of aminoglycosides that ultimately results in microbial death (Allison et al. 2011b). Persister sensitization may occur independently of proton motive force such as fluoroquinolones, which work by generating reactive oxygen species (Duan et al. 2016). Daptomycin, which is a lipopeptide, can convert aminoglycoside into lytic antibiotic in *S. aureus*. In *P. aeruginosa*, quorum-sensing inhibitor BF8 can sensitize the persister using the additional targets without inhibiting the quorum sensing (Pan et al. 2012; Pan et al. 2013). Anti-TB drugs like PA-824 also work by the same mechanism by inducing reactive nitrogen species to reduce persister levels (Nuermberger et al. 2006; Zhang et al. 2012). Despite various convincing outcomes, a deeper understanding of persister metabolism is essential in designing an effective strategy for this category.

2.10.2.3 Repurposing of Drugs.

Repurposing of some existing drugs, which are already in use to treat some other diseases, is a viable option to treat persisters. Whole-cell screening-based approaches might pick some ideal candidates which show well in vitro activity against slow growth models (Kumar et al. 2019). They could be some anticancer

Table 2.2 Some effective anti-persister strategies against some clinically relevant pathogens, which are already discussed in Sect. 2.7

Target bacteria	Compound/category	Remarks	Clinical relevance	References
Gram-negative bacteria				
P. aeruginosa	• Cadaverine. • Drug adjuvant. • Known target.	• Increased carbenicillin and ticarcillin killing. • Loss of lysine decarboxylase activity.		Manuel et al. (2010)
	• ECCP (electrochemical control of persisters).	• Synergistic effects exist, e.g., with tobramycin. • Effective against both planktonic persisters and biofilms.	• Potential application with medical implants.	Niepa et al. (2012)
	• Antifungal. • 5-Fluorocytosine	• Inhibition of virulence factor production.	• Suppresses pathogenicity in a mouse model of lung infection.	Imperi et al. (2013b)
	• Antihelminthic – Niclosamide.	• Inhibition of QS systems.	• Reduce pathogenicity in an insect model.	Imperi et al. (2013a)
	• Antibiotic – Azithromycin.	• Inhibition of QS systems.	• Tested in clinical settings – Efficacy in the treatment of chronic pulmonary infections.	Imperi et al. (2014)
	• Anti-virulence – Pentetic acid.	• Reduction of virulence factor production.	• Effective to alleviate mouse airway infection.	Gi et al. (2014)
Gram-positive bacteria				
VRE, S. epidermidis	Cocktail of antibiotics tedizolid, telavancin, dalbavancin, and oritavancin	• More prolonged exposure to vancomycin and ciprofloxacin. Inhibit biofilm persister.	• Effective against *MRSE (methicillin-resistant Staphylococcus epidermidis)*.	O'Driscoll and Crank (2015)
Staphylococcus aureus	• CD437 and CD1530 (retinoids (vitamin A	• Induced rapid membrane permeabilization.	• Completely eradicate stationary-phase persister cells.	Kim et al. (2018)

Organism				
	analogs))/target-independent direct killing. • NH125 (1-hexadecyl-2-methyl-3-(phenylmethyl)-1H-imidazolium iodide.	• Target – Histidine kinase, WalK. • Showed relatively low selectivity between bacterial and mammalian membrane.	• High membrane selectivity. • No hemolytic activity. • Killed over 99% of MRSA biofilm persisters. • Benzalkonium chloride, a widely used topical antiseptic and an eye drop preservative. • Potential as a topical antimicrobial to treat chronic skin infections.	Wang et al. (2013), Liu et al. (2018b)
	• Target dependent. • ADEP4 – Direct killing.	• Targeting protease – clpP. • Nonspecific protein degradation.	• ADEP4 with rifampicin eradicates biofilm in a mouse model.	Jarchum (2014)
	• Anti-virulence: Diflunisal. • Toxin suppression.	• Small molecule: Biaryl compounds. • Fused/separated by a short linker. • Useful in prophylaxis and as adjuvants in antibiotic therapy.	• Against MRSA. • USA300 FDA-approved nonsteroidal anti-inflammatory.	Khodaverdian et al. (2013)
Listeria monocytogenes	• Curvacin. • Nitrofurantoin.	• Inhibit biofilm formation.		Komp Lindgren et al. (2015)
M. Tuberculosis	• Target dependent. • TCA1 – An adjuvant combination with rifampicin or isoniazid.	• Target – Decaprenyl-phosphoryl-β-D-ribofuranose oxidoreductase. • DprE1 and MoeW enzymes involved in the cell wall and molybdenum cofactor biosynthesis.	• Efficacious in mouse models.	Wang et al. (2013)
	• Suramin.	• Inhibit RecA/SOS.		Nautiyal et al. (2014)

(continued)

Table 2.2 (continued)

Target bacteria	Compound/category	Remarks	Clinical relevance	References
	• PA-824 – Nitroimidazole prodrug.	• It is activated by a reductase resulting in the generation of reactive nitrogen species like nitric oxide.		Singh et al. (2008)
	• Rhodamine.	• Inhibit dihydrolipoamide acyltransferase.		Bryk et al. (2008)
Broad spectrum				
Gram (+) and gram (−) bacterial species	• Anticancer drug – 5-fluorouracil.	• Treatment of actinic keratosis and Bowen's disease. • Inhibition of biofilm formation, repression of QS in *P. aeruginosa*. • Growth inhibition.	• Successful in clinical trials in humans, as an external coating of central venous catheters.	Walz et al. (2010), Ueda et al. (2009)
P. aeruginosa *A. baumannii* *M. Tuberculosis*	• Gallium compounds: Gallium nitrate.	• Treatment of hypercalcemia of malignancy.	• Effective against acute and chronic mouse infections.	Hijazi et al. (2018), DeLeon et al. (2009), de Leseleuc et al. (2012), Kaneko et al. (2007)
E. coli *S. aureus* *P. aeruginosa*	• Awakening of persisters. • Basic amino acids (L–Arg and L–Lys).	• Potentiate aminoglycosides by modifying the membrane, pH gradient.	• In vitro. • In biofilms. • In vivo in a mouse model.	Van den Bergh et al. (2017)
E. coli *E. tarda* *L. Monocytogenes* *P. aeruginosa* *S. aureus*	• Mannitol and other sugars.	• Generation of proton motive force. • Improved uptake of aminoglycosides.	• Effective in biofilm and in chronic mouse urinary infection model.	Allison et al. (2011a), Peng et al. (2015), Barraud et al. (2013), Meena et al. (2015)
A. baumannii, P. aeruginosa	• Artilysin®. • Art-175; chimeric product – KZ144	• Kills by osmotic rupture.	• No resistance development or hemolysis. • No cross-resistance.	Defraine et al. (2016)

	endolysin and lipopolysaccharide.		
E. coli, P. aeruginosa	• Persister awakening: Cis-2-decenoic acid.	• Fatty acid signaling molecule that converts persisters to a metabolically active state.	Marques et al. (2014)
Pseudomonas aeruginosa, Acinetobacter baumannii	• ZY4 – Cyclic peptide.	• Permeabilizing bacterial membrane.	• High stability in vivo. • Therapeutic potentials in MDR.	Mwangi et al. (2019)
Listeria monocytogenes, Staphylococcus aureus, methicillin-resistant Staphylococcus aureus	• Target independent direct killing. • Enterocidin B3A–B3B (bacteriocin).	• Membrane permeabilization consecutively to pore formation.	• Impeded biofilm formation. • MRSA.	Al-Seraih et al. (2017)
S. aureus S. epidermidis	• An enzyme isolated from bacteriophage: LysH5.	• Inhibit biofilm formation. • Kill persister cells.		Yang et al. (2015)
E. coli, S aureus	• Hypotonic treatment. • Proton motive force. • Independent improvement of aminoglycoside killing.	• Probably by increased uptake via mechanosensitive channels.		Jiafeng et al. (2015)
E. coli, S. aureus	• Pentobra: Tobramycin with 12-amino-acid-long peptide	• Membrane crossing and killing.	• Non-cytotoxic to eukaryotic cell.	Schmidt et al. (2014)
MRSA (*S. aureus*) MDR (*A. baumannii*)	• SAAPs (synthetic antimicrobial and antibiofilm peptides).	• Membrane killing. • Effective against an *E. coli* isolates resistant to colistin.	• SAAP-148 kills ESKAPE pathogens without resistance selection. • Long-term exposure: No bacterial resistance. • SAAP-148 ointments are highly effective against	de Breij et al. (2018)

(continued)

Table 2.2 (continued)

Target bacteria	Compound/category	Remarks	Clinical relevance	References
S. aureus *M. Tuberculosis*	• Antihistamine – Terfenadine.	• Studying the structure-activity relationship (SAR) of *S. aureus* antimicrobial activity. • Inhibition of the bacterial type II topoisomerases.	biofilm-associated skin infections. • Clinical use of its active metabolite fexofenadine (Allegra) is discontinued as it can cause cardiac arrhythmia, attributed to. • Due to inhibition of the human-related gene (hERG) potassium channel. • But can improve by SAR.	Perlmutter et al. (2014)
Mycobacterium tuberculosis, Streptococcus pyogenes, Streptococcus agalactiae, Propionibacterium acnes	• HT61 (analog of quinolone-like compounds).	• The depolarization and physical disruption of the bacterial cell membrane.	• Hydrophobicity and toxicity – Limited to topical applications. • Is in a clinical trial. • 1% HT61 in gel formulation – Against MRSA in a mouse skin infection model • *103 clinical MRSA isolates.*	Amison et al. (2020), Hubbard et al. (2017)
MRSA/MRSE, VRE, M. tuberculosis	• Halogenated phenazines.	• Inspired on 2-bromo-1-hydroxyphenazine that was identified from marine sources.	• *Most potent antibiofilm agents will not do hemolysis.*	Garrison et al. (2015)
Anti-mycobacteria	• Zafirlukast.	• Treatment of asthma.	• Not yet tested.	Pinault et al. (2013)
E. coli, M. tuberculosis	• Pyrazinamide, KKL-35.	• *Trans*-translation rescues stalled ribosomes from toxic protein products.		Shi et al. (2011), Mukherjee et al. (2016), Ramadoss et al. (2013)
M. bovis, M. tuberculosis	• AM-0016 – Amphiphilic derivative of α-mangostin.	• Induce cell envelope damage and rapidly collapsing membrane potential.	• The low propensity for resistance development.	Mukherjee et al. (2016)

(Chowdhury et al. 2016), anti-inflammatory drugs (Jiang et al. 2020), and antihelminthic (Imperi et al. 2013a), which are already in the market. As they are already proved to be safe for human use, any lead compound will possess the potential to save resources and time (Rangel-Vega et al. 2015).

2.10.2.4 Novel Approaches

Phage therapy is an emerging novel approach as the anti-persister strategy. For instance, M13mp18 phage in *E. coli* and Sb-1 in *S. aureus*, respectively, can effectively kill persister cells (Tkhilaishvili et al. 2018; Lu and Collins 2009; Tkhilaishvili et al. 2020). Recently, it has also been seen to be effective against dual-species biofilm of *S. aureus/P. aeruginosa*, when combined with ciprofloxacin (Tkhilaishvili et al. 2020). Phage can lyse/kill bacterial cells by activating the expression of the lethal genes. Nevertheless, over time, due to evolutionary pressure, bacterial hosts can develop resistance to the lethal phage, so in such cases, we can engineer phages to target non-essential genes or those not directly targeted by the antibiotics (Lu and Collins 2009).

2.11   Conclusion and Perspectives

Here in this book chapter, we try to capture different aspects of bacterial persisters, which is an ever-evolving field of research with significant clinical importance. Different sections of the book chapter are articulated in such a way that readers can have a comprehensive view of all the research done so far as well as the future challenges. We have highlighted the basic concept of persister formation, the complex mechanism associated with it, and the technical limitations to study this complex characteristic. Next, we have discussed the clinical complications of recurrent infections due to the formation of drug-tolerant persisters for several important pathogens. As we know, persister formation is a stepping stone of evolution for virulent strains to become antibiotic resistant, and there is an utmost need to design novel treatment strategies to eliminate persisters specifically. Containing the persisters and biofilm formation in the first place and effectively killing them will be a significant advancement toward defeating multidrug resistance (MDR), which is a huge threat to human civilization. As of now, the persister research community has successfully gathered vast in vitro data for specific mechanisms, but that knowledge does not always translate well in in vivo models from a treatment point of view. Despite so many years of effort to design effective anti-persister treatment strategies, there is no single anti-persister drug that has passed clinical trial so far. Hence, the standard route of antimicrobial drug designing based on target needs to be re-visited, and screening of the potential inhibitors should be done more effectively in relevant non-replicating and slow growth infection models for a better outcome.

References

Adams KN et al (2011) Drug tolerance in replicating mycobacteria mediated by a macrophage-induced efflux mechanism. Cell 145:39–53

Allison KR, Brynildsen MP, Collins JJ (2011a) Heterogeneous bacterial persisters and engineering approaches to eliminate them. Curr Opin Microbiol 14:593–598

Allison KR, Brynildsen MP, Collins JJ (2011b) Metabolite-enabled eradication of bacterial persisters by aminoglycosides. Nature 473:216–220

Al-Seraih A et al (2017) Enterocin B3A-B3B produced by LAB collected from infant faeces: potential utilization in the food industry for *Listeria monocytogenes* biofilm management. Antonie Van Leeuwenhoek 110:205–219

Amato SM, Orman MA, Brynildsen MP (2013) Metabolic control of persister formation in *Escherichia coli*. Mol Cell 50:475–487

Amison RT et al (2020) The small quinolone derived compound HT61 enhances the effect of tobramycin against *Pseudomonas aeruginosa* in vitro and in vivo. Pulm Pharmacol Ther 61:101884

Arnion H et al (2017) Mechanistic insights into type I toxin antitoxin systems in *Helicobacter pylori*: the importance of mRNA folding in controlling toxin expression. Nucleic Acids Res 45:4782–4795

Balaban NQ, Merrin J, Chait R, Kowalik L, Leibler S (2004) Bacterial persistence as a phenotypic switch. Science 305:1622–1625

Balaban NQ et al (2019) Definitions and guidelines for research on antibiotic persistence. Nat Rev Microbiol 17:441–448

Barak Z, Gallant J, Lindsley D, Kwieciszewki B, Heidel D (1996) Enhanced ribosome frameshifting in stationary phase cells. J Mol Biol 263:140–148

Barraud N, Buson A, Jarolimek W, Rice SA (2013) Mannitol enhances antibiotic sensitivity of persister bacteria in *Pseudomonas aeruginosa* biofilms. PLoS One 8:e84220

Battesti A, Bouveret E (2006) Acyl carrier protein/SpoT interaction, the switch linking SpoT-dependent stress response to fatty acid metabolism. Mol Microbiol 62:1048–1063

Beaber JW, Hochhut B, Waldor MK (2004) SOS response promotes horizontal dissemination of antibiotic resistance genes. Nature 427:72–74

Berghoff BA, Hoekzema M, Aulbach L, Wagner EG (2017) Two regulatory RNA elements affect TisB-dependent depolarization and persister formation. Mol Microbiol 103:1020–1033

Bigger J (1944) Treatment of *Staphylococcal* infections with penicillin by intermittent sterilisation. The Lancet 244:497–500

Bougdour A, Cunning C, Baptiste PJ, Elliott T, Gottesman S (2008) Multiple pathways for regulation of sigmaS (RpoS) stability in *Escherichia coli* via the action of multiple anti-adaptors. Mol Microbiol 68:298–313

Boutte CC, Crosson S (2013) Bacterial lifestyle shapes stringent response activation. Trends Microbiol 21:174–180

Brauner A, Fridman O, Gefen O, Balaban NQ (2016) Distinguishing between resistance, tolerance and persistence to antibiotic treatment. Nat Rev Microbiol 14:320–330

Brauner A, Shoresh N, Fridman O, Balaban NQ (2017) An experimental framework for quantifying bacterial tolerance. Biophys J 112:2664–2671

Brileya KA, Camilleri LB, Fields MW (2014) 3D-fluorescence in situ hybridization of intact, anaerobic biofilm. Methods Mol Biol 1151:189–197

Bryk R et al (2008) Selective killing of nonreplicating mycobacteria. Cell Host Microbe 3:137–145

Butt A et al (2014) The HicA toxin from *Burkholderia pseudomallei* has a role in persister cell formation. Biochem J 459:333–344

Cabral DJ, Wurster JI, Belenky P (2018) Antibiotic persistence as a metabolic adaptation: stress, metabolism, the host, and new directions. Pharmaceuticals (Basel) 11(**1**):**14**

Cameron DR, Shan Y, Zalis EA, Isabella V, Lewis K (2018) A genetic determinant of Persister cell formation in bacterial pathogens. J Bacteriol 200(**17**):**e00303–e00318**

Chan K et al (2002) Complex pattern of *Mycobacterium marinum* gene expression during long-term granulomatous infection. Proc Natl Acad Sci U S A 99:3920–3925

Chen X, Zhang M, Zhou C, Kallenbach NR, Ren D (2011) Control of bacterial persister cells by Trp/Arg-containing antimicrobial peptides. Appl Environ Microbiol 77:4878–4885

Cheverton AM et al (2016) A *Salmonella* toxin promotes Persister formation through acetylation of tRNA. Mol Cell 63:86–96

Chowdhury N, Wood TL, Martinez-Vazquez M, Garcia-Contreras R, Wood TK (2016) DNA-crosslinker cisplatin eradicates bacterial persister cells. Biotechnol Bioeng 113:1984–1992

Christensen LD et al (2007) Impact of *Pseudomonas aeruginosa* quorum sensing on biofilm persistence in an in vivo intraperitoneal foreign-body infection model. Microbiology 153:2312–2320

Chuang YM et al (2015) Deficiency of the novel exopolyphosphatase Rv1026/PPX2 leads to metabolic downshift and altered cell wall permeability in *Mycobacterium tuberculosis*. mBio 6:e02428

Chubukov V, Sauer U (2014) Environmental dependence of stationary-phase metabolism in *Bacillus subtilis* and *Escherichia coli*. Appl Environ Microbiol 80:2901–2909

Claudi B et al (2014) Phenotypic variation of *Salmonella* in host tissues delays eradication by antimicrobial chemotherapy. Cell 158:722–733

Cohen NR, Lobritz MA, Collins JJ (2013) Microbial persistence and the road to drug resistance. Cell Host Microbe 13:632–642

Conlon BP et al (2013) Activated ClpP kills persisters and eradicates a chronic biofilm infection. Nature 503:365–370

Conlon BP et al (2016) Persister formation in *Staphylococcus aureus* is associated with ATP depletion. Nat Microbiol 1:**16051**

Corona F, Martinez JL (2013) Phenotypic resistance to antibiotics. Antibiotics 2:237–255

de Breij A et al (2018) The antimicrobial peptide SAAP-148 combats drug-resistant bacteria and biofilms. Sci Transl Med 10(**423**):**eaan4044**

De Groote VN et al (2009) Novel persistence genes in *Pseudomonas aeruginosa* identified by high-throughput screening. FEMS Microbiol Lett 297:73–79

de Jong IG, Haccou P, Kuipers OP (2011) Bet hedging or not? a guide to proper classification of microbial survival strategies. Bioessays 33:215–223

De Leenheer P, Cogan NG (2009) Failure of antibiotic treatment in microbial populations. J Math Biol 59:563–579

de Leseleuc L, Harris G, KuoLee R, Chen W (2012) In vitro and in vivo biological activities of iron chelators and gallium nitrate against *Acinetobacter baumannii*. Antimicrob Agents Chemother 56:5397–5400

Defraine V et al (2016) Efficacy of Artilysin Art-175 against resistant and persistent *Acinetobacter baumannii*. Antimicrob Agents Chemother 60:3480–3488

DeLeon K et al (2009) Gallium maltolate treatment eradicates *Pseudomonas aeruginosa* infection in thermally injured mice. Antimicrob Agents Chemother 53:1331–1337

Deng W et al (2020) L-lysine potentiates aminoglycosides against *Acinetobacter baumannii* via regulation of proton motive force and antibiotics uptake. Emerg Microbes Infect 9:639–650

Dhar N, McKinney JD (2010) *Mycobacterium tuberculosis* persistence mutants identified by screening in isoniazid-treated mice. Proc Natl Acad Sci U S A 107:12275–12280

Diard M et al (2017) Inflammation boosts bacteriophage transfer between *Salmonella* spp. Science 355:1211–1215

Dorr T, Lewis K, Vulic M (2009) SOS response induces persistence to fluoroquinolones in *Escherichia coli*. PLoS Genet 5:e1000760

Dorr T, Vulic M, Lewis K (2010) Ciprofloxacin causes persister formation by inducing the TisB toxin in *Escherichia coli*. PLoS Biol 8:e1000317

Duan X et al (2016) l-Serine potentiates fluoroquinolone activity against *Escherichia coli* by enhancing endogenous reactive oxygen species production. *J Antimicrob Chemother* 71:2192–2199

Eisenreich W, Dandekar T, Heesemann J, Goebel W (2010) Carbon metabolism of intracellular bacterial pathogens and possible links to virulence. Nat Rev Microbiol 8:401–412

Ethapa T et al (2013) Multiple factors modulate biofilm formation by the anaerobic pathogen Clostridium difficile. J Bacteriol 195:545–555

Fauvart M, De Groote VN, Michiels J (2011) Role of persister cells in chronic infections: clinical relevance and perspectives on anti-persister therapies. J Med Microbiol 60:699–709

Feng J, Zhang S, Shi W, Zhang Y (2016) Ceftriaxone pulse dosing fails to eradicate biofilm-like microcolony *B. burgdorferi* persisters which are sterilized by daptomycin/doxycycline/cefuroxime without pulse dosing. Front Microbiol 7:1744

Ford CB et al (2011) Use of whole genome sequencing to estimate the mutation rate of *Mycobacterium tuberculosis* during latent infection. Nat Genet 43:482–486

Foster TJ (2005) Immune evasion by *staphylococci*. Nat Rev Microbiol 3:948–958

Fridman O, Goldberg A, Ronin I, Shoresh N, Balaban NQ (2014) Optimization of lag time underlies antibiotic tolerance in evolved bacterial populations. Nature 513:418–421

Fung DK, Chan EW, Chin ML, Chan RC (2010) Delineation of a bacterial starvation stress response network which can mediate antibiotic tolerance development. Antimicrob Agents Chemother 54:1082–1093

Garrison AT et al (2015) Halogenated Phenazines that potently eradicate biofilms, MRSA Persister cells in non-biofilm cultures, and *Mycobacterium tuberculosis*. Angew Chem Int Ed Engl 54:14819–14823

Gefen O, Chekol B, Strahilevitz J, Balaban NQ (2017) TDtest: easy detection of bacterial tolerance and persistence in clinical isolates by a modified disk-diffusion assay. Sci Rep 7:41284

Germain E, Castro-Roa D, Zenkin N, Gerdes K (2013) Molecular mechanism of bacterial persistence by HipA. Mol Cell 52:248–254

Ghosh A et al (2018) Contact-dependent growth inhibition induces high levels of antibiotic-tolerant persister cells in clonal bacterial populations. EMBO J 37(9):e98026

Gi M et al (2014) A drug-repositioning screening identifies pentetic acid as a potential therapeutic agent for suppressing the elastase-mediated virulence of *Pseudomonas aeruginosa*. Antimicrob Agents Chemother 58:7205–7214

Giard JC et al (2001) The stress proteome of *Enterococcus faecalis*. Electrophoresis 22:2947–2954

Gobert AP et al (2001) *Helicobacter pylori* arginase inhibits nitric oxide production by eukaryotic cells: a strategy for bacterial survival. Proc Natl Acad Sci U S A 98:13844–13849

Goneau LW et al (2014) Selective target inactivation rather than global metabolic dormancy causes antibiotic tolerance in uropathogens. Antimicrob Agents Chemother 58:2089–2097

Grassi L, Maisetta G, Esin S, Batoni G (2017) Combination strategies to enhance the efficacy of antimicrobial peptides against bacterial biofilms. Front Microbiol 8:2409

Gutierrez A et al (2013) Beta-Lactam antibiotics promote bacterial mutagenesis via an RpoS-mediated reduction in replication fidelity. Nat Commun 4:1610

Hansen S, Lewis K, Vulic M (2008) Role of global regulators and nucleotide metabolism in antibiotic tolerance in *Escherichia coli*. Antimicrob Agents Chemother 52:2718–2726

Harrison JJ et al (2009) The chromosomal toxin gene yafQ is a determinant of multidrug tolerance for *Escherichia coli* growing in a biofilm. Antimicrob Agents Chemother 53:2253–2258

Haseltine WA, Block R (1973) Synthesis of guanosine tetra- and pentaphosphate requires the presence of a codon-specific, uncharged transfer ribonucleic acid in the acceptor site of ribosomes. Proc Natl Acad Sci U S A 70:1564–1568

He Z et al (2012) Use of the quorum sensing inhibitor furanone C-30 to interfere with biofilm formation by *Streptococcus mutans* and its luxS mutant strain. Int J Antimicrob Agents 40:30–35

Helaine S et al (2014) Internalization of *Salmonella* by macrophages induces formation of nonreplicating persisters. Science 343:204–208

Henry TC, Brynildsen MP (2016) Development of Persister-FACSeq: a method to massively parallelize quantification of persister physiology and its heterogeneity. Sci Rep 6:25100

Hijazi S et al (2018) Antimicrobial activity of gallium compounds on ESKAPE pathogens. Front Cell Infect Microbiol 8:316

Hong SH, Wang X, O'Connor HF, Benedik MJ, Wood TK (2012) Bacterial persistence increases as environmental fitness decreases. Microb Biotechnol 5:509–522

Honsa ES et al (2017) RelA mutant Enterococcus faecium with multiantibiotic tolerance arising in an immunocompromised host. mBio 8(1):e02124

Hubbard AT et al (2017) Mechanism of action of a membrane-active Quinoline-based antimicrobial on natural and model bacterial membranes. Biochemistry 56:1163–1174

Iino R, Matsumoto Y, Nishino K, Yamaguchi A, Noji H (2013) Design of a large-scale femtoliter droplet array for single-cell analysis of drug-tolerant and drug-resistant bacteria. Front Microbiol 4:300

Imperi F, Leoni L, Visca P (2014) Antivirulence activity of azithromycin in *Pseudomonas aeruginosa*. Front Microbiol 5:178

Imperi F et al (2013a) New life for an old drug: the anthelmintic drug niclosamide inhibits *Pseudomonas aeruginosa* quorum sensing. Antimicrob Agents Chemother 57:996–1005

Imperi F et al (2013b) Repurposing the antimycotic drug flucytosine for suppression of *Pseudomonas aeruginosa* pathogenicity. Proc Natl Acad Sci U S A 110:7458–7463

Jarchum I (2014) A one-two punch knocks out biofilms. Nat Biotechnol 32:142

Javid B et al (2014) Mycobacterial mistranslation is necessary and sufficient for rifampicin phenotypic resistance. Proc Natl Acad Sci U S A 111:1132–1137

Jiafeng L, Fu X, Chang Z (2015) Hypoionic shock treatment enables aminoglycosides antibiotics to eradicate bacterial persisters. Sci Rep 5:14247

Jiang M et al (2020) Antimicrobial activities of peptide Cbf-K16 against drug-resistant *Helicobacter pylori* infection in vitro and in vivo. Microb Pathog 138:103847

Joers A, Kaldalu N, Tenson T (2010) The frequency of persisters in *Escherichia coli* reflects the kinetics of awakening from dormancy. J Bacteriol 192:3379–3384

Joers A, Tenson T (2016) Growth resumption from stationary phase reveals memory in *Escherichia coli* cultures. Sci Rep 6:24055

Jubair M, Morris JG Jr, Ali A (2012) Survival of *Vibrio cholerae* in nutrient-poor environments is associated with a novel "persister" phenotype. PLoS One 7:e45187

Kalan L, Wright GD (2011) Antibiotic adjuvants: multicomponent anti-infective strategies. Expert Rev Mol Med 13:e5

Kaldalu N, Joers A, Ingelman H, Tenson T (2016) A general method for measuring Persister levels in *Escherichia coli* cultures. Methods Mol Biol 1333:29–42

Kaneko Y, Thoendel M, Olakanmi O, Britigan BE, Singh PK (2007) The transition metal gallium disrupts *Pseudomonas aeruginosa* iron metabolism and has antimicrobial and antibiofilm activity. J Clin Invest 117:877–888

Keren I, Minami S, Rubin E, Lewis K (2011) Characterization and transcriptome analysis of *Mycobacterium tuberculosis* persisters. mBio 2:e00100–e00111

Keren I, Shah D, Spoering A, Kaldalu N, Lewis K (2004) Specialized persister cells and the mechanism of multidrug tolerance in *Escherichia coli*. J Bacteriol 186:8172–8180

Keseler IM et al (2017) The EcoCyc database: reflecting new knowledge about *Escherichia coli* K-12. Nucleic Acids Res 45:D543–D550

Khan F, Pham DTN, Tabassum N, Oloketuyi SF, Kim YM (2020) Treatment strategies targeting persister cell formation in bacterial pathogens. Crit Rev Microbiol 46:665–688

Khodaverdian V et al (2013) Discovery of antivirulence agents against methicillin-resistant *Staphylococcus aureus*. Antimicrob Agents Chemother 57:3645–3652

Kieser KJ, Rubin EJ (2014) How sisters grow apart: mycobacterial growth and division. Nat Rev Microbiol 12:550–562

Kim JH et al (2017) Synergistic antibacterial effects of chitosan-Caffeic acid conjugate against antibiotic-resistant acne-related Bacteria. Mar Drugs 15(6):167

Kim W et al (2018) A new class of synthetic retinoid antibiotics effective against bacterial persisters. Nature 556:103–107

Kim Y, Wood TK (2010) Toxins Hha and CspD and small RNA regulator Hfq are involved in persister cell formation through MqsR in *Escherichia coli*. Biochem Biophys Res Commun 391:209–213

Komp Lindgren P, Klockars O, Malmberg C, Cars O (2015) Pharmacodynamic studies of nitrofurantoin against common uropathogens. J Antimicrob Chemother 70:1076–1082

Koo H, Allan RN, Howlin RP, Stoodley P, Hall-Stoodley L (2017) Targeting microbial biofilms: current and prospective therapeutic strategies. Nat Rev Microbiol 15:740–755

Korch SB, Henderson TA, Hill TM (2003) Characterization of the hipA7 allele of *Escherichia coli* and evidence that high persistence is governed by (p)ppGpp synthesis. Mol Microbiol 50:1199–1213

Kostakioti M, Hadjifrangiskou M, Pinkner JS, Hultgren SJ (2009) QseC-mediated dephosphorylation of QseB is required for expression of genes associated with virulence in uropathogenic *Escherichia coli*. Mol Microbiol 73:1020–1031

Kotte O, Volkmer B, Radzikowski JL, Heinemann M (2014) Phenotypic bistability in *Escherichia coli*'s central carbon metabolism. Mol Syst Biol 10:736

Kumar R et al (2019) Exploring the new horizons of drug repurposing: a vital tool for turning hard work into smart work. Eur J Med Chem 182:111602

Kwan BW, Chowdhury N, Wood TK (2015) Combatting bacterial infections by killing persister cells with mitomycin C. Environ Microbiol 17:4406–4414

Lehar SM et al (2015) Novel antibody-antibiotic conjugate eliminates intracellular *S. aureus*. *Nature* 527:323–328

Lei J et al (2019) The antimicrobial peptides and their potential clinical applications. Am J Transl Res 11:3919–3931

Leszczynska D, Matuszewska E, Kuczynska-Wisnik D, Furmanek-Blaszk B, Laskowska E (2013) The formation of persister cells in stationary-phase cultures of *Escherichia coli* is associated with the aggregation of endogenous proteins. PLoS One 8:e54737

Leung V, Levesque CM (2012) A stress-inducible quorum-sensing peptide mediates the formation of persister cells with noninherited multidrug tolerance. J Bacteriol 194:2265–2274

Levin BR, Concepcion-Acevedo J, Udekwu KI (2014) Persistence: a copacetic and parsimonious hypothesis for the existence of non-inherited resistance to antibiotics. Curr Opin Microbiol 21:18–21

Levin-Reisman I et al (2010) Automated imaging with ScanLag reveals previously undetectable bacterial growth phenotypes. Nat Methods 7:737–739

Levin-Reisman I et al (2017) Antibiotic tolerance facilitates the evolution of resistance. Science 355:826–830

Lewis K (2008) Multidrug tolerance of biofilms and persister cells. Curr Top Microbiol Immunol 322:107–131

Li C et al (2019) Two optimized antimicrobial peptides with therapeutic potential for clinical antibiotic-resistant *Staphylococcus aureus*. Eur J Med Chem 183:111686

Li F et al (2018) In vitro activity of beta-lactams in combination with beta-lactamase inhibitors against *Mycobacterium tuberculosis* clinical isolates. Biomed Res Int 2018:3579832

Li T, Yin N, Liu H, Pei J, Lai L (2016) Novel inhibitors of toxin HipA reduce multidrug tolerant Persisters. ACS Med Chem Lett 7:449–453

Li XZ, Nikaido H (2009) Efflux-mediated drug resistance in bacteria: an update. Drugs 69:1555–1623

Li Y, Zhang Y (2007) PhoU is a persistence switch involved in persister formation and tolerance to multiple antibiotics and stresses in *Escherichia coli*. Antimicrob Agents Chemother 51:2092–2099

Liebens V et al (2014) A putative de-N-acetylase of the PIG-L superfamily affects fluoroquinolone tolerance in *Pseudomonas aeruginosa*. Pathog Dis 71:39–54

Liu Q, Zheng Z, Kim W, Burgwyn Fuchs B, Mylonakis E (2018b) Influence of subinhibitory concentrations of NH125 on biofilm formation & virulence factors of *Staphylococcus aureus*. Future Med Chem 10:1319–1331

Liu S et al (2017) Variable Persister gene interactions with (p)ppGpp for Persister formation in *Escherichia coli*. Front Microbiol 8:1795

Liu Y, Kyle S, Straight PD (2018a) Antibiotic stimulation of a *Bacillus subtilis* migratory response. mSphere 3:00586

Lu TK, Collins JJ (2009) Engineered bacteriophage targeting gene networks as adjuvants for antibiotic therapy. Proc Natl Acad Sci U S A 106:4629–4634

Luidalepp H, Joers A, Kaldalu N, Tenson T (2011) Age of inoculum strongly influences persister frequency and can mask effects of mutations implicated in altered persistence. J Bacteriol 193:3598–3605

Ma D et al (2019) The toxin-antitoxin MazEF drives *Staphylococcus aureus* biofilm formation, antibiotic tolerance, and chronic infection. mBio 10(6):e01658

MacGilvary NJ, Tan S (2018) Fluorescent *Mycobacterium tuberculosis* reporters: illuminating host-pathogen interactions. Pathog Dis 76(**3**):**fty017**

Maiden MM et al (2018) Triclosan is an aminoglycoside adjuvant for eradication of *Pseudomonas aeruginosa* biofilms. Antimicrob Agents Chemother 62(6):e00146

Manina G, Dhar N, McKinney JD (2015) Stress and host immunity amplify *Mycobacterium tuberculosis* phenotypic heterogeneity and induce nongrowing metabolically active forms. Cell Host Microbe 17:32–46

Manuel J, Zhanel GG, de Kievit T (2010) Cadaverine suppresses persistence to carboxypenicillins in *Pseudomonas aeruginosa* PAO1. Antimicrob Agents Chemother 54:5173–5179

Marques CN, Morozov A, Planzos P, Zelaya HM (2014) The fatty acid signaling molecule cis-2-decenoic acid increases metabolic activity and reverts persister cells to an antimicrobial-susceptible state. Appl Environ Microbiol 80:6976–6991

Marr AK, Gooderham WJ, Hancock RE (2006) Antibacterial peptides for therapeutic use: obstacles and realistic outlook. Curr Opin Pharmacol 6:468–472

Meena M, Prasad V, Zehra A, Gupta VK, Upadhyay RS (2015) Mannitol metabolism during pathogenic fungal-host interactions under stressed conditions. Front Microbiol 6:1019

Meyer KJ, Taylor HB, Seidel J, Gates MF, Lewis K (2020) Pulse dosing of antibiotic enhances killing of a *Staphylococcus aureus* biofilm. Front Microbiol 11:596227

Michiels JE, Van den Bergh B, Verstraeten N, Michiels J (2016) Molecular mechanisms and clinical implications of bacterial persistence. Drug Resist Updat 29:76–89

Mohamed MF, Brezden A, Mohammad H, Chmielewski J, Seleem MN (2017) Targeting biofilms and persisters of ESKAPE pathogens with P14KanS, a kanamycin peptide conjugate. Biochim Biophys Acta Gen Subj 1861:848–859

Moker N, Dean CR, Tao J (2010) *Pseudomonas aeruginosa* increases formation of multidrug-tolerant persister cells in response to quorum-sensing signaling molecules. J Bacteriol 192:1946–1955

Mukamolova GV, Kaprelyants AS, Young DI, Young M, Kell DB (1998) A bacterial cytokine. Proc Natl Acad Sci U S A 95:8916–8921

Mukamolova GV et al (2006) Muralytic activity of Micrococcus luteus Rpf and its relationship to physiological activity in promoting bacterial growth and resuscitation. Mol Microbiol 59:84–98

Mukherjee D, Zou H, Liu S, Beuerman R, Dick T (2016) Membrane-targeting AM-0016 kills mycobacterial persisters and shows low propensity for resistance development. Future Microbiol 11:643–650

Mulcahy LR, Burns JL, Lory S, Lewis K (2010) Emergence of *Pseudomonas aeruginosa* strains producing high levels of persister cells in patients with cystic fibrosis. J Bacteriol 192:6191–6199

Mwangi J et al (2019) The antimicrobial peptide ZY4 combats multidrug-resistant *Pseudomonas aeruginosa* and *Acinetobacter baumannii* infection. Proc Natl Acad Sci U S A 116 (52):26516–26522

Namugenyi SB, Aagesen AM, Elliott SR, Tischler AD (2017) *Mycobacterium tuberculosis* PhoY proteins promote persister formation by mediating Pst/SenX3-RegX3 phosphate sensing. mBio 8(4):e00494

Nathan C (2012) Fresh approaches to anti-infective therapies. Sci Transl Med 4:140–142

Nautiyal A, Patil KN, Muniyappa K (2014) Suramin is a potent and selective inhibitor of *Mycobacterium tuberculosis* RecA protein and the SOS response: RecA as a potential target for antibacterial drug discovery. J Antimicrob Chemother 69:1834–1843

Niepa TH, Gilbert JL, Ren D (2012) Controlling *Pseudomonas aeruginosa* persister cells by weak electrochemical currents and synergistic effects with tobramycin. Biomaterials 33:7356–7365

Nikitushkin VD et al (2015) A product of RpfB and RipA joint enzymatic action promotes the resuscitation of dormant mycobacteria. FEBS J 282:2500–2511

Nistico L et al (2009) Fluorescence "in situ" hybridization for the detection of biofilm in the middle ear and upper respiratory tract mucosa. Methods Mol Biol 493:191–213

Norton JP, Mulvey MA (2012) Toxin-antitoxin systems are important for niche-specific colonization and stress resistance of uropathogenic *Escherichia coli*. PLoS Pathog 8:e1002954

Nuermberger E et al (2006) Combination chemotherapy with the nitroimidazopyran PA-824 and first-line drugs in a murine model of tuberculosis. Antimicrob Agents Chemother 50:2621–2625

O'Driscoll T, Crank CW (2015) Vancomycin-resistant enterococcal infections: epidemiology, clinical manifestations, and optimal management. Infect Drug Resist 8:217–230

Orman MA, Brynildsen MP (2013a) Dormancy is not necessary or sufficient for bacterial persistence. Antimicrob Agents Chemother 57:3230–3239

Orman MA, Brynildsen MP (2013b) Establishment of a method to rapidly assay bacterial persister metabolism. Antimicrob Agents Chemother 57:4398–4409

Orman MA, Brynildsen MP (2015) Inhibition of stationary phase respiration impairs persister formation in *E. coli*. Nat Commun 6:7983

Pan J, Bahar AA, Syed H, Ren D (2012) Reverting antibiotic tolerance of *Pseudomonas aeruginosa* PAO1 persister cells by (Z)-4-bromo-5-(bromomethylene)-3-methylfuran-2(5H)-one. PLoS One 7:e45778

Pan J, Song F, Ren D (2013) Controlling persister cells of *Pseudomonas aeruginosa* PDO300 by (Z)-4-bromo-5-(bromomethylene)-3-methylfuran-2(5H)-one. Bioorg Med Chem Lett 23:4648–4651

Paranjape SS, Shashidhar R (2020) Glucose sensitizes the stationary and persistent population of *Vibrio cholerae* to ciprofloxacin. Arch Microbiol 202:343–349

Pedersen K et al (2003) The bacterial toxin RelE displays codon-specific cleavage of mRNAs in the ribosomal a site. Cell 112:131–140

Peng B et al (2015) Exogenous alanine and/or glucose plus kanamycin kills antibiotic-resistant bacteria. Cell Metab 21:249–262

Perlmutter JI et al (2014) Repurposing the antihistamine terfenadine for antimicrobial activity against *Staphylococcus aureus*. J Med Chem 57:8540–8562

Perry JA, Jones MB, Peterson SN, Cvitkovitch DG, Levesque CM (2009) Peptide alarmone signalling triggers an auto-active bacteriocin necessary for genetic competence. Mol Microbiol 72:905–917

Petchiappan A, Chatterji D (2017) Antibiotic resistance: current perspectives. ACS Omega 2:7400–7409

Pinault L, Han JS, Kang CM, Franco J, Ronning DR (2013) Zafirlukast inhibits complexation of Lsr2 with DNA and growth of *Mycobacterium tuberculosis*. Antimicrob Agents Chemother 57:2134–2140

Portela CA, Smart KF, Tumanov S, Cook GM, Villas-Boas SG (2014) Global metabolic response of *Enterococcus faecalis* to oxygen. J Bacteriol 196:2012–2022

Potrykus K, Cashel M (2008) (p)ppGpp: still magical? Annu Rev Microbiol 62:35–51

Prax M, Bertram R (2014) Metabolic aspects of bacterial persisters. Front Cell Infect Microbiol 4:148

Prysak MH et al (2009) Bacterial toxin YafQ is an endoribonuclease that associates with the ribosome and blocks translation elongation through sequence-specific and frame-dependent mRNA cleavage. Mol Microbiol 71:1071–1087

Pu Y et al (2016) Enhanced efflux activity facilitates drug tolerance in dormant bacterial cells. Mol Cell 62:284–294

Que YA et al (2013) A quorum sensing small volatile molecule promotes antibiotic tolerance in bacteria. PLoS One 8:e80140

Ragheb MN et al (2019) Inhibiting the evolution of antibiotic resistance. Mol Cell 73:157–165

Ramadoss NS et al (2013) Small molecule inhibitors of trans-translation have broad-spectrum antibiotic activity. Proc Natl Acad Sci U S A 110:10282–10287

Ramirez M et al (2016) Diverse drug-resistance mechanisms can emerge from drug-tolerant cancer persister cells. Nat Commun 7:10690

Rangel-Vega A, Bernstein LR, Mandujano-Tinoco EA, Garcia-Contreras SJ, Garcia-Contreras R (2015) Drug repurposing as an alternative for the treatment of recalcitrant bacterial infections. Front Microbiol 6:282

Rao NN, Liu S, Kornberg A (1998) Inorganic polyphosphate in *Escherichia coli*: the phosphate regulon and the stringent response. J Bacteriol 180:2186–2193

Ratcliff WC, Denison RF (2011) Bacterial persistence and bet hedging in *Sinorhizobium meliloti*. Commun Integr Biol 4:98–100

Ravagnani A, Finan CL, Young M (2005) A novel firmicute protein family related to the actinobacterial resuscitation-promoting factors by non-orthologous domain displacement. BMC Genom 6:39

Reffuveille F, de la Fuente-Nunez C, Mansour S, Hancock RE (2014) A broad-spectrum antibiofilm peptide enhances antibiotic action against bacterial biofilms. Antimicrob Agents Chemother 58:5363–5371

Rego EH, Audette RE, Rubin EJ (2017) Deletion of a mycobacterial divisome factor collapses single-cell phenotypic heterogeneity. Nature 546:153–157

Roostalu J, Joers A, Luidalepp H, Kaldalu N, Tenson T (2008) Cell division in *Escherichia coli* cultures monitored at single cell resolution. BMC Microbiol 8:68

Rosser A, Stover C, Pareek M, Mukamolova GV (2017) Resuscitation-promoting factors are important determinants of the pathophysiology in *Mycobacterium tuberculosis* infection. Crit Rev Microbiol 43:621–630

Rowe SE, Conlon BP, Keren I, Lewis K (2016) Persisters: methods for isolation and identifying contributing factors--a review. Methods Mol Biol 1333:17–28

Sakatos A et al (2018) Posttranslational modification of a histone-like protein regulates phenotypic resistance to isoniazid in mycobacteria. Sci Adv 4:eaao1478

Saliba AE et al (2016) Single-cell RNA-seq ties macrophage polarization to growth rate of intracellular *Salmonella*. Nat Microbiol 2:16206

Scherrer R, Moyed HS (1988) Conditional impairment of cell division and altered lethality in hipA mutants of *Escherichia coli* K-12. J Bacteriol 170:3321–3326

Schmidt NW et al (2014) Engineering persister-specific antibiotics with synergistic antimicrobial functions. ACS Nano 8:8786–8793

Schmidt NW et al (2015) Pentobra: a potent antibiotic with multiple layers of selective antimicrobial mechanisms against Propionibacterium acnes. J Invest Dermatol 135:1581–1589

Shah D et al (2006) Persisters: a distinct physiological state of E. coli. BMC Microbiol 6:53

Shan Y, Lazinski D, Rowe S, Camilli A, Lewis K (2015) Genetic basis of persister tolerance to aminoglycosides in *Escherichia coli*. mBio 6(2):e0007

Shapiro JA, Nguyen VL, Chamberlain NR (2011) Evidence for persisters in *Staphylococcus epidermidis* RP62a planktonic cultures and biofilms. J Med Microbiol 60:950–960

Sharma B, Brown AV, Matluck NE, Hu LT, Lewis K (2015) *Borrelia burgdorferi*, the causative agent of Lyme disease, forms drug-tolerant persister cells. Antimicrob Agents Chemother 59:4616–4624

Shi W et al (2011) Pyrazinamide inhibits trans-translation in *Mycobacterium tuberculosis*. Science 333:1630–1632

Singh R et al (2008) PA-824 kills nonreplicating *Mycobacterium tuberculosis* by intracellular NO release. Science 322:1392–1395

Spoering AL, Vulic M, Lewis K (2006) GlpD and PlsB participate in persister cell formation in *Escherichia coli*. J Bacteriol 188:5136–5144

Stapels DAC et al (2018) *Salmonella* persisters undermine host immune defenses during antibiotic treatment. Science 362:1156–1160

Tang HJ et al (2019) Colistin-sparing regimens against *Klebsiella pneumoniae* carbapenemase-producing *K. pneumoniae* isolates: Combination of tigecycline or doxycycline and gentamicin or amikacin. *J Microbiol Immunol Infect* **52**:273–281

Tashiro Y et al (2012) RelE-mediated dormancy is enhanced at high cell density in *Escherichia coli*. J Bacteriol 194:1169–1176

Theodore A, Lewis K, Vulic M (2013) Tolerance of *Escherichia coli* to fluoroquinolone antibiotics depends on specific components of the SOS response pathway. Genetics 195:1265–1276

Tiwari P et al (2015) MazF ribonucleases promote *Mycobacterium tuberculosis* drug tolerance and virulence in Guinea pigs. Nat Commun 6:6059

Tkhilaishvili T, Lombardi L, Klatt AB, Trampuz A, Di Luca M (2018) Bacteriophage Sb-1 enhances antibiotic activity against biofilm, degrades exopolysaccharide matrix and targets persisters of *Staphylococcus aureus*. Int J Antimicrob Agents 52:842–853

Tkhilaishvili T, Wang L, Perka C, Trampuz A, Gonzalez Moreno M (2020) Using bacteriophages as a Trojan horse to the killing of dual-species biofilm formed by *Pseudomonas aeruginosa* and methicillin resistant *Staphylococcus aureus*. Front Microbiol 11:695

Torrey HL, Keren I, Via LE, Lee JS, Lewis K (2016) High Persister mutants in *Mycobacterium tuberculosis*. PLoS One 11:e0155127

Tripathi A, Dewan PC, Siddique SA, Varadarajan R (2014) MazF-induced growth inhibition and persister generation in *Escherichia coli*. J Biol Chem 289:4191–4205

Ueda A, Attila C, Whiteley M, Wood TK (2009) Uracil influences quorum sensing and biofilm formation in *Pseudomonas aeruginosa* and fluorouracil is an antagonist. Microb Biotechnol 2:62–74

Ueta M et al (2008) Role of HPF (hibernation promoting factor) in translational activity in *Escherichia coli*. J Biochem 143:425–433

Van Acker H, Sass A, Dhondt I, Nelis HJ, Coenye T (2014) Involvement of toxin-antitoxin modules in *Burkholderia cenocepacia* biofilm persistence. Pathog Dis 71:326–335

Van den Bergh B, Fauvart M, Michiels J (2017) Formation, physiology, ecology, evolution and clinical importance of bacterial persisters. FEMS Microbiol Rev 41:219–251

Van den Bergh B et al (2016) Frequency of antibiotic application drives rapid evolutionary adaptation of *Escherichia coli* persistence. Nat Microbiol 1:16020

van Opijnen T, Bodi KL, Camilli A (2009) Tn-seq: high-throughput parallel sequencing for fitness and genetic interaction studies in microorganisms. Nat Methods 6:767–772

Vazquez-Laslop N, Lee H, Neyfakh AA (2006) Increased persistence in *Escherichia coli* caused by controlled expression of toxins or other unrelated proteins. J Bacteriol 188:3494–3497

Vega NM, Allison KR, Khalil AS, Collins JJ (2012) Signaling-mediated bacterial persister formation. Nat Chem Biol 8:431–433

Vega NM, Allison KR, Samuels AN, Klempner MS, Collins JJ (2013) *Salmonella typhimurium* intercepts *Escherichia coli* signaling to enhance antibiotic tolerance. Proc Natl Acad Sci U S A 110:14420–14425

Verstraeten N et al (2015) Obg and membrane depolarization are part of a microbial bet-hedging strategy that leads to antibiotic tolerance. Mol Cell 59:9–21

Walz JM et al (2010) Anti-infective external coating of central venous catheters: a randomized, noninferiority trial comparing 5-fluorouracil with chlorhexidine/silver sulfadiazine in preventing catheter colonization. Crit Care Med 38:2095–2102

Wang F et al (2013) Identification of a small molecule with activity against drug-resistant and persistent tuberculosis. Proc Natl Acad Sci U S A 110:E2510–E2517

Wenthzel AM, Stancek M, Isaksson LA (1998) Growth phase dependent stop codon readthrough and shift of translation reading frame in *Escherichia coli*. FEBS Lett 421:237–242

Westermann AJ, Barquist L, Vogel J (2017) Resolving host-pathogen interactions by dual RNA-seq. PLoS Pathog 13:e1006033

Wilmaerts D, Windels EM, Verstraeten N, Michiels J (2019) General mechanisms leading to Persister formation and awakening. Trends Genet 35:401–411

Wilmaerts D et al (2018) The persistence-inducing toxin HokB forms dynamic pores that cause ATP leakage. mBio 9(4):e00744

Windels EM et al (2019) Bacterial persistence promotes the evolution of antibiotic resistance by increasing survival and mutation rates. ISME J 13:1239–1251

Wood DN, Chaussee MA, Chaussee MS, Buttaro BA (2005) Persistence of *Streptococcus pyogenes* in stationary-phase cultures. J Bacteriol 187:3319–3328

Wu N et al (2015) Ranking of persister genes in the same *Escherichia coli* genetic background demonstrates varying importance of individual persister genes in tolerance to different antibiotics. Front Microbiol 6:1003

Wu S, Yu P-L, Flint S (2017) Persister cell formation of *Listeria monocytogenes* in response to natural antimicrobial agent nisin. Food Control 77:243–250

Wu Y, Vulic M, Keren I, Lewis K (2012) Role of oxidative stress in persister tolerance. Antimicrob Agents Chemother 56:4922–4926

Wuyts J, Van Dijck P, Holtappels M (2018) Fungal persister cells: the basis for recalcitrant infections? PLoS Pathog 14:e1007301

Xiao H et al (1991) Residual guanosine 3′,5′-bispyrophosphate synthetic activity of *relA* null mutants can be eliminated by *spoT* null mutations. J Biol Chem 266:5980–5990

Yamaguchi Y, Park JH, Inouye M (2009) MqsR, a crucial regulator for quorum sensing and biofilm formation, is a GCU-specific mRNA interferase in *Escherichia coli*. J Biol Chem 284:28746–28753

Yang S et al (2015) Antibiotic regimen based on population analysis of residing persister cells eradicates *Staphylococcus epidermidis* biofilms. Sci Rep 5:18578

Yuksel M, Power JJ, Ribbe J, Volkmann T, Maier B (2016) Fitness trade-offs in competence differentiation of *Bacillus subtilis*. Front Microbiol 7:888

Zalis EA et al (2019) Stochastic variation in expression of the tricarboxylic acid cycle produces persister cells. mBio 10(5):e01930

Zhang Y (2014) Persisters, persistent infections and the Yin-Yang model. Emerg Microbes Infect 3:e3

Zhang Y, Yew WW, Barer MR (2012) Targeting persisters for tuberculosis control. Antimicrob Agents Chemother 56:2223–2230

Zhang Y, Zhang J, Hara H, Kato I, Inouye M (2005) Insights into the mRNA cleavage mechanism by MazF, an mRNA interferase. J Biol Chem 280:3143–3150

Microbial Pathogenesis: Mechanism and Recent Updates on Microbial Diversity of Pathogens

3

Swasti Dhagat and Satya Eswari Jujjavarapu

Contents

S. Dhagat · S. E. Jujjavarapu (✉)
Department of Biotechnology, National Institute of Technology Raipur, Raipur, India
e-mail: satyaeswarij.bt@nitrr.ac.in

© The Author(s), under exclusive license to Springer Nature Singapore Pte
Ltd. 2022
V. Kumar et al. (eds.), *Antimicrobial Resistance*,
https://doi.org/10.1007/978-981-16-3120-7_3

71

Abstract

Microorganisms or microbes are microscopic organisms that are found all around the globe and are too small to be seen by naked eyes. Microorganisms have a wide range of applications. They help to digest food, fasten the decay and decomposition processes, and produce macromolecules of industrial and commercial importance. They also help in environmental bioremediation and production of biofuels. Microorganisms are highly diversified in the environment ranging from single-celled to multicellular microscopic beings. They are classified as bacteria, archaea, fungi, protozoa, algae, and virus. Some of these also negatively impact our lives. They are pathogens and cause various diseases in humans, animals, and plants and also damage products and surfaces. Microorganisms produce various virulence factors, such as exoenzymes and toxins, which are responsible for causing diseases and invading host immune system. The pathogens enter into human body through various entry portals in the body after which they adhere to the receptors present on host cells. They damage host cells by either secreting biochemicals or by changing the morphology of host cells. This results in abnormal functioning of host cells which might even lead to death of cell or sometimes death of host too. Among these various classes of microbial diversity, pathogenic microorganisms constitute only a small portion, but they have high genetic diversity. A deep study of diversity of pathogens helps in understanding evolution and taxonomy of pathogens better along with their pathogenicity. This chapter focuses on the diversity of microorganisms and their subdivisions which affect plants, animals, and humans both positively and negatively. Moreover, the mechanisms by which various microorganisms cause pathogenesis in humans are also explained in this chapter along with the current scenario on different species of microbial pathogens.

Keywords

Pathogenesis · Microbial diversity · Bacteria · Fungi · Viruses

3.1   Diversity of Microorganisms

Microorganisms or microbes are microscopic organisms that are found all around the globe. They are too small to be seen by naked eyes. Microorganisms have a wide range of applications. They help to digest food, fasten the decay and decomposition processes, and produce macromolecules of industrial and commercial importance. They also help in environmental bioremediation and production of biofuels. Some of the microorganisms also negatively impact our lives. They are pathogens and cause various diseases in humans, animals, and plants and also damage products and surfaces. Microbes are classified as bacteria, archaea, fungi, protozoa, algae, and virus.

3.2   Bacteria

All bacteria are unicellular organisms. They are classified as prokaryotes as they lack well-defined nucleus. They are found in varied environments including inside human gut, ocean, and soil. Relationship of human and bacteria is complex. They not only help humans in metabolism of food or fermentation of various food products, but they also cause diseases which at some cases may be lethal too. Typically, they have a length of 0.5–5 micrometers. According to their cellular shapes, they can be classified as bacilli (rod shaped, e.g., *Bacillus*, *Clostridium*, *Propionibacterium*, *Corynebacterium*), cocci (spherical or round shaped, e.g., *Streptococcus*, *Staphylococcus*, *Micrococcus*), spirilla (spiral shaped, e.g., *Spirillum*, *Campylobacter*, *Helicobacter*, *Borrelia*), and vibrio (comma or curved rod shaped). Unlike eukaryotic cells, they lack nucleus and membrane-bound organelles, such as mitochondria and endoplasmic reticulum. They have a cell wall made up of peptidoglycan and contain a circular chromosome in the cytoplasm. They can be classified as Gram-positive and Gram-negative on the basis of their cell wall composition. The presence of peptidoglycan layer in the bacterial cell wall binds strongly to a violet dye (crystal violet) and stains the cell purple. This dye is not retained in the Gram-negative bacterial cells, and hence their staining depends upon the secondary red-colored dye (safranin) used. *Bacillus*, *Clostridium*, *Lactobacillus*, *Mycobacterium*, *Staphylococcus*, *Streptococcus*, and *Streptomyces* are few genera of Gram-positive bacteria. The Gram-negative bacteria include *Acinetobacter*, *Campylobacter*, *Enterobacter*, *Escherichia*, *Pseudomonas*, and *Shigella*, most of which are pathogenic. On the basis of utilization of gaseous oxygen, they are categorized as aerobic (that live in the presence of oxygen), anaerobic (that live without oxygen), and facultative (that live with or without oxygen) and autotrophs or heterotrophs according to the way they obtain energy.

Multiplication of most bacteria is by binary fission wherein the parent bacterial cell grows in size, copies its DNA, and splits into two identical daughter cells. Another form of asexual reproduction occurring in bacteria is budding. Here, an organism develops as a bud which is pinched off as it matures. The process of horizontal gene transfer introduces variations in the bacterial genome. This occurs by transformation, transduction, and conjugation. In bacterial transformation, the cells take up short fragments of DNA from surrounding environment and integrate into their genome. Transduction also integrates DNA from environment by viruses called bacteriophage and transfers the DNA. Conjugation involves physical contact between two bacteria (known as donor and recipient) for transfer of genetic material.

Apart from their pathogenicity and contaminant in food, bacteria have a varied range of applications. In food industries, they are used for the fermentation of bread and production of yoghurt, cheese, and organic acids in pickles and vinegar. They are also exploited for the production of organic acids, enzymes, alcohols, acetones, and pharmaceutical products like human insulin, growth hormones, and vitamins. Bacteria present in the gut of cattle secrete cellulase that helps in the digestion of cellulose. Because of their specificity, few species of bacteria have also been used as an alternative to pesticides in pest control. Apart from these applications, they have the ability to digest pollutants and recycle them to energy and nutrients or to less toxic products.

3.2.1 Bacterial Pathogenesis

Pathogenic bacteria attach themselves to receptors present on host cells by adhesin. Adhesin is a protein or glycoprotein and is present on the surface of pathogens. Adhesins are present on the surface of pathogenic bacteria, fungi, virus, and protozoa. Examples of bacterial adhesins are type I fimbriae present on enterotoxigenic *Escherichia coli* (causing traveler's disease) that help in attachment of bacteria to intestinal epithelial cells of host. Type IV pili on *Neisseria gonorrhoeae* help in its attachment to host urethral epithelial cell and cause gonorrhea. *Streptococcus mutans* produce adhesin P1 and attached to host teeth causing dental caries. *Streptococcus pyogenes* (causative agent of strep throat) produces F-protein and attaches to respiratory epithelial cells. N-Methylphenylalanine pili are produced by *Vibrio cholerae* and attach to intestinal epithelial cells causing cholera.

Patients who have bacteria present and multiplying in blood can get septic shock where the blood of the person decreases. This results in insufficient supply of oxygen and nutrients to cells and organism. Some bacteria also produce toxins which results in low blood pressure. Bacteria can also cause excessive swelling by release of many pro-inflammatory molecules from immune cells. They cause increased permeability of blood vessels which allows fluid to escape the bloodstream and enter tissues causing edema.

Exoenzymes are extracellular enzymes that help pathogen invade host cells and tissues. For example, Hyaluronidase S produced by *Staphylococcus aureus*, *Clostridium perfringens*, and *Streptococcus pyogenes* degrade hyaluronic acid present as intracellular cement in tissues. This increases the permeability of pathogens inside the tissue layers. Phospholipase C is also an exoenzyme produced by *Bacillus anthracis*. This enzyme degrades membrane of phagosomes which leads to escape of bacteria into cytoplasm. Collagenase produced by *Clostridium perfringens* degrades collagen present in connective tissue and promotes its spread to other cells. *C. perfringens* then multiply in blood and use toxins and phospholipase to cause lysis and necrosis. After the death of host cells, the organism ferments carbohydrates of muscles to produce gas. This results in necrosis of tissue with gas known as gas gangrene.

Pathogenic bacteria also produce biological poison called as toxins which invade the host by damaging tissues. They can be classified as endotoxins or exotoxins. An example of bacterial endotoxin is lipopolysaccharide (LPS) present on the outer membrane of Gram-negative bacteria. Lipid A is the lipid component of endotoxin and is highly conserved among Gram-negative pathogen. The Gram-negative pathogenic bacteria release endotoxin when the cells die during infection by pathogens. This results in disruption of host cell membrane. Lipid A triggers inflammatory response in host. The inflammatory response caused by low concentration of endotoxin in body is effective to defend against pathogens, whereas excessive inflammatory response due to high concentration of endotoxin can result in drop in blood pressure, multiple organ failure, and death of host. They are stable at high temperature (121 °C for 15 min) to get inactivated.

Exotoxins are produced by both Gram-positive and Gram-negative bacteria. Exotoxins are target specific and interact with specific receptors present on specific cells and cause cellular damage. They are inactivated at temperatures above 41 °C due to the presence of protein in them. Even small concentration of exotoxins is lethal. Exotoxins can be classified as intracellular targeting, membrane disrupting, and superantigens. The intracellular-targeting exotoxins consist of two subunits: A and B. The B subunit binds to specific receptors present on host cell, whereas the A toxin is responsible for interfering with the specific cellular activity. Examples of intracellular-targeting exotoxins are tetanus, botulinum, diphtheria, and cholera toxins. Membrane-disrupting toxins disrupt cell membrane by either degrading the phospholipid bilayer of cell membrane in host cells or by forming pores. Examples of membrane-disrupting toxins are hemolysins, leukocidins, and streptolysins. Superantigens trigger an excessive stimulation of immune cells to secrete cytokines known as cytokine storm. This results in a strong immune response and inflammation that may cause low blood pressure, high fever, multiple organ failure, and death. Examples of superantigen exotoxins are toxic shock syndrome toxin, streptococcal mitogenic exotoxins, and streptococcal pyrogenic toxins.

Bacteria produce a number of virulence factors to evade the immune system of host cells, especially phagocytosis. Capsules produced by many bacteria are used for the adhesion of bacteria to host cells. They also prevent ingestion of bacteria by phagocytes. In addition to this, the capsules increase the size of bacterial cells making them difficult to be engulfed by immune cells. Example of capsule-producing bacterium is *Streptococcus pneumoniae*. Pathogenic bacteria also produce protease. These proteases cleave and digest antibodies, thereby preventing phagocytosis and killing of bacteria. Some other virulence factors are M-protein and mycolic acid.

M-protein is present in fimbriae of *Streptococcus* sp. This protein alters the surface of *Streptococcus* and blocks binding of complement molecules. As complement molecules assist in phagocytosis, the binding of M-protein inhibits the process of phagocytosis, thereby evading the host immune system. Mycolic acid is a waxy substance present on the cell envelope of *Mycobacterium tuberculosis*. When the pathogen is engulfed by the phagocytes present in lungs, the mycolic acid coat helps the bacteria in resisting the killing of pathogen in phagolysosome.

Some bacteria also exploit the natural mechanism of the host cells by producing certain virulence factors. Coagulase, produced by *Staphylococcus aureus*, exploits the mechanism of blood clotting in host. The release of coagulase into the bloodstream results in clot formation without damaging blood vessels. This clot coats bacteria in fibrin and prevents the bacteria from getting exposed to immune cells in bloodstream. Kinases, another virulence factor, have effects opposite to coagulase. Kinases convert plasminogen to plasmin which digests fibrin clot. Kinases digest the clot and release trapped pathogens in clot to escape. Examples of kinases as virulence factors are streptokinase and staphylokinase produced by *Streptococcus pyogenes* and *Staphylococcus aureus*, respectively.

Another mechanism by which pathogens evade the immune system of host is antigenic variation. In this method, the surface proteins of pathogens are altered so that they are not recognized by host immune system, for example, variation of VlSE protein in *Borrelia burgdorferi* (causative agent of Lyme disease) and type IV pili in *Neisseria gonorrhoeae* (causative agent of gonorrhea).

3.2.2 Diseases Caused by Pathogenic Bacteria in Humans

Human body contains more number of bacterial cells than human cells themselves. Many species of bacteria are either harmless or beneficial to human being, whereas only a few species of pathogenic bacteria exist. The pathogenic bacteria are troublesome due to the generation of strains which show resistance to antimicrobial agents. Generally, the infections from these bacteria are nosocomial, that is, the infections of these multidrug-resistant strains originate from hospitals (Fair and Tor 2014). Few of such pathogenic bacterial strains are discussed below.

3.2.2.1 *Enterococcus faecalis*

Enterococcus faecalis is normally present in the gut of human beings but can cause serious human infections such as bacteremia, endocarditis, urinary tract infections, and wound infections. Among these infections, urinary tract infections are the most common and cause around 110,000 cases every year. The infections by *E. faecalis* are hospital-acquired infections and affect hospitalized people as they have weak immune system. The people at high risk of *E. faecalis* infections are those that have undergone surgery or undergoing cancer treatment, dialysis or HIV/AIDS patients and those who have received organ transplant or had root canal. The bacteria cause infections when they enter human body through wounds, blood, or urine. These infections are difficult to treat as the organism is resistant to multiple antibiotics. The intravascular and urinary tract catheter devices harbor *E. faecalis*, and hence their use also leads to spread of infection. The common infections caused by *E. faecalis* are bacteremia (infection of blood), endocarditis (infection of endocardium, heart's inner lining), urinary tract infections (infections of the bladder, urethra, and kidneys), wound infections (infection through open cut, especially during surgery), periodontitis (infection of gum found in people who had root canal), and meningitis (inflammation of membranes surrounding the brain and spinal cord) (Kau et al. 2005).

3.2.2.2 *Enterococcus faecium*

Enterococcus faecium is a Gram-positive bacterium inhabiting the intestinal tract of humans (Fisher and Phillips 2009). It causes infections of the skin, urinary tract, and endocardium in humans (Arias and Murray 2012) and has been found to be associated with infections in ventilators, urinary drainage catheters, and central lines of medical intensive care units. It is one of the leading cause of multidrug-resistant enterococcal infections (Hidron et al. 2008). The vancomycin-resistant enterococci (VRE) were first detected in the late 1980s in hospitals in the USA, and since then, they have spread to Europe and worldwide making their treatment difficult (Werner et al. 2008). The World Health Organization has included vancomycin-resistant *E. faecium* in a list of antibiotic-resistant bacteria as a high-priority pathogen (Tacconelli et al. 2018). Germany experienced increased rates of VRE infections of the blood and urinary tract in intensive care units between 2007 and 2016 (Higuita and Huycke 2014; Remschmidt et al. 2018; Markwart et al. 2019).

3.2.2.3 *Escherichia coli*

Escherichia coli colonizes the gastrointestinal tract of humans in few hours after birth where they coexist with the host as commensals. They generally do not cause any disease but infect immunocompromised individuals or when the normal gastrointestinal barriers are breached. *E. coli* contains different disease-causing pathogens. They affect cell signaling, mitochondrial function, cytoskeletal function, mitosis, and protein synthesis in eukaryotes. They cause diverse disease in humans such as diarrhea to severe dysentery, cystitis or pyelonephritis in the urinary tract, hemorrhagic colitis, meningitis, and septicemia (Donnenberg and Whittam 2001; Kaper et al. 2004).

3.2.2.4   *Pseudomonas aeruginosa*

Pseudomonas aeruginosa is an opportunistic pathogenic bacterium that affects immunocompromised individuals and causes severe pulmonary disease. *P. aeruginosa* also results in neutropenia, severe burns, or cystic fibrosis (Lyczak et al. 2000). It is a leading pathogen for nosocomial infection. *P. aeruginosa* has been found in cystic fibrosis, pulmonary infections, burn wounds, urinary tract infections, and medical equipment, such as ventilators, inhalers, respirators, dialysis equipment, vaporizers, and anesthesiology equipment, and from toilet and sinks (Gales et al. 2001). Along with these infections, it also causes gastrointestinal infections; skin infections such as external otitis, folliculitis, and dermatitis; bacteremia; soft tissue infections; respiratory system infections in patients with cystic fibrosis; and bone and joint infections (Tacconelli et al. 2002; Gellatly and Hancock 2013; Alhazmi 2015; Azam and Khan 2019).

3.2.2.5   *Staphylococcus aureus*

Staphylococcus aureus causes skin infections, pneumonia, bone infections, heart valve infections, bacteremia, infective endocarditis, osteoarticular infections, soft tissue infections, pleuropulmonary infections, and device-related infections. These infections can be from mild to serious. These bacteria infect catheters that are inserted through the skin into the blood vessel or medical implants. The bacteria cause bacteremia by traveling through the bloodstream and infect heart valves and bones. *Staphylococcus aureus* also produce toxins that cause staphylococcal food poisoning, scalded skin syndrome, or toxic shock syndrome. Toxic shock syndrome progresses rapidly and causes rash, fever, low blood pressure, and multiple organ failure (Tong et al. 2015; Eswari and Yadav 2019).

3.2.2.6   *Clostridioides difficile*

Clostridioides difficile is a bacterium that causes diarrhea and colitis which can lead to serious bowel problems. *C. difficile* was discovered in 1935, but the antibiotic-associated diarrhea was not recognized until 1978. The widespread use of antibiotic clindamycin in the 1970s led to a rise in the infections of *C. difficile*. The broad-spectrum antibiotics increased the *C. difficile* epidemic in the next 20 years. The feces of patients infected with *C. difficile* contain bacterial spores, which when transferred to another person's gastrointestinal tract through contaminated food or hands come to life. Normally, the bacteria remain dormant and do not develop any illness. But when good bacteria of the stomach are knocked down by antibiotics, *C. difficile* grows and multiplies. The bacteria produce toxins in the colon which injures the lining of the colon and causes diarrhea and inflammation. Generally, they affect those who have been taking broad-spectrum antibiotics or medication (proton pump inhibitor) to reduce amount of stomach acid, had surgery of the digestive tract, have weak immune system, are 65 years old, or have inflammatory bowel disease, kidney disease, or cancer. It causes around half a million infections in the USA every year with the recurrence of 1 in 6 patients in 2–8 weeks.

3.2.2.7 *Streptococcus pneumoniae*

Streptococcus pneumoniae causes pneumococcal disease which causes pneumonia, otitis, sinusitis, meningitis, and bacteremia. The bacteria mainly colonize mucosal surfaces of the upper respiratory tract of humans. Normally 27–65% of children and < 10% of adults are carriers of *S. pneumoniae*, which causes invasive inflammatory disease when it enters the bloodstream. Pneumococcal disease is common in low- and middle-income countries where availability of pneumococcal vaccine is less. Pneumococcal disease is common in dry season with tropical climate, whereas in temperate climate, it is more common during winter and early spring. In 2017, *S. pneumoniae* was included in the list of 12 priority pathogens by WHO (Dion and Ashurst 2020; Weiser et al. 2018).

3.2.2.8 *Acinetobacter baumannii*

Acinetobacter baumannii is the leading pathogen responsible for nosocomial infections among patients in intensive care units. *A. baumannii* leads to approximately 9% of hospital-acquired infections and can cause post-surgical urinary tract and respiratory tract infections in hospitalized patients. The major route of transmission of bacteria is via the hands of hospital staff. *Acinetobacter* results in mild-to-severe infections and can also be fatal. The severity of disease also depends on the site of infection and immune response of the individual (Joly-Guillou 2005; Lee et al. 2017).

3.2.2.9 *Klebsiella pneumoniae*

Klebsiella pneumoniae is an opportunistic pathogen which is found in the skin, mouth, and intestines. *Klebsiella* spp. affect immunocompromised individuals who suffer from diabetes mellitus or chronic pulmonary obstruction. *K. pneumoniae* is the most important infection-causing pathogen of *Klebsiella* spp. *K. pneumoniae* causes pyogenic liver abscess, necrotizing fasciitis, meningitis, severe pneumonia, endophthalmitis, soft tissue infections, septicemia, and urinary tract infections. The transmission of *K. pneumoniae* is via the hands of hospital staff and medical devices as *K. pneumoniae* forms biofilms on endotracheal tubes and catheters which causes infections in patients (Schroll et al. 2010). The hospital-acquired outbreaks of *K. pneumoniae* are generally caused by multidrug-resistant *Klebsiella* spp., the extended-spectrum β-lactamase (ESBL) producers. *Klebsiella* spp. account for approximately 8% of hospital-acquired infections in Europe and the USA. In the USA, *Klebsiella* is the eighth most important nosocomial infectious pathogen and causes around 3–7% of all hospital-acquired infections (Horan et al. 1988; Schaberg et al. 1991; Podschun and Ullmann 1998; Li et al. 2014).

3.2.2.10 *Neisseria gonorrhoeae*

Neisseria gonorrhoeae is an obligate pathogen that infects mucosal surfaces of the pharynx, female and male reproductive tracts, conjunctiva, and rectum. Asymptomatic infections in female reproductive tract can also result in serious infections which might affect the fallopian tube. The infections of reproductive tracts can also lead to inflammatory response causing pelvic inflammatory disease (PID). The infection causes scarring of tubes, occlusion of oviduct, and loss of ciliated cells. This can lead to infertility and problems during pregnancy and can also cause chronic pelvic pain (Lenz and Dillard 2018).

3.2.2.11 *Mycobacterium tuberculosis*

Even though the vaccine for tuberculosis has been used worldwide, it accounts for one of the highest mortalities among infectious diseases. Multidrug-resistant *Mycobacterium tuberculosis* is resistant to treatment of first-line anti-TB drugs such as rifampin and isoniazid, while resistance to second-line medications results in generation of extensively drug-resistant TB (XDR-TB). Almost 25% of people in the world are infected with *Mycobacterium tuberculosis*, but the disease results when the bacteria become active in individuals. The individuals with compromised immune system, old people, diabetics, and HIV patients are more prone to activation of *M. tuberculosis*. The year 2014 experienced around 9.6 million new TB cases with 2.96 cases per 100,000 people in the USA, whereas 1.5 million people died of it. Year 2016 resulted in 600,000 new cases of TB with 240,000 deaths, out of which multidrug-resistant strains accounted for 4.1% of all new TB cases. The major cases of multidrug-resistant *Mycobacterium tuberculosis* occurred in India, China, South America, former Soviet Union, and southern region of Africa (Smith 2003).

3.2.2.12 *Helicobacter pylori*

Helicobacter pylori infect approximately half of the world's population. *H. pylori* can colonize human body lifelong, if not treated. Medical conditions associated with *H. pylori* include chronic active gastritis and peptic ulcers and gastrointestinal diseases, duodenal ulcer, gastric mucosa-associated lymphoid tissue lymphoma, and gastric adenocarcinoma. Hence *H. pylori* can also be referred to as bacterial carcinogen. *H. pylori* infections prevail highly in Latin American countries and have low prevalence in the USA and Japan (Kusters et al. 2006; Calvet et al. 2013; Kao et al. 2016).

3.2.2.13 *Campylobacter* spp.

Campylobacter jejuni is transmitted via contaminated food or water and causes foodborne diarrhea and acute human enterocolitis even at low doses. The organism has also been linked with the development of Guillain-Barré syndrome (GBS), a neurological disorder. *C. jejuni* adheres to enterocytes in the gut and induces diarrhea by production of toxins (cytotoxins and enterotoxins) (Wallis 1994; Van Vliet and Ketley 2001).

3.2.2.14 *Salmonella*

Salmonella is one of the most common foodborne pathogens. In severe cases, it can also lead to chronic enterocolitis. *Salmonella enterica* is highly pathogenic among all strains of *Salmonella*. It accounts for approximately 93.8 million foodborne illness with 155,000 deaths every year. The mortality rate of *Salmonella* infections is due to the prevalence of multidrug-resistant strains of *Salmonella* (D'Aoust 1991; Eng et al. 2015).

3.2.2.15 *Haemophilus influenzae*

The problem with *Haemophilus influenzae*, causal organism of *H. influenzae* type B (Hib) disease, is an increase in the number of antibiotic-resistant strains of the organism. The nontypeable strains of *H. influenzae* (NTHi) live within biofilms and hence induce otitis media which is a bacterial infection of the middle ear (Harrison and Mason 2015).

3.2.2.16 *Shigella* spp.

Shigella is a common bacterial pathogen isolated from patients suffering with diarrhea. Shigellosis is an acute intestinal infection which can result in watery diarrhea to severe inflammatory bacillary dysentery. This can be accompanied by fever, abdominal cramps, and stools containing blood and mucus. These diseases can severely affect immunocompromised patients. The world encounters about 5–15% of diarrheal cases due to *Shigella* with approximately 1.1 million deaths, majority of which are children below the age of 5 years. The continuous increase in the incidence of shigellosis has become a global health problem which is yet to be solved (Schroeder and Hilbi 2008; Organization 2017).

3.3   Archaea

The domain Archaea was not recognized until the twentieth century, but in the late 1970s, a new group of microorganisms were discovered based on their differences with bacteria. Archaea are prokaryotic cells that are different from bacteria as they lack peptidoglycans in their cell wall but consist of ether-linked lipids termed as pseudopeptidoglycan. They comprise approximately 20% of the earth's biomass. They inhabit extreme environmental conditions such as hot springs and deep-sea well with temperatures over 100 °C and in extremely acidic or alkaline water; digestive tracts of cattle producing methane; anoxic conditions, such as muds of marshes and bottom of ocean; and underground petroleum deposits.

The archaeans use different inorganic substances as energy sources like sulfur, hydrogen gas, and carbon dioxide. Some of them also use sunlight by absorbing it using their light-sensitive membrane pigment, bacteriorhodopsin. The pigment reacts with sunlight and pumps protons out of the membrane. These protons flow back inside and are used in the synthesis of adenosine triphosphate (ATP) which is the energy source of cells. Bacteriorhodopsin exhibits similarity to rhodopsin, a light-detecting pigment found in vertebral retina.

3.3.1   Classification of Archaea

Archaea are classified as crenarchaeota, euryarchaeota, and korarchaeota. Crenarchaeota consists of microorganisms that can tolerate extreme temperature and acidic conditions. They can be further divided as thermophiles, mesophiles, and psychrophiles. Thermophiles inhabit extremely hot temperatures and can grow and reproduce at 100 °C or higher. They inhabit hot springs, volcanic openings, and acidic soils (Eswari et al. 2019; Dhagat and Jujjavarapu 2020). Mesophiles live in neither hot nor cold conditions with a temperature range from 20 °C to 40 °C. They are mostly used in fermentation processes requiring room temperature. Psychrophiles grow and reproduce at extremely low temperatures ranging from −20 °C to 10 °C. They are found in deep ocean water, high altitude, snowfields, and Arctic and alpine soil.

Euryarchaeota are methane-producing and salt-loving archaea and are categorized into methanogens and halophiles. They are the only life forms that can utilize carbon as electron acceptor and perform cellular respiration. Methanogens break down complex carbon molecules into methane during digestion, which leads to their degradation. They play an important role in sewage treatment plants and carbon cycle and are present in the stomach of cows where they help in the breakdown of sugars in grass which are generally not digested by eukaryotes. Halophiles, on the other hand, thrive in extremely salty or saline environments. They generally inhabit places such as Dead Sea and Great Salt Lake where salt concentration is five times higher than oceans. Korarchaeota is the oldest lineage of archaebacteria. They consist of genes found in both crenarchaeota and euryarchaeota and also genes different from both groups. They are not abundantly present in nature, and their presence is restricted to high-temperature hydrothermal vents.

Two of the other minor subdivisions of archaea are thaumarchaeota and nanoarchaeota. Thaumarchaeota are the most abundant and unique archaea. Initially, they were classified as "mesophilic crenarchaeota." They are ammonia-oxidizing microorganisms present in soil, hot springs, acidic soils, and marine water. They are capable of surviving under low concentrations of ammonia. They aerobically oxidize ammonia to nitrate in a process termed as nitrification. They are autotrophic and help in carbon dioxide fixation. They consist of ammonia monooxygenases which belong to a family of copper-containing membrane-bound monooxygenases with a wide range of substrates. Some thaumarchaeota are also dependent on other bacteria or organic material for their survival. The strains belonging to thaumarchaeota have tetraether lipids with crenarchaeol which is a thaumarchaeota-specific core lipid (Pester et al. 2011; Stieglmeier et al. 2014).

Nanoarchaeota are very small obligate parasites or symbionts belonging to the Archaea domain. They inhabit terrestrial hot springs, marine thermal vents, and mesophilic hypersaline environments (Hohn et al. 2002; Casanueva et al. 2008), and thus these microorganisms are capable of surviving in a wide range of temperature and geochemical environments (Munson-McGee et al. 2015). The only genus belonging to this phylum is *Nanoarchaeum*. The optimal environmental conditions of these small coccoids are pH 6.0 and salinity of 2%. *Nanoarchaeum equitans* is a hyperthermophile which can survive temperatures up to 90 °C. It is found to be associated with its host *Ignicoccus hospitalis* which is a marine hyperthermophilic crenarchaeon. It is incapable of synthesizing its own amino acids, lipids, and nucleotides but obtains them from its host cells. It is the smallest known archaeal parasite with a diameter of 400 nm without any contribution to the environment. Due to the lack of its ability to metabolize inorganic compounds, it inhabits places rich in carbon dioxide, hydrogen, and sulfur.

3.3.2   Pathogenicity in Archaea

The ability of archaea to act as a pathogen is still not clear, but the mechanism of pathogenesis of archaea may be linked to the following factors. Archaea are present in the gut of human body, and they have high access opportunity to colonize hosts. Methanogens are present in human colon (Miller et al. 1982), vagina (Belay et al. 1990), and subgingival area (Bonelo et al. 1984; Belay et al. 1988; Kulik et al. 2001). They require anaerobic environments for their growth and hence exist in sites in human body where anaerobic bacteria flourish.

Some archaea consist of unique flagella which show structural similarity to type IV pili (adhesin) of bacteria rather than the bacterial flagella (Thomas et al. 2001). Archaea also produce Tad-like proteins which are involved in fibril formation and help the archaea adhere to the surface of host cells. For example, *Pyrodictium* produces a network of flagellum-like filaments which helps connect other *Pyrodictium* cells. The role of these structures is not clear but might help in host-microbe adhesion or microbe-microbe interaction within a host.

Archaea, especially methanogens, have also been shown to cause disease in syntropy. Syntropy involves cooperation of two or more microorganisms for the consumption of a substrate which cannot be catabolized alone by either of the microbial species (Madigan et al. 2000). This also makes the process energy-efficient. Syntropy between anaerobic microbial community of deep periodontal pockets leads to periodontal disease. In this condition, the hydrogen atoms produced by secondary fermenters are consumed by methanogens (Carlsson 2000). Lack of methanogenic population compared to sulfate-reducing bacteria in human intestine leads to increase in toxic levels of hydrogen sulfide. This is the main reason for ulcerative colitis, a kind of inflammatory bowel disease in humans (Levine et al. 1996).

The structure of cell wall of archaea differs from bacteria and eukarya. The cell walls of archaea lack murein, and their lipids are made up of branched phytanyl chains which are attached to glycerol backbones via ether bonds. This is in contrast to bacteria and eukarya where the fatty acyl chains are ester-linked. They also lack lipopolysaccharides. The unique polar lipids of archaea can be incorporated into liposomes, forming archaeosomes which act as potent immune adjuvants in vitro and in vivo (Krishnan et al. 2000). Archaeosomes activate antigen-presenting cells by increasing expression of major histocompatibility complex class II and costimulatory molecules and induce strong antigen-specific response similar to lipopolysaccharides (Krishnan et al. 2001). This shows that the archaeal polar lipids are specifically recognized by the human immune system.

It is not known till date whether archaea have specific mechanism to evade human immune system. The paracrystalline cell surface S-layer present in archaea may play a role in evading immune response. It is proposed that the acquisition of virulence factors by archaea might be due to lateral gene transfer between bacteria and archaea over evolutionary period, but the evidence for the same is not available.

Methanogens are present in anaerobic environments in human body where other disease-causing endogenous anaerobic microflorae are present. These anaerobic microflorae are responsible for bite wounds, genitourinary and gastrointestinal surgery, malignancy, and aspiration. As methanogens colonize alimentary canal in humans, they may be attributed to anaerobic microbial infection. *Methanobrevibacter smithii* is a dominant methanogen present in human colon and has been found to be more in the fecal samples of patients with diverticulosis than normal humans.

The excretion of breath methane can be linked to the methane production by methanogenic bacteria present inside a host (Bond Jr et al. 1971). The concentration of breath methane increases in patients with precancerous condition than in healthy patients (Haines et al. 1977; Piqué et al. 1984). This might be due to the use of laxatives and enemas (Karlin et al. 1982). It is still not clear whether the excretion of methane increases after the development of diseases or the increase in methanogenic population leads to disease development (Eckburg et al. 2003).

3.3.3   Pathogenic Archaea

Not many reports have been available till date that focus on pathogenic species of archaea. Due to the presence of archaea in subgingival area, they have been mostly linked with diseases of oral cavity. The most common disease of oral cavity caused by archaea is periodontitis which is an infection and inflammation of periodontal tissue. It is caused by a mixed culture of oral microorganisms consisting of Gram-negative anaerobes, Gram-positive bacteria, spirochetes, and methanogenic archaea (Socransky 2002; Wade 2011). *Methanobrevibacter oralis* and species of *Methanobacterium*, *Methanosarcina*, and *Thermoplasmata* are some of the archaeal species that inhabit periodontal pocket and gingival sulcus and cause periodontitis, whereas *M. oralis* and *Synergistes* sp. are responsible for causing apical periodontitis (Maeda et al. 2013).

3.4   Fungi

Fungi are eukaryotic, multicellular organisms with cell wall made up of chitin. They consist of mushroom, molds, mildews, rusts, smuts, and yeasts. They may be free-living present in soil or water, have symbiotic relationship with plants, or are parasites. They are decomposers, that is, they absorb organic matter from dead material present in environment and thus play an important part in recycling the nutrients back to the soil. Some fungi, known as mycorrhizae, are present as symbionts with plants. They live in the roots of plants and affect their growth by supplying essential nutrients to them. Fungi are also source of some antibiotic drugs and foods such as morels, mushrooms, and truffles and are also required in the fermentation of beer, champagne, and bread. Some fungi can even harm hosts and are responsible for causing many diseases in plants and animals. Fungi, being eukaryotes, share chemical and genetic similarity with animals which makes the fungal diseases difficult to treat. Fungal diseases in plants might lead to severe damage to crops.

Fungi can be single-celled, for example, yeasts, or multicellular, wherein cells are arranged in the form of characteristic filaments called hyphae. These filamentous tubes called hyphae help fungi in absorbing nutrients. A group of hyphae is called as mycelium. They can either be well structured, for example, in mushroom, or tangled and unstructured, e.g., in molds. Dimorphic fungi exist in the form of yeast and hyphae. The reproduction of fungi is by the release of spores. They can be found in any habitat but live mostly in soil or on plants rather than in aquatic bodies.

3.4.1 Fungal Classification

Fungi are further classified into five phyla, namely, Chytridiomycota, Zygomycota, Glomeromycota, Ascomycota, and Basidiomycota. Chytridiomycota or chytrids are aquatic and microscopic organisms. They move with the help of flagella, are asexual, and produce motile spores. They are present in various environments, such as in the Arctic and high-elevation soils. They have the ability to grow under anaerobic conditions with a variety of pH ranges. Their morphology is different from other groups as they have zoospores with single flagellum at the posterior end. These zoospores are produced in zoosporangia. The zoosporangium of Chytridiomycota is a sac-like structure which results in the production of zoospores. One of the examples of chytrids, *Batrachochytrium dendrobatidis*, has the ability to cause fungal infection in frogs. They burrow under the skin of frogs and have been found to kill two-thirds of frogs in South and Central America.

Zygomycetes (singular Zygomycota) are terrestrial microorganisms that consume dead and decaying animal matter and plant detritus. They are also responsible for contaminating food, such as *Rhizopus stolonifer*, which is a bread mold. The phylum Zygomycota consists of approximately 900 species which consists of 1% of the true fungi. The mycelium of zygomycetes is a large cell with many nuclei as their hyphae are not separated by septa. Their reproduction is asexual through non-motile endospores formed in sporangia, sporangiola, or merosporangia and sexual by forming zygospores in zygosporangia. Zygomycetous fungi are saprotrophs or parasites or cause diseases in plants, fungi, animals, and humans. Some zygomycetous fungi, known as mycorrhizae, also form mutualistic relation with plants.

Glomeromycota consists of almost half of the fungi present in soil and is often associated with plants as mycorrhizae. They obtain sugar from plants and help the plants by dissolving minerals in soil which the plants take up as nutrients. This is done with the help of specialized structures called as arbuscules which help in the exchange of nutrients. Glomeromycota also reproduce asexually.

Ascomycota or ascomycetes comprise the largest phylum of fungi. They consist of both single-celled and multicellular fungi. Their reproduction is majorly asexual with the help of conidia, but some of them produce sexual spores in reproductive sacs known as asci. Some of the microorganisms belonging to this group are *Aspergillus nidulans, Laboulbeniales, Monascus, Pezizomycotina, Verticillium,* and yeast. Ascomycetes have commercial importance. *Aspergillus oryzae* are used for the fermentation of rice. Morels and truffles are used as gourmet delicacies. Yeasts are used in fermentation of wine, brewing, and baking. They are also pathogens of plants and animals and cause various infections like ringworm, ergotism, and athlete's foot, which may also lead to death. Pneumonia caused by fungi also poses threat to AIDS patients. Ascomycetes may also live inside humans, such as *Candida albicans*, which exists in the respiratory, gastrointestinal, and female reproductive tracts. They are capable of producing poisonous secondary metabolites making the crops unfit for consumption. They also infest and destroy the crops.

Basidiomycota produce sexual spores called basidiospores in club-shaped basidia. Because of these club-shaped bodies, basidiomycote or basidiomycetes are called as club fungi. Mushrooms, rusts, and smuts are some of the most common examples of basidiomycetes. Basidiomycota consists of edible fungi as well as some deadly toxins, such as *Cryptococcus neoformans*, which cause severe respiratory illness. The Basidiomycota also contributed to the ecosystem by degrading components of wood, especially lignin.

3.4.2   Mechanism of Fungal Pathogenesis

The virulence factors produced by pathogenic fungi are similar to the ones produced by pathogenic bacteria. But certain strains of fungi produce specific virulence factors. *Candida albicans* produce surface glycoproteins as adhesins which bind to phospholipids of endothelial and epithelial cells. *Candida* also produces exoenzymes, proteases, and phospholipases that help the pathogen to invade tissues. The proteases degrade keratin which is a structural protein found on epithelial cells. This increases the ability of fungi to invade host tissue. The phospholipase disrupts the host cell membrane leading to killing of host cells or pathogenic invasion of host tissue.

Claviceps purpurea is a pathogenic fungus that causes ergotism and grows on grains, such as rye. The fungus produces a mycotoxin (fungal toxin) called as ergot toxin, an alkaloid. The ergotism caused by ergot toxin is of two types: gangrenous and convulsive. The gangrenous ergotism occurs when the ergot toxin causes vasoconstriction. This results in improper blood flow to the extreme sites of the body leading to gangrene. The convulsive ergotism occurs when the ergot toxin targets the central nervous system. This results in mania and hallucinations. *Cryptococcus* which causes pneumonia and meningitis produces capsule as virulence factor. The capsule is made up of a polysaccharide, glucuronoxylomannan, and helps the pathogen resist phagocytosis than the non-capsulated strains.

The virulence factor produced by *Aspergillus* is a mycotoxin, aflatoxin. Aflatoxin can act both as a mutagen and as a carcinogen in host cells. It is responsible for development of liver cancer, and it can also cross blood-placental barrier (Wild et al. 1991). *Aspergillus* also produces gliotoxin which promotes virulence by inhibiting the function of phagocytic cells and pro-inflammatory response as well as by inducing host cells for self-destruction. *Aspergillus* produces two protease enzymes which help the pathogen to evade the immune system. Elastase degrades elastin protein present in the connective tissue of lungs. This leads to development of lung disease. Catalase, produced by *Aspergillus*, protects the pathogen from hydrogen peroxide which is produced by host immune system as a mechanism to destroy pathogens.

3.4.3   Diseases Caused by Pathogenic Fungi in Humans

Many fungal species cause disease in humans, animals, and plants. Fungal diseases have a serious impact on human health. Some of the most common fungal diseases affecting humans are enlisted here.

3.4.3.1   Adiaspiromycosis
It is a rare pulmonary infection caused by *Chrysosporium parvum* var. *crescens* (*Emmonsia parva*) due to inhalation of fungal spores (Buyuksirin et al. 2011; Anstead et al. 2012).

3.4.3.2   Aspergillosis
It is caused by various species of *Aspergillus*, especially *A. fumigatus* and *A. flavus*. It occurs in humans, animals, and birds. It might result in acute or chronic infections. Acute aspergillosis is associated with severely compromised immune patients. Patients with respiratory illness such as cystic fibrosis, asthma, tuberculosis, sarcoidosis, and chronic obstructive pulmonary disease are more susceptible to chronic infections of aspergillosis (Smith and Denning 2011; Denning et al. 2013; Denning et al. 2014; Warris et al. 2019).

3.4.3.3   Candidiasis
It is caused by different types of *Candida* with the most prominent being *Candida albicans*. It affects the mouth where it causes white patches on tongue, mouth, and throat and may lead to soreness. It also causes infections in vagina. Rare candidial infections might also become invasive and spread to other parts of the body (Andrews and Domonkos 1963).

3.4.3.4   Entomophthoromycosis
It is a rare fungal infection affecting immunocompromised patients and is dominant in tropical and subtropical regions. Depending on the affected tissue, it can be classified as basidiobolomycosis and conidiobolomycosis. Basidiobolomycosis is caused by *Basidiobolus ranarum* and occurs due to implantation of fungus into

subcutaneous tissues of gluteal muscles, thighs, or torso. Conidiobolomycosis is caused by *Conidiobolus coronatus* and *C. incongruous*. It occurs when fungal spores enter nasal tissue, facial soft tissue, and paranasal sinuses (Gugnani 1992; Ribes et al. 2000). It is generally limited to rhinofacial area (Costa et al. 1991; El-Shabrawi et al. 2014).

3.4.3.5   Fungal Keratitis
It is an inflammation of eye's cornea caused by species of *Fusarium*, *Aspergillus*, *Candida*, basidiomycetes, or Mucorales (Bongomin et al. 2017).

3.4.3.6   Lobomycosis
It is a chronic skin and subcutaneous mycosis caused by *Lacazia loboi*. It is endemic to the Americas and Amazon basin (Francesconi et al. 2014). This results in appearance of cutaneous lesions which might be ulcerated, keloid-like, nodular, or plaque-like and develop slowly (Rodríguez-Toro 1993; Elsayed et al. 2004).

3.4.3.7   Pneumocystis Pneumonia
It is a fungal pneumonia caused by *Pneumocystis jirovecii*. *P. jirovecii* inhabit lungs of healthy individuals and make them susceptible to other lung infections. They are seen in people with weak immune system (Aliouat-Denis et al. 2008).

3.4.3.8   Pythiosis
It is a difficult-to-treat and life-threatening infectious disease caused by a water mold *Pythium insidiosum*. This disease is found in temperate, tropical, and subtropical regions of the world. Depending on the site of infection, they can be classified as systemic or vascular, cutaneous, and ocular. Vascular pythiosis is the most common and lethal form of pythiosis in humans (Sudjaritruk and Sirisanthana 2011).

3.4.3.9   Tinea Capitis
Tinea capitis is a fungal infection of the scalp and is also known as scalp ringworm, ringworm of the scalp, ringworm of hair, herpes tonsurans, and tinea tonsurans. It is caused by *Trichophyton* and *Microsporum* sp. that colonize hair shaft. It is mainly seen in prepubertal children and occurs more in boys than girls (Bongomin et al. 2017).

3.5   Protozoa

Protozoa, also called as protists, are unicellular eukaryotic organisms. They have complex intracellular structures and perform complex metabolic activities. They have vesicular nucleus (except ciliates) in which the chromatin is scattered. They are aerobic organisms, that is, they require oxygen for their survival. Protozoan forms a link between plants, animals, and fungi. In terms of number, diversity, and biomass, they constitute the largest group of organisms. They consist of nucleus, complex cell organelles, and cellulosic cell wall. Their mode of nourishment is by absorption or ingestion of organic compounds through specialized structures.

3.5.1   Various Classes of Protozoa

Earlier the classification of protozoa was based on their mode of locomotion or the presence of different locomotory organs which are as follows:

- Class I: Mastigophora/Flagellata.
  They consist of a whip-like structure called flagella which helps the protozoan to propel forward. Their body is covered with chitin, silica, or cellulose. They are generally free-living or parasitic and reproduce sexually by longitudinal fission. Examples of this class are *Euglena*, *Giardia*, and *Trypanosoma*.
- Class II: Ciliata.
  They have tiny hair that beat to produce movement. Their body is covered by pellicle and their locomotion is by cilia. The asexual reproduction is by binary fission, whereas sexual reproduction is by conjugation. They consist of two types of nuclei, micronucleus and macronucleus. Examples of Ciliata are *Balantidium*, *Paramecium*, and *Vorticella*.
- Class III: Sarcodina/Amoeboids.
  Amoeboids have pseudopodia, false feet, which are used for feeding and locomotion. They are free-living and reproduce sexually by syngamy and asexually by binary fission. Some examples include *Amoeba* and *Entamoeba*.
- Class IV: Sporozoa.
  Sporozoans are non-motile. They are endoparasites, and their body is covered with pellicle. Their asexual reproduction is by fission and sexual reproduction is by spores. Examples of this class are *Monocystis* and *Plasmodium*.

In 1985, the Society of Protozoologists classified protozoa into six phyla: Sarcomastigophora, Apicomplexa, Entamoebidae, Trichomonadida, Naegleria, and *Balantidium*. Sarcomastigophora and Apicomplexa are the protozoan species which cause diseases in humans. Sarcomastigophora consists of a wide range of protozoa that either consists of flagella (Class Mastigophora) or pseudopods (Class Amoeboids). This phylum has both free-living and parasitic protozoans with reproduction by closed mitosis. *Trichomonas*, *Giardia*, *Leishmania*, and *Trypanosoma* are the disease-causing protozoa under this phylum.

Apicomplexa are parasitic protozoans comprising of ciliates and dinoflagellates. This phylum consists of cortical alveolae which are flattened vesicle-like structures found beneath the plasma membrane. Initially, this phylum was a part of Class Sporozoa. These cause serious illness in birds, animals, and humans, such as leucocytòzoonosis in birds, babesiosis in cattle and dogs, and malaria in humans. The species under this phylum consist of a group of secretory organelles called apicoplast. This helps in invasion of host cells by parasitic cells.

Entamoebidae is a family of Archamoeba and comprise of *Entamoeba*, *Endolimax*, and *Iodamoeba*. They lack mitochondria and perform anaerobic respiration. They consist of vesicular nucleus with a central endosome. The microorganisms belonging to this phylum are responsible for causing diseases of digestive systems in animals and humans. Trichomonadida consists of four to six flagella and one or two nuclei. They reproduce asexually by binary fission. They are further subdivided into Monocercomonadidae and Trichomonadidae. They consist of both pathogenic and non-pathogenic strains. The non-pathogenic strains are found in alimentary canal and reproductive tract. The pathogenic Trichomonadida belong to the genus *Giardia*, *Hexamita*, *Trichomonas*, and *Tritrichomonas*.

Naegleria is a free-living protozoan which is found in soil and aquatic environments. The life cycle of *Naegleria* consists of three stages, amoeboid, cyst, and flagellated stages, and it can easily change from amoeboid to flagellated stage. *Naegleria fowleri*, also called as brain-eating amoeba, is a human pathogenic strain responsible for causing primary amoebic meningoencephalitis (PAM). *Balantidium coli*, belonging to the phylum *Balantidium*, is a large ciliated protozoan and causes infection in humans. It is responsible for causing balantidiasis in humans. It also infects domestic and wild mammals, such as pigs and monkeys.

According to the means of nutrition, they are grouped as autotrophs or heterotrophs. *Euglena* contains chloroplasts and synthesizes their own food, and hence they are classified as autotrophs. *Amoeba*, on the other hand, cannot synthesize their food, and hence, they are heterotrophs. Protozoans can either be free-living or parasitic, unicellular, or colonial. Protozoa reproduce either asexually, as in amoeba and flagellates, or both asexually and sexually as in Apicomplexa. Binary fission is the most common type of asexual reproduction seen in protozoans. Protozoans help in controlling biomass as they consume bacteria. The diseases caused by protozoans can be mild or life-threatening depending upon the species and strain of the parasite.

3.5.2   Pathogenesis of Protozoa

Even though protozoa are eukaryotic pathogens, they also produce adhesins and toxins analogous to bacterial pathogens. They undergo antigenic variation and are able to survive inside phagocytic vesicles. Different strains of protozoa have different mechanism for invasion of host cells.

Giardia lamblia have a large adhesive disk of microtubules which helps in their attachment to intestinal mucosa of host cells. When the protozoa adhere to the host cells, the protozoan flagella move in a way that removes fluid from under the disk. This results in generation of an area of low pressure which helps in its adhesion to epithelial cells. *Giardia* causes inflammation through the release of cytopathic substances and shortens the intestinal villi. This inhibits the absorption of nutrients.

Plasmodium falciparum resides inside the red blood cells and produces PfEMP1, an adhesin membrane protein. This protein is present on the surface of infected red blood cells and causes the blood cells to stick to each other and to the walls of blood vessels. This leads to reduced blood flow and, in severe cases, anemia, jaundice, organ failure, and death. PfEMP1 is recognized by host immune system, but antigenic variation in its structure prevents the protein from getting easily recognized by host immune cell.

Trypanosoma brucei produces capsules made up of a dense glycoprotein (similar to bacterial capsule) to prevent phagocytosis by host immune system. The antibodies produced by host immune system can recognize the coat, but the pathogen alters the structure of capsules by antigenic variations and thereby evades its recognition by immune system.

3.5.3 Current Scenario of Pathogenic Protozoa

Infections caused by pathogenic protozoa cause around 58 million cases of diarrhea every year with 1.8 million deaths per year (Putignani and Menichella 2010; Thompson and Ash 2016). It has also been reported by WHO in 2004 that diarrhea affects more individuals worldwide than any other disease (Press and Geneva 2008). Most intestinal parasites result in malnutrition, iron deficiencies, and long-term adverse effects (Organization 2002), while many species of enteric protozoa are linked with diarrhea in humans. Some of the pathogenic protozoa that cause diarrhea are discussed below.

3.5.3.1 *Cryptosporidium* Species
These species are responsible for causing infections in AIDS patients (Sterling and Adam 2006). *Cryptosporidium* causes approximately 20% of diarrheal cases in children in developing countries and 9% in developed countries (Xiao 2010; Rimšelienė et al. 2011).

3.5.3.2 *Giardia intestinalis*
Giardia intestinalis is another common protozoan parasite that causes giardiasis. Giardiasis infects approximately 280 million people every year and results in 2.5 million deaths annually which consists of 2–7% in developed countries and 20–30% in developing countries (Franzen et al. 2009; Jerlström-Hultqvist et al. 2010). Hence, this becomes the second largest protozoan infection causing morbidity and mortality next to malaria (Menkir and Mengestie 2014).

3.5.3.3 *Entamoeba* Species
Six species of *Entamoeba* have been found in human: *E. histolytica*, *E. coli*, *E. dispar*, *E. hartmanni*, *E. moshkovskii*, and *E. polecki*. Only *E. histolytica* is pathogenic among these, and as per WHO reports, it infects approximately 500 million people annually with around 100,000 deaths worldwide every year (Jackson 1998; World Health Organization 2004; Lebbad 2010; Santos et al. 2010).

3.5.3.4 *Balantidium coli*

Balantidium coli is the largest protozoan infecting humans (Farthing and Kelly 2005; Solaymani-Mohammadi and Petri Jr. 2006) and causes balantidiasis. This leads to perforation in the intestine. It has a mortality rate of 30% (Ferry et al. 2004; Schuster and Ramirez-Avila 2008). Balantidiasis has been reported in Finland, Sweden, northern Russia, rural South America, Southeast Asia, and Western Pacific Islands (Esteban et al. 1998; Ferry et al. 2004; Schuster and Visvesvara 2004; Solaymani-Mohammadi and Petri Jr. 2006; Schuster and Ramirez-Avila 2008).

Cyclospora cayetanensis also causes epidemic or endemic diarrhea in children and adults (Chacín-Bonilla 2010). Other protozoans causing diarrhea in humans are *Dientamoeba fragilis*, *Blastocystis* species, and *Cystoisospora belli* (Fletcher et al. 2012).

Some of the other common diseases caused by protozoa are:

- Malaria caused by *Plasmodium falciparum*, *P. malariae*, *P. vivax*, and *P. ovales*.
- Trypanosomiasis (African sleeping sickness) caused by *Trypanosoma brucei gambiense* (TbG) and *Trypanosoma brucei rhodesiense* (TbR).
- Chagas disease (American trypanosomiasis) caused by *Trypanosoma cruzi*.
- Lambliasis (beaver fever) caused by *Giardia lamblia*.
- Babesiosis caused by *Babesia microti*.
- *Sappinia* amoebic encephalitis (SAE) caused by *Sappinia pedata*.
- Blastocystosis caused by *Blastocystis hominis*.
- Trichomoniasis caused by *Trichomonas vaginalis*.
- Toxoplasmosis caused by *Toxoplasma gondii*.
- Schistosomiasis caused by *Schistosoma* species.

3.6 Algae

Algae are unicellular or multicellular eukaryotes. They are also called cyanobacteria or blue-green algae. These organisms obtain nourishment by photosynthesis and produce carbohydrates and oxygen. Their habitats include damp soil, water, and rocks. Cyanobacteria are aquatic organisms that are unicellular or colonial.

Cyanobacteria are thought to be the precursors of green land plants. Millions of years ago, chloroplast, the organelle of the plant which produces chlorophyll, was believed to be free-living cyanobacteria. In the late Proterozoic era, cyanobacteria began to reside within eukaryotic cells, and these cyanobacteria generate energy for host cells through a process called endosymbiosis. This theory of endosymbiosis is supported by the structural and genetic similarities between cyanobacteria and chloroplasts.

Some of the algae reproduce by spores which are motile reproductive cells. Algae reproduce both sexually and asexually. The sexual reproduction of algae is by the formation of genetically diverse gametes by meiosis. Two gametes from different individuals join to form a new individual. Some types of algae undergo simple sexual reproduction where the algae themselves act as gametes. In other types of algae, the process involves egg- and sperm-like cells and sex-attractant pheromones. Algae are believed to be one of the first organisms to undergo sexual reproduction seen in plants and animals today.

Based on the estimates by Guiry (2012), around 37,000 species of algae and 4000 species of cyanobacteria have been identified, and 30,000 species are yet to be discovered. Algae have existed for more than two billion years now. Algae, being autotrophs, are the key producers of the food in the aquatic ecosystem. They are the major source of food for fishes which are indirectly the food for many animals. Hence, they are an energy source which powers the entire ecosystem. Apart from being a food source, they supply oxygen to marine animals. The kelps (kombu) and red algae *Porphyra* (nori) have been cultivated and harvested in the Pacific Basin to be used as a source of food in Asia for hundreds of years. Kelp is also used as fertilizer. Kelp ash has been industrially used for its potassium and sodium salts. Agar and carragee are some of the other algal products which are used as stabilizer in foods, cosmetics, and paints.

3.6.1 Algal Classification

Fritsch, in 1938, provided the classification of algae in his book *The Structure and Reproduction of Algae*. He classified algae into 11 classes based on types of pigments present, types of flagella, mode of reproduction, structure of thallus, and assimilatory products.

Class I: Chlorophyceae.

These are green algae and contain chlorophyll a and b, carotenoids, and xanthophylls as pigments. The cell wall is made up of cellulose and chloroplast has pyrenoids. The flagella are of equal lengths. Reproduction in Chlorophyceae is by vegetative, sexual, and asexual means. Chlorophyceae is subdivided into nine orders:

Order 1: Volvovales, for example, *Volvox*.
Order 2: Chlorococcales, for example, *Chlorella*.
Order 3: Ulotrichales, for example, *Ulothrix*.
Order 4: Cladophorales, for example, *Cladophora*.
Order 5: Chaetophorales, for example, *Fritschiella*.
Order 6: Oedogoniales, for example, *Oedogonium*.
Order 7: Conjugales, for example, *Zygnema*.
Order 8: Siphonales, for example, *Vaucheria*.
Order 9: Charales, for example, *Chara*.

Class II: Xanthophyceae.

These are yellow-green algae and have chlorophyll a and e, β-carotene, and xanthophylls. They have plastids without pyrenoids. Their cell wall is made up of pectic substance and cellulose. They store reserve food material in the form of oil. They have two unequal flagella inserted anteriorly with short whiplash and longer tinsel type. Reproduction is by vegetative, sexual, and asexual means. Xanthophyceae is divided into four orders:

Order 1: Heterochloridales, for example, *Heterochloris* and *Chloramoeba.*
Order 2: Heterococcales, for example, *Myxochloris* and *Halosphaera.*
Order 3: Heterotrichales, for example, *Tribonema* and *Microspora.*
Order 4: Heterosiphonales, for example, *Botrydium.*
Class III: Chrysophyceae.

The dominant pigment present in this class is phycochrysin which gives it brown or orange color. Chromatophores have naked pyrenoid-like bodies. Their cell wall consists of silica or calcium. They store reserve food in the form of chrysolaminarin and leucosin. They are motile cells with two equal or unequal flagella inserted anteriorly. Sexual reproduction is rare in this class but, if occurs, is isogamous. This class is further classified into three orders:

Order 1: Chrysomonadales, for example, *Chrysococcus*, *Chrysodendron*, and *Chromulina.*

Order 2: Chrysosphaerales, for example, *Chrysosphaera* and *Echinochrysis.*

Order 3: Chrysotrichales, for example, *Nematochrysis* and *Chrysoclonium.*
Class IV: Bacillariophyceae.

They are diatoms and yellow or golden-brown algae. The dominating pigments in this class are golden-brown pigments, fucoxanthin, diatoxanthin, and diadinoxanthin. The chromatophores have pyrenoids, and their cell wall is made up of pectin and silica. They produce fats and volutin as a result of photosynthesis. The mitotic cells are flagellated with a single flagellum. The reproduction is sexual by fusion of gametes. This class is subdivided into two orders:

Order 1: Centrales, for example, *Cyclotella* and *Chaetoceras.*

Order 2: Pennales, for example, *Grammatophora*, *Navicula*, and *Pinnularia.*
Class V: Cryptophyceae.

The main pigment in this class is xanthophylls which gives the algae red or brown color. They have pyrenoid-like bodies which are independent of chromatophores. The cells are motile with anteriorly inserted unequal flagella. Sexual reproduction in this class is rare. Cryptophyceae is further subdivided into two orders:

Order 1: Cryptomonadales, for example, *Cryptomonas*, *Rhodomonas*, and *Cyanomonas.*

Order 2: Cryptococcales, for example, *Tetragonidium.*
Class VI: Dinophyceae (Peridineae).

The main pigment is xanthophylls which gives the algae red or brown color. Many species of this class are colorless saprophytes. The cell wall is made up of cellulose. The cells are motile with two flagella. Food is stored in the form of starch or fats. Sexual reproduction is rare in this class but, if present, is of isogamous type. This class can be classified into six orders:

Order 1: Desmonadales, for example, *Desmocapsa*, *Pleromonas*, and *Desmomastix.*

Order 2: Thecatales, for example, *Exuviaella* and *Prorocentrum.*

Order 3: Dinophyceales, for example, *Dinophysis*, *Ornithocercus*, and *Phalacroma.*

Order 4: Dinoflagellata, for example, *Amphidinium, Blastidinium, Ceratium,* and *Heterocapsa.*

Order 5: Dinococcales, for example, *Dinastridium, Cystodinium*, and *Dissodinium.*

Order 6: Dinotrichales, for example, *Dinothria* and *Dinoclonium.*

Class VII: Chloromonadineae.

They have bright green tint due to the presence of excess of xanthophylls. They do not have pyrenoids but have numerous disk-shaped chromatophores. They are motile cells with equal flagella. Their reserve food material as fats and oil. Chloromonadineae is subdivided into only one order:

Order 1: Chloromonadales, for example, *Trentonia* and *Vacuolaria.*

Class VIII: Euglenineae.

They are unicellular green algae with many chromatophores in all cells and pyrenoid-like bodies in some. They have one or two flagella that arise from an invagination at the anterior end of the cell. The reserve food material is a polysaccharide, paramylon. This class is subdivided into three orders:

Order 1: Euglenaceae, for example, *Euglena.*

Order 2: Astasiaceae, for example, *Astasia.*

Order 3: Peranemacea, for example, *Anisonema.*

Class IX: Phaeophyceae.

They are brown algae and contain the pigment, fucoxanthin, in the chromatophores. The lower forms of Phaeophyceae have naked pyrenoid-like bodies. The cell wall is made up of cellulose with alginic acid and fucinic acid. The cells have two lateral or subapical flagella. They reserve food material as laminarin (polysaccharide) and mannitol (alcohol). The reproduction is sexual from isogamous to oogamous type. They are divided into nine orders:

Order 1: Ectocarpales, for example, *Ectocarpus, Punctaria*, and *Holothrix.*

Order 2: Tilopteridales, for example, *Tilopteris.*

Order 3: Cutleriales, for example, *Cutleria.*

Order 4: Sporochnales, for example, *Sporochnus.*

Order 5: Desmarestiales, for example, *Desmarestia.*

Order 6: Laminariales, for example, *Laminaria, Chorda*, and *Alaria.*

Order 7: Sphacelariales, for example, S*phacelaria* and *Halopteria.*

Order 8: Dictyotales, for example, *Dictyota, Hormosira*, and *Ascoseira.*

Order 9: Fucales, for example, *Fucus* and *Sargassum.*

Class X: Rhodophyceae.

These are mostly marine algae with uniaxial or multiaxial thalli. The chromatophores contain pigments such as r-phycoerythrin and r-phycocyanin which impart red color. The outer cell wall is made up of pectin, whereas the inner cell wall is made of cellulose. The reserve food is floridean starch. They are non-motile and sexual reproduction is oogamous type. This class consists of seven orders:

Order 1: Bangiales, for example, *Bangia* and *Porphyra.*

Order 2: Nemationales, for example, *Batrachospermum.*

Order 3: Gelidiales, for example, *Gelidium.*

Order 4: Cryptonemiales, for example, *Corallina* and *Gloiopeltis.*

Order 5: Gigartinales, for example, *Chondrus, Gigartina* and *Gracilaria.*

Order 6: Rhodomeniales, for example, *Rhodymenia.*
Order 7: Ceramiales, for example, *Polysiphonia, Ceramium,* and *Lophosiphonia.*
Class XI: Myxophyceae.
The pigments present in Myxophyceae are chlorophyll a, β-carotene, and c-phycocyanin. They have rudimentary nucleus with no chromatophores. Their cell wall is made up of mucopolymers. Sexual reproduction is absent in this class. The reserve food is in the form of cyanophycean starch. They are subdivided into five orders:
Order 1: Chlorococcales, for example, *Chlorococcus, Gloeocapsa,* and *Microcystis.*
Order 2: Chaemaesiphonales, for example, *Chamesiphon* and *Dermocarpa.*
Order 3: Pleurocapsales, for example, *Pleurocapsa.*
Order 4: Nostocales, for example, *Nostoc, Oscillatoria,* and *Spirulina.*
Order 5: Stigonematales, for example, *Stigonema*

3.6.2   Microscopic Algae as Pathogens: Mechanism of Pathogenicity

Although algae are non-pathogenic, some might produce toxins. For example, algal blooms produce high concentrations of toxins that hampers with functions of the nervous system and liver in humans and animals. Neurotoxins produced by some dinoflagellates cause paralysis in fish and human. The neurotoxins can be taken up by feeding on the organisms that have consumed dinoflagellates or by coming in contact with water contaminated with toxins of dinoflagellates.

Pfiesteria piscicida produces toxins in its life cycle that kill fish. The exposure to water containing *P. piscicida* can cause memory loss and confusion in humans. *Desmodesmus armatus* is a green alga which has found to cause swelling and redness of injured knee and foot along with fever and leukocytosis in patients who suffered injuries in freshwater. *Prototheca* species of algae grow in soil and sewage water. They do not contain chlorophyll and hence do not undergo photosynthesis and cause infections in immunocompromised humans and animals by entering through wounds. They cause skin infections such as discharging ulcers and protothecosis in cats, dogs, cattle, and humans. *P. cutis* causes human chronic skin ulcers, whereas *P. wickerhamii* causes septicemia or meningitis. *Prototheca* sp. also causes infections in cattle, such as bovine mastitis, an inflammatory disease of the udder (Satoh et al. 2010).

The mechanism by which algae negatively affect humans and animals is unknown. The biofilms formed by *Prototheca* sp. consist of surface-attached cells which are linked together by matrix containing DNA and polysaccharides. These biofilms decrease the release of IL-6 by mononuclear immune cells, which renders them less susceptible to antimicrobial agents (Kwiecinski 2015).

3.6.3   Disease-Causing Algae

Infections caused by algae in humans are very uncommon. Protothecosis is an infection caused by achlorophyllous algae, *Prototheca wickerhamii* and *P. zopfii*, in humans and animals with majority being caused by *P. wickerhamii* (Krcmery Jr. 2000, Consuelo Quinet Leimann et al. 2004). *Prototheca* sp. colonizes the fingernails, skin, digestive system, and respiratory tracts in humans (Huerre et al. 1993; Wirth et al. 1999). The introduction of *Prototheca* inside humans is via traumatic inoculation. Protothecosis is reported in all continents except Antarctica (Nelson et al. 1987).

Protothecosis are of three forms: cutaneous, disseminated, and olecranon bursitis (Krcmery Jr. 2000; Torres et al. 2003; Consuelo Quinet Leimann et al. 2004). Simple cutaneous infections account for almost half of protothecosis cases (Krcmery Jr. 2000; Consuelo Quinet Leimann et al. 2004). These occur in immunocompromised individuals undergoing treatment for AIDS, cancer, renal or hepatic diseases, or autoimmune disorders (Woolrich et al. 1994; Carey et al. 1997; Wirth et al. 1999; Torres et al. 2003; Consuelo Quinet Leimann et al. 2004). Lesions occur at the site of algal inoculation. Disseminated protothecosis is rare and also occurs in immunocompromised patients with cancer, AIDS, and organ transplant (Heney et al. 1991; Kunova et al. 1996; Wirth et al. 1999; Torres et al. 2003). The infected organs are subcutaneous tissue (Torres et al. 2003). In some cases, this might also result in central venous catheter-related algaemia along with fever and sepsis syndrome (Kunova et al. 1996; Torres et al. 2003). On the other hand, individuals with olecranon bursitis are not immunocompromised but have penetrating/non-penetrating trauma to the affected elbow (Nosanchuk and Greenberg 1973; de Montclos et al. 1995; Pfaller and Diekema 2005).

3.7   Virus

Viruses are non-cellular entities, that is, even though they are considered as microorganisms, they are non-living organisms. This is because they cannot reproduce outside a host and cannot perform metabolism on their own. Viruses are pathogens causing diseases in prokaryotes and eukaryotes. They are smaller than most of the microbes. They are made up of a nucleic acid core (DNA or RNA) surrounded with a protein coat called capsid.

Viruses are classified based on symmetry of capsid, dimensions of virion and capsid, presence or absence of envelope, and nature of nucleic acid. On the basis of morphology and symmetry of capsid, they can have helical or icosahedral symmetry. A virus is considered to have helical symmetry if the capsid is shaped into a filamentous or rod-shaped structure. The viruses with helical symmetry exhibit flexibility which is dependent upon the arrangement of capsomeres, subunits of capsid, in the capsid. Viruses with icosahedral symmetry have identical subunits making equilateral triangles and arranged symmetrically. Icosahedral symmetry is found in many animal viruses. This shape provides the virus a very stable shape and a lot of space inside to store nucleic acid.

3.7.1   Classification of Virus

Depending upon the nature of nucleic acid, David Baltimore classified viruses into seven groups in the early 1970s:

- Group I: Double-stranded DNA.
  These viruses have double-stranded DNA (dsDNA) as their genetic material. They replicate either in the nucleus using cellular proteins (adenoviruses) or in the cytoplasm by making their own enzymes for replication of nucleic acid (poxviruses). The transcription for the production of viral mRNA is similar to its cellular DNA. Adenoviruses, herpes viruses, papovaviruses, and poxviruses fall into this category.
- Group II: Single-stranded sense (+) DNA.
  Their genetic material is single-stranded DNA (ssDNA) which gets converted to double-stranded DNA intermediate before transcription. Replication takes place in nucleus. This involves the formation of a (−) sense strand, which serves as a template for (+) strand DNA and RNA synthesis. This group includes parvoviruses, circoviruses, anelloviruses, nanoviruses, and geminiviruses.
- Group III: Double-stranded RNA.
  This group has double-stranded RNA as the genetic material. The genomes of these viruses are segmented, and each genomic segment is separately transcribed to produce monocistronic mRNAs using RNA-dependent RNA polymerase encoded by virus. Examples of this group of viruses include reoviruses and birnaviruses.
- Group IV: Single-stranded sense (+) RNA.
  In this group, the genetic material is single-stranded RNA with positive polarity. The RNA can be directly accessed by host ribosomes to form proteins. They can be polycistronic mRNA where the genomic RNA is directly used as mRNA for protein translation which results in mature protein after cleavage or have a complex transcription process. They form replicative intermediates or intermediates of dsRNA while replicating genomic RNA. These intermediates result in the formation of RNA strands with negative polarity which acts as a template for the production of single-stranded RNA with positive polarity. This group includes picornaviruses, hepatitis A, togaviruses, astroviruses, coronaviruses, flaviviruses, calciviruses, and arteriviruses.
- Group V: Single-stranded (−) sense RNA.
  These viruses have single-stranded RNA with negative polarity. Here, the RNA and genes cannot be directly accessed by host ribosomes to form proteins and so must be transcribed by viral polymerases. Hence, in this group also, double-stranded RNA intermediates are formed to produce mRNA. The positive RNA strands act as template for the synthesis of negative RNA strands. The site of

replication of viruses containing non-segmented genomes is cytoplasm, whereas replication occurs in nucleus for viruses with segmented genomes. In both of the cases, viral RNA-dependent RNA polymerases produce monocistronic mRNAs. Examples of this group are orthomyxoviruses, rhabdoviruses, filoviruses, bunyaviruses, arenaviruses, and paramyxoviruses.

- Group VI: Single-stranded (+) sense RNA with DNA intermediate in life cycle.
  The viruses that belong to this group have two copies of single-stranded RNA genomes. This group uses the enzyme reverse transcriptase to convert RNA to DNA which is then transported to the nucleus of host cell. Instead of using RNA as templates for protein synthesis, they use DNA. The DNA is spliced in the host genome using integrase. Replication occurs with the help of host cell's polymerases. The viral DNA, integrated into the host genome, is transcribed into mRNA which is later translated into proteins. The viruses belonging to this group are retroviruses.
- Group VII: Double-stranded DNA with RNA intermediate.
  They consist of partial double-stranded DNA genome which makes single-stranded RNA intermediates acting as mRNA. They can also be converted to double-stranded DNA by undergoing reverse transcription, but this process occurs inside the virus particle upon maturation. The double-stranded genome is gapped which is filled to form covalently closed circular DNA to be used as template for mRNA production. This group includes hepadnaviruses.

Viruses are further classified based on many factors, such as the morphology of the virus, the presence of envelope, the type of nucleic acid present, the type of host, the mode of replication, etc. Their hosts range from single-celled bacteria, fungi, and protozoa to multicellular fungi, plants, and animals.

On the basis of the type of genetic material present, viruses can be classified into three types:

1. DNA viruses.
   These viruses use DNA as the genetic material and affect animals and humans. Their effects range from benign symptoms to serious health issues. Example: herpesvirus, papillomavirus, and parvovirus
2. RNA viruses.
   They have RNA as genetic material. Example: dengue virus, Ebola virus, hepatitis C virus, influenza virus, measles virus, poliovirus, rabies virus, rotavirus, and yellow fever virus
3. DNA-RNA viruses.
   This type contains both DNA and RNA as genetic material. Example: leukovirus and Rous's virus.

On the basis of the number of strands of genetic material present, viruses are classified into four types:

1. Double-stranded DNA.

Example: adenovirus; bacteriophages T2, T3, T4, T6, T7, and lambda; herpesvirus, and pox virus
2. Single-stranded DNA.
   Example: bacteriophages X, 74, and Φ
3. Double-stranded RNA.
   Example: reovirus of animals, rice dwarf virus of plants, and wound tumor virus
4. Single-stranded RNA.
   Example: avian leukemia virus, bacteriophage MS-2, influenza virus, poliomyelitis, and tobacco mosaic virus.

Viral envelope is the lipid-containing membrane which surrounds some virus particles. They are obtained during the process of viral maturation by budding through a cellular membrane. Glycoproteins encoded by virus are present on the surfaces of viral envelope as projections called as peplomers. On the basis of the presence of envelope, viruses are classified as:

1. Enveloped virus.
   Example:

- DNA viruses: hepadnavirus, herpesvirus, and poxvirus.
- RNA viruses: bunyavirus, coronavirus, filovirus, flavivirus, hepatitis D, orthomyxovirus, paramyxovirus, rhabdovirus, and togavirus.
- Retroviruses.

2. Non-enveloped virus.
   Example:

- DNA viruses: adenovirus, papovavirus, and parvovirus.
- RNA viruses: hepatitis A virus, hepatitis E virus, and picornavirus.

Mostly animal viruses are roughly spherical in shape except for few viruses. On the basis of the shape of viruses, they are classified as:

1. Bullet shaped.
   Example: rabies virus
2. Filamentous shaped.
   Example: Ebola virus
3. Brick shaped.
   Example: poxvirus
4. Space vehicle shaped.
   Example: adenovirus.

On the basis of structure, viruses are classified into the following four types:

1. Cubical virus.

They are also termed as icosahedral symmetry virus. Example: picornavirus and reovirus
2. Spiral virus.
   They are also known as helical symmetry virus. Example: orthomyxovirus and paramyxovirus
3. Radial symmetry virus.
   Example: bacteriophage
4. Complex virus.
   Example: poxvirus.

On the basis of capsid structure, viruses are classified as:

1. Naked icosahedral.
   Example: hepatitis A virus and poliovirus
2. Enveloped icosahedral.
   Example: Epstein-Barr virus, herpes simplex virus, HIV-1, rubella virus, and yellow fever virus
3. Enveloped helical.
   Example: influenza virus, measles virus, mumps virus, and rabies virus
4. Naked helical.
   Example: tobacco mosaic virus
5. Complex with many proteins.

They have a combination of icosahedral and helical capsid structures. Example: hepatitis B virus, herpesvirus, smallpox virus, and T4 bacteriophage.
   On the basis of the type of host, viruses are classified into three types:

1. Animal viruses.
   These viruses infect and live inside animal cells. Their genetic material is either RNA or DNA. Example: influenza virus, mumps virus, poliovirus, and rabies virus
2. Plant viruses.
   These viruses infect plants and their genetic material is RNA enclosed in a protein coat. Example: beet yellows virus, potato virus, tobacco mosaic virus, and turnip yellows virus
3. Bacteriophages.

They infect bacterial cells. Their genetic material is DNA, and each bacteriophage will infect only a particular strain or species of bacteria. Example: lambda and T4.
   On the basis of mode of transmission, viruses are classified as:

1. Virus transmitted through respiratory route.
   Example: rhinovirus and swine flu virus
2. Virus transmitted through fecal-oral route.
   Example: hepatitis A virus, poliovirus, and rotavirus

3. Virus transmitted through sexual contacts.
   Example: retrovirus
4. Viruses transmitted through blood transfusion.
   Example: hepatitis B virus and HIV
5. Zoonotic viruses transmitting through the bite of infected animals.
   Example: alphavirus, flavivirus, and rabies virus.

On the basis of site of replication and replication properties, viruses are classified into the following five types:

1. Replication and assembly in cytoplasm of host.
   All RNA viruses replicate and assemble in cytoplasm of host cell except influenza virus.
2. Replication in nucleus and assembly in cytoplasm of host.
   Example: influenza virus and poxvirus.
3. Replication and assembly in nucleus of host.
   All DNA viruses replicate and assemble in nucleus of host cell except poxvirus.
4. Replication through double-stranded DNA intermediate.
   Example: all DNA viruses, retroviruses, and some tumor-causing RNA virus.
5. Replication through single-stranded RNA intermediate.
   Example: all RNA viruses except reovirus and tumor-causing RNA viruses.

3.7.2 Mechanism of Pathogenesis of Viruses

Similar to bacteria, viruses also use adhesins to bind to host cells. Antigenic variation is a mechanism used by some enveloped viruses to evade host immune system. Adhesins are present on viral capsid or membrane envelope. The viral adhesin interacts with specific cell receptors of certain cells, tissues, and organs of hosts, for example, hemagglutinin on influenza virus. Hemagglutinin is a spike protein and helps in attachment of virus to sialic acid present on the membrane of intestinal and respiratory cells of host. Glycoprotein gp120 is also an adhesin which is found on HIV. Infection of HIV requires its interaction with two receptors present on host cells. In the first interaction, gp120 is bound to CD4 marker present on T helper cells. The second interaction of HIV with the host occurs before entry of virus into the cell. This interaction is between gp120 and either CCR5 and CXCR4 (chemokine receptors). The glycoproteins gB, gC, and gD are the adhesin molecules of herpes simplex virus I or II. These adhesins attach to heparan sulfate present on mucosal surface of the mouth and genitals.

Antigenic variation occurs in enveloped viruses with the most common being in influenza virus. Antigenic variation occurs by antigenic drift or antigenic shift. Antigenic drift results from mutations in genes involved for the synthesis of spike proteins, hemagglutinin, and/or neuraminidase. These variations are a result of antigenic changes occurring over time. Antigenic shift occurs by a process of gene reassortment of spike proteins between two influenza viruses which infect the same host. In this process, infection of a host cell with two different influenza viruses simultaneously leads to mixing of genes. As a result, the new virus consists of a mixture of proteins from both of the unmutated viruses. The antigenic variation in influenza virus occurs at a high rate, and hence the immune system cannot recognize the different strains of influenza virus leading to frequent outbreaks of flu.

3.7.3   Deadliest Diseases Caused by Various Strains of Virus in Humans

Viruses are responsible for causing many diseases in humans, animals, and plants. Some of them also infect bacteria called as bacteriophages. The most deadly diseases caused by viruses are mentioned below.

1. Marburg virus.
   Marburg virus was identified in 1967. Small outbreaks of this virus were seen in lab workers in Germany when they came in contact with infected monkeys from Uganda. It causes hemorrhagic fever that leads to high fever and bleeding throughout the body. This results in shock, organ failure, and death. According to the reports of the World Health Organization, the mortality rate during the outbreak was 25% which increased to 80% in the Democratic Republic of the Congo during 1998–2000 and in Angola during 2005 outbreak (Harding 2020).
2. Rabies.
   Due to the development of rabies vaccines in the 1920s, the disease has become rare in the developed world. But still the disease is present on all continents except Antarctica and is a serious problem in India and parts of Africa with over 95% of deaths in Asia and Africa. It generally affects populations living in remote areas. It causes inflammation of the brain and leads to death once symptoms appear (World Health Organization 2020a).
3. Ebola virus.
   Ebola virus causes Ebola virus disease (EVD) which was formerly known as Ebola hemorrhagic fever. The outbreak of Ebola was first struck simultaneously in 1976 in the Democratic Republic of the Congo and the Republic of Sudan. Ebola virus spreads through blood or other body fluids. The virus is transmitted from wild animals to humans and then spreads through contact with blood or other body fluids. The outbreak of Ebola in West Africa during 2014–2016 was largest since the discovery of the virus. The average fatality rate of EVD is around 50%. EVD also leads to external bleeding, diarrhea, and multiple organ failure along with fever, sore throat, muscle pain, and headache (World Health Organization 2020b).
4. Hantavirus.
   Hantavirus causes hantavirus pulmonary syndrome (HPS) in humans. The first case of HPS was reported in the USA in 1993 with the death of two people. The virus is transmitted by the exposure to droppings of those infected with HPS in

the USA with a mortality rate of 36% (Centers for Disease Control and Prevention 2017).

5. Influenza.

Approximately 500,000 people die of influenza worldwide during a flu season. The pandemic spreads even faster when a new flu strain emerges with a high mortality rate. Spanish flu pandemic, the deadliest flu pandemic, began in 1918 and killed around 50 million people (Harding 2020).

6. HIV.

HIV is the deadliest virus of all viruses. According to WHO reports, HIV has claimed almost 33 million lives so far. By the end of 2019, a total of approximately 38 million people were living with the virus. In 2019 alone, 1.7 million people were newly infected, whereas 690,000 died of the virus. The disease mainly affects middle- to low-income countries from where 95% of new cases are reported. It affects 1 in 25 adults in Africa which accounts for two-thirds of the infected patients (World Health Organization 2020c).

7. Dengue.

The cases of dengue virus were first reported in the 1950s in the Philippines and Thailand, and since then, it has grown dramatically across the globe. It is estimated that almost 390 million people are infected with dengue virus every year (Bhatt et al. 2013). The infections of dengue virus occurred in 129 countries (Brady et al. 2012) with 70% of the cases in Asia itself (Tjaden et al. 2013). The total number of dengue cases, as reported by WHO, was 505,430 in 2000 which increased to 2.4 million in 2010 and 4.2 million in 2019 indicating an eightfold increase. The deaths increased from 960 in 2000 to 4032 in 2015.

Year 2019 experienced the largest number of dengue cases globally with 3.1 million cases in the American region, 131,000 in Malaysia, 101,000 in Bangladesh, 320,000 in Vietnam, and 420,000 in the Philippines. Although the mortality rate of dengue fever is low, it can still cause dengue hemorrhagic fever and can increase the mortality to 20% if not treated. Even though the vaccine for dengue was approved by FDA in 2019, it can only be administered to patients who have previously contracted the disease; otherwise, there is a serious risk of developing severe dengue in unaffected people (World Health Organization 2020d).

8. Rotavirus.

Rotavirus causes severe diarrheal disease in children. According to WHO in 2013, approximately 215,000 children under the age of 5 years die every year from rotavirus infections. Children in developing and low-income group countries are most severely affected by rotavirus. Due to the availability of vaccines against rotavirus, the total cases and mortality rates have declined sharply (World Health Organization 2018).

9. SARS-CoV.

SARS-associated coronavirus (SARS-CoV) is a causative agent of viral respiratory disease, severe acute respiratory syndrome (SARS). It was first identified during an outbreak in China on February 2003 from where it spread to four other countries. Over the course of 2 years, the virus spread to 26 countries, infected more than 8000 people, and killed 770 people. The virus has a zoonotic origin which emerged in bats and was transmitted to nocturnal animals before infecting humans. During the progression of disease, the lungs of infected individuals become inflamed and filled with pus. The mortality rate associated with SARS was 3%. No new cases of SARS have been reported by WHO since May 2004 (World Health Organization 2004; Harding 2020).

10. MERS-CoV.

MERS-CoV causes Middle East Respiratory Syndrome (MERS). The outbreak of MERS was seen in Saudi Arabia in 2012 and in South Korea in 2015. This virus belongs to a family of coronaviruses, similar to SARS, and also has a zoonotic origin. The virus was transmitted from bats to camels before reaching humans. It causes fever, cough, and shortness of breath and might even progress to severe pneumonia in certain cases. It is the most lethal coronavirus with a mortality rate between 30 and 40%. Till 31 January 2020, the total number of MERS-CoV cases globally was 2519 with 866 deaths. From December 2019 to January 2020 alone, 19 additional cases of MERS-CoV were reported including 8 deaths (Harding 2020, World Health Organization 2020e).

11. Zika virus.

Zika virus disease is caused by mosquito-borne transmission of Zika virus. The outbreak of Zika virus was reported in Brazil in 2015 from where it spread to the Americas, Asia, and Africa (Lan et al. 2017). As of 2019, a total of 87 countries have reported cases of Zika. In 2018, the American region had 3473 of the laboratory-confirmed cases of Zika virus. In Brazil, the total cases were 19,020, and the highest cases were reported in Panama and Bolivia (World Health Organization 2019).

12. SARS-CoV-2.

Since its outbreak in late December 2019 in China, SARS-CoV-2 has created havoc throughout the world. Similar to SARS-CoV and MERS-CoV, this virus is also a zoonotic virus and has thought to be originated in bats from where it was transmitted to pangolins before humans. The disease caused by SARS-CoV-2, termed as COVID-19, also results in pneumonia and in extreme cases leads to multiple organ failure in infected individuals. As per the reports of WHO on 24 July 2020, a total of 15,296,926 individuals have been infected with the virus globally, and 628,903 individuals have succumbed to death making its mortality rate as 4% which is still increasing day by day. The highest cases and deaths have been reported in the American region (World Health Organization 2020f).

References

Alhazmi A (2015) Pseudomonas aeruginosa-pathogenesis and pathogenic mechanisms. Int J Biol 7 (2):44

Aliouat-Denis C-M, Chabé M, Demanche C, Viscogliosi E, Guillot J, Delhaes L, Dei-Cas E (2008) Pneumocystis species, co-evolution and pathogenic power. Infect Genet Evol 8(5):708–726

Andrews GC, Domonkos AN (1963) Diseases of the skin: clinical dermatology. Saunders, Philadelphia, PA

Anstead GM, Sutton DA, Graybill JR (2012) Adiaspiromycosis causing respiratory failure and a review of human infections due to Emmonsia and Chrysosporium spp. J Clin Microbiol 50 (4):1346–1354

Arias CA, Murray BE (2012) The rise of the Enterococcus: beyond vancomycin resistance. Nat Rev Microbiol 10(4):266–278

Azam MW, Khan AU (2019) Updates on the pathogenicity status of Pseudomonas aeruginosa. Drug Discov Today 24(1):350–359

Belay N, Johnson R, Rajagopal B, De Macario EC, Daniels L (1988) Methanogenic bacteria from human dental plaque. Appl Environ Microbiol 54(2):600–603

Belay N, Mukhopadhyay B, De Macario EC, Galask R, Daniels L (1990) Methanogenic bacteria in human vaginal samples. J Clin Microbiol 28(7):1666–1668

Bhatt S, Gething PW, Brady OJ, Messina JP, Farlow AW, Moyes CL, Drake JM, Brownstein JS, Hoen AG, Sankoh O (2013) The global distribution and burden of dengue. Nature 496 (7446):504–507

Bond JH Jr, Engel RR, Levitt MD (1971) Factors influencing pulmonary methane excretion in man: an indirect method of studying the in situ metabolism of the methane-producing colonic bacteria. J Exp Med 133(3):572–588

Bonelo G, Ventosa A, Megás M, Ruiz-Berraquero F (1984) The sensitivity of halobacteria to antibiotics. FEMS Microbiol Lett 21(3):341–345

Bongomin F, Gago S, Oladele RO, Denning DW (2017) Global and multi-national prevalence of fungal diseases—estimate precision. J Fungi 3(4):57

Brady OJ, Gething PW, Bhatt S, Messina JP, Brownstein JS, Hoen AG, Moyes CL, Farlow AW, Scott TW, Hay SI (2012) Refining the global spatial limits of dengue virus transmission by evidence-based consensus. PLoS Negl Trop Dis 6(8):e1760

Buyuksirin M, Ozkaya S, Yucel N, Guldaval F, Ceylan K, Polat GE (2011) Pulmonary adiaspiromycosis: the first reported case in Turkey. Respiratory Med CME 4(4):166–169

Calvet X, Ramírez Lázaro MJ, Lehours P, Mégraud F (2013) Diagnosis and epidemiology of helicobacter pylori infection. Helicobacter 18:5–11

Carey WP, Kaykova Y, Bandres JC, Sidhu GS, Bräu N (1997) Cutaneous protothecosis in a patient with AIDS and a severe functional neutrophil defect: successful therapy with amphotericin B. Clin Infect Dis 25(5):1265–1266

Carlsson J (2000) Growth and nutrition as ecological factors. In: Kuramitsu HK, Ellen RP (eds) Oral bacterial ecology: the molecular basis. Horizon Scientific Press, Wymondham, UK, pp 67–130

Casanueva A, Galada N, Baker GC, Grant WD, Heaphy S, Jones B, Yanhe M, Ventosa A, Blamey J, Cowan DA (2008) Nanoarchaeal 16S rRNA gene sequences are widely dispersed in hyperthermophilic and mesophilic halophilic environments. Extremophiles 12(5):651–656

Centers for Disease Control and Prevention (2017) Hantavirus disease, by state of reporting. https://www.cdc.gov/hantavirus/surveillance/reporting-state.html Accessed 25 July 2020

Chacín-Bonilla L (2010) Epidemiology of Cyclospora cayetanensis: a review focusing in endemic areas. Acta Trop 115(3):181–193

Consuelo Quinet Leimann B, Monteiro PCF, Lazéra M, Ulloa Candanoza ER, Wanke B (2004) Protothecosis. Med Mycol 42(2):95–106

Costa AR, Porto E, Pegas JRP, dos Reis VMS, Pires MC, da Silva Lacaz C, Rodrigues MC, Müller H, Cucé LC (1991) Rhinofacial zygomycosis caused by Conidiobolum coronatus. Mycopathologia 115(1):1–8

D'Aoust J-Y (1991) Pathogenicity of foodborne Salmonella. Int J Food Microbiol 12(1):17–40

de Montclos M, Chatte G, Perrin-Fayolle M, Flandrois J-P (1995) Olecranon bursitis due to Prototheca wickerhamii, an algal opportunistic pathogen. Eur J Clin Microbiol Infect Dis 14 (6):561–562

Denning DW, Pashley C, Hartl D, Wardlaw A, Godet C, Del Giacco S, Delhaes L, Sergejeva S (2014) Fungal allergy in asthma–state of the art and research needs. Clin Transl Allergy 4(1):14

Denning DW, Pleuvry A, Cole DC (2013) Global burden of chronic pulmonary aspergillosis complicating sarcoidosis. Eur Respir J 41(3):621–626

Dhagat S, Jujjavarapu SE (2020) Isolation of a novel thermophilic bacterium capable of producing high-yield bioemulsifier and its kinetic modelling aspects along with proposed metabolic pathway. Braz J Microbiol 51(1):135–143

Dion CF, Ashurst JV (2020) Streptococcus Pneumoniae. In: StatPearls. StatPearls Publishing LLC, Treasure Island, FL

Donnenberg MS, Whittam TS (2001) Pathogenesis and evolution of virulence in enteropathogenic and enterohemorrhagic *Escherichia coli*. J Clin Invest 107(5):539–548

Eckburg PB, Lepp PW, Relman DA (2003) Archaea and their potential role in human disease. Infect Immun 71(2):591–596

Elsayed S, Kuhn SM, Barber D, Church DL, Adams S, Kasper R (2004) Human case of lobomycosis. Emerg Infect Dis 10(4):715

El-Shabrawi MH, Arnaout H, Madkour L, Kamal NM (2014) Entomophthoromycosis: a challenging emerging disease. Mycoses 57:132–137

Eng S-K, Pusparajah P, Ab Mutalib N-S, Ser H-L, Chan K-G, Lee L-H (2015) Salmonella: a review on pathogenesis, epidemiology and antibiotic resistance. Front Life Sci 8(3):284–293

Esteban J-G, Aguirre C, Angles R, Ash LR, Mas-Coma S (1998) Balantidiasis in Aymara children from the northern Bolivian Altiplano. Am J Trop Med Hygiene 59(6):922–927

Eswari J, Yadav M (2019) New perspective of drug discovery from herbal medicinal plants: Andrographis paniculata and Bacopa monnieri (terpenoids) and novel target identification against Staphylococcus aureus. S Afr J Bot 124:188–198

Eswari JS, Dhagat S, Sen R (2019) Introduction. In: Thermophiles for biotech industry: a bioprocess technology perspective. Singapore, Springer, pp 1–30

Fair RJ, Tor Y (2014) Antibiotics and bacterial resistance in the 21st century. Perspect Med Chem 6: S14459

Farthing MJ, Kelly P (2005) Protozoal gastrointestinal infections. Medicine 33(4):81–83

Ferry T, Bouhour D, De Monbrison F, Laurent F, Dumouchel-Champagne H, Picot S, Piens M, Granier P (2004) Severe peritonitis due to Balantidium coli acquired in France. Eur J Clin Microbiol Infect Dis 23(5):393–395

Fisher K, Phillips C (2009) The ecology, epidemiology and virulence of Enterococcus. Microbiology 155(6):1749–1757

Fletcher SM, Stark D, Harkness J, Ellis J (2012) Enteric protozoa in the developed world: a public health perspective. Clin Microbiol Rev 25(3):420–449

Francesconi VA, Klein AP, Santos APBG, Ramasawmy R, Francesconi F (2014) Lobomycosis: epidemiology, clinical presentation, and management options. Ther Clin Risk Manag 10:851

Franzen O, Jerlström-Hultqvist J, Castro E, Sherwood E, Ankarklev J, Reiner DS, Palm D, Andersson JO, Andersson B, Svärd SG (2009) Draft genome sequencing of Giardia intestinalis assemblage B isolate GS: is human giardiasis caused by two different species? PLoS Pathog 5 (8):e1000560

Gales A, Jones R, Turnidge J, Rennie R, Ramphal R (2001) Characterization of Pseudomonas aeruginosa isolates: occurrence rates, antimicrobial susceptibility patterns, and molecular typing in the global SENTRY Antimicrobial Surveillance Program, 1997–1999. Clin Infect Dis 32 (Suppl_2):S146–S155

Gellatly SL, Hancock RE (2013) Pseudomonas aeruginosa: new insights into pathogenesis and host defenses. Pathogens Dis 67(3):159–173

Gugnani H (1992) Entomophthoromycosis due to conidiobolus. Eur J Epidemiol 8(3):391–396

Guiry MD (2012) How many species of algae are there? J Phycol 48(5):1057–1063

Haines A, Dilawari J, Metz G, Blendis L, Wiggins H (1977) Breath-methane in patients with cancer of the large bowel. Lancet 310(8036):481–483

Harding, A. (2020) The 12 deadliest viruses on earth. https://www.livescience.com/56598-deadliest-viruses-on-earth.html Accessed 25 July 2020

Harrison A, Mason KM (2015) Pathogenesis of Haemophilus influenzae in humans. In: Human emerging and re-emerging infections: viral and parasitic infections. Wiley, Hoboken, NJ, pp 517–533

Heney C, Greeff M, Davis V (1991) Hickman catheter-related protothecal algaemia in an immuno-compromised child. J Infect Dis 163(4):930–931

Hidron AI, Edwards JR, Patel J, Horan TC, Sievert DM, Pollock DA, Fridkin SK, Team NHSN, Facilities PNHSN (2008) Antimicrobial-resistant pathogens associated with healthcare-associated infections: annual summary of data reported to the National Healthcare Safety Network at the Centers for Disease Control and Prevention, 2006–2007. Infect Control Hosp Epidemiol 29(11):996–1011

Higuita NIA, Huycke MM (2014) Enterococcal disease, epidemiology, and implications for treatment. Enterococci: from commensals to leading causes of drug resistant infection [internet], Massachusetts Eye and Ear Infirmary

Hohn MJ, Hedlund BP, Huber H (2002) Detection of 16S rDNA sequences representing the novel phylum "Nanoarchaeota": indication for a wide distribution in high temperature biotopes. Syst Appl Microbiol 25(4):551–554

Horan T, Culver D, Jarvis W, Emori G, Banerjee S, Martone W, Thornsberry C (1988) Pathogens causing nosocomial infections preliminary data from the national nosocomial infections surveillance system. Antimicrob Newsl 5(9):65–67

Huerre M, Ravisse P, Solomon H, Ave P, Briquelet N, Maurin S, Wuscher N (1993) Human protothecosis and environment. Bull Soc Pathol Exot 86(5 Pt 2):484–488

Jackson T (1998) Entamoeba histolytica and Entamoeba dispar are distinct species; clinical, epidemiological and serological evidence. Int J Parasitol 28(1):181–186

Jerlström-Hultqvist J, Ankarklev J, Svärd SG (2010) Is human giardiasis caused by two different Giardia species? Gut Microbes 1(6):379–382

Joly-Guillou M-L (2005) Clinical impact and pathogenicity of Acinetobacter. Clin Microbiol Infect 11(11):868–873

Kao C-Y, Sheu B-S, Wu J-J (2016) Helicobacter pylori infection: an overview of bacterial virulence factors and pathogenesis. Biom J 39(1):14–23

Kaper JB, Nataro JP, Mobley HL (2004) Pathogenic escherichia coli. Nat Rev Microbiol 2 (2):123–140

Karlin DA, Jones R, Stroehlein JR, Mastromarino AJ, Potter GD (1982) Breath methane excretion in patients with unresected colorectal cancer. J Natl Cancer Inst 69(3):573–576

Kau AL, Martin SM, Lyon W, Hayes E, Caparon MG, Hultgren SJ (2005) Enterococcus faecalis tropism for the kidneys in the urinary tract of C57BL/6J mice. Infect Immun 73(4):2461–2468

Krcmery V Jr (2000) Systemic chlorellosis, an emerging infection in humans caused by algae. Int J Antimicrob Agents 15(3):235–237

Krishnan L, Dicaire CJ, Patel GB, Sprott GD (2000) Archaeosome vaccine adjuvants induce strong humoral, cell-mediated, and memory responses: comparison to conventional liposomes and alum. Infect Immun 68(1):54–63

Krishnan L, Sad S, Patel GB, Sprott GD (2001) The potent adjuvant activity of archaeosomes correlates to the recruitment and activation of macrophages and dendritic cells in vivo. J Immunol 166(3):1885–1893

Kulik EM, Sandmeier H, Hinni K, Meyer J (2001) Identification of archaeal rDNA from subgingival dental plaque by PCR amplification and sequence analysis. FEMS Microbiol Lett 196(2):129–133

Kunova A, Kollar T, Spanik S, Krcméry V (1996) First report of Prototheca wickerhami algaemia in an adult leukemic patient. J Chemother 8(2):166–167

Kusters JG, Van Vliet AH, Kuipers EJ (2006) Pathogenesis of helicobacter pylori infection. Clin Microbiol Rev 19(3):449–490

Kwiecinski J (2015) Biofilm formation by pathogenic Prototheca algae. Lett Appl Microbiol 61 (6):511–517

Lan PT, Quang LC, Huong VTQ, Thuong NV, Hung PC, Huong TTLN, Thao HP, Thao NTT, Mounts AW, Nolen LD (2017) Fetal Zika virus infection in Vietnam. PLoS Curr 9:5693598

Lebbad M (2010) Molecular diagnosis and characterization of two intestinal Protozoa: Entamoeba histolytica & Giardia intestinalis, Institutionen för mikrobiologi, tumör-och cellbiologi/department of . . .

Lee C-R, Lee JH, Park M, Park KS, Bae IK, Kim YB, Cha C-J, Jeong BC, Lee SH (2017) Biology of Acinetobacter baumannii: pathogenesis, antibiotic resistance mechanisms, and prospective treatment options. Front Cell Infect Microbiol 7:55

Lenz JD, Dillard JP (2018) Pathogenesis of Neisseria gonorrhoeae and the host defense in ascending infections of human fallopian tube. Front Immunol 9:2710

Levine J, Furne JK, Levitt MD (1996) Ashkenazi Jews, sulfur gases, and ulcerative colitis. J Clin Gastroenterol 22(4):288–291

Li B, Zhao Y, Liu C, Chen Z, Zhou D (2014) Molecular pathogenesis of Klebsiella pneumoniae. Future Microbiol 9(9):1071–1081

Lyczak JB, Cannon CL, Pier GB (2000) Establishment of Pseudomonas aeruginosa infection: lessons from a versatile opportunist. Microbes Infect 2(9):1051–1060

Madigan M, Martinko J, Parker J (2000) Prokaryotic diversity: the archaea. In: Brock biology of microorganisms. Prentice-Hall, Upper Saddle River, NJ, pp 546–572

Maeda H, Hirai K, Mineshiba J, Yamamoto T, Kokeguchi S, Takashiba S (2013) Medical microbiological approach to archaea in oral infectious diseases. Jpn Dental Sci Rev 49(2):72–78

Markwart R, Willrich N, Haller S, Noll I, Koppe U, Werner G, Eckmanns T, Reuss A (2019) The rise in vancomycin-resistant Enterococcus faecium in Germany: data from the German antimicrobial resistance surveillance (ARS). Antimicrob Resist Infect Control 8(1):147

Menkir S, Mengestie H (2014) Prevalence of intestinal parasitic infections among people with and without Hiv infection and their association with diarrhea in Debre Markos town, East Gojjam Zone, Ethiopia, Haramaya University

Miller TL, Wolin M, de Macario EC, Macario A (1982) Isolation of Methanobrevibacter smithii from human feces. Appl Environ Microbiol 43(1):227–232

Munson-McGee JH, Field EK, Bateson M, Rooney C, Stepanauskas R, Young MJ (2015) Nanoarchaeota, their Sulfolobales host, and Nanoarchaeota virus distribution across Yellowstone National Park hot springs. Appl Environ Microbiol 81(22):7860–7868

Nelson AM, Neafie RC, Connor DH (1987) 11 cutaneous protothecosis and chlorellosis, extraordinary "aquatic-borne" algal infections. Clin Dermatol 5(3):76–87

Nosanchuk JS, Greenberg RD (1973) Prototheecosis of the olecranon bursa caused by achloric algae. Am J Clin Pathol 59(4):567–573

Organization (2002) Prevention and control of schistosomiasis and soil-transmitted helminthiasis: report of a WHO expert committee. World Health Organization, Geneva

Organization (2017) WHO publishes list of bacteria for which new antibiotics are urgently needed. World Health Organization, Geneva

Pester M, Schleper C, Wagner M (2011) The Thaumarchaeota: an emerging view of their phylogeny and ecophysiology. Curr Opin Microbiol 14(3):300–306

Pfaller M, Diekema D (2005) Unusual fungal and pseudofungal infections of humans. J Clin Microbiol 43(4):1495–1504

Piqué JM, Pallarés M, Cusó E, Vilar-Bonet J, Gassull MA (1984) Methane production and colon cancer. Gastroenterology 87(3):601–605

Podschun R, Ullmann U (1998) Klebsiella spp. as nosocomial pathogens: epidemiology, taxonomy, typing methods, and pathogenicity factors. Clin Microbiol Rev 11(4):589–603

Press W, Geneva S (2008) The global burden of disease: 2004 update. World Health Organization, Geneva

Putignani L, Menichella D (2010) Global distribution, public health and clinical impact of the protozoan pathogen Cryptosporidium. Interdiscip Perspect Infect Dis 2010:753512

Remschmidt C, Schröder C, Behnke M, Gastmeier P, Geffers C, Kramer TS (2018) Continuous increase of vancomycin resistance in enterococci causing nosocomial infections in Germany— 10 years of surveillance. Antimicrob Resist Infect Control 7(1):54

Ribes JA, Vanover-Sams CL, Baker DJ (2000) Zygomycetes in human disease. Clin Microbiol Rev 13(2):236–301

Rimšelienė G, Vold L, Robertson L, Nelke C, Søli K, Johansen ØH, Thrana FS, Nygård K (2011) An outbreak of gastroenteritis among schoolchildren staying in a wildlife reserve: thorough investigation reveals Norway's largest cryptosporidiosis outbreak. Scand J Public Health 39 (3):287–295

Rodríguez-Toro G (1993) Lobomycosis. Int J Dermatol 32(5):324–332

Santos HLC, Bandea R, Martins LAF, De Macedo HW, Peralta RHS, Peralta JM, Ndubuisi MI, Da Silva AJ (2010) Differential identification of Entamoeba spp. based on the analysis of 18S rRNA. Parasitol Res 106(4):883–888

Satoh K, Ooe K, Nagayama H, Makimura K (2010) Prototheca cutis sp. nov., a newly discovered pathogen of prototliecosis isolated from inflamed human skin. Int J Syst Evol Microbiol 60 (5):1236–1240

Schaberg DR, Culver DH, Gaynes RP (1991) Major trends in the microbial etiology of nosocomial infection. Am J Med 91(3):S72–S75

Schroeder GN, Hilbi H (2008) Molecular pathogenesis of Shigella spp.: controlling host cell signaling, invasion, and death by type III secretion. Clin Microbiol Rev 21(1):134–156

Schroll C, Barken KB, Krogfelt KA, Struve C (2010) Role of type 1 and type 3 fimbriae in Klebsiella pneumoniae biofilm formation. BMC Microbiol 10(1):179

Schuster FL, Ramirez-Avila L (2008) Current world status of Balantidium coli. Clin Microbiol Rev 21(4):626–638

Schuster FL, Visvesvara GS (2004) Amebae and ciliated protozoa as causal agents of waterborne zoonotic disease. Vet Parasitol 126(1-2):91–120

Smith I (2003) Mycobacterium tuberculosis pathogenesis and molecular determinants of virulence. Clin Microbiol Rev 16(3):463–496

Smith N, Denning D (2011) Underlying conditions in chronic pulmonary aspergillosis including simple aspergilloma. Eur Respir J 37(4):865–872

Socransky SS (2002) Dental biofilms: difficult therapeutic targets. Periodontol 2000(28):12–55

Solaymani-Mohammadi S, Petri W Jr (2006) Zoonotic implications of the swine-transmitted protozoal infections. Vet Parasitol 140(3-4):189–203

Sterling CR, Adam RD (2006) The pathogenic enteric protozoa: giardia, entamoeba, cryptosporidium and cyclospora. Springer, Cham

Stieglmeier M, Alves RJ, Schleper C (2014) The phylum thaumarchaeota. In: Prokaryotes. Springer, Cham, pp 347–362

Sudjaritruk T, Sirisanthana V (2011) Successful treatment of a child with vascular pythiosis. BMC Infect Dis 11(1):33

Tacconelli E, Carrara E, Savoldi A, Harbarth S, Mendelson M, Monnet DL, Pulcini C, Kahlmeter G, Kluytmans J, Carmeli Y (2018) Discovery, research, and development of new antibiotics: the WHO priority list of antibiotic-resistant bacteria and tuberculosis. Lancet Infect Dis 18(3):318–327

Tacconelli E, Tumbarello M, Bertagnolio S, Citton R, Spanu T, Fadda G, Cauda R (2002) Multidrug-resistant Pseudomonas aeruginosa bloodstream infections: analysis of trends in prevalence and epidemiology. Emerg Infect Dis 8(2):220

Thomas NA, Bardy SL, Jarrell KF (2001) The archaeal flagellum: a different kind of prokaryotic motility structure. FEMS Microbiol Rev 25(2):147–174

Thompson R, Ash A (2016) Molecular epidemiology of Giardia and Cryptosporidium infections. Infect Genet Evol 40:315–323

Tjaden NB, Thomas SM, Fischer D, Beierkuhnlein C (2013) Extrinsic incubation period of dengue: knowledge, backlog, and applications of temperature dependence. PLoS Negl Trop Dis 7(6): e2207

Tong SY, Davis JS, Eichenberger E, Holland TL, Fowler VG (2015) Staphylococcus aureus infections: epidemiology, pathophysiology, clinical manifestations, and management. Clin Microbiol Rev 28(3):603–661

Torres HA, Bodey G, Tarrand JJ, Kontoyiannis DP (2003) Prototsis in patients with cancer: case series and literature review. Clin Microbiol Infect 9(8):786–792

Van Vliet A, Ketley J (2001) Pathogenesis of enteric campylobacter infection. J Appl Microbiol 90 (S6):45S–56S

Wade WG (2011) Has the use of molecular methods for the characterization of the human oral microbiome changed our understanding of the role of bacteria in the pathogenesis of periodontal disease? J Clin Periodontol 38:7–16

Wallis M (1994) The pathogenesis of campylobacter jejuni. Br J Biomed Sci 51(1):57

Warris A, Bercusson A, Armstrong-James D (2019) Aspergillus colonization and antifungal immunity in cystic fibrosis patients. Med Mycol 57(Suppl_2):S118–S126

Weiser JN, Ferreira DM, Paton JC (2018) Streptococcus pneumoniae: transmission, colonization and invasion. Nat Rev Microbiol 16(6):355–367

Werner G, Coque T, Hammerum A, Hope R, Hryniewicz W, Johnson A, Klare I, Kristinsson K, Leclercq R, Lester C (2008) Emergence and spread of vancomycin resistance among enterococci in Europe. Eur Secur 13(47):19046

Wild C, Rasheed F, Jawla M, Hall A, Jansen L, Montesano R (1991) In-utero exposure to aflatoxin in West Africa. Lancet 337(8757):1602

Wirth F, Passalacqua J, Kao G (1999) Disseminated cutaneous protothecosis in an immunocompromised host: a case report and literature review. Cutis 63(3):185–188

Woolrich A, Koestenblatt E, Don P, Szaniawski W (1994) Cutaneous protothecosis and AIDS. J Am Acad Dermatol 31(5):920–924

World Health Organization (2004) China's latest SARS outbreak has been contained, but biosafety concerns remain—Update 7. https://www.who.int/csr/don/2004_05_18a/en/ Accessed 25 July 2020

World Health Organization (2018) Rotavirus. https://www.who.int/immunization/diseases/rotavi rus/en/ Accessed 25 July 2020

World Health Organization (2019) Zika epidemiology update. https://www.who.int/emergencies/ diseases/zika/zika-epidemiology-update-july-2019.pdf?ua=1 Accessed 25 July 2020

World Health Organization (2020a) Rabies. https://www.who.int/news-room/fact-sheets/detail/ rabies Accessed 25 July 2020

World Health Organization (2020b) Ebola virus disease. https://www.who.int/health-topics/ebola/ #tab=tab_1 Accessed 25 July 2020

World Health Organization (2020c) HIV/AIDS. https://www.who.int/news-room/fact-sheets/detail/ hiv-aids Accessed 25 July 2020

World Health Organization (2020d) Dengue and severe dengue. https://www.who.int/news-room/ fact-sheets/detail/dengue-and-severe-dengue Accessed 25 July 2020

World Health Organization (2020e) MERS situation update, January 2020. http://www.emro.who. int/pandemic-epidemic-diseases/mers-cov/mers-situation-update-january-2020.html Accessed 25 July 2020

World Health Organization (2020f) Coronavirus disease (COVID-19) Situation Report – 186. https://www.who.int/docs/default-source/coronaviruse/situation-reports/20200724-covid-19-sitrep-186.pdf?sfvrsn=4da7b586_2 Accessed 25 July 2020

Xiao L (2010) Molecular epidemiology of cryptosporidiosis: an update. Exp Parasitol 124(1):80–89

Pseudomonas aeruginosa: Pathogenic Adapter Bacteria

4

Swati Sagarika Panda, Khusbu Singh, Sanghamitra Pati, Rajeev Singh, Rajni Kant, and Gaurav Raj Dwivedi

Contents

S. S. Panda · K. Singh · S. Pati
Microbiology Department, ICMR-Regional Medical Research Centre Bhubaneswar, Bhubaneswar, Odisha, India

R. Singh · R. Kant · G. R. Dwivedi (✉)
Microbiology Department, ICMR-Regional Medical Research Centre, Gorakhpur, Uttar Pradesh, India
e-mail: grd.rmrcb@gov.in

Abstract

The origin of life was one of the best revolutions in the history of evolution. Bacteria which are supposed to be the earliest life forms developed on earth 3.7 million years ago. In evolutionary history, humans are considered as the world's greatest evolutionary force (omnipotent). Bacteria are unicellular; however, they understand the power of many; that's why they adopt multicellularity which birth new capabilities. *Pseudomonas aeruginosa* is one of the best adapters which is widespread in different environments such as lakes, plants, and animals. *P. aeruginosa* was placed in the critical category of pathogens which needs new treatment options. *Pseudomonas* infection is one of the best examples of fight between the bacteria (omnipresent) and humans (omnipotent). In this book chapter, human pathogenic *P. aeruginosa* have been studied. Efforts have also been made to better understand the classification, morphology, genetics and molecular biology, treatment option, drug resistance, etc.

Keywords

Evolution · Omnipotent · Omnipresent · Cystic fibrosis · Drug resistance

4.1 Introduction

A stable and comparative study of disease and its associated risk factors are necessary for deadliest diseases in the world (Bygbjerg 2012). They are probably moving toward the fast-acting, incurable ones that grab headlines from time to time. Measuring disease and burden of injury on populations depends upon a composite metric that covers both premature mortality and the predominance and harshness of ill-health (Murray et al. 2012). The transmission of the pathogens is of great interest across the host and has importance in the view of community health and finance (Cleaveland et al. 2001). The successful establishment of microorganisms at the site of infection in the host causes disease. After establishment, causative agents start to spread from the original site of infection by secreting toxins which help to reach other parts of the body (Janeway 2001). Although our immune system is an effective barrier against infectious agents, these pathogens bypass the immune system either via virulence factors or higher growth rate of pathogens.

Even after the presence of antibiotics, vaccines, and many successful anti-infectives, the spreading of infectious disease has been occupying an alarming position for public health in recent decades. Microorganisms (bacteria, virus, fungi, parasite, etc.) are the key source of communicable diseases and the cause of death in developing and underdeveloped countries (Dwivedi et al. 2016). In lower-income countries, the infectious disease condition is more alarming than the middle- and upper-income country. According to the World Health Organization (WHO), a catalogue of microbes (DIRTY DOZEN) which desperately require novel drugs was released (Tacconelli et al. 2018). The list of WHO is mainly dominated by Gram-negative bacteria (GNB), and among GNB, *P. aeruginosa* placed on second position which is responsible for hospital-acquired infection (Bassetti et al. 2018). *P. aeruginosa* range of niches such as air, water, plant, and animal, as well as it is metabolically versatile and characterized by biofilm formation (Dwivedi et al. 2016; Morales et al. 2012).

The rampant use of the available set of antibiotics for infectious disease as well as the limiting drug target leads to the development of resistance in clinical isolates. When any organism/strain/isolates develop resistance toward two or more structurally unrelated classes of drugs, it is termed as multidrug resistant (MDR), and when these clinical isolates become resistant to one agent in all but two or fewer antimicrobial classes, it is called extensively drug resistant (XDR) (Perez and Van Duin 2013). Similarly, when any organism/strain/isolates develop resistance to all agents in all antimicrobial classes, it is known as pandrug resistant (PDR) (Basak et al. 2016). Bacteria developed multidrug resistance by certain mechanisms such as overexpression of biofilm, efflux pump, and β-lactamase enzyme and downexpression of porins (Dwivedi et al. 2018; Harmsen et al. 2010). Currently, Gram-positive bacteria (GPB) have various options for treatment, and the introduction of the teixobactin also strengthens the hopes without generation of MDR strain (Ling et al. 2015; Piddock 2015). In the case of GNB, we have only limited treatment options (carbapenems and third-generation cephalosporins). Earlier carbapenems were supposed to be the medical miracle against MDR GNB, but the origin of the metallo-β-lactamases strengthened the defiance of *P. aeruginosa*. This is the need of the hour to find out novel antibiotics as well as alternative therapies, and rejuvenation of the old antibiotics for these bad bugs is required.

In view of these problems and absence of effective therapy, this book chapter deals with (a) *P. aeruginosa* in detail. The literature is reviewed, concerning all aspects of *P. aeruginosa* (history, classification, morphology, biochemical test, molecular identification, genome, pathogenicity, antibiotic option and combination coverage) (b) also explore the different mechanisms of MDR and treatment options.

4.2 Disease: An Introduction

Any destructive abnormality in the normal structural or functional condition of an organism is called disease. It is generally linked with certain signs and symptoms which are contradicted to the physical injury. The successful infection can leads to the diseases condition with the cooperation of susceptible host, environmental support, and presence of virulence factors (Fig. 4.1). This triad affects human, animal, and environmental health known as one health. According to WHO, one health is an approach to designing and implementing programs, policies, legislations, and research in which multiple sectors communicate and work together to achieve better public health outcomes. According to CDC, one health is an approach which encourages collaborative effort of many experts working across human, animal, and environmental health to improve the health of people, animals, and wildlife.

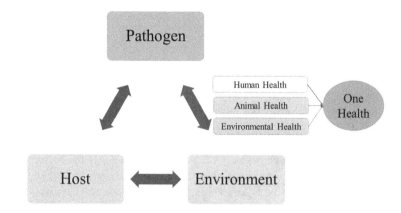

Fig. 4.1 Disease triad showing relationship with one health

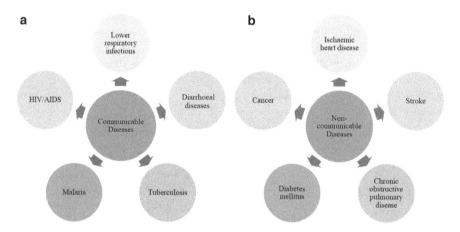

Fig. 4.2 Diseases: (**a**) communicable and (**b**) non-communicable

When a microorganism enters the human host and makes itself comfortable, it leads to infection. In the case where bacteria are well established in our body and start dividing at an exponential rate, then it is known as bacterial infectious disease (CDC 2018). Infectious diseases are transmitted from one person to another person via various routes such as air, water, food, vector, etc. In infectious diseases, the major cause of death is lower respiratory infection, diarrheal disease, tuberculosis, etc. (Fig. 4.2a). However, non-communicable diseases (NCDs) are non-transmissible and can be classified into rapid and chronic diseases. In NCDs, the major cause of mortality includes ischemic heart disease, stroke, cancer, etc. (Fig. 4.2b). The double burden of communicable diseases and NCDs is trending upward in low- and middle-income countries (Bygbjerg 2012). In both the low- and high-income countries, the alliance between communicable and NCD is also developing. The adults who are affected by communicable disease (immunodeficiency and tuberculosis) are more sensitive toward the NCDs (cardiovascular disease and diabetes mellitus) (Harries et al. 2015; Kaye and Pogue 2015).

From the last two decades, the number of incidence and severity of infection are in increasing order (Bouza and Finch 2001). In recent years, it has also been observed that there is quite increase in the number of GNB infection in many ways: (a) negligence toward GNB, (b) attention toward discovery of inhibitor against MRSA and MDR TB, and (c) GNB that are 1000 times sensitive toward natural compound (Vincent et al. 2009). During this time gap, GNB acquire resistance toward all available sets of antibiotics. Spreading of antimicrobial resistance among strains either occurs by horizontal gene transfer or by genetic transfer. Nowadays, some of the GNB require more attention for the development of effective therapies, and WHO classified these into three categories: critical (*Acinetobacter baumannii*, *P. aeruginosa*, *Enterobacteriaceae*), high (*Enterococcus faecium*, *Staphylococcus aureus*, *Helicobacter pylori*, *Neisseria gonorrhoeae*), and medium (*Streptococcus pneumoniae*, *Haemophilus influenzae*, *Shigella*).

Since 1962, there was no novel drug for MDR bacteria, but in 2015, there was the introduction of a new group of antibiotics, i.e., teixobactin (Ling et al. 2015; Piddock 2015). This drug has shown effectiveness against MDR GPB (Ling et al. 2015). At the same time, GNB established high-degree resistance against used antibiotics (Dwivedi et al. 2015). Reports from the United States, the United Kingdom, India, and Greece emphasized that *Escherichia coli*, *Klebsiella pneumoniae*, *P. aeruginosa*, and *A. baumannii* isolates were found to have resistance to a latest reported group of antibiotic, i.e., carbapenems (Gaynes et al. 2005; Kumarasamy et al. 2010; Peleg and Hooper 2010; Souli et al. 2008; Valencia et al. 2009). Many organizations such as the National Healthcare Safety Network, Centers for Disease Control and Prevention (CDC), and World Health Organization (WHO) used their operational programs on antimicrobial resistance for bacilli, i.e., *Enterobacteriaceae* (Peleg and Hooper 2010; Tokars et al. 2004). GNB infection is one of the causes of death particularly in nosocomial infections caused by carbapenem-resistant *Enterobacteriaceae* (CRE), extended-spectrum β-lactamase (ESBL)-producing *Enterobacteriaceae*, MDR *P. aeruginosa*, and MDR *A. baumannii* (Kaye and Pogue 2015).

Being the commander of nosocomial infection among GNB and diverse infection in any part of the body makes this *P. aeruginosa* a greatest concern forcing us to study in detail (Nayar et al. 2018).

4.3   *P. aeruginosa.*

4.3.1   History

P. aeruginosa was first recognized by Carle Gessard, a French pharmacist, in 1882. However, for the first time, it was discovered as a water-soluble pigment-producing organism which under exposure to ultraviolet light enlightened green-blue color

(Young et al., 2001). Carle Gessard studied the pathogenicity and infectious nature of *P. aeruginosa*.

4.3.2 Classification

Initially, taxonomic studies of the genus *Pseudomonas* follow basic classification methods. These methods were organized to depict and classify the organisms. At the end of the twentieth century, the introduction of several molecular biology-based biomarkers changes the classical taxonomy and introduced authenticated identification. The genus *Pseudomonas* was classified into mainly five groups based on rRNA homology. Presently, the number of species has dramatically increased. The new methodology for the taxonomy is based on the macromolecules other than rRNA component which helps to reduce the *Pseudomonas* nomenclature to a practicable size (Young et al., 2001). Currently, genus *Pseudomonas* has more than 200 species. This genus consists of a diverse group of bacteria which are responsible for different human ailments (Fig. 4.3).

4.3.3 Morphology

Morphologically, *P. aeruginosa* is a rod-shaped bacterium which has an incredible nutritional adaptability. It is about 1.5–3.0 micron in length and 0.4–0.8 micron in width, and majority of them have a single polar flagellum (Zhao et al. 2018). *P. aeruginosa* lives in a wide range of temperature (20–42°C). Naturally, this opportunistic pathogen has a broad range of hosts like plants, insects, nematodes, and mammals. It also relates with amoebae and yeasts (Cosson et al. 2002;

Fig. 4.3 Classification of *P. aeruginosa*

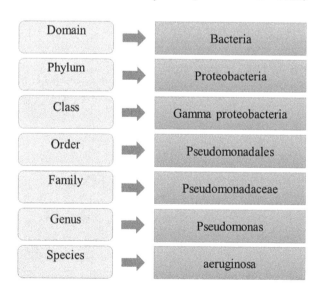

Domain	Bacteria
Phylum	Proteobacteria
Class	Gamma proteobacteria
Order	Pseudomonadales
Family	Pseudomonadaceae
Genus	Pseudomonas
Species	aeruginosa

D'Argenio et al. 2001; Hogan and Kolter 2002; Mahajan-Miklos et al. 1999; Pukatzki et al. 2002). The diverse types of virulence factors and regulatory systems regulate the virulence and pathogenesis to a wide range of host (Cao et al. 2001; Chugani et al. 2001; Cosson et al. 2002; Gallagher and Manoil 2001; Mahajan-Miklos et al. 1999; Pukatzki et al. 2002). Alteration in the colony architecture of *P. aeruginosa* largely depends upon the environmental pressure. Due to this environmental pressure, the colony of *P. aeruginosa* differs from each other. Few of these different colonies are also responsible for biofilm-related phenomenon (Kirisits et al. 2005).

   P. aeruginosa may produce three types of colonies. Generally, the strains which are confined from natural surroundings show a morphology of rough and small colonies. The specimens, which are collected from clinical settings, produce the colony which appears as smooth with an unbroken edge. Similarly, the samples obtained from urinary and respiratory infection give a mucoid appearance with production of alginate slime. The smooth and mucoid colonies are mainly responsible for colonization and virulence (Sader et al. 2017; Yang et al. 2017).

   In general, *P. aeruginosa* produces two types of pigments, i.e., pyoverdine and pyocyanin. Pyoverdine is fluorescent in nature, while pyocyanin is a blue color pigment. This pyocyanin is formed when there is a low amount of iron in the media. These pigments enhance the possibility of the infection caused by *P. aeruginosa* (Ben-David et al. 2018).

   A polar flagellum of *P. aeruginosa* is used in movement and displaying chemotaxis to useful molecules. Depending upon the size and antigenicity of the flagellin subunit, the flagellum is classified as either a type or b type. Initially, this flagellum helps in attachment and invasion of pathogens in the host's tissue at the time of infection (Park et al. 2005). Likewise, pili of *P. aeruginosa* also support it to hold to the mucosal surfaces and epithelial cells of the host. The pili are defined by polar filaments made up of homopolymers from the protein pilin, encoded by the piliA gene.

4.4 Identification of *P. aeruginosa*

4.4.1 Biochemical Test and Molecular Identification

As discussed above, *P. aeruginosa* has a distinctive action for the production of water-dissolved, green-blue pigment, but some exceptional strain does not produce any such pigments (O'Reilly et al. 2018; Sutter 1968). Specially, the primary cultures of this microbe, confined from patients under antibiotic treatment, do not show any pigmentation (Arai et al. 1970a). So, for confirmation and identification of *P. aeruginosa*, a set of biochemical tests are needed. These include the cytochrome oxidase test and fermentative behavior (Hugh and Leifson 1953). For distinction from other Gram-negative bacilli, gluconate oxidation test is required (Arai et al. 1970b). Even this gluconate oxidation test is supposed to be the ideal biochemical test for identification of *P. aeruginosa*, although there are some strains which show negativity for this test (Arai et al. 1970b). Amidase an enzyme efficient in hydrolyzing acetamide/propionamide was identified by Kelly and Clarke in 1960. Acylamidase activity was detected by using acetamide (substrate) and Nessler's reagent (Arai et al. 1970b). In 1965, Jessen reported that strains of *P. aeruginosa* can break Tween 80 and exploit it as an energy source. Synthetic medium consisting of surface-active agents of Tween derivatives which is the source of carbon, suitable for *P. aeruginosa*. This character makes it distinct from other GNB (Arai et al. 1970a).

In order to validate the biochemical test, different molecular markers are also used for identification of *P. aeruginosa*. Generally, microscopic and biochemical tests are performed for the identification of bacteria (Capuzzo et al. 2005). But these two methods do not give a confirmatory result. So, to support the previous identification, genotypic characterization becomes essential for classifying *P. aeruginosa*. 16S rRNA, gyrB, toxA, and 16S–23S rRNA genes are the molecular marker which becomes the powerful tools to recognize *P. aeruginosa* (Xu et al. 2004). However, in some cases, outer membrane protein-encoding gene (*opr*L and *opr*I)-based PCR is also done for confirmation (Douraghi et al. 2014).

4.4.2 Genetics and Molecular Biology

The whole genome sequencing is a powerful tool for identification of functional characterization of an organism. For the first time, the whole genome sequencing of *P. aeruginosa* was done in 2000 (Stover et al. 2000). The genome is a compression of a large repertoire of transporters and transcriptional regulators. The whole genome contains a total of 5697 genes, among which 5572 genes encode proteins (Table 4.1) (Dwivedi et al. 2018; Stover et al. 2000). The size of the genome is approximately 5.5–7 Mbp, and the G + C content is around 65–67% (Klockgether et al. 2011). There is only 0.5–0.7% interclonal sequence diversity among the different strains of *P. aeruginosa*. Mainly the proteome is dominated by hypothetical/unknown proteins followed by proteins involved in metabolism. Nutritional diversity of the *P. aeruginosa* is due to the presence of two components in the regulatory system. The protein which helps in transportation (primary transporters, secondary transporters, phosphotransferase system, and ion channels) is about 8% (Dwivedi et al. 2018). Accessory DNA is responsible for the genomic diversity which is dispersed around the genome.

Table 4.1 Distribution of different categories of genes in *P. aeruginosa* PAO1

Gene category	Corresponding number
Total number of genes present in *P. aeruginosa*	5697
Total number of genes encoding proteins	5572
Total number of genes encoding rRNA	13
Total number of genes encoding tRNA	63
Genes for other RNA	30
Total number of genes encoding pseudogene	19

4.5   Pathogenicity, Morbidity, and Mortality of *P. aeruginosa*

P. aeruginosa stands second in hospital-acquired pneumonia, third in urinary tract infections (UTI), fourth in infections after surgery, globally is the fifth most frequent pathogen, and ranked seventh for sepsis (Lister et al. 2009). In the patient of cystic fibrosis and airway infection, it is dominant due to the production of many virulence factors (VF), i.e., phenazines, adhesins, exotoxins, and proteases (Fricks-Lima et al. 2011). Multidrug resistance (MDR) mechanism in *P. aeruginosa* infection is more challenging for eradication of the expression of virulence factors, linked to antibiotic resistance, while other reports concluded that the effect is antagonistic (Finlayson and Brown 2011; Fricks-Lima et al. 2011; Jeannot et al. 2008; Ramisse et al. 2000). One of the main causes of pathogenesis by *P. aeruginosa* is the formation of biofilm. This biofilm helps the microorganisms to live in a community and make them 1000 times more resistant than their sensitive strains (Baugh et al. 2014; Dwivedi et al. 2016). The role of biofilm in the formation of MDR has been discussed in detail in the "Drug Resistance" section. There are two hypotheses: according to one, antibiotic resistance is correlated with expression of virulence factor, while the second opposes this hypothesis. *P. aeruginosa* causes airway tansmissible infection which can be both acute or chronic in nature (Gellatly and Hancock 2013). Various types of cells play an important role in the host defense mechanism. In *P. aeruginosa* which is a metabolically adaptable ubiquitous bacterium that inhabitant in different environments and grows on various surfaces such as animals, plants, and humans. A variety of life-threatening infections are caused by *P. aeruginosa* (Klockgether et al. 2011). The first *Pseudomonas* infection case was reported in the year 1890. Nowadays, *P. aeruginosa* is responsible for more than 15% cases of bacteremia caused by GNB (Iglewski 1996). The United States accounts for about 51,000 cases of nosocomial infection caused by *P. aeruginosa* every year (CDC 2018). Dendritic cells, T cells, macrophages, and neutrophils play a major role in immunological defense against airway infection in host (Gellatly and Hancock 2013). Appropriate response of the host and bacterial virulence factors are the two major reasons that determine the degree of symptoms and outcomes of *P. aeruginosa* infection (Gellatly and Hancock 2013).

4.6   Prevention

P. aeruginosa is a common occupant in soil, water, and vegetation. The hospital environment is most suitable for it. The pool of *P. aeruginosa* is normally found inside respiratory equipment, food, sinks, taps, toilets, showers, etc. In view of the above facts, we have to take prevention accordingly. According to the recent report of the Centers for Disease Control and Prevention (CDC 2018), hand washing is an utmost important process for both patients and their care takers to avoid getting and spreading the infection. Patients' rooms and shared equipment should be sanitized to reduce the risk of spreading of infection. Healthcare facilities should have well-established water management system to avoid the exposure to the bacteria.

Accurate sterilization and handling of respirators, catheters, and other instruments should be done to avoid any contamination. The hospital surrounding should be properly treated with disinfectants. The burn wounds should be dealt with antibacterial agents such as silver sulfadiazine, coupled with surgical debridement. These above methods adequately cut down the frequency of *P. aeruginosa* sepsis in burn patients (Peña et al. 2013).

4.7 Treatment Option

4.7.1 Antibiotic Option

From the serendipity discovery of penicillin till date, antibiotics are responsible to save enormous life. The golden era of antibiotics started after discovery of many novel groups of antibiotics. From the discovery of the first antibiotics till date, antibiotics are the first treatment options for *P. aeruginosa* infections. The correct doses of antibiotics should be prescribed immediately after isolation of *P. aeruginosa* from blood of a patient (Bassetti et al. 2018).

Normally, cephalosporins (ceftazidime, cefepime), aminoglycosides (gentamicin, tobramycin, amikacin, netilmicin), carbapenems (imipenem, meropenem), fluoroquinolones (levofloxacin, ciprofloxacin), penicillin with β-lactamase inhibitors (BLI) (piperacillin and ticarcillin in combination with clavulanic acid or tazobactam), monobactams (aztreonam), fosfomycin, and polymyxins (polymyxin B, colistin) are eight major categories for the treatment of *P. aeruginosa*, out of which cephalosporin, or a carbapenem, or an antipseudomonal β-lactam/BLI is more efficient for monotherapy than others (Bassetti et al. 2018). Aminoglycosides should not be prescribed for monotherapy due to its less efficiency (Kalil et al. 2016).

One recent report has suggested that delafloxacin is another antibiotic under fluoroquinolone group, which can be used as an antipseudomonal agent mainly for skin-related infection. Here the administration can be done either intravenously or orally. Murepavadin is one more antipseudomonal agent, which has different mode of action. It is a synthetic peptide developed by a pharma company in Switzerland. It blocks the lipopolysaccharide transport to the cell by binding with N-terminal of beta-barrel protein and hence causes the death of bacterial cells. Although it has an innovative mode of action, still its role is controversial due to its side effect in nosocomial infection. This is the main reason for terminating its application since July 2019.

Urinary tract infection (UTI) caused by *P. aeruginosa* is generally reported in hypertension, cognitive impairment, and diabetes mellitus patients (Lamas Ferreiro et al. 2017). Combination with a single antibiotic therapy that is standard for the treatment of UTI in absence of septic shock (Bassetti et al. 2018; Lamas Ferreiro et al. 2017).

Ecthyma gangrenosum (a cutaneous infection in immunocompromised patients) and burn wound infections are the most sensitive cases among skin and soft tissue infection caused by *P. aeruginosa* (Bassetti et al. 2018). Ecthyma gangrenosum is seen in neutropenic patients (Sarkar et al. 2016). Initially it is associated with painless erythematous areas with papules and/or bullae. It gradually becomes painful and develops into gangrenous ulcers (Stevens et al. 2014). For treatment, we can use the combination coverage of β-lactams and aminoglycosides (Azzopardi et al. 2014; Bassetti et al. 2018; Mayhall 2003). In case of cystic fibrosis, nebulized antibiotics, alone or in combination with oral antibiotics, are better than any other option.

Sometimes unnecessary use of antibiotics can lead to antibiotic resistance; hence, we should focus more on novel therapeutic approaches. These may include the approaches that target quorum sensing, uses of nanoparticles, phage therapy, and most importantly developing the vaccine strategy against *P. aeruginosa* Besides therapeutic treatment, we should also be conscious about early diagnosis, appropriate dosing, interval of drug administration, and duration of therapy (Bassetti et al. 2018). Ventilator-associated pneumonia (VAP) which includes prolonged mechanical ventilation, linked with very high risk factors of cancer and shock, is caused by *P. aeruginosa* (Fernández-Barat et al. 2017; Venier et al. 2011). So, in this case, antipseudomonal cephalosporin, or a carbapenem, or an antipseudomonal β-lactam/ BLI is an effective antibiotic option for the treatment (Bassetti et al. 2018). Here instead of aerosol therapy, intravenous therapy with combination coverage may be more beneficial (Bassetti et al. 2018).

4.7.2 Combination Coverage

The antibiotic inefficiency results from the continuous increase in antibiotic resistance among *P. aeruginosa*. This leads to the negative impact on the patient survival rate (Garnacho-Montero et al. 2007). In the view of the above problem, antibiotics used for *P. aeruginosa* infection should be the combination of two different class agents. The double coverage is preferred to promote the activity of antibiotics during observational therapy against *P. aeruginosa* infection (Peña et al. 2013). So, combination coverage has been suggested if the patient is suspected to have *P. aeruginosa* infection. Common prescription of an antipseudomonal β-lactam (piperacillin/ tazobactam, ceftolozane/tazobactam, ceftazidime, cefepime, or a carbapenem) plus a second antipseudomonal agent (aminoglycoside or a fluoroquinolones) is advised for the patient. The treatment rate was enhanced up to 93% when treated with amoxicillin-clavulanate of patients suffering with cystitis (Rodríguez-Baño et al. 2008). This study is signifying that amoxicillin-clavulanate may be fruitful in the treatment of simple cystitis. Similarly, piperacillin-tazobactam may be positively used in the treatment of UTI (Gavin et al. 2006). In addition to this standard antibiotic therapy, now some novel drugs like ceftolozane-tazobactam, ceftazidime-avibactam, imipenem-cilastatin-relebactam, etc. and advanced clinical treatment are available in the European market for the treatment of *P. aeruginosa* (Bassetti et al. 2018). Some study also showed that this new way of treatments is showing more efficiency than the traditional antibiotic therapy (Wright et al., 2017).

4.8   Drug Resistance

Antibiotics are there to eliminate the pathogenic bacteria. In contrast there is some natural process in bacteria that encourages resistance against antibiotics, for example, gene level mutation. The degree of antibiotic consumption was noticed to strongly correspond to the level of antibiotic-resistant infection (Goossens et al. 2005). The full course of recommended antibiotic treatment is crucial if an inappropriate dose is taken by the patients, and the possibility of development of antibiotic resistance increases many folds. Bacteria become resistant to multiple classes of antibiotics by collecting multiple resistance traits over time (Avorn 2001). Microorganisms contain many resistance genes. these genes developed resistance before the antibiotic started functioning for treatment purposes (Chadwick 1997)

When a drug unclasps its capacity to inhibit bacterial growth successfully, it is mentioned as antibiotic resistance. Bacteria become resistant and increase in the presence of therapeutic levels of the antibiotics (Zaman et al. 2017). However, due to the production of β-lactamase and aminoglycosides by *P. aeruginosa*, active efflux pump, membrane impermeability and porin alteration, target modification, and biofilm formation, it develops drug resistance capacity (Fig. 4.4) (Bassetti et al. 2018; Dias et al. 2018; Llano-Sotelo et al. 2002) (Table 4.2).

P. aeruginosa is responsible for various diseases. In UTI, the presence of *P. aeruginosa* is more common. Patients with cystic fibrosis are repeatedly colonized with *P. aeruginosa*. Because of *P. aeruginosa* antibodies, cystic fibrosis patients rarely have *P. aeruginosa* bacteria. But, localization of *P. aeruginosa* infections results in the death of cystic fibrosis patients. By use of contaminated respirators, there is a chance of necrotizing *P. aeruginosa* pneumonia in other patients. In case of eye surgery or injury especially in children with middle ear infections, *P. aeruginosa*

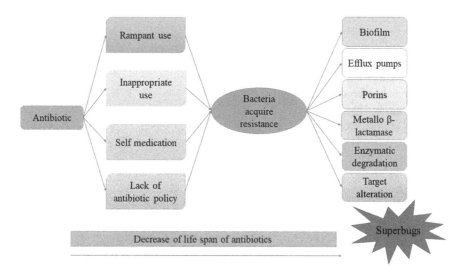

Fig. 4.4   Consequence of events responsible for MDR and development of pre-antibiotic era

Table 4.2 Antibiotics: class, introduction, and mechanism of resistance pattern

Class of antibiotics	Name of antibiotic	Year of introduction	Year of resistance	Modes of resistance	References
β-Lactam	Penicillin (penicillin G)	1942	1942	Efflux/altered target/hydrolysis	CDC (2013), Decuyper et al. (2018), Lobanovska and Pilla (2017)
	Imipenem (carbapenem)	1985	1998		
	Ceftazidime (third-generation cephalosporin)	1985	1987		
	Ceftaroline (fifth-generation cephalosporin)	2010	2011		
Aminoglycosides	Streptomycin	1944	1950	Efflux/altered target/acetylation/phosphorylation/nucleotidylation	CDC (2013), Davies and Davies (2010), Zaman et al. (2017)
	Gentamicin	1967	1979		
Tetracyclines	Tetracycline	1950	1959	Monooxygenation/efflux/altered target	CDC (2013), Demerec (1949), Nelson and Levy (2011)
	Chlortetracycline	1948	1949		
Glycopeptides	Vancomycin	1972	1988	Reprogramming peptidoglycan biosynthesis	CDC (2013)
Macrolides	Erythromycin	1953	1968	Hydrolysis/glycosylation/phosphorylation/efflux/altered target	CDC (2013), Zaman et al. (2017)
Quinolones	Nalidixic acid	1962	1969	Acetylation/efflux/altered target	Emmerson and Jones (2003)
Fluoroquinolones	Levofloxacin	1996	1996	Efflux/altered target	CDC (2013)
Sulfonamides	Methicillin	1960	1962	Efflux/altered target	Davies and Davies (2010), Zaman et al. (2017)

can cause severe infections. It also occasionally causes meningitis following lumbar puncture and cardiac surgery (Iglewski 1996).

Even after the development of different types of resistance, still antibiotics are the main treatment option for *P. aeruginosa* (Hancock 1998; Henwood et al. 2001). The resistance properties developed mainly due to overexpression of efflux pump, MBL production, outer membrane impermeability, and biofilm formation (Hancock 1998). Excluding this, there are some other factors which influence the drug resistance, i.e.:

- Due to low permeability of cell walls, it is resistant to antimicrobial agents.
- Through the mutation, these isolates develop the genetic capacity to show a broad range of resistance.
- It also gets additional resistance mechanism through horizontal gene transfer (HGT) (Lambert 2002).

4.8.1 Porins

This organism has an outer membrane with a low level of permeability. The outer membrane of GNB is a permeability hurdle, depending on the different size of solute. So, it plays a major role in antibiotic resistance. The outer membrane contains a certain group of protein called porins that allows only hydrophobic substances below a certain molecular size (Dwivedi et al. 2016; Yoshimura and Nikaido 1982). From the studies, it has been revealed that this porin also exists in different isolates of *P. aeruginosa*. From research, it has been shown that pure form of porin in the *P. aeruginosa* produced a significantly larger pore than those of enteric bacteria such as *E. coli* and *S. typhimurium* (Hancock and Nikaido 1978). Overall, we can say that the outer membrane comprises a semipermeable barrier to the uptake of antibiotics and substrate molecules. Hydrophilic antibiotic uptake is regulated by a water-filled channel composed of porin proteins, and the outer membrane restricts the movement of such molecules into the cell. This phenomenon is valid for all GNB but is especially true in the case of *P. aeruginosa*, which has an overall outer membrane permeability that is 100-fold lower than that of *E. coli* (Angus et al. 1982). There is a logical proof that OprF is the important porin which attributes to the limitation of *P. aeruginosa* (Bellido et al. 1992). OprF constitutes an inefficient uptake route for antibiotics (Woodruff et al. 1986). OprD is a crucial outer membrane porin through which antibiotics such as carbapenem (imipenem and meropenem) enter (Huang and Hancock 1993). However, loss of oprD results in the development of carbapenem resistance by *P. aeruginosa*. Other small channel porins like OprC and OprE may be available in wild type (Yamano et al. 1993), but the triple mutant OprC, OprD, and OprE are defenseless against antibiotics (Yoneyama et al. 1995).

4.8.2   Efflux

Primary level of resistance is governed by several types of permeability agents in *P. aeruginosa* outer membrane, and secondary level of resistance is provided by efflux system which also plays a major role in multidrug resistance mechanism (Poole and Srikumar 2001). Efflux pumps are localized in the cytoplasmic membranes with the combination of cell walls (Saier et al. 1998). Limited drug influx and active drug efflux perform simultaneously to maintain this resistance (Dwivedi et al. 2016; Yoshihara et al. 1999). The major resistance is controlled by four multidrug efflux systems in *P. aeruginosa*, i.e., MexXY-OprM, MexAB-OprM, MexCD-OprJ, and MexEF-OprN. The first two are responsible for inherent MDR, while the last two are responsible for acquired MDR. MexCD-OprJ and MexEF-OprN are observed only in mutant strains of *P. aeruginosa* (Poole and Srikumar 2001).

4.8.3   Biofilm

Biofilms have been the main focus of the researchers during the last decades due to two reasons. Firstly, the microorganisms are able to sustain in those multicellular communities. Secondly, as it can resist antibiotic treatment, host immune responses, and biocide treatment, its formation makes noticeable problems in the medical and industrial environment (Harmsen et al. 2010). Biofilm is a developmental process in which microorganisms attach to a solid surface followed by formation of micro-colony and maturation of biofilm (Ramsey and Whiteley 2004). Transportation of *P. aeruginosa* bacteria to a surface before attachment is assumed to involve disseminative, convective, and active flagellum-driven transport (Harmsen et al. 2010; van Loosdrecht et al. 1990). Previous studies have shown that the various types of components like flagella (O'Toole and Kolter 1998; Sauer et al. 2002) and type IV pili (Chiang and Burrows 2003; Déziel et al. 2001; O'Toole and Kolter 1998)-dependent twitching motility are important for attachment to abiotic surfaces (O'Toole and Kolter 1998; Ramsey and Whiteley 2004). After *P. aeruginosa* has attached to a surface, it either detaches, remains attached at the position of attachment, or moves with the help of type IV pili or flagella (Harmsen et al. 2010; Klausen et al. 2003; Shrout et al. 2006; Singh 2004; Singh et al. 2002), but the mutations in flagella and type IV pili assembly result in the reduction of the attachment in comparison to wild types (O'Toole et al. 2000; O'Toole and Kolter 1998). Except these, Cup fimbria (Vallet et al. 2001), extracellular DNA (eDNA) (Whitchurch et al. 2002), and Psl polysaccharide (Ma et al. 2009) also play an important role in the attachment of *P. aeruginosa* to surfaces in microtiter trays and flow chambers (Harmsen et al. 2010). It also has been reported that *sad*B gene regulates the attachment and detachment of *P. aeruginosa* from solid surfaces (Caiazza and O'Toole 2004). This SadB is synthesized by regulating the intracellular level of the second messenger molecule, i.e., cyclic diguanylate-guanosine monophosphate (c-di-GMP) (Harmsen et al. 2010; Merritt et al. 2007).

4.8.4 Metallo-β-Lactamases

Bacteria are becoming resistant to β-lactam antibiotics by producing an enzyme, i.e., extended-spectrum β-lactamase. This enzyme provides resistance by hydrolyzing β-lactam antibiotics such as penicillin, cephalosporins, cephamycins, etc. (Dwivedi et al. 2016). Carbapenem resistance in *P. aeruginosa* is mediated by five genes, namely, IMP, VIM, SPM, GIM, and AIM (Dwivedi et al. 2016; Zubair et al. 2011). New Delhi metallo-β-lactamase 1 (NDM-1) is an enzyme resistant to carbapenem which was discovered in India and Pakistan (Kumarasamy et al. 2010). Strains containing NDM-1 gene producing carbapenemases are highly resistant and difficult to treat (Dwivedi et al. 2016; Muir and Weinbren 2010).

4.8.5 Quorum Sensing

Quorum sensing is another factor responsible for drug resistance. It is a process in which bacterial cells communicate with each other and exchange information about the density of cell and gene expression (Rutherford and Bassler 2012). In *P. aeruginosa*, quorum sensing is generally regulated by three interconnected systems, i.e., Las system, Rhl system, and Pqs system (Harmsen et al. 2010). From various studies of microarray analysis, it has been found that quorum sensing is regulated by several hundred genes in *P. aeruginosa* (Harmsen et al. 2010; Schuster et al. 2003; Wagner et al. 2003). Except the production of rhamnolipid and eDNA, quorum sensing also clearly regulates a number of other factors involved in *P. aeruginosa* biofilm formation. The central metabolism in the bacteria is also related to the effect of quorum sensing on *P. aeruginosa* biofilm development. Anaerobic nitrate respiration may play an important role in *P. aeruginosa* biofilm development in clinical settings, and the rhlRI system is necessary to prevent accumulation of toxic nitric oxide during the process (Worlitzsch et al. 2002; Yoon et al. 2002). In agreement, the *nir*CMSQ and *nap*EF genes, which are required for respiratory nitrate reduction, were found to be strongly upregulated in *P. aeruginosa* during biofilm growth.

4.9 Summary

P. aeruginosa is considered as the most formidable pathogen found frequently in the hospital settings and responsible for various human infections. This pathogen is the cause of about 17% of healthcare-associated pneumonia, an enhancer of cystic fibrosis, and the most common organism of bloodstream infections. The onset of the different resistance mechanisms and presence of a multitude of virulence factors made *P. aeruginosa* the worst nightmare. So, this is the need of the hour to find the way to tackle the problem caused by *P. aeruginosa*. This chapter might be useful for easy understanding of negative aspects of *P. aeruginosa* and their possible treatment options from different resources.

Acknowledgments Authors acknowledge the project with project file no. SB/YS/LS-77/2014 funded by Science and Engineering Research Board (SERB-DBT). Thanks are also due to scientists and colleagues for providing supports for this study.

References

Angus BL, Carey AM, Caron DA, Kropinski AM, Hancock RE (1982) Outer membrane permeability in Pseudomonas aeruginosa: comparison of a wild-type with an antibiotic-supersusceptible mutant. Antimicrob Agents Chemother 21:299–309

Arai T, Enomoto S, Goto S (1970a) Determination of Pseudomonas aeruginosa by biochemical test methods. 3. Utilization of tween 80 by Pseudomonas aeruginosa. Jpn J Microbiol 14:285–290

Arai T, Enomoto S, Kuwahara S (1970b) Determination of Pseudomonas aeruginosa by biochemical test methods I an improved method for gluconate oxidation test. Jpn J Microbiol 14:49–56

Avorn J (2001) Improving drug use in elderly patients: getting to the next level. JAMA 286:2866–2868

Azzopardi EA, Azzopardi E, Camilleri L, Villapalos J, Boyce DE, Dziewulski P, Dickson WA, Whitaker IS (2014) Gram negative wound infection in hospitalised adult burn patients--systematic review and metanalysis. PLoS One 9:e95042

Basak S, Singh P, Rajurkar M (2016) Multidrug resistant and extensively drug resistant Bacteria: a study. J Pathog 2016:4065603

Bassetti M, Vena A, Croxatto A, Righi E, Guery B (2018) How to manage Pseudomonas aeruginosa infections. Drugs Context 7:212527

Baugh S, Phillips CR, Ekanayaka AS, Piddock LJV, Webber MA (2014) Inhibition of multidrug efflux as a strategy to prevent biofilm formation. J Antimicrob Chemother 69:673–681

Bellido F, Martin NL, Siehnel RJ, Hancock RE (1992) Reevaluation, using intact cells, of the exclusion limit and role of porin OprF in Pseudomonas aeruginosa outer membrane permeability. J Bacteriol 174:5196–5203

Ben-David Y, Zlotnik E, Zander I, Yerushalmi G, Shoshani S, Banin E (2018) SawR a new regulator controlling pyomelanin synthesis in Pseudomonas aeruginosa. Microbiol Res 206:91–98

Bouza E, Finch R (2001) Infections caused by gram-positive bacteria: situation and challenges of treatment. Clin Microbiol Infect 7(Suppl 4):3

Bygbjerg IC (2012) Double burden of noncommunicable and infectious diseases in developing countries. Science 337:1499–1501

Caiazza NC, O'Toole GA (2004) SadB is required for the transition from reversible to irreversible attachment during biofilm formation by Pseudomonas aeruginosa PA14. J Bacteriol 186:4476–4485

Cao H, Krishnan G, Goumnerov B, Tsongalis J, Tompkins R, Rahme LG (2001) A quorum sensing-associated virulence gene of Pseudomonas aeruginosa encodes a LysR-like transcription regulator with a unique self-regulatory mechanism. Proc Natl Acad Sci U S A 98:14613–14618

Capuzzo C, Firrao G, Mazzon L, Squartini A, Girolami V (2005) "Candidatus Erwinia dacicola", a coevolved symbiotic bacterium of the olive fly Bactrocera oleae (Gmelin). Int J Syst Evol Microbiol 55:1641–1647

CDC (2013) Antibiotic resistance threats in the United States

CDC (2018) Pseudomonas aeruginosa in healthcare settings

Chadwick D (1997) Monotherapy clinical trials of new antiepileptic drugs: design, indications, and controversies. Epilepsia 38(Suppl 9):S16–S20

Chiang P, Burrows LL (2003) Biofilm formation by hyperpiliated mutants of Pseudomonas aeruginosa. J Bacteriol 185:2374–2378

Chugani SA, Whiteley M, Lee KM, D'Argenio D, Manoil C, Greenberg EP (2001) QscR, a modulator of quorum-sensing signal synthesis and virulence in Pseudomonas aeruginosa. Proc Natl Acad Sci U S A 98:2752–2757

Cleaveland S, Laurenson MK, Taylor LH (2001) Diseases of humans and their domestic mammals: pathogen characteristics, host range and the risk of emergence. Philos Trans R Soc London B Biol Sci 356:991–999

Cosson P, Zulianello L, Join-Lambert O, Faurisson F, Gebbie L, Benghezal M, Van Delden C, Curty LK, Köhler T (2002) Pseudomonas aeruginosa virulence analyzed in a Dictyostelium discoideum host system. J Bacteriol 184:3027–3033

D'Argenio DA, Gallagher LA, Berg CA, Manoil C (2001) Drosophila as a model host for Pseudomonas aeruginosa infection. J Bacteriol 183:1466–1471

Davies J, Davies D (2010) Origins and evolution of antibiotic resistance. Microbiol Mol Biol Rev 74:417–433

Decuyper L, Jukič M, Sosič I, Žula A, D'hooghe M, Gobec S (2018) Antibacterial and β-lactamase inhibitory activity of monocyclic β-lactams. Med Res Rev 38:426–503

Demerec M (1949) Patterns of bacterial resistance to penicillin, Aureomycin, and streptomycin. J Clin Invest 28:891–893

Déziel E, Comeau Y, Villemur R (2001) Initiation of biofilm formation by Pseudomonas aeruginosa 57RP correlates with emergence of hyperpiliated and highly adherent phenotypic variants deficient in swimming, swarming, and twitching motilities. J Bacteriol 183:1195–1204

Dias C, Borges A, Oliveira D, Martinez-Murcia A, Saavedra MJ, Simões M (2018) Biofilms and antibiotic susceptibility of multidrug-resistant bacteria from wild animals. PeerJ 6:e4974

Douraghi M, Ghasemi F, Dallal MMS, Rahbar M, Rahimiforoushani A (2014) Molecular identification of Pseudomonas aeruginosa recovered from cystic fibrosis patients. J Prev Med Hyg 55:50–53

Dwivedi GR, Gupta S, Maurya A, Tripathi S, Sharma A, Darokar MP, Srivastava SK (2015) Synergy potential of indole alkaloids and its derivative against drug-resistant Escherichia coli. Chem Biol Drug Des 86:1471–1481

Dwivedi GR, Sanchita, Singh DP, Sharma A, Darokar MP, Srivastava SK (2016) Nano particles: emerging warheads against bacterial superbugs. Curr Top Med Chem 16:1963–1975

Dwivedi GR, Tyagi R, Sanchita, Tripathi S, Pati S, Srivastava SK, Darokar MP, Sharma A (2018) Antibiotics potentiating potential of catharanthine against superbug Pseudomonas aeruginosa. J Biomol Struct Dyn 36(16):4270–4284

Emmerson AM, Jones AM (2003) The quinolones: decades of development and use. J Antimicrob Chemother 51(Suppl 1):13–20

Fernández-Barat L, Ferrer M, De Rosa F, Gabarrús A, Esperatti M, Terraneo S, Rinaudo M, Li Bassi G, Torres A (2017) Intensive care unit-acquired pneumonia due to Pseudomonas aeruginosa with and without multidrug resistance. J Infect 74:142–152

Finlayson EA, Brown PD (2011) Comparison of antibiotic resistance and virulence factors in pigmented and non-pigmented Pseudomonas aeruginosa. West Indian Med J 60:24–32

Fricks-Lima J, Hendrickson CM, Allgaier M, Zhuo H, Wiener-Kronish JP, Lynch SV, Yang K (2011) Differences in biofilm formation and antimicrobial resistance of Pseudomonas aeruginosa isolated from airways of mechanically ventilated patients and cystic fibrosis patients. Int J Antimicrob Agents 37:309–315

Gallagher LA, Manoil C (2001) Pseudomonas aeruginosa PAO1 kills Caenorhabditis elegans by cyanide poisoning. J Bacteriol 183:6207–6214

Garnacho-Montero J, Sa-Borges M, Sole-Violan J, Barcenilla F, Escoresca-Ortega A, Ochoa M, Cayuela A, Rello J (2007) Optimal management therapy for Pseudomonas aeruginosa ventilator-associated pneumonia: an observational, multicenter study comparing monotherapy with combination antibiotic therapy. Crit Care Med 35:1888–1895

Gavin PJ, Suseno MT, Thomson RB, Gaydos JM, Pierson CL, Halstead DC, Aslanzadeh J, Brecher S, Rotstein C, Brossette SE, Peterson LR (2006) Clinical correlation of the CLSI susceptibility breakpoint for piperacillin-tazobactam against extended-spectrum-beta-lactamase-producing Escherichia coli and Klebsiella species. Antimicrob Agents Chemother 50:2244–2247

Gaynes R, Edwards JR, National Nosocomial Infections Surveillance System (2005) Overview of nosocomial infections caused by gram-negative bacilli. Clin Infect Dis 41:848–854

Gellatly SL, Hancock REW (2013) Pseudomonas aeruginosa: new insights into pathogenesis and host defenses. Pathog Dis 67:159–173

Goossens H, Ferech M, Vander Stichele R, Elseviers M, ESAC Project Group (2005) Outpatient antibiotic use in Europe and association with resistance: a cross-national database study. Lancet 365:579–587

Hancock RE (1998) Resistance mechanisms in Pseudomonas aeruginosa and other nonfermentative gram-negative bacteria. Clin Infect Dis 27(Suppl 1):S93–S99

Hancock RE, Nikaido H (1978) Outer membranes of gram-negative bacteria. XIX. Isolation from Pseudomonas aeruginosa PAO1 and use in reconstitution and definition of the permeability barrier. J Bacteriol 136:381–390

Harmsen M, Yang L, Pamp SJ, Tolker-Nielsen T (2010) An update on Pseudomonas aeruginosa biofilm formation, tolerance, and dispersal. FEMS Immunol Med Microbiol 59:253–268

Harries AD, Kumar AMV, Satyanarayana S, Lin Y, Takarinda KC, Tweya H, Reid AJ, Zachariah R (2015) Communicable and non-communicable diseases: connections, synergies and benefits of integrating care. Public Health Action 5:156–157

Henwood CJ, Livermore DM, James D, Warner M, Pseudomonas Study Group (2001) Antimicrobial susceptibility of Pseudomonas aeruginosa: results of a UK survey and evaluation of the British Society for Antimicrobial Chemotherapy disc susceptibility test. J Antimicrob Chemother 47:789–799

Hogan DA, Kolter R (2002) Pseudomonas-Candida interactions: an ecological role for virulence factors. Science 296:2229–2232

Huang H, Hancock RE (1993) Genetic definition of the substrate selectivity of outer membrane porin protein OprD of Pseudomonas aeruginosa. J Bacteriol 175:7793–7800

Hugh R, Leifson E (1953) The taxonomic significance of fermentative versus oxidative metabolism of carbohydrates by various gram negative bacteria. J Bacteriol 66:24–26

Iglewski BH (1996) Pseudomonas. In: Baron S (ed) Medical Microbiology. University of Texas Medical Branch at Galveston, Galveston, TX

Janeway CA (2001) How the immune system protects the host from infection. Microbes Infect 3:1167–1171

Jeannot K, Elsen S, Köhler T, Attree I, Van Delden C, Plésiat P (2008) Resistance and virulence of Pseudomonas aeruginosa clinical strains overproducing the MexCD-OprJ efflux pump. Antimicrob Agents Chemother 52:2455–2462

Kalil AC, Metersky ML, Klompas M, Muscedere J, Sweeney DA, Palmer LB, Napolitano LM, O'Grady NP, Bartlett JG, Carratalà J, El Solh AA, Ewig S, Fey PD, File TM, Restrepo MI, Roberts JA, Waterer GW, Cruse P, Knight SL, Brozek JL (2016) Management of Adults with Hospital-acquired and Ventilator-associated Pneumonia: 2016 clinical practice guidelines by the Infectious Diseases Society of America and the American Thoracic Society. Clin Infect Dis 63: e61–e111

Kaye KS, Pogue JM (2015) Infections caused by resistant gram-negative Bacteria: epidemiology and management. Pharmacotherapy 35:949–962

Kirisits MJ, Prost L, Starkey M, Parsek MR (2005) Characterization of colony morphology variants isolated from Pseudomonas aeruginosa biofilms. Appl Environ Microbiol 71:4809–4821

Klausen M, Aaes-Jørgensen A, Molin S, Tolker-Nielsen T (2003) Involvement of bacterial migration in the development of complex multicellular structures in Pseudomonas aeruginosa biofilms. Mol Microbiol 50:61–68

Klockgether J, Cramer N, Wiehlmann L, Davenport CF, Tümmler B (2011) Pseudomonas aeruginosa genomic structure and diversity. Front Microbiol 2:150

Kumarasamy KK, Toleman MA, Walsh TR, Bagaria J, Butt F, Balakrishnan R, Chaudhary U, Doumith M, Giske CG, Irfan S, Krishnan P, Kumar AV, Maharjan S, Mushtaq S, Noorie T,

Paterson DL, Pearson A, Perry C, Pike R, Rao B, Ray U, Sarma JB, Sharma M, Sheridan E, Thirunarayan MA, Turton J, Upadhyay S, Warner M, Welfare W, Livermore DM, Woodford N (2010) Emergence of a new antibiotic resistance mechanism in India, Pakistan, and the UK: a molecular, biological, and epidemiological study. Lancet Infect Dis 10:597–602

Lamas Ferreiro JL, Álvarez Otero J, González González L, Novoa Lamazares L, Arca Blanco A, Bermúdez Sanjurjo JR, Rodríguez Conde I, Fernández Soneira M, De la Fuente Aguado J (2017) Pseudomonas aeruginosa urinary tract infections in hospitalized patients: mortality and prognostic factors. PLoS One 12:e0178178

Lambert PA (2002) Mechanisms of antibiotic resistance in Pseudomonas aeruginosa. J R Soc Med 95(Suppl 41):22–26

Ling LL, Schneider T, Peoples AJ, Spoering AL, Engels I, Conlon BP, Mueller A, Schäberle TF, Hughes DE, Epstein S, Jones M, Lazarides L, Steadman VA, Cohen DR, Felix CR, Fetterman KA, Millett WP, Nitti AG, Zullo AM, Chen C, Lewis K (2015) A new antibiotic kills pathogens without detectable resistance. Nature 517:455–459

Lister PD, Wolter DJ, Hanson ND (2009) Antibacterial-resistant Pseudomonas aeruginosa: clinical impact and complex regulation of chromosomally encoded resistance mechanisms. Clin Microbiol Rev 22:582–610

Llano-Sotelo B, Azucena EF, Kotra LP, Mobashery S, Chow CS (2002) Aminoglycosides modified by resistance enzymes display diminished binding to the bacterial ribosomal aminoacyl-tRNA site. Chem Biol 9:455–463

Lobanovska M, Pilla G (2017) Penicillin's discovery and antibiotic resistance: lessons for the future? Yale J Biol Med 90:135–145

Ma L, Conover M, Lu H, Parsek MR, Bayles K, Wozniak DJ (2009) Assembly and development of the Pseudomonas aeruginosa biofilm matrix. PLoS Pathog 5:e1000354

Mahajan-Miklos S, Tan MW, Rahme LG, Ausubel FM (1999) Molecular mechanisms of bacterial virulence elucidated using a Pseudomonas aeruginosa-Caenorhabditis elegans pathogenesis model. Cell 96:47–56

Mayhall CG (2003) The epidemiology of burn wound infections: then and now. Clin Infect Dis 37:543–550

Merritt JH, Brothers KM, Kuchma SL, O'Toole GA (2007) SadC reciprocally influences biofilm formation and swarming motility via modulation of exopolysaccharide production and flagellar function. J Bacteriol 189:8154–8164

Morales E, Cots F, Sala M, Comas M, Belvis F, Riu M, Salvadó M, Grau S, Horcajada JP, Montero MM, Castells X (2012) Hospital costs of nosocomial multi-drug resistant Pseudomonas aeruginosa acquisition. BMC Health Serv Res 12:122

Muir A, Weinbren MJ (2010) New Delhi metallo-beta-lactamase: a cautionary tale. J Hosp Infect 75:239–240

Murray CJL, Vos T, Lozano R, Naghavi M, Flaxman AD et al (2012) Disability-adjusted life years (DALYs) for 291 diseases and injuries in 21 regions, 1990-2010: a systematic analysis for the global burden of disease study 2010. Lancet 380:2197–2223

Nayar G, Darley ESR, Hammond F, Matthews S, Turton J, Wach R (2018) Does screening neonates in the neonatal intensive care unit for Pseudomonas aeruginosa colonisation help prevent infection? J Hosp Infect 16(18):1963–1975

Nelson ML, Levy SB (2011) The history of the tetracyclines. Ann N Y Acad Sci 1241:17–32

O'Reilly MC, Dong S-H, Rossi FM, Karlen KM, Kumar RS, Nair SK, Blackwell HE (2018) Structural and biochemical studies of non-native agonists of the LasR quorum-sensing receptor reveal an L3 loop "out" conformation for LasR. Cell Chem Biol 25(9):1128–1139

O'Toole G, Kaplan HB, Kolter R (2000) Biofilm formation as microbial development. Annu Rev Microbiol 54:49–79

O'Toole GA, Kolter R (1998) Flagellar and twitching motility are necessary for Pseudomonas aeruginosa biofilm development. Mol Microbiol 30:295–304

Park S-Y, Heo Y-J, Choi Y-S, Déziel E, Cho Y-H (2005) Conserved virulence factors of Pseudomonas aeruginosa are required for killing Bacillus subtilis. J Microbiol 43:443–450

Peleg AY, Hooper DC (2010) Hospital-acquired infections due to gram-negative bacteria. N Engl J Med 362:1804–1813

Peña C, Suarez C, Ocampo-Sosa A, Murillas J, Almirante B, Pomar V, Aguilar M, Granados A, Calbo E, Rodríguez-Baño J, Rodríguez F, Tubau F, Oliver A, Martínez-Martínez L, Spanish Network for Research in Infectious Diseases (REIPI) (2013) Effect of adequate single-drug vs combination antimicrobial therapy on mortality in Pseudomonas aeruginosa bloodstream infections: a post hoc analysis of a prospective cohort. Clin Infect Dis 57:208–216

Perez F, Van Duin D (2013) Carbapenem-resistant Enterobacteriaceae: a menace to our most vulnerable patients. Cleve Clin J Med 80:225–233

Piddock LJV (2015) Teixobactin, the first of a new class of antibiotics discovered by iChip technology? J Antimicrob Chemother 70:2679–2680

Poole K, Srikumar R (2001) Multidrug efflux in Pseudomonas aeruginosa: components, mechanisms and clinical significance. Curr Top Med Chem 1:59–71

Pukatzki S, Kessin RH, Mekalanos JJ (2002) The human pathogen Pseudomonas aeruginosa utilizes conserved virulence pathways to infect the social amoeba Dictyostelium discoideum. Proc Natl Acad Sci U S A 99:3159–3164

Ramisse F, Van Delden C, Gidenne S, Cavallo J, Hernandez E (2000) Decreased virulence of a strain of Pseudomonas aeruginosa O12 overexpressing a chromosomal type 1 beta-lactamase could be due to reduced expression of cell-to-cell signaling dependent virulence factors. FEMS Immunol Med Microbiol 28:241–245

Ramsey MM, Whiteley M (2004) Pseudomonas aeruginosa attachment and biofilm development in dynamic environments. Mol Microbiol 53:1075–1087

Rodríguez-Baño J, López-Cerero L, Navarro MD, Díaz de Alba P, Pascual A (2008) Faecal carriage of extended-spectrum beta-lactamase-producing Escherichia coli: prevalence, risk factors and molecular epidemiology. J Antimicrob Chemother 62:1142–1149

Rutherford ST, Bassler BL (2012) Bacterial quorum sensing: its role in virulence and possibilities for its control. Cold Spring Harb Perspect Med 2(11):a012427

Sader HS, Huband MD, Castanheira M, Flamm RK (2017) Pseudomonas aeruginosa antimicrobial susceptibility results from four years (2012 to 2015) of the international network for optimal resistance monitoring program in the United States. Antimicrob Agents Chemother 61(3): e02252

Saier MH, Paulsen IT, Sliwinski MK, Pao SS, Skurray RA, Nikaido H (1998) Evolutionary origins of multidrug and drug-specific efflux pumps in bacteria. FASEB J 12:265–274

Sarkar S, Patra AK, Mondal M (2016) Ecthyma gangrenosum in the periorbital region in a previously healthy immunocompetent woman without bacteremia. Indian Dermatol Online J 7:36–39

Sauer K, Camper AK, Ehrlich GD, Costerton JW, Davies DG (2002) Pseudomonas aeruginosa displays multiple phenotypes during development as a biofilm. J Bacteriol 184:1140–1154

Schuster M, Lostroh CP, Ogi T, Greenberg EP (2003) Identification, timing, and signal specificity of Pseudomonas aeruginosa quorum-controlled genes: a transcriptome analysis. J Bacteriol 185:2066–2079

Shrout JD, Chopp DL, Just CL, Hentzer M, Givskov M, Parsek MR (2006) The impact of quorum sensing and swarming motility on Pseudomonas aeruginosa biofilm formation is nutritionally conditional. Mol Microbiol 62:1264–1277

Singh PK (2004) Iron sequestration by human lactoferrin stimulates P. aeruginosa surface motility and blocks biofilm formation. Biometals 17:267–270

Singh PK, Parsek MR, Greenberg EP, Welsh MJ (2002) A component of innate immunity prevents bacterial biofilm development. Nature 417:552–555

Souli M, Galani I, Giamarellou H (2008) Emergence of extensively drug-resistant and pandrug-resistant gram-negative bacilli in Europe. Euro Surveill 13(47):19045

Stevens DL, Bisno AL, Chambers HF, Dellinger EP, Goldstein EJC, Gorbach SL, Hirschmann JV, Kaplan SL, Montoya JG, Wade JC, Infectious Diseases Society of America (2014) Practice guidelines for the diagnosis and management of skin and soft tissue infections: 2014 update by the Infectious Diseases Society of America. Clin Infect Dis 59:e10–e52

Stover CK, Pham XQ, Erwin AL, Mizoguchi SD, Warrener P, Hickey MJ et al (2000) Complete genome sequence of Pseudomonas aeruginosa PAO1, an opportunistic pathogen. Nature 406:959–964

Sutter VL (1968) Identification of Pseudomonas species isolated from hospital environment and human sources. Appl Microbiol 16:1532–1538

Tacconelli E, Carrara E, Savoldi A, Harbarth S, Mendelson M, Monnet DL, Pulcini C, Kahlmeter G, Kluytmans J, Carmeli Y, Ouellette M, Outterson K, Patel J, Cavaleri M, Cox EM, Houchens CR, Grayson ML, Hansen P, Singh N, Theuretzbacher U, Magrini N, WHO Pathogens Priority List Working Group (2018) Discovery, research, and development of new antibiotics: the WHO priority list of antibiotic-resistant bacteria and tuberculosis. Lancet Infect Dis 18:318–327

Tokars JI, Richards C, Andrus M, Klevens M, Curtis A, Horan T, Jernigan J, Cardo D (2004) The changing face of surveillance for health care-associated infections. Clin Infect Dis 39:1347–1352

Valencia R, Arroyo LA, Conde M, Aldana JM, Torres M-J, Fernández-Cuenca F, Garnacho-Montero J, Cisneros JM, Ortíz C, Pachón J, Aznar J (2009) Nosocomial outbreak of infection with pan-drug-resistant Acinetobacter baumannii in a tertiary care university hospital. Infect Control Hosp Epidemiol 30:257–263

Vallet I, Olson JW, Lory S, Lazdunski A, Filloux A (2001) The chaperone/usher pathways of Pseudomonas aeruginosa: identification of fimbrial gene clusters (cup) and their involvement in biofilm formation. Proc Natl Acad Sci U S A 98:6911–6916

Van Loosdrecht MC, Norde W, Zehnder AJ (1990) Physical chemical description of bacterial adhesion. J Biomater Appl 5:91–106

Venier AG, Gruson D, Lavigne T, Jarno P, L'hériteau F, Coignard B, Savey A, Rogues AM, REA-RAISIN group (2011) Identifying new risk factors for Pseudomonas aeruginosa pneumonia in intensive care units: experience of the French national surveillance. REA-RAISIN J Hosp Infect 79:44–48

Vincent J-L, Rello J, Marshall J, Silva E, Anzueto A, Martin CD, Moreno R, Lipman J, Gomersall C, Sakr Y, Reinhart K, EPIC II Group of Investigators (2009) International study of the prevalence and outcomes of infection in intensive care units. JAMA 302:2323–2329

Wagner VE, Bushnell D, Passador L, Brooks AI, Iglewski BH (2003) Microarray analysis of Pseudomonas aeruginosa quorum-sensing regulons: effects of growth phase and environment. J Bacteriol 185:2080–2095

Whitchurch CB, Tolker-Nielsen T, Ragas PC, Mattick JS (2002) Extracellular DNA required for bacterial biofilm formation. Science 295:1487

Woodruff WA, Parr TR, Hancock RE, Hanne LF, Nicas TI, Iglewski BH (1986) Expression in Escherichia coli and function of Pseudomonas aeruginosa outer membrane porin protein F. J Bacteriol 167:473–479

Worlitzsch D, Tarran R, Ulrich M, Schwab U, Cekici A, Meyer KC, Birrer P, Bellon G, Berger J, Weiss T, Botzenhart K, Yankaskas JR, Randell S, Boucher RC, Döring G (2002) Effects of reduced mucus oxygen concentration in airway Pseudomonas infections of cystic fibrosis patients. J Clin Invest 109:317–325

Wright H, Bonomo RA, Paterson DL (2017) New agents for the treatment of infections with gram-negative bacteria: restoring the miracle or false dawn? Clin Microbiol Infect 23:704–712

Xu J, Moore JE, Murphy PG, Millar BC, Elborn JS (2004) Early detection of Pseudomonas aeruginosa--comparison of conventional versus molecular (PCR) detection directly from adult patients with cystic fibrosis (CF). Ann Clin Microbiol Antimicrob 3:21

Yamano Y, Nishikawa T, Komatsu Y (1993) Cloning and nucleotide sequence of anaerobically induced porin protein E1 (OprE) of Pseudomonas aeruginosa PAO1. Mol Microbiol 8:993–1004

Yang Q, Zhang H, Wang Y, Xu Z, Zhang G, Chen X, Xu Y, Cao B, Kong H, Ni Y, Yu Y, Sun Z, Hu B, Huang W, Wang Y, Wu A, Feng X, Liao K, Luo Y, Hu Z, Chu Y, Lu J, Su J, Gui B, Duan Q, Zhang S, Shao H, Badal RE (2017) Antimicrobial susceptibilities of aerobic and facultative gram-negative bacilli isolated from Chinese patients with urinary tract infections between 2010 and 2014. BMC Infect Dis 17:192

Yoneyama H, Yamano Y, Nakae T (1995) Role of porins in the antibiotic susceptibility of Pseudomonas aeruginosa: construction of mutants with deletions in the multiple porin genes. Biochem Biophys Res Commun 213:88–95

Yoon SS, Hennigan RF, Hilliard GM, Ochsner UA, Parvatiyar K, Kamani MC, Allen HL, DeKievit TR, Gardner PR, Schwab U, Rowe JJ, Iglewski BH, McDermott TR, Mason RP, Wozniak DJ, Hancock REW, Parsek MR, Noah TL, Boucher RC, Hassett DJ (2002) Pseudomonas aeruginosa anaerobic respiration in biofilms: relationships to cystic fibrosis pathogenesis. Dev Cell 3:593–603

Yoshihara Y, Mizuno T, Nakahira M, Kawasaki M, Watanabe Y, Kagamiyama H, Jishage K, Ueda O, Suzuki H, Tabuchi K, Sawamoto K, Okano H, Noda T, Mori K (1999) A genetic approach to visualization of multisynaptic neural pathways using plant lectin transgene. Neuron 22:33–41

Yoshimura F, Nikaido H (1982) Permeability of Pseudomonas aeruginosa outer membrane to hydrophilic solutes. J Bacteriol 152:636–642

Young JM, Bull CT, De Boer SH, Firrao G, Gardan L, Saddler GE, Stead DE, Takikawa Y (2001) Classification, nomenclature, and plant pathogenic bacteria - a clarification. Phytopathology 91:617–620

Zaman SB, Hussain MA, Nye R, Mehta V, Mamun KT, Hossain N (2017) A review on antibiotic resistance: alarm bells are ringing. Cureus 9:e1403

Zhao J, Cheng W, He X, Liu Y, Li J, Sun J, Li J, Wang F, Gao Y (2018) Association of furanone C-30 with biofilm formation & antibiotic resistance in Pseudomonas aeruginosa. Indian J Med Res 147:400–406

Zubair M, Malik A, Ahmad J (2011) Prevalence of metallo-β-lactamase-producing Pseudomonas aeruginosa isolated from diabetic foot ulcer patients. Diabetes Metab Syndr 5:90–92

Impact of Antibiotic Resistance of Bacteria in Biofilms and Microbial Fuel Cell: Confronting the Dark Box for Global Health Threat

5

Sweta Naik, Srilakshmi Mutyala, and J. Satya Eswari

Contents

Abstract

Increased population activities, a growing market for animal nutrition, and extensive usage of antibiotics were responsible for the continued presence in the ecosystem of antibiotic contamination. Such contamination has gained enhanced scrutiny because it can worsen the presence of antibiotic-resistant bacteria and antibiotic resistance genes. Consequently, the successful removal of antibiotic toxins has been a popular subject of environmental science.

S. Naik · J. S. Eswari (✉)
Department of Biotechnology, National Institute of Technology, Raipur, Raipur, Chhattisgarh, India
e-mail: satyaeswarij.bt@nitrr.ac.in

S. Mutyala
Department of Chemical Engineering, National Institute of Technology, Raipur, Raipur, Chhattisgarh, India

V. Kumar et al. (eds.), *Antimicrobial Resistance*,
https://doi.org/10.1007/978-981-16-3120-7_5

137

Microbial fuel cells (MFCs) coupled with microbial metabolisms and electrochemical redox reactions are deemed interesting alternatives for antibiotic contaminant degradation. This chapter discusses state-of-the-art MFCs for improved antibiotic treatment and examines processes for the elimination of antibiotics dependent on MFCs. The results of standard parameters, such as salinity, temperature, source of carbon, material used as electrode, applied potential, primary concentration of antibiotic, and electrochemical characteristics, are expanded on the overall efficiency of these systems. There are also records of degradation mechanisms and metabolic pathways through antibiotic products linked to MFC processes. Also, the impact on the fate of antibiotic resistance genes from the temperature, salinity, initial antibiotic concentration, and potential added in MFCs is revealed. Ultimately, an overview is given on potential technologies and problems which are beneficial to the creation of antibiotic-containing wastewater treatment MFCs.

Keywords

Antibiotic resistance · Biofilm · Microbial fuel cell · Toxicant · Wastewater

5.1 Introduction

Once bacteria developed on a surface as a biofilm, they are covered from toxins, biocides, and certain chemical or physical challenges. This trend is increasingly recognized as a key factor in varied infection persistence. Chronic infections involving biofilms include cystic fibrosis pneumonia, periodontitis, and various infections associated with residential products, such as heart valves, prostheses, and catheters. Bacteria resistance to antibiotics in biofilms is easily replicated in vitro, demonstrating that host factors are not necessary for this biofilm protection manifestation (Khatoon et al. 2018). Although biofilms of the common opportunistic pathogens *Staphylococcus epidermidis* and *Pseudomonas aeruginosa* are well known for their resistance to antibiotics, biofilms are produced by numerous other microbes that provide defence against many antibiotics. The yeast *Candida albicans* and mandatory anaerobic *Porphyromonas gingivalis* (Hansen and Velschow 2000) were shown to be less susceptible to antibiotics when grown in biofilms than free-floating cells. Various types of antibiotics have been investigated against microbial biofilms, and when grown in free aqueous suspension, microbes in the growth model of biofilm are found to be less susceptible to inhibit than the same strain. Such findings inform us that the capacity to build protective biofilms is distributed widely amongst microbes and that the mechanisms of resistance used in biofilms are strong protection of the spectrum (Sharma et al. 2019).

5.1.1 Antibiotic Resistance

This chapter summarizes the outcome of antibiotic resistance that leads to worldwide health problem and emphasizes attempts to recover and resolve this critical problem.

Educational programmes, developmental therapies, political agendas, and regulations are important to reduce the increase in antibiotic resistance. The dynamic mission of reducing antibiotic resistance in an era while novel therapies for microbial infections are inadequate is entrusted to prescribers, policymakers, and researchers. Researchers propose that surveillance, surveillance practice, regulation, and novel therapies offer solutions to antibiotic resistance in both the mankind and agronomic sectors. This report also discusses the importance of antibiotic resistance and emphasizes the necessity for a multifaceted strategy for improving healthcare outcomes (Aslam et al. 2018).

Resistance to antimicrobials in microbial pathogens is a global problem attributed to great mortality and morbidity. Multidrug-resistant variations in Gram-positive and Gram-negative microbes have led to infections with traditional antimicrobials which are very challenging to treat or even not curable (Buchy et al. 2020; Andam et al. 2017). Since, in many healthcare settings, there is a lack of early detection of causative microbes and their patterns of antibacterial susceptivity in patients with bacteraemia and other severe diseases, wide-spectrum antibiotics are used substantially and often gratuitously (Akova 2016). The substantial rise in the emergence of resistance occurs, and bacterial species can easily be transmitted to other patients and also to the environment when combined with inadequate infection management practices. The availability of validated epidemiologic study evidence on antimicrobial resistance in commonly experienced microbial infections would be beneficial not only in choosing treatment approaches but also in the production of an efficient antimicrobial management system in hospitals (Frieri et al. 2020). Resistance to familiar antibacterial treatments and the development of multidrug-resistant microbes by significant bacterial pathogens are growing very fast, which is very harmful for the environment. There are challenges in the battle against microbial diseases and associated infections, and the existing lack of effective medications, deficiency of active preventive strategies, and just a few novel antibiotics in the clinical pipeline would need for new therapeutic options and alternative antimicrobial therapies to be established (Prestinaci et al. 2015). The researchers indicated that increasing understanding of bacterial virulence techniques and triggered infectious disease molecular pathways offers great opportunities for targeting and interfering with critical pathogenic organism factors or microorganism-associated virulence traits while bypassing the evolutionary pressure on the microorganism for the development of resistance (Laws et al. 2019). Infection occurred from antibiotic-resistant microbes is a global problem for human health as well as for researchers who are involved in developing alternative strategies to resolve this problem. In the last few decades, human pathogenic microbes have significantly increased worldwide, and these microbes have immunity to multiple antibiotics. Various biofilm-based infections were not cured by conventional therapies due to their antibiotic-resistant properties (Annunziato 2019). Furthermore, market supplies for the production of novel antibiotics have been running dry in recent decades. A recent WHO Global Health Day under the slogan "Drug Resistance: No intervention today means no solution tomorrow" has prompted an uptick in the clinical effort, and many innovative approaches have been created to improve treatment options for resistant bacterial infection pathogens. It was observed that not only antibiotic resistance genes are important but also the bacteria shape, genetic element, and repository of antibiotic resistance genes (ARGs) through which pathogenic bacteria can gain resistance by horizontal gene transfer (HGT) are important to study. As has been shown for some clinically significant ARGs, HGT induced antibiotic resistance to spread from commensal and environmental organisms to pathogenic ones (Von Wintersdorff et al. 2016). Even when transformation and transduction are considered less significant, recent studies indicate that their function could be greater than previously assumed. Knowing the nature of the resistance and how it mobilizes against pathogenic bacteria is important for attempts to monitor the spread of certain genes. In the past decades, multidrug-resistant bacteria have emerged at an unprecedented pace and created severe problems. The advent of resistant infections triggered by these bacteria has contributed to mortality and morbidity, and the ways to counter bacterial resistance desperately need to be sought (Bello-López et al. 2019).

5.1.2   Antibiotics as an Organic Pollutant in Wastewater

Nowadays, organic pollutants in wastewater are increasing very fast, which is a major concern worldwide. Sources of these organic pollutants are pharmaceuticals, pesticides, domestic wastes, chemicals, industrial effluents, sewages, and so on. Pharmaceutical pollutants are highly lipophilic or accumulate in fat, semi volatility, bioaccumulation, highly toxic, it can migrate in to the food chains, and persistence in the environment (Jayaraj et al. 2016).

As organic matter in a body of water increases, the number of decomposers increases. These decomposers are steadily growing, and they use a lot of oxygen during development. As decomposition gradually occurs, the levels of dissolved oxygen (DO) decrease. Oxygen deficiency can kill aquatic species. Though marine species die, decomposers break them down, which further decreases DO levels. Naturally, the water quality is mainly influenced by the presence of organic compounds. Contaminated water can have unhealthy colour, odour, pesticides, taste, turbidity, toxic chemicals, industrial waste products, high levels of TDS, alkaline, acids, and other domestic sewage material (Watkinson et al. 2007).

Aldrin – This antibiotic can easily metabolized to dieldrin by plants and animals alike. As a consequence, traces of aldrin are seldom contained in plants and livestock, even only in limited concentrations. Aldrin is known to be bioconcentrate due to its persistent nature and hydrophobicity.

Dieldrin – Strongly attaches to soil seeds. It is constant in existence and hydrophobicity, and bioconcentrate is related to dieldrin.

Endrin – It was used as a rodenticide as well. This antibiotic can easily metabolized by livestock and does not retain in fat as much as other chemicals with identical structures. This antibiotic can also penetrate the environment by volatilization, which may contaminate surface water from soil runoff.

Chlordane – Chlordane is extremely water-insoluble and is soluble in organic solvents. It is semi-volatile, and as a consequence, partitioning into the environment may be predicted. Because of its strong partition coefficient, it attaches readily to marine sediments and bioconcentrates in the fat of animals.

Heptachlor – It is extremely water-insoluble and soluble in organic solvents. It is very dynamic and can, therefore, be forced to partition into the environment. In the fat of living creatures, it attaches readily to marine sediments and bioconcentrates.

HCB – Hexachlorobenzene is also a by-product of industrial chemical manufacturing, namely, carbon tetrachloride, perchloroethylene, trichloroethylene, and pentachlorobenzene. HCB is extremely water-insoluble and soluble in organic solvents. It is very dynamic and can, therefore, be forced to partition into the environment. It is particularly immune to degradation and has a high partition coefficient (KOW = 3.03–6.42), and as a consequence, bioconcentrate is found in the fat of living organisms.

Mirex – Mirex is very immune to degradation and is extremely insoluble in soil, and bioaccumulation and biomagnification have been reported. Mirex attaches tightly to marine sediments, owing to its insolubility.

Toxaphene – This is also one type of antibiotic, which cannot dissolve in water. It can sustain in soil up to 12 years (i.e. half-life of toxaphene). Bioconcentration in marine animals has been demonstrated, and air transmission is likely to occur.

DDT – Like toxaphene, DDT also cannot dissolve in water, but it can easily dissolve in most organic solvents. It is semi-volatile, and as a consequence, partitioning into the environment may be predicted. Their existence in the atmosphere is widespread, and traces were also found in the Arctic. It is lipophilic and divides readily into the fat of all living species, and bioconcentration and biomagnification have been demonstrated.

PCB – In the environment, polychlorinated biphenyls were known to be correlated with the organic materials of biological tissues, sediments, and soils or with organic matter dissolved in marine environments, instead of in water solution. In spite of their low vapour pressure, and partly because of their hydrophobicity, PCBs volatilize from water surfaces; thus, atmospheric transport may be a major mechanism for the delivery of PCBs in the atmosphere.

Dioxins and furans – They are usually very water-insoluble, lipophilic, and very permanent. None of them are commercially produced and have no known use. They are by-products of the production of other chemicals.

5.1.3   Biofilm and Antibiotics

Pathogenic biofilm is one of the major concerns worldwide due to its inherent antibiotic resistance properties. Trying to combat this cell organization typically entails high doses of antibiotic for a long period, and this strategy often fails to contribute to the persistence of infection (Singh et al. 2017). Worldwide, now scientists are focusing on their research to develop an alternative process to control the pathogenic microbial biofilm. Although treatment of pathogenic microbial biofilm-based infections is like a puzzle for researchers, still they are trying hard to find the appropriate solutions. In the recent past, some researchers have studied two antimicrobial agents, namely, pharmacokinetic (PK) and pharmacodynamic (PD), to minimize the antibiotic resistance and antibiotic tolerance in the pathogenic microbial biofilm. But the major issue that arises from this study is they were tested on planktonic cells, which are far more different in all aspects from the pathogenic microbial biofilm in terms of metabolism, rate of growth, expression of gene, etc. These researchers have established several procedures for antibiotic PK/PD studies in biofilm infections of *P. aeruginosa* in vitro and in vivo. It should be noted that none of the biofilm procedures have so far been certified for clinical use or useful for antibiotic therapy guidance (Sharma et al. 2019). Nowadays, an increase in multidrug-resistant infections from pathogenic biofilm is a worldwide problem, especially biofilm with *P. aeruginosa*. The researchers who investigated on *P. aeruginosa* biofilms have answered all unsolved problems concerning with the impact of the widely used calcium channel blocker (CCB) diltiazem on the growth of biofilms. To test the overall growth, antibiotic tolerance, and antibiotic resistance of diltiazem on *P. aeruginosa*-based biofilm, several tests have been conducted during fluoroquinolone therapy in both the conditions, i.e. in the presence and absence of diltiazem. The authors observed that the widely used calcium channel blocker diltiazem induces resistance to primary-line fluoroquinolones for *P. aeruginosa* biofilms. Throughout the treatment of planktonic *P. aeruginosa*, the traditional in vitro models simulate antibiotic pharmacodynamics (Noreddin and Elkhatib 2009).

5.2   Molecular Mechanism of Antibiotic Resistance

Antibiotic resistance can be defined as the ability of the microorganisms to tolerate and resist the antibiotic, which means microbial growth is not stopped and also these microorganisms can't be killed. The molecular mechanisms behind the antibiotic resistance of microbes have been extensively studied. Generally, there are two types of antibiotic resistance: type 1, intrinsic antibiotic resistance, and type 2, acquired antibiotic resistance. Microbes have some inherent structural and functional properties, so they can intrinsically resist some antibiotics. There are few antibiotics are present, due to their structure they can't able to penetrate the microbial outer cell membrane and those antibiotic who can able enter the cell membrane can be removed by efflux pumps. Microorganisms, meanwhile, may have gained resistance that can be produced by chromosome mutations or, more generally, by the acquisition of an antibiotic-resistant gene from other bacteria by transposons or mobile plasmids (also called as horizontal gene transfer) (Li and Webster 2018).

As per Riedl et al., there are currently five key targets for antibiotics (i.e. membrane structure, folate mechanism, DNA gyrase and RNA polymerase, protein synthesis, and cell wall synthesis), and antibiotic resistance can be assimilated via four types of pathways (i.e. mutation, conjugation, transduction, and transformation) and expressed through four different mechanisms. For example, MRSA is resistant to many penicillin-like β-lactam antibiotics primarily because it expresses the mecA gene encoding the low-affinity penicillin-binding protein 2a (PBP 2a) (Chellat et al. 2016).

Recently, the Antibiotic Resistance Genes Database has reported that there are 23,000 antibiotic-resistant genes that are present in 380 types of different forms from accessible sequences of bacterial genomes. Fortunately, the amount of genes with adaptive tolerance is far lower. The development of biofilm in orthopaedic implant-associated infections is often considered to be one of the main forms of antibiotic resistance. As we already discussed, pathogenic microbial biofilm has inherent ability to resist and tolerate the specific antibiotics because (1) some antibiotics generally are not able to penetrate the outer cell membrane of microbes; (2) due to nutrient limitation, some biofilm microbial cells are very slow in their growth rate; and (3) some biofilm may adopt a protected phenotype.

For example, the biofilm-forming *S. epidermidis* strains (126 out of 342) displayed a slightly higher prevalence of tolerance to ciprofloxacin and sulfamethoxazole as well as to the four aminoglycosides relative to certain non-producing isolates (Kapoor et al. 2017; Fair and Tor 2014; Arciola et al. 2005). The existence of antibiotic-resistant genes that are widely spread in natural environment may be affected by several factors like human activities. Numerous forms of anthropogenic behaviour (such as the usage of antibiotics in livestock and the use of antibiotics in waste disposal) have produced large reservoirs for antibiotic-resistant genes. Water, wind, animals, etc. may spread antibiotic-resistant genes throughout the natural habitat. As a result, treatment plants for wastewater are abundant in genes for antibiotic-resistant microbes, and reservoirs of antibiotic resistance genes are bacteriophages (viruses that infect bacteria) found in wastewater (Kraemer et al. 2019).

It is worthy to note that mechanisms of antibiotic resistance have long existed. Antibiotic encoding integrons and antibiotic-resistant genes can be found in the intestine of gut flora who are isolated from modern civilization and are not exposed to any antibiotics.

5.3   Biofilm Resistance and Tolerance

In the presence of any bacteriostatic antimicrobial agent or bactericidal agent, resistant microorganisms can grow at a concentration that is usually inhibitory to microbial growth. Generally, resistivity can be investigated by applying minimum inhibitory concentration (MIC) in planktonic cells, and this antimicrobial agent concentration is the lowermost amount which can inhibit the growth of microorganisms. Usually, it was thought that resistance is responsible for mutations or alteration of antibiotic resistance genetic component, while resistance might be intrinsic, and therefore it is dependent on instinctive properties of the cell and wild-type genes. For example, intrinsically, the antibiotic resistance of Gram-negative microbes is more than Gram-positive microbes like vancomycin because the outer membrane of the Gram-negative bacteria is relatively impermeable (Li et al. 2017). Resistance mechanisms were reported by Lewis (2008) through which the interaction of an antimicrobial agent with the target can be prevented. The resistance of microorganisms to an antimicrobial agent can be defined as the ability of that microorganism to survive in the presence of (bactericidal) antimicrobial agents. Minimum bactericidal concentration (MBC) is the lowest concentration, which is a measure of tolerance that kills 99.9% of cells in culture (Lewis 2008). Brauner described the different alternative measures of tolerance. It was thought that the tolerance mechanism will inhibit the bactericidal agent from employing its toxic effects even though the agent has bound to its target. Significantly, the pathogenic microbes apply both the resistance and tolerance mechanisms to resist the antimicrobial problems in the biofilm. Since decades, microscopy has been used to research microbes, and it's appropriate that it has now found its place in biofilm susceptibility testing as a visual way of corroborating certain test methods (Brauner et al. 2016). Biofilms that are exposed to a given antimicrobial agent can be labelled to differentiate between living and dead cells. By using microscopy, the structural architecture of biofilm and the effect of antimicrobial agent on biofilm can be interpreted. A concern with microscopy as a means to determine the sensitivity of biofilms is that this is not appropriate for all methods of biofilm production (Brauner et al. 2016). Furthermore, Müsken et al. (2010) identified a microscopic approach for quantitatively and qualitatively evaluating the antibiotic susceptibility of *P. aeruginosa* biofilms grown in 96-well microtiter plates, thus opening the ability for microscopy to be used in high-throughput biofilm susceptibility assays (Müsken et al. 2010). While clinical research of biofilm antibiotic susceptibility is not greater to normal (planktonic cells) antibiotic resistance study in directing antibiotic treatment choices that increase results for CF patients (Waters and Ratjen 2015), it is reasonable to conclude that biofilm susceptibility testing could be clinically important, since most bacterial infections are based on biofilm.

5.4   Bio-Electrochemical Concepts in Wastewater (Microbial Fuel Cell)

Microbial metabolism energy production for both catabolism and anabolism is a mixture of oxidation and reduction of substrate. This cycle includes a source of electron (substratum) that drops inside the microbe's metabolic flux (can be used by the microbes) and a powerful/weak electron sink (acceptor) to complete the electron transport chain. The isolation of these two processes (oxidation and reduction) by an ion-permeable membrane (optional) in an electrode-equipped device (artificial electron acceptors) provides an atmosphere for harnessing the energy produced by the microbe in the form of current density, against the possible differences created between them. The bacteria use the accessible substrate (fermentation) producing the reduction equivalents at anode [protons (H+) and electrons (e-)]. Protons are transferred to the cathode through the ion-selective membrane via the electrode interface of the solution, creating a theoretical gap between anode and cathode through which the electrons migrate across the circuit (current) through the external load. The reduction equivalents produced during the activity of BES have several applications in the areas of energy generation and waste remediation. Broadly speaking, BES technology may be categorized as a power generator, wastewater treatment facility, and value-added substance recovery device. Alternatively, whether the waste/wastewater acts as a donor or acceptor of electrons, its remediation is represented either by anodic oxidation or by cathodic reduction under specified conditions (Pant et al. 2010). Recently, declines in certain substrates or carbon dioxide (CO_2) as electron acceptors are also recorded during bioelectrochemical systems (BES) activity, raising its commercial viability (Pant et al. 2015). The present chapter reflects extensively on the remediation dimensions of BES including multiple drainage, different toxins, and desalination. Microbial fuel cell-coupled constructed wetlands (CW-MFCs) utilize electrochemical, biological, and ecological roles for wastewater treatment. However, few studies have studied the threats of antibiotic resistance genes (ARGs) by utilizing such antibiotic elimination mechanisms. Thus, Zhang et al. designed three CW-MFCs in 2017 to investigate the antibiotic resistance of biofilm. In their experiment, they studied the dynamics of antibiotic resistance gene in the biofilm filler, and the experiment was operated till 5000 h. Their experimental result showed that relatively high steady voltages were obtained, i.e. 605.8 mV, 613.7 mV, and 541.4 mV, with effluent antibiotic concentrations of 400, 1000, and 1600µg L^{-1}, respectively. They investigated the ARGs in both the electrode by 16 s rRNA sequencing, and they found that cathode layer has higher amount as compared to anode and middle layers, but the opposite pattern for the sul and tet genes was observed. The relative abundance of the three sul genes measured was in the order sulI > sulII > sulIII and those of the five teta genes was in the order tetA > tetC > tetW > tetO > tetQ. Compared to the filler biofilm, the effluent water had comparatively small abundances of sul and tet genes. No changes were observed throughout the treatment time for most ARGs, and no major differences were found between the ARGs and the copy numbers of 16S rRNA genes, except for sulI and tetW in the effluent. For several of the ARG copy numbers, however, major similarities have been found (Zhang et al. 2017).

5.5 Potential for New Therapies

In future, researchers should focus on developing new techniques and treatment strategies to inhibit the microbial growth of biofilm and also investigate the detailed mechanism behind the antibiotic resistance. There are various mechanisms of antibiotic resistance, which can act together. To be clinically effective, anti-biofilm treatments should have to frustrate more than one mechanism at the same time. Heterogeneity is a prevalent feature in these processes in resistance; in a biofilm, microorganisms live in a wide range of environments (Fair and Tor 2014).

Firstly, based on their spatial location, the microbial cells may be exposed to various antibiotic concentrations. Secondly, gradients in microbial nutrient concentration and waste products crisscross the biofilm and alter the local environment, leading to a wide range of growth rates of individual microbial cells. Thirdly, a limited proportion of cells in a bacterial biofilm may differentiate themselves into a strongly safe phenotypic state and coexist with antibiotic-prone neighbours. The proliferation of states that occur when these three forms of heterogeneity are crossed indicates that any provided antimicrobial agent may destroy some of the cells in a biofilm but is unlikely to target all effectively. The majority of all of the antibiotics in current usage has been established against the growing culture of individual cell based on their behaviour. Novel screens of current and potential antibiotics screened for action against non-growing or biofilm cells may produce clinically successful antimicrobial agents against biofilm infections (Stewart 2002). Scientists need to discover the gene that is responsible for antibiotic resistance of biofilm, and this gene of interest should be the targets for chemotherapeutic adjuvants to increase the efficiency of current antibiotics against pathogenic biofilm. Since biofilm resistance depends on bacterial aggregation in multicellular cultures, one approach may be to create therapies that destroy the biofilm's multicellular structure. If the biofilm's multicellularity is destroyed, the host defences will overcome the infection, and the antibiotic effectiveness could be recovered. Promising therapy involves enzymes that remove the biofilm's matrix polymers, chemical reactions that inhibit the formation of the biofilm matrix, and analogy of microbial signalling molecules that interfere with cell-to-cell contact needed to shape natural biofilms. The gene of interest involved in the antibiotic resistivity of biofilm is one of the potential targets for cancer therapies because it requires the gene product for formation of multicellular colony. In other words, we agree that therapeutic approaches are directed at constructing multicellular systems rather than basic human cell functions. When we grasp the multicellular complexity of microbial existence, we'll know how to handle the recurrent infections associated with biofilms (Koo et al. 2017).

5.6 Antibiotic Removal Mechanisms Based on Microbial Fuel Cell

The antibiotic removal mechanisms are divided into three groups, taking into account the various MFC operating modes used to degrade antibiotics:

Most MFCs are served in group A (Fig. 5.1a). This figure represents a dual chambered MFC which consists of an anode and a cathode. The anode is biologically active as biofilm grows over the anode surface. The cathode is abiotic in nature, and the cathodic chamber consists of potassium ferricyanide, which acts as an electron acceptor (Yuan et al. 2016). Antibiotics function as electron donors and origins of carbon in biological anodes. Exoelectrogenic microbes and decaying antibiotic bacteria that bind biofilms inside extracellular polymeric substances to the anodes are liable for growing the overpotential of biorefractory parents and their metabolites. In MFCs, anaerobic digestion of organic compound is an important factor as this digestion mechanism is associated with the electrical signal which leads to antibiotic mineralization in MFCs. Nonstop electrical signal will help in stimulating the microbial metabolism with direct or indirect electron transfer to the microbial cell by providing a microenvironment with electron (Dominguez-Benetton

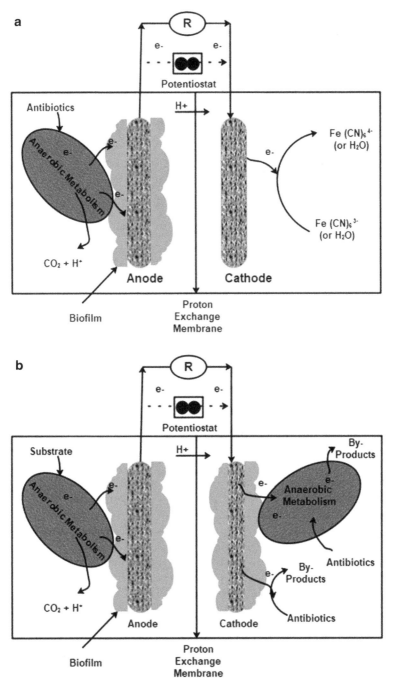

Fig. 5.1 (**a**) Bio-anode on oxidation of antibiotic coupled with abiotic cathode, (**b**) bio-anode coupled to bio-cathode for biodegradation of antibiotics, and (**c**) bio-anode coupled to a cathode compartment with antibiotic reduction by advanced oxidation processes

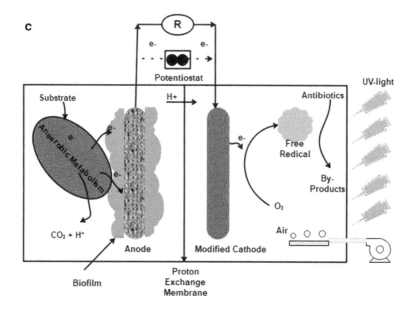

Fig. 5.1 (continued)

et al. 2018). The activated microorganisms quickly metabolize antibiotics through the secretion of enzymes, such as SMX and its 3A5MI by-product (Wang et al. 2016).

In group B (Fig. 5.1b), the devices are typically run as bio-cathodes, i.e. a Microbial electrolysis cells (MEC) with an external energy source, since at the biological anode, the capacity for removal of antibiotics is also greater (Jafary et al. 2015). In this type of devices, generally antibiotic degradation happened with respect to bio-electrochemical reaction. That means the antibiotics take electrons from cathode, and when the electrochemical reaction reduces, it leads to biodegradation of antibiotics inside the system. Consequently, microbes thriving as biocatalysts in cathode electrodes speed up antibiotic removal by growing their overpotential. Examples of these include the chloramphenicol (CAP) and nitrofurazone (NFZ) (Kong et al. 2017; Liang et al. 2013). If the antibiotics have more tendency for reduction to bio-anode, then the MFC device can perform as bio-cathode without any external power supply.

Updated materials are also used in group C (Fig. 5.1c) to produce radical species in the Bioelectrochemical systems (BES) cathode, which strike and degrade goal antibiotics. This method includes the following reactions: (1) bio-anode-generated electrons transferred to the cathodic surface material by external circuit; (2) cathodic material which accepts electrons for the production of H_2O_2 to reduce dissolved oxygen or air from pumps, i.e. $O_2 + 2e + 2H + H_2O_2$; and (3) H_2O_2 that can produce radical species under ultraviolet light or cathodic content, such as hydroxyl radicals (percent OH), which oxidize and kill parents of antibiotics, i.e. $H_2O_2 + hv \rightarrow 2OH\%$ and antibiotic+$OH\% \rightarrow$ degradation products.

5.7   Overall Performance/Discussion of Antibiotic Resistance in Biofilm and Fuel Cells

Biofilm-based pathogens were hard to eliminate owing to their ability to tolerate concentrations of antibiotics which would usually destroy free planktonic aquatic cells. Bacterial biofilms, therefore, add greatly to patient morbidity and mortality. As described in this study, several pathways have been found that lead to biofilm antibiotic tolerance and antibiotic resistance, like heterogeneity of biofilm, stress responses, and persistent cells. It has now been apparent that the fundamental processes of tolerance and susceptibility to antibiotics in biofilms are, in certain instances, genetically dependent. From the various reports, it was found that *P. aeruginosa* was the most widely researched to study the antibiotic resistance mechanism (Verderosa et al. 2019). Because this analysis should not be deemed exhaustive, pathogenic microbial biofilm-forming microbes other than *P. aeruginosa* there is an apparent shortage of articles explicitly identifying distinct processes of antibiotic resistance and tolerance (Ciofu and Tolker-Nielsen 2019). Going forward, we must continue to recognize and characterize the pathways that encourage antibiotic recalcitrance in bacterial biofilms. Advanced technologies and creative laboratory setups and methods, such as work into pathogenic bacterial biofilm metabolomes (Zhang and Powers 2012; Stipetic et al. 2016).

The research done by Amini et al. (2011) (Amini et al. 2011) is very much important for scientific world as they summarized the concept that biofilms largely depend upon the different genetic factors than planktonic cells to survive from antibiotic treatment. Significantly, it is crucial that biofilm recalcitrance determinants continue to drive therapeutic approaches that would inhibit such defences (Taylor et al. 2014). The medical therapy strategies should improve in order to cure patients from biofilm-based infection with the application of high doses of antibiotics. One of the important issues related to antibiotic resistance which is very negligibly reported is the effect of commonly used medicine, which is non-antimicrobial in nature, to cure biofilm-based infections. So, the scientific community should also focus on this topic to investigate consequences of those medicines on patients susceptible to biofilm-based infections (Gebreyohannes et al. 2019). Impressively, some calcium channel blockers (usually recommended for patients with arrhythmias, coronary heart disease, or asthma) were shown to have an effect on the microbial biofilm response to fluoroquinolones. Although the precise medical strategy of advanced drug-to-drug interaction needs to be studied, it is very much significant that medications used for some comorbidities may theoretically have an impact on antibiotic treatment results at the microbial stage (Santos-Lopez 2019).

5.8 Conclusions

Pathogenic microbial biofilm uses multiple pathways to survive from antibiotic. The significance of involving pathways for antibiotic resistance differs from each other which mainly depends upon the growth condition of biofilm and antimicrobial agent. As per recent reports, the multifactorial nature of biofilm-specific antibiotic resistance and tolerance was investigated, which means the scientific world is one step closer to improving clinical outcomes for patients suffering from chronic, treatment-refractory biofilm-based infections.

References

Akova M (2016) Epidemiology of antimicrobial resistance in bloodstream infections. Virulence 7 (3):252–266

Amini S, Hottes AK, Smith LE, Tavazoie S (2011) Fitness landscape of antibiotic tolerance in *pseudomonas aeruginosa* biofilms. PLoS Pathog 7(10):e1002298

Andam CP, Worby CJ, Gierke R, McGee L, Pilishvili T, Hanage WP (2017) Penicillin resistance of nonvaccine type pneumococcus before and after PCV13 introduction, United States. Emerg Infect Dis 23(6):1012–1015

Annunziato G (2019) Strategies to overcome antimicrobial resistance (AMR) making use of non-essential target inhibitors: a review. Int J Mol Sci 20(23):5844

Arciola CR et al (2005) Antibiotic resistance in exopolysaccharide-forming Staphylococcus epidermidis clinical isolates from orthopaedic implant infections. Biomaterials 26 (33):6530–6535

Aslam B et al (2018) Antibiotic resistance: a rundown of a global crisis. Infect Drug Resist 11:1645–1658

Bello-López JM et al (2019) Horizontal gene transfer and its association with antibiotic resistance in the genus aeromonas spp. Microorganisms 7(9):363

Brauner A, Fridman O, Gefen O, Balaban NQ (2016) Distinguishing between resistance, tolerance and persistence to antibiotic treatment. Nat Rev Microbiol 14(5):320–330

Buchy P et al (2020) Impact of vaccines on antimicrobial resistance. Int J Infect Dis 90:188–196

Chellat MF, Raguž L, Riedl R (2016) Targeting antibiotic resistance. Angew Chemie Int Ed 55 (23):6600–6626

Ciofu O, Tolker-Nielsen T (2019) Tolerance and resistance of pseudomonas aeruginosa biofilms to antimicrobial agents-how P. aeruginosa can escape antibiotics. Front Microbiol 10:913

Dominguez-Benetton X et al (2018) Metal recovery by microbial electro-metallurgy. Prog Mater Sci 94:435–461

Fair RJ, Tor Y (2014) Antibiotics and bacterial resistance in the 21st century. Perspect Med Chem 6 (6):25–64

Frieri M, Kumar K, Boutin A (2020) Antibiotic resistance. J Infect Public Health 10(4):369–378

Gebreyohannes G, Nyerere A, Bii C, Sbhatu DB (2019) Challenges of intervention, treatment, and antibiotic resistance of biofilm-forming microorganisms. Heliyon 5(8):e02192

Hansen AK, Velschow S (2000) Antibiotic resistance in bacterial isolates from laboratory animal colonies naive to antibiotic treatment. Lab Anim 34(4):413–422

Jafary T et al (2015) Biocathode in microbial electrolysis cell; present status and future prospects. Renewab Sustain Energy Rev 47:23–33

Jayaraj R, Megha P, Sreedev P (2016) Organochlorine pesticides, their toxic effects on living organisms and their fate in the environment. Interdiscip Toxicol 9(3–4):90–100

Kapoor G, Saigal S, Elongavan A (2017) Action and resistance mechanisms of antibiotics: A guide for clinicians. J Anaesthesiol Clin Pharmacol 33(3):300–305

Khatoon Z, McTiernan CD, Suuronen EJ, Mah TF, Alarcon EI (2018) Bacterial biofilm formation on implantable devices and approaches to its treatment and prevention. Heliyon 4(12):e01067

Kong D et al (2017) Response of antimicrobial nitrofurazone-degrading biocathode communities to different cathode potentials. Bioresour Technol 241:951–958

Koo H, Allan RN, Howlin RP, Stoodley P, Hall-Stoodley L (2017) Targeting microbial biofilms: current and prospective therapeutic strategies. Nat Rev Microbiol 15(12):740–755

Kraemer SA, Ramachandran A, Perron GG (2019) Antibiotic pollution in the environment: from microbial ecology to public policy. Microorganisms 7(6):180

Laws M et al (2019) Antibiotic resistance breakers: current approaches and future directions. FEMS Microbiol Rev 43(5):490–516

Lewis K (2008) Multidrug tolerance of biofilms and persister cells. In: Current topics in microbiology and immunology, vol 322. Springer, Berlin, pp 107–131

Li B, Webster TJ (2018) Bacteria antibiotic resistance: new challenges and opportunities for implant-associated orthopedic infections. J Orthop Res 36(1):22–32

Li J et al (2017) Antimicrobial activity and resistance: influencing factors. Front Pharmacol 8:364

Liang B et al (2013) Accelerated reduction of chlorinated nitroaromatic antibiotic chloramphenicol by biocathode. Environ Sci Technol 47(10):5353–5361

Müsken M, Di Fiore S, Römling U, Häussler S (2010) A 96-well-plate-based optical method for the quantitative and qualitative evaluation of Pseudomonas aeruginosa biofilm formation and its application to susceptibility testing. Nat Protoc 5(8):1460–1469

Noreddin AM, Elkhatib WF (2009) Novel in vitro pharmacodynamic model simulating ofloxacin pharmacokinetics in the treatment of Pseudomonas aeruginosa biofilm-associated infections. J Infect Public Health 2(3):120–128

Pant D, Mohanakrishna G, Srikanth S (2015) Bioelectrochemical systems (Bes) for microbial electroremediation: an advanced wastewater treatment technology. In: Applied environmental biotechnology: present scenario and future trends. Springer, New Delhi, India, pp 145–167

Pant D, Van Bogaert G, Diels L, Vanbroekhoven K (2010) A review of the substrates used in microbial fuel cells (MFCs) for sustainable energy production. Bioresour Technol 101 (6):1533–1543

Prestinaci F, Pezzotti P, Pantosti A (2015) Antimicrobial resistance: a global multifaceted phenomenon. Pathogens Global Health 109(7):309–318

Alfonso Santos-Lopez (2019) "Biofilm-dependent evolutionary pathways to antibiotic resistance | bioRxiv." https://www.biorxiv.org/content/10.1101/581611v1.full. Accessed 19 May 2020

Sharma D, Misba L, Khan AU (2019) Antibiotics versus biofilm: An emerging battleground in microbial communities. Antimicrob Resist Infect Control 8(1):1–10

Singh S, Singh SK, Chowdhury I, Singh R (2017) Understanding the mechanism of bacterial biofilms resistance to antimicrobial agents. Open Microbiol J 11(1):53–62

Stewart PS (2002) Mechanisms of antibiotic resistance in bacterial biofilms. Int J Med Microbiol 292(2):107–113

Stipetic LH, Dalby MJ, Davies RL, Morton FR, Ramage G, Burgess KEV (2016) A novel metabolomic approach used for the comparison of Staphylococcus aureus planktonic cells and biofilm samples. Metabolomics 12(4):75

Taylor PK, Yeung ATY, Hancock REW (2014) Antibiotic resistance in Pseudomonas aeruginosa biofilms: towards the development of novel anti-biofilm therapies. J Biotechnol 191:121–130

Verderosa AD, Totsika M, Fairfull-Smith KE (2019) Bacterial biofilm eradication agents: a current review. Front Chem 7:824

Von Wintersdorff CJH et al (2016) Dissemination of antimicrobial resistance in microbial ecosystems through horizontal gene transfer. Front Microbiol 7:173

Wang L, Liu Y, Ma J, Zhao F (2016) Rapid degradation of sulphamethoxazole and the further transformation of 3-amino-5-methylisoxazole in a microbial fuel cell. Water Res 88:322–328

Waters V, Ratjen F (2015) Pulmonary exacerbations in children with cystic fibrosis. Ann Am Thorac Soc 12:S200–S206

Watkinson AJ, Murby EJ, Costanzo SD (2007) Removal of antibiotics in conventional and advanced wastewater treatment: implications for environmental discharge and wastewater recycling. Water Res 41(18):4164–4176

Yuan H, Hou Y, Abu-Reesh IM, Chen J, He Z (2016) Oxygen reduction reaction catalysts used in microbial fuel cells for e1nergy-efficient wastewater treatment: a review. Mater Horizons 3 (5):382–401

Zhang B, Powers R (2012) Analysis of bacterial biofilms using NMR-based metabolomics. Future Med Chem 4(10):1273–1306

Zhang S et al (2017) Dynamics of antibiotic resistance genes in microbial fuel cell-coupled constructed wetlands treating antibiotic-polluted water. Chemosphere 178:548–555

Plant Secondary Metabolites for Tackling Antimicrobial Resistance: A Pharmacological Perspective

6

Sathiya Maran, Wendy Wai Yeng Yeo, Swee-Hua Erin Lim, and Kok-Song Lai

Contents

S. Maran
School of Pharmacy, Monash University Malaysia, Jalan Lagoon Selatan, Subang Jaya, Malaysia

W. W. Y. Yeo
Perdana University Graduate School of Medicine, Perdana University, Wisma Chase Perdana, Changkat Semantan, Damansara Heights, Kuala Lumpur, Malaysia

S.-H. E. Lim · K.-S. Lai (✉)
Health Sciences Division, Abu Dhabi Women's College, Higher Colleges of Technology, Abu Dhabi, United Arab Emirates
e-mail: lkoksong@hct.ac.ae

Abstract

The practice of using medicinal plants in treating various diseases and ailments dates back to over 60,000 years ago. These medicinal plants produce a diversity of secondary metabolites which are natural sources of biologically active compounds that can be classified into phenolics, alkaloids, saponins, terpenes, glycosides and flavonoids. These active compounds exert significant antimicrobial resistance (AMR) and pharmacological activity. In the current scenario, more attention is being focussed towards plant-based drug discovery in treating diseases, especially in fulfilling therapeutic requirements of dreaded diseases such as cancer, HIV and neurodegenerative diseases. This review aims to elucidate and understand different classifications of plant secondary metabolites towards understanding their mechanisms of action in tackling antimicrobial resistance (AMR) as well as their benefits and usage in pharmacology.

Keywords

Antimicrobial resistance · Plant secondary metabolites · Pharmacology · DNA replication · Quorum sensing

6.1   Introduction

Antimicrobial resistance (AMR) is reported as one of the most pressing public health problems globally; AMR has caused infections in at least 2.8 million people with more than 35,000 deaths in the USA alone in 2019 (Mahizan et al. 2019). AMR is defined as inefficacious infection-associated treatment with an antimicrobial agent that used to be effective (Barbieri et al. 2017) and threatens the effective prevention and treatment of infections caused by bacteria, parasites, viruses and fungi (Prestinaci et al. 2015; Moo et al. 2019). Initial measures by WHO via the Global Strategy for Containment of Antimicrobial Resistance provided a framework of intervention to slow the development and reduce the spread of antimicrobial-resistant microorganisms (WHO 2001). The Evolving Threat of Antimicrobial Resistance – Options for Action (2012) then proposed a combination of interventions comprising of health systems and surveillance in improving the use of antimicrobials in hospitals and community, via implementation of infection prevention and control. This is in addition to the development of appropriate new drugs and vaccines in tandem with political commitment (Podolsky 2018). In 2014, WHO published the first global report on surveillance of AMR which showed that surveillance data is important for orienting treatment choices, understanding AMR trends, identifying priority areas for interventions and monitoring the impact of interventions to contain resistance (WHO 2014).

Lack of compliance by both the consumers and healthcare practitioners has been reported as the main cause of emergence of antibiotic resistance bacteria (Yang et al. 2018a). Alongside these, other factors that have been associated with the increase in AMR are the limited effectiveness of the antibiotics' life spans, the excessive and unmonitored use in agriculture and the slow rate in release of new antimicrobial agents (Othman et al. 2019). These issues have pivoted the focus towards natural products as a possible remedy for AMR. In the current scenario, plant-based drug discovery and its applications in pharmacology have been seeing a surge since 2008 (Newman et al. 2008; Madhumitha and Saral 2009; Velu et al. 2018). Numerous factors contribute to this surge, thus leading to the possibility of using plant secondary metabolites as antimicrobial agents especially against causative human pathogens which can enhance the conventional antibiotic actions. Besides these, the abundance of fauna (400,000–500,000 species) (Othman et al. 2019) on earth leading to richness in secondary metabolites, such as phenolic and polyphenols, alkaloids, terpenoids and essential oils, lectins and others, provides a sea of opportunity for identifications of novel compounds. Furthermore, the advancement of modern technology that enables large screening and isolation of different compounds using high-performance tools alleviates the laborious and time-consuming procedures of isolation and extraction of natural product compounds (Gorlenko et al. 2020).

This chapter intends to examine the different groups of antimicrobial plant secondary metabolites, their structural diversity and mechanisms of antimicrobial resistance, towards understanding their pharmacological significance in medicine.

6.2   Groups of Antimicrobial Plant Secondary Metabolite

Secondary plant metabolites are chemical compounds produced by plant cells through metabolic pathways derived from the primary metabolic pathways (Hussein and El-Anssary 2018). About <10% of plant chemical compounds are secondary metabolites, and these are generally used as defence mechanisms against insects, herbivores and microorganisms. These metabolites harbour medical applications such as antibacterial, antifungal and antiviral activities which make them attractive in traditional medications (Othman et al. 2019). Plant secondary metabolites can be classified into phenolics, alkaloids, saponins, terpenes, glycosides and flavonoids according to their chemical structures (Velu et al. 2018).

6.2.1   Phenolics

Phenolic compounds contain benzene rings, with one or more hydroxyl substituents, which are categorized from simple phenolic molecules to highly polymerized compounds (Velderrain-Rodríguez et al. 2014). They are produced in the shikimic acid of plants and pentose phosphate pathway through the phenylpropanoid metabolization (Randhir et al. 2004). Phenolics are found abundantly in plants and contain numerous compounds such as simple flavonoids, phenolic acids, complex flavonoids and coloured anthocyanins (Babbar et al. 2014). Phenolic compounds have been reported as an anti-aging, anti-inflammatory, antioxidant and antiproliferative agents and have been known to promote health benefits by reducing the risk of metabolic syndrome and the related complications of type 2 diabetes and long-term diabetes complications, including cardiovascular disease, neuropathy, nephropathy and retinopathy (Lin et al. 2016). Numerous clinical studies mainly reporting the benefits of phenols when tested in human trials have also been conducted in determining the potential health benefits of phenolic compounds (Table 6.1).

6.2.2 Alkaloids

Alkaloids are large and structurally diverse groups of natural products, consisting of one to five nitrogen atoms, a primary amine (RNH_2), a secondary amine (R_2NH) or a tertiary amine (R_3N) (Robbers et al. 1996). Alkaloids can be found in 300 plant families, and to date, about 18,000 alkaloids have been discovered (Othman et al. 2019; Dembitsky 2005) in different parts of plants (Robbers et al. 1996), insects (Lusebrink et al. 2008; Savitzky et al. 2012), amphibians (Toledo and Jared 1995; Stynoski et al. 2014), reptiles (Lainson et al. 2003), birds (Clark 2010), mammals and marine animals (Dembitsky 2005; Pawlik 2011; McClintock and Baker 2013; Baker and Alvi 2004; Chadwick and Morrow 2011; Moore et al. 1993).

Alkaloids have been extensively utilized in the development of antibacterial drugs (Cushnie et al. 2014), largely due to their beneficial pharmacological properties such as analgesic (e.g. codeine), central nervous stimulant (e.g. brucine), central nervous depressant (e.g. morphine), antihypotensive (e.g. ephedrine), antihypertensive (e.g. reserpine), antipyretic (e.g. quinine), anticholinergic (e.g. atropine), antiemetic (e.g. scopolamine) (Robbers et al. 1996), oxytocic and vasoconstrictor (e.g. ergometrine) (Dewick 1997), antitumour (e.g. vinblastine) and antimalarial (e.g. quinine) (Evans 2009) activities. Despite their beneficial properties, some alkaloids have also been reported to be highly toxic, contributing to incidents of human poisoning (Dewick 1997; Evans 2009; Stegelmeier et al. 1999). Furthermore, certain alkaloids such as the belladonna (containing atropine) used in medicine have been reported to be toxic (Robbers et al. 1996; Dewick 1997).

The antibacterial mechanism of action of alkaloids has been tested in the indolizidine, isoquinoline, quinolone, agelasine and polyamine classes (Cushnie et al. 2014). Studies have reported the effectiveness of these alkaloids in perturbing the Z-ring composed of filamenting temperature-sensitive mutant Z (FtsZ) subunits and inhibiting bacterial cell division (Beuria et al. 2005). Furthermore, evidences have also showed the mechanisms of alkaloid in binding to FtsZ; inhibiting FtsZ GTPase activity (Domadia et al. 2008); inhibiting Z-ring formation and inducing cell elongation, without affecting DNA replication, nucleoid segregation and membrane structure; and inducing the SOS response (Beuria et al. 2005; Domadia et al. 2008; Boberek et al. 2010).

Table 6.1 Clinical studies using plant secondary metabolites

NCT number	Study design	Number of participants enrolled	Condition/disease	Intervention	Outcomes	References
NCT04317079	Interventional, randomized, parallel assignment	37	Obesity	5 milligrams and 15 milligrams of an extra virgin olive oil phenolic compound, versus placebo, combined with diet	Suggested phenols as a good lead compound for future drug development in obesity treatments	Cotrim et al. (2012)
NCT01290250	Interventional, randomized, crossover assignment, quadruple masking	210	Overweight/obesity	Orange juice-based beverage enriched in polyphenols given 2 daily doses (250 ml each) for 3 months	Normal or high concentration of polyphenols protects against DNA damage and lipid peroxidation and reduced body weight in overweight or obese non-smoking adults	Rangel-Huerta et al. (2015)
NCT01813981	Interventional, randomized, crossover assignment, single masking	19	Cardiovascular disease	Impact of coffee intake, matched for caffeine but differing in CGA content (89 and 310 mg) on flow-mediated dilatation (FMD), was assessed	Coffee intake acutely improves human vascular function, an effect, in part, mediated by 5-CQA and its physiological metabolites	Mills et al. (2017)
NCT04025281	Interventional, randomized, crossover assignment	164	Cardiovascular disease	Participants at high risk for cardiovascular disease randomized into group: Receiving Mediterranean diet supplemented with 50 mL/d of extra virgin olive oil (MD + EVOO) or 30 g/d of nuts (MD + nuts) and a low-fat diet	MD + EVOO and MD + nuts showed a higher decrease in systolic (6 mmHg) and diastolic (3 mmHg) blood pressure	Casas et al. (2014)

(continued)

Table 6.1 (continued)

NCT number	Study design	Number of participants enrolled	Condition/disease	Intervention	Outcomes	References
NCT04164446	Cardiovascular diseases Hyperlipidaemias Hypercholesterolaemia	100	Randomized, parallel assignment, triple masking	Volunteers were supplemented with 450 mg gallic acid equivalent (GAE)/day of oil palm phenolics or control treatments for a 60-day period	Improved total cholesterol and LDL-C levels were observed	Fairus et al. (2018)

6.2.3 Saponins

Saponins are naturally occurring bio-organic compounds that have at least one glycosidic linkage (C-O-sugar bond) at C-3 between a glycone and a sugar chain (El Aziz et al. 2019). Saponins are widely distributed in higher plants; however, studies have also reported it to be found in marine invertebrates. They can be divided into two groups, namely, steroidal saponins and triterpenoid saponins (Tagousop et al. 2018). Over the years, their biological activities have been reported to show antibacterial, antifungal, antiviral, anti-inflammatory, anti-ulcer, haemolytic and hepatoprotective properties (Sonfack et al. 2019; Yang et al. 2018b; Pu et al. 2015; Dong et al. 2019; Vieira Júnior et al. 2015; Michel et al. 2011). In addition, the potential synergistic act of saponin with antibiotics has led to new treatment options especially in the field of infectious disease (Tagousop et al. 2018; Krstin et al. 2015).

6.2.4 Terpenes

Terpenes are made up of large hydrocarbon groups consisting of 5-carbon isoprene (C5H8) units (Mahizan et al. 2019). Terpenes are synthesized via two pathways: the non-mevalonate pathway and the mevalonate pathway from acetyl-CoA precursor. Over the years, about 40,000 structural varieties with a few classes reported as pharmaceutical agents have been reported (Roberts 2007; Cho et al. 2017). In terms of the different classes of terpenes, to date, a total of eight different classes of terpenoids have been classified: hemiterpenoids, monoterpenoids, sesquiterpenoids, diterpenoids, sesterpenoids, triterpenoids, tetrapenoids and polyterpenoids; these classes differ in the number of isoprene (C5H8) units (Mahizan et al. 2019).

Functional groups of terpenoids determine their antimicrobial activity, and most terpenoids possess the ability to uptake oxygen and undergo oxidative phosphorylation (Griffin et al. 1999). The presence of hydroxyl groups plays an important role towards contributing to terpenes' antiseptic and antibacterial action. The antiseptic and antibacterial potentials of terpenes are classified according to their solubility in water and the lipophilicity and/or hydrophobicity and presence of hydroxyl groups (Zengin and Baysal 2014).

6.2.5 Flavonoids

Flavonoids are composed of carbon, hydrogen, oxygen, sulphur and nitrogen. Flavonoids can be classified according to biosynthetic origins: chalcones, flavanones, flavan-3-ols and flavan-3,4-diols (Cushnie and Lamb 2005). Flavonoids have been reported to provide attractive colours to the flowers in aiding plant pollination and to promote physiological survival of the plant from fungal pathogens and UV-B radiation (Middleton Jr 1993; Harborne and Williams 2000). An increasing interest in flavonoids is being focussed on in medical research, largely due to their anti-inflammatory activity, oestrogenic activity, enzyme inhibition, antimicrobial activity, antiallergic activity, antioxidant activity, vascular activity and cytotoxic antitumour activity (Harborne and Williams 2000; Havsteen 1983).

6.3   Mechanisms/Mode of Action of Plant Secondary Metabolites

6.3.1   Disruption of Plasma Membrane

Basically, the mechanism of plant secondary metabolites against microorganisms relies on their chemical structure and properties, which affects the microbial cells in several ways. The bacterial plasma membrane which is a dynamic structure has several functions including biosynthesis of lipids, osmoregulation, respiration and transport processes as well as biosynthesis and cross-linking of peptidoglycan (Górniak et al. 2019). If there were any disruption of cytoplasmic membrane structure or function, leakage of ions or proteins leading to cell death will occur (Yang et al. 2020).

Various essential oils have been reported to disrupt cytoplasmic membrane structure that resulted in cell death (Yang et al. 2017; Yang et al. 2019; Yang et al. 2018c). The isothiocyanates (ITCs), plant secondary metabolites found in cruciferous vegetable such as broccoli, cabbage and kale, also demonstrated the potential to disrupt the bacterial cell membranes of *Escherichia coli*, *Pseudomonas aeruginosa*, *Staphylococcus aureus* and *Listeria monocytogenes* (Borges et al. 2015). Lipophilic plant secondary metabolites such as steroids, mustard oils, phenylpropanoids and mono-, sesqui-, di- and triterpenes which are trapped inside the membrane will change the organism's membrane fluidity and permeability (Wink 2018).

The carvacrol (5-isopropyl-2-methylphenol), from the class of terpene, can be found in the leaves of herbs such as thyme, wild bergamot, pepperwort and oregano (Marchese et al. 2018). Due to its hydrophobic characteristic, the hydroxyl group of carvacrol will interact with the lipid bilayer of the bacterial membrane and align between fatty acid chains (Marchese et al. 2018). This arrangement destabilizes the membrane structure causing an increase of the membrane's fluidity and permeability of proton and ions. Loss of the ion gradient eventually leads to bacterial cell death which has been observed in different species such as *L. monocytogenes*, *S. enteritidis*, *E. coli* and *S. aureus* (Siroli et al. 2015; Wang et al. 2016).

A study conducted by Xu et al. (Xu et al. 2019) demonstrated that kaempferol, quercetin dihydrate and catechin, the most predominant flavonoids from *Sedum aizoon* L., worked against *Aeromonas* in vitro. These flavonoids exhibited antimicrobial activities by targeting the bacterial cell surface and internal ultrastructure resulting in proteins and reducing sugar leakage.

Similarly, the cell wall is an essential component in fungal homeostasis, which is composed of different layers and primarily consists of β-glucans. Plant secondary metabolites functioning as defence molecules are not only limited to bacteria, but they may also function as antifungal agents (Freiesleben and Jäger 2014; Lima et al. 2019). It is speculated that the terpenoid phenols have potent antifungal activity against *Saccharomyces cerevisiae* by affecting membrane expansion and fluidity correlating with cytosolic calcium surge and are defective in ion homeostasis (Rao et al. 2010). In addition, many efforts have been made to find alternative treatments for viral infections including using plant secondary metabolites (Kapoor et al. 2017).

6.3.2   Inhibition of DNA Replication

With potential medical uses, there is a growing demand for plant secondary metabolites for medicine, diagnostics and industry. Plant secondary metabolites include lipophilic, aromatic and planar compounds that can act as intercalating or alkylating agents by intercalate between the stacking DNA base pairs or directly bind to nucleotide bases, respectively (Wink 2018). Thereby, it interfers with DNA replication, transcription and repair, which may even lead to mutations and genotoxicity of the microorganisms.

For example, berbamine, berberine, harmine, harman alkaloids and sanguinarine are strong DNA intercalators which inhibit DNA replication and ribosomal protein biosynthesis, while aristolochic acid, furanoquinoline alkaloids, safrole, senecionine and other phenylpropanoids are alkylating agents (Wink 2018). A previous study suggested that berberine, which is an isoquinoline derivative alkaloid, may inhibit the DNA synthesis of *Streptococcus agalactiae* by interfering with the activity of DNA topoisomerase, which is associated with DNA replication of the bacteria (Peng et al. 2015).

6.3.3   Interference of Quorum Sensing

Quorum sensing (QS) or bacterial pheromone which is known as autoinducer is secreted by the bacteria to regulate cell activity such as antibiotic resistance in order to adapt to changing environmental conditions (Bouyahya et al. 2017). Thus, inhibition of quorum sensing by the medicinal plants containing bioactive compounds to counteract the bacteria resistance by targeting their quorum sensing signalling pathways opens up the possibility of using these plant secondary metabolites as novel anti-QS agents.

The different structure and chemical composition of secondary metabolites result in differences in their QS inhibitory action (Asfour 2018). It has been reported that flavonoid-rich quorum quenching fraction of *Glycyrrhiza glabra* attenuated *Acinetobacter baumannii*, which is an opportunistic pathogen of hospital-derived infection (Bhargava et al. 2015). The results of this study highlight that downregulation of the autoinducer synthase by *Glycyrrhiza glabra* can be used as an alternative quorum quenching therapy against multidrug-resistant *Acinetobacter baumannii*.

6.3.4   Inhibition of Protein Synthesis

A study from Peng and colleagues (Peng et al. 2015) has shown that berberine, a derivative from alkaloid, had effect on proteins of *Streptococcus agalactiae*, a Gram-positive bacterium causing postpartum infection and neonatal sepsis. It was demonstrated that the berberine destroyed *S. agalactiae*'s proteins or partially degraded the proteins leading to the death of the bacteria.

The bioactive secondary metabolites such as tannins, terpenoids, saponins, alkaloids and flavonoids which can be found in higher plants have in vitro antifungal properties which may present constitutively or be synthesized de novo in response to pathogen infections (Ribera and Zuñiga 2012). Berberine, an isoquinoline alkaloid which can be extracted from different plants in *Berberis* genus, interferes viral immediate-early 2 (IE2) protein, transactivating activity in human cytomegalovirus prior replication (Luganini et al. 2019). Berberine has been used as a traditional component of Chinese and Ayurvedic medicine (Neag et al. 2018).

Bacterial cell walls are made up of peptidoglycan, which is consisting of a long chain of sugar polymers. It was reported that D-alanine-D-alanine ligase (Ddl) is an important enzyme in the D-Ala branch of bacterial cell wall peptidoglycan, which can be an attractive antimicrobial drug target (Wu et al. 2008). Also, Ddl, which catalyses the dimerization of two D-alanine molecules, is the potential target for drug development to fight against *Mycobacterium tuberculosis*, which causes tuberculosis (Yang et al. 2018b).

It was reported that quercetin (3,3′,4′,5,7-pentahydroxyflavone) and apigenin (4′,5,7-trihydroxyflavone), two representative flavonoids, function as reversible inhibitors of Ddl against both *Helicobacter pylori* and *E. coli* (Wu et al. 2008). In another study by Rao and Venkatachalam (Rao and Venkatachalam 2000), phenanthroindolizidine plant alkaloids pergularinine and tylophorinidine inhibited the activity of dihydrofolate reductase, which has an essential role in production of purine and pyrimidine precursors for DNA, RNA and amino acid biosynthesis.

6.3.5   Combinatorial Effect

Nevertheless, secondary metabolite alone may not be sufficient to fight against pathogen infections. Thus, an alternative way to combat infections caused by resistant pathogens can be enhanced by combination or having synergistic actions of various phytoactive components at multiple targets in bacterial cell. Various studies have been explored to understand the structure, biological function, biosynthesis pathway and possible modifications of the plant secondary metabolites to which they can be used in medicine (Kowalczyk et al. 2020).

A study was performed using the combination of nanoliposomal formulations piperine, a piperidine-type alkaloid, which has shown the inhibition of bacterial efflux pumps, and gentamicin, a type of aminoglycoside against methicillin-resistant *Staphylococcus aureus* (MRSA) (Khameneh et al. 2015). Similarly, piperine in combination with mupirocin had been reported to show better in vivo efficacy against *Staphylococcus aureus* strains of MRSA 15187 in a Swiss albino mouse dermal infection model as compared to the formulation of 2% mupirocin, which is commercially available (Mirza et al. 2011). Another study by Tegos and colleague (Tegos et al. 2002) demonstrated that the *Berberis* plant which produces berberine, a potent molecule in pharmacology and medicinal chemistry, synthesizes multidrug resistance pump inhibitors, which increases the penetration level of berberine into the cells of Gram-negative bacteria.

6.4   Mechanisms of Antimicrobial Resistance

The overuse and inappropriate usage of antibiotics have led to the emergence of multidrug-resistant pathogens such as *Staphylococcus aureus* (MRSA), multidrug-resistant (MDR) *Mycobacterium tuberculosis* and the extended-spectrum beta-lactamase (ESBL)-producing bacteria (Medina and Pieper 2016). As most of the antimicrobial compounds are derived from natural molecules such as the secondary metabolites, due to overuse and inappropriate use, the co-resident microorganism in bacterial pathogens have evolved mechanisms to overcome these antimicrobial agents (Munita and Arias 2016).

Generally, the susceptibility of Gram-negative bacteria to antibiotics is variable due to the presence of the outer membrane and multidrug resistance pumps (Tegos et al. 2002; Zgurskaya et al. 2015). Antimicrobial activity can be achieved by destroying the pathogenic microbial cell wall. Thus, one of the common mechanisms used by microorganisms to overcome the accumulation of antimicrobial agents inside the cells is by changing their outer membrane permeability in order to limit the influx of substances from the external milieu. The classical example is the influx of β-lactam and fluoroquinolone antibiotics which are reduced via changes in permeability of the outer membrane of the bacteria through the porin, an open channel that allows the passive transportation of molecules across lipid bilayer membranes (Pagès et al. 2008).

Likewise, microorganisms are capable of overexpression of efflux pumps that expel the antimicrobial agents from the cells. The multidrug efflux systems are ancient elements which are believed to have been existing in the bacterial genomes long time ago before the use of antibiotic therapy (Davies and Davies 2010; Blanco et al. 2016). These efflux pumps are able to expel various substrates including the organic pollutants, plant-produced compounds, heavy metals, quorum sensing signals or bacterial metabolites besides the antibiotics (Blanco et al. 2016). *Pseudomonas aeruginosa* is one of the most prevalent opportunistic pathogens which uses the antibiotic efflux pump mechanism, efficiently extruding the antibiotics through its cell wall (Issa et al. 2018). Other antibiotic classes extruded include phenicols, quinolones, sulfoamides and oxazolidinones which use the similar mode of resistance mechanism as reviewed by Davies and Davies in 2010.

The other mechanism of resistance is the modification of target sites in which antimicrobial agents such as tetracyclines, beta-lactams and glycopeptide are no longer able to react with, thus decreasing efficacy of antimicrobials (Othman et al.

2019). Currently, colistin has been an effective option for most multidrug-resistant Gram-negative bacteria; it is used as the last-line drug for infections caused by severe Gram-negative bacteria such as *Enterobacteriaceae* (Aghapour et al. 2019). Unfortunately, colistin resistance has occurred with lipopolysaccharide modification at the outer cell membrane of Gram-negative bacteria, which is the main site of action for colistin (Aghapour et al. 2019).

Some microorganisms have the capability of developing resistance to antimicrobial agents by direct modification or inactivation of the antibiotics. There are three main enzymes which inactivate antibiotics that include the β-lactamases, aminoglycoside-modifying enzymes and chloramphenicol acetyltransferases (Zeng and Lin 2013; Tang et al. 2014). These bacteria employ the modifying enzymes' effects by catalysing the common biochemical reactions as follows: (1) acetylation (aminoglycosides, chloramphenicol, streptogramins), (2) phosphorylation (aminoglycosides, chloramphenicol) and (3) adenylation (aminoglycosides, lincosamides) (Munita and Arias 2016). The bleomycin family members produced by *Streptomyces verticillus* are subject to acetylation that leads to the disruption of the metal-binding domain of the antibiotics required for activity (Peterson and Kaur 2018).

6.5 Pharmacological Significance of Plant Secondary Metabolites in Medicine

Plant secondary metabolites play an important role in self-defence of plant cells. Pharmacological significance of plant secondary metabolites has been significantly recognized since the early ages, and its applications is currently expanded towards plant-based drug discovery research and natural products (Madhumitha and Saral 2009; Newman and Cragg 2007; Anupama et al. 2014). Table 6.2 summarizes recent findings in the application of plant secondary metabolites towards pharmacological benefits. The World Health Organization (WHO) documented the guidance for strategic methods for standardization and drug discovery development process from the medicinal plant materials (Manila 1993).

The pharmacological significance of flavonoids has been recognized to have anti-obesity, anti-inflammatory and vasodilator effects and antioxidant, immunostimulant, antidiabetic, antihypertensive, anti-atherosclerosis and anti-hypercholesterolemic activities (Salvamani et al. 2014). It is also being manufactured as supplements in the form of capsules and powders (Batiha et al. 2020). Quercetin, a bioflavonoid compound found in vegetables, fruits and grains, has been widely used as a pharmacological compound. The *Moringa oleifera* (drumstick tree), *Centella asiatica* (Indian pennywort), *Hypericum perforatum* (St. John's wort) and *Brassica oleracea* (wild cabbage) are known for their antihypertensive, anticancer and anti-depressive activities; they have also been found to reduce the risk of stroke (Batiha et al. 2020). Quercetin has also been reported to protect the brain cells from oxidative stress especially due to tissue damage among the Alzheimer patients (Lakhanpal and Rai 2007). Quercetin combined with fish oil

Table 6.2 List of plant secondary metabolites and pharmacological benefits and antimicrobial activity

Compounds	Plant source	Pharmacological benefits	References
Alkaloids and phenols	*Uncaria tomentosa*	Potential for the treatment of periodontitis	Lima et al. (2020)
Alkaloids, flavonoids, phenols and tannins	*Vernonia amygdalina*	Exhibited inhibitory activity on bacteria	Dumas et al. (2020)
Alkaloids, flavonoids, phenolics, terpenoids, cardiac glycosides, saponins, steroids and tannins	*Centella asiatica*	Treatment of symptoms related to bacterial infections	Sieberi et al. (2020)
Curcumin	*Curcuma longa*	Anticancer, antioxidant, antimalarial, anti-inflammatory	Razali et al. (2018)
Flavonoids	*Aristolochia indica*	Antibacterial activity against multidrug-resistant bacteria	Bartha et al. (2019)
Linalyl anthranilate	*Lavender, thyme*	Antibacterial activity against multidrug-resistant *Klebsiella pneumoniae*	Yang et al. (2020)
Phenolic and flavonoid	*Maclura tricuspidata*	Improve therapeutic efficiency for hepatocellular carcinoma	Park et al. (2020)
Phenolics and terpenoids	*Tanacetum sonbolii* Mozaff *Moringa oleifera* lam.	Inhibitory activity against *T. b. rhodesiense* increases cytotoxicity against growth of colon cancer cells	Mofidi Tabatabaei et al. (2020), Shousha et al. (2019)
Polyphenol	*Thymbra sintenisii* subsp. *isaurica*	Anticancer (breast carcinoma)	Hepokur et al. (2020)
Quercetin	*T. Spruneriana*	Healing of hyperpigmentation, Alzheimer and diabetes	Aylanc et al. (2020)
Tannins, saponins, flavonoids, terpenoids, glycosides, alkaloids and phenols	*Rhus vulgaris* Meikle	Antimicrobial activity against MRSA	Mutuku et al. (2020)
Triterpenoid saponin	*Ardisia gigantifolia* Stapf.	Exhibits high antiangiogenic potency	Mu et al. (2020)
β-Caryophyllene	*Oregano, black pepper, cinnamon bark*	Antibacterial activity against *Bacillus cereus*	Moo et al. (2020)
Phenolic	*Zingiber officinale*	Induces apoptosis and inhibits cancer signalling	Qian et al. (2020)

has demonstrated beneficial effects against neurodegenerative diseases (Denny Joseph and Muralidhara 2015). A recent study by Zhang et al. (2020) also demonstrated that quercetin can be developed into a promising nanotherapeutic for the effective treatment of breast cancer and lung metastasis (Zhang et al. 2020).

Besides flavonoids, other polyphenols which are widely studied for their secondary metabolites' properties are the *Cannabis*. The pharmacokinetics of cannabinoids has stimulated significant interest in pharmacology research, and it was supported for legalization in terms of medical purposes (Vergara et al. 2020; Gregus and Buczynski 2020). Cannabinoids contain more than 545 secondary metabolites and possess a great structural diversity of non-nitrogen compounds which are highly capable of interfering with the central nervous system (Seca and Pinto 2019). Over the last few years, interest has been generated towards the beneficial effects of cannabis products for the treatment of drug-resistant or refractory epilepsy (Gonçalves et al. 2019).

In terms of infectious disease, malaria represents the most common worldwide health issue. Chloroquine and artesunate, a derivative of artemisinin from the plant *Artemisia annua*, or also known as mugworts, have been widely used as antimalarial medication. Although therapeutic effectiveness has been observed, the side effects due to drug resistance are inevitable. A recent study by Avitabile and colleagues (2020) showed that the "green" silver nanoparticles (AgNPs) from two *Artemisia* species (*A. abrotanum* and *A. arborescens*) lowered levels of parasitaemia, thus signifying the greatest level of antimalarial activity of artemisinin plant secondary metabolite.

Plant secondary metabolites have also been recognized in treating diabetes mellitus (Tran et al. 2020). Saponins isolated from ginseng have shown significant antidiabetic activity by moderating enzyme activity in influencing glucose metabolism and controlling insulin secretion (Dans et al. 2007; Attele et al. 2002; Chen et al. 2019; Hyun et al. 2020). Saponins which have been studied for their antimicrobial activity and the crude leaf extract of *A. carambola*, or also commonly known as star fruit, presented a broad spectrum of action against *S. aureus* ATCC 29213 MRSA, *S. aureus* 6 MRSA, *S. aureus* 10 MRSA, *S. aureus* 12 MRSA, *E. faecalis* ATCC 29212, *K. pneumoniae* 8 ESBL and *A. baumannii* 2 MBL (Silva et al. 2020).

The Chinese rhubarb (*Rheum officinale*) showed effective inhibitory effect against *Campylobacter* (Yosri et al. 2020), a serious global cause of diarrhoeal disease worldwide, which has been reported to be fatal among young children, elderly and immunosuppressed individuals (Yosri et al. 2020). Additionally, the *Rheum* spp. have also been reported to show antimicrobial activity against a wide range of pathogens such as the *Candida albicans* DSMZ 1386, *Enterococcus durans, Enterococcus faecalis* ATCC 29212, *Escherichia coli* ATCC 25922, *Klebsiella pneumoniae, Listeria monocytogenes ATCC 7644, Pseudomonas fluorescens P1, Salmonella enteritidis* ATCC 13075, *Salmonella infantis, Salmonella typhimurium* SL 1344 and *Staphylococcus epidermidis* DSMZ 20044 (Canli et al. 2016).

6.6   Future Perspectives and Concluding Remarks

It is clear that plant secondary metabolites are emerging as a possible alternative compound for management and treatment of many diseases. Their benefits leading towards plant-based drug development with minimal or no side effects are being highly sought after. Undoubtedly, despite the increase in demand, limited clinical trials in determining the efficacy remain vague. This might be due to challenges faced in analysing and documenting toxicological, epidemiological and other pharmacognosy-based data (Ahmad Khan and Ahmad 2019). Another perspective that needs to be addressed is the standardization of usage of plant secondary metabolites in pharmacology. The complexity and diversity of its compounds has been postulated to lead to intense differences in pharmacological activity. However, recent advancements, especially in high-throughput screening and big data analysis, have shed some light towards its vast application in modern medicine. Furthermore, the omics technology has led to diverse information and insights into application and benefits of plant secondary metabolites and their application in pharmacology and their ability to compete in the mainstream biomedical science (Ahmad Khan and Ahmad 2019).

An additional future prospect of plant secondary metabolites in tackling AMR and their application in pharmacology is the commercial income. Countries like China, India and other developing countries could be the main source of exporter of plant compounds leading to a surge in commercial income.

From this review, it is evident that plant secondary metabolite-based drug research has been yielding promising and beneficial results, especially in AMR and pharmacology. Therefore, further identification and isolation of similar compounds using high-end technologies could be useful to significantly impact the field of antimicrobial resistance especially in pharmacology. Moreover, incorporation of improved quality control and regulatory measures by WHO foresees the application and usage of plant secondary metabolites into conventional medical systems.

Acknowledgement The authors would like to thank the Higher Colleges of Technology, United Arab Emirates, and UCSI University Research Excellence and Innovation Grant (REIG-FAS-2020/032) for the support.

References

Aghapour Z et al (2019) Molecular mechanisms related to colistin resistance in Enterobacteriaceae. Infect Drug Resist 12:965

Ahmad Khan MS, Ahmad I (2019) Chapter 1 - herbal medicine: current trends and future prospects. In: Khan MSA, Ahmad I, Chattopadhyay D (eds) New look to phytomedicine. Academic Press, Cambridge, MA, pp 3–13

Anupama N, Madhumitha G, Rajesh K (2014) Role of dried fruits of Carissa carandas as anti-inflammatory agents and the analysis of phytochemical constituents by GC-MS. Biomed Res Int 2014:512369

Asfour HZ (2018) Anti-quorum sensing natural compounds. J Microscopy Ultrastruct 6(1):1

Attele AS et al (2002) Antidiabetic effects of Panax ginseng berry extract and the identification of an effective component. Diabetes 51(6):1851–1858

Aylanc V et al (2020) In vitro studies on different extracts of fenugreek (Trigonella spruneriana BOISS.): Phytochemical profile, antioxidant activity, and enzyme inhibition potential. J Food Biochem 2020:e13463

Babbar N et al (2014) Influence of different solvents in extraction of phenolic compounds from vegetable residues and their evaluation as natural sources of antioxidants. J Food Sci Technol 51 (10):2568–2575

Baker DD, Alvi KA (2004) Small-molecule natural products: new structures, new activities. Curr Opin Biotechnol 15(6):576–583

Barbieri R et al (2017) Phytochemicals for human disease: an update on plant-derived compounds antibacterial activity. Microbiol Res 196:44–68

Bartha GS et al (2019) Analysis of aristolochlic acids and evaluation of antibacterial activity of Aristolochia clematitis L. Biol Futura 70(4):323

Batiha GE-S et al (2020) The pharmacological activity, biochemical properties, and pharmacokinetics of the major natural polyphenolic flavonoid: quercetin. Foods 9(3):374

Beuria TK, Santra MK, Panda D (2005) Sanguinarine blocks cytokinesis in Bacteria by inhibiting FtsZ assembly and bundling. Biochemistry 44(50):16584–16593

Bhargava N et al (2015) Attenuation of quorum sensing-mediated virulence of Acinetobacter baumannii by Glycyrrhiza glabra flavonoids. Future Microbiol 10(12):1953–1968

Blanco P et al (2016) Bacterial multidrug efflux pumps: much more than antibiotic resistance determinants. Microorganisms 4(1):14

Boberek JM, Stach J, Good L (2010) Genetic evidence for inhibition of bacterial division protein FtsZ by berberine. PLoS One 5(10):e13745

Borges A et al (2015) Antibacterial activity and mode of action of selected glucosinolate hydrolysis products against bacterial pathogens. J Food Sci Technol 52(8):4737–4748

Bouyahya A et al (2017) Medicinal plant products targeting quorum sensing for combating bacterial infections. Asian Pac J Trop Med 10(8):729–743

Canli K et al (2016) In vitro antimicrobial activity screening of Rheum rhabarbarum roots. Int J Pharm Sci Invent 5(2):1–4

Casas R et al (2014) The effects of the mediterranean diet on biomarkers of vascular wall inflammation and plaque vulnerability in subjects with high risk for cardiovascular disease. A randomized trial. PLoS One 9(6):e100084

Chadwick NE, Morrow KM (2011) Competition among sessile organisms on coral reefs, in Coral Reefs: an ecosystem in transition. Springer, pp 347–371

Chen W, Balan P, Popovich DG (2019) Review of ginseng anti-diabetic studies. Molecules 24 (24):4501

Cho KS et al (2017) Terpenes from forests and human health. Toxicol Res 33(2):97–106

Clark VC (2010) Collecting arthropod and amphibian secretions for chemical analyses. In: Behavioral and chemical ecology, 1st edn. Nova Science Publication, New York, pp 1–46

Cotrim BA et al (2012) Unsaturated fatty alcohol derivatives of olive oil phenolic compounds with potential low-density lipoprotein (LDL) antioxidant and antiobesity properties. J Agric Food Chem 60(4):1067–1074

Cushnie TPT, Lamb AJ (2005) Antimicrobial activity of flavonoids. Int J Antimicrob Agents 26 (5):343–356

Cushnie TT, Cushnie B, Lamb AJ (2014) Alkaloids: an overview of their antibacterial, antibiotic-enhancing and antivirulence activities. Int J Antimicrob Agents 44(5):377–386

Dans AML et al (2007) The effect of Momordica charantia capsule preparation on glycemic control in type 2 diabetes mellitus needs further studies. J Clin Epidemiol 60(6):554–559

Davies J, Davies D (2010) Origins and evolution of antibiotic resistance. Microbiol Mol Biol Rev 74(3):417–433

Dembitsky VM (2005) Astonishing diversity of natural surfactants: 6. Biologically active marine and terrestrial alkaloid glycosides. Lipids 40(11):1081

Denny Joseph KM, Muralidhara (2015) Combined oral supplementation of fish oil and quercetin enhances neuroprotection in a chronic rotenone rat model: relevance to Parkinson's disease. Neurochem Res 40(5):894–905

Dewick P (1997) Medicinal natural products. Wiley, Chichester, UK

Domadia PN et al (2008) Berberine targets assembly of Escherichia coli cell division protein FtsZ. Biochemistry 47(10):3225–3234

Dong J et al (2019) Saponins regulate intestinal inflammation in colon cancer and IBD. Pharmacol Res 144:66–72

Dumas NGE, Anderson NTY, Godswill NN, Thiruvengadam M, Ana-Maria G, Ramona P, Crisan GC, Laurian V, Shariati MA, Tokhtarov Z, Emmanuel Y (2020) Secondary metabolite contents and antimicrobial activity of leaf extracts reveal genetic variability of Vernonia amygdalina and Vernonia calvoana morphotypes. Biotechnol Appl Biochem. https://doi.org/10.1002/bab.2017. Epub ahead of print. PMID: 32881085

El Aziz M, Ashour A, Melad A (2019) A review on saponins from medicinal plants: chemistry, isolation, and determination. J Nanomed Res 8(1):6–12

Evans WC (2009) Trease and evans' pharmacognosy E-book. Elsevier, Amsterdam

Fairus S et al (2018) A phase I single-blind clinical trial to evaluate the safety of oil palm phenolics (OPP) supplementation in healthy volunteers. Sci Rep 8(1):8217

Freiesleben S, Jäger A (2014) Correlation between plant secondary metabolites and their antifungal mechanisms–a review. Med Aromat Plants 3(2):1–6

Gonçalves J et al (2019) Cannabis and its secondary metabolites: their use as therapeutic drugs, toxicological aspects, and analytical determination. Medicines 6(1):31

Gorlenko CL et al (2020) Plant secondary metabolites in the battle of drugs and drug-resistant bacteria: new heroes or worse clones of antibiotics. Antibiotics 9(4):170

Górniak I, Bartoszewski R, Króliczewski J (2019) Comprehensive review of antimicrobial activities of plant flavonoids. Phytochem Rev 18(1):241–272

Gregus AM, Buczynski MW (2020) Druggable targets in endocannabinoid signaling. Adv Exp Med Biol 1274:177–201

Griffin SG et al (1999) The role of structure and molecular properties of terpenoids in determining their antimicrobial activity. Flavour Fragr J 14(5):322–332

Harborne JB, Williams CA (2000) Advances in flavonoid research since 1992. Phytochemistry 55 (6):481–504

Havsteen B (1983) Flavonoids, a class of natural products of high pharmacological potency. Biochem Pharmacol 32(7):1141–1148

Hepokur C et al (2020) Evaluation of antioxidant and anticancer effects of Thymbra sintenisii subsp. isaurica extract. J Cancer Res Ther 16(4):822–827

Hussein RA, El-Anssary AA (2018) In: Builders PF (ed) Plants secondary metabolites: the key drivers of the pharmacological actions of medicinal plants, in herbal medicine. IntechOpen

Hyun SH et al (2020) Physiological and pharmacological features of the non-saponin components in Korean red ginseng. J Ginseng Res 44:527–537

Issa HB, Phan G, Broutin I (2018) Functional mechanism of the efflux pumps transcription regulators from Pseudomonas aeruginosa based on 3D structures. Front Mol Biosci 5:57

Kapoor R, Sharma B, Kanwar S (2017) Antiviral phytochemicals: an overview. Biochem Physiol 6 (2):7

Khameneh B et al (2015) Investigation of the antibacterial activity and efflux pump inhibitory effect of co-loaded piperine and gentamicin nanoliposomes in methicillin-resistant Staphylococcus aureus. Drug Dev Ind Pharm 41(6):989–994

Kowalczyk T et al (2020) Transgenesis as a tool for the efficient production of selected secondary metabolites from in vitro plant cultures. Plan Theory 9(2):132

Krstin S, Peixoto HS, Wink M (2015) Combinations of alkaloids affecting different molecular targets with the saponin digitonin can synergistically enhance trypanocidal activity against Trypanosoma brucei brucei. Antimicrob Agents Chemother 59(11):7011–7017

Lainson R, de Souza MC, Franco CM (2003) Haematozoan parasites of the lizard Ameiva ameiva (Teiidae) from Amazonian Brazil: a preliminary note. Mem Inst Oswaldo Cruz 98 (8):1067–1070

Lakhanpal P, Rai DK (2007) Quercetin: a versatile flavonoid. Int J Med Update 2(2):22–37

Lima SL, Colombo AL, de Almeida Junior JN (2019) Fungal cell wall: emerging antifungals and drug resistance. Front Microbiol 10:2573

Lima V et al (2020) Uncaria tomentosa reduces osteoclastic bone loss in vivo. Phytomedicine 79:153327

Lin D et al (2016) An overview of plant phenolic compounds and their importance in human nutrition and management of type 2 diabetes. Molecules 21(10):1374

Luganini A et al (2019) The isoquinoline alkaloid berberine inhibits human cytomegalovirus replication by interfering with the viral immediate Early-2 (IE2) protein transactivating activity. Antivir Res 164:52–60

Lusebrink I, Dettner K, Seifert K (2008) Stenusine, an antimicrobial agent in the rove beetle genus Stenus (Coleoptera, Staphylinidae). Naturwissenschaften 95(8):751–755

Madhumitha G, Saral AM (2009) Free radical scavenging assay of Thevetia neriifolia leaf extracts. Asian J Chem 21(3):2468–2470

Mahizan NA et al (2019) Terpene derivatives as a potential agent against antimicrobial resistance (AMR) pathogens. Molecules 24(14):2631

Manila S (1993) Research guidelines for evaluating the safety and efficacy of herbal medicine. WHO publications, Philippines, PA

Marchese A et al (2018) The natural plant compound carvacrol as an antimicrobial and anti-biofilm agent: mechanisms, synergies and bio-inspired anti-infective materials. Biofouling 34 (6):630–656

McClintock JB, Baker BI (2013) Chemistry and ecological role of starfish secondary metabolites. In: Starfish: biology and ecology of the Asteroidea. The Johns Hopkins University Press, Baltimore, MD, p 81

Medina E, Pieper DH (2016) Tackling threats and future problems of multidrug-resistant bacteria. In: How to overcome the antibiotic crisis. Springer, Cham, pp 3–33

Michel CG et al (2011) Phytochemical and biological investigation of the extracts of Nigella sativa L. seed waste. Drug Test Anal 3(4):245–254

Middleton E Jr (1993) The impact of plant flavonoids on mammalian biology: implications for immunity, inflammation and cancer. In: The flavonoids: advances in research since 1986. Routledge, Abingdon, UK, pp 337–370

Mills CE et al (2017) Mediation of coffee-induced improvements in human vascular function by chlorogenic acids and its metabolites: two randomized, controlled, crossover intervention trials. Clin Nutr 36(6):1520–1529

Mirza ZM et al (2011) Piperine as an inhibitor of the MdeA efflux pump of Staphylococcus aureus. J Med Microbiol 60(10):1472–1478

Mofidi Tabatabaei S et al (2020) Phytochemical study of Tanacetum Sonbolii aerial parts and the antiprotozoal activity of its components. Iran J Pharm Res 19(1):77–83

Moo CL et al (2020) Antibacterial activity and mode of action of β-caryophyllene on Bacillus cereus. Pol J Microbiol 69(1):1–6

Moo C-L et al (2019) Mechanisms of antimicrobial resistance (AMR) and alternative approaches to overcome AMR. Curr Drug Discov Technol 16:430–447

Moore KS et al (1993) Squalamine: an aminosterol antibiotic from the shark. Proc Natl Acad Sci 90 (4):1354–1358

Mu LH et al (2020) Antiangiogenic effects of AG36, a triterpenoid saponin from Ardisia gigantifolia stapf. J Nat Med 74(4):732–740

Munita JM, Arias CA (2016) Mechanisms of antibiotic resistance. In: Virulence mechanisms of bacterial pathogens. Wiley, pp 481–511

Mutuku A et al (2020) Evaluation of the antimicrobial activity and safety of Rhus vulgaris (Anacardiaceae) extracts. BMC Compl Med Ther 20(1):272

Neag MA et al (2018) Berberine: botanical occurrence, traditional uses, extraction methods, and relevance in cardiovascular, metabolic, hepatic, and renal disorders. Front Pharmacol 9:557

Newman DJ, Cragg GM (2007) Natural products as sources of new drugs over the last 25 years. J Nat Prod 70(3):461–477

Newman RA et al (2008) Cardiac glycosides as novel cancer therapeutic agents. Mol Interv 8(1):36

Othman L, Sleiman A, Abdel-Massih RM (2019) Antimicrobial activity of polyphenols and alkaloids in middle eastern plants. Front Microbiol 10:911–911

Pagès J-M, James CE, Winterhalter M (2008) The porin and the permeating antibiotic: a selective diffusion barrier in gram-negative bacteria. Nat Rev Microbiol 6(12):893–903

Park SY et al (2020) Anti-metastatic effect of gold nanoparticle-conjugated Maclura tricuspidata extract on human hepatocellular carcinoma cells. Int J Nanomed 15:5317–5331

Pawlik JR (2011) The chemical ecology of sponges on Caribbean reefs: natural products shape natural systems. Bioscience 61(11):888–898

Peng L et al (2015) Antibacterial activity and mechanism of berberine against Streptococcus agalactiae. Int J Clin Exp Pathol 8(5):5217

Peterson E, Kaur P (2018) Antibiotic resistance mechanisms in bacteria: relationships between resistance determinants of antibiotic producers, environmental bacteria, and clinical pathogens. Front Microbiol 9:2928

Podolsky SH (2018) The evolving response to antibiotic resistance (1945–2018). Palgrave Commun 4(1):1–8

Prestinaci F, Pezzotti P, Pantosti A (2015) Antimicrobial resistance: a global multifaceted phenomenon. Pathog Global Health 109(7):309–318

Pu X et al (2015) Polyphylla saponin I has antiviral activity against influenza a virus. Int J Clin Exp Med 8(10):18963–18971

Qian S et al (2020) Zingerone suppresses cell proliferation via inducing cellular apoptosis and inhibition of the PI3K/AKT/mTOR signaling pathway in human prostate cancer PC-3 cells. J Biochem Mol Toxicol 35:e22611

Randhir R, Lin Y-T, Shetty K (2004) Stimulation of phenolics, antioxidant and antimicrobial activities in dark germinated mung bean sprouts in response to peptide and phytochemical elicitors. Process Biochem 39(5):637–646

Rangel-Huerta OD et al (2015) Normal or high polyphenol concentration in Orange juice affects antioxidant activity, blood pressure, and body weight in obese or overweight adults. J Nutr 145 (8):1808–1816

Rao A et al (2010) Mechanism of antifungal activity of terpenoid phenols resembles calcium stress and inhibition of the TOR pathway. Antimicrob Agents Chemother 54(12):5062–5069

Rao KN, Venkatachalam S (2000) Inhibition of dihydrofolate reductase and cell growth activity by the phenanthroindolizidine alkaloids pergularinine and tylophorinidine: the in vitro cytotoxicity of these plant alkaloids and their potential as antimicrobial and anticancer agents. Toxicol In Vitro 14(1):53–59

Razali NA et al (2018) Curcumin derivative, 2,6-bis(2-fluorobenzylidene)cyclohexanone (MS65) inhibits interleukin-6 production through suppression of NF-κB and MAPK pathways in histamine-induced human keratinocytes cell (HaCaT). BMC Complement Altern Med 18 (1):217

Ribera A, Zuñiga G (2012) Induced plant secondary metabolites for phytopatogenic fungi control: a review. J Soil Sci Plant Nutr 12(4):893–911

Robbers JE, Speedie MK, Tyler VE (1996) Pharmacognosy and pharmacobiotechnology. Williams & Wilkins, Philadelphia, PA

Roberts SC (2007) Production and engineering of terpenoids in plant cell culture. Nat Chem Biol 3 (7):387–395

Salvamani S et al (2014) Antiartherosclerotic effects of plant flavonoids. Biomed Res Int 2014:480258

Savitzky AH et al (2012) Sequestered defensive toxins in tetrapod vertebrates: principles, patterns, and prospects for future studies. Chemoecology 22(3):141–158

Seca AML, Pinto DCGA (2019) Biological potential and medical use of secondary metabolites. Medicines 6(2):66

Shousha WG et al (2019) Evaluation of the biological activity of Moringa oleifera leaves extract after incorporating silver nanoparticles, in vitro study. Bull Nat Res Centre 43(1):212

Sieberi BM et al (2020) Screening of the dichloromethane: Methanolic extract of Centella asiatica for antibacterial activities against Salmonella typhi, Escherichia coli, Shigella sonnei, Bacillus subtilis, and Staphylococcus aureus. Sci World J 2020:6378712

Silva KB et al (2020) Phytochemical characterization, antioxidant potential and antimicrobial activity of *Averrhoa carambola L. (Oxalidaceae)* against multiresistant pathogens. Braz J Biol 81(3):509–515

Siroli L et al (2015) Effects of sub-lethal concentrations of thyme and oregano essential oils, carvacrol, thymol, citral and trans-2-hexenal on membrane fatty acid composition and volatile molecule profile of Listeria monocytogenes, Escherichia coli and Salmonella enteritidis. Food Chem 182:185–192

Sonfack G et al (2019) Saponin with antibacterial activity from the roots of *Albizia adianthifolia*. Nat Prod Res 2019:1–9

Stegelmeier B et al (1999) Pyrrolizidine alkaloid plants, metabolism and toxicity. J Nat Toxins 8 (1):95

Stynoski JL et al (2014) Evidence of maternal provisioning of alkaloid-based chemical defenses in the strawberry poison frog *Oophaga pumilio*. Ecology 95(3):587–593

Tagousop CN et al (2018) Antimicrobial activities of saponins from Melanthera elliptica and their synergistic effects with antibiotics against pathogenic phenotypes. Chem Cent J 12(1):97

Tang SS, Apisarnthanarak A, Hsu LY (2014) Mechanisms of β-lactam antimicrobial resistance and epidemiology of major community-and healthcare-associated multidrug-resistant bacteria. Adv Drug Deliv Rev 78:3–13

Tegos G et al (2002) Multidrug pump inhibitors uncover remarkable activity of plant antimicrobials. Antimicrob Agents Chemother 46(10):3133–3141

Toledo Rd, Jared C (1995) Cutaneous granular glands and amphibian venoms. Comp Biochem Physiol A Physiol 111(1):1–29

Tran N, Pham B, Le L (2020) Bioactive compounds in anti-diabetic plants: from herbal medicine to modern drug discovery. Biology 9(9):252

Velderrain-Rodríguez G et al (2014) Phenolic compounds: their journey after intake. Food Funct 5 (2):189–197

Velu G, Palanichamy V, Rajan AP (2018) Phytochemical and pharmacological importance of plant secondary metabolites in modern medicine. In: Roopan SM, Madhumitha G (eds) Bioorganic phase in natural food: an overview. Springer, Cham, pp 135–156

Vergara D et al (2020) Modeling cannabinoids from a large-scale sample of Cannabis sativa chemotypes. PLoS One 15(9):e0236878

Vieira Júnior GM et al (2015) New steroidal saponin and antiulcer activity from Solanum paniculatum L. Food Chem 186:160–167

Wang L-H et al (2016) Membrane destruction and DNA binding of Staphylococcus aureus cells induced by carvacrol and its combined effect with a pulsed electric field. J Agric Food Chem 64 (32):6355–6363

WHO (2001) WHO global strategy for containment of antimicrobial resistance. World Health Organization, Geneva

WHO (2014) Antimicrobial resistance: global report on surveillance. World Health Organization, Geneva

Wink M (2018) Plant secondary metabolites modulate insect behavior-steps toward addiction? Front Physiol 9:364

Wu D et al (2008) Enzymatic characterization and crystal structure analysis of the D-alanine-D-alanine ligase from helicobacter pylori. Proteins: Struct Funct Bioinform 72(4):1148–1160

Xu F et al (2019) Antimicrobial activity of flavonoids from Sedum aizoon L. against Aeromonas in culture medium and in frozen pork. Food Sci Nutr 7(10):3224–3232

Yang L et al (2018a) Antifungal effects of Saponin extract from rhizomes of Dioscorea panthaica Prain et Burk against Candida albicans. Evid Based Complement Alternat Med 2018:6095307

Yang S et al (2018b) The biological properties and potential interacting proteins of d-Alanyl-d-alanine ligase a from Mycobacterium tuberculosis. Molecules 23(2):324

Yang S-K et al (2017) Additivity vs synergism: investigation of the additive interaction of cinnamon bark oil and meropenem in combinatory therapy. Molecules 22(11):1733

Yang S-K et al (2018c) Plant-derived antimicrobials: insights into mitigation of antimicrobial resistance. Rec Nat Prod 12(4):295–316

Yang S-K et al (2019) Disruption of KPC-producing Klebsiella pneumoniae membrane via induction of oxidative stress by cinnamon bark (Cinnamomum verum J. Presl) essential oil. PloS One 14(4):e0214326

Yang S-K et al (2020) Antimicrobial activity and mode of action of terpene linalyl anthranilate against carbapenemase-producing Klebsiella pneumoniae. J Pharm Anal 11(2):210–219

Yosri M et al (2020) Identification of novel bioactive compound derived from Rheum officinalis against campylobacter jejuni NCTC11168. Sci World J 2020:3591276

Zeng X, Lin J (2013) Beta-lactamase induction and cell wall metabolism in gram-negative bacteria. Front Microbiol 4:128

Zengin H, Baysal AH (2014) Antibacterial and antioxidant activity of essential oil terpenes against pathogenic and spoilage-forming bacteria and cell structure-activity relationships evaluated by SEM microscopy. Molecules 19(11):17773–17798

Zgurskaya HI, Lopez CA, Gnanakaran S (2015) Permeability barrier of gram-negative cell envelopes and approaches to bypass it. ACS Infect Dis 1(11):512–522

Zhang X et al (2020) Inhibition of growth and lung metastasis of breast cancer by tumor-homing triple-bioresponsive nanotherapeutics. J Control Release 328:454–469

Can Nanoparticles Help in the Battle against Drug-Resistant Bacterial Infections in "Post-Antibiotic Era"?

7

Niranjana Sri Sundaramoorthy and Saisubramanian Nagarajan

Contents

Abstract

Antimicrobial resistance (AMR) is gaining a foothold as a major public health crisis. AMR has been partly fueled by human's negligence (misuse/abuse) of antibiotics in aquaculture, poultry, animal farming, etc., which ultimately results in increased prevalence of antimicrobial agents in the environment (predominantly in water and soil), which, in turn, propels most microbes to acquire drug-resistant phenotype in order to survive. Most of the "marvel drugs" that were once effective have started to become ineffective against drug-resistant microbes, ultimately deserting human race with negligible therapeutic options left out to

N. S. Sundaramoorthy · S. Nagarajan (✉)
Center for Research on Infectious Diseases, School of Chemical and Biotechnology, SASTRA deemed University, Thanjavur, Tamil Nadu, India
e-mail: sai@scbt.sastra.edu

© The Author(s), under exclusive license to Springer Nature Singapore Pte Ltd. 2022
V. Kumar et al. (eds.), *Antimicrobial Resistance*,
https://doi.org/10.1007/978-981-16-3120-7_7

combat the multidrug-resistant (MDR) superbugs. The announcement of list of priority pathogens a couple of years ago by WHO has created a strong urge among researchers to identify new antimicrobial agents that would help in resolving AMR crisis. Although novel antimicrobial agents might turn out to be effective, due to evolutionary selection pressures, there is an increased propensity for microbes to gain resistance for these new agents too. Meanwhile, numerous reports depicting the antimicrobial nature of nanoparticles have thrown light on an alternative approach for tackling the situation. Nanomaterials, owing to flexibility in their engineering, have been explored to target various bacterial virulence factors and have been observed to be successful in inhibiting biofilm formation, efflux pumps, and quorum sensing and for plasmid curing which is responsible for dissemination of antibiotic-resistant genes (ARGs) among microbial populations/communities. In addition, nanoparticles have also been used to deliver antibiotics by conjugating the nanoparticles with targeting molecules that could ensure the specificity of the system. Nanoparticles have also been reported to detect bacteria responsible for infections in samples in a short duration with increased accuracy relative to the conventional procedures. This chapter will attempt to cover all possible modes of action by which nanoparticles can be employed to combat drug-resistant bacteria. In a retrospective view, chances for development of bacterial resistance to nanoparticles, question of enhancement of drug resistance due to nanoparticles, and other shortcomings in the use of nanoparticles will also be addressed.

Keywords

Nanoparticles · Multidrug-resistant bacteria · Biofilm inhibition · Drug delivery · Efflux pump inhibition

7.1   Introduction

There is a rise in threat of multidrug-resistant (MDR) bacteria globally on public health due to the occurrence and dissemination of acquired antibiotic resistance in both pathobionts and environmental isolates coupled with lack of treatment options available. Use of antibiotics has increased by 36% globally from 2000 to 2010, and the consumption of the last-resort antibiotics carbapenems and polymyxins has increased by 45% and 13%, respectively (Van Boeckel et al. 2014). In 2030, antimicrobials used in food animals have been estimated to increase by 200,235 tons (Van Boeckel et al. 2017). Evidence clearly points out the causal link between increased use of antimicrobials and selection of drug-resistant microbes (Holmberg et al. 1984; Read and Woods 2014; Nature 2013). It has been estimated that if the present antimicrobial resistance (AMR) state continues unabated, ten million deaths would occur in 2050 pushing deaths due to cancer to the second place worldwide (Neill 2014). Hence, the World Health Organization (WHO) has listed out the priority antimicrobial-resistant (AMR) pathogens (12 classes of bacteria) in 2017, to realize the ground scenario and to encourage the medical research community to

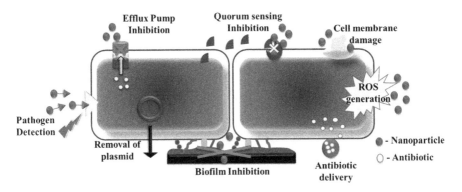

Fig. 7.1 Schematic representation of different actions/uses of nanoparticles on bacteria

develop alternative or novel approaches to treat these AMR pathogens (Tacconelli et al. 2017). Currently, we have 50 new antibiotics in developmental stage, out of which 32 target the list of AMR pathogens mentioned above, but the impact of these new antimicrobials in preventing evolution of drug resistance remains to be awaited further. The Global Antibiotic Research and Development Partnership (GARDP), a nonprofit research and development organization, has planned to deliver five new treatments by 2025 to tackle drug-resistant infections (Balasegaram and Piddock 2020).

To overcome this grim AMR condition is a mandate to ensure infection-free healthy life span for human race. Non-orthodox approaches are of current interest to mitigate these infectious agents. Nano-based platforms can be one potential alternative to deal with the pathogens. Studies have shown that metal nano-constructs possess antibacterial activity and arrest infectious diseases (Huh and Kwon 2011). Nanoparticles can be either designed or modified to reduce acute toxicity and ensure sustained release, target specificity, and increased bioavailability (Sharma et al. 2016; Wang et al. 2017). Apart from the well-known antibacterial nature of nanoparticles, they can also interrupt with other drug resistance mechanisms like biofilm formation, overexpression of efflux pumps, and acquisition of resistant genes through horizontal gene transfer (HGT); hence, in combination with existing antibiotics, nanoparticles have the potential to restore drug sensitivity in drug-resistant pathogens (Gupta et al. 2017). Nanoparticles can also be engineered to specifically target and deliver antibiotics. They can also detect pathogens when conjugated with targeting molecules with high sensitivity and with appropriate design can also act as theranostics (both diagnostic and therapeutic). Action of nanoparticles on bacteria is given as a schematic representation in Fig. 7.1. This chapter focuses on the different possible roles of nanomaterials in combating the MDR pathogens. Recent reports on the function of various types of nanoparticles in curtailing microbes are compiled in Table 7.1. In a retrospective manner, the possible shortcomings in the usage of nanoparticles in clinical therapy against infectious agents will also be discussed.

Table 7.1 Recent reports on various functions of different nanoparticles

S. no.	Nanoparticle type	Dedicated property of NPs	Effect on pathogens	Pathogens targeted	References
1	AgNPs	PVP-coated	Antibacterial	NDM containing *A. baumannii*	Shi et al. (2019)
		Gallic acid functionalization	Antibacterial by photothermal lysis	*S. aureus* and *E. coli*	Liu et al. (2020)
		Mycosynthesis	Biofilm inhibition	UPEC	Rodríguez-Serrano et al. (2020)
		pH sensitive	Biofilm inhibition	*S. aureus*	Wu et al. (2019)
		Thiosemicarbazide functionalization with glutamic acid	Downregulation of *icaA* and *icaD*	MRSA	Montazeri et al. (2020)
		Seed lectin functionalized	Biofilm inhibition and eradication	UPEC	Bala Subramaniyan et al. (2020)
		N-acyl-homoserine lactonase stabilized	Quorum sensing and biofilm inhibition	*H. pylori*	Gopalakrishnan et al. (2020)
		Biosynthesis	Alters expression of *mrkA* and *luxS*	MDR *K. pneumoniae*	Foroohimanjili et al. (2020)
		Biosynthesis	Growth inhibition and interference in QS	*S. Typhi*	Balakrishnan et al. (2020)
		Nanocomposite	Downregulation of *norA*	*S. aureus*	Kahzad and Salehzadeh (2020)
		—	Removal of plasmids	*E. coli, P. aeruginosa, S. aureus,* and *S. pyogenes*	Mhawesh et al. (2019)
		Biosynthesis of CuFe$_2$O$_4$@ag composite	Efflux pump inhibition by downregulating *norA*	*S. aureus*	Kahzad and Salehzadeh (2020)
2	AuNPs	Bacterial synthesis	Inhibition of pyocyanin production	*P. aeruginosa*	Palpperumal et al. (2016)
		Ethnobotanical crude extract synthesis	Downregulation of AhyR (involved in production of NAHL)	*A. hydrophila*	Fernando and Judan Cruz (2020)

		Modified with antibody/antibiotic	Bacterial detection	Carbapenem-resistant Enterobacteriaceae	Wang et al. (2019a)
		Chitosan mediated	Antibacterial	Few gram-negatives and gram-positives	Kalaivani et al. (2020)
		Display P9b bacteriophage	Pathogen detection	P. aeruginosa	Franco et al. (2020)
		Flower shaped	Detect pathogens simultaneously from PCR product	Salmonella typhimurium, Listeria monocytogenes, and E. coli	Du et al. (2020)
		Nanorods with DNA aptamer functionalization	Antibacterial by photothermal lysis	MRSA	Ocsoy et al. (2017)
3	PtNPs	Green synthesis	Antibacterial	Resistant P. aeruginosa and B. subtilis	Tahir et al. (2017)
		Green synthesis	Antibacterial	S. mutans, E. faecalis, Porphyromonas gingivalis	Itohiya et al. (2019)
		Phytoprotein stabilized, self-assembled	Biofilm eradication	S. Typhi	Subramaniyan et al. (2018)
		Pectin capped	Reduction in biofilm formation due to removal of plasmid	ESBL producing MDR UPEC	Bharathan et al. (2019)
4	ZnO NPs	Green synthesis	Antibacterial	P. vulgaris, P. aeruginosa, E. coli, and K. pneumoniae	Fahimmunisha et al. (2020)
			Antibacterial	M. Smegmatis	Mistry et al. (2020)
		Nanocomposite of chitosan and chitosan-ZnO	Downregulation of LasI and RhlI	P. aeruginosa	Badawy et al. (2020)
		Thiolated chitosan coated and doped with cobalt	Efflux pump inhibition	MRSA	Iqbal et al. (2019)
		Cysteamine functionalized	Detection by sensing NAHLs	P. aeruginosa	Vasudevan et al. (2020)

(continued)

Table 7.1 (continued)

S. no.	Nanoparticle type	Dedicated property of NPs	Effect on pathogens	Pathogens targeted	References
5	Se NPs	Synthesized by pulsed laser ablation in liquids	Antibacterial	MDR *E. coli*, MRSA, *P. aeruginosa*, and *S. epidermidis*	Geoffrion et al. (2020)
		Green synthesis and coated with bovine serum albumin	Antibacterial	*S. aureus*	Chung et al. (2020)
		Integrated with lysozyme	Antibacterial	*S. aureus* and *E. coli*	Vahdati and Tohidi Moghadam (2020)
		Brooms by biogenic synthesis	Antibacterial	*B. subtilis*	Sadalage et al. (2020)
		Nanocomposite with iron oxide NPs	Antibiofilm activity in the presence of magnetic field	*S. aureus*	Li et al. (2020b)
			Inhibited violacein production and biofilm	*C. violaceum* and *P. aeruginosa*	Gómez-Gómez et al. (2019a)
6	Ceria NPs	Biosynthesis and encapsulated with nano-chitosan	Biofilm disruption	*S. aureus* and *P. aeruginosa*	Bushra et al. (2020)
7	TiO$_2$ NPs	Green synthesis	Antibacterial	*E. coli, K. pneumoniae, B. subtilis, S. aureus,* and *S. typhi*	Thakur et al. (2019)
		Nanocomposite film formed with cellulose and polypyrrole	Bacterial detection	*E. coli, S. epidermidis, A. hydrophila, S. aureus*	Ghasemi et al. (2020)
8	Magnetic NPs	Pseudobactin coated	Antibacterial	*P. aeruginosa, E. coli, S. typhimurium,* and *S. aureus*	Kotb et al. (2020)
		Alginate coated and conjugated with tobramycin	Growth and biofilm inhibition	*P. aeruginosa*	Armijo et al. (2020)
		Chitosan coated	Bacterial detection	*E. coli* and *S. aureus*	Le et al. (2020)

#		Description	Application	Bacteria	Reference
9	CuNPs	Laser ablation synthesis	Antibacterial	*Aggregatibacter actinomycetemcomitans*	Fernández-Arias et al. (2020)
		Sulfide NPs coated with vancomycin	Photolysis	Vancomycin-resistant *Enterococcus*	Zou et al. (2020)
		Self-assembled sulfide NPs	Antibacterial	*E. coli* and *S. aureus*	Gargioni et al. (2020)
10	Chitosan NPs	Cationic and fluorescein labelled and conjugated with DNA aptamer	Bacterial detection	*E. coli*	Zhao et al. (2020)
11	CuO/ZnO	5-Nitroindole capped	Antibacterial and antibiofilm	*E. coli, P. aeruginosa,* and *S. aureus*	Manoharan et al. (2020)
12	Cu-cysteamine NPs	Action with potassium iodide	Photodynamic therapy-mediated killing	MRSA	Zhen et al. (2020)
13	Ag-PtNPs	Dendrimer shaped	Antibacterial	*S. aureus, P. aeruginosa,* and MDR *E. coli*	Ruiz et al. (2020)

7.2   Antibacterial Activity of Nanoparticles

Prevalence and spread of antibiotic resistance and lack of efficient treatment options have called for innovative approaches for the development of novel antibiotics. Reduced economic gains and challenging regulatory norms coupled with the ability of bacteria to evolve resistant trait quickly have made the pharmaceutical industries highly reluctant to venture into new antibiotic production (Committee on New Directions in the Study of Antimicrobiol Therapeutics: New Classes of Antimicrobials 2006). As an alternative to novel antibiotics, nanoparticles were widely explored. Nanoparticles (NPs) are generally of size 1–100 nm and are widely known for their enhanced biological activity due to their large surface area to volume ratio relative to their bulk particles. This property has also made them to possess antimicrobial activity as they can interact easily with bacterial cells and interfere with metabolic pathways in microbes (Huh and Kwon 2011). Among the nanoparticles, metal nanoparticles have been found to inhibit infectious bacteria effectively. Silver nanoparticles (AgNPs) are widely explored for their antimicrobial activity. AgNPs are capable of inhibiting the growth and multiplication of important infectious candidates like *Staphylococcus aureus, Pseudomonas aeruginosa, Salmonella typhi, Escherichia coli, Klebsiella pneumoniae*, and few others (Li et al. 2011; Prakash et al. 2013; Rai et al. 2014; Siddiqi and Husen 2016; Siddiqi et al. 2018). Mycosynthesis of AgNPs showed significant efficacy against *S. aureus, Enterococcus faecalis*, and *Salmonella enterica* (Devi and Joshi 2012) and has shown to enhance the activity of erythromycin, methicillin, chloramphenicol, and ciprofloxacin against *Klebsiella pneumoniae* and *Enterobacter aerogenes* (Bawaskar et al. 2010). The mechanism of AgNPs' bactericidal activity has been found to involve light-activated protein oxidation catalyzed by AgNPs and is independent of silver ion release or ROS generation as proposed in earlier studies (Shi et al. 2019). Polyvinylpyrrolidone (PVP)-coated AgNPs were able to inhibit the growth of NDM-1 containing pan-drug-resistant *Acinetobacter baumannii* at 0.9 µg/ml, relative to conventional antibiotics like gentamicin, amikacin, and carbapenems, which failed to curtail the growth of this strain (Shi et al. 2019). Gallic acid-functionalized AgNPs were embedded in a biocompatible polysaccharide to obtain a hydrogel with antibacterial activity and ensured sustained release and, when irradiated with NIR (Near Infrared), exhibited photothermal therapy against *S. aureus* and *E. coli* by killing 94% of bacteria (Liu et al. 2020). In vivo *S. aureus* wound infection in mice treated with the hydrogel and NIR radiation showed reduced ulceration on day 1 and decreased size by day 3 relative to the other groups (Liu et al. 2020). PEGylated dendronized AgNPs were able to permeabilize the outer membrane of Gram-negative *P. aeruginosa* and allowed phage endolysin to reach peptidoglycan overcoming LPS (Lipopolysaccharide) barrier at a lower concentration of 20 µg/ml (Ciepluch et al. 2019). AgNPs functionalized with gold (Au) have shown to increase the stability of nanoparticles without compromising their antibacterial activity (Baptista et al. 2018).

Au-PtNPs possessed an MIC of 5 µg/ml against MDR *E. coli*, whereas levofloxacin was able to inhibit microbial growth at 32 µg/ml (Zhao et al. 2014).

2 nm core cationic monolayer-protected AuNPs interacted with cell membrane of Gram-positive and Gram-negative bacteria, forming aggregation patterns and resulting in lysis of bacterial cell (Rai et al. 2010). Water-soluble AuNPs with n-decane end group (NP 3) suppressed the growth of uropathogenic strains of *E. coli* and inhibited the MDR pathogens at 16 nM. NP 3 also damaged cell membrane at 500 nM in resistant *E. coli* and MRSA as evidenced by propidium iodide assay (Li et al. 2014). AuNPs possess a remarkable property of transforming light into heat under laser irradiation (Mocan et al. 2017). Organic dyes can be used to activate and photosensitize AuNPs of size less than 5 nm (Hu et al. 2006). Laser radiation (660 nm) for 5 min of polysiloxane polymers with methylene blue (MB) and AuNPs resulted in significant killing of *E. coli* and MRSA (Methicillin Resistant *Staphylococcus aureus*). 2 nM of AuNPs enhanced the activity of MB to kill bacteria (Perni et al. 2009). Monoclonal antibody-modified popcorn-shaped gold nanoparticle upon 670 nm laser radiation was able to kill 100% of drug-resistant *Salmonella typhimurium* by photothermal lysis (Khan et al. 2011). DNA aptamer-functionalized gold nanorods (AuNRs) were able to bind specifically to MRSA by targeted photothermal lysis and inactivated 95% of bacteria, whereas aptamer-functionalized AuNPs were able to inactivate only 5% of bacterial cells. This difference was due to high longitudinal absorption of NIR and efficient conversion of light to heat energy by AuNRs (Ocsoy et al. 2017). Pectin-capped platinum nanoparticles (PtNPs) of size 2–5 nm had an MIC of 31.2 μM and 62.5 μM against *B. subtilis*, *S. aureus*, and *P. aeruginosa* and *E. coli*. Triple dose of PtNPs was able to cure infections of *E. coli* and *Aeromonas hydrophila* in zebrafish infection model by generating ROS coupled with loss of membrane integrity (Ayaz Ahmed et al. 2016). Green synthesis of PtNPs from plant extract of *Taraxacum laevigatum*, of size 2–7 nm, exhibited strong antibacterial activity against resistant *P. aeruginosa* and *B. subtilis* (Tahir et al. 2017). PtNPs at concentration > 5 ppm suppressed growth of bacteria involved in dental infections – *Streptococcus mutans*, *Enterococcus faecalis*, and *Porphyromonas gingivalis* – and at >125 pico/mL, it resulted in significant amount of decomposition and elimination of LPS (Itohiya et al. 2019). PtNPs synthesized from marine actinobacteria exhibited broad-spectrum activity against *Proteus vulgaris*, *Staphylococcus aureus*, *E. cloacae*, and *E. coli* (Dev Sharma 2017). Dendrimer-shaped silver-platinum (AgPt) nanoparticles inhibited the growth of pathogenic *S. aureus*, *P. aeruginosa*, and MDR *E. coli* at 10–50 μg/ml with no fibroblast cytotoxicity relative to AgNPs (Ruiz et al. 2020). A recent study showed that chitosan-mediated AuNPs showed antibacterial activity against few Gram-negative and Gram-positive isolates (Kalaivani et al. 2020). In addition, the nanoparticles also induced apoptosis-mediated cytotoxicity against MCF-7 (breast cancer cell lines) at 250 μg/ml (Kalaivani et al. 2020).

Chemically synthesized zinc oxide nanoparticles (ZnO NPs) had an IC_{50} value of 2 mM against carbapenem-resistant *A. baumannii* strain due to fourfold increase in generation of ROS upon treatment with nanoparticles, ultimately leading to bacterial cell destruction (Tiwari et al. 2018). Another study revealed a contradictory result that antibacterial activity of ZnO NPs is not due to ROS generation but due to the reason that ZnO NP treatment led to upregulation of biosynthesis and carbohydrate degradation using gene transcription microarray analysis. They also observed a significant 3-log reduction in colonies of MRSA with negligible increase in ROS generation (Kadiyala et al. 2018). Ciprofloxacin-conjugated ZnO NPs of size 18–20 nm exhibited antibacterial activity against multidrug-resistant clinical isolates of *E. coli*, *S. aureus*, and *Klebsiella* sp. at 10 µg/ml (Patra et al. 2014). A recent study showed that green synthesis of ZnO NPs using *Aloe socotrina* leaf extract of size 15–50 nm had antibacterial activity at 50 µg/ml and 75 µg/ml against *P. vulgaris*, *P. aeruginosa*, and *E. coli* and *K. pneumoniae* (Fahimmunisha et al. 2020). Nanoparticles of ZnO and hydroxyapatite (nHA) and combination of (ZnO NPs + nHA) were coated onto Ti disks (Abdulkareem et al. 2015). Biofilms were allowed to form on coated disks for 96 h in fermenter under aerobic conditions with artificial saliva and peri-implant sulcular fluid, and it was found that ZnO NPs and composite coatings were effective in reducing the thickness of biofilm to 24 µM relative to untreated group (91 µM) (Abdulkareem et al. 2015). Another recent study revealed that ZnO NPs, at its sub-MIC (32 µg/ml), established synergistic relationship with rifampicin and caused a fourfold reduction in MIC of rifampicin against *Mycobacterium smegmatis*, which can be contribute to increased membrane permeability upon treatment with ZnO NPs, rather than ROS generation (Mistry et al. 2020).

Selenium nanoparticles (Se NPs) synthesized by a novel green process called pulsed laser ablation in liquids (PLAL) showed a dose-dependent antibacterial activity toward MDR *E. coli*, MRSA, *P. aeruginosa*, and *S. epidermidis* at concentrations 0.05 and 25 ppm (Geoffrion et al. 2020). These nanoparticles also showed anticancer properties in malignant melanoma and glioblastoma cells. Se NPs coated with bovine serum albumin (BSA) achieved a significant tenfold reduction in CFU/ml of *S. aureus* but not against MDR *E. coli* (Chung et al. 2020). A recent study has shown the enhanced antibacterial efficacy of a nanohybrid system – Se NPs – integrated with lysozyme against *S. aureus* and *E. coli,* relative to the individual components (Vahdati and Tohidi Moghadam 2020). Chitosan-based Se NPs had a MIC of 0.068, 0.137, and 0.274 mg/ml against *S. sanguinis*, *S. aureus*, and *E. faecalis*, respectively. At 0.274 mg/ml, the bacteria were completely eliminated in vitro after 1, 2, and 6 h of treatment, respectively (Abdolrasoul et al. 2019). Another study showed that chitosan-stabilized Se NPs inhibited the growth of *M. smegmatis* and *M. tuberculosis* by damaging their cell wall integrity (Estevez et al. 2020). A nanohybrid system was prepared using chitosan-cetyltrimethyl ammonium bromide (CTAB)-based hydrogel, which would entrap mupirocin and Se NPs. The nanohybrid was able to reduce the MIC of mupirocin by threefold, and an in vivo study on rat diabetic wound infection model was performed (Golmohammadi et al. 2020). The rats were infected by mupirocin- and methicillin-resistant *S. aureus* and were treated with the nanohybrid system for 21 days, which showed significant improvement in the treated group relative to untreated as evident from histopathological and wound contraction analysis (Golmohammadi et al. 2020). Selenium brooms (Se Brs), formed by biogenic synthesis using almond skin extract, showed selective antibacterial activity against *B. subtilis* at 7.8 µg/ml (Sadalage et al. 2020). Se Brs coated on cotton fabrics were found to retard the growth of bacteria, thus can be applied in coating equipment's surface and hand gloves to prevent *B. subtilis* bacteremia from hospitals (Sadalage et al. 2020). Another study showed a synergistic nanocomposite formed by conjugating quercetin and acetylcholine to the surface of Se NPs exhibited enhanced antibacterial activity against MDR *E. coli* and *S. aureus* by reducing the viability of cells to 90% at 25 μg/ml (Huang et al. 2016).

Dextran-coated nano-ceria was found to be effective against *P. aeruginosa* and *S. epidermidis* at basic pH (9.0) causing a 2-log reduction after 6 h of treatment at a concentration of 500 μg/ml (Alpaslan et al. 2017). Polyacrylic acid-coated or dextran-coated ceria nanoparticles inhibited the growth of *P. aeruginosa* by 55.1% after 24 h relative to untreated group (Wang et al. 2013). Ceria nanoparticle, itself, was not able to inhibit the growth of carbapenem-resistant *K. pneumoniae* but was able to act synergistically with imipenem, cefotaxime, and amoxicillin, which was attributed to the membrane-permeabilizing ability of the nanoparticles, that allowed passive diffusion of antibiotics (Bellio et al. 2018).

Green synthesis of titanium dioxide nanoparticles (TiO_2 NPs) by *Azadirachta indica* extract yielded spherical NPs of size ranging from 15 to 45 nm and exhibited antibacterial activity against *E. coli*, *K. pneumoniae*, *B. subtilis*, *S. aureus*, and *S. typhi* with MIC of 10–25 μg/ml (Thakur et al. 2019). In order to prevent implant-associated infections, TiO_2 nanotubes incorporated with AgNPs were fabricated on Ti implants. These were able to kill *S. aureus* in the first few days and were able to prevent bacterial adhesion to implants for 30 days (Zhao et al. 2011). Iron oxide nanoparticle (Fe_3O_4 NP) was able to exhibit antibacterial activity against ampicillin- and kanamycin-resistant *E. coli* strains by causing growth reduction at 250 μg/ml, indicating its bacteriostatic nature. It also inhibited proton flux through membrane and altered the membrane permeability (Gabrielyan et al. 2019). Pseudobactin is one of the main siderophores of *P. aeruginosa* and is known for antimicrobial activity against foreign bacteria. Pseudobactin-coated iron nanoparticles showed antibacterial activity against resistant *P. aeruginosa*, *E. coli*, and *S. typhimurium* and to a lesser extent against *S. aureus* (Kotb et al. 2020). A nanocomposite of Fe/Ni oxide of varying molar concentrations of nickel was prepared along with Fe_2O_3 and NiO nanoparticles (Bhushan et al. 2019). On testing the samples against *B. subtilis*, *S. aureus*, *E. coli*, and *S. typhi* for antibacterial activity by Kirby-Bauer disk diffusion assay, the authors observed that the nanomaterial exhibited bactericidal activity in dose-dependent manner with order of susceptibility being *E. coli* > *B. subtilis* > *S. aureus* > *S. typhi*. One of the composites, FeN4, was able to eradicate *S. typhi* growth completely and almost abolished *B. subtilis* and *E. coli* at 120 mg/dl (Bhushan et al. 2019). Vancomycin and Fe_3O_4 nanoparticles were bound to lanthanum hexaboride (LaB_6) nanoparticles with a silica coating and carboxyl functionalization to develop a nanomaterial (Van-LaB_6@SiO_2/Fe_3O_4) for NIR thermal ablation of bacteria (Lai and Chen 2013). The composite nanomaterials were able to achieve magnetic separation upon external field and were able to decrease the survival fractions of *S. aureus* and *E. coli* to 51% and 65%, respectively, whereas nanocomposite devoid of vancomycin (LaB_6@SiO_2/Fe_3O_4) did not decrease the survival fractions greater than 5%. Treatment of the nanocomposite Van-LaB$_6$@SiO$_2$/Fe$_3$O$_4$ for 5 min along with NIR irradiation reduced the survival fraction to 1% depicting composite nanoparticles' excellent photothermal ablation of bacteria (Lai and Chen 2013). Magnetic nanoparticles functionalized with CPQN (quaternarized N-halamine-based cationic polymer) were able to successfully eliminate 100% of *S. aureus* and 99.9% of *E. coli* in 5 min at 10^{7-8} CFU/mg nanoparticles and can be used as water disinfectant (Chen et al. 2016). A pH-responsive nanocarrier (Amp-MSN@FA@CaP@FA) carrying ampicillin was constructed that covered double folic acid and calcium phosphate on mesoporous silica surface which reduced the mortality due to drug-resistant *E. coli* infection and promoted wound healing by resisting *S. aureus* infection in vivo (Chen et al. 2018).

Zero-valent copper nanoparticles of 12 nm size inhibited the growth of *E. coli* at 60 μg/ml (Raffi et al. 2010). SEM analysis revealed that the nanoparticles created pits/cavities in bacterial cell wall leading to cell death. 2–5-nm-sized CuNPs embedded in sepiolite (Mg$_8$Si$_{12}$O$_{30}$(OH)$_4$(H$_2$O)$_4$·8H$_2$O) were able to decrease the bacterial concentration of *S. aureus* and *E. coli* by 99.9% (Raffi et al. 2010). Growth of *Aggregatibacter actinomycetemcomitans*, a Gram-negative pathogen responsible for periodontitis, peri-implantitis, and non-oral infections, was inhibited by CuNPs synthesized by laser ablation. The CuNPs were also found to be cyto-compatible with human periodontal ligament stem cells (Fernández-Arias et al. 2020). The researchers have also shown that the bactericidal activity was not due to Cu ion release but due to the influence of size and crystallographic structure (Fernández-Arias et al. 2020). Copper sulfide (CuS) nanoparticles were synthesized using vancomycin as reducing and capping agent. CuS@Van did not exhibit antibactericidal activity against vancomycin-resistant *Enterococcus* (VRE) in dark but resulted in 99% death of bacterial cells upon NIR radiation for 10 min due to photothermal effect at 64 μg/ml of CuS@Van (Zou et al. 2020). BALB/C mice were infected with VRE and treated with CuS@Van + NIR radiation which eradicated the infection completely in the second day and showed reduction in inflammatory cells as evidenced by histopathology analysis. Copper-cysteamine (Cu-Cy) nanoparticles along with potassium iodide (KI) under UV light produced singlet oxygen, H$_2$O$_2$, and triiodide ions and significantly inactivated MRSA and *E. coli* (Zhen et al. 2020). The role of KI was to enhance the photodynamic therapy (PDT)-mediated killing of MRSA and *E. coli*. Self-assembled monolayer of CuS NPs was grafted on glass eliminated 95% of *E. coli* and *S. aureus* after 5 h of contact time and 99.9% of bacteria after 24 h (Gargioni et al. 2020). Another interesting study showed the nanomechanical action of a nanosystem to drill the cell membrane of carbapenemase-producing multidrug-resistant *K. pneumoniae* (Galbadage et al. 2019). Molecular nanomachines (MNMs) were designed with a rotor and a stator with functionalization ability. Upon exposure to UV light of 365 nm, the MNM is activated, and due to the rotational activity of the rotor, MNM can permeabilize the cell membrane. This enhanced the susceptibility to meropenem, by exposing the penicillin-binding proteins (PBPs), and effectively killed both resistant and sensitive *K. pneumoniae* (Galbadage et al. 2019).

7.3   In Biofilm Prevention and Disruption

Gene mutation or resistance gene acquisition of planktonic cells is not the sole reason of antibiotic resistance. Biofilm formation, especially in MDR pathogens, is a major challenge and a hurdle in the treatment of pathogens. Bacterial cells adhere to a surface and pack themselves in a self-produced extracellular polymeric substance (EPS), composed of exopolysaccharides, eDNA, proteins, and enzymes (Costerton et al. 1999). These serve as barricade for the entry and access to antibiotics and help in evading host immune responses (Davies 2003; Yang et al. 2012). Hence, it is essential to take measures that can eradicate or prevent the biofilm formation for eliminating pathogens successfully. Fungal biosynthesized AgNPs using extracellular metabolites from *Fusarium scirpi* inhibited 97% biofilm formation and disrupted 80% mature biofilm in vitro formed by uropathogenic *E. coli* (UPEC) clinical isolate that possesses *fimH* and curli fimbriae, at its sub-MIC concentration of 7.5 mg/L, depicting its potential to prevent and cure UPEC infections (Rodríguez-Serrano et al. 2020). AgNPs functionalized with *Butea monosperma* seed lectin (BAgNPs) were able to disturb biofilm formation of UPEC and eradicate preformed biofilm at 37.5 μM in vitro, which can be attributed to the sugar-binding site of BMSL (Bala Subramaniyan et al. 2020). Ag-Au hybrid nanoparticles were synthesized using quercetin as reducing agent. 40-nm-sized hybrid nanoparticles had MBC at 10 μg/ml and 20 μg/ml against Gram-positive and Gram-negative bacteria, respectively, and mixed infections were affected at 20 μg/ml. Biofilm formed by polymicrobes was also reduced, and the intracellular infection was suppressed by 70% to 90% in monocyte and fibroblast cell lines (Bhatia and Banerjee 2020). AgNPs immobilized on titanium by silver plasma immersion ion implantation possessed antibiofilm activity in vitro and in vivo against *S. epidermidis* by inhibiting bacterial adhesion and transcription of *icaAD* which was independent of Ag ion release. Hence, these AgNPs immobilized on titanium can be utilized in preventing periprosthetic infection that typically poses danger post-orthopedic surgery (Qin et al. 2014). Green synthesis of silver nanoparticles capped with semi-synthetic polysaccharide-based biopolymer (carboxymethyl tamarind polysaccharide) yielded nanoparticles of size 20–40 nm, which inhibited biofilm formation of *B. subtilis*, *E. coli*, and *S. typhimurium* at sub-MIC with no toxicity to mammalian cells and at 175 μM completely eradicated and prevented regrowth of planktonic cells even after 48 h. The nanoparticles altered the expression and positioning of bacterial cytoskeletal proteins – FtsZ and FtsA – leading to membrane damage and blocking cell division (Sanyasi et al. 2016). Green synthesized AgNPs using leaf extract of *Allophylus cobbe* in combination with antibiotics – ampicillin and vancomycin – inhibited biofilm activity in Gram-negative (*P. aeruginosa*, *S. flexneri*) and Gram-positive (*S. aureus*, *S. pneumoniae*) bacteria by 55% and 75% and showed increased ROS generation in presence of antibiotics (Gurunathan et al. 2014). Chemically synthesized AgNPs in combination with polymyxin B exhibited synergistic effect and enhanced the antibiofilm activity against *P. aeruginosa* (Salman et al. 2019). Biofilm-responsive silver nanoantibiotics (rAgNAs) made up of self-assembled nanoclusters and pH-sensitive charge reversal ligands show boosted bactericidal activity in biofilm environment (Wu et al. 2019). Bactericidal activity of rAgNAs remains quenched under neutral physiological conditions, and upon entry to acidic MRSA biofilm microenvironment, rAgNAs exhibited charge reversal, thus facilitating local accumulation, which got split into small Ag nanoclusters, enabling increased penetration and Ag release, which in turn amplifies bactericidal activity. Pathological examination of *S. aureus*-infected and rAgNA-treated mouse confirmed the disruption of biofilm and inhibition of further infection progression in vivo. The rAgNA-treated group had preserved tight junctions and intact structure of the muscle, evidencing its non-toxic nature (Wu et al. 2019). Biosynthesized AgNPs prepared using extract of marine sponges inhibited biofilm formation against 16 biofilm-forming bacterial strains isolated from ship hull at 50 µg/ml and thus AgNP coating on marine industrial surfaces to control biofouling (Inbakandan et al. 2013). AgNPs conjugated with thiosemicarbazide functionalized by glutamic acid (Ag@Glu/Tsc NPs) inhibited biofilm formation of MRSA strains up to 76.7% and downregulated the expression of biofilm-associated genes *icaA* and *icaD* by 66.7% and 60.3% at sub-MIC (Montazeri et al. 2020).

Mixed charged zwitterion-modified AuNPs were able to transit faster from negative to positive charge (Hu et al. 2017). This enabled the AuNPs to disperse well in healthy tissues (pH ~ 7.4) and then immediately change their affinity to negatively charged bacterial surfaces in a MRSA biofilm (pH ~5.5), which upon NIR irradiation caused thermal ablation of MRSA biofilm leaving the surrounding healthy tissues undamaged in vivo (Hu et al. 2017). Laser radiation of monolayers of gold nanostars at 808 nm resulted in efficient killing of MRSA biofilm by local hyperthermia (Pallavicini et al. 2014). PtNPs stabilized by polyacrylic acid, of size ~4 nm, exhibited antibiofilm activity at 400 µg/ml against *Streptococcus mutans*, an important causative of oral biofilms (Hashimoto et al. 2017). Phytoprotein-stabilized self-assembled Pt nanoclustes (PtNCs), synthesized using proteins from spinach leaves, exhibited antibiofilm activity and disrupted preformed biofilm of *S. typhi* at 6.8 µM (Subramaniyan et al. 2018). Citrate-coated 6 nm PtNPs, PEG-coated 11 nm AuNPs, and PEG-coated 8 nm iron oxide NPs reduced the bio-volume and biofilm formation of *Legionella pneumophila*, a primary pathogen that forms biofilms in cooling towers, spas, and dental lines and causes disease outbreaks (Raftery et al. 2014). These nanoparticles exhibited remarkable antibiofilm activity and increased roughness coefficient at smaller size and at lower concentration of 1 µg/ml but not when size or concentration is increased. Interestingly, AgNPs, known for their antibacterial activity, did not alter biofilm formation (citrate-coated AgNPs) (Raftery et al. 2014). Biologically synthesized Se NPs were able to inhibit biofilm formation of *S. aureus*, *P. aeruginosa*, and *P. mirabilis* by 42%, 34.3%, and 53.4%, respectively, in selected biofilm-forming clinical isolates (Shakibaie et al. 2015). A nanocomposite containing Se NPs and iron oxide nanoparticles (IONPs) was formed using chitosan-coated pentasodium triphosphate as a cross-linking agent. This nanocomposite exhibited excellent antibiofilm effect against *S. aureus* biofilm in the presence of external magnetic field with a relative fraction of dead to live bacteria of ~400% (Li et al. 2020b). This efficacy of nanocomposite relative to individual components can be attributed to magnetic field, which aided the nanoparticles to penetrate the biofilm (Li et al. 2020b).

Alginate-coated iron oxide nanoparticles (magnetite) were conjugated with tobramycin and were investigated against in vitro cystic fibrosis model by growing *P. aeruginosa* (PaO1) biofilms for 60 days (Armijo et al. 2020). The conjugate was able to inhibit the bacterial growth and biofilm formation, which would be otherwise impossible. Susceptibility to tobramycin was found to reduce with increasing culture time, whereas there was no decrease to susceptibility of the nanoparticle (Armijo et al. 2020). Superparamagnetic iron oxide nanoparticles (SPIONs) were evaluated for their antibiofilm activity against *P. aeruginosa* biofilms (Ramezani Ali Akbari and Abdi Ali 2017). Among 20 isolates that were drug resistant and form strong biofilms, SPIONs were able to reduce biofilm biomass in 11 isolates at 30 µg/ml. Contrary to this observation, it stimulated biofilm formation in nine isolates. SPIONs exhibited synergistic relationship with imipenem in few of these isolates (Ramezani Ali Akbari and Abdi Ali 2017). Ceragenin (a synthetic mimic of antibacterial peptides)-coated magnetic nanoparticles exhibited strong antibacterial activity and prevented formation of *P. aeruginosa* biofilm in different body fluids – saliva, cerebrospinal fluid, abdominal fluid, serum, plasma, and urine – at 100 µg/ml (Niemirowicz et al. 2015). ZnO NP-coated glass slides restricted biofilm formation of *E. coli* and *S. aureus* by generation of hydroxyl radicals from the coated surface (Applerot et al. 2012). Uncoated glass surface allowed a biofilm formation of 10^{10}–10^{11} CFU/cm^2, whereas ZnO-coated glass surfaces did not permit any colonization of *S. aureus* and 20 CFU/cm^2 for *E. coli* at the end of 10 days, which establishes the potential use of the nanoparticles in medical and environmental applications (Applerot et al. 2012).

Biogenic CuNPs and ZnO NPs were synthesized from non-pathogenic *E. faecalis*. Both nanoparticles had antibacterial activity against clinical isolates of *E. coli*, *K. pneumoniae*, and MRSA and standard strains of *S. flexneri*, *E. faecalis*, *E. coli*, *K. pneumoniae*, and *S. aureus*, except *P. aeruginosa*, at concentrations ranging from 10 to 64 µg/ml (Ashajyothi et al. 2016). CuNPs also exerted enhanced antibiofilm activity against these strains relative to ZnO NPs and antibiotics by inhibiting ~80% biofilm formation alone and ~ 90% in combination with antibiotics (Ashajyothi et al. 2016). Casein-capped CuNPs efficiently inhibited biofilm biomass formed by *P. aeruginosa* and *S. aureus* by 88.6% and 90.4% at 1X MIC (Christena et al. 2015). 5-Nitroindole (5 N)-capped CuO/ZnO bimetal nanoparticles (5NNP) were synthesized to improve the antibacterial and antibiofilm activities of 5 N against *E. coli*, *P. aeruginosa*, and *S. aureus* (Manoharan et al. 2020). 5NNP containing 1 mM of 5 N reduced biofilm by 97% against strong biofilm-forming MDR isolate from membrane bioreactor (Manoharan et al. 2020). Ceria nanoparticles synthesized using root extract of *Arctium lappa L.* encapsulated with nano-chitosan effectively disrupted biofilm of *S. aureus* and *P. aeruginosa* (Bushra et al. 2020).

7.4   Quorum Sensing Inhibitors

Quorum sensing (QS) is a density dependent cell-cell signaling that induces alter-ation in behavior when a critical population density has been attained (Fuqua et al. 1994; Abisado et al. 2018). Bacteria generate a particular signal (autoinducers – AI), which reaches a threshold concentration as the population increases. The signal then interacts with a particular receptor protein causing an alteration in gene expression in the associated population that is often associated with increased virulence (Abisado et al. 2018). Autoinducing peptides (AIPs) and N-acyl-homoserine lactonase (NAHLs) are the widely reported QS systems in Gram-positive and Gram-negative bacteria, respectively (Rémy et al. 2018). Autoinducer-2, another QS system, is produced and sensed by both Gram-positive and Gram-negative bacteria (Pereira et al. 2013). Bacteria utilize quorum sensing inhibitors (QSIs) to negate the action of AIs and use quorum-quenching enzymes to destroy the signaling molecules. Researchers surmised this as a potential target and started to explore a similar strategy to reduce bacterial pathogenicity and virulence to tackle drug-resistant bacteria. N-acyl-homoserine lactonase-stabilized silver nanoparticles (AiiA-AgNPs) inhibited quorum sensing by degradation of QS molecules at 1–5 μM. This in turn reduced biofilm formation, urease production and altered cell surface hydrophobicity of *Helicobacter pylori*, thus sensitizing the bacteria to treatments, which would other-wise fail due to biofilm formation (Gopalakrishnan et al. 2020). Interestingly, AiiA-AgNPs at higher concentrations exhibited cytotoxicity (100 μM) against carcinoma cells, realizing their potential in cancer therapy. Honey polyphenols were introduced in scaffold of selenium nanovectors (SeNPs@HP) that inhibited 60% of protease activity, decreased 49.6% of pyocyanin content, reduced 52.7% of elastase activity, and suppressed 59.6% of rhamnolipid production in *P. aeruginosa* at 4.5 μg/ml (Prateeksha et al. 2017). Nanoparticles containing chitosan, sulfo-butyl-ether-β-cyclodextrin, and pentasodium triphosphate integrated with quercetin inhibited bacterial quorum sensing up to 61.12% in *E. coli* (Thanh Nguyen and Goycoolea 2017). Chitosan and chitosan-zinc oxide (CH/ZnO) nanocomposite downregulated the expression of QS-dependent virulence factors, LasI and RhlI, in *P. aeruginosa* (Badawy et al. 2020). CH/ZnO nanocomposite alone reduced the expression of RhlI by 1240-fold in PaO1 and by 1778- and 627-fold in clinical isolates (Badawy et al. 2020). Gold nanoparticles (AuNPs) synthesized by *Lysinibacillus* sp. and *Pseudomonas stutzeri* inhibited pyocyanin production of *P. aeruginosa* during 72 h incubation period (Palpperumal et al. 2016). AgNPs synthesized using *Mespilus germanica* extract inhibited the expression of *mrkA* (type 3 fimbriae) and *luxS* (QS system) genes significantly in MDR strains of *K. pneumoniae* (Foroohimanjili et al. 2020). AgNPs synthesized from seed extracts of *Myristica fragrans* inhibited the growth of MDR *S. typhi* strains and the pigment production of indicator bacteria *Chromobacterium violaceum*, revealing their quorum sensing inhibition potential (Balakrishnan et al. 2020). *AhyR* involved in the produc-tion of pentasodium triphosphate in *A. hydrophila* was downregulated by AuNPs synthesized by ethnobotanical crude extracts, thus inhibiting the bacterial biofilms by either interrupting QS system or by deregulating the synthesis of pentasodium triphosphate (Fernando and Judan Cruz 2020). AgNPs synthesized from extract of *Carum copticum* (Ag@CC-NPs) inhibited ~75% of violacein against *C. violaceum* and inhibited 77, 49, 71, 53, 89, and 60% production of pyocyanin and pyoverdine, exoprotease activity, elastase, swimming motility and rhamnolipid production in *P. aeruginosa* (Qais et al. 2020). AgCl-TiO$_2$ reduced violacein production in bacterial model *C. violaceum*, and the anti-QS activity was confirmed by the absence of signaling molecule, oxo-octanoyl Pentasodium triphosphate, and can be used as efficient model for controlling food spoilage in food packaging materials (Naik and Kowshik 2014). ZnO NPs decreased elastase, pyocyanin, and biofilm formation in most tested strains of *P. aeruginosa* that included six strains from cystic fibrosis patients, two resistant strains, two gallium-resistant strains, and four environmental isolates (García-Lara et al. 2015). Se NPs and TeNPs inhibited significant amount of violacein production in *C. violaceum* and biofilm architecture of *P. aeruginosa*, which are important QS signaling systems (Gómez-Gómez et al. 2019a). Another study showed that monophasic tin dioxide nanoflowers (TONFs) inhibited QS-regulated virulence in *C. violaceum*, *P. aeruginosa*, and *Serratia marcescens* (Al-Shabib et al. 2018). Mycofabricated AgNPs, synthesized from metabolites of soil fungus *Rhizopus arrhizus*; inhibited biofilm formation and production of pyocyanin, pyoverdine, pyochelin, and rhamnolipid; reduced NAHL production; and downregulated QS-regulated genes that encoded virulence factors in *P. aeruginosa* (Singh et al. 2015). Kaempferol-loaded chitosan/TPP nanoparticles significantly inhibited the production of violacein pigment in *C. violaceum* during 30-day storage (Ilk et al. 2017). Nanocapsules of chitosan, containing an oil core and stabilized by surfactant, inhibited quorum sensing of *E. coli* (Qin et al. 2017).

7.5   Role on Efflux Pumps

Efflux pumps are a predominant reason for drug resistance in bacteria. Efflux was first described as a mechanism for tetracycline resistance in *E. coli* (McMurry et al. 1980), and later, numerous reports on chromosome- and plasmid-mediated efflux were reported. Five major classes of efflux systems have been widely reported in bacteria and are mostly drug specific (Poole 2007). Hence, efforts to identify agents that can inhibit efflux pumps or downregulate gene corresponding to efflux gene expression could be of help to combat drug-resistant strains. Despite the potential in targeting efflux pumps and multitasking ability of nanoparticles, there are very few reports that have identified nanoparticle candidates that can alter antibiotic resistance contributed by efflux pumps. ZnO NPs conjugated to thiosemicarbazide and functionalized with glutamic acid (ZnO@Glu-TSC), in combination with ciprofloxacin, downregulated the expression of *norA*, *norB*, *norC*, and *tet38* by 5.4-, 3.8-, 2.1-, and 3.4-fold, respectively, as evidenced from qPCR assay, relative to ciprofloxacin alone in *S. aureus* strains (Nejabatdoust et al. 2019). 65 µM casein-stabilized CuNPs effectively inhibited efflux pumps of *P. aeruginosa* and *S. aureus* similar to standard EPI inhibitor verapamil (Christena et al. 2015). In addition, the nanoparticles also enhanced the permeability of membrane in both Gram-positive and Gram-negative bacteria as evidenced by propidium iodide assay and N-phenyl-naphthylamine (NPN) assay at 1X MIC, respectively. Another recent study showed that biosynthesized $CuFe_2O_4$@Ag nanocomposite using extract from *Chlorella vulgaris*, in combination with ciprofloxacin, reduced the expression of *norA*, a major contributor of drug efflux in *S. aureus*, by 59% and 65% in resistant clinical and standard strains of *S. aureus*, thus inhibiting efflux pump genes and increasing the efficacy of ciprofloxacin (Kahzad and Salehzadeh 2020). Thiolated chitosan-coated cobalt-doped zinc oxide (Co-ZnO) nanoparticles exhibited 100% antibacterial activity against MRSA at 10 µg/ml in visible light (Iqbal et al. 2019). Cartwheel assay revealed that at 2.5 µg/ml, Co-ZnO NPs showed enhanced fluorescence due to increased ethidium bromide (EtBr) accumulation, which depicts the efflux pump inhibitory activity of the nanoparticles, though the efflux pump involved was not identified (Iqbal et al. 2019). Citric acid-coated iron oxide (magnetite) nanoparticles (CA-MNP) increased the membrane permeability and ROS generation in *Mycobacterium smegmatis*, thus increasing the susceptibility of the strain to isoniazid and rifampicin, which belong to first-line anti-tuberculosis drugs. Intracellular accumulation of the drug was found to be enhanced in the presence of nanoparticles, which was evident from EtBr accumulation (Padwal et al. 2015).

7.6   Action on Plasmids

Plasmids are extrachromosomal DNA that can replicate independently and can be transferred to other bacteria by means of horizontal gene transfer (HGT) (San Millan 2018). Plasmids form a major contributor in acquisition of antibiotic-resistant genes (ARG) and disseminate to other bacteria, ultimately creating superbugs. Removal of plasmids from bacterial population is termed as plasmid curing. This strategy can be employed for the removal of ARGs, without leading to dysbiosis. Acridine orange, ethidium bromide, plumbagin, and sodium dodecyl sulfate are few curing agents that have successfully eliminated plasmids in vitro from various Gram-negative and Gram-positive bacteria (Michelle et al. 2018). Owing to their toxic nature, the use of these agents has been restricted to research purposes. For therapeutic interventions, non-toxic curing agents that eliminate plasmid both in vitro and in vivo are the need of the hour. Few studies have reported the ability of nanoparticles in reducing plasmid-mediated antibiotic resistance in bacteria. Chitosan nanoparticles were evaluated against plasmid-mediated multidrug resistance of *E. faecalis*, *S. aureus*, and *Bacillus* sp. isolated from saliva of chronic periodontitis patients (Subbiah et al. 2019). Chitosan nanoparticles were able to cure plasmids, and the efficiency of curing was 88% in tetracycline-resistant *E. faecalis*. Metal oxide nanoparticles (ZnO, Al_2O_3, and TiO_2) at concentration < 50 mg/L inhibited lateral transfer of pUC19 plasmid carrying ampicillin-resistant genes to *E. coli* in a decreasing order of Al_2O_3 NPs > ZnO NPs > TiO_2 NPs (Hu et al. 2019). Metal oxide nanoparticles interact with phosphate and base groups of pUC19, forming aggregates, which inhibited lateral transfer of ARG. 1.5 ppm of AgNPs was able to cure plasmids of resistant *E. coli*, *P. aeruginosa*, *S. aureus*, and

S. pyogenes with efficiency ranging from 50% to 100%, respectively, in vitro. 3 ppm of AgNPs completely cured plasmids in all these strains (Mhawesh et al. 2019). AgNPs synthesized using extract of *Zingiber officinale* exhibited plasmid curing in *E. coli*, *K. pneumoniae*, and *S. aureus* at 3%, but the mechanism of curing was not explored (Hashim 2017). Pectin-capped PtNPs at sublethal concentration of 20 μM caused loss of extended-spectrum beta-lactamase (ESBL) containing plasmid in carbapenem-resistant clinical isolate of *E. coli* and resulted in small colony variant of cured strain relative to its wild type (Bharathan et al. 2019). Plasmid loss caused a 16–64-fold reduction in MIC of meropenem and ceftriaxone, respectively, and in addition, it reduced biofilm formation by 50%. The mechanism by which PtNPs eliminate plasmid might be due to DNA cleavage and by affecting membrane integrity. Moreover, PtNPs acted as an adjuvant along with meropenem and caused a 5-log decline in colony counts relative to the untreated group in a zebrafish infection model (Bharathan et al. 2019).

7.7   As delivery Systems to Combat Infections

Owing to increase in drug solubility, modulation of drug release, ability to deliver multiple drugs simultaneously, and evading immune system, nanoparticles prove themselves as promising candidates to be used in drug delivery systems (Gao et al. 2018). Nanoparticle-based delivery can promote antibiotic localization to pathogen, can alter interaction between drug and pathogen, and enables drug-free anti-viru-lence therapy. Self-assembled nanoparticles of cationic peptides crossed the blood-brain barrier (BBB), which can be used to treat brain diseases such as meningitis or encephalitis caused by *B. anthrax*, *B. subtilis*, or *S. aureus* (Liu et al. 2009). A lipid-polymer hybrid nanoparticle, with poly(lactic-co-glycolic acid) as a polymeric nanoparticle core and phosphatidylcholine as the lipid coat, encapsulating class III compounds of the Biopharmaceutical Classification Systems (BCS) was able to encounter rhamnolipids, an essential component of biofilms of *P. aeruginosa* (Cheow and Hadinoto 2012). BCS class III compounds have poor lipid membrane permeability, whose release from the hybrid is triggered by the rhamnolipids, whereas compounds having high lipid membrane permeability are released irrespective of the presence or absence of rhamnolipids. Self-assembled nanoparticles formed with diblock copolymers composed of 2-(dimethylamino)ethyl methacrylate, butyl methacrylate (BMA), and 2-propylacrylic acid (PAA) (p(DMAEMA)-b-p(DMAEMA-co-BMA-co-PAA)) exhibited good adsorption efficiencies to negatively charged hydroxyapatite (HA), saliva-coated HA (sHA), and exopolysaccharide-coated sHA (Horev et al. 2015). Farnesol loaded in the hydrophobic core was released depending on pH and disrupted *S. mutans* biofilms fourfold greater than free farnesol upon topical appli-cation (Horev et al. 2015). About >99% of biofilm cells formed by *P. aeruginosa*, *E. coli*, *S. aureus*, and *S. epidermidis* were killed by nitric oxide-releasing silica nanoparticles in vitro, and the nanoparticles were found to be less cytotoxic than clinical concentrations of currently administered antiseptics (Hetrick et al. 2009).

DNase I-functionalized ciprofloxacin-loaded poly(lactic-co-glycolic acid) (PLGA) nanoparticles made a notable shift in biofilm treatment, by two modes of action: (1) release of ciprofloxacin in controlled fashion and (2) disassembly of biofilm by degrading the extracellular matrix in both young and established biofilms of *P. aeruginosa* (Baelo et al. 2015). DNase I-loaded silver-doped mesoporous silica nanoparticles enhanced the biofilm dispersion in both Gram-positive and Gram-negative bacteria (Tasia et al. 2020). Another recent study showed that co-immobilization of cellobiose dehydrogenase and DNase I on chitosan nanoparticles targeted both polymicrobial biofilm matrix and the cells of *S. aureus* and *C. albicans* by degrading eDNA, by reducing biofilm thickness, and by killing cells on silicone substrate (Tan et al. 2020). Clarithromycin (CLR)-loaded PLGA nanoparticles possessed a loading efficiency of 58–80% and were found to have an initial burst of drug followed by a plateau during a period of 24 h. Moreover, the CLR-loaded nanoparticles showed equal antibacterial effect against *S. aureus* at 1/8 concentration relative to free drug (Mohammadi et al. 2011). Levofloxacin was conjugated to hyaluronic acid to prepare nitric oxide-sensitive nanomicelles (HA-NO-LF) (Lu et al. 2020a). These nanomicelles can enter host cells by endocytosis (CD44 mediated) and release the drug on exposure to nitric oxide, owing to which the therapeutic effect of the nanosystem was observed in upgrading inflammatory levels in pneumonia mouse model relative to free levofloxacin (Lu et al. 2020a). Antimicrobial peptides linked to cathepsin B can be delivered by vitamin C lipid nanoparticles in macrophage lysosomes, which is the center of bactericidal activities (Hou et al. 2020). Adaptive transfer of macrophages containing the nanoparticles enabled elimination of MDR *E. coli* and *S. aureus* leading to recovery of immunocompromised septic mice. N'-((5-nitrofuran-2-yl)methylene)-2-benzohydrazide was incorporated in polysorbate 20 micelles, which was then loaded in chitosan nanoparticles (CH-5-NFB-NP) and was able to inhibit the growth of oxacillin-resistant *S. aureus* and vancomycin intermediate-resistant *S. aureus* strains (de Andrade et al. 2020). Due to the combined properties of tissue regeneration, protective biofilm property, and antibacterial activity, the nanosystem can be used to treat MDR in patients with burn wounds. Kappa-carrageenan-wrapped ZnO NPs inhibited biofilm growth of MRSA at 100 µg/ml and were non-toxic to fibroblast cell lines with good biocompatibility (Vijayakumar et al. 2020). In addition, the nanoparticles did not show mortality of *Artemia salina*, an aquatic crustacean, at 500 µg/ml after 48 h, highlighting their ecosafety. Cationic liposomes adsorbed with AuNPs interacted with bacteria only at acidic pH, which make them suitable to treat skin pathogens such as *Propionibacterium acnes* and *S. aureus* (Pornpattananangkul et al. 2010). Nanocarriers comprising hydrophobic SPIONs and hydrophilic antibiotic methicillin were developed to treat medical device-associated infections by *S. aureus* (Geilich et al. 2017). This possessed high magneticity and was able to penetrate 20- µm-thick biofilm upon application of external magnetic field and eradicated all bacteria in the biofilm at 40 µg/ml SPION with 20 µg/ml methicillin. In addition, the formulation showed selective toxicity toward MRSA and not to mammalian cells (Geilich et al. 2017). Anti-tuberculosis drugs – isoniazid, rifampin, and streptomycin – were encapsulated in the nanoparticle and were evaluated against

M. tuberculosis in human monocytes. Antimicrobial activity and intracellular drug accumulation of isoniazid and streptomycin were enhanced, but the effect on antimicrobial activity of rifampin on intracellular bacteria was not augmented (Anisimova et al. 2000). QSI (Precirol, Compritol, and Dynasan)-loaded ultra-small solid lipid nanoparticles (QSI-us-SLNs) released 60–95% QSI in controlled manner over 8 h and inhibited pyocyanin (*pqs* QS system in *P. aeruginosa*), resulting in reduced virulence (Nafee et al. 2014). Recent years have stimulated the research of employing polymeric nanoparticles as alternative for viral delivery systems of CRISPR/Cas9 to treat or inhibit microbial infections. Polymer-derivatized Cas9 was covalently modified with cationic polymer and further complexed with single-guide RNA that targeted *mecA* gene, a major gene involved in methicillin resistance in *S. aureus* (Kang et al. 2017). This nanocomplex was able to retain the endonuclease activity of inducing double-strand DNA cleavage and allowed the editing of the resistant gene in bacterial genome, thus successfully mitigating the resistance. Moxifloxacin (MXF) was encapsulated in ROS-responsive material 4-(hydroxymethyl) phenylboronic acid pinacol ester-modified a-cyclodextrin and formulated as ROS-responsive MXF containing nanodelivery system (Wang et al. 2019b). These were coated with 1,2-distearoyl-sn-glycero-3-phosphoethanolamine-*N*-methoxy(polyethylene glycol) (DSPE-PEG$_{2000}$) and conjugated with folic acid (DSPE-PEG$_{3400}$-FA) to penetrate the sputum secreted from *P. aeruginosa*-infected lung, thus enabling active targeting of macrophages in inflammatory tissues (Wang et al. 2019b). In vitro cell line studies showed that DSPE-PEG-FA-modified nanodelivery system was effectively internalized by the infected macrophages and eradicated resistant bacteria, which was reflected in vivo in mouse pulmonary infection model. Polycaprolactone was electrospun and loaded with antibiotics – 12.5% colistin and 1.4% vancomycin – and cationic/anionic gold nanoparticles to develop antibiotic-loaded nanomesh (Fuller et al. 2019). Colistin-loaded nanomesh had the highest sustained release for 14 days for a 4 mg, 1.5 cm^2 nanomesh in vitro. In addition, colistin-loaded nanomesh with cationic AuNPs showed a relatively larger zone of inhibition against *E. coli*. Lipid-polymer hybrid nanoparticle (LPN) was synthesized with PLGA as polymer core and DOTAP (Dioleoyl-3-trimethylammonium propane) as the cationic lipid shell to release antimicrobial agents or bioimaging agents to biofilms (Baek et al. 2018). DOTAP enabled binding to surface of Gram-positive and Gram-negative pathogens. The formulation reduced 95% biofilm at 8–32-fold concentration lower than free antibiotics.

7.8   Nanoparticles in Detection and Diagnosis of Infection

Detection of infection and resistance is an important step to proceed with successful treatment. Most of the conventional methods are time-consuming and compromise sensitivity and specificity. Hence, in search of a more promising alternative, researchers have resorted to the use of nanoparticles, owing to their multitude abilities. As expected, nanoparticles have been successful in detecting infections, which when done at an early stage could reduce treatment failure and help in approaching proper cure. Anti-DH5α strain polyclonal antibody-modified AuNPs were used as a probe to detect the *E. coli* DH5α strain (Xu et al. 2012). The NPs were able to bind 77% of the target strain from a mixture of other *E. coli* strains, which can be detected by dark-field imaging with a detection limit of 2×10^4 CFU/ml and duration of 15–30 min (Xu et al. 2012). Cationic AuNPs with quaternary amine head groups were electrostatically bound to β-galactosidase, thus inhibiting the enzyme activity (Miranda et al. 2011). When the nanoparticle binds with bacteria, the enzyme gets released enabling colorimetric readout. AuNPs are widely known to quench fluorescence of any molecule linked with it. Studies have reported to conjugate fluorophores to AuNPs, which upon binding to bacteria release the fluorophores, thus enhancing fluorescence of the molecule. The displacement is due to the hydrophobic spots on the bacterial cell surface. The fluorescence patterns were able to discern difference in the bacterial strains due to change in surface chemistries (Phillips et al. 2008; Bunz and Rotello 2010). Stabilized flower-shaped AuNP-assisted multiplex PCR assay was able to simultaneously detect *Salmonella typhimurium*, *Listeria monocytogenes*, and *E. coli* (Du et al. 2020). PCR product obtained, when mixed with AuNPs and NaCl, can form a colorimetric product within 10 min, which can be visualized by the naked eye, and the detection limit was 3.125 ng/ μl. A fluorescent probe, poly[9,9′-bis(6''- (N,N,N-trimethylammonium) hexyl)fluorine-co-alt-4,7-(2,1,3-benzothiadiazole)dibromide], was encapsulated in maleimide-functionalized NPs. Thiolated vancomycin was attached to the NPs to target MRSA (Norouz Dizaji et al. 2020). The NP was able to successfully bind to its target bacteria in mice model, but not to its non-target, *E. coli*, and hence can be used to detect bacterial infections with specificity. Vancomycin-modified $Fe_3O_4@Au$ NPs were able to enrich carbapenem-resistant *Enterobacteriaceae* (CRE) in urine, and the interaction of vancomycin-cell wall was exploited to separate CRE. Carbapenemases in urine were quantified by the change in pH due to alteration in concentration of hydrogen ions caused by imipenem hydrolysis (Wang et al. 2019a). This can determine the concentration of CRE in urine with a detection limit of 1×10^3 CFU/ml in 3.5 h. Magnetic nanoparticles modified with 3-aminopropyltriethoxysilane were covalently linked to a β-lactam antibiotic, amoxicillin. The modified antibiotic binds to the penicillin-binding proteins (PBPs) of bacteria, which can be separated by external magnetic field (Hasan et al. 2016). High molecular PBPs were detected by MALDI MS (Matrix assisted laser desorption/ionization Mass spectrometry), which contained 10^4 and 10^3 CFU/ ml of *S. aureus* and *E. coli*, respectively. Iron oxide magnetic nanoparticles functionalized with chlorin e6 and species-specific aptamers were able to identify and enrich bacteria of different bacterial species (*S. aureus* and *E. coli*) in blood of mice, which can be detected under fluorescent microscope (Wang et al. 2018a). Analysis of amine, methyl, and carboxylic acid functionalization on T4 bacterio-phage adsorbed to silane-functionalized Fe_3O_4 revealed that bare iron oxide and amine-functionalized nanoparticles were able to capture threefold more *E. coli* (Liana et al. 2017). Nevertheless, an incubation temperature of 37 °C and tryptophan-rich tryptone in the media are mandatory for long, irreversible binding of phage to bacteria. *P. aeruginosa*-specific virulent bacteriophage was isolated and functionalized to tosyl-activated magnetic beads (He et al. 2017). The phage was able to attach to the target and lysed the bacteria, releasing ATP. Luciferase-adenosine triphosphate bioluminescence was used to quantify the captured bacteria with a detection limit of 2×10^2 CFU/ml, and the duration of the entire process was 2 h, including the 100 min replication cycle of the phage. Streptavidin-modified magnetic nanoparticles were functionalized with biotinylated antibodies and further coated with AgNPs (Fargašová et al. 2017). These nanoparticles successfully detected *S. aureus* and *S. pyogenes*, the major pathogens in prosthetic joint infections, detected by magnetically assisted surface-enhanced Raman spectroscopy (MA-SERS). Some nanoparticles have shown to exhibit theranostic property – diagnosis with therapy. Gold nanorods (68 nm $\times$ 18 nm) selectively killed *P. aeruginosa* by labelling them with polyclonal IgG antibodies (obtained from rabbit). Upon NIR radiation of 50 mW for 10 min, which penetrated for 15 cm, 80% of bacterial cells were compromised (Sean Norman et al. 2008). Another study showed that antimicrobial peptide-functionalized magnetic NPs and Au-coated Ag-decorated graphene oxides modified with 4-mercaptophenylboronic acid (4-MPBA) can be used as probes for bacterial isolation and SERS tags, respectively, to develop a sandwich-structured biosensor (Yuan et al. 2018). On interacting with different pathogens – *E. coli*, *S. aureus*, and *P. aeruginosa* – the fingerprint of 4-MPBA alters due to difference in interaction between the SERS tag and the pathogen. The biosensor was able to classify 97.3% of the blood samples from 39 patients and can be used to isolate, discriminate, and kill the bacteria. Plasmonic gold colloids were assembled with P9b bacteriophage that displayed specific peptide to bind *P. aeruginosa* and can be detected based on the differences in Raman spectra within 1 h (Franco et al. 2020). Positively charged fluorescein-labelled trimethyl chitosan nanoparticles were conjugated with DNA aptamer specific to *E. coli* DH5α and were found to selectively bind *E. coli* in vitro, which was detected by fluorescence emission at 520 nm (Zhao et al. 2020). Chitosan-coated magnetic nanoparticles (CS-MNPs), when incubated with *E. coli* or *S. aureus*, bind to the negatively charged cell membrane and caused a hindrance in accessibility of negatively charged substrate – 2,2′-azino-bis(3-ethylbenzothiazoline-6-sulfonic acid) diammonium salts – to positively charged CS-MNPs, thus reducing their peroxidase-like activity (Le et al. 2020). This was able to detect 10^4 CFU/ml by the naked eye and 10^2 CFU/ml by spectrophotometry within a span of 10 min. Another recent study showed that nanocomposite film formed with bacterial cellulose, polypyrrole, and TiO_2-Ag was able to detect and measure growth of five pathogenic bacteria – *E. coli*, *S. epidermidis*, *A. hydrophila*, and two strains of *S. aureus* (Ghasemi et al. 2020). Increasing bacterial concentration would decrease the electrical resistance of the sensor, and the change varies with different bacterial types based on the size, Gram type, and environment. The sensitivity of the sensor can be increased by increasing the concentration of polypyrrole and TiO_2-Ag. Photoluminescence-based ZnO NPs, functionalized with cysteamine, were able to sense NAHLs produced by *P. aeruginosa*, and the linear detection range was 10–120 nM in artificial urine medium, with 97% sensitivity (Vasudevan et al.

2020). Quantum dots have been widely reported for their ability to detect pathogens. CdSe/ZnS core/shell quantum dots (QDs) conjugated with streptavidin exhibited 2 orders of magnitude more sensitivity relative to conventional dye-based detection of *E. coli* (Hahn et al. 2005). Upon excitation, QDs retain their fluorescence for a long time leading to accurate detection of single cell. Another group designed a convective PCR (cPCR) in combination with nucleic acid lateral flow assay to detect MRSA (Rajendran et al. 2019). Multiplex PCR was performed using fluorescein- and digoxigenin-modified primers of *femA* and *mecA*, respectively, to differentiate MRSA from methicillin-sensitive strains. Amplification was achieved in less than 30 min, the amplicons can be analyzed using duplex lateral flow assay using QD-labelled reporter probes, and the fluorescence signal can be acquired in a smartphone camera. This detected MRSA with a limit of 4.7×10^3 copies of DNA and can be used to rapidly identify the pathogen (Rajendran et al. 2019).

7.9   Other Contributions of Nanoparticles to Mitigate Drug Resistance

Se NPs at concentration of 0.9 mg/kg improved the gut health in chicken by increasing beneficial bacteria such as *Lactobacillus* and *Faecalibacterium* and by increasing the short-chain fatty acids (SCFA), which are used as energy source for colonic cells (Gangadoo et al. 2018). This can reduce the use of antibiotics in poultry which in turn can mitigate the indirect overuse of antibiotics. Antibiotic residues in aquatic environment might lead to toxic effects on aquatic ecosystem and become a major source for spread of antibiotic resistance. Ceria nanoparticles were able to exhibit photocatalytic activity and degrade 98% of norfloxacin after 30 min upon UV irradiation (Remani and Binitha 2020). Norfloxacin can interfere in DNA replication, which can lead to mutations and may ultimately cause resistance to drugs. It was found that photocatalytic degradation of norfloxacin using ceria nanoparticles did not lead to any toxic byproducts and the catalyst could also be reused. TiO_2 NPs were modified with zinc phthalocyanine (Znpc-TiO_2) to improve their photocatalytic activity, and degradation of erythromycin was tested to eliminate residual antibiotic in aquatic bodies (Vignesh et al. 2014). Znpc-TiO_2 at 0.4 g/L was able to degrade 74.21% of erythromycin, whereas TiO_2 achieved only 31.57% at 180 min in the presence of visible light. Regenerated photocatalyst was able to degrade ~69% of erythromycin after five recycles. Outer membrane vesicles (OMVs) from carbapenem-resistant *K. pneumoniae* (CRKP) were reinforced by BSA nanoparticles (Wu et al. 2020). Subcutaneous vaccination with these nanoparticles yielded higher CRKP-specific titers, and the survival rate of mice infected with CRKP was higher after immunization. *Vibrio cholerae*, upon accumulation of quorum sensing autoinducer CAI-1, transits to a non-biofilm-forming, avirulent state and produces protease that detaches *V. cholerae* from the intestine (Lu et al. 2015). CAI-1-loaded NPs with PEG as stabilizer and vitamin E as core component were developed using flash nanoprecipitation (FNP), which was able to activate *V. cholerae* QS responses 5 orders of magnitude and was able to diffuse delivery barrier such as intestinal mucus in vivo (Lu et al. 2015). Magnetic core-shell nanoparticles, which are positively charged, trapped multidrug-resistant *S. aureus* and uropathogenic *E. coli* by electrostatic interaction and completely killed them upon exposure to radiofrequency current for 30 min due to loss of membrane potential and dysfunction of membrane-associated complexes (Chaurasia et al. 2016). Such physical antibacterial strategies reduce the possibility of triggering resistance and can be applied to ensure safety in hospitals and among communities.

7.10   Have Bacteria Developed Resistance to Nanoparticles?

Bacteria have excellent adaptation ability in response to their external factors. Some known mechanisms of resistance to nanoparticles are biofilm formation, altering surface charge, ion efflux pumps, mutations, and expression of extracellular proteins or matrix (Salas Orozco et al. 2019). Most reported studies have revolved around resistance development against AgNPs, due to the interest it has gained relative to other nanoparticles. A study showed that bacteria *E. coli* and *P. aeruginosa*, when repeatedly cultured in sub-inhibitory concentration of AgNPs, developed resistance and reduced their antimicrobial activity by increasing their MIC from 1.69–13.5 mg/ L to 432 mg/L after 20th step of sub-culturing, but no resistance was observed on exposure to silver ions (Panáček et al. 2018). The researchers also found aggregation of AgNPs, when exposed to the AgNP-resistant bacteria relative to the sensitive (AgNPs unexposed) strain, leads to resistance for the nanoparticles. This aggregation was not reversed by increasing the concentration of stabilizing agent, and experiments revealed that the aggregation was due to the secretion of flagellin protein which can be reversed by the use of inhibitors of flagellin production (pomegranate rind extract) (Panáček et al. 2018). Production of extracellular substances was seen upon continuous exposure of AgNPs for 180 generations in *E. coli*, which caused agglomeration of nanoparticles and may decrease AgNPs' bactericidal activity (Faghihzadeh et al. 2018). Another study was performed to determine the ability of *P. aeruginosa*, *S. aureus*, and *A. baumannii* to develop resistance to AgNPs relative to ciprofloxacin and silver nitrate (Ellis et al. 2019). It was observed that only *P. aeruginosa* was able to resist AgNPs by passaging them at increasing concentrations of nanoparticles, but all three strains developed resistance against ciprofloxacin, and no resistance was observed for silver nitrate. Presence of citrate-capped AgNPs increased the growth fitness of *E. coli* MG1655 strains after 225 generations (Graves et al. 2015). Genome analysis revealed changes toward AgNPs' resistance started from 100th generation and at 200th generation, 3 mutations were observed, which might have caused resistance to AgNPs. Survival rates of *E. coli* with extracellular polymeric substances (EPS) in the presence of ZnO NPs and SiO_2 NPs were 65% and 79%, respectively (Wang et al. 2016). Removal of EPS by sonication leads to the survival rate of 11% and 63% in the presence of ZnO NPs and SiO_2 NPs, respectively. TEM and FTIR analysis showed protein-like substances and C=O groups were the binding sites for sequestering nanoparticles (Wang et al. 2016). Copper oxide nanoparticle-activated lysogenic bacteriophage accumulated nitrite and increased N_2O emissions in *P. aeruginosa* (Guo et al. 2017). In addition, exposure to CuO NPs upregulated genes responsible for copper resistance, RND pumps, P-type ATPase efflux, and cation diffusion facilitator transporters. Hence, the combined and ultimate fate of using nanoparticle against resistant traits is variable and has to be analyzed thoroughly with caution before claiming it to be a therapy. It, thus, becomes unavoidable to evaluate the risk of resistance development after prolonged exposure to nanoparticles, or we may revert to the same cycle of searching a potential agent to tackle resistant pathogens.

7.11 Every Rose Has Its Thorn: NPs Worsen AMR Condition

Any therapy would have its downside, besides its extraordinary bustle. Nanomaterials are in no way an exception. Few studies have shown that exposure to nanoparticles has increased the virulence of pathogens, and substantial reports have shown enhancement in transfer of ARGs upon exposure to NPs, leading to a paradoxical state. Most nanoparticles induce ROS generation, which, at sublethal concentrations, can stimulate defense mechanisms in bacteria due to hormesis effect (Iavicoli et al. 2010). Combination of cerium oxide and iron oxide (maghemite) nanoparticles was tested against a panel of Gram-positive and Gram-negative strains that included resistant strains too. Contrary to the well-known antibacterial activities of nanoparticles, ceria NPs and iron oxide NPs did not have any inhibitory activity on any of the strains in both planktonic and biofilm states. Unexpectedly, these nanoparticles, when combined with ciprofloxacin, reduced the antibacterial activity of the antibiotic against the strains (Masadeh et al. 2015). Another study reported that sublethal concentrations (0.5 and 2 mg/L) of CeO_2 NPs enhanced the biofilm biomass after 36 h (Xu et al. 2019). In addition, bacterial diversity of biofilm formation was affected upon NP exposure, with *Citrobacter* and *Pseudomonas* dominating the biofilm. NPs also upregulated QS signals – NAHL and AI-2 – by causing an increase in ROS and stimulated inherent resistance after 4 h with reduced ATP content (Xu et al. 2019). ZnO NPs disrupted signal perception and response in QS system, and TiO_2 NPs and AgNPs affected autoinducer biosynthesis in two different strains of *C. violaceum* (Gómez-Gómez et al. 2019b). These metal-based NPs when released in the environment can disturb the QS-based communication in microbial communities that plays a major role in environmental and technological process. Nanoalumina promoted conjugative transfer of RP4 plasmid from *E. coli* to *Salmonella* spp. by 200-fold (Qiu et al. 2012). Oxidative stress, cell membrane damage, enhancing expression of mating pair formation genes, and downregulating the expression of conjugative transfer regulators are the possible mechanisms for conjugation enhancement by nanoalumina when they are released to the environment, specifically in water treatment systems. A recent study revealed that AgNPs and silver ions have shown to facilitate horizontal transfer of plasmid-mediated antibiotic resistance genes from *E. coli* to *Pseudomonas putida* at sublethal concentrations due to ROS generation, SOS response, and membrane damage (Lu et al. 2020b). CuO NPs and copper ions, at sub-inhibitory and environmentally relevant concentrations (1–100 μmol/L), significantly enhanced plasmid-encoded resistance from *E. coli* to *P. putida* (Zhang et al. 2019). ROS generation, cell membrane damage, and upregulation of pilus generation genes were observed on exposure to Cu^+ ions and CuO NPs. ZnO NPs at sublethal concentrations (1 to 10 mg/L) increased conjugative frequency up to 24.3-fold in pure *E. coli* culture and 8.3-fold in mixed aquatic microbiota in 24 h (Wang et al. 2018b). Transformation efficiency of naked plasmid pGEX4T-1 was enhanced by threefold in *E. coli* on exposure to ZnO NPs. Natural sphalerite nanoparticles promoted HGT of ARG among different *E. coli* strains, which increased with increasing concentration, though the significance in aquatic system was not fully understood (Li et al. 2020a). Contrary to the aforementioned reports, a recent study has shown that Cu, Zn, CuO, and ZnO NPs attenuated ARG transfer, with metal oxide nanoparticles producing pronounced activity (Su et al. 2019). The nanoparticles were able to enter the bacterial cell, induce ROS generation, and increase membrane permeability. Dissolved metal ions and growth inhibition by NPs decreased transfer of ARGs by exerting metal stress and reduction in population density (Su et al. 2019). In short, nanoparticles can either induce ROS generation and increase membrane permeability, thus causing an increase in transfer of ARGs, or decrease HGT by causing damage to the DNA.

Apart from the downside of nanoparticles in involvement of bacterial resistance, there are equal reports on common toxic effects in therapeutic usage of nanoparticles leading to nano-toxicology, which, in itself, is a huge topic of debate against its immense beneficial activities. Most metal nanoparticles have shown cytotoxicity owing to inflammatory response and ROS generation, perforate cell membrane, damage cytoskeletal components, affect transcription by damaging DNA, disturb metabolism of mitochondria, and interfere in lysosome formation, thus impeding autophagy (Sukhanova et al. 2018). Multiple factors are involved in toxic fate of the nanoparticles. Few of them are capping or stabilizing agents, shape/structure of nanoparticles, concentration, solubility, size, mode of synthesis, and surface functionalization (Hossain et al. 2011). There exists a plethora of studies to exhibit the antibacterial nature of nanoparticles, but only a handful of reports focus on long-term toxicity induced by persistence and accumulation of nanoparticles in the environment. To hone the beneficial effect of nanoparticles, while reducing nanoparticle-induced toxicity, it becomes essential to optimize various aspects like size, shape, and capping and stabilizing agents such that nanoparticles attain inverse quasi-stable state when compared with phages/viruses. In other words, if nanoparticles can be designed such that they are stable while interacting with microbes and attain a less stable/less persistent state upon release into the environment, that will be a highly desirous goal. If the above said goal can be attained by fine-tuning size, shape, capping agent, and thus stability of NPs, use of NPs in therapy is not a far-fetched goal.

7.12   Conclusion

Nanomaterials are one of the promising candidates that can be hired to combat infections caused by multidrug-resistant bacteria. Enormous reports provide convincing proofs about the multimodal action of nanoparticles – as antibacterial agents; as inhibitors of biofilms, efflux pumps, and quorum sensing; and ability to cure plasmids. Detection of infectious pathogens and ability to deliver antibiotics in sustained manner are its added assets. Nevertheless, possibilities of being put to therapeutic use are hindered due to the toxic nature of some nanoparticles. Reports also show that nanoparticles can enhance antibiotic resistance under some conditions by enhancement of horizontal gene transfer. In some cases, bacteria have also developed mechanisms to resist the activity of nanoparticles. These shortcomings have to be addressed to ensure use of NPs as therapeutic interventions. In addition, if stability of nanoparticles can be engineered such that they maintain stability while interacting with microbes and concomitantly display reduced stability/persistence in the environment, then engineered NPs will become a boon to healthcare sector to deal with microbial infections caused by both planktonic cells and biofilms.

References

Abdolrasoul R, Bagheri H, Ghazvini K, Boruziniat A, Darroudi M (2019) Synthesis and antibacterial activity of colloidal selenium nanoparticles in chitosan solution: a new antibacterial agent. Mater Res Express 6:105356. https://doi.org/10.1088/2053-1591

Abdulkareem EH, Memarzadeh K, Allaker RP et al (2015) Anti-biofilm activity of zinc oxide and hydroxyapatite nanoparticles as dental implant coating materials. J Dent 43:1462–1469. https://doi.org/10.1016/j.jdent.2015.10.010

Abisado RG, Benomar S, Klaus JR et al (2018) Bacterial quorum sensing and microbial community interactions. MBio 9:e02331

Alpaslan E, Geilich BM, Yazici H, Webster TJ (2017) PH-controlled cerium oxide nanoparticle inhibition of both gram-positive and gram-negative Bacteria growth. Sci Rep 7:1–12. https://doi.org/10.1038/srep45859

Al-Shabib NA, Husain FM, Ahmad N et al (2018) Facile synthesis of tin oxide hollow Nanoflowers interfering with quorum sensing-regulated functions and bacterial biofilms. J Nanomater 2018:1–11. https://doi.org/10.1155/2018/6845026

Anisimova YV, Gelperina SI, Peloquin CA, Heifets LB (2000) Nanoparticles as antituberculosis drugs carriers: effect on activity against Mycobacterium tuberculosis in human monocyte-derived macrophages. J Nanopart Res 2:165–171. https://doi.org/10.1023/A:1010061013365

Applerot G, Lellouche J, Perkas N et al (2012) ZnO nanoparticle-coated surfaces inhibit bacterial biofilm formation and increase antibiotic susceptibility. RSC Adv 2:2314–2321. https://doi.org/10.1039/c2ra00602b

Armijo LM, Wawrzyniec SJ, Kopciuch M et al (2020) Antibacterial activity of iron oxide, iron nitride, and tobramycin conjugated nanoparticles against Pseudomonas aeruginosa biofilms. J Nanobiotechnol 18:1–27. https://doi.org/10.1186/s12951-020-0588-6

Ashajyothi C, Harish KH, Dubey N, Chandrakanth RK (2016) Antibiofilm activity of biogenic copper and zinc oxide nanoparticles-antimicrobials collegiate against multiple drug resistant bacteria: a nanoscale approach. J Nanostruct Chem 6:329–341. https://doi.org/10.1007/s40097-016-0205-2

Ayaz Ahmed KB, Raman T, Anbazhagan V (2016) Platinum nanoparticles inhibit bacteria prolif-eration and rescue zebrafish from bacterial infection. RSC Adv 6:44415–44424. https://doi.org/10.1039/c6ra03732a

Badawy MSEM, Riad OKM, Taher FA, Zaki SA (2020) Chitosan and chitosan-zinc oxide nanocomposite inhibit expression of LasI and RhlI genes and quorum sensing dependent virulence factors of Pseudomonas aeruginosa. Int J Biol Macromol 149:1109–1117. https://doi.org/10.1016/j.ijbiomac.2020.02.019

Baek JS, Tan CH, Ng NKJ et al (2018) A programmable lipid-polymer hybrid nanoparticle system for localized, sustained antibiotic delivery to gram-positive and gram-negative bacterial biofilms. Nanoscale Horizons 3:305–311. https://doi.org/10.1039/c7nh00167c

Baelo A, Levato R, Julián E et al (2015) Disassembling bacterial extracellular matrix with DNase-coated nanoparticles to enhance antibiotic delivery in biofilm infections. J Control Release 209:150–158. https://doi.org/10.1016/j.jconrel.2015.04.028

Bala Subramaniyan S, Senthilnathan R, Arunachalam J, Anbazhagan V (2020) Revealing the significance of the glycan binding property of Butea monosperma seed lectin for enhancing the Antibiofilm activity of silver nanoparticles against Uropathogenic Escherichia coli. Bioconjug Chem 31:139–148. https://doi.org/10.1021/acs.bioconjchem.9b00821

Balakrishnan S, Ibrahim KS, Duraisamy S et al (2020) Antiquorum sensing and antibiofilm potential of biosynthesized silver nanoparticles of Myristica fragrans seed extract against MDR Salmonella enterica serovar Typhi isolates from asymptomatic typhoid carriers and typhoid patients. Environ Sci Pollut Res 27:2844–2856. https://doi.org/10.1007/s11356-019-07169-5

Balasegaram M, Piddock LJV (2020) The global antibiotic Research and Development Partnership (GARDP) not-for-profit model of antibiotic development. ACS Infect Dis 6:1295–1298

Baptista PV, McCusker MP, Carvalho A et al (2018) Nano-strategies to fight multidrug resistant bacteria-"a Battle of the titans". Front Microbiol 9:1441

Bawaskar M, Gaikwad S, Ingle A et al (2010) A new report on Mycosynthesis of silver nanoparticles by Fusarium culmorum. Curr Nanosci 6:376–380. https://doi.org/10.2174/157341310791658919

Bellio P, Luzi C, Mancini A et al (2018) Cerium oxide nanoparticles as potential antibiotic adjuvant. Effects of CeO_2 nanoparticles on bacterial outer membrane permeability. Biochim Biophys Acta Biomembr 1860:2428–2435. https://doi.org/10.1016/j.bbamem.2018.07.002

Bharathan S, Sundaramoorthy NS, Chandrasekaran H et al (2019) Sub lethal levels of platinum nanoparticle cures plasmid and in combination with carbapenem, curtails carbapenem resistant Escherichia coli. Sci Rep 9:5305. https://doi.org/10.1038/s41598-019-41489-3

Bhatia E, Banerjee R (2020) Hybrid silver-gold nanoparticles suppress drug resistant polymicrobial biofilm formation and intracellular infection. J Mater Chem B 8:4890–4898. https://doi.org/10.1039/D0TB00158A

Bhushan M, Kumar Y, Periyasamy L, Viswanath AK (2019) Fabrication and a detailed study of antibacterial properties of αFe_2O_3/NiO nanocomposites along with their structural, optical, thermal, magnetic and cytotoxic features nanotechnology. Nanotechnology 30(18):185101. https://doi.org/10.1088/1361-6528

Bunz UHF, Rotello VM (2010) Gold nanoparticle-fluorophore complexes: sensitive and discerning "noses" for biosystems sensing. Angew Chemie Int Ed 49:3268–3279

Bushra U, Nousheen A, Shamaila S et al (2020) Targeting microbial biofilms: by Arctium lappa l. synthesised biocompatible CeO2-NPs encapsulated in nano-chitosan. IET Nanobiotechnol 14 (3):217–223

Chaurasia AK, Thorat ND, Tandon A et al (2016) Coupling of radiofrequency with magnetic nanoparticles treatment as an alternative physical antibacterial strategy against multiple drug resistant bacteria. Sci Rep 6:1–13. https://doi.org/10.1038/srep33662

Chen X, Hu B, Xiang Q et al (2016) Magnetic nanoparticles modified with quaternarized N-halamine based polymer and their antibacterial properties. J Biomater Sci Polym Ed 27:1187–1199. https://doi.org/10.1080/09205063.2016.1188471

Chen X, Liu Y, Lin A et al (2018) Folic acid-modified mesoporous silica nanoparticles with pH-responsiveness loaded with amp for an enhanced effect against anti-drug-resistant bacteria by overcoming efflux pump systems. Biomater Sci 6:1923–1935. https://doi.org/10.1039/c8bm00262b

Cheow WS, Hadinoto K (2012) Lipid-polymer hybrid nanoparticles with rhamnolipid-triggered release capabilities as anti-biofilm drug delivery vehicles. Particuology 10:327–333. https://doi.org/10.1016/j.partic.2011.08.007

Christena LR, Mangalagowri V, Pradheeba P et al (2015) Copper nanoparticles as an efflux pump inhibitor to tackle drug resistant bacteria. RSC Adv 5:12899–12909. https://doi.org/10.1039/C4RA15382K

Chung S, Zhou R, Webster TJ (2020) Green synthesized BSA-coated selenium nanoparticles inhibit bacterial growth while promoting mammalian cell growth. Int J Nanomed 15:115–124. https://doi.org/10.2147/IJN.S193886

Ciepluch K, Skrzyniarz K, Barrios-Gumiel A et al (2019) Dendronized silver nanoparticles as bacterial membrane Permeabilizers and their interactions with P. aeruginosa lipopolysaccharides, lysozymes, and phage-derived Endolysins. Front Microbiol 10:2771. https://doi.org/10.3389/fmicb.2019.02771

Committee on New Directions in the Study of Antimicrobiol Therapeutics: New Classes of Antimicrobials (2006) Challenges for the development of new antimicrobials-rethinking the approaches. National Academic Press, Washington, DC

Costerton JW, Stewart PS, Greenberg EP (1999) Bacterial biofilms: a common cause of persistent infections. Science 284:1318–1322

Davies D (2003) Understanding biofilm resistance to antibacterial agents. Nat Rev Drug Discov 2:114–122

de Andrade LF, Apolinário AC, Rangel-Yagui CO et al (2020) Chitosan nanoparticles for the delivery of a new compound active against multidrug-resistant Staphylococcus aureus. J Drug Deliv Sci Technol 55:101363. https://doi.org/10.1016/j.jddst.2019.101363

Dev Sharma K (2017) Antibacterial activity of biogenic platinum nanoparticles: an in vitro study. Int J Curr Microbiol App Sci 6:801–808. https://doi.org/10.20546/ijcmas.2017.602.089

Devi LS, Joshi SR (2012) Antimicrobial and synergistic effects of silver nanoparticles synthesized using: soil fungi of high altitudes of eastern Himalaya. Mycobiology 40:27–34. https://doi.org/10.5941/MYCO.2012.40.1.027

Du J, Wu S, Niu L et al (2020) A gold nanoparticles-assisted multiplex PCR assay for simultaneous detection of: Salmonella typhimurium, listeria monocytogenes and Escherichia coli O157:H7. Anal Methods 12:212–217. https://doi.org/10.1039/c9ay02282a

Ellis DH, Maurer-Gardner EI, Sulentic CEW, Hussain SM (2019) Silver nanoparticle antibacterial efficacy and resistance development in key bacterial species. Biomed Phys Eng Express 5:015013. https://doi.org/10.1088/2057-1976/AAD5A7

Estevez H, Palacios A, Gil D et al (2020) Antimycobacterial effect of selenium nanoparticles on Mycobacterium tuberculosis. Front Microbiol 11:800. https://doi.org/10.3389/FMICB.2020.00800

Faghihzadeh F, Anaya NM, Astudillo-Castro C, Oyanedel-Craver V (2018) Kinetic, metabolic and macromolecular response of bacteria to chronic nanoparticle exposure in continuous culture. Environ Sci Nano 5:1386–1396. https://doi.org/10.1039/c8en00325d

Fahimmunisha BA, Ishwarya R, AlSalhi MS et al (2020) Green fabrication, characterization and antibacterial potential of zinc oxide nanoparticles using Aloe socotrina leaf extract: a novel drug delivery approach. J Drug Deliv Sci Technol 55:101465. https://doi.org/10.1016/j.jddst.2019.101465

Fargašová A, Balzerová A, Prucek R et al (2017) Detection of prosthetic joint infection based on magnetically assisted surface enhanced Raman spectroscopy. Anal Chem 89:6598–6607. https://doi.org/10.1021/acs.analchem.7b00759

Fernández-Arias M, Boutinguiza M, Del Val J et al (2020) Copper nanoparticles obtained by laser ablation in liquids as bactericidal agent for dental applications. Appl Surf Sci 507:145032. https://doi.org/10.1016/j.apsusc.2019.145032

Fernando SID, Judan Cruz KG (2020) Ethnobotanical biosynthesis of gold nanoparticles and its downregulation of quorum sensing-linked AhyR gene in Aeromonas hydrophila. SN Appl Sci 2:570. https://doi.org/10.1007/s42452-020-2368-1

Foroohimanjili F, Mirzaie A, Hamdi SMM et al (2020) Antibacterial, antibiofilm, and antiquorum sensing activities of phytosynthesized silver nanoparticles fabricated from *Mespilus germanica* extract against multidrug resistance of *Klebsiella pneumoniae* clinical strains. J Basic Microbiol 60:216–230. https://doi.org/10.1002/jobm.201900511

Franco D, De Plano LM, Rizzo MG et al (2020) Bio-hybrid gold nanoparticles as SERS probe for rapid bacteria cell identification. Spectrochim Acta - Part A Mol Biomol Spectrosc 224:117394. https://doi.org/10.1016/j.saa.2019.117394

Fuller MA, Carey A, Whiley H et al (2019) Nanoparticles in an antibiotic-loaded nanomesh for drug delivery. RSC Adv 9:30064–30070. https://doi.org/10.1039/c9ra06398f

Fuqua WC, Winans SC, Greenberg EP (1994) Quorum sensing in bacteria: the LuxR-LuxI family of cell density-responsive transcriptional regulators. J Bacteriol 176:269–275

Gabrielyan L, Hakobyan L, Hovhannisyan A, Trchounian A (2019) Effects of iron oxide (Fe_3O_4) nanoparticles on Escherichia coli antibiotic-resistant strains. J Appl Microbiol 126:1108–1116. https://doi.org/10.1111/jam.14214

Galbadage T, Liu D, Alemany LB et al (2019) Molecular Nanomachines disrupt bacterial Cell Wall, increasing sensitivity of extensively drug-resistant Klebsiella pneumoniae to Meropenem. ACS Nano 13:14377–14387. https://doi.org/10.1021/acsnano.9b07836

Gangadoo S, Dinev I, Chapman J et al (2018) Selenium nanoparticles in poultry feed modify gut microbiota and increase abundance of Faecalibacterium prausnitzii. Appl Microbiol Biotechnol 102:1455–1466. https://doi.org/10.1007/s00253-017-8688-4

Gao W, Chen Y, Zhang Y et al (2018) Nanoparticle-based local antimicrobial drug delivery. Adv Drug Deliv Rev 127:46–57

García-Lara B, Saucedo-Mora MÁ, Roldán-Sánchez JA et al (2015) Inhibition of quorum-sensing-dependent virulence factors and biofilm formation of clinical and environmental *Pseudomonas aeruginosa* strains by ZnO nanoparticles. Lett Appl Microbiol 61:299–305. https://doi.org/10.1111/lam.12456

Gargioni C, Borzenkov M, D'Alfonso L et al (2020) Self-assembled monolayers of copper sulfide nanoparticles on glass as antibacterial coatings. Nano 10:352. https://doi.org/10.3390/nano10020352

Geilich BM, Gelfat I, Sridhar S et al (2017) Superparamagnetic iron oxide-encapsulating polymersome nanocarriers for biofilm eradication. Biomaterials 119:78–85. https://doi.org/10.1016/j.biomaterials.2016.12.011

Geoffrion LD, Hesabizadeh T, Medina-Cruz D et al (2020) Naked selenium nanoparticles for antibacterial and anticancer treatments. ACS Omega 5(6):2660–2669. https://doi.org/10.1021/acsomega.9b03172

Ghasemi S, Bari MR, Pirsa S, Amiri S (2020) Use of bacterial cellulose film modified by polypyrrole/TiO2-ag nanocomposite for detecting and measuring the growth of pathogenic bacteria. Carbohydr Polym 232:115801. https://doi.org/10.1016/j.carbpol.2019.115801

Golmohammadi R, Najar-Peerayeh S, Tohidi Moghadam T, Hosseini SMJ (2020) Synergistic antibacterial activity and wound healing properties of selenium-chitosan-mupirocin Nanohybrid system: an in vivo study on rat diabetic Staphylococcus aureus wound infection model. Sci Rep 10:1–10. https://doi.org/10.1038/s41598-020-59510-5

Gómez-Gómez B, Arregui L, Serrano S et al (2019a) Selenium and tellurium-based nanoparticles as interfering factors in quorum sensing-regulated processes: Violacein production and bacterial biofilm formation. Metallomics 11:1104–1111. https://doi.org/10.1039/c9mt00044e

Gómez-Gómez B, Arregui L, Serrano S et al (2019b) Unravelling mechanisms of bacterial quorum sensing disruption by metal-based nanoparticles. Sci Total Environ 696:133869. https://doi.org/10.1016/j.scitotenv.2019.133869

Gopalakrishnan V, Masanam E, Ramkumar VS et al (2020) Influence of *N*-acylhomoserine lactonase silver nanoparticles on the quorum sensing system of *Helicobacter pylori*: A potential strategy to combat biofilm formation. J Basic Microbiol 60:207–215. https://doi.org/10.1002/jobm.201900537

Graves JL, Tajkarimi M, Cunningham Q et al (2015) Rapid evolution of silver nanoparticle resistance in Escherichia coli. Front Genet 5:42. https://doi.org/10.3389/fgene.2015.00042

Guo J, Gao SH, Lu J et al (2017) Copper oxide nanoparticles induce lysogenic bacteriophage and metal-resistance genes in Pseudomonas aeruginosa PAO1. ACS Appl Mater Interfaces 9:22298–22307. https://doi.org/10.1021/acsami.7b06433

Gupta D, Singh A, Khan AU (2017) Nanoparticles as efflux pump and biofilm inhibitor to rejuvenate bactericidal effect of conventional antibiotics. Nanoscale Res Lett 12(1):454

Gurunathan S, Han JW, Kwon DN, Kim JH (2014) Enhanced antibacterial and anti-biofilm activities of silver nanoparticles against gram-negative and gram-positive bacteria. Nanoscale Res Lett 9:1–17. https://doi.org/10.1186/1556-276X-9-373

Hahn MA, Tabb JS, Krauss TD (2005) Detection of single bacterial pathogens with semiconductor quantum dots. Anal Chem 77:4861–4869. https://doi.org/10.1021/ac050641i

Hasan N, Guo Z, Wu HF (2016) Large protein analysis of Staphylococcus aureus and Escherichia coli by MALDI TOF mass spectrometry using amoxicillin functionalized magnetic nanoparticles. Anal Bioanal Chem 408:6269–6281. https://doi.org/10.1007/s00216-016-9730-6

Hashim FJ (2017) Characterization and biological effect of silver nanoparticles synthesized by Zingiber officinale aqueous extract. Pharmacogn Mag 13:S201–S208

He Y, Wang M, Fan E et al (2017) Highly specific bacteriophage-affinity strategy for rapid separation and sensitive detection of viable Pseudomonas aeruginosa. Anal Chem 89:1916–1921. https://doi.org/10.1021/acs.analchem.6b04389

Hetrick EM, Shin JH, Paul HS, Schoenfisch MH (2009) Anti-biofilm efficacy of nitric oxide-releasing silica nanoparticles. Biomaterials 30:2782–2789. https://doi.org/10.1016/j.biomaterials.2009.01.052

Holmberg SD, Osterholm MT, Senger KA, Cohen ML (1984) Drug-resistant Salmonella from animals fed Antimicrobials. N Engl J Med 311:617–622. https://doi.org/10.1056/NEJM198409063111001

Horev B, Klein MI, Hwang G et al (2015) PH-activated nanoparticles for controlled topical delivery of farnesol to disrupt oral biofilm virulence. ACS Nano 9:2390–2404. https://doi.org/10.1021/nn507170s

Hossain S, Chowdhury EH, Akaike T (2011) Nanoparticles and toxicity in therapeutic delivery: the ongoing debate. Ther Deliv 2:125–132. https://doi.org/10.4155/tde.10.109

Hou X, Zhang X, Zhao W et al (2020) Vitamin lipid nanoparticles enable adoptive macrophage transfer for the treatment of multidrug-resistant bacterial sepsis. Nat Nanotechnol 15:41–46

Hu D, Li H, Wang B et al (2017) Surface-adaptive gold nanoparticles with effective adherence and enhanced Photothermal ablation of methicillin-resistant Staphylococcus aureus biofilm. ACS Nano 11:9330–9339. https://doi.org/10.1021/acsnano.7b04731

Hu M, Chen J, Li ZY et al (2006) Gold nanostructures: engineering their plasmonic properties for biomedical applications. Chem Soc Rev 35:1084–1094

Hu X, Yang B, Zhang W et al (2019) Plasmid binding to metal oxide nanoparticles inhibited lateral transfer of antibiotic resistance genes. Environ Sci Nano 6:1310–1322. https://doi.org/10.1039/c8en01447g

Huang X, Chen X, Chen Q et al (2016) Investigation of functional selenium nanoparticles as potent antimicrobial agents against superbugs. Acta Biomater 30:397–407. https://doi.org/10.1016/j.actbio.2015.10.041

Huh AJ, Kwon YJ (2011) "Nanoantibiotics": a new paradigm for treating infectious diseases using nanomaterials in the antibiotics resistant era. J Control Release 156:128–145

Iavicoli I, Calabrese EJ, Nascarella MA (2010) Exposure to nanoparticles and hormesis. Dose-Response 8:501–517. https://doi.org/10.2203/dose-response.10-016.Iavicoli

Ilk S, Sağlam N, Özgen M, Korkusuz F (2017) Chitosan nanoparticles enhances the anti-quorum sensing activity of kaempferol. Int J Biol Macromol 94:653–662. https://doi.org/10.1016/j.ijbiomac.2016.10.068

Inbakandan D, Kumar C, Abraham LS et al (2013) Silver nanoparticles with anti microfouling effect: a study against marine biofilm forming bacteria. Colloids Surf B Biointerfaces 111:636–643. https://doi.org/10.1016/j.colsurfb.2013.06.048

Iqbal G, Faisal S, Khan S et al (2019) Photo-inactivation and efflux pump inhibition of methicillin resistant Staphylococcus aureus using thiolated cobalt doped ZnO nanoparticles. J Photochem Photobiol B Biol 192:141–146. https://doi.org/10.1016/j.jphotobiol.2019.01.021

Itohiya H, Matsushima Y, Shirakawa S et al (2019) Organic resolution function and effects of platinum nanoparticles on bacteria and organic matter. PLoS One 14:e0222634. https://doi.org/10.1371/journal.pone.0222634

Kadiyala U, Turali-Emre ES, Bahng JH et al (2018) Unexpected insights into antibacterial activity of zinc oxide nanoparticles against methicillin resistant: Staphylococcus aureus (MRSA). Nanoscale 10:4927–4939. https://doi.org/10.1039/c7nr08499d

Kahzad N, Salehzadeh A (2020) Green synthesis of CuFe2O4@ag nanocomposite using the Chlorella vulgaris and evaluation of its effect on the expression of norA efflux pump Gene among Staphylococcus aureus strains. Biol Trace Elem Res 198:1–12. https://doi.org/10.1007/s12011-020-02055-5

Kalaivani R, Maruthupandy M, Muneeswaran T et al (2020) Chitosan mediated gold nanoparticles against pathogenic bacteria, fungal strains and MCF-7 cancer cells. Int J Biol Macromol 146:560–568. https://doi.org/10.1016/j.ijbiomac.2020.01.037

Kang YK, Kwon K, Ryu JS et al (2017) Nonviral genome editing based on a polymer-Derivatized CRISPR Nanocomplex for targeting bacterial pathogens and antibiotic resistance. Bioconjug Chem 28:957–967. https://doi.org/10.1021/acs.bioconjchem.6b00676

Khan SA, Singh AK, Senapati D et al (2011) Bio-conjugated popcorn shaped gold nanoparticles for targeted photothermal killing of multiple drug resistant Salmonella DT104. J Mater Chem 21:17705–17709. https://doi.org/10.1039/c1jm13320a

Kotb E, Ahmed AA, Saleh TA et al (2020) Pseudobactins bounded iron nanoparticles for control of an antibiotic-resistant Pseudomonas aeruginosa ryn32. Biotechnol Prog 36(1):e2907. https://doi.org/10.1002/btpr.2907

Lai BH, Chen DH (2013) Vancomycin-modified LaB6@SiO2/Fe3O4 composite nanoparticles for near-infrared photothermal ablation of bacteria. Acta Biomater 9:7573–7579. https://doi.org/10.1016/j.actbio.2013.03.023

Le TN, Tran TD, Il Kim M (2020) A convenient colorimetric Bacteria detection method utilizing chitosan-coated magnetic nanoparticles. Nano 10:92. https://doi.org/10.3390/nano10010092

Li G, Chen X, Yin H et al (2020a) Natural sphalerite nanoparticles can accelerate horizontal transfer of plasmid-mediated antibiotic-resistance genes. Environ Int 136:105497. https://doi.org/10.1016/j.envint.2020.105497

Li S, Chang R, Chen J et al (2020b) Novel magnetic nanocomposites combining selenium and iron oxide with excellent anti-biofilm properties. J Mater Sci 55:1012–1022. https://doi.org/10.1007/s10853-019-04019-0

Li WR, Xie XB, Shi QS et al (2011) Antibacterial effect of silver nanoparticles on Staphylococcus aureus. Biometals 24:135–141. https://doi.org/10.1007/s10534-010-9381-6

Li X, Robinson SM, Gupta A et al (2014) Functional gold nanoparticles as potent antimicrobial agents against multi-drug-resistant bacteria. ACS Nano 8:10682–10686. https://doi.org/10.1021/nn5042625

Liana AE, Marquis CP, Gunawan C et al (2017) T4 bacteriophage conjugated magnetic particles for E. coli capturing: influence of bacteriophage loading, temperature and tryptone. Colloids Surf B Biointerfaces 151:47–57. https://doi.org/10.1016/j.colsurfb.2016.12.009

Liu L, Xu K, Wang H et al (2009) Self-assembled cationic peptide nanoparticles as an efficient antimicrobial agent. Nat Nanotechnol 4:457–463. https://doi.org/10.1038/nnano.2009.153

Liu Y, Li F, Guo Z et al (2020) Silver nanoparticle-embedded hydrogel as a photothermal platform for combating bacterial infections. Chem Eng J 382:122990. https://doi.org/10.1016/j.cej.2019.122990

Lu C, Xiao Y, Liu Y et al (2020a) Hyaluronic acid-based levofloxacin nanomicelles for nitric oxide-triggered drug delivery to treat bacterial infections. Carbohydr Polym 229:115479. https://doi.org/10.1016/j.carbpol.2019.115479

Lu HD, Spiegel AC, Hurley A et al (2015) Modulating Vibrio cholerae quorum-sensing-controlled communication using autoinducer-loaded nanoparticles. Nano Lett 15:2235–2241. https://doi.org/10.1021/acs.nanolett.5b00151

Lu J, Wang Y, Jin M et al (2020b) Both silver ions and silver nanoparticles facilitate the horizontal transfer of plasmid-mediated antibiotic resistance genes. Water Res 169:115229. https://doi.org/10.1016/j.watres.2019.115229

Manoharan RK, Mahalingam S, Gangadaran P, Ahn YH (2020) Antibacterial and photocatalytic activities of 5-nitroindole capped bimetal nanoparticles against multidrug resistant bacteria. Colloids Surf B Biointerfaces 188:110825. https://doi.org/10.1016/j.colsurfb.2020.110825

Masadeh MM, Karasneh GA, Al-Akhras MA et al (2015) Cerium oxide and iron oxide nanoparticles abolish the antibacterial activity of ciprofloxacin against gram positive and gram negative biofilm bacteria. Cytotechnology 67:427–435. https://doi.org/10.1007/s10616-014-9701-8

Hashimoto M, Yanagiuchi H, Kitagawa H, Honda Y (2017) Inhibitory effect of platinum nanoparticles on biofilm formation of oral bacteria. Nano Biomed 9:77–82

McMurry L, Petrucci RE, Levy SB (1980) Active efflux of tetracycline encoded by four genetically different tetracyline resistance determinants in Escherichia coli. Proc Natl Acad Sci U S A 77:3974–3977. https://doi.org/10.1073/pnas.77.7.3974

Mhawesh AA, Aal Owaif HA, Abdulateef SA (2019) In vitro experimental research for using the silver nanoparticles as plasmid curing agent in some types of multi-antibiotic resistant pathogenic bacteria. Indian J Public Heal Res Dev 10:2448–2454. https://doi.org/10.5958/0976-5506.2019.02233.2

Michelle MCB, Maria Laura C, Laura JVP (2018) Strategies to combat antimicrobial resistance: anti-plasmid and plasmid curing. FEMS Microbiol Rev 42:781–804. https://doi.org/10.1093/FEMSRE

Miranda OR, Li X, Garcia-Gonzalez L et al (2011) Colorimetric bacteria sensing using a supramolecular enzyme-nanoparticle biosensor. J Am Chem Soc 133:9650–9653. https://doi.org/10.1021/ja2021729

Mistry N, Bandyopadhyaya R, Mehra S (2020) ZnO nanoparticles and rifampicin synergistically damage the membrane of mycobacteria. ACS Appl Nano Mater 3(4):3174–3184. https://doi.org/10.1021/acsanm.9b02089

Mocan L, Tabaran FA, Mocan T et al (2017) Laser thermal ablation of multidrug-resistant bacteria using functionalized gold nanoparticles. Int J Nanomed 12:2255–2263. https://doi.org/10.2147/IJN.S124778

Mohammadi G, Nokhodchi A, Barzegar-Jalali M et al (2011) Physicochemical and anti-bacterial performance characterization of clarithromycin nanoparticles as colloidal drug delivery system. Colloids Surf B Biointerfaces 88:39–44. https://doi.org/10.1016/j.colsurfb.2011.05.050

Montazeri A, Salehzadeh A, Zamani H (2020) Effect of silver nanoparticles conjugated to thiosemicarbazide on biofilm formation and expression of intercellular adhesion molecule genes, icaAD, in Staphylococcus aureus. Folia Microbiol 65:153–160. https://doi.org/10.1007/s12223-019-00715-1

Nafee N, Husari A, Maurer CK et al (2014) Antibiotic-free nanotherapeutics: ultra-small, mucus-penetrating solid lipid nanoparticles enhance the pulmonary delivery and anti-virulence efficacy of novel quorum sensing inhibitors. J Control Release 192:131–140. https://doi.org/10.1016/j.jconrel.2014.06.055

Naik K, Kowshik M (2014) Anti-quorum sensing activity of AgCl-TiO$_2$ nanoparticles with potential use as active food packaging material. J Appl Microbiol 117:972–983. https://doi. org/10.1111/jam.12589

Nature (2013) The antibiotic alarm. Nature 495:141

Neill JO (2014) Antimicrobial Resistance: Tackling a crisis for the health and wealth of nations. Wellcome Trust

Nejabatdoust A, Zamani H, Salehzadeh A (2019) Functionalization of ZnO nanoparticles by glutamic acid and conjugation with Thiosemicarbazide alters expression of efflux pump genes in multiple drug-resistant Staphylococcus aureus strains. Microb Drug Resist 25:966–974. https://doi.org/10.1089/mdr.2018.0304

Niemirowicz K, Surel U, Wilczewska AZ et al (2015) Bactericidal activity and biocompatibility of ceragenin-coated magnetic nanoparticles. J Nanobiotechnol 13:32. https://doi.org/10.1186/s12951-015-0093-5

Norouz Dizaji A, Ding D, Kutsal T et al (2020) In vivo imaging/detection of MRSA bacterial infections in mice using fluorescence labelled polymeric nanoparticles carrying vancomycin as the targeting agent. J Biomater Sci Polym Ed 31:293–309. https://doi.org/10.1080/09205063.2019.1692631

Ocsoy I, Yusufbeyoglu S, Yılmaz V et al (2017) DNA aptamer functionalized gold nanostructures for molecular recognition and photothermal inactivation of methicillin-resistant Staphylococcus aureus. Colloids Surf B Biointerfaces 159:16–22. https://doi.org/10.1016/j.colsurfb.2017.07.056

Padwal P, Bandyopadhyaya R, Mehra S (2015) Biocompatible citric acid-coated iron oxide nanoparticles to enhance the activity of first-line anti-TB drugs in Mycobacterium smegmatis. J Chem Technol Biotechnol 90:1773–1781. https://doi.org/10.1002/jctb.4766

Pallavicini P, Donà A, Taglietti A et al (2014) Self-assembled monolayers of gold nanostars: a convenient tool for near-IR photothermal biofilm eradication. Chem Commun 50:1969–1971. https://doi.org/10.1039/c3cc48667b

Palpperumal S, Sankaralingam S, Kathiresan D et al (2016) Partial purification and characterization of a Haloalkaline protease from Pseudomonas aeruginosa. Br Microbiol Res J 15:1–7. https://doi.org/10.9734/bmrj/2016/26289

Panáček A, Kvítek L, Smékalová M et al (2018) Bacterial resistance to silver nanoparticles and how to overcome it. Nat Nanotechnol 13:65–71. https://doi.org/10.1038/s41565-017-0013-y

Patra P, Mitra S, Debnath N et al (2014) Ciprofloxacin conjugated zinc oxide nanoparticle: a camouflage towards multidrug resistant bacteria. Bull Mater Sci 37:199–206

Pereira CS, Thompson JA, Xavier KB (2013) AI-2-mediated signalling in bacteria. FEMS Microbiol Rev 37:156–181. https://doi.org/10.1111/j.1574-6976.2012.00345.x

Perni S, Piccirillo C, Pratten J et al (2009) The antimicrobial properties of light-activated polymers containing methylene blue and gold nanoparticles. Biomaterials 30:89–93. https://doi.org/10.1016/j.biomaterials.2008.09.020

Phillips RL, Miranda OR, You CC et al (2008) Rapid and efficient identification of bacteria using gold-nanoparticle-poly(para-phenyleneethynylene) constructs. Angew Chemie Int Ed 47:2590–2594. https://doi.org/10.1002/anie.200703369

Poole K (2007) Efflux pumps as antimicrobial resistance mechanisms. Ann Med 39:162–176

Pornpattananangkul D, Olson S, Aryal S et al (2010) Stimuli-responsive liposome fusion mediated by gold nanoparticles. ACS Nano 4:1935–1942. https://doi.org/10.1021/nn9018587

Prakash P, Gnanaprakasam P, Emmanuel R et al (2013) Green synthesis of silver nanoparticles from leaf extract of Mimusops elengi, Linn. for enhanced antibacterial activity against multidrug resistant clinical isolates. Colloids Surf B Biointerfaces 108:255–259. https://doi.org/10.1016/j.colsurfb.2013.03.017

Prateeksha SBR, Shoeb M et al (2017) Scaffold of selenium Nanovectors and honey phytochemicals for inhibition of Pseudomonas aeruginosa quorum sensing and biofilm formation. Front Cell Infect Microbiol 7:93. https://doi.org/10.3389/fcimb.2017.00093

Qais FA, Shafiq A, Ahmad I et al (2020) Green synthesis of silver nanoparticles using Carum copticum: assessment of its quorum sensing and biofilm inhibitory potential against gram negative bacterial pathogens. Microb Pathog 2020:104172. https://doi.org/10.1016/j.micpath. 2020.104172

Qin H, Cao H, Zhao Y et al (2014) In vitro and in vivo anti-biofilm effects of silver nanoparticles immobilized on titanium. Biomaterials 35:9114–9125. https://doi.org/10.1016/j.biomaterials. 2014.07.040

Qin X, Engwer C, Desai S et al (2017) An investigation of the interactions between an E. coli bacterial quorum sensing biosensor and chitosan-based nanocapsules. Colloids Surf B Biointerfaces 149:358–368. https://doi.org/10.1016/j.colsurfb.2016.10.031

Qiu Z, Yu Y, Chen Z et al (2012) Nanoalumina promotes the horizontal transfer of multiresistance genes mediated by plasmids across genera. Proc Natl Acad Sci U S A 109:4944–4949. https:// doi.org/10.1073/pnas.1107254109

Raffi M, Mehrwan S, Bhatti TM et al (2010) Investigations into the antibacterial behavior of copper nanoparticles against Escherichia coli. Ann Microbiol 60:75–80. https://doi.org/10.1007/ s13213-010-0015-6

Raftery TD, Kerscher P, Hart AE et al (2014) Discrete nanoparticles induce loss of Legionella pneumophila biofilms from surfaces. Nanotoxicology 8:477–484. https://doi.org/10.3109/ 17435390.2013.796537

Rai A, Prabhune A, Perry CC (2010) Antibiotic mediated synthesis of gold nanoparticles with potent antimicrobial activity and their application in antimicrobial coatings. J Mater Chem 20:6789–6798. https://doi.org/10.1039/c0jm00817f

Rai M, Kon K, Ingle A et al (2014) Broad-spectrum bioactivities of silver nanoparticles: the emerging trends and future prospects. Appl Microbiol Biotechnol 98:1951–1961

Rajendran VK, Bakthavathsalam P, Bergquist PL, Sunna A (2019) Smartphone detection of antibiotic resistance using convective PCR and a lateral flow assay. Sens Actuators B Chem 298:126849. https://doi.org/10.1016/j.snb.2019.126849

Ramezani Ali Akbari K, Abdi Ali A (2017) Study of antimicrobial effects of several antibiotics and iron oxide nanoparticles on biofilm producing pseudomonas aeruginosa. Nanomed J 4:37–43. https://doi.org/10.22038/nmj.2017.8051

Read AF, Woods RJ (2014) Antibiotic resistance management. Evol Med Public Heal 2014:147. https://doi.org/10.1093/emph/eou024

Remani KC, Binitha NN (2020) Photocatalytic degradation of norfloxacin under UV, visible and solar light using ceria nanoparticles. Mater Today Proc 25:246–251. https://doi.org/10.1016/j. matpr.2020.01.212

Rémy B, Mion S, Plener L et al (2018) Interference in bacterial quorum sensing: a biopharmaceutical perspective. Front Pharmacol 9:203

Rodríguez-Serrano C, Guzmán-Moreno J, Ángeles-Chávez C et al (2020) Biosynthesis of silver nanoparticles by Fusarium scirpi and its potential as antimicrobial agent against uropathogenic Escherichia coli biofilms. PLoS One 15:e0230275. https://doi.org/10.1371/journal.pone. 0230275

Ruiz AL, Garcia CB, Gallón SN, Webster TJ (2020) Novel silver-platinum nanoparticles for anticancer and antimicrobial applications. Int J Nanomed 15:169–179. https://doi.org/10. 2147/IJN.S176737

Sadalage PS, Nimbalkar MS, Sharma KKK et al (2020) Sustainable approach to almond skin mediated synthesis of tunable selenium microstructures for coating cotton fabric to impart specific antibacterial activity. J Colloid Interface Sci 569:346–357. https://doi.org/10.1016/j. jcis.2020.02.094

Salas Orozco MF, Niño-Martínez N, Martínez-Castañón GA et al (2019) Molecular mechanisms of bacterial resistance to metal and metal oxide nanoparticles. Int J Mol Sci 20(11):2808

Salman M, Rizwana R, Khan H et al (2019) Synergistic effect of silver nanoparticles and polymyxin B against biofilm produced by *Pseudomonas aeruginosa* isolates of pus samples *in vitro*. Artif Cells Nanomed Biotechnol 47:2465–2472. https://doi.org/10.1080/21691401.2019.1626864

San Millan A (2018) Evolution of plasmid-mediated antibiotic resistance in the clinical context. Trends Microbiol 26:978–985

Sanyasi S, Majhi RK, Kumar S et al (2016) Polysaccharide-capped silver nanoparticles inhibit biofilm formation and eliminate multi-drug-resistant bacteria by disrupting bacterial cytoskeleton with reduced cytotoxicity towards mammalian cells. Sci Rep 6:1–16. https://doi.org/10.1038/srep24929

Sean Norman R, Stone JW, Gole A et al (2008) Targeted photothermal lysis of the pathogenic bacteria, pseudomonas aeruginosa, with gold nanorods. Nano Lett 8:302–306. https://doi.org/10.1021/nl0727056

Shakibaie M, Forootanfar H, Golkari Y et al (2015) Anti-biofilm activity of biogenic selenium nanoparticles and selenium dioxide against clinical isolates of Staphylococcus aureus, Pseudomonas aeruginosa, and Proteus mirabilis. J Trace Elem Med Biol 29:235–241. https://doi.org/10.1016/j.jtemb.2014.07.020

Sharma M, Sharma R, Jain DK (2016) Nanotechnology based approaches for enhancing Oral bioavailability of poorly water soluble antihypertensive drugs. Scientifica 2016:8525679. https://doi.org/10.1155/2016/8525679

Shi T, Wei Q, Wang Z et al (2019) Photocatalytic protein damage by silver nanoparticles circumvents bacterial stress response and multidrug resistance. mSphere 4(3):e00175. https://doi.org/10.1128/msphere.00175-19

Siddiqi KS, Husen A (2016) Fabrication of metal nanoparticles from Fungi and metal salts: scope and application. Nanoscale Res Lett 11:1–15

Siddiqi KS, Husen A, Rao RAK (2018) A review on biosynthesis of silver nanoparticles and their biocidal properties. J Nanobiotechnol 16:1–28

Singh BR, Singh BN, Singh A et al (2015) Mycofabricated biosilver nanoparticles interrupt Pseudomonas aeruginosa quorum sensing systems. Sci Rep 5:1–14. https://doi.org/10.1038/srep13719

Su Y, Wu D, Xia H et al (2019) Metallic nanoparticles induced antibiotic resistance genes attenuation of leachate culturable microbiota: the combined roles of growth inhibition, ion dissolution and oxidative stress. Environ Int 128:407–416. https://doi.org/10.1016/j.envint.2019.05.007

Subbiah U, Elayaperumal G, Elango S et al (2019) Effect of chitosan, chitosan nanoparticle, Anacyclus pyrethrum and Cyperus rotundus in combating plasmid mediated resistance in periodontitis. Anti Infective Agents 18:43–53. https://doi.org/10.2174/2211352517666190221150743

Subramaniyan SB, Ramani A, Ganapathy V, Anbazhagan V (2018) Preparation of self-assembled platinum nanoclusters to combat Salmonella typhi infection and inhibit biofilm formation. Colloids Surfaces B Biointerfaces 171:75–84. https://doi.org/10.1016/j.colsurfb.2018.07.023

Sukhanova A, Bozrova S, Sokolov P et al (2018) Dependence of nanoparticle toxicity on their physical and chemical properties. Nanoscale Res Lett 13:1–21

Tacconelli E, Carrara E, Savoldi A et al (2017) Global priority list of antibiotic-resistant bacteria to guide research, discovery, and development of new antibiotics. WHO Press, Geneva, pp 1–7

Tahir K, Nazir S, Ahmad A et al (2017) Facile and green synthesis of phytochemicals capped platinum nanoparticles and in vitro their superior antibacterial activity. J Photochem Photobiol B Biol 166:246–251. https://doi.org/10.1016/j.jphotobiol.2016.12.016

Tan Y, Ma S, Leonhard M et al (2020) Co-immobilization of cellobiose dehydrogenase and deoxyribonuclease I on chitosan nanoparticles against fungal/bacterial polymicrobial biofilms targeting both biofilm matrix and microorganisms. Mater Sci Eng C 108:110499. https://doi.org/10.1016/j.msec.2019.110499

Tasia W, Lei C, Cao Y et al (2020) Enhanced eradication of bacterial biofilms with DNase I-loaded silver-doped mesoporous silica nanoparticles. Nanoscale 12:2328–2332. https://doi.org/10.1039/c9nr08467c

Thakur BK, Kumar A, Kumar D (2019) Green synthesis of titanium dioxide nanoparticles using Azadirachta indica leaf extract and evaluation of their antibacterial activity. South Afr J Bot 124:223–227. https://doi.org/10.1016/j.sajb.2019.05.024

Thanh Nguyen H, Goycoolea F (2017) Chitosan/Cyclodextrin/TPP nanoparticles loaded with quercetin as novel bacterial quorum sensing inhibitors. Molecules 22:1975. https://doi.org/10.3390/molecules22111975

Tiwari V, Mishra N, Gadani K et al (2018) Mechanism of anti-bacterial activity of zinc oxide nanoparticle against Carbapenem-resistant Acinetobacter baumannii. Front Microbiol 9:1218. https://doi.org/10.3389/fmicb.2018.01218

Vahdati M, Tohidi Moghadam T (2020) Synthesis and characterization of selenium nanoparticles-lysozyme Nanohybrid system with synergistic antibacterial properties. Sci Rep 10:1–10. https://doi.org/10.1038/s41598-019-57333-7

Van Boeckel TP, Gandra S, Ashok A et al (2014) Global antibiotic consumption 2000 to 2010: an analysis of national pharmaceutical sales data. Lancet Infect Dis 14:742–750. https://doi.org/10.1016/S1473-3099(14)70780-7

Van Boeckel TP, Glennon EE, Chen D et al (2017) Reducing antimicrobial use in food animals. Science 357(6358):1350–1352. https://doi.org/10.1126/science.aao1495

Vasudevan S, Srinivasan P, Rayappan JBB, Solomon AP (2020) Photoluminescence biosensor for the detection of N-acyl Homoserine lactone using Cysteamine functionalized ZnO nanoparticles for the early diagnosis of urinary tract infection. J Mater Chem B 8:4228–4236. https://doi.org/10.1039/C9TB02243K

Vignesh K, Rajarajan M, Suganthi A (2014) Photocatalytic degradation of erythromycin under visible light by zinc phthalocyanine-modified titania nanoparticles. Mater Sci Semicond Process 23:98–103. https://doi.org/10.1016/j.mssp.2014.02.050

Vijayakumar S, Saravanakumar K, Malaikozhundan B et al (2020) Biopolymer K-carrageenan wrapped ZnO nanoparticles as drug delivery vehicles for anti MRSA therapy. Int J Biol Macromol 144:9–18. https://doi.org/10.1016/j.ijbiomac.2019.12.030

Wang J, Wu H, Yang Y et al (2018a) Bacterial species-identifiable magnetic nanosystems for early sepsis diagnosis and extracorporeal photodynamic blood disinfection. Nanoscale 10:132–141. https://doi.org/10.1039/c7nr06373c

Wang J, Yang W, Peng Q et al (2019a) Rapid detection of carbapenem-resistant Enterobacteriaceae using pH response based on vancomycin-modified Fe3O4@au nanoparticle enrichment and the carbapenemase hydrolysis reaction. Anal Methods 12:104–111. https://doi.org/10.1039/c9ay02196e

Wang L, Hu C, Shao L (2017) The antimicrobial activity of nanoparticles: present situation and prospects for the future. Int J Nanomedicine 12:1227–1249

Wang Q, Kang F, Gao Y et al (2016) Sequestration of nanoparticles by an EPS matrix reduces the particle-specific bactericidal activity. Sci Rep 6:1–10. https://doi.org/10.1038/srep21379

Wang Q, Perez JM, Webster TJ (2013) Inhibited growth of Pseudomonas aeruginosa by dextran- and polyacrylic acid-coated ceria nanoparticles. Int J Nanomedicine 8:3395–3399. https://doi.org/10.2147/IJN.S50292

Wang X, Yang F, Zhao J et al (2018b) Bacterial exposure to ZnO nanoparticles facilitates horizontal transfer of antibiotic resistance genes. Nano Impact 10:61–67. https://doi.org/10.1016/j.impact.2017.11.006

Wang Y, Yuan Q, Feng W et al (2019b) Targeted delivery of antibiotics to the infected pulmonary tissues using ROS-responsive nanoparticles. J Nanobiotechnol 17:103. https://doi.org/10.1186/s12951-019-0537-4

Wu G, Ji H, Guo X et al (2020) Nanoparticle reinforced bacterial outer-membrane vesicles effectively prevent fatal infection of carbapenem-resistant Klebsiella pneumoniae. Nanomed Nanotechnol Biol Med 24:102148. https://doi.org/10.1016/j.nano.2019.102148

Wu J, Li F, Hu X et al (2019) Responsive assembly of silver nanoclusters with a biofilm locally amplified bactericidal effect to enhance treatments against multi-drug-resistant bacterial infections. ACS Cent Sci 5:1366–1376. https://doi.org/10.1021/acscentsci.9b00359

Xu X, Chen Y, Wei H et al (2012) Counting Bacteria using functionalized gold nanoparticles as the light-scattering reporter. Anal Chem 84:9721–9728. https://doi.org/10.1021/ac302471c

Xu Y, Wang C, Hou J et al (2019) Effects of cerium oxide nanoparticles on bacterial growth and behaviors: induction of biofilm formation and stress response. Environ Sci Pollut Res 26:9293–9304. https://doi.org/10.1007/s11356-019-04340-w

Yang L, Liu Y, Wu H et al (2012) Combating biofilms. FEMS Immunol Med Microbiol 65:146–157

Yuan K, Mei Q, Guo X et al (2018) Antimicrobial peptide based magnetic recognition elements and au@ag-GO SERS tags with stable internal standards: a three in one biosensor for isolation, discrimination and killing of multiple bacteria in whole blood. Chem Sci 9:8781–8795. https://doi.org/10.1039/c8sc04637a

Zhang S, Wang Y, Song H et al (2019) Copper nanoparticles and copper ions promote horizontal transfer of plasmid-mediated multi-antibiotic resistance genes across bacterial genera. Environ Int 129:478–487. https://doi.org/10.1016/j.envint.2019.05.054

Zhao L, Wang H, Huo K et al (2011) Antibacterial nano-structured titania coating incorporated with silver nanoparticles. Biomaterials 32:5706–5716. https://doi.org/10.1016/j.biomaterials.2011.04.040

Zhao X, Lu C, Yang S, Zhang J (2020) Bioconjugation of aptamer to fluorescent trimethyl chitosan nanoparticles for bacterial detection. Mater Lett 264:127330. https://doi.org/10.1016/j.matlet.2020.127330

Zhao Y, Ye C, Liu W et al (2014) Tuning the composition of AuPt bimetallic nanoparticles for antibacterial application. Angew Chemie Int Ed 53:8127–8131. https://doi.org/10.1002/anie.201401035

Zhen X, Chudal L, Pandey NK et al (2020) A powerful combination of copper-cysteamine nanoparticles with potassium iodide for bacterial destruction. Mater Sci Eng C 110:110659. https://doi.org/10.1016/j.msec.2020.110659

Zou Z, Sun J, Li Q et al (2020) Vancomycin modified copper sulfide nanoparticles for photokilling of vancomycin-resistant enterococci bacteria. Colloids Surfaces B Biointerfaces 189:110875. https://doi.org/10.1016/j.colsurfb.2020.110875

Precise Sequence-Specific Antimicrobials Based on CRISPR: Toward Prevailing Over Bacterial Antibiotic Resistance

8

Vida Ebrahimi and Atieh Hashemi

Contents

Abstract

Antibiotic resistance is one of the major obstacles of modern medicine in treating common infectious diseases. One of the common recently known genome editing tools, CRISPR (clustered regularly interspaced short palindromic repeats), can be utilized to selectively remove the genes associated with antibiotic resistance. These sequence-specific antimicrobials can be delivered to target organisms using phage-based vehicles. CRISPR technology's clinical application is still

V. Ebrahimi · A. Hashemi (✉)
Department of Pharmaceutical Biotechnology, School of Pharmacy, Shahid Beheshti University of Medical Sciences, Tehran, Iran
e-mail: at_hashemi@sbmu.ac.ir

© The Author(s), under exclusive license to Springer Nature Singapore Pte
Ltd. 2022
V. Kumar et al. (eds.), *Antimicrobial Resistance*,
https://doi.org/10.1007/978-981-16-3120-7_8

questionable since some major barriers, derived from delivery systems, pose a threat to the wide utilization of this solution to solve the antimicrobial resistance problem. Also, social issues of the therapeutic use of the CRISPR method need to be solved. Ideal CRISPR/Cas antimicrobial activity is obtained when there is no tolerance against CRISPR machine components, so the resistance factors against the CRISPR system's gene-editing roll need to be identified and minimized as possible. There are some examples of the in vivo use of CRISPR antimicrobials, but still a gap needs to be filled before the clinical application of these antimicrobials.

Keywords

Antimicrobial resistance · Genome editing · CRISPR · CRISPR/Cas, CRISPR antimicrobials

8.1   Introduction

Prokaryotic organisms have developed various protection mechanisms, both innate and adaptive, to defend themselves against the invader genetic elements such as parasitic plasmids and bacteriophages (Hampton et al. 2020). Each of these multiple bacterial immune strategies obstructs the various phases of the phage life cycle (van Houte et al. 2016a). Among these systems, clustered regularly interspaced short palindromic repeats (CRISPRs) are exceptional in that they can continuously update and customize their immune responses to match the invaders' particular nucleic acid sequences (Hashemi 2018). CRISPR, together with the CRISPR-associated proteins (Cas endonucleases), protect their hosts against foreign invader nucleic acids through an adaptive immune system (Hashemi 2020). Protection begins with an acquisition stage in which, following infection, tiny pieces of the invader's genome, known as spacers, are collected and integrated into the CRISPR locus, embedded in the bacterial genome. Then, spacers are transcribed into CRISPR RNAs (crRNAs) to identify and inactivate the cognate targets during the targeting step, directing Cas endonucleases to kill the DNA or RNA of the invader (Ebrahimi and Hashemi 2020) (Fig. 8.1).

The adaptive CRISPR/Cas systems are classified into two distinct classes based on the effector modules, class 1 with multi-subunit effector complex comprising four to seven Cas endonuclease subunits and class 2 with a single-subunit large protein as the effector, e.g., Cas9 or Cas12 or Cas13 (Koonin and Makarova 2019) (Fig. 8.2). Each of the two classes of the CRISPR/Cas system is subdivided into three types and several subtypes. The effector complexes of class 1 systems (comprising types I, III, and IV) encompass RNA-binding RRMs (RNA recognition motifs) domain. In class 2 systems (containing types II, V, and VI), the best-characterized effector module is the Cas9 (type II) molecule. The Cas9 endonuclease consists of two different nuclease domains, HNH and RuvC, responsible for the target DNA strands cut in the crRNA–target DNA complex. A trans-acting CRISPR RNA (tracrRNA), crucial for pre-crRNA processing and target identification, is also encoded from the type II

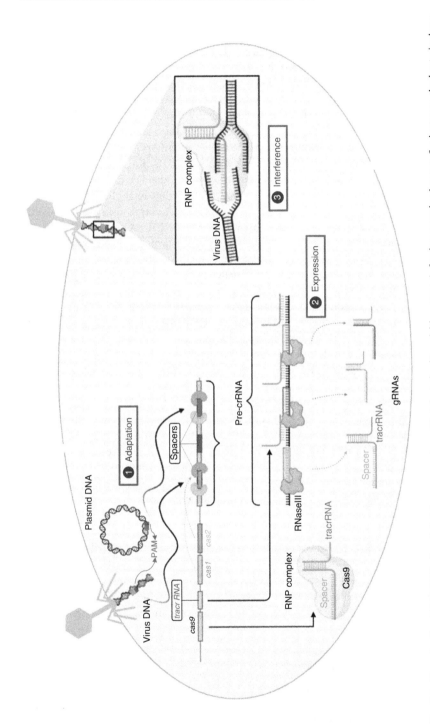

Fig. 8.1 The adaptive CRISPR/Cas9 antiviral immune system. (1) Adaptation/Acquisition step: A foreign genetic element, for instance, the bacteriophage DNA or a plasmid attacks the bacteria and a part of its sequence, adjacent to PAM site, named a protospacer is kept in CRISPR array as cell memory.

loci (Mohanraju et al. 2016) (Fig. 8.3). CRISPR technology has been used to create efficient genome manipulation in microorganisms, animals, and plants. First, a conserved 2- to 4-bp sequence called the protospacer-adjacent motif (PAM) that flanks a target DNA site is recognized by the Cas endonuclease under the RNA guidance. Next, the Cas9 protein inquires the flanking DNA sequences for base-pairing complementarity to a single guide RNA (sgRNA) when binding to the PAM, resulting in DNA cleavage (Barrangou and Doudna 2016). Genome-wide screening is the other application of the CRISPR tool. These studies are utilized in identifying genes occupied in the development and metastasis of tumors and characterization of noncoding and regulatory sequences (Chen et al. 2015; Rajagopal et al. 2016). CRISPR-based technologies can also be used for cell therapy that aims to correct defective genotypes. CRISPR-based genome engineering is applicable in agricultural, food, and industrial biotechnology, biological control, and medical fields (Li et al. 2015; Selle and Barrangou 2015; Gantz et al. 2015; Long et al. 2014).

The homeostasis of eukaryotic cells is maintained through several robust DNA repair mechanisms. By contrast, bacterial cells readily obtain genetic material and keep mutation levels high adequate to ensure evolutionary suppleness since they have limited DNA repair processes. This property has paved the way for the lethal effect of CRISPR-mediated DNA cleavage for prokaryotic cells (Beisel et al. 2014). So, either the native or engineered self-targeting CRISPR approach can be considered as an antimicrobial by generating a double-stranded break (DSB) within the target sequence. The DSB creation can be programmed for any bacterial species by sequence-specific antibiotics, which are able to eliminate bacterial pathogens (Gomaa et al. 2014).

In this book chapter, the antibiotic application of CRISPR/Cas technology is aimed to be introduced. First, the advantages of sequence-specific antimicrobials in comparison with conventional antibiotics are mentioned. Second, the classification of known CRISPR antimicrobials is introduced. Next, different delivery strategies for these antibiotics are discussed. Then, the existing obstacles to the application of CRISPR antimicrobials are explained. Finally, examples of these agents' utilization are presented, and the future prospect in this field is pictured.

8.2   Problem(s) Associated with Conventional Antibiotics

Antibiotics have long been considered as one of the twentieth century's most important discoveries with history of over 70 years. Sadly, the dawn of the antibiotic period has mirrored the emergence of the antimicrobial resistance phenomenon.

Fig. 8.1 (continued) (2) Expression step: The CRISPR/Cas9-related genes containing the Cas9 and CRISPR array are transcribed. The CRISPR locus generates a long pre-crRNA then the pre-crRNA is diced into small crRNAs by RNaseIII. (3) Interference step: In the next invading of the viral DNA, the ribonucleoprotein (RNP) complex assembles and the Cas9 is guided towards the foreign DNA by the crRNA homologous to the invading sequence. Finally, the viral DNA is cleaved.

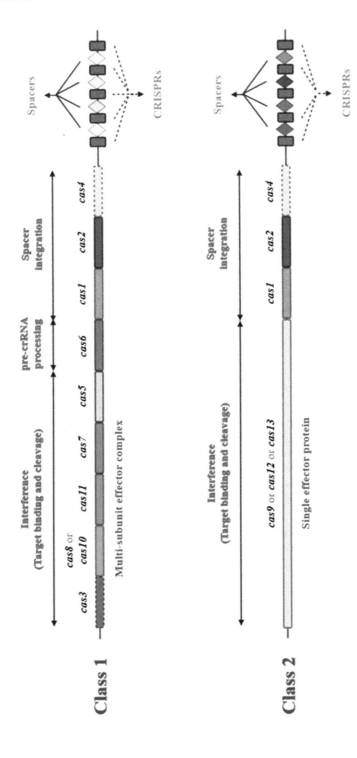

Fig. 8.2 Class 1 and class 2 CRISPR/Cas networks with their and 13 core gene families. The overall architectures of class 1 (multi-subunit effector complexes) and class 2 (single effector proteins) CRISPR/Cas systems. Genes are drawn as rectangular; homologous genes are displayed with the same color. Unnecessary genes are shown by a dashed outline

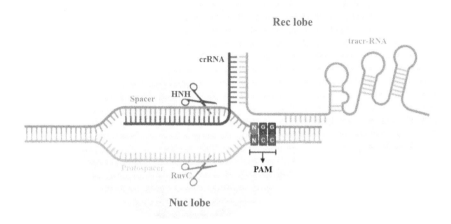

Fig. 8.3 Roll of SpCas9 in target sequence cleavage. The sgRNA leads the Cas9 towards the target sequence with nearly 20 crRNA-homologous nucleotides. The Cas9 attaches reversibly to the PAM site, adjacent to the target sequence, via recognition (Rec) lobe and is stabilized through tracr-RNA:: crRNA duplex. The crRNA binds to the target sequence via base-pairing. Finally, each strands of the target DNA are cleaved by separate domains, named HNH and RuvC domains, in nuclease (Nuc) lobe of the Cas9 protein

Antibiotics have significantly enabled us to fight against various pathogens; however, promiscuous AMR exchange among bacterial populations has appeared a situation in which a limited number of efficacious antimicrobials are left. The emerging of some pan-drug-resistant strains is quite a formidable obstacle (Kothari 2020). The yearly fatalities from drug-resistant microbes surpass 50,000 worldwide. Currently, the antimicrobial resistance (AMR) is one of the most severe defects of commonly used antibiotics in long-term application which is estimated to be the leading cause of death for almost ten million people at 2050. The antimicrobial resistance is a common occurrence in which bacteria develop to survive antibiotic action, rendering them seemingly inactive. Nowadays, more than 150 various antibiotics and their derivatives are utilized to treat different infectious diseases which in turn established a modern and severe artificial selection way for bacterial strains to acclimate by becoming antibiotic resistance (Ghosh et al. 2019). There are four main mechanisms for antibiotic misuse- and/or overuse-related AMR. The first strategy for loss of susceptibility is based on preventing the access of the antibiotic to its target site. The other way is the alteration of target site by mutation. The third mechanism is target protection against antimicrobial agent, and the last strategy is direct antimicrobial modification by enzymatic inactivation (Moo et al. 2020).

Three major causes of AMR are yet known. The use of antimicrobials, which has risen strongly in the last few decades, is believed to be the first cause. The second cause lies in the fact that patients are not able to correctly obey the counseling orders in most cases. The third reason is that there are very small quantities of new medications under development within a given class of antibiotics to replace those made obsolete by increasing drug resistance. In addition, study on pharmacodynamic (PD) and pharmacokinetic (PK) of antibiotics that increases the possibility of a good clinical outcome is required. It is conceivable that the patient received subtherapeutic drug concentration for causes that may be due to unsuitable dosing, altered renal and/or hepatic activity, and protein binding or volume of distribution heterogeneity (Admassie 2018).

Antimicrobial resistance globally is now known to be one of the major risks to public health. The yearly fatalities from drug-resistant microbes surpass 50,000 worldwide. So, the advent and propagation of antimicrobial resistance demand immediate action from both health providers and government officials. The complicated and multifactorial nature of antimicrobial resistance, especially in terms of human, animal, and environmental interactions, is not well recognized. The condition becomes further complicated by the absence of accurate knowledge, the time-consuming process of new antimicrobials' development, and the high prevalence of horizontal gene transfer (Singh et al. 2020).

The abovementioned serious drawbacks as well as the potential of bacteria to rapidly adapt with media condition has made it imperative to search for new and further potent antimicrobials and to establish antibiotics with different scaffolding and novel mechanisms of action. A great struggle is being made, especially in academia, to explore the potential of the exploited critical mechanism in bacteria and to develop new interfering molecules in this process, as well as to investigate the biochemical details of these targets (Wright 2016). Several techniques were used to combat antibiotic-resistant bacteria, including the development of novel antibiotics, bacteriophage utilization lysing these bacteria, and the application of natural and artificial enzymes or peptides targeting genomes or essential proteins of these bacteria (Matsuzaki et al. 2003; Narenji et al. 2017).

8.3   CRISPR/Cas Potential to Serve as Programmable Sequence-Specific Antimicrobials

The microbial community has been influenced by the progress of drug development and success in creating new types of antimicrobial drugs. After all, the microorganisms have been very intelligent in improving themselves to resist these modern antimicrobial weapons and have evolved, making most of the routinely used antibacterial agents less effective. The world is looking at an issue of how the successful performance of microbial pathogens versus antimicrobial agents emerged. One of the highest priorities in medicine is to devise novel antibiotics to combat resistant pathogens, but a certain solution to this issue has yet to be discovered. The sequence-specific and titratable elimination of distinct bacterial strains using CRISPR/Cas method has emerged as a potential approach for combating antimicrobial resistance (Sarma et al. 2020). Sequence-specific targeting, the capability to differentiate between commensal and pathogenic bacterial organisms, is one of the highly essential key properties of this method (Cui and Bikard 2016).

The programmability of Cas9 arises from its nuclease activity. This activity helps more cytotoxic effect on resistant cells. A CRISPR-guide RNA may then be precisely engineered to target resistance or virulence-related genes, causing a cut in resistant bacteria's dsDNA, returning them to antibiotic-sensitive bacteria (Aslam et al. 2020). Taken together, RNA-based spacers flanked by partial repeat sequences guide CRISPR-associated proteins to directly target and cut DNA encoding corresponding protospacers. The CRISPR/Cas machinery can therefore be designed to precisely target and break any in vivo DNA according to the information given in the CRISPR array used to target bacterial populations that bear unique genes responsible for antibiotic resistance (Bikard et al. 2014; Hale et al. 2012). However, deliberate or unintended targeting of the bacterial genome sequence by the CRISPR/ Cas method is cytotoxic and may contribute to cell death due to the creation of permanent chromosomal lesions (Citorik et al. 2014; Vercoe et al. 2013).

8.4 Benefits and Advantages of CRISPR-Based Antimicrobials

In all medicine, biotechnology, and environmental aspects, modifying the composition of microbial species is a crucial point of view. Although various antimicrobial methods, such as antibiotics, antimicrobial peptides, and bacteriophages, give partial remedies and constrained opportunities, what is left indefinable is a generalized and programmable approach that can discriminate even between the two almost similar microorganisms that allow for fine control of the composition of the microbial community (Koskella and Meaden 2013). It's confirmed that CRISPR/Cas strategy could offer such an approach with high specificity (Gomaa et al. 2014) (Fig. 8.4). The CRISPR/Cas antimicrobial agents provide several possible benefits over conventional antimicrobials. Besides selective killing, provided with an effective conveyance method, the multiplex gene targeting potential of CRISPR/Cas systems may be used to attack multiple organisms at the same time and/or multiple sequences of the same bacterium to avoid resistant mutants from growing (Bikard et al. 2014).

Due to the extensively conserved bacterial and archaeal CRISPR/Cas systems, the isolation, calibration, and evolvement of delivery vectors of the CRISPR/Cas machinery will be required for the development of RNA-guided nucleases proficient of targeting further strains, as well as multidrug-resistant (MDR) bacteria (Citorik et al. 2014). Moreover, RNA-guided Cas proteins could help us to temper the incidence of exact genes, for instance, antibiotic resistance genes and virulence elements, in wild-type populations due to delivery vehicles capable of acting in higher organisms (Esvelt et al. 2014).

The gain of programmed Cas-mediated antimicrobial treatment is the possibility to target more than one genetic element by a nuclease with two or more sgRNAs, which could reduce resistant clones that reject phagemid-related DNA cleavage via the formation of target mutations, along with increase in the spectrum of targeted cells. Furthermore, sequence-specific Cas protein dissemination and simple reprogramming to attack various sequences decrease the plasmid content of a

Fig. 8.4 Selective elimination of a specific bacterial strain. A strategy is required which is able to selectively omit discrete constituents (green) but not the rest within a miscellaneous microbial population

bacterial community without destroying the cells immunizing them against plasmids containing to antibiotic resistance or virulence genes (Gholizadeh et al. 2020).

8.5 Different Types of Available Nucleases for CRISPR Antimicrobials

The Cas9 endonuclease as a compartment of CRISPR/Cas type II system can be utilized to design CRISPR-based antimicrobials. For example, Jinek et al. used this system to program antimicrobial activity against resistant *E. coli* (Jinek et al. 2012). As an example, this system was used to combat against carbapenem-resistant *Enterobacteriaceae* (CRE) (Hao et al. 2020). Bikard et al. used the Cas9 nuclease with phagemid delivery system, a plasmid that is intended to be wrapped in phage capsids, to eliminate methicillin-resistant *Staphylococcus aureus* (MRSA) strains from a mixed bacterial community. The Cas9, reprogrammed for targeting virulence genes, destroyed virulent *S. aureus*, despite being avirulent. Reprogramming of Cas9 to target genes of antibiotic resistance extinguishes staphylococcal plasmids that host genes of antibiotic resistance and immunizes avirulent staphylococci to avoid the dissemination of plasmid-borne resistance genes (Bikard et al. 2014). Gomma et al. (2014) utilized the *E. coli* type I–E CRISPR/Cas system, containing six *cas* genes located in two separate operons, namely, *casABCDE* and *cas3*, for targeting and selective elimination of individual bacterial strains. Cas3 protein contains multi-domains acting as helicase and nuclease (Sinkunas et al. 2011). They determined that powerful elimination can be accomplished by the targeting several and varied locations in the genome, including *nusB*, *ftsA*, *msbA*, and *asd*, and also the elimination ranges seen by simultaneous targeting of multiple locations were close to those targeting only one of the locations (Brouns et al. 2008). The delivery method in that investigation was based on transformation, and the chromosomal genes were targeted to be eliminated (Gomaa et al. 2014). The *E. coli* type I–E system, delivered as temperate phage, was also used to sensitize different strains of antibiotic-resistant *E. coli* (Yosef et al. 2015). In another example, the endogenous *C. difficile* type I–B CISPR system was used as self-targeting antimicrobial agent (Selle et al. 2020). The Cas13a single-stranded RNA nuclease as member of type VI class 2 system was also used as antibacterial agent against carbapenem-resistant *E. coli* (mediated by bla_{IMP-1}, bla_{VIM-2}, bla_{OXA-48}, bla_{KPC-2}, and bla_{NDM-1} genes) and methicillin-resistant *S. aureus* (mediated by *mecA* gene) by Kiga et al. (2020).

8.6 CRISPR Antimicrobial Delivery Systems for Targeted Killing or Antibiotic Sensitivity

One of the main criteria required to be tested for the effective application of CRISPR in reducing antimicrobial resistance is the delivery of the CRISPR/Cas components in vivo. Through multiple approaches, the CRISPR/Cas9 machinery may be delivered into the target cell (Sarma et al. 2020). The Cas9 endonuclease targeting various bacterial pathogens and antibiotic resistance gene sequences is conveyed to microbial communities using polymer-derivatized CRISPR nanocomplexes (Kang et al. 2017), conjugate-transmitted plasmid-carrying bacteria (Citorik et al. 2014; Kim et al. 2016), and bacteriophages (Bikard et al. 2014; Citorik et al. 2014).

As an example, the phage-based delivery system was used in Yosef et al. investigation. Researchers engineered a λ-prophage carrier, targeting the bla_{NDM-1} and $bla_{CTX-M-15}$ genes. This strategy ruined both antibiotic resistance plasmids and genetically engineered lytic phages (Yosef et al. 2015). A plasmid with genes encoding Cas9 protein and sgRNA may be used as delivery vector. The main benefit of utilization just one vector coding for both nuclease and sgRNA is to reduce steps of separated vectors' transformation comprising these genes. This vector's high stability allows it being an effective but easy approach (Sarma et al. 2020). For instance, Citorik et al. utilized M13-phagemids to deliver CRISPR/Cas machinery. The authors made a genetic construct comprising two spacers in order to target the bla_{SHV-18} and bla_{NDM-1} genes and detected almost 2 to 3 log10 decrease in viable *E. coli* count containing plasmids with bla_{NDM-1} or bla_{SHV-18} genes, but no changes are observed for the wild-type strain (Citorik et al. 2014).

Conveyance of Cas protein and sgRNA as a complex, known as RNP complex, using a transfer vehicle is another broadly used CRISPR delivery technique. The bonuses of exploitation of this complex are fast performance, provision of high precision, and decreased off-target reactions. Furthermore, there is no need for codon optimization. The complex can be transferred using liposomes, polymeric nanocomplexes, metallic nanoparticles, etc. (Liu et al. 2017).

8.7 Existing Challenges for Sufficient Antimicrobial Activity and Ways of Efficacy Enhancement

The use of CRISPR network as a gene-editing method has attracted the science community owing to its future uses. With the advent of this instrument, overcoming antibiotic resistance using CRSIPR, along with its share of difficulties, has provided new perspectives.

8.7.1 Resistance Against CRISPR/Cas Antimicrobials

Target site modification is one of the mechanisms of failure in CRISPR antimicrobial activity. Point mutation in the target sequence of the CRISPR/Cas antibiotics, which is responsible for antimicrobial resistance, or complete removal of target gene is a defense mechanism against CRISPR-associated DNA cleavage in bacterial cells. This phenomenon can hinder the CRISPR-based antimicrobial resistance prevention in bacteria (Cui and Bikard 2016; Leenay et al. 2019; Selle and Barrangou 2015; Standage-Beier et al. 2015; Zerbini et al. 2017). In addition to point mutation, insertion or deletion mutations through either sgRNA- or Cas-related genes can also protect the bacterial cell against CRISPR-related gene editing (Vento et al. 2019; Ebrahimi and Hashemi 2020). The sgRNA sequence can be altered by homologous recombination process between the repeat sequence section of sgRNA and the homologue outside-present repeats (Gomaa et al. 2014; Zerbini et al. 2017). This mutation can deactivate the leading role of sgRNA toward the target sequence and prevent the CRISPR-based antibiotic from nuclease activity. The bacterial resistance toward CRISPR antimicrobials generates obstacles to deliver CRISPR machinery. Besides resistance due to mutations, another resistance mechanism is the production of anti-CRISPR (Acrs) small protein molecules which is responsible to attach the target DNA of CRISPR network and to disable the whole Cas-related cleavage of the target sequence (Vercoe et al. 2013). Intensified specificity and complex diversity of sequences for Arcs demonstrates that they are omnipresent and can be carried by mobile gene elements such as phages and extrachromosomal DNA pieces (Pawluk et al. 2018; Borges et al. 2017; van Houte et al. 2016b; Jiang et al. 2013; Pawluk et al. 2016; Harrington et al. 2017).

There are some introduced solutions to overcome bacterial tolerance against CRISPR antibiotic activity. The optimized expression level of Cas nuclease and sgRNA can prevent antibiotic failure. Perhaps, exploitation of stronger promotors upstream of the sgRNA gene or transporting CRISPR components through high-copy number plasmids or codon optimization can efficiently reduce bacterial tolerance (Guo et al. 2019; Li et al. 2019; Song et al. 2017; Wu et al. 2019).

8.7.2 CRISPR/Cas Common Delivery Vehicles and Associated Challenges

Because of the large molecular weight of the CRISPR complex (almost 160KDa), the delivery of the components into the cell is a complex process. The tools used to cell delivery of this machinery are diverse. For instance, bacteriophages, nanocages, and nano-sized RNP complexes are some introduced methods. A study in 2014 recommended that the CRISPR system can be packaged inside phage capsid as a transfer approach into the bacterial cell (Citorik et al. 2014). Since being wrapped in capsid, special plasmids named phagemids are utilized. Phagemid transduction results in successful cleavage of the target gene (Citorik et al. 2014; Bikard et al. 2014). There are two main limitations in the application of phagemids as delivery vehicles. First, following administration, the phagemid does not develop further phages, which implies that the phagemid dose necessary for treatment is much greater than the target cell population. Second, their widespread usage could be precluded by the narrow-host compatibility (34, 60).

The low conjugation rate for plasmids is one of the restrictive factors in the utilization of plasmid as a carrier of CRISPR components. The disadvantages of plasmids as delivery system include the time-consuming activity since the plasmid must first code the Cas protein and sgRNA, and next the CRISPR performance begins (Liu et al. 2017), increased off-target activity (Fu et al. 2013; Cradick et al. 2013), and inaccessibility to nucleus (Sarma et al. 2020). However, due to independence of conjugative plasmids from any cellular receptor, easy to be engineered with wide coding capacities, resistance against restriction-modification systems, and compatibility with wide-spectrum hosts, conjugative plasmid-based nuclease transmission remains a fascinating choice. Because of increased cell-to-cell communication, conjugative plasmids that are enrolled in biofilm formation could increase rates of conjugative plasmid transmission (Pérez-Mendoza and de la Cruz 2009; Smillie et al. 2010; Oliveira et al. 2014; Jain and Srivastava 2013; Hausner and Wuertz 1999).

A key to above-mentioned complications could be the usage of mRNA insisted of the DNA which subsequently translates into the nuclease and sgRNA. This will agree for the construction of CRISPR/Cas active complex in a rather quicker time with low off-target activity. In addition, since the RNA places in the cytoplasm, the requirement to access nucleus is mitigated, though the low stability of mRNA is still a hurdle (Liu et al. 2017).

One of the most important difficulties in exploitation of Cas9 is its innate toxicity as seen in *Corynebacterium glutamicum* and *Synechococcus elongatus* (Naduthodi et al. 2018; Jiang et al. 2017). A substitute to this could be the application of Cas12a instead of Cas9 which is less toxic (Hatoum-Aslan et al. 2011; Yosef et al. 2012; Swarts et al. 2012).

8.7.3 Social Issues of Using CRISPR Antimicrobials

The usage of the CRISPR/Cas network in order to degrade and eradicate antimicrobial resistance genes from the bacterial community present in nature will have to meet various regulatory and judicial hurdles. There are important ethical issues that need to be considered before using any gene-editing approaches as well as CRISPR antibiotics (Carter and Friedman 2016; Adelman et al. 2017).

8.8   In Vivo Application of CRISPR Antimicrobials

CRISPR-based antimicrobials as novel antibiotic agents are moving toward being applied in vivo. Some recent studies in this field are summarized in this section. The *C. difficile* infection (CDI) causes yearly thousands of deaths in the United States.

The main underlying risk factor for CDI is the administration of broad-spectrum antibiotics, damaging the normal gut microflora. The routine CDI treatment is vancomycin, perturbating the gut microbiota as an unintended effect. This side effect leads to CDI recurrency, so novel therapeutic methods need to be developed. In a study at 2020, the endogenous CRISPR system was repurposed to combat *C. difficile*. For this aim, the self-targeting Cas3 was generated by phage-delivered spacers against bacterial chromosome. The efficacy of this novel antimicrobial method was assessed in mouse model of CDI (Selle et al. 2020).

In another in vivo investigation, the CRISPR/Cas9 system, targeting bla_{NDM-1} resistance gene, was used to reduce phage-delivered resistance in *E. coli*. In this study, the CRISPR machinery was delivered to bacterial cells with PDC (phage-delivered CRISPR/Cas) method. This strategy sensitized the resistance plasmid-harboring bacteria. Next, the phage-delivered resistance eradication with subsequent antibiotic treatment (PRESA) method was performed by kanamycin. The efficacy of combinational CRISPR antimicrobial and kanamycin administration was assessed in vitro and in vivo. The in vivo analysis was performed on mouse intestine or skin infection models revealing continuous repressive impact against resistant bacteria (Liu et al. 2020).

The other example for CRISPR-based antimicrobial in vivo analysis is Bikard et al. investigation. In this study, the capsule genes of *S. pneumoniae*, as a critical virulence factor, was targeted by CRISPR/Cas9 system. The inhibiting CRISPR roll against capsule switching from heat-killed encapsulated strain to live nonencapsulated strain was tested through in vivo analysis on infected mice (Bikard et al. 2012).

8.9 Future Prospects

Recently, most studies are designed based on a sole CRISPR delivery system selected on the basis of organism type and with the aim of highest killing performance for cells containing target genes. Several studies indicate that the application of mixture of delivery systems such as bacteriophages, plasmids, and polymeric nanocomplex vehicles could be ideal in efficacy (Roach et al. 2017). Furthermore, the full extinction of the intended organism(s) and the absence of special methods to combat pathogenic organisms or to change genetic content of microbial populations are a key problem in microbiology. The relative quantity of pathogenic targeted bacteria may be decreased by CRISPR-based antibiotics as sequence-specific nucleases, but the optimization of a wide-spectrum and robust delivery mechanism remains a crucial issue. CRISPR/Cas transmission vectors or vehicles and their design, diverse bacterial communities, various pathways of tolerance against antibiotics in different organisms, probability of mutations in target sequences, legislation, and social issues of CRISPR/Cas-based antibiotics may therefore be the obstacles ahead toward CRISPR-based antimicrobial studies.

References

Adelman Z, Akbari O, Bauer J, Bier E, Bloss C, Carter SR, Callender C, Costero-Saint Denis A, Cowhey P, Dass B (2017) Rules of the road for insect gene drive research and testing. Nat Biotechnol 35(8):716–718. https://doi.org/10.1038/nbt.3926

Admassie M (2018) Current review on molecular and phenotypic mechanism of bacterial resistance to antibiotic. Sci J Clin Med 7:13. https://doi.org/10.11648/j.sjcm.20180702.11

Aslam B, Rasool M, Idris A, Muzammil S, Alvi RF, Khurshid M, Rasool MH, Zhang D, Ma Z, Baloch Z (2020) CRISPR-Cas system: a potential alternative tool to cope antibiotic resistance. Antimicrob Resist Infect 9(1):1–3. https://doi.org/10.1186/s13756-020-00795-6

Barrangou R, Doudna JA (2016) Applications of CRISPR technologies in research and beyond. Nat Biotechnol 34(9):933–941. https://doi.org/10.1038/nbt.3659

Beisel CL, Gomaa AA, Barrangou R (2014) A CRISPR design for next-generation antimicrobials. Genome Biol 15(11):516. https://doi.org/10.1186/s13059-014-0516-x

Bikard D, Euler CW, Jiang W, Nussenzweig PM, Goldberg GW, Duportet X, Fischetti VA, Marraffini LA (2014) Exploiting CRISPR-Cas nucleases to produce sequence-specific antimicrobials. Nat Biotechnol 32(11):1146–1150. https://doi.org/10.1038/nbt.3043

Bikard D, Hatoum-Aslan A, Mucida D, Marraffini LA (2012) CRISPR interference can prevent natural transformation and virulence acquisition during in vivo bacterial infection. Cell Host Microbe 12(2):177–186. https://doi.org/10.1016/j.chom.2012.06.003

Borges AL, Davidson AR, Bondy-Denomy J (2017) The discovery, mechanisms, and evolutionary impact of anti-CRISPRs. Annu Rev Virol 4(1):37–59. https://doi.org/10.1146/annurev-virology-101416-041616

Brouns SJ, Jore MM, Lundgren M, Westra ER, Slijkhuis RJ, Snijders AP, Dickman MJ, Makarova KS, Koonin EV, Van Der Oost J (2008) Small CRISPR RNAs guide antiviral defense in prokaryotes. Science 321(5891):960–964. https://doi.org/10.1126/science.1159689

Carter S, Friedman R (2016) Policy and regulatory issues for gene drives in insects. In: Workshop report

Chen S, Sanjana NE, Zheng K, Shalem O, Lee K, Shi X, Scott DA, Song J, Pan JQ, Weissleder R (2015) Genome-wide CRISPR screen in a mouse model of tumor growth and metastasis. Cell 160(6):1246–1260. https://doi.org/10.1016/j.cell.2015.02.038

Citorik RJ, Mimee M, Lu TK (2014) Sequence-specific antimicrobials using efficiently delivered RNA-guided nucleases. Nat Biotechnol 32(11):1141–1145. https://doi.org/10.1038/nbt.3011

Cradick TJ, Fine EJ, Antico CJ, Bao G (2013) CRISPR/Cas9 systems targeting β-globin and CCR5 genes have substantial off-target activity. Nucleic Acids Res 41(20):9584–9592. https://doi.org/10.1093/nar/gkt714

Cui L, Bikard D (2016) Consequences of Cas9 cleavage in the chromosome of Escherichia coli. Nucleic Acids Res 44(9):4243–4251. https://doi.org/10.1093/nar/gkw223

Ebrahimi V, Hashemi A (2020) Challenges of in vitro genome editing with CRISPR/Cas9 and possible solutions: a review. Gene 753:144813. https://doi.org/10.1016/j.gene.2020.144813

Esvelt KM, Smidler AL, Catteruccia F, Church GM (2014) Emerging technology: concerning RNA-guided gene drives for the alteration of wild populations. Elife 3:e03401. https://doi.org/10.7554/eLife.03401

Fu Y, Foden JA, Khayter C, Maeder ML, Reyon D, Joung JK, Sander JD (2013) High-frequency off-target mutagenesis induced by CRISPR-Cas nucleases in human cells. Nat Biotechnol 31(9):822–826. https://doi.org/10.1038/nbt.2623

Gantz VM, Jasinskiene N, Tatarenkova O, Fazekas A, Macias VM, Bier E, James AA (2015) Highly efficient Cas9-mediated gene drive for population modification of the malaria vector mosquito Anopheles stephensi. Proc Natl Acad Sci 112(49):E6736–E6743. https://doi.org/10.1073/pnas.1521077112

Gholizadeh P, Köse Ş, Dao S, Ganbarov K, Tanomand A, Dal T, Aghazadeh M, Ghotaslou R, Rezaee MA, Yousefi B (2020) How CRISPR-Cas system could be used to combat antimicrobial resistance. Infect Drug Resist 13:1111. https://doi.org/10.2147/IDR.S247271

Ghosh C, Sarkar P, Issa R, Haldar J (2019) Alternatives to conventional antibiotics in the era of antimicrobial resistance. Trends Microbiol 27(4):323–338. https://doi.org/10.1016/j.tim.2018. 12.010

Gomaa AA, Klumpe HE, Luo ML, Selle K, Barrangou R, Beisel CL (2014) Programmable removal of bacterial strains by use of genome-targeting CRISPR-Cas systems. MBio 5:1. https://doi.org/ 10.1128/mBio.00928-13

Guo T, Xin Y, Zhang Y, Gu X, Kong J (2019) A rapid and versatile tool for genomic engineering in Lactococcus lactis. Microb Cell Fact 18(1):22. https://doi.org/10.1186/s12934-019-1075-3

Hale CR, Majumdar S, Elmore J, Pfister N, Compton M, Olson S, Resch AM, Glover CV III, Graveley BR, Terns RM (2012) Essential features and rational design of CRISPR RNAs that function with the Cas RAMP module complex to cleave RNAs. Mol Cell 45(3):292–302. https://doi.org/10.1016/j.molcel.2011.10.023

Hampton HG, Watson BN, Fineran PC (2020) The arms race between bacteria and their phage foes. Nature 577(7790):327–336. https://doi.org/10.1038/s41586-019-1894-8

Hao M, He Y, Zhang H, Liao XP, Liu YH, Sun J, Du H, Kreiswirth BN, Chen L (2020) CRISPR-Cas9-mediated Carbapenemase gene and Plasmid curing in Carbapenem-resistant Enterobacteriaceae. Antimicrob Agents Chemother 64:9. https://doi.org/10.1128/aac.00843-20

Harrington LB, Doxzen KW, Ma E, Liu J-J, Knott GJ, Edraki A, Garcia B, Amrani N, Chen JS, Cofsky JC (2017) A broad-spectrum inhibitor of CRISPR-Cas9. Cell 170(6):1224–1233. https://doi.org/10.1016/j.cell.2017.07.037

Hashemi A (2018) CRISPR-Cas system as a genome engineering platform: applications in biomedicine and biotechnology. Curr Gene Ther 18(2):115–124. https://doi.org/10.2174/ 1566523218666180221110627

Hashemi A (2020) CRISPR–Cas9/CRISPRi tools for cell factory construction in E. coli. World J Microbiol Biotechnol 36(7):1–13. https://doi.org/10.1007/s11274-020-02872-9

Hatoum-Aslan A, Maniv I, Marraffini LA (2011) Mature clustered, regularly interspaced, short palindromic repeats RNA (crRNA) length is measured by a ruler mechanism anchored at the precursor processing site. Proc Natl Acad Sci 108(52):21218–21222. https://doi.org/10.1073/ pnas.1112832108

Hausner M, Wuertz S (1999) High rates of conjugation in bacterial biofilms as determined by quantitative in situ analysis. Appl Environ Microbiol 65(8):3710–3713. https://doi.org/10.1128/ AEM.65.8.3710-3713.1999

Jain A, Srivastava P (2013) Broad host range plasmids. FEMS Microbiol Lett 348(2):87–96. https:// doi.org/10.1111/1574-6968.12241

Jiang W, Maniv I, Arain F, Wang Y, Levin BR, Marraffini LA (2013) Dealing with the evolutionary downside of CRISPR immunity: bacteria and beneficial plasmids. PLoS Genet 9(9):e1003844. https://doi.org/10.1371/journal.pgen.1003844

Jiang Y, Qian F, Yang J, Liu Y, Dong F, Xu C, Sun B, Chen B, Xu X, Li Y (2017) CRISPR-Cpf 1 assisted genome editing of Corynebacterium glutamicum. Nat Commun 8:15179. https://doi. org/10.1038/ncomms15179

Jinek M, Chylinski K, Fonfara I, Hauer M, Doudna JA, Charpentier E (2012) A programmable dual-RNA–guided DNA endonuclease in adaptive bacterial immunity. Science 337 (6096):816–821. https://doi.org/10.1126/science.1225829

Kang YK, Kwon K, Ryu JS, Lee HN, Park C, Chung HJ (2017) Nonviral genome editing based on a polymer-derivatized CRISPR nanocomplex for targeting bacterial pathogens and antibiotic resistance. Bioconjug Chem 28(4):957–967. https://doi.org/10.1021/acs.bioconjchem.6b00676

Kiga K, Tan XE, Ibarra-Chávez R, Watanabe S, Aiba Y, Sato'o Y, Li FY, Sasahara T, Cui B, Kawauchi M, Boonsiri T, Thitiananpakorn K, Taki Y, Azam AH, Suzuki M, Penadés JR, Cui L (2020) Development of CRISPR-Cas13a-based antimicrobials capable of sequence-specific killing of target bacteria. Nat Commun 11(1):2934. https://doi.org/10.1038/s41467-020-16731-6

Kim J-S, Cho D-H, Park M, Chung W-J, Shin D, Ko KS, Kweon D-H (2016) CRISPR/Cas9-mediated re-sensitization of antibiotic-resistant Escherichia coli harboring extended-spectrum beta-lactamases. J Microbiol Biotechnol 26(2):394–401. https://doi.org/10.4014/jmb.1508. 08080

Koonin EV, Makarova KS (2019) Origins and evolution of CRISPR-Cas systems. Philos Trans R Soc B 374(1772):20180087. https://doi.org/10.1098/rstb.2018.0087

Koskella B, Meaden S (2013) Understanding bacteriophage specificity in natural microbial communities. Viruses 5(3):806–823. https://doi.org/10.3390/v5030806

Kothari V (2020) Discovery and development of new anti-infectives: perspectives and prospective. Curr Drug Discov Technol 17(4):414–414. https://doi.org/10.2174/157016381704200908145734

Leenay RT, Vento JM, Shah M, Martino ME, Leulier F, Beisel CL (2019) Genome editing with CRISPR-Cas9 in Lactobacillus plantarum revealed that editing outcomes can vary across strains and between methods. Biotechnol J 14(3):1700583. https://doi.org/10.1002/biot.201700583

Li Q, Seys FM, Minton NP, Yang J, Jiang Y, Jiang W, Yang S (2019) CRISPR–Cas9^{D10A} nickase-assisted base editing in the solvent producer *Clostridium beijerinckii*. Biotechnol Bioeng 116 (6):1475–1483. https://doi.org/10.1002/bit.26949

Li Z, Liu Z-B, Xing A, Moon BP, Koellhoffer JP, Huang L, Ward RT, Clifton E, Falco SC, Cigan AM (2015) Cas9-guide RNA directed genome editing in soybean. Plant Physiol 169 (2):960–970. https://doi.org/10.1104/pp.15.00783

Liu C, Zhang L, Liu H, Cheng K (2017) Delivery strategies of the CRISPR-Cas9 gene-editing system for therapeutic applications. J Control Release 266:17–26. https://doi.org/10.1016/j.jconrel.2017.09.012

Liu H, Li H, Liang Y, Du X, Yang C, Yang L, Xie J, Zhao R, Tong Y, Qiu S, Song H (2020) Phage-delivered sensitisation with subsequent antibiotic treatment reveals sustained effect against antimicrobial resistant bacteria. Theranostics 10(14):6310–6321. https://doi.org/10.7150/thno.42573

Long C, McAnally JR, Shelton JM, Mireault AA, Bassel-Duby R, Olson EN (2014) Prevention of muscular dystrophy in mice by CRISPR/Cas9–mediated editing of germline DNA. Science 345 (6201):1184–1188. https://doi.org/10.1126/science.1254445

Matsuzaki S, Yasuda M, Nishikawa H, Kuroda M, Ujihara T, Shuin T, Shen Y, Jin Z, Fujimoto S, Nasimuzzaman M (2003) Experimental protection of mice against lethal Staphylococcus aureus infection by novel bacteriophage φMR11. J Infect Dis 187(4):613–624. https://doi.org/10.1086/374001

Mohanraju P, Makarova KS, Zetsche B, Zhang F, Koonin EV, Van der Oost J (2016) Diverse evolutionary roots and mechanistic variations of the CRISPR-Cas systems. Science 353:6299. https://doi.org/10.1126/science.aad5147

Moo C-L, Yang S-K, Yusoff K, Ajat M, Thomas W, Abushelaibi A, Lim S-H-E, Lai K-S (2020) Mechanisms of antimicrobial resistance (AMR) and alternative approaches to overcome AMR. Curr Drug Discov Technol 17(4):430–447. https://doi.org/10.2174/1570163816666190304122219

Naduthodi MIS, Barbosa MJ, van der Oost J (2018) Progress of CRISPR-Cas based genome editing in photosynthetic microbes. Biotechnol J 13(9):1700591. https://doi.org/10.1002/biot.201700591

Narenji H, Gholizadeh P, Aghazadeh M, Rezaee MA, Asgharzadeh M, Kafil HS (2017) Peptide nucleic acids (PNAs): currently potential bactericidal agents. Biomed Pharmacother 93:580–588. https://doi.org/10.1016/j.biopha.2017.06.092

Oliveira PH, Touchon M, Rocha EP (2014) The interplay of restriction-modification systems with mobile genetic elements and their prokaryotic hosts. Nucleic Acids Res 42(16):10618–10631. https://doi.org/10.1093/nar/gku734

Pawluk A, Davidson AR, Maxwell KL (2018) Anti-CRISPR: discovery, mechanism and function. Nat Rev Microbiol 16(1):12. https://doi.org/10.1038/nrmicro.2017.120

Pawluk A, Staals R, Taylor C, Watson B, Saha S, Fineran P (2016) Inactivation of CRISPR-Cas systems by anti-CRISPR proteins in diverse bacterial species. Nat Microbiol 1(8):16085. https://doi.org/10.1038/nmicrobiol.2016.85

Pérez-Mendoza D, de la Cruz F (2009) Escherichia coli genes affecting recipient ability in plasmid conjugation: are there any? BMC Genomics 10(1):71. https://doi.org/10.1186/1471-2164-10-71

Rajagopal N, Srinivasan S, Kooshesh K, Guo Y, Edwards MD, Banerjee B, Syed T, Emons BJ, Gifford DK, Sherwood RI (2016) High-throughput mapping of regulatory DNA. Nat Biotechnol 34(2):167–174. https://doi.org/10.1038/nbt.3468

Roach DR, Leung CY, Henry M, Morello E, Singh D, Di Santo JP, Weitz JS, Debarbieux L (2017) Synergy between the host immune system and bacteriophage is essential for successful phage therapy against an acute respiratory pathogen. Cell Host Microbe 22(1):38–47. https://doi.org/10.1016/j.chom.2017.06.018

Sarma AP, Jain C, Solanki M, Ghangal R, Patnaik S (2020) Role of gene editing tool CRISPR-Cas in the Management of Antimicrobial Resistance. In: Panwar H, Sharma C, Lichtfouse E (eds) Sustainable agriculture reviews 46: mitigation of antimicrobial resistance. Tools and targets, vol 1. Springer, Cham, pp 129–146. https://doi.org/10.1007/978-3-030-53024-2_6

Selle K, Barrangou R (2015) CRISPR-based technologies and the future of food science. J Food Sci 80(11):R2367–R2372. https://doi.org/10.1111/1750-3841.13094

Selle K, Fletcher JR, Tuson H, Schmitt DS, McMillan L, Vridhambal GS, Rivera AJ, Montgomery SA, Fortier LC, Barrangou R, Theriot CM, Ousterout DG (2020) In vivo targeting of Clostridioides difficile using phage-delivered CRISPR-Cas3 antimicrobials. MBio 11:2. https://doi.org/10.1128/mBio.00019-20

Singh KS, Anand S, Dholpuria S, Sharma JK, Shouche Y (2020) Antimicrobial resistance paradigm and one-health approach. In: Sustainable agriculture reviews. Springer, Cham, pp 1–32

Sinkunas T, Gasiunas G, Fremaux C, Barrangou R, Horvath P, Siksnys V (2011) Cas3 is a single-stranded DNA nuclease and ATP-dependent helicase in the CRISPR/Cas immune system. EMBO J 30(7):1335–1342. https://doi.org/10.1038/emboj.2011.41

Smillie C, Garcillán-Barcia MP, Francia MV, Rocha EP, de la Cruz F (2010) Mobility of plasmids. Microbiol Mol Biol Rev 74(3):434–452. https://doi.org/10.1128/MMBR.00020-10

Song X, Huang H, Xiong Z, Ai L, Yang S (2017) CRISPR-Cas9^{D10A} nickase-assisted genome editing in *Lactobacillus casei*. Appl Environ Microbiol 83(22):e01259–e01217. https://doi.org/10.1128/aem.01259-17

Standage-Beier K, Zhang Q, Wang X (2015) Targeted large-scale deletion of bacterial genomes using CRISPR-nickases. ACS Synth Biol 4(11):1217–1225. https://doi.org/10.1021/acssynbio.5b00132

Swarts DC, Mosterd C, Van Passel MW, Brouns SJ (2012) CRISPR interference directs strand specific spacer acquisition. PLoS One 7(4):e35888. https://doi.org/10.1371/journal.pone.0035888

van Houte S, Buckling A, Westra ER (2016a) Evolutionary ecology of prokaryotic immune mechanisms. Microbiol Mol Biol Rev 80(3):745–763. https://doi.org/10.1128/MMBR.00011-16

van Houte S, Ekroth AK, Broniewski JM, Chabas H, Ashby B, Bondy-Denomy J, Gandon S, Boots M, Paterson S, Buckling A (2016b) The diversity-generating benefits of a prokaryotic adaptive immune system. Nature 532(7599):385–388. https://doi.org/10.1038/nature17436

Vento JM, Crook N, Beisel CL (2019) Barriers to genome editing with CRISPR in bacteria. J Ind Microbiol Biotechnol 46(9–10):1327–1341. https://doi.org/10.1007/s10295-019-02195-1

Vercoe RB, Chang JT, Dy RL, Taylor C, Gristwood T, Clulow JS, Richter C, Przybilski R, Pitman AR, Fineran PC (2013) Cytotoxic chromosomal targeting by CRISPR/Cas systems can reshape bacterial genomes and expel or remodel pathogenicity islands. PLoS Genet 9(4):e1003454. https://doi.org/10.1371/journal.pgen.1003454

Wright GD (2016) Antibiotic adjuvants: rescuing antibiotics from resistance. Trends Microbiol 24 (11):862–871. https://doi.org/10.1016/j.tim.2016.06.009

Wu Z, Chen Z, Gao X, Li J, Shang G (2019) Combination of ssDNA recombineering and CRISPR-Cas9 for Pseudomonas putida KT2440 genome editing. Appl Microbiol Biotechnol 103 (6):2783–2795. https://doi.org/10.1007/s00253-019-09654-w

Yosef I, Goren MG, Qimron U (2012) Proteins and DNA elements essential for the CRISPR adaptation process in Escherichia coli. Nucleic Acids Res 40(12):5569–5576. https://doi.org/10.1093/nar/gks216

Yosef I, Manor M, Kiro R, Qimron U (2015) Temperate and lytic bacteriophages programmed to sensitize and kill antibiotic-resistant bacteria. Proc Natl Acad Sci 112(23):7267–7272. https://doi.org/10.1073/pnas.1500107112

Zerbini F, Zanella I, Fraccascia D, König E, Irene C, Frattini LF, Tomasi M, Fantappiè L, Ganfini L, Caproni E (2017) Large scale validation of an efficient CRISPR/Cas-based multi gene editing protocol in Escherichia coli. Microb Cell Fact 16(1):68. https://doi.org/10.1186/s12934-017-0681-1

Antibiotic-Resistant *Klebsiella pneumoniae* and Targeted Therapy

9

Ishika Verma, Rika Semalty, and Reema Gabrani

Contents

Abstract

Rapid progression of antimicrobial resistance is a major concern in current scenario. *Klebsiella pneumoniae* is one of the ESKAPE pathogens that exhibits multidrug resistance. *K. pneumoniae* is linked to pneumonia, urinary tract infections and septicaemia. The development of antimicrobial resistance in *K. pneumoniae* has led to decreased efficacy of conventional therapeutics against

I. Verma · R. Semalty · R. Gabrani (✉)
Department of Biotechnology, Jaypee Institute of Information Technology, Noida, Uttar Pradesh, India
e-mail: reema.gabrani@jiit.ac.in

© The Author(s), under exclusive license to Springer Nature Singapore Pte Ltd. 2022
V. Kumar et al. (eds.), *Antimicrobial Resistance*,
https://doi.org/10.1007/978-981-16-3120-7_9

233

the pathogen. The resistance could occur due to drug inactivation, increased efflux or altered binding to the target site. The resistance is further compounded by extended spectrum β-lactamase (ESBL) production or biofilm formation by many strains of *K. pneumoniae*. Constant increase in mortality rate owing to antimicrobial resistance makes it imperative to look out for alternative approaches to therapy. Many therapies like drugs in combination, bacteriophages, phytochemicals, nanoparticles, antimicrobial peptides and photodynamic light therapy are being used to tackle the issues due to antimicrobial resistance. The chapter discusses about mechanisms of resistance and various possible therapeutic approaches to overcome antimicrobial resistance in *K. pneumoniae*.

Keywords

Antimicrobial peptides · Bacteriophage · Drug resistance · ESKAPE pathogen · Nanoparticle formulations

9.1   Introduction

Klebsiella pneumoniae is a type of Gram-negative, nonmotile and encapsulated bacteria that belongs to the *Enterobacteriaceae* family (Ahmed et al. 2016). The polysaccharide capsule, which allows the bacteria to evade phagocytosis- and antibody-mediated killing by the host organism, is the most important virulence factor of the bacterium. Another virulence factor is the outer surface of the bacteria, which is coated with lipopolysaccharides. A release of an inflammatory cascade in the host organism due to the sensing of lipopolysaccharides is responsible for sepsis and septic shock. Fimbriae and siderophores also are virulence factors for the bacterium. Fimbriae help the bacterium in attaching itself to the host cell, and siderophores allow propagation of infection in the host by acquiring iron from the host (Ashurst and Dawson 2020; Highsmith and Jarvis 1985). *K. pneumoniae* can live as a saprophyte in the gastrointestinal tract and nasopharynx and are also present in the human stool (Ashurst and Dawson 2020).

   K. pneumoniae causes both community-acquired and hospital-acquired pneumonia, but the latter is more frequent (Ashurst and Dawson 2020). Usually, immunocompromised individuals who are taking long courses of antibiotics and those who require care devices like intravenous catheters or ventilators are at a high risk of *K. pneumoniae* infections (Paczosa and Mecsas 2016). Patients in healthcare settings can acquire *K. pneumoniae* infection because of being in contact with an infected person, by the contamination of the environment or via ventilators, intravenous catheters or wounds during surgery (Monegro et al. 2020). *K. pneumoniae* can cause a variety of infections such as urinary tract infections (UTIs), intra-abdominal infection, liver abscesses, bloodstream infection (BSI) and pneumonia (Vading et al. 2018).

9.2   Drug Resistance in *K. pneumoniae*

The term "ESKAPE" is an acronym for six pathogens which exhibit multidrug resistance. The group consists of *Enterococcus faecium, Staphylococcus aureus, K. pneumoniae, Acinetobacter baumannii, Pseudomonas aeruginosa,* and *Enterobacter species.* ESKAPE pathogens are a major cause of lethal nosocomial infections (Mulani et al. 2019; De Oliveira et al. 2020). Negligent exploitation of antimicrobial drugs is the primary cause of rapid development of resistance against these agents in *K. pneumoniae*-correlated infections (Li et al. 2019). Emergence of multidrug-resistant *K. pneumoniae* is a worldwide concern leading to a high mortality rate (Sikarwar and Batra 2011). *K. pneumoniae* shows resistance against the four major antibiotic classes: carbapenems, cephalosporins, aminoglycosides and fluoroquinolones leading to the therapeutic failure of these agents (Ferreira et al. 2019). Various resistant strategies are used by *K. pneumoniae* to exhibit multidrug resistance against these agents. Enzymatic drug inactivation and modification, alteration of drug targets, increased efflux of the drug and porin loss are the major strategies (Sikarwar and Batra 2011; Mulani et al. 2019) (Fig. 9.1). Pathogen-induced biofilm formation inhibits the immune response of the host and the antibiotics by protecting specialised dormant cells called persister cells (Mulani et al. 2019).

Fig. 9.1 Various mechanisms conferring drug resistance to *K. pneumoniae*

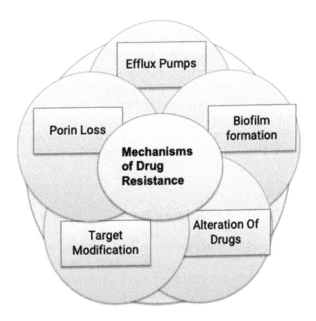

9.3   Mechanisms of Drug Resistance

9.3.1   Drug Resistance Due to Efflux Pumps

Efflux pumps are membrane proteins associated with the expulsion of substances across the cell membrane, without altering or degrading the compound. Efflux pumps decrease the intracellular drug concentrations by releasing the antimicrobial cells outside the cells, leading to reduced susceptibility to a broad range of antibiotics. On the basis of structure and substrate specificity, efflux pumps can be classified into five major families: the major facilitator superfamily (MFS), the small multidrug resistance (SMR) family, the multidrug and toxic compound extrusion (MATE) family, the ATP binding cassette (ABC) family and the resistance nodulation-cell division (RND) family (Willers et al. 2016). Frequent use of antibiotics has led to microorganisms overexpressing chromosomally encoded efflux pumps to induce cross-resistance to β-lactams (Willers et al. 2016; Türkel et al. 2018). The AcrAB pump, from RND family, is the most common overexpressed system in the case of *K. pneumoniae*. Overexpression of acrAB is responsible for the development of resistance against a broad range of drugs like chloramphenicol, β-lactams, erythromycin, fluoroquinolones, tetracycline, aminoglycosides and tigecycline (Türkel et al. 2018). According to a study, overexpression of the efflux pumps AcrAB and quinolone/olaquindox efflux pump (OqxAB) confers tigecycline and quinolone resistance in *K. pneumoniae*. Mutations in ramR, acrR and rpsJ genes, which encode efflux pumps transcriptional regulators, also lead to tigecycline non-susceptibility (Chiu et al. 2017; Zheng et al. 2018). Another study revealed that the upregulation of AcrAB, OqxAB efflux pumps and RamA, SoxS transcriptional activators is responsible for eravacycline resistance (Zheng et al. 2018).

Conjugative plasmids transmit antimicrobial resistance genes among pathogens. A study has reported carbapenem resistance by carbapenem-resistant plasmids (blaKPC-2 or blaNDM-1 plasmids) and colistin resistance by colistin-resistant plasmid (mcr plasmids). Novel plasmid containing tmexCD1-toprJ1 and mcr8.5 genes leads to simultaneous transmission of colistin and tigecycline resistance (Sun et al. 2020).

9.3.2   Drug Resistance Due to Porin Loss

Alterations in membrane permeability affect the drug influx and prevent interaction between the drug and its intracellular target leading to drug resistance. Outer membrane protein (OMP) or porins are trimeric transmembrane proteins expressed in large amounts in the outer membrane of Gram-negative bacteria. Porins form channels, which allow nonspecific diffusion of molecules across the lipid bilayer membranes. Porin channels majorly control the influx of hydrophilic and charged drugs like β-lactams and fluoroquinolones. In *K. pneumoniae*, OmpK35 and OmpK36 are two major nonspecific porins associated with antimicrobial resistance. LamB, OmpK26, PhoE and KpnO porins also contribute towards the intrinsic resistance. Insertional interruption is the most observed type of change in OmpK36-deficient strains. Insertion sequences (IS), simple transposable elements responsible for genetic variability, vary from highly specific sequences to random insertions (Wu et al. 2020). The lower expression of OmpK35 due to insertional disruption in the promoter region is associated with resistance against chloramphenicol, cephalosporins, ciprofloxacin and meropenem. It has been reported that insertional inactivation by IS903 of ompK36 is the major cause of porin alterations. The coding region of ompK36 gene is disturbed by the insertion of ISEc68. The downregulation of OmpK36 expression is related to cefoxitin, fluoroquinolone and cephalosporin resistance (Wu et al. 2020; Pulzova et al. 2017). The expression of alternative porins has been observed with the downregulation of OmpK35 and OmpK36. OmpK26 and LamB are less permeable to β-lactams as compared to OmpK35 and OmpK36. A study showed a significant increase in ompK26 gene expression in OmpK35-deficient strains. The overexpression of the lamB gene has been correlated with the loss of OmpK36 (Brunson et al. 2019). OmpK26, which is less permeable to carbapenem, is expressed in the absence of LamB. This results in carbapenem resistance due to under-expression of LamB porin. The disruption of KpnO porin renders resistance against β-lactams, aminoglycosides and tetracycline (Pulzova et al. 2017).

9.3.3 Drug Resistance Due to Target Modification

Drug target modification is simply the alteration in the drug binding site so that the drug cannot bind to its specific site anymore. Hence, the organism is rendered resistant to that drug.

Polymyxin binds to the negatively charged lipopolysaccharides (LPS) and displaces the cations in the outer membrane finally leading to cell lysis (Trimble et al. 2016). Lipid A, a part of conserved LPS structure, embedded in the outer membrane of Gram-negative bacteria, is an important target for polymyxin (Velkov et al. 2014). The mechanism of resistance in *K. pneumoniae* against polymyxin commonly involves modification of lipid A. This modification is achieved by the incorporation of positively charged moieties such as 4-amino-4-deoxy-L-arabinopyranose and palmitoyl (Velkov et al. 2014). These additions neutralise lipid A, hence reducing the binding affinity of polymyxin (Haeili et al. 2017).

Fosfomycin, another antibiotic used to treat *K. pneumoniae* infections, inhibits biogenesis of cell wall by targeting MurA enzyme. The mechanism of resistance against fosfomycin involves the modification of the target MurA (Liu et al. 2020; Lu et al. 2016).

Fluoroquinolones inhibit the activity of enzymes important for DNA replication and transcription – topoisomerase IV and DNA gyrase. Topoisomerase IV has two subunits ParC and ParE. DNA gyrase also has two subunits GyrA and GyrB (Mirzaii et al. 2018). Mutations in these subunits lead to the modification of the target enzymes, hence making *K. pneumoniae* resistant to fluoroquinolones (Azargun et al. 2019).

9.3.4 Drug Resistance Due to Alteration of the Drug

Drug alteration is a major mechanism of resistance against antibiotics in *K. pneumoniae*. This alteration is done by drug inactivating enzymes, which are produced by the bacteria (Santajit and Indrawattana 2016).

One of the major examples of drug alteration is the presence of enzymes called ®-lactamases which hydrolyses the beta rings of β-lactams. Hence, the expression of these enzymes in *Klebsiella pneumoniae* makes it resistant to penicillins, cephalosporins and carbapenems. As per the Ambler classification, there are four classes of β-lactamases—A, B, C and D (Hall and Barlow 2005).

Class A, C and D enzymes, collectively known as serine-beta-lactamases, form an acyl enzyme to hydrolyses their targets through an active site serine. Class B enzymes are metallo-β-lactamases which utilise zinc in their active site to hydrolyse their substrate (Swain and Padhy 2016; Bush and Jacoby 2010).

In *K. pneumoniae* class A, β-lactamases are sulfhydryl variable β-lactamases (SHV), TEM, cefotaxime β-lactamases (CTX-M) and *K. pneumoniae* carbapenemase (KPC).

Verona integron-encoded metallo-β-lactamase (VIM), imipenemase (IMP) and new discovered New Delhi metallo-β-lactamases (NDM) belong to class B. The OXA and AmpC belong to classes C and D, respectively (Santajit and Indrawattana 2016; Navon-Venezia et al. 2017).

KPC, which is encoded by bla_{KPC} gene, confers resistance against carbapenems. *K. pneumoniae* infections caused by strains carrying bla_{KPC} gene are challenging to treat. The occurrence of NDM-1 and OXA-48 in *K. pneumoniae* strains is common in India (Remya et al. 2018; Shankar et al. 2018).

Resistance against aminoglycosides is caused by modifying enzymes that alter hydroxyl or amine groups of 2-deoxystreptamine nucleus or the sugar moieties of aminoglycoside molecules (El-Badawy et al. 2017). Acetyltransferases, nucleotidyl transferases and phosphotransferases are the types of aminoglycoside-modifying enzymes (Soleimani et al. 2014; Navon-Venezia et al. 2017).

9.3.5 Drug Resistance Due to Biofilm Formation

Biofilm is the aggregation of microorganisms attached to an inert or living surface, coated by extracellular matrix, such as polysaccharides, proteins and extracellular DNA. The restricted drug penetration across the biofilm is one of the major causes of antibiotic resistance (Naparstek et al. 2014). *K. pneumoniae* possess capsule polysaccharide (CPS) at the surface which inhibits opsonisation and phagocytosis. Two types of fimbriae produced by *K. pneumoniae* are associated with the biofilm formation. Type 1 fimbriae is encoded by fim gene, whereas type 3 fimbriae is encoded by mrk gene (Desai et al. 2019). MrkA type 3 fimbrial protein is responsible for the growth of *K. pneumoniae* on abiotic surfaces. MrkD type 3 protein facilitates fimbrial adhesion on the extracellular matrix. TreC and sugE genes modulate CPS production, thereby affecting the biofilm synthesis. The luxS and pgaABCD operons enhance biofilm formation by promoting intercellular adhesion and abiotic surface binding (Vuotto et al. 2017). KPF-28 adhesin leads to the colonisation of *K. pneumoniae*. A study showed *K. pneumoniae* biofilms had reduced susceptibility to gentamicin, ampicillin and ciprofloxacin (Chung et al. 2016). Another study showed resistance of *K. pneumoniae* biofilms to ceftazidime, cefepime, meropenem, ciprofloxacin and piperacillin-tazobactam (Vuotto et al. 2014). Colistin resistance has also been associated with the formation of biofilms (Cepas et al. 2019).

9.4   Novel Therapies to Overcome Drug Resistance in *K. pneumoniae* Infections

For combating drug resistance in *K. pneumoniae* infections, several therapies are currently being developed (Fig. 9.2). These therapies aim either to kill/limit the growth of resistant strains or increase the efficiency of existing antimicrobials. Also, few of these therapies decelerate the development of further resistance.

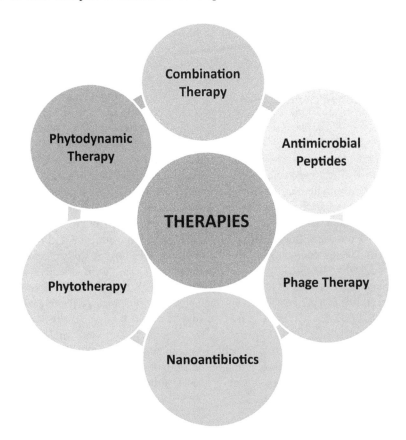

Fig. 9.2 Therapies used to treat infections caused by drug-resistant *K. pneumoniae*

9.4.1 Phage Therapy

Bacteriophages are the most abundant and diverse life forms on the Earth and are found in every ecological niche that supports bacterial proliferation. Lytic bacteriophages infect the host cells by attaching to specific surface epitopes and inject the phage genetic material to hijack the host machinery, thereby inducing replication, assembly and release of infectious progeny by bursting the host cell (Principi et al. 2019). This reduces the development cost of phage therapy compared to antibiotics (Szeloch et al. 2013). Bacteriophages display low natural toxicity and high strain-specific selectivity. The effect of bacteriophage is limited to the site of infection due to their specificity, which prevents the inherent microbiota from damage (Moghadam et al. 2020). Lack of cross-resistance with antibiotics in case of bacteriophages provides them an edge over conventional antibiotics. Bacteriophages overcome the host resistance mechanisms by inhibiting host restriction enzymes, hydrolysing cofactors and simultaneous injection of viral DNA and resistance mechanism inhibitors. According to a study, phage KP36, a member of the Tuna virus, has developed various mechanisms to overcome host restriction (Szeloch et al. 2013). Another study demonstrated that bacteriophage ZCKP1, of the Caudovirales order, had a broad lytic profile and was highly effective against antimicrobial-resistant *K. pneumoniae*. Phage ZCKP1 inhibited the formation of *K. pneumoniae* biofilms by disrupting the bacterial capsule owing to the exopolysaccharide depolymerase activity, thereby rendering the host cell vulnerable to antibacterial agents. ZCKP1 was stable at high temperatures along with a wide pH range (Taha et al. 2018). Phage 0507-KN2–1 was reported to be specific to KN2 capsular-type *K. pneumoniae* due to KN2 capsule recognising polysaccharide depolymerase (Hsu et al. 2013). Phage KPO1K2, of the Podoviridae family, has been reported to infect *K. pneumoniae* B5055 and a few more isolates. KPO1K2 had a large burst size and was effective over a wide range of pH and temperature (Verma et al. 2009). The lytic phage φBO1E targets KPC-producing *K. pneumoniae* due to specific polysaccharide recognition in the host capsule (Herridge et al. 2020).

Phage cocktail consists of a combination of phages exhibiting different but overlapping host specificities. They possess a broad lytic spectrum and reduce the chances of development of resistance within the bacterial host. A study proposed that phage cocktails, BFC-1, Pyophage and PhageBioDerm, were effective against *K. pneumoniae* (Haddad et al. 2019). It was shown that *K. pneumoniae* strain K7 was susceptible to a phage cocktail consisting of three lytic phages, namely, GH-K1, GH-K2 and GH-K3 (Gu et al. 2012). A study showed that a cocktail of six lytic phages (HD001, Kp152, Kp154, Kp155, Kp164 and Kp6377) in combination with trimethoprim-sulfamethoxazole effectively inhibited drug-resistant *K. pneumoniae* (Bao et al. 2020). Hence, a combination of antibiotics with lytic phage is another approach to prevent the development of resistance in the host cell.

9.4.2 Nanoantibiotics

Nanoantibiotics refer to the nanoparticles that possess antimicrobial activity. Application of nanoparticles (NPs) is an alternative therapy used to overcome drug resistance by regulating cell membrane disruption, free radical generation leading to enhanced oxidative stress, denaturation of proteins, inhibition of DNA replication and hampering metabolic pathways (Khana et al. 2016; Lee et al. 2019). NPs coupled with antibiotics exhibit synergistic antimicrobial effects. Zinc oxide (ZnO) NPs inhibited the development of *K. pneumoniae* in the lag phase by disrupting the lipopolysaccharide membrane and suppression of porins. ZnO NPs were also responsible for increased oxidative stress due to deregulation in synthesis of antioxidant enzyme catalase (Reddy et al. 2014). A study conjugated histidine (H-)-rich peptides, known to have antimicrobial activity, with silver nanoparticles (AgNPs) to treat MDR *K. pneumoniae*. H-AgNPs alone and in combination with gentamicin showed significant deduction in biofilm production. The decrease in bacterial count was attributed to the DNA fragmentation due to H-AgNPs (Chhibber et al. 2017). Bactericidal effects of AgNPs were also linked with alteration of protein expression and induction of apoptosis (Rai et al. 2012). According to a study, extract from *Mukia scabrella*, herbal plant effective against respiratory tract ailments, when combined with AgNPs was successful in inhibiting drug-resistant *K. pneumoniae* (Prabakar et al. 2013). The combination of AgNPs and imipenem has been administered to overcome imipenem resistance in the pathogen (Aziz et al. 2017). Chitosan nanoparticles (CSNPs) are nontoxic, biodegradable polymers highly effective against MDR *K. pneumoniae* (Jamil et al. 2016). Induction of reactive oxygen species (ROS) along with the release of Zn ions from zinc oxide (ZnO) and neodymium (Nd)-doped ZnO NPs has shown to inhibit the growth of *K. pneumoniae* (Aziz et al. 2017). PF-TiO$_2$ (phosphorus and fluorine co-doped titanium dioxide) NPs exhibited antibacterial potential by production of ROS which damaged the DNA and other metabolites of the bacterial cell (Kőrösi et al. 2018). Chlorhexidine (CHX), an antibacterial agent, has been coupled with gold (Au) NPs to form Au-CHX NPs. Au-CHX NPs disrupted the biofilm formation and hampered the growth of *K. pneumoniae* (Ahmed et al. 2016). Another study showed a successful approach to overcome tigecycline-resistant *K. pneumonia* (TRKP) by loading tigecycline (TIG) inside I-TPGS/Ga2O3 NPs. These formulations reduced the overexpression of efflux pump genes, thereby preventing expulsion of antibiotics from the bacterial cells and rendered the cells sensitive to the antibiotic (Kang et al. 2019).

9.4.3 Phytotherapy

The use of bioactive components of plants, like secondary metabolites, containing antimicrobial properties is an alternative approach to overcome antimicrobial resistance. Plant antimicrobials act synergistically with conventional antibiotics to enhance the antibiotic efficacy by inhibiting metabolic pathways, disrupting cell membrane and hampering the bacterial enzymes (Haroun et al. 2016). Increased efficiency, accessibility and minimum side effects make phytotherapy a favourable approach (Salou et al. 2019). *Plectranthus amboinicus, Lamiaceae* family, contain essential oils that change the membrane composition resulting in increase of membrane fluidity. *P. amboinicus* leaf extract exhibited antioxidant properties that damaged and lysed the methicillin-resistant strains (Ismayil and Nimila 2019). A study reported that *Thymbra spicata* of the Lamiaceae family enhanced the activity of amikacin, ampicillin and cefotaxime against resistant *K. pneumoniae* by more than four times. *T. spicata* extract inhibited protein and DNA synthesis and depolarised the cell membrane to facilitate the influx of antibiotics (Haroun et al. 2016). Another study showed that methanol extract from *Nauclea pobeguinii* roots increased the efficacy of ampicillin and amoxicillin when used in combination (Seukep et al. 2016). Essential oils like thyme oil, peppermint oil, cinnamon oil and eucalyptus oil exhibited strong biofilm inhibition potential. These oils when used in combination with ciprofloxacin reduced the MIC of the drug against *K. pneumoniae* (Mohamed et al. 2018). A study reported ethanol extract of *Punica granatum* showed synergistic antibacterial properties with ciprofloxacin, ceftazidime, meropenem, gentamicin and norfloxacin. Constituents of butanol fraction were suggested to be efflux pump inhibitors (Rafiq et al. 2017). *K. pneumoniae* was shown to be sensitive to alcoholic extracts of propolis (Taher 2015). Lavender essential oil in combination with meropenem disrupted the cell wall by inducing oxidative stress. This allowed a significant influx of meropenem inside the cell resulting in reduced MIC (Yang et al. 2020a, b). *K. pneumoniae* was shown to be highly sensitive to *Ocimum gratissimum* extracts (Salou et al. 2019). OmpK35, OmpK36 porins were overexpressed along with downregulation of efflux pumps when cinnamaldehyde was administered with cefotaxime and ciprofloxacin. This combination resulted in damage to the cell wall, organelles ensuing apoptosis (Dhara and Tripathi 2020). Linalyl anthranilate (LNA) exhibited bactericidal action by inducing lipid peroxidation that led to loss of the cell membrane. LNA also resulted in a significant increase in oxidative stress due to ROS generation, thereby rendering *K. pneumoniae* cells susceptible to damage (Yang et al. 2020a, b).

9.4.4 Combination Therapy

Combination therapy means targeting multiple sites of a pathogen so that there are fewer chances for drug resistance development. The benefit of combination therapy is the potential delay in the emergence of resistance because multiple mechanisms of action are used simultaneously, and the pharmacodynamic killing activity of the antibiotic agents increases (Jacobs et al. 2017). The process of development of new antimicrobials is an extremely time-consuming and expensive process having many potential ramifications. Hence, employing a combination of antimicrobial agents instead of a single antimicrobial agent may be a feasible treatment option (Yu et al. 2019). Depending on the patient population's disease profile and age, the mortality rate due to infections caused by carbapenem-resistant *K. pneumoniae* strains can go up to 75%; hence, treating such infections is very important (Vivas et al. 2020).

Treating infections caused by carbapenemase-producing *K. pneumoniae* (CKP) strains is a global challenge. KPC, a class A carbapenemase, discovered in *K. pneumoniae,* is one of the significant carbapenemases associated with multidrug resistance (Vivas et al. 2020). High mortality rates and association with complex comorbidities in infections caused by KPC-producing *K. pneumoniae* (KKP) strains make them a serious problem. A study showed that the mortality of patients with KKP bloodstream infections was significantly lower when they were treated with a combination of polymyxin B and amikacin compared to when a single in vitro active monotherapy agent was administered. In the same study, the combination of polymyxin B and amikacin was shown to be beneficial over other combinations or single agents (Medeiros et al. 2019). A combination of cefepime and amoxicillin/clavulanic acid was found to be more promising and efficient than tigecycline-based therapy against *K. pneumoniae* (Ji et al. 2015).

In a study, ceftazidime/avibactam-based combination therapy showed extremely positive results with 91.7% patients recovered from KKP gut colonisation (Bassetti et al. 2019). Souli et al. stated the importance of double carbapenem combination therapy in treating untreatable KKP infections (Souli et al. 2017). A study, published in 2019, reported a combination of colistin with either meropenem or amikacin to be an option for targeting colistin-resistant CKP infections (Yu et al. 2019).

9.4.5 Antimicrobial Peptides

Antimicrobial peptides (AMPs) are a group of natural compounds having low molecular weight (<10 kDa) and can be isolated from plants and animals. They are of a wide variety depending on the sequence length (12–50 amino acids), amino acid residues and structure (Wang et al. 2016).

AMP antibacterial mechanisms include the formation of transmembrane pores leading to lysis, interference with cell wall synthesis, disruption of biochemical processes and activation of immune response (Swedan et al. 2019). Due to rising antimicrobial resistance, AMPs could be used as an alternative therapeutic agent (Dias et al. 2020). Bacteria are less likely to develop and transfer resistant genes against peptides; hence, AMPs provide a good treatment option over conventionally used antimicrobial agents. AMPs are mostly cationic in nature and possess an amphipathic structure, which can interact electrostatically with bacterial membrane targets (Swedan et al. 2019).

An AMP thanatin displaced divalent cations in the outer membrane of *K. pneumoniae* strain that produced NDM (New Delhi metallo-beta-lactamase-1). Also, it inactivated NDM-1 by displacing zinc ions from its active site, thereby reversing carbapenem resistance in NDM-1-producing *K. pneumoniae* (Ma et al. 2019).

A study reported that WLBU2, a 24-residue engineered cationic peptide containing only tryptophan, valine and arginine, can be used in combination with amoxicillin-clavulanate or ciprofloxacin for treating multidrug-resistant *K. pneumoniae* (Swedan et al. 2019). A combination of hLF1–11 (a synthetic AMP) with antibiotics was reported to have a great potential against multidrug-resistant *K. pneumoniae* infections (Morici et al. 2017).

A small peptide, RcAlb-PepII, was designed on the basis of the primary structure of Rc-2S-Alb, a 2S albumin family protein isolated from the seed cake of *Ricinus communis*. It was suggested to have a great potential to be developed as an antibacterial agent to kill or inhibit *K. pneumoniae*. RcAlb-PepII at 11 mM concentration showed bacteriostatic activity, whereas at 42 mM it showed bactericidal activity (Dias et al. 2020). Another antimicrobial peptide, cathelicidin, which was present in the serum of *Alligator mississippiensis*, showed antibacterial activity against multidrug-resistant *K. pneumoniae*. Moreover, cathelicidin peptides were not significantly toxic towards mammalian cells (Barksdale et al. 2017). A novel AMP Cm38, which was purified from the venom of *Centruroides margaritatus*, exhibited antibacterial activity against *K. pneumoniae* (Dueñas-Cuellar et al. 2015).

Biofilm-forming ability of multidrug-resistant *K. pneumoniae* was a critical issue as it allowed the bacteria to become resistant to a number of antibiotics. Thus, higher concentrations of antibiotics were required to eradicate bacteria. The R5F5, a cationic AMP, had selective anti-biofilm activity against CKP (Cardoso et al. 2019). Another anti-biofilm activity exhibiting AMP, DJK-6, has been reported to enhance the activity of beta-lactams particularly meropenem (Ribeiro et al. 2015). Polyionenes were demonstrated to be effective in treating lung infections caused by multidrug-resistant *K. pneumoniae* (Lou et al. 2018).

9.4.6 Photodynamic Therapy

Photodynamic therapy (PT) involves the use of photosensitiser (PS) and a type of light. These PS molecules are activated after being exposed to light of a certain wavelength and get excited from the ground state to a triplet state. The PS in triplet state reacts with oxygen close to the bacteria leading to the formation of ROS such as superoxide anion and hydroxyl radicals (type I reactions) or singlet oxygen ($^1O_2$, type II reactions). The ROS formed kill the bacteria nonspecifically by interacting with their cellular components and membranes (Tosato et al. 2020). The singlet oxygen or ROS are effective against drug-resistant bacteria including those producing carbapenemases. The oxidative damage caused by PT could cause carbapenemases to lose their function. Many evidences have suggested that PT can kill drug-resistant bacteria at a clinically acceptable dosage without inducing resistance (Feng et al. 2020).

The cationic PS have been demonstrated to be more effective against *K. pneumoniae* than anionic PS possibly due to its greater interaction with negatively charged membrane of the bacteria (Valenzuela-Valderrama et al. 2020a, b). A study reported photodynamic inactivation of extended-spectrum beta-lactamase (ESBL)-producing *K. pneumoniae* by 5-ALA (5-aminolevulinic acid) and its derivative MAL (5-aminolevulinic acid methyl ester). 5-ALA or MAL induced damage to the genomic DNA and envelope of ESBL-producing *K. pneumoniae*. It denatured the cytoplasmic components, released intracellular biopolymers and inactivated the biofilm too (Liu et al. 2016). Methylene blue-based PT was reported to have a great potential to inactivate hypervirulent and hypermucoviscous *K. pneumoniae* (dos Anjos et al. 2020). Another study showed that methylene blue-based PT can reverse carbapenem resistance by inactivating carbapenemases and consequently made bacteria susceptible to carbapenems again (Feng et al. 2020). Toluidine blue and rose Bengal have also been reported to be used as PS in PT against *K. pneumoniae* infections (Liu et al. 2016).

PSIR-3, Ir(III)-based cationic compound, was reported to produce an effective photodynamic effect against carbapenemase-producing *K. pneumoniae* strains and also increased its susceptibility when used in combination with imipenem (Valenzuela-Valderrama et al. 2020b). Other Ir(III)-based cationic compounds PSIR-1 and PSIR-2 were also reported to have bactericidal effects on being used as PS and could be developed as an alternative to conventional antibacterial agents to treat multidrug-resistant bacteria (Valenzuela-Valderrama et al. 2020a). A PT using a combination of PSs methylene blue and 6-carboxypterin was reported to be efficient in controlling and eradicating multidrug-resistant *K. pneumoniae* mature biofilms (Tosato et al. 2020).

9.5 Conclusion

In conclusion, increasing antibiotic resistance in *K. pneumoniae* is a global burden. *K. pneumoniae,* being a nosocomial pathogen, is continuously exposed to multiple antibiotics, which favour the selection of its resistant strains. The infections caused by the resistant *K. pneumoniae* strains are associated with limited treatment options.

Combination therapies increase the existing antimicrobial spectrum targeting important drug resistance mechanisms, such as drug efflux, biofilm formation, enzymatic inactivation of drug, alteration of drug targets and reduced permeability due to porin loss or modification. Antimicrobial combinations are an efficient way of treating multidrug-resistant *K. pneumoniae* infections, but thorough investigations are required to study the drug interactions to minimise the side effects and ultimately find the best and safest combination regime. There is a dearth of production of new antimicrobials in the post-antibiotic era. Thus, there is an urgency for the development of alternative treatment options. In this context, we have discussed the potential of some alternatives like nanoantibiotics, phytotherapy, photodynamic therapy, antimicrobial peptides and phage therapy, to be developed as antimicrobials. Extensive research and new product development are earnestly required to minimise the possible side effects and formulate the most effective regime.

Acknowledgement The authors acknowledge Jaypee Institute of Information Technology for all the support.

Conflict of Interest The authors declare no conflict of interest.

References

Ahmed A, Khan AK, Anwar A, Ali SA, Shah MR (2016) Biofilm inhibitory effect of chlorhexidine conjugated gold nanoparticles against *Klebsiella pneumoniae*. Microb Pathog 98:50–56. https://doi.org/10.1016/j.micpath.2016.06.016

Ashurst JV, Dawson A (2020) Klebsiella Pneumonia. StatPearls Publishing, Treasure Island

Azargun R, Barhaghi MHS, Kafil HS, Oskouee MA, Sadeghi V, Memar MY, Ghotaslou R (2019) Frequency of DNA gyrase and topoisomerase IV mutations and plasmid-mediated quinolone resistance genes among Escherichia coli and *Klebsiella pneumoniae* isolated from urinary tract infections in Azerbaijan, Iran. J Glob Antimicrob Resist 17:39–43. https://doi.org/10.1016/j.jgar.2018.11.003

Aziz MMA, Yosri M, Amin BH (2017) Control of imipenem resistant-*Klebsiella pneumoniae* pulmonary infection by oral treatment using a combination of mycosynthesized ag-nanoparticles and imipenem. J Radiat Res Appl Sci 10(4):353–360. https://doi.org/10.1016/j.jrras.2017.09.002

Bao J, Wu N, Zeng Y, Chen L, Li L, Yang L, Zhang Y, Guo M, Li L, Li J, Tan D, Cheng M, Gu J, Qin J, Liu J, Li S, Pan Q, Jin X, Yao B, Guo X, Zhu T, Le S (2020) Non-active antibiotic and bacteriophage synergism to successfully treat recurrent urinary tract infection caused by extensively drug-resistant *Klebsiella pneumoniae*. Emerg Microbes Infect 9(1):771–774. https://doi.org/10.1080/22221751.2020.1747950

Barksdale SM, Hrifko EJ, van Hoek ML (2017) Cathelicidin antimicrobial peptide from Alligator mississippiensis has antibacterial activity against multi-drug resistant Acinetobacter baumannii and *Klebsiella pneumoniae*. Dev Comp Immunol 70:135–144. https://doi.org/10.1016/j.dci.2017.01.011

Bassetti M, Carannante N, Pallottob C, Righia E, CaprioGD BM, Sodano G, Mallardo E, Francisci D, Sartor A, Graziano E, Tascini C (2019) KPC-producing *Klebsiella pneumoniae* gut decolonisation following ceftazidime/avibactam-based combination therapy: a retrospective observational study. J Glob Antimicrob Resist 17:109–111. https://doi.org/10.1016/j.jgar.2018.11.014

Brunson DN, Maldosevic E, Velez A, Figgins E, Ellis TN (2019) Porin loss in *Klebsiella pneumoniae* clinical isolates impacts production of virulence factors and survival within macrophages. Int J Med Microbiol 309(3–4):213–224. https://doi.org/10.1016/j.ijmm.2019.04.001

Bush K, Jacoby GA (2010) Updated functional classification of b-lactamases. Antimicrob Agents Chemother 54(3):969–976. https://doi.org/10.1128/AAC.01009-09

Cardoso MH, Santos VPM, Costa BO, Buccini DF, Rezende SB, Porto WF, Santos MJ, Silva ON, Ribeiro SM, Franco OL (2019) A short peptide with selective anti-biofilm activity against Pseudomonas aeruginosa and *Klebsiella pneumoniae* carbapenemase-producing bacteria. Microb Pathog 135:103605. https://doi.org/10.1016/j.micpath.2019.103605

Cepas V, Lopez Y, Munoz E, Rolo D, Ardanuy C, Marti S, Xercavins M, Horcajada JP, Bosch J, Soto SM (2019) Relationship between biofilm formation and antimicrobial resistance in gram-negative bacteria. Microb Drug Resist 25(1):72–79. https://doi.org/10.1089/mdr.2018.0027

Chhibber S, Gondil VS, Sharma S, Kumar M, Wangoo N, Sharma RK (2017) A novel approach for combating *Klebsiella pneumoniae* biofilm using histidine functionalized silver nanoparticles. Front Microbiol 8:1104. https://doi.org/10.3389/fmicb.2017.01104

Chiu SK, Chan MC, Huang LY, Lin YT, Lin JC, Lu PL, Siu LK, Chang FY, Yeh KM (2017) Tigecycline resistance among carbapenem-resistant *Klebsiella pneumoniae*: clinical characteristics and expression levels of efflux pump genes. PLoS One 12(4):e0175140. https://doi.org/10.1371/journal.pone.0175140

Chung PY (2016) The emerging problems of *Klebsiella pneumoniae* infections: carbapenem resistance and biofilm formation. FEMS Microbiol Lett 363(20):fnw219. https://doi.org/10.1093/femsle/fnw219

De Oliveira DMP, Forde BM, Kidd TJ, Harris PNA, Schembri MA, Beatson SA, Paterson DL, Walker MJ (2020) Antimicrobial resistance in ESKAPE pathogens. Clin Microbiol Rev 33(3): e00181–e00119. https://doi.org/10.1128/cmr.00181-19

Desai S, Sanghrajka K, Gajjar D (2019) High adhesion and increased cell death contribute to strong biofilm formation in *Klebsiella pneumoniae*. Pathogens 8(4):277. https://doi.org/10.3390/pathogens8040277

Dhara L, Tripathi A (2020) Cinnamaldehyde: a compound with antimicrobial and synergistic activity against ESBL-producing quinolone-resistant pathogenic Enterobacteriaceae. Eur J Clin Microbiol Infect Dis 39(1):65–73. https://doi.org/10.1007/s10096-019-03692-y

Dias LP, Souza PFN, Oliveira JTA, Vasconcelos IM, Araújo NMS, Tilburg MFV, Guedes MIF, Carneiro RF, Lopes JLS, Sousa DOB (2020) RcAlb-PepII, a synthetic small peptide bioinspired in the 2S albumin from the seed cake of Ricinus communis, is a potent antimicrobial agent against *Klebsiella pneumoniae* and Candida parapsilosis. Biochim Biophys Acta Biomembr 1862(2):183092. https://doi.org/10.1016/j.bbamem.2019.183092

dos Anjos C, Sabino CP, Sellera FP, Esposito F, Pogliani FC, Lincopan N (2020) Hypervirulent and hypermucoviscous strains of *Klebsiella pneumoniae* challenged by antimicrobial strategies using visible light. Int J Antimicrob Agents 56(1):106025. https://doi.org/10.1016/j.ijantimicag.2020.106025

Dueñas-Cuellar RA, Kushmerick C, Naves LA, Batista IFC, Guerrero-Vargas JA, Pires OR Jr, Fontes W, Mariana SC (2015) Cm38: a new antimicrobial peptide active against *Klebsiella pneumoniae* is homologous to Cn11. Protein Pept Lett 22(2):164–172. https://doi.org/10.2174/0929866652202150128143048

El-Badawy MF, Tawakol WM, El-Far SW, Maghrabi IA, Al-Ghamdi SA, Mansy MS, Ashour MS, Shohayeb MM (2017) Molecular identification of aminoglycoside-modifying enzymes and plasmid-mediated quinolone resistance genes among *Klebsiella pneumoniae* clinical isolates recovered from Egyptian patients. Int J Microbiol 2017:8050432. https://doi.org/10.1155/2017/8050432

Feng Y, Palanisami A, Ashraf S, Bhayana B, Hasan T (2020) Photodynamic inactivation of bacterial carbapenemases restores bacterial carbapenem susceptibility and enhances carbapenem antibiotic effectiveness. Photodiagnosis Photodyn Ther 30:101693. https://doi.org/10.1016/j.pdpdt.2020.101693

Ferreira RL, da Silva BCM, Rezende GS, Nakamura-Silva R, Pitondo-Silva A, Campanini EB, Brito MCA, da Silva EML, Freire CC d M, da Cunha AF, Pranchevicius MC d S (2019) High prevalence of multidrug-resistant *Klebsiella pneumoniae* harboring several virulence and β-lactamase encoding genes in a Brazilian intensive care unit. Front Microbiol 9:3198. https://doi.org/10.3389/fmicb.2018.03198

Gu J, Liu X, Li Y, Han W, Lei L, Yang Y, Zhao H, Gao Y, Song J, Lu R, Sun C, Feng X (2012) A method for generation phage cocktail with great therapeutic potential. PLoS One 7(3):e31698. https://doi.org/10.1371/journal.pone.0031698

Haddad LE, Harb CP, Gebara MA, Stibich MA, Chemaly RF (2019) A systematic and critical review of bacteriophage therapy against multidrug-resistant ESKAPE organisms in humans. Clin Infect Dis 69(1):167–178. https://doi.org/10.1093/cid/ciy947

Haeili M, Javani A, Moradi J, Jafari Z, Feizabadi MM, Babaei E (2017) MgrB alterations mediate Colistin resistance in *Klebsiella pneumoniae* isolates from Iran. Front Microbiol 8:2470. https://doi.org/10.3389/fmicb.2017.02470

Hall BG, Barlow M (2005) Revised Ambler classification of b-lactamases. J Antimicrob Chemother 55(6):1050–1051. https://doi.org/10.1093/jac/dki130

Haroun MF, Al-Kayali RS (2016) Synergistic effect of Thymbra spicata L. extracts with antibiotics against multidrug- resistant Staphylococcus aureus and *Klebsiella pneumoniae* strains. Iran J Basic Med Sci 19(11):1193–1200

Herridge WP, Shibu P, O'Shea J, Brook TC, Hoyles L (2020) Bacteriophages of *Klebsiella spp.*, their diversity and potential therapeutic uses. J Med Microbiol 69:176–194. https://doi.org/10.1099/jmm.0.001141

Highsmith AK, Jarvis WR (1985) *Klebsiella pneumoniae*: selected virulence factors that contribute to pathogenicity. Infect Control 6(2):75–77. https://doi.org/10.1017/S0195941700062640

Hsu CR, Lin TL, Pan YJ, Hsieh PF, Wang JT (2013) Isolation of a bacteriophage specific for a new capsular type of *Klebsiella pneumoniae* and characterization of its polysaccharide depolymerase. PLoS One 8(8):e70092. https://doi.org/10.1371/journal.pone.0070092

Ismayil S, Nimila PJ (2019) Antimicrobial activity of *Plectranthus amboinicus* (Lour.) against gram negative bacteria *Klebsiella pneumoniae* and *Shigella flexneri* and their phytochemical tests. Int J Health Sci Res 9(5):304–311

Jacobs DM, Safir MC, Huang D, Minhaj F, Parker A, Rao GG (2017) Triple combination antibiotic therapy for carbapenemase-producing *Klebsiella pneumoniae*: a systematic review. Ann Clin Microbiol Antimicrob 16:76. https://doi.org/10.1186/s12941-017-0249-2

Jamil B, Habibb H, Abbasi S, Nasir H, Rahmane A, Rehmanf A, Bokhari H, Imrana M (2016) Cefazolin loaded chitosan nanoparticles to cure multi drug resistant gram-negative pathogens. Carbohydr Polym 136:682–691. https://doi.org/10.1016/j.carbpol.2015.09.078

Ji S, Lv F, Du X, Wei Z, Fu Y, Mu X, Jiang Y, Yu Y (2015) Cefepime combined with amoxicillin/clavulanic acid: a new choice for the KPC-producing K. pneumoniae infection. Int J Infect Dis 38:108–114. https://doi.org/10.1016/j.ijid.2015.07.024

Kang XQ, Shu GF, Jiang SP, Xu XL, Qi J, Jin FY, Liu D, Xiao YH, Lu XY, Du YZ (2019) Effective targeted therapy for drug-resistant infection by ICAM-1 antibody-conjugated TPGS modified β-Ga$_2$O$_3$:Cr^{3+} nanoparticles. Theranostics 9(10):2739–2753. https://doi.org/10.7150/thno.33452

Khana ST, Musarrat J, Al-Khedhairya AA (2016) Countering drug resistance, infectious diseases, and sepsis using metal and metal oxides nanoparticles: current status. Colloids Surf B Biointerfaces 146:70–83. https://doi.org/10.1016/j.colsurfb.2016.05.046

Kőrösi L, Bognár B, Horváth M, Schneider G, Kovács J, Scarpellini A, Castelli A, Colombo M, Prato M (2018) Hydrothermal evolution of PF-co-doped TiO2 nanoparticles and their antibacterial activity against carbapenem-resistant *Klebsiella pneumoniae*. Appl Catal B 231:115–122. https://doi.org/10.1016/j.apcatb.2018.03.012

Lee NY, Ko WC, Hsueh PR (2019) Nanoparticles in the treatment of infections caused by multidrug resistant organisms. Front Microbiol 10:1153. https://doi.org/10.3389/fphar.2019.01153

Li G, Zhao S, Wang S, Sun Y, Zhou Y, Pan X (2019) A 7-year surveillance of the drug resistance in *Klebsiella pneumoniae* from a primary health care center. Ann Clin Microbiol Antimicrob 18:34. https://doi.org/10.1186/s12941-019-0335-8

Liu C, Zhou Y, Wang L, Han L, Lei J, Ishaq HM, Nair SP, Xu J (2016) Photodynamic inactivation of *Klebsiella pneumoniae* biofilms and planktonic cells by 5-aminolevulinic acid and 5-aminolevulinic acid methyl ester. Lasers Med Sci 31(3):557–565. https://doi.org/10.1007/s10103-016-1891-1

Liu P, Chen S, Wu ZY, Qi M, Li XY, Liu CX (2020) Mechanisms of fosfomycin resistance in clinical isolates of carbapenem-resistant *Klebsiella pneumoniae*. J Glob Antimicrob Resist 22:238–243. https://doi.org/10.1016/j.jgar.2019.12.019

Lou W, Venkataraman S, Zhong G, Ding B, Tan JPK, Xu L, Fan W, Yang YY (2018) Antimicrobial polymers as therapeutics for treatment of multidrug-resistant *Klebsiella pneumoniae* lung infection. Acta Biomater 78:78–88. https://doi.org/10.1016/j.actbio.2018.07.038

Lu PL, Hsieh YJ, Lin JE, Huang JW, Yang TY, Lin L, Tseng SP (2016) Characterisation of fosfomycin resistance mechanisms and molecular epidemiology in extended-spectrum β-lactamase-producing *Klebsiella pneumoniae* isolates. Int J Antimicrob Agents 48 (5):564–568. https://doi.org/10.1016/j.ijantimicag.2016.08.013

Ma B, Fang C, Lu L, Wang M, Xue X, Zhou Y, Li M, Hu Y, Luo X, Hou Z (2019) The antimicrobial peptide thanatin disrupts the bacterial outer membrane and inactivates the NDM-1 metallo-β-lactamase. Nat Commun 10:3517. https://doi.org/10.1038/s41467-019-11503-3

Medeiros GS, Rigatto MH, Falci DR, Zavascki AP (2019) Combination therapy with polymyxin B for carbapenemase-producing *Klebsiella pneumoniae* bloodstream infection. Int J Antimicrob Agents 53(2):152–157. https://doi.org/10.1016/j.ijantimicag.2018.10.010

Mirzaii M, Jamshidi S, Zamanzadeh M, Marashifard M, Malek Hosseini SAA, Haeili M, Jahanbin F, Mansouri F, Darban-Sarokhalil D, Khoramrooz SS (2018) Determination of gyrA and parC mutations and prevalence of plasmid-mediated quinolone resistance genes in Escherichia coli and *Klebsiella pneumoniae* isolated from patients with urinary tract infection in Iran. J Glob Antimicrob Resist 13:197–200. https://doi.org/10.1016/j.jgar.2018.04.017

Moghadam MT, Amirmozafari N, Shariati A, Hallajzadeh M, Mirkalantari S, Khoshbayan A, Jazi FM (2020) How phages overcome the challenges of drug resistant bacteria in clinical infections. Infect Drug Resist 7(13):45–61. https://doi.org/10.2147/IDR.S234353

Mohamed SH, Mohamed MSM, Khalil MS, Azmy M, Mabrouk MI (2018) Combination of essential oil and ciprofloxacin to inhibit/eradicate biofilms in multidrug-resistant *Klebsiella pneumoniae*. J Appl Microbiol 125(1):84–95. https://doi.org/10.1111/jam.13755

Monegro AF, Muppidi V, Regunath H (2020) Hospital acquired infections. StatPearls Publishing, Treasure Island

Morici P, Florio W, Rizzato C, Ghelardi E, Tavanti A, Rossolini GM, Lupetti A (2017) Synergistic activity of synthetic N-terminal peptide of human lactoferrin in combination with various antibiotics against carbapenem-resistant *Klebsiella pneumoniae* strains. Eur J Clin Microbiol Infect Dis 36:1739–1748. https://doi.org/10.1007/s10096-017-2987-7

Mulani MS, Kamble EE, Kumkar SN, Tawre MS, Pardesi KR (2019) Emerging strategies to combat ESKAPE pathogens in the era of antimicrobial resistance: a review. Front Microbiol 10:539. https://doi.org/10.3389/fmicb.2019.00539

Naparstek L, Carmeli Y, Venezia SN, Banin E (2014) Biofilm formation and susceptibility to gentamicin and colistin of extremely drug-resistant KPC-producing *Klebsiella pneumoniae*. J Antimicrob Chemother 69:1027–1034. https://doi.org/10.1093/jac/dkt487

Navon-Venezia S, Kondratyeva K, Carattoli A (2017) *Klebsiella pneumoniae*: a major worldwide source and shuttle for antibiotic resistance. FEMS Microbiol Rev 41(3):252–272. https://doi.org/10.1093/femsre/fux013

Paczosa MK, Mecsas J (2016) *Klebsiella pneumoniae*: going on the offense with a strong defense. Microbiol Mol Biol Rev 80(3):629–661. https://doi.org/10.1128/MMBR.00078-15

Prabakar K, Sivalingamb P, Rabeeka SIM, Muthuselvamb M, Devarajana N, Arjunanb A, Karthicka R, Sureshd MM, Wembonyamac JP (2013) Evaluation of antibacterial efficacy of phyto fabricated silver nanoparticles using Mukia scabrella (Musumusukkai) against drug resistance nosocomial gram negative bacterial pathogens. Colloids Surf B Biointerfaces 104:282–288. https://doi.org/10.1016/j.colsurfb.2012.11.041

Principi N, Silvestri E, Esposito S (2019) Advantages and limitations of bacteriophages for the treatment of bacterial infections. Front Pharmacol 10:513. https://doi.org/10.3389/fphar.2019.00513

Pulzova L, Navratilova L, Comor L (2017) Alterations in outer membrane permeability favor drug-resistant phenotype of *Klebsiella pneumoniae*. Microb Drug Resist 23(4):413–420. https://doi.org/10.1089/mdr.2016.0017

Rafiq Z, Narasimhan S, Haridoss M, Vennila R, Vaidyanathan R (2017) Punica Granatum rind extract: antibiotic Potentiator and efflux pump inhibitor of multidrug resistant *Klebsiella Pneumoniae* clinical isolates. Asian J Pharm Clin Res 10(3):1–5. https://doi.org/10.22159/ajpcr.2017.v10i3.16000

Rai MK, Deshmukh SD, Ingle AP, Gade AK (2012) Silver nanoparticles: the powerful nanoweapon against multidrug-resistant bacteria. J Appl Microbiol 112(5):841–852. https://doi.org/10.1111/j.1365-2672.2012.05253.x

Reddy LS, Nisha MM, Joice M, Shilpa PN (2014) Antimicrobial activity of zinc oxide (ZnO) nanoparticle against *Klebsiella pneumoniae*. Pharm Biol 52(11):1388–1397. https://doi.org/10.3109/13880209.2014.893001

Remya P, Shanthi M, Sekar U (2018) Prevalence of blaKPC and its occurrence with other beta-lactamases in *Klebsiella pneumoniae*. J Lab Physicians 10(4):387–391. https://doi.org/10.4103/JLP.JLP_29_18

Ribeiro SM, de la Fuente-Núñez C, Baquir B, Faria-Junior C, Franco OL, Hancock REW (2015) Anti-biofilm peptides increase the susceptibility of carbapenemase- producing *Klebsiella pneumoniae* clinical isolates to β-lactam antibiotics. Antimicrob Agents Chemother 59 (7):3906–3912. https://doi.org/10.1128/aac.00092-15

Salou M, Toulan DEE, Dossim S, Agbonon A (2019) In vitro activities of aqueous and hydro-ethanolic extracts of Ocimum gratissimum on Escherichia coli ESBL, *Klebsiella pneumoniae* ESBL and methicillin resistant Staphylococcus aureus. Afr J Microbiol Res 13(3):55–59. https://doi.org/10.5897/AJMR2018.9004

Santajit S, Indrawattana N (2016) Mechanisms of antimicrobial resistance in ESKAPE pathogens. Biomed Res Int 2016:2475067. https://doi.org/10.1155/2016/2475067

Seukep JA, Sandjo LP, Ngadjui BT, Kuete V (2016) Antibacterial and antibiotic-resistance modifying activity of the extracts and compounds from Nauclea pobeguinii against gram-negative multi-drug resistant phenotypes. BMC Complement Altern Med 16:193. https://doi.org/10.1186/s12906-016-1173-2

Shankar C, Shankar BA, Manesh A, Veeraraghavan B (2018) KPC-2 producing ST101 *Klebsiella pneumoniae* from bloodstream infection in India. J Med Microbiol 67(7):927–930. https://doi.org/10.1099/jmm.0.000767

Sikarwar AS, Batra HV (2011) Prevalence of antimicrobial drug resistance of *Klebsiella pneumoniae* in India. Int J Biosci Biochem Bioinform 1:3. https://doi.org/10.7763/IJBBB.2011.V1.38

Soleimani N, Aganj M, Ali L, Shokoohizadeh L, Sakinc T (2014) Frequency distribution of genes encoding aminoglycoside modifying enzymes in uropathogenic *E. coli* isolated from Iranian hospital. BMC Res Notes 7:842. https://doi.org/10.1186/1756-0500-7-842

Souli M, Karaiskos I, Masgala A, Galani L, Barmpouti E, Giamarellou H (2017) Double-carbapenem combination as salvage therapy for untreatable infections by KPC-2-producing *Klebsiella pneumoniae*. Eur J Clin Microbiol Infect Dis 36(7):1305–1315. https://doi.org/10.1007/s10096-017-2936-5

Sun S, Gao H, Liu Y, Jin L, Wang R, Wang X, Wang Q, Yin Y, Zhang Y, Wang H (2020) Co-existence of a novel plasmid mediated efflux pump with colistin resistance gene mcr in one plasmid confers transferable multidrug resistance in *Klebsiella pneumoniae*. Emerg Microbes Infect 9(1):1102–1113. https://doi.org/10.1080/22221751.2020.1768805

Swain SS, Padhy RN (2016) Isolation of ESBL-producing gram-negative bacteria and in silico inhibition of ESBLs by flavonoids. J Taibah Univ Sci 11(3):217–229. https://doi.org/10.1016/j.jtumed.2016.03.007

Swedan S, Shubair Z, Almaaytah A (2019) Synergism of cationic antimicrobial peptide WLBU2 with antibacterial agents against biofilms of multi-drug resistant Acinetobacter baumannii and *Klebsiella pneumoniae*. Infect Drug Resist 12:2019–2030. https://doi.org/10.2147/IDR.S215084

Szeloch AK, Kawa ZD, Dąbrowska BW, Kassner J, Skrobek GM, Augustyniak D, Szelachowska MŁ, Żaczek M, Górski A, Kropinski AM (2013) Characterising the biology of novel lytic bacteriophages infecting multidrug resistant *Klebsiella pneumoniae*. Virol J 10:100. https://doi.org/10.1186/1743-422X-10-100

Taha OA, Connerton PL, Connerton IF, Shibiny AE (2018) Bacteriophage ZCKP1: a potential treatment for *Klebsiella pneumoniae* isolated from diabetic foot patients. Front Microbiol 9:2127. https://doi.org/10.3389/fmicb.2018.02127

Taher NM (2015) Synergistic effect of Propolis extract and antibiotics on multi-resist *Klebsiella pneumoniae* strain isolated from wound. Adv Life Sci Technol 43:23–28. https://doi.org/10.7176/ALST

Tosato MG, Schilardi P, Lorenzo de Mele MF, Thomas AH, Lorente C, Miñán A (2020) Synergistic effect of carboxypterin and methylene blue applied to antimicrobial photodynamic therapy against mature biofilm of *Klebsiella pneumoniae*. Heliyon 6(3):e03522. https://doi.org/10.1016/j.heliyon.2020.e03522

Trimble MJ, Mlynárčik P, Kolář M, Hancock RE (2016) Polymyxin: alternative mechanisms of action and resistance. Cold Spring Harb Perspect Med 6(10):a025288. https://doi.org/10.1101/cshperspect.a025288

Türkel İ, Yıldırım T, Yazgan B, Bilgin M, Başbulut E (2018) Relationship between antibiotic resistance, efflux pumps, and biofilm formation in extended-spectrum β-lactamase producing *Klebsiella pneumoniae*. J Chemother 30(6–8):354–363. https://doi.org/10.1080/1120009X.2018.1521773

Vading M, Nauclér P, Kalin M, Giske CG (2018) Invasive infection caused by *Klebsiella pneumoniae* is a disease affecting patients with high comorbidity and associated with high long-term mortality. PLoS One 13(4):e0195258. https://doi.org/10.1371/journal.pone.0195258

Valenzuela-Valderrama M, Bustamantea V, Carrascoa N, Gonzálezb IA, Dreysec P, Palavecino CE (2020a) Photodynamic treatment with cationic Ir(III) complexes induces a synergistic antimicrobial effect with imipenem over carbapenem-resistant *Klebsiella pneumoniae*. Photodiagnosis Photodyn Ther 30:101662. https://doi.org/10.1016/j.pdpdt.2020.101662

Valenzuela-Valderrama M, Carrasco-Veliz N, Gonzalez IA, Dreyse P, Palavecino CE (2020b) Synergistic effect of combined imipenem and photodynamic treatment with the cationic Ir(III) complexed to polypyridine ligand on carbapenem-resistant *Klebsiella pneumoniae*. Photodiagnosis Photodyn Ther 30:101882. https://doi.org/10.1016/j.pdpdt.2020.101882

Velkov T, Deris ZZ, Huang JX, Azad MA, Butler M, Sivanesan S, Kaminskas LM, Dong YD, Boyd B, Baker MA, Cooper MA, Nation RL, Li J (2014) Surface changes and polymyxin interactions with a resistant strain of *Klebsiella pneumoniae*. Innate Immun 20(4):350–363. https://doi.org/10.1177/1753425913493337

Verma V, Harjai K, Chhibber S (2009) Characterization of a T7-like lytic bacteriophage of *Klebsiella pneumoniae* B5055: a potential therapeutic agent. Curr Microbiol 59(3):274–281. https://doi.org/10.1007/s00284-009-9430-y

Vivas R, Dolabella SS, Barbosa AAT, Jain S (2020) Prevalence of *Klebsiella pneumoniae* carbapenemase - and New Delhi metallo-beta-lactamase-positive K. pneumoniae in Sergipe, Brazil, and combination therapy as a potential treatment option. Rev Soc Bras Med Trop 53:e20200064. https://doi.org/10.1590/0037-8682-0064-2020

Vuotto C, Longo F, Balice MP, Donelli G, Varaldo PE (2014) Antibiotic resistance related to biofilm formation in *Klebsiella pneumoniae*. Pathogens 3:743–758. https://doi.org/10.3390/pathogens3030743

Vuotto C, Longo F, Pascolini C, Donelli G, Balice MP, Libori MF, Tiracchia V, Salvia A, Varaldo PE (2017) Biofilm formation and antibiotic resistance in *Klebsiella pneumoniae* urinary strains. J Appl Microbiol 123:1003–1018. https://doi.org/10.1111/jam.13533

Wang S, Zeng X, Yang Q, Qiao S (2016) Antimicrobial peptides as potential alternatives to antibiotics in food animal industry. Int J Mol Sci 17(5):603. https://doi.org/10.3390/ijms17050603

Willers C, Wentzel JF, du Plessis LH, Gouws C, Hamman JH (2016) Efflux as a mechanism of antimicrobial drug resistance in clinical relevant microorganisms: the role of efflux inhibitors. Expert Opin Ther Targets 21(1):23–36. https://doi.org/10.1080/14728222.2017.1265105

Wu LT, Guo MK, Ke SC, Lin YP, Pang YC, Nguyen HTV, Chen CM (2020) Characterization of the genetic background of KPC-2-producing *Klebsiella pneumoniae* with insertion elements disrupting the ompK36 Porin gene. Microb Drug Resist 26(9):1050–1057. https://doi.org/10.1089/mdr.2019.0410

Yang SK, Yusof K, Thomas W, Akseer R, Alhosani MS, Abushelaibi A, Lim SHE, Lai KS (2020a) Lavender essential oil induces oxidative stress which modifies the bacterial membrane permeability of carbapenemase producing *Klebsiella pneumoniae*. Sci Rep 10:819. https://doi.org/10. 1038/s41598-019-55601-0

Yang SK, Yusoff K, Ajat M, Yap WS, Lim SHE, Lai KS (2020b) Antimicrobial activity and mode of action of terpene linalyl anthranilate against carbapenemase-producing *Klebsiella pneumoniae*. J Pharm Anal. https://doi.org/10.1016/j.jpha.2020.05.014

Yu L, Zhang J, Fu Y, Zhao Y, Wang Y, Zhao J, Guo Y, Li C, Zhang X (2019) Synergetic effects of combined treatment of Colistin with Meropenem or amikacin on Carbapenem-resistant *Klebsiella pneumoniae* in vitro. Front Cell Infect Microbiol 9:422. https://doi.org/10.3389/fcimb. 2019.00422

Zheng JX, Lin XW, Sun X, Lin WH, Chen Z, Wu Y, Qi GB, Deng QW, Qu D, Yu ZJ (2018) Overexpression of OqxAB and MacAB efflux pumps contributes to eravacycline resistance and heteroresistance in clinical isolates of *Klebsiella pneumoniae*. Emerg Microbes Infect 7 (1):1–11. https://doi.org/10.1038/s41426-018-0141-y

Plant-Associated Endophytic Fungi and Its Secondary Metabolites Against Drug-Resistant Pathogenic Microbes

10

Sriravali K., Anil B. Jindal, Bhim Pratap Singh, and Atish T. Paul

Contents

Abstract

Antimicrobial resistance (AMR) is now one of the major concern in the public health globally, as it is increasing the cost of the treatment as well as time of hospitalization. The resistant microbes like bacteria, fungi, protozoa, and virus are responsible for causing simple skin infections to severe infectious diseases like tuberculosis (TB). The major resistant microbes includes methicillin-resistant *Staphylococcus aureus* (MRSA), vancomycin-resistant *Staphylococcus aureus*

S. K. · A. B. Jindal · A. T. Paul (✉)
Laboratory of Natural Products Chemistry, Department of Pharmacy, Birla Institute of Technology and Sciences Pilani (BITS Pilani), Pilani, Rajasthan, India
e-mail: atish.paul@pilani.bits-pilani.ac.in

B. P. Singh (✉)
Department of Agriculture & Environmental Sciences (AES), National Institute of Food Technology Entrepreneurship & Management (NIFTEM), Sonepat, Haryana, India

V. Kumar et al. (eds.), *Antimicrobial Resistance*,
https://doi.org/10.1007/978-981-16-3120-7_10

(VRSA), vancomycin-resistant *Enterococcus faecalis* (VREF), resistant *Pseudomonas* sp., resistant *Mycobacterium* sp., etc. The World Health Organization (WHO) has also taken many initiatives to combat AMR, one of which is the Global Antimicrobial Resistance and Use Surveillance System (GLASS). The pathogenic microbes are developing resistance to most of the existing antibiotics by different mechanisms, thereby decreasing the efficiency of molecules. New drug molecule identification and development is essential to tackle the AMR. Nature gives the best remedies for various diseases; plant-endophytic fungi symbiotic relation is one of the important sources having high diversity within the various plant tissues and produces therapeutically active secondary metabolites. This chapter mainly focused on compounds isolated from plant associated fungal endophytes that are active against various drug-resistant microbes and may also exhibits synergistic activitywith existing antibiotics. Further investigation is needed for the identification of more active compounds and to bring them for use in therapeutic level.

Keywords

Antimicrobial resistance · Fungal endophytes · Secondary metabolites

Abbreviations

ABR	Antibacterial resistance
AMR	Antimicrobial resistance
GARDP	Global Antibiotic Research and Development Partnership
GIT	Gastrointestinal tract
GLASS	Global Antimicrobial Resistance and Use Surveillance System
HIV	Human immunodeficiency virus
IACG	Interagency Coordination Group on Antimicrobial Resistance
ICMR	Indian Council of Medical Research
MDR-TB	Multidrug-resistant tuberculosis
MRSA	Methicillin-resistant *Staphylococcus aureus*
MSSA	Methicillin-susceptible *Staphylococcus aureus*
NCDC	National AMR Surveillance Network
SSSIs	Skin and skin structure infections
SSTIs	Skin and soft tissue infections (SSTIs)
VREF	Vancomycin-resistant *Enterococcus faecalis*
VRSA	Vancomycin-resistant *Staphylococcus aureus*
WHO	World Health Organization
WWTPs	Wastewater treatment plants
XDR-TB	Extensively drug-resistant tuberculosis

10.1   Introduction

Antimicrobial resistance (AMR) to various antibiotics is one of the most threatening conditions that results in the reduction of efficacy of antibiotics against various pathogenic microbes such as bacteria (*Mycobacterium* sp., *Staphylococcus* sp., *Enterococcus* sp., *Salmonella* sp., *Escherichia sp.*), fungi (*Candida* sp., *Aspergillus* sp.), protozoan (*Plasmodium* sp., *Trypanosoma* sp., *Leishmania* sp.), and virus (McKeegan et al. 2002). AMR is spreading across the globe that increases the severity of infection and prolongs the illness, thereby affecting the socioeconomic status of the patient compared to nonresistant infected patients (Ventola 2015). The pathogenic microbes are responsible for causing various infectious diseases like tuberculosis, meningitis, pneumonia, malaria, influenza, trichomoniasis, toxoplasmosis, etc. (Singh et al. 2014) affecting various parts of human body including the skin, respiratory tract, systemic circulation, gastrointestinal tract (GIT), urinary tract, and reproductive organs. Without effective treatment with antibiotics, there will be difficulty in treating various medical surgeries such as organ transplantation, cholecystectomy, biopsy, carotid endarterectomy, joint replacements, coronary artery bypass, craniectomy, caesarean, appendectomy, etc. (Ventola 2015). AMR further complicates the treatment for these conditions. The World Health Organization (WHO) defined the resistant microbes as "superbugs." In 2014, the WHO published AMR global report on surveillance, and it mainly focused on the antibacterial resistance (ABR). The resistance was observed in various human pathogenic microbes including *Escherichia coli* (*E. coli*), *Staphylococcus aureus* (*S. aureus*), *Klebsiella pneumoniae* (*K. pneumoniae*), and *Pseudomonas aeruginosa* (*P. aeruginosa*). These are majorly responsible for common healthcare-associated and community-acquired infections. In 2015, the WHO launched the Global Antimicrobial Resistance and Use Surveillance System (GLASS), and it will collect, analyze, and interpret the information regarding AMR provided by different countries. It published its early implementation reports for the years 2016–2017 and 2017–2018 and recent report in May 2020. At present, 92 countries were enrolled in the GLASS by April 2020. According to GLASS, in India, three AMR surveillance networks participated, which includes the National AMR Surveillance Network (NCDC), Antimicrobial Surveillance and Research Network (Indian Council of Medical Research, ICMR), and Gonococcal Antimicrobial Resistance Surveillance Network, Safdarjung Hospital. Other initiatives which deal with the AMR by the WHO include Global Antibiotic Research and Development Partnership (GARDP) and Interagency Coordination Group on Antimicrobial Resistance (IACG).

10.1.1 Factors Causing the Resistance in Pathogenic Microbes

Various factors triggered the AMR in pathogenic microbes, and it continues to spread across the globe through mutation of genes that usually occurred over the time naturally. The major factors that were responsible for contribution of AMR include:

1. Overuse of antibiotics: Due to inappropriate prescription by the healthcare professionals as well as purchasing the antibiotics without prescription by the physician (self-medication), people are misusing the antibiotics for simple ailments.
2. Prolonged use of antibiotics: In the hospitals and intensive care units, prolonged use of antibiotics is mandatory especially in immunosuppressed patients (e.g., human immunodeficiency virus (HIV), cancer, organ transplantation) and elderly patients.
3. Environment: Soil and water are the main reservoirs for the resistant microorganisms. Soil is the main source for various microorganisms that produces antibiotics, acquired resistance intrinsically to antibiotics. Water is another major source which was contaminated by large amounts of antibiotics due to disposal of drugs, industrial waste, and excretory compounds by humans. Several studies also reported about the resistant bacteria found in wastewater treatment plants (WWTPs).
4. The use of antibiotics in livestock and agriculture: The use of antibiotics for growth promotion as well as treatment is one of the major contributions of antibiotic resistance. The use of large amounts of antibiotics for disease control in plants also leads to antibiotic resistance (Prestinaci et al. 2015).

10.1.2 Mechanism of Antibiotic Resistance

A large number of antibiotics are available for the treatment of various infectious diseases that inhibit the microbial growth by different mechanisms. The major mechanisms include the following: inhibition of cell wall synthesis (β-lactam antibiotics and glycopeptides), depolarization of the cell membrane (lipopeptides), inhibition of protein synthesis (aminoglycosides, tetracyclines, macrolides), inhibition of nucleic acid synthesis (fluoroquinolones), inhibition of metabolic pathways (sulfonamides), etc. After subsequent development of new molecules for treating infectious diseases, microbes are acquiring resistance to developed antibiotics by different mechanisms. Major mechanisms include modification of cell membranes which limits the uptake of drug, effluxing the drug molecules by using special transporter proteins, production of antibiotic altering and degrading enzymes, overproduction of drug targets, and alteration of the target site. All these alterations may be inherent to microbes, which already have resistant genes or obtained by transfer of resistant genetic material from one pathogen to another. The microorganisms may use both of these mechanisms for acquiring resistance (Reygaert 2013).

ABR is one of the most prevailing conditions, and people are at higher risk of getting infected by drug-resistant bacteria. The WHO published the global priority list of antibiotic-resistant human pathogenic microbes that categorize the resistant bacteria into priority 1 (critical), priority 2 (high), and priority 3 (medium) pathogens. The examples of most common pathogenic resistant bacteria include *P. aeruginosa, S. aureus, K. pneumoniae, E. coli, Acinetobacter baumannii* (*A. baumannii*), *Enterococcus faecium* (*E. faecium*), *Helicobacter pylori* (*H. pylori*), *Salmonella* spp., *Campylobacter* spp., etc. Apart from these bacterial infections, multidrug-resistant tuberculosis (MDR-TB) is caused by *Mycobacterium tuberculosis (M. tuberculosis)*, one of the most typical conditions which is getting complicated gradually and also associated with comorbidities like HIV. This MDR-TB is resistant to two of the most prescribed anti-TB drugs (rifampicin and isoniazid). Extensively drug-resistant tuberculosis (XDR-TB) is another form of tuberculosis that is resistant to four anti-TB drugs (Seung et al. 2015). Resistance in HIV is also reported after starting antiretroviral therapy (Kuritzkes 2011). *Plasmodium falciparum* (*P. falciparum*) is a protozoa that causes malaria, which also acquired resistant to first-line therapeutic medicines. Drug-resistant strains for influenza, *Candida albicans* (*C. albicans*), *P. aeruginosa, Enterococcus faecalis* (*E. faecalis*), *K. pneumoniae, Salmonella* species, etc., were also reported. Due to the AMR, the efficacy of the antibiotics decreases that leads to failure of treatment. So, new molecule lead identification or development is necessary to effectively act against drug-resistant pathogenic microbes.

10.1.3 Endophytes as a Source of Therapeutic Compounds

Plant-associated endophytes are one of the major reservoirs for novel secondary metabolites that exhibit various activities such as anti-inflammatory, antimicrobial, anticancer, antidiabetic, antioxidant activity, etc. The endophytes live inside the plant tissues in symbiotic association without causing any disease. The various categories of bioactive compounds produced by endophytes include alkaloids, glycosides, terpenoids, quinones, phenols, xanthones, polyketides, isoprenoids, steroids, isocoumarins, perylene derivatives, furandiones, peptides, depsipeptides, proteins, lipids, shikimates, cytochalasines, etc. In the nineteenth century, the existence of fungal endophytes was identified. The secondary metabolites produced by endophytic fungi not only exhibit bioactivity but also help the plant to fight against external biotic and abiotic stress by enhancing the immune response to invade pathogenic microbes. The "endophytic continuum" model explains that the interaction between fungi and plant may be mutualism or parasitism depending on the fungal species and the environment. The endophytic fungi associated with plants were classified into four types, Ascomycota, Basidiomycota, Zygomycota, and Oomycota (Rajamanikyam et al. 2017). Some endophytic fungi associated with plants include *Penicillium* sp., *Fusarium* sp., *Colletotrichum* sp., *Cytonaema* sp., *Gliocladium* sp., *Xylaria* sp., *Cladosporium* sp., *Phomopsis* sp., *Alternaria* sp., *Pestalotiopsis* sp., etc. Pandey et al. (2014) reported about the various bioactive compounds isolated from fungal endophytes and its application in pharmaceutical science. Some examples of the antibiotics produced by fungal endophytes include Griseofulvin, Brefeldin A, colletotric acid, torreyanic acid, etc., that were isolated from the endophytic fungus of *Xylaria* sp. F0010, *Cladosporium* sp., *Colletotrichum gloeosporioides,* and *Pestalotiopsis microspore*, respectively. Various antimicrobial compounds produced by endophytic fungus include 7-amino-4 methylcoumarin, hypericin, emodin, ambuic acid, citrinin, pestalachlorides A–C, butanedioic acid and cytochalasin D, phenylpropanoid amides, etc. The compounds with anticancer properties include paclitaxel, vinblastine and vincristine, podophyllotoxin, and camptothecin and its analogues that were also isolated from the endophytic fungi from various plants. Antioxidants (exopolysaccharides, phenylpropanoid amides, graphislactone A), anti-insect agents (naphthalene), immunosuppressive agents (subglutinols A and B), and acetylcholinesterase inhibitors were also reported to be isolated from various species of endophytic fungus (Pandey et al. 2014).

Deshmukh et al. (2014) reported about the compounds produced by endophytic fungi with antibacterial activity. The compounds exhibited activity against *S. aureus, E. faecalis, B. subtilis, K. pneumoniae, E. coli, M. luteus,* and *M. tuberculosis* including drug-resistant strains of *S. aureus* (MRSA and MDR *S. aureus*), vancomycin-resistant *E. faecium,* and MDR-TB. Many researchers also reported about the activity of fungal endophytes against AMR. The present topic will mainly focus on the compounds isolated from the endophytic fungi and its activity against various resistant strains of microbes.

Synergistic activity of fungal endophytic extracts with antibiotics is also reported. Iswarya and Ramesh (2019) reported about the synergistic interaction between endophytic fungus *Penicillium daleae* EF 4 extract with penicillin combination and antagonistic with tetracycline combination. Synergistic activity is also observed against *S. aureus* with the combination of *Colletotrichum gloeosporioides* fungal extract with antibiotics such as penicillin, methicillin, and vancomycin. Significant decrease in MIC values was observed (Arivudainambi et al. 2011). The compounds produced by fungal endophytes alone or in combination with existing antibiotics are potential source to combat AMR.

10.2 Fungal Endophytes Against MRSA

Bacterial infection caused by the Methicillin-resistant *S.aureus* (MRSA) acquired resistance to most of the antibiotics such as methicillin, penicillin, oxacillin, vancomycin, and cephalosporin. It is the major cause of the nosocomial infection across the world and most frequently the common cause of wound infection after surgery. It was first identified in 1960, and the prevalence was increased to 80.89% in 1999 from 12% in 1992 (Kaur and Chate 2015). MRSA became epidemic, as it is spreading globally with increasing its pathological potential. Based on the antimicrobial resistance and severity, the WHO categorizes the MRSA bacteria as priority 2 pathogen in 2017. Types of MRSA include hospital-acquired, community-acquired, and livestock-associated infections. Community-acquired MRSA is the most commonly occurring infection in the public health, and it is reported to be endemic in Southeast Asia, Sri Lanka, Taiwan, China, Vietnam, Australia, Greece, United Kingdom (UK), and the United States (Chatterjee and Otto 2013). The death rates in United States are higher for MRSA infection compared to other infectious diseases like tuberculosis, HIV, and viral hepatitis for the year 2002–2004 (Boucher and Corey 2008).

S. aureus is a facultative anaerobe, gram-positive, nonmotile, and nonspore-forming spherical bacterium that is responsible for various skin and skin structure infections (SSSIs), also known as skin and soft tissue infections (SSTIs), and also affects the other parts of the body, mostly upper respiratory tract. It belongs to the family Staphylococcaceae, in the order Bacillales and class Bacilli. Various infectious diseases caused by *S. aureus* include furuncles, folliculitis, impetigo, carbuncles, and mastitis. *S. aureus* can also cause serious infections like endocarditis, pneumonia, bacteremia, osteomyelitis, and septic arthritis. In the earlier days, penicillin G was used for the treatments for these infectious diseases that was introduced in the early 1940. But resistant strains were identified after few years. The mechanism of resistance is due to the inactivation of drug by the penicillinase/β-lactamase enzyme that hydrolyzes the ring of β-lactam antibiotics. Then, methicillin, a semisynthetic penicillinase-resistant β-lactam antibiotic, was developed. Later, the resistant strains for methicillin are also identified (Peacock and Paterson 2015). These MRSA now acquired resistance to most of the available antibiotics, further reducing the efficacy and thereby complicating the treatment.

10.2.1 Mechanism of Resistance in *S. aureus*

The cell wall of most of the gram-positive bacteria is made up of peptidoglycan layer, and it is very thick and the major constituent of the cell wall of gram-positive bacteria, whereas this layer is very thin in the gram-negative bacteria. As *S. aureus* is a gram-positive bacteria, most of the commercially available antibiotics act by inhibiting the cell wall synthesis. Peptidoglycan layer is made up of repeated subunits of N-acetylglucosamine and N-acetylmuramic acid that are cross-linked by peptide bond. In cell wall formation, long glycan cross-linked chains are formed by transglycosylation and transpeptidation. The glycan chains are formed by transglycosylation reaction and cross-linking between the peptide subunits taking place by transpeptidation reaction. The peptide chain in gram-positive bacteria is composed of L-Ala–γ-D-Glu–L-Lys–D-Ala–D-Ala subunit. For the formation of cell wall, cross-linking occurs between the 4 D-Ala of one stand of peptidoglycan subunit and 3 L-Lys of another peptide subunit. In bacteria, transglycosylation and transpeptidation reactions are carried out by enzyme transpeptidases known as penicillin-binding proteins (PBPs) (Peacock and Paterson 2015).

The β-lactam antibiotics include penicillin, cephalosporins, monobactams, and carbapenems that work by inhibiting the bacterial cell wall synthesis especially by targeting the PBPs. These drugs will bind to the PBPs and disturb its normal cellular function that leads to the formation of defective cell wall. Later the resistance to β-lactam antibiotics was developed due to the production of penicillinase enzyme. The penicillinase enzyme inactivates the antibiotics by hydrolyzing the amide bond of β-lactam ring (a four-membered cyclic amide) and further inhibits the binding of drugs to PBPs. The production of penicillinase enzyme was carried out by the expression of structural gene *blaZ*. That was further controlled by regulatory genes *blaI* and *blaR1*. The plasmids encoded for the production of penicillinase enzyme are also responsible for resistance to other drugs, e.g., erythromycin, aminoglycosides, and fusidic acid. These plasmids also carry a gene that acquired resistance to heavy metals, disinfectants, and dyes (Pantosti et al. 2007).

Modified β-lactam antibiotics such as methicillin, ampicillin, and oxacillin were developed to combat resistance developed by penicillins. These drugs protect the β-lactam ring by preventing the attack from penicillinase enzyme. A few years after treatment with methicillin, the resistance was developed by the bacteria. The acquisition of resistance to methicillin is due to the production of new penicillin-binding proteins named as PBP2. This has very less affinity to bind with β-lactam antibiotics as well as toward methicillin (Reygaert 2013). The production PBP2 is due to the expression of *mec*A gene; this is also further controlled by two regulatory genes *mec*I and *mec*R1. The sequence and organization of *mec*A gene is similar to that of *bla* gene complex (Pantosti et al. 2007). Staphylococcal chromosome cassette (SCCmec) is a mobile genetic element that mediates the resistance in *S. aureus* to methicillin by horizontal transfer of this *mec*A gene that was probably absent in methicillin-susceptible *S. aureus* (MSSA). SCCmec contain *mec* gene complex and the cassette chromosome recombinase (ccr) gene complex (contains site-specific recombinase genes). Reygaert et al. (2013) reported that SCCmec elements contain eight subtypes I–VIII and that resistance genes are only carried by types II, III, and VIII and the remaining types do not carry resistance genes. This is also helpful in determining the difference between hospital-acquired and community-acquired MRSA, in which types IV, V, and VII were carried by community-acquired MRSA, whereas types I, II, III, VI, and VIII were carried by hospital-acquired MRSA.

As MRSA strains were identified and characterized, it acquired resistance to another antibiotics like vancomycin, teicoplanin, linezolid, and daptomycin. Later, it was categorized as MDR strain, and it was continuously evolving as paradigmatic pathogen (Kaur and Chate 2015). Fungal endophytes are one of the major sources to combat resistance against MRSA. Many researches are still ongoing for the identification and isolation of compounds from various fungal endophytes with anti-MRSA activity. Here some compounds isolated from various fungal sources are mentioned.

3-Hydroxy-1-(1,3,8-trihydroxy-6-methoxynaphthalen-2-yl)propan-1-one; 3-hydroxy-1-(1,8-dihydroxy-3,5-dimethoxynaphthalen-2-yl)propan-1-one and 3-hydroxy-1-(1,8-dihydroxy-3,6-dimethoxynaphthalen-2-yl)propan-1-one (Fig. 10.1) were isolated from the *Phomopsis fukushii*, an endophyte of plant *Paris polyphylla* (*Melanthiaceae*), and evaluated for its antimicrobial activity. The three compounds exhibited anti-MRSA activity with MIC values of 4, 4, and 8 mg/mL, respectively (H. Y. Yang et al. 2017).

Oxysporone and xylitol (Fig. 10.1) are extracted from the endophytic fungus *Pestalotia* sp. from the cladodes and leaves of mangrove plant *Heritiera fomes* (Malvaceae). The antimicrobial activity was tested against various strains of MRSA, i.e., XU212, ATCC 25923, SA-1199B, EMRSA-15, MRSA340702. The MIC values were found to be 128,128, 64, 64, and 128 μM, respectively, for xylitol

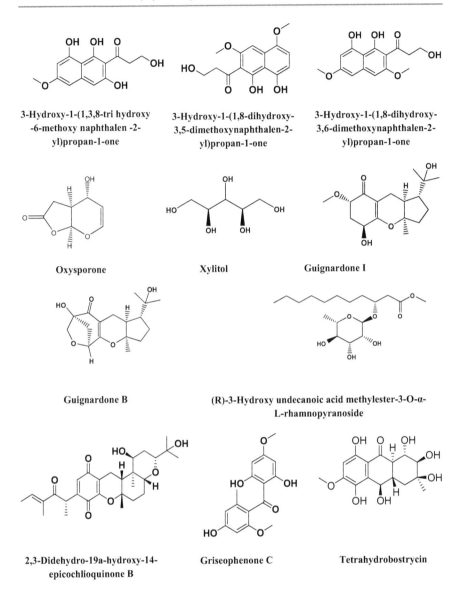

Fig. 10.1 Structure of compounds isolated from fungal endophytes against MRSA

and 128, 64, 32, 32, and 64 μM, respectively, for oxysporone (Nurunnabi et al. 2018).

Debbab et al. (2013) reported about the compounds isolated from the endophytic fungi of the mangrove plants with anti-MRSA activity. Guignardones F–I and guignardones A and B were isolated from the endophytic fungus *Guignardia* sp. from the mangrove plant *Scyphiphora hydrophyllacea* (Rubiaceae). Among all isolated compounds, guignardone I (a new monoterpene) and guignardone B

(Fig. 10.1) exhibited zone of inhibition of 9.0 mm and 8.0 mm, respectively, against MRSA at 65 μM concentration (Mei et al. 2012). A new fatty acid glycoside, (R)-3-hydroxy undecanoic acid methylester-3-O-α-L-rhamnopyranoside (Fig. 10.1) was also isolated from the *Guignardia* sp. from same plant, and zone of inhibition was found to be 10.7 mm against MRSA (Zeng et al. 2012).

2,3-Didehydro-19a-hydroxy-14-epicochlioquinone B, griseophenone C, and tetrahydrobostrycin (Fig. 10.1) were isolated from the *Nigrospora* sp. MA75 from the mangrove plant *Pongamia pinnata* (Fabaceae) that exhibited anti-MRSA activity with MIC values of 16.5, 1.6, and 5.9 μM, respectively (Xu 2015).

Two new succinic acid derivatives along with seven known compounds were isolated from the fungus *Xylaria cubensis* PSU-MA34, an endophyte of mangrove plant *Bruguiera parviflora* (Rhizophoraceae). Among all, 2-chloro-5-methoxy-3-methylcyclohexa-2,5-diene-1,4-dione (Fig. 10.2) exhibited activity against MRSA with MIC value of 128 μg/mL (Klaiklay et al. 2012).

Thirteen compounds were isolated from the *Alternaria* sp., an endophyte of mangrove plant *Sonneratia alba* (Sonneratiaceae). Among all, xanalteric acid I and xanalteric acid II and altenusin (Fig. 10.2) exhibited antimicrobial activity against MRSA with MIC values of 125, 250, and 31.25 μg/mL, respectively (Kjer et al. 2009).

Eight endophytic fungi were isolated from the plant *Opuntia dillenii*, a well-established invasive plant in southeastern arid zone of Sri Lanka that belongs to the family Cactaceae. Among all, *Fusarium* sp. fungal extract showed antimicrobial activity. A bioactive compound named Equisetin (Fig. 10.2) was isolated from the active *Fusarium* sp., and it exhibited anti-MRSA activity with MIC value 16 μg/mL (Ratnaweera et al. 2015).

A nortriterpenoid helvolic acid was isolated from the organic extract of endophyte *Xylaria* sp. The fungus was isolated from the leaves of *Anoectochilus setaceus*, an orchid plant endemic to Sri Lanka. The antimicrobial activity of the Helvolic acid (Fig. 10.2) was tested against various microorganisms including MRSA. The MIC was found to be 4 μg/mL against MRSA (Macías-Rubalcava et al. 2017).

(22E,24R)-Stigmasta-5,7,22-trien-3-b-ol, stigmast-4-ene-3-one, stigmasta-4,6,8,22-tetraen-3-one, terretonin, terretonin A, butyrolactone VI, aspernolide F, and aspernolide G were isolated from the endophytic fungus *Aspergillus terreus* from the roots of *Carthamus lanatus* belonging to the family Asteraceae. Their structure was determined by various spectrophotometric techniques, and the isolated compounds were evaluated for various biological activities such as antileishmanial, antimicrobial, antimalarial, and cytotoxic activities. All compounds exhibited anti-MRSA activity with IC $_{50}$ values in the range of 0.94 to <20 mg/mL. Among all, (22E,24R)-stigmasta-5,7,22-trien-3-b-ol (Fig. 10.2) exhibited potential antimicrobial activity against MRSA with IC_{50} value of 0.96 mg/mL, and aspernolides F (Fig. 10.2) exhibited mild activity against MRSA with IC_{50} value of 6.39 mg/mL (Ibrahim et al. 2015).

Botryosphaeria mamane PSU-M76, an endophytic fungus isolated from the leaves of *Garcinia mangostana* (Guttiferae), exhibited antibacterial activity against *S. aureus* and MRSA. Botryomaman, 2,4-dimethoxy-6-pentylphenol, (R)-(−)-

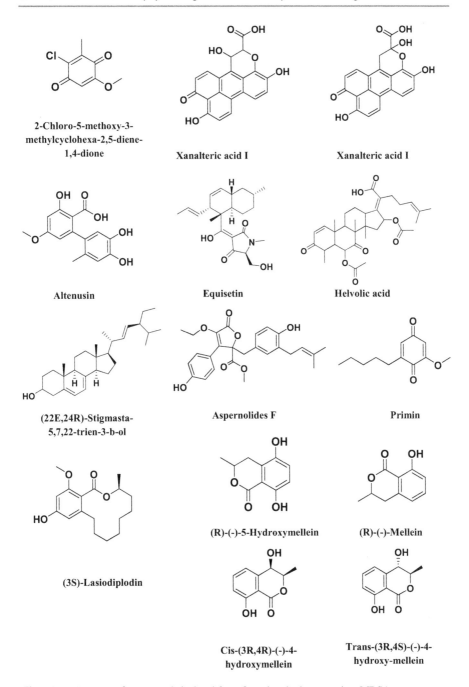

Fig. 10.2   Structure of compounds isolated from fungal endophytes against MRSA

mellein, primin, cis-4-hydroxymellein, trans-4-hydroxymellein, and 4,5-dihydroxy-2-hexenoic acid were isolated from the endophyte. Among all isolated compounds, Primin (Fig. 10.2) exhibited antimicrobial activity with MIC value of 8 mg/ mL, while other compounds exhibited less activity with MIC values equal to 128 mg/ mL against MRSA (Pongcharoen et al. 2007).

Two endophytic fungi *Botryosphaeria rhodina* PSU-M35 and PSU-M11 were isolated from the leaves of *Garcinia mangostana* (Guttiferae). A total of 19 compounds were extracted from two fungal strains. Antimicrobial activity of 12 compounds (3S)-lasiodiplodin, (R)-(−)-5-hydroxymellein, (R)-(−)-mellein, cis-(3R,4R)-(−)-4-hydroxymellein, trans-(3R,4S)-(−)-4-hydroxy-mellein, (R)-(−)-2-octeno-d-lactone,tetrahydro-4-hydroxy-6-propylpyran-2-one, dihydro-4-(hydroxymethyl)-3,5-dimethyl-2(3H)-fur-anone, (3S,4S)-(−)-4-acetyl-3-methyl-2 (3H)-dihydrofur-anone, cyclopentanone, and two dimeric α-lactones was tested against *S. aureus* and MRSA strains. Among all, (3S)-lasiodiplodin exhibited the best antimicrobial activity with MIC values of 64 and 128 mg/mL against *S. aureus* and MRSA, respectively. (R)-(−)-5-Hydroxymellein also exhibited antimicrobial activity with MIC value >128 mg/mL against both *S. aureus* and MRSA. (R)-(−)-Mellein, cis-(3R,4R)-(−)-4-hydroxymellein, and trans-(3R,4S)-(−)-4-hydroxy-mellein (Fig. 10.2) were also active against both strains. Other compounds did not exhibit antimicrobial activity at 200 mg/mL concentration (Rukachaisirikul et al. 2009).

Seven compounds were isolated from the *Aspergillus terreus*, an endophyte of *Carthamus lanatus* (Asteraceae). Among all isolated compounds, (22E,24R)-Stigmasta-5,7,22-trien-3-β-ol (Fig. 10.3) exhibited anti-MRSA activity with IC_{50} value of 2.29 µM (Elkhayat et al. 2016).

Neosartorin, (Fig. 10.3) a yellow pigment that was isolated from the fungus *Aspergillus fumigatiaffinis*, showed antibacterial activity against various gram-positive human pathogens including *Enterococci*, *B. subtilis*, *Staphylococci*, and *Streptococci*. The fungus was isolated from the plant *Tribulus terrestris* belonging to the family Zygophyllaceae. The MIC value was found to be 8 µg/mL against MDR strains of *S. aureus*, i.e., *S. aureus* Mu50 and *S. aureus* 25,697 (Ola et al. 2013a, b).

Twenty fungal strains were isolated from the plant *Nymphaea nouchali* (Nymphaeaceae). Among all, 8 out of 20 fungal extracts showed antimicrobial activity on tested human pathogens such as *S. aureus, Bacillus cereus* (*B. cereus*), *P. aeruginosa*, and *E. coli*. The fungal strains were identified by using different morphological and molecular techniques. In further investigation, two cytochalasans, chaetoglobosin A and C, were isolated from the crude extract of fungi *Chaetomium globosum*. Chaetoglobosin A and chaetoglobosin C (Fig. 10.3) exhibited antimicrobial activity with MIC values of 32 µg/mL and >64 µg/mL, respectively (Dissanayake et al. 2016a).

Nine endophytic fungal strains were isolated from the mangrove plant *Premna serratifolia* (Lamiaceae) and evaluated for its antimicrobial activity. *Hypocrea virens* fungal extract showed potent antimicrobial activity out of nine fungal extracts. Two compounds, gliotoxin and bisdethiobis(methylthio)gliotoxin, were isolated from the organic extract of fungus *Hypocrea virens*. Gliotoxin (Fig. 10.3) exhibited

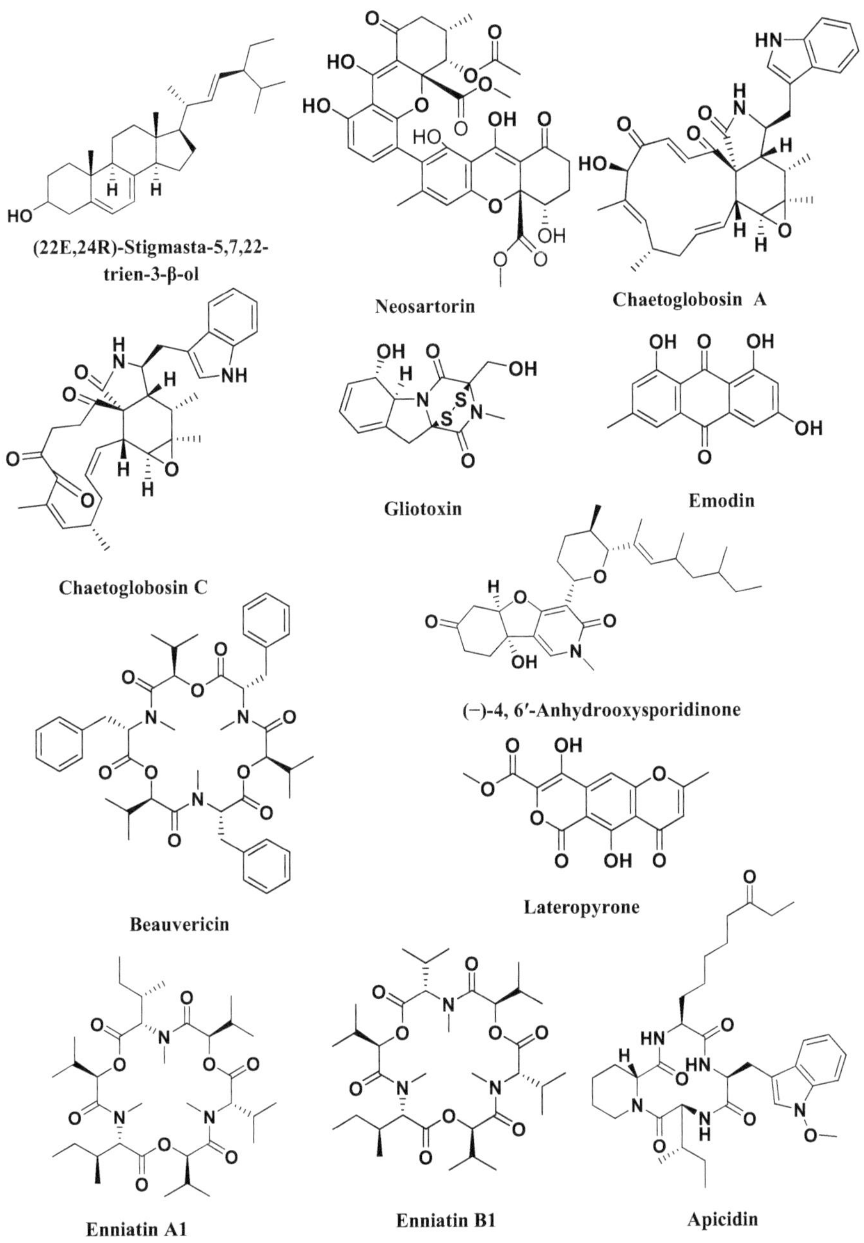

Fig. 10.3 Structure of compounds isolated from fungal endophytes against MRSA

anti-MRSA activity with MIC value of 32 μg/mL, whereas bisdethiobis(methylthio)-gliotoxin was inactive against MRSA (Ratnaweera et al. 2016).

Eurotium chevalieri KUFA, a mangrove-derived endophytic fungus 0006, was isolated from the inner twig of plant *Rhizophora mucronata Poir* (Rhizophoraceae). Nineteen compounds were isolated from the ethyl acetate extract of fungus. Ten compounds were tested for its antibiofilm activity against *E. coli, E. faecalis,* and *S. aureus*. Among all tested compounds, Emodin (Fig. 10.3) exhibited potent activity against tested pathogens, and the MIC values were determined against various pathogenic microbes including MRSA (>64 μg/mL). Emodin also produced synergistic activity with oxacillin against MRSA determined by disk diffusion method (Zin et al. 2017).

Fusarium oxysporum an endophytic fungus was isolated from the bark of *Cinnamomum kanehirae* belonging to the family Lauraceae endemic to Taiwan. Nine compounds were isolated from the fungus *Fusarium oxysporum* and evaluated for its cytotoxicity and antibacterial activity. Beauvericin (Fig. 10.3) exhibited potent antibacterial activity against MRSA with MIC value of t 3.125 μg/mL. Moderate anti-MRSA activity was exhibited by (−)-4, 6′-anhydrooxysporidinone (Fig. 10.3) with MIC value of 100 μg/mL. Remaining all compounds did not exhibit any antibacterial activity at MIC value of 100 μg/mL (Wang et al. 2011).

An endophytic fungus *Fusarium tricinctum* is generally endosymbiont of the plant *Aristolochia paucinervis* (Aristolochiaceae). The fungi *Fusarium tricinctum* when cocultured with *B. subtilis* on solid rice medium increase the secondary metabolite production by 78-fold, i.e., increase in concentration of lateropyrone, lipopeptide fusaristatin A, and three cyclic depsipeptides of the enniatin type (enniatin B, enniatin B1, enniatin A1). Apart from that, four metabolites include macrocarpon C, 2-(carboxymethylamino)benzoic acid, and d (−)-citreoisocoumarinol that were only identified in the coculture of *Fusarium tricinctum* with *B. subtilis*. The antibacterial activity of nine compounds was tested against various pathogenic microbes including MDR *S. aureus* (*S. aureus* 25,697 strain). The MIC values of nine compounds were found to be in the range of 2–64 μg/mL. The highest antimicrobial activity was shown by lateropyrone with MIC value of 2–4 μg/mL, followed by Enniatin A1 with MIC value of 4–8 μg/mL and Enniatin B1 (Fig. 10.3) that showed activity at MIC of 8 μg/mL (Ola et al. 2013a, b).

Apicidin (Fig. 10.3) was isolated from the endophytic fungi *Fusarium* sp. from the plant *Anemopsis californica* belonging to the family Saururaceae. Anti-quorum-sensing inhibition of apicidin was tested against MRSA. Apicidin inhibited the MRSA function by targeting *Agr*A plasmid. In vivo studies also conducted on C57BL/6 and BALB/c mice to check the impact of apicidin on skin tissue infected with MRSA. The decreased bacterial burden was observed with apicidin treatment. The attenuation of MRSA pathogenesis by apicidin corresponding with the quorum-sensing inhibition both in vivo and in vitro (Parlet et al. 2019).

3-(2-Hydroxypropyl)benzene-1,2-diol and desoxybostrycin along with ten other compounds were isolated from the PSU-N24, an endophyte of the plant *Garcinia nigrolineata* (Clusiaceae). The fungal extract was tested for antimycobacterial, antimicrobial, and antimalarial activities. At 128 μg/mL, 3-(2-hydroxypropyl) benzene-1,2-diol and desoxybostrycin (Fig. 10.4) exhibited antibacterial activity against *S. aureus* (MRSA). Remaining compounds exhibited very weak antibacterial activity >128 μg/mL (Sommart et al. 2008).

Twenty-one endophytic fungal strains were isolated from the plant *Calamus thwaitesii* (Arecaceae). 7 out of 21 fungal strains such as *Calonectria pteridis, Lasiodiplodia theobromae, Mycoleptodiscus* sp., *Dendriphiella* sp., *Macrophomina phaseolina, Phomopsis* sp., and *Aspergillus terreus* showed the antimicrobial activity against tested pathogenic microbes. Among all, *Mycoleptodiscus* sp. isolated from the leaves exhbited best antimicrobial activity. This fungus was further grown in large scale on sterile PDA and incubated for 14 days, then extracted with ethanol. Mycoleptodiscin B (Fig. 10.4), an alkaloid was isolated from the ethanolic extract and exhibited high activity against *B. subtilis* and *S. aureus* and showed anti-MRSA activity with MIC value of 32 μg/mL (Dissanayake et al. 2016a, b).

An endophytic fungi Fusarium sp. TP-G1 was isolated from the roots of Dendrobium *officinale Kimura* (Orchidaceae). Nine compounds were isolated from the fungal extract and checked for their antimicrobial activity against *S. aureus*, MRSA, and A. baumannii. Best activity was exhibited by trichosetin, beauvericin A (Fig. 10.4), beauvericin (Fig. 10.3), enniatin MK1688, enniatin I, and enniatin H with MIC values of 2, 2, 4, 8, 16, and 32, respectively, against MRSA. Enniatin B, fusaric acid, and dehydrofusaric acid (Fig. 10.4) exhibited anti-MRSA activity with MIC value >128 μg/mL (Shi et al. 2018).

Nine compounds were isolated from the Talaromyces wortmannii, an endophyte of aloe vera, and tested for its antibacterial activity. Among all, skyrin and rugulosin A (Fig. 10.4) exhibited anti-MRSA activity with MIC values of 4 and 16 μg/mL, respectively. Biemodin (Fig. 10.4) also exhibited anti-MRSA activity but less potent compared to skyrin and rugulosin A (Bara et al. 2013).

Piliformic acid (Fig. 10.5) was isolated from the Xylaria cubensis BCRC 09F 0035, an endophyte of Litsea akoensis (Lauraceae). Piliformic acid exhibited antimicrobial activity with MIC value of 200 μg/mL against *S. aureus* and MRSA (Macías-Rubalcava and Sánchez-Fernández 2017).

Aspermerodione and Andiconin C (Fig. 10.5) isolated from the Aspergillus sp. TJ23, an endophyte of Hypericum perforatum (Hypericaceae), exhibited anti-MRSA activity with MIC values of 32 and >100 μg/mL, respectively. Aspermerodione also showed synergistic activity along with oxacillin and piperacillin, with $\sum$ FIC values of 0.25 and 0.375 (Qiao et al. 2018).

Alternariol and 3,7-dihydroxy-9-methoxy-2-methyl-6Hbenzo[c]chromen-6-one (Fig. 10.5) (structural isomer of alternariol) are isolated from the endophytic fungus Alternaria alternata resident of plant Grewia asiatica (Malvaceae). They exhibited anti-MRSA activity with MIC values of 8 and 64 μg/mL, respectively, with ciprofloxacin as a positive control that also exhibited anti-MRSA activity with MIC value 8 μg/mL (Deshidi et al. 2017).

Seven compounds were isolated from the Penicillium sp. GD6, an endophyte of Bruguiera gymnorrhiza (Rhizophoraceae). Among all, 2-deoxy-sohirnone C (Fig. 10.5) showed anti-MRSA activity with MIC value of 80 μg/mL (Jiang et al. 2018).

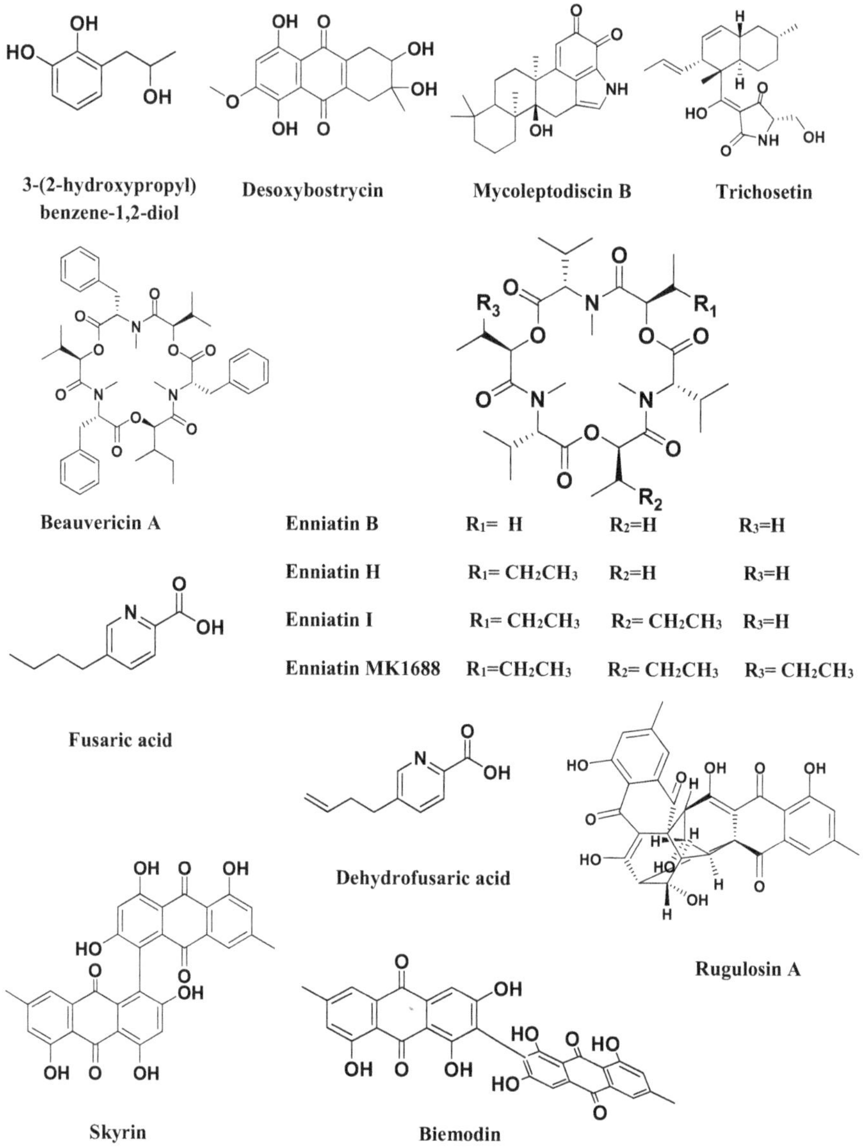

Fig. 10.4 Structure of compounds isolated from fungal endophytes against MRSA

Three compounds were isolated from the Penicillium sp., resident of Cerbera manghas (Apocynaceae). Among all, 4-(3-hydroxybutan-2-yl)-3,6-dimethylbenzene-1,2-diol and 3,4,5-trimethyl-1,2-benzenediol (Fig. 10.5) exhibited anti-MRSA activity (Cui et al. 2008).

Guanacastepene A (Fig. 10.5), a diterpene, was isolated from the unidentified fungal extract from the plant Daphnopsis americana (Thymelaeaceae). It exhibited

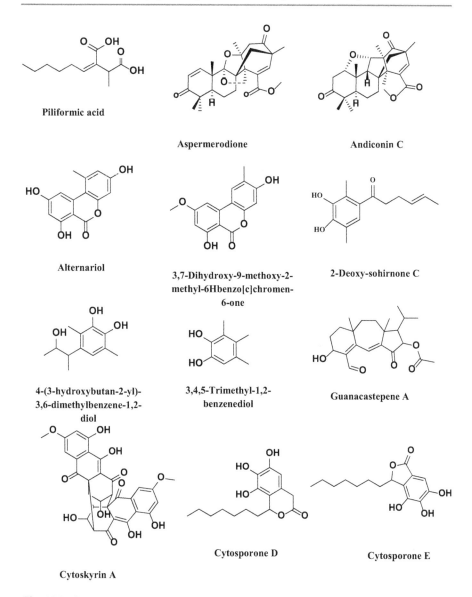

Fig. 10.5 Structure of compounds isolated from fungal endophytes against MRSA

antimicrobial activity against MRSA, MSSA, and vancomycin-resistant E. faecalis (VREF) by disk diffusion method (Radić and Štrukelj 2012).

Cytosporone D, cytosporone E, and cytoskyrin A (Fig. 10.5) were isolated from the Cytospora sp. CR200, an endophyte collected from the branches of Conocarpus erectus belonging to the family *Combretaceae* (buttonwood tree). The antibacterial activity was tested against various bacteria including MRSA. Cytoskyrin A exhibited activity with MIC value in the range of 0.03–0.25 µg/mL against tested pathogens. Cytosporone D and E exhibited moderate activity with MIC in the range of 8–64 µg/mL (Deshmukh et al. 2014).

10.3 Fungal Endophytes Against Drug-Resistant *Plasmodium* Sp.

Parasites are the causative agents for malaria, transmitted by the infected female *Anopheles* mosquito bite. Humans can be affected by five species of plasmodium that includes *P. falciparum*, *Plasmodium malariae* (*P. malariae*), *Plasmodium vivax* (*P. vivax*), *Plasmodium ovale* (*P. ovale*), and *Plasmodium knowlesi* (*P. knowlesi*). Among all, malaria caused by *P. falciparum* is dangerous form and life-threatening if left untreated. In severe cases, it may cause liver and kidney damage, convulsions, and coma that often leads to death. According to the WHO, there are 228 million cases reported globally and 405,000 deaths that occurred in 2018. Children under 5 years of age are mostly affected by malaria, and it accounts for 67% of all malaria deaths worldwide. In 2018, 99.7%, 50%, 71%, and 65% of cases are caused by the *P. falciparum* in African region, Southeast Asian region, and Eastern Mediterranean and Western Pacific areas, respectively.

The malarial parasite proceeds with two stages, asexual and sexual stages, in its life cycle. Asexual stage proceeds in any host (human) body, and sexual stage takes place in mosquito. Both phases occurred as continuous cyclic form. By using insecticides and other precautionary measures, transmission by vectors can be controlled. The antimalarial drugs mainly target the asexual stage of malarial parasite. The drugs which can inhibit the growth of parasite include quinolines (chloroquine, quinine, and mefloquine), the antifolates (sulfadoxine, pyrimethamine), the naphthoquinone, and the artemisinin derivatives. Subsequently after the introduction of medicine for the treatment of malaria, the resistance strains were developed. Quinine, chloroquine, sulfadoxine, and pyrimethamine were introduced in the year 1632, 1945, and 1967; its resistance strain was identified in the year 1910, 1957, and 1967, respectively. Later, drug-resistant strains were identified for mefloquine and artemisinin in the year 1982 and 2006, respectively (Rout and Mahapatra 2019).

Chloroquine is a basic drug molecule; after reaching the target site in the food vacuoles of the parasite, it will become protonated due to the acidic environment in the food vacuoles and get accumulated inside the vacuoles, as it is protonated and cannot cross the cell membrane. Then the chloroquine binds with the free heme (Fe (II)-protoporphyrin IX (FP) moiety to form chloroquine-FP complex, which is toxic to the parasite compared to heme moiety. The chloroquine-FP complex will disturb the membrane function that leads to lysis of the cell.

After 20 years of successful treatment with the chloroquine, resistant strain was identified. The chloroquine resistance was due the mutation in the *P. falciparum* chloroquine-resistant transporter gene (PfCRT), *P. falciparum* multidrug-resistant gene (PfMDR1), and *P. falciparum* Na+/H+ exchanger (pfnhe-1). The chloroquine resistance was primarily carried by the PfCRT gene. This mutation will lead to less drug concentration inside the vacuoles due to the leakage. Multidrug-resistant gene was carried by PfMDR1, which encodes P-glycoprotein homolog-1 (Pgh-1). Pgh-1 regulates MDR role in many organisms belonging to ATP-binding cassette transporter (ABC transporter). The mutation in PfMDR1 gene may affect the uptake and transportation of chloroquine by parasite. PfMDR1 gene mutation not only confers resistance to chloroquine but also to other drugs like quinine, lumefantrine, and mefloquine (Rout and Mahapatra 2019).

A number of researchers worked on fungal endophytes to combat resistance against *P. falciparum*. Here the compounds isolated from the fungal endophytes active against *P. falciparum* resistant strain were mentioned.

Pestalotiopsis sp., an endophytic fungus, was isolated from the ethyl acetate extract from the stem of *Melaleuca quinquenervia* (Myrtaceae). Three compounds named pestalactams A, B, and C were isolated from the endophytic fungi. Pestalactam A (Fig. 10.6) exhibited activity against chloroquine-sensitive and chloroquine-resistant strain of *P. falciparum* with IC_{50} values of 16.2 and 41.3 µM, respectively. Pestalactam B (Fig. 10.6) exhibited activity against chloroquine-sensitive and chloroquine-resistant strain of *P. falciparum* with IC_{50} values of 20.7 and 36.3 µM, respectively (Ferreira et al. 2019).

An endophytic fungus *Diaporthe miriciae* was isolated from the plant *Vellozia gigantea* (Velloziaceae). The crude endophytic extract showed antiplasmodial, antibacterial, and antifungal activities. Epoxycytochalasin H (Fig. 10.6) was isolated from the endophytic extract of *Diaporthe miriciae* and exhibited activity with IC_{50} values of 105.34 and 79 nM against chloroquine-sensitive and chloroquine-resistant strains of *P. falciparum*, respectively (Ferreira et al. 2019).

(+)-Phomalactone, 6-(1-propenyl)-3,4,5,6-tetrahydro-5-hydroxy-4Hpyran-2-one, and 5-hydroxymellein were isolated from the endophytic fungi *Xylaria* sp. The fungus was isolated from the leaf of *Siparuna* sp. belonging to the family Siparunaceae collected from the region of Altos de Campana National Park, Panama. (+)-Phomalactone, 6-(1-propenyl)-3,4,5,6-tetrahydro-5-hydroxy-4Hpyran-2-one (Fig. 10.6), and 5-hydroxymellein (Fig. 10.2) exhibited activity with IC_{50} values of 13, >50, and 19 µg/mL, respectively, against chloroquine-resistant strain of *P. falciparum* (Jiménez-Romero et al. 2008).

3-(2-Hydroxypropyl)benzene-1,2-diol was isolated along with 11 other compounds from the fungi PSU-N24. The fungus was isolated from the *Garcinia nigrolineata* (Clusiaceae). 3-(2-Hydroxypropyl)benzene-1,2-diol (Fig. 10.4) exhibited the best antimalarial activity against *P. falciparum* (K1, multidrug-resistant strain) with an IC_{50} of 6.68 mg/ml, while 9α-hydroxhalorosellinia A and desoxybostrycin exhibited less activity with the IC_{50} values of 7.94 and 10 mg/mL, respectively (Sommart et al. 2008).

Geotrichum sp., an endophytic fungus, was isolated from the plant *Crassocephalum crepidioides* belonging to the family Asteraceae. Three dihydroisocoumarin derivatives named 7-butyl-6,8-dihydroxy3(R)-pent-11-enyliso chroman-1-one, 7-butyl-15-enyl-6,8-dihydroxy-3(R)-pent-11-enylisochroman-1-one, and 7-butyl-6,8-dihydroxy-3(R)-pentylisochroman-1-one were isolated from the

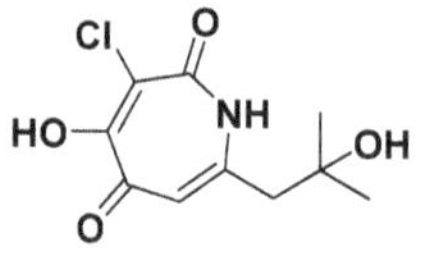

Fig. 10.6 Structure of compounds isolated from fungal endophytes against. *Plasmodium* sp.

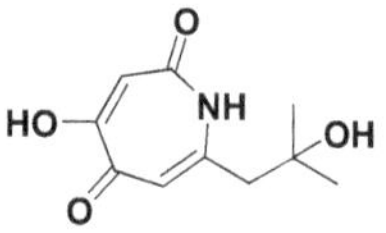

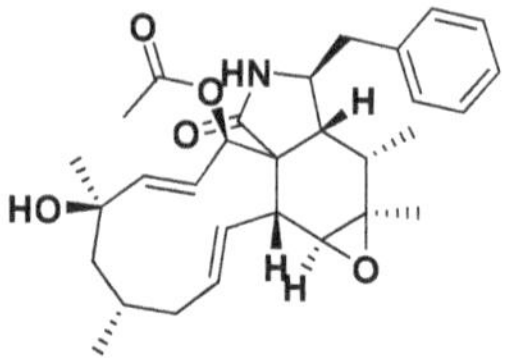

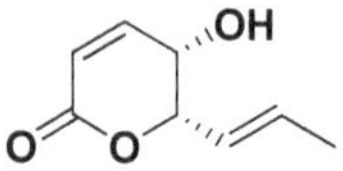

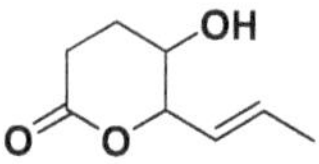

endophytic fungus, and its antimalarial, antituberculosis, and antifungal activities were tested against *P. falciparum* K1 (multidrug-resistant strain), *M. tuberculosis* H27Ra, and *C. albicans*, respectively. 7-Butyl-6,8-dihydroxy3(R)-pent-11-enylisochroman-1-

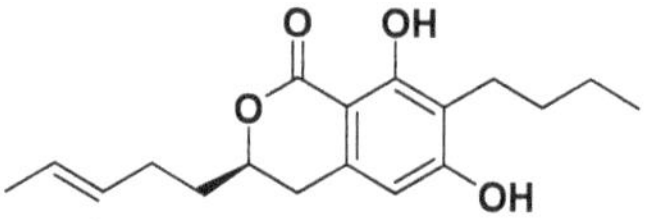

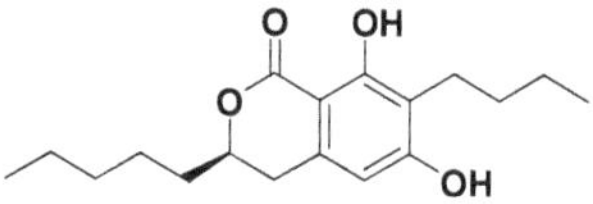

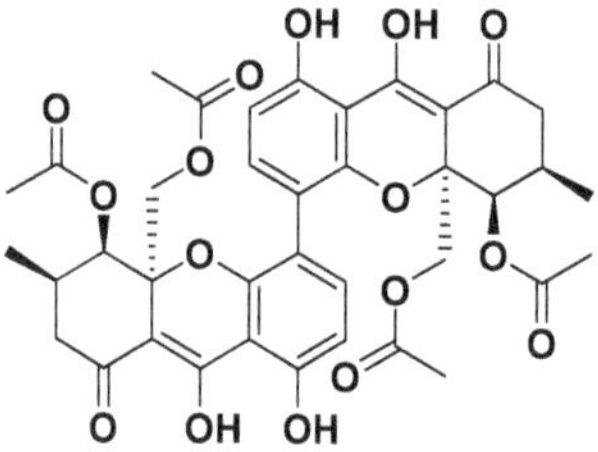

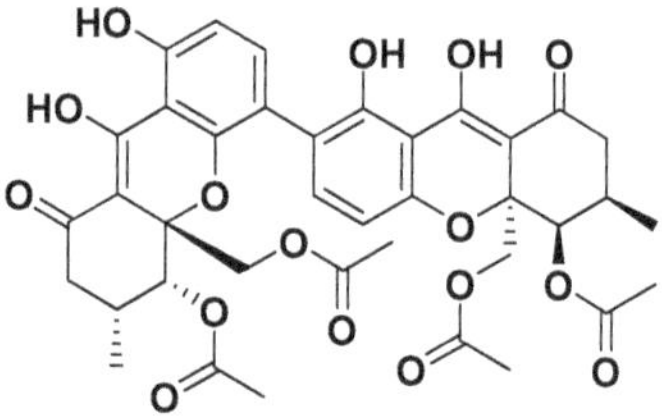

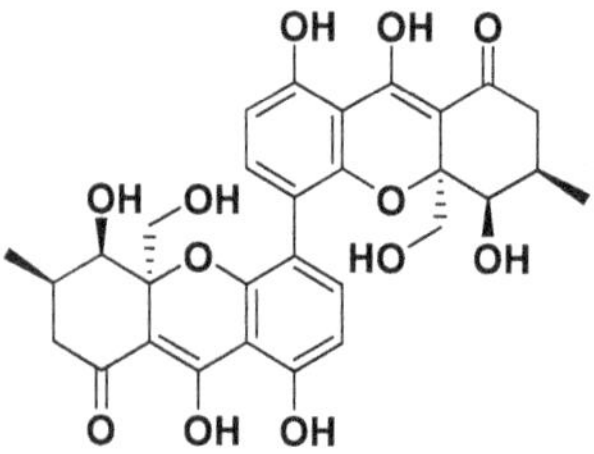

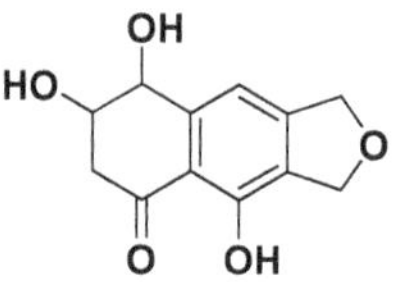

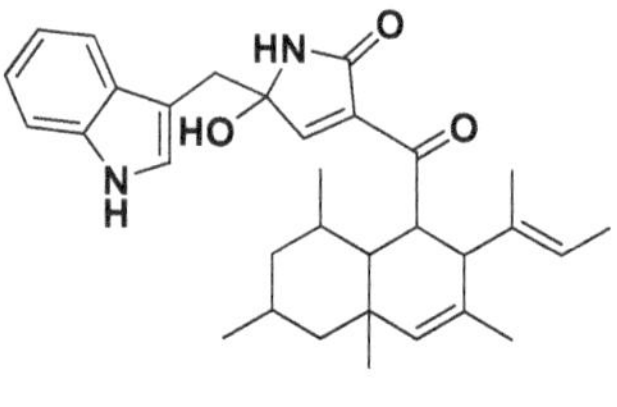

one and 7-butyl-6,8-dihydroxy-3(R)-pentylisochroman-1-one (Fig. 10.6) exhibited activity against *P. falciparum* K1 (multidrug-resistant strain) with IC_{50} values of 4.7 and 2.6 µg/mL, respectively (Kongsaeree et al. 2003).

Two xanthone dimmers Phomoxanthone A and Phomoxanthone B (Fig. 10.6) were isolated from the fungus *Phomopsis* sp. BCC 1323 from the leaf of *Tectona grandis* (Lamiaceae) collected from Northern Thailand. The antimalarial and antitubercular activity of isolated compounds was tested against *P. falciparum* (K1, multidrug-resistant strain) and *M. tuberculosis* (H37Ra strains. Phomoxanthone A and phomoxanthone B exhibited activity against *P. falciparum* (K1, multidrug-resistant strain) with IC_{50} values of 0.11 and 0.33 µg/mL, respectively, while deacetyl-phomoxanthone A (Fig. 10.6) (deacetylated product of phomoxanthones A) exhibited activity with IC_{50} value of >20 µg/mL (M. Isaka et al. 2001).

Nodulisporium sp., an endophytic fungus, was isolated from the *Antidesma ghaesembilla* (Phyllanthaceae). The crude ethyl acetate showed activity against *M. tuberculosis* with IC_{50} value of 100 mg/mL. Further six compounds were isolated from *Nodulisporium* sp. Antimalarial activity was performed using the microculture radioisotope technique against *P. falciparum* K1, the multidrug-resistant strain. Among all, isofuranonaphthalenone (a dihydronaphthalenones) (Fig. 10.6) exhibited antimalarial activity with MIC value of IC_{50} value 11.3 mg/mL (Prabpai et al. 2015).

An antimalarial polyketide known as codinaeopsin was isolated from the endophytic fungus of plant *Vochysia guatemalensis* (vochysia), a white yemeri tree. The endophytic fungal isolate CR127A was found to be 98% identical to *Codinaeopsis gonytrichoides*. Codinaeopsin (Fig. 10.6) exhibited antimalarial activity against *P. falciparum* 3D7 (sensitive to all drugs), HB3 (resistant to pyrimethamine and mefloquine) and Dd2 (resistant to chloroquine, pyrimethamine, and mefloquine) strains with IC_{50} values of 4.66 µM, 4.73 µM, and 4.84 µM, respectively (Kontnik and Clardy 2008).

5-Carboxylmellein, Cytochalasin Q, and Halorosellinic acid (Fig. 10.7) were isolated from different fungal endophytes of *Xylaria* sp., i.e., 5-carboxylmellein isolated from *Xylaria* sp. SNB-GTC2501 (an endophyte of *Bisboecklera microcephala*), cytochalasin Q isolated from *Xylaria* sp. NC1214 (an endophyte isolated from *Hypnum* sp.), and halorosellinic isolated from *Xylaria* sp. YC-10 (an endophyte isolated from the stem of *Azadirachta indica*) (Macías-Rubalcava et al. 2017). For these three compounds, antimalarial activity was previously reported by Chinworrungsee et al. (2001). 5-Carboxylmellein, cytochalasin Q, and halorosellinic exhibited antimalarial activity with IC_{50} values of 4, 17, and 13 µg/mL against a multidrug-resistant strain of *P. falciparum* K1.

Pullularia sp. BCC 8613, an endophytic fungi, was isolated from the leaf of *Calophyllum* sp. collected from Narathiwat Province, Hala-Bala Wildlife Sanctuary. Four compounds named pullularins A–D (cyclohexadepsipeptides) were isolated from the fungus. Deprenylation of pullularin A leads to the formation of another compound known as deprenylpullularin A. The compounds pullularin A, Pullularin B, Pullularin C, and Deprenylpullularin A (Fig. 10.7) were tested for its antimycobacterial and antimalarial activities. The antimalarial activity of these four compounds exhibited activity against *P. falciparum* (K1, multidrug-resistant strain)

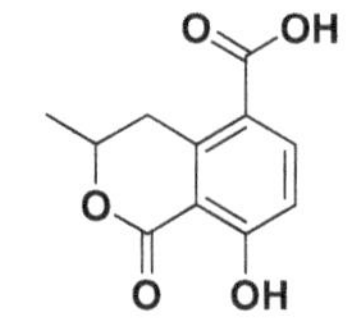

5-Carboxylmellein

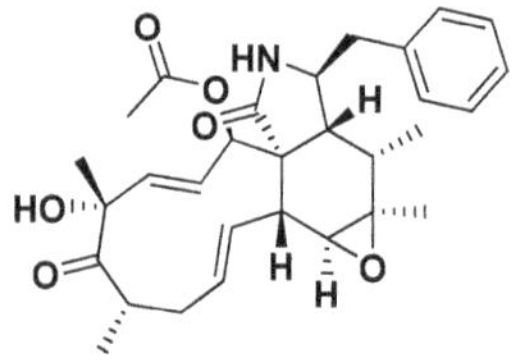

Cytochalasin Q

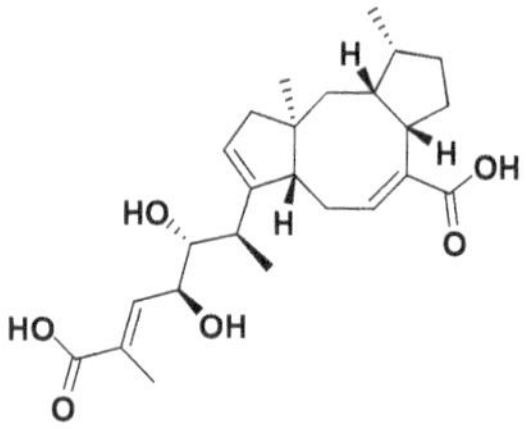

Halorosellinic acid

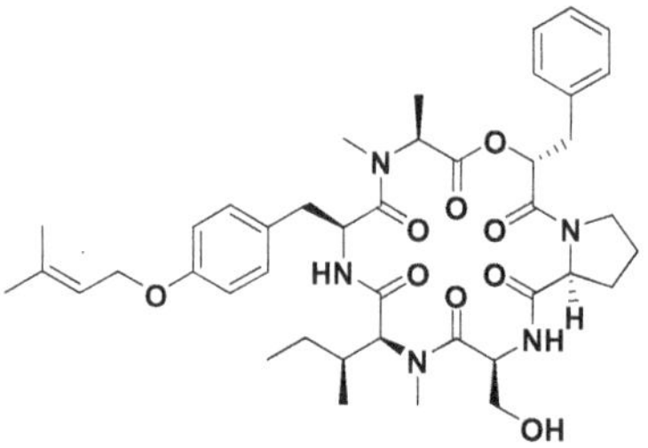

Pullularin A

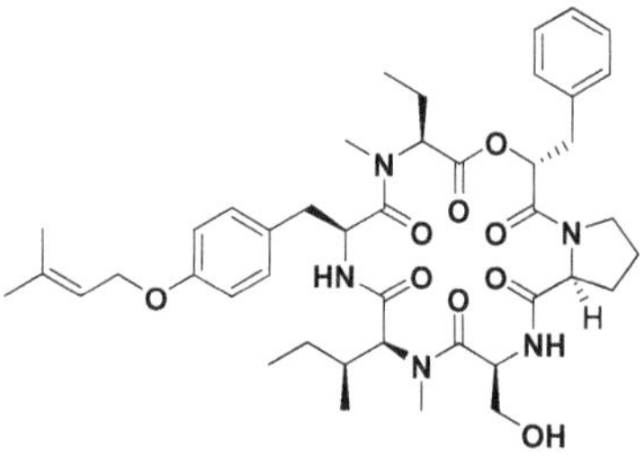

Pullularin B

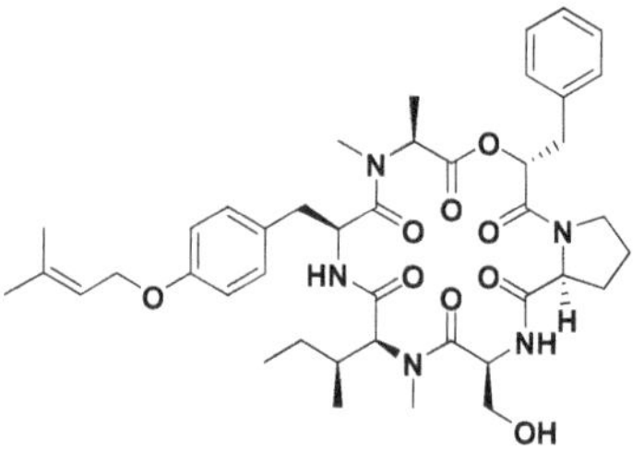

Pullularin C

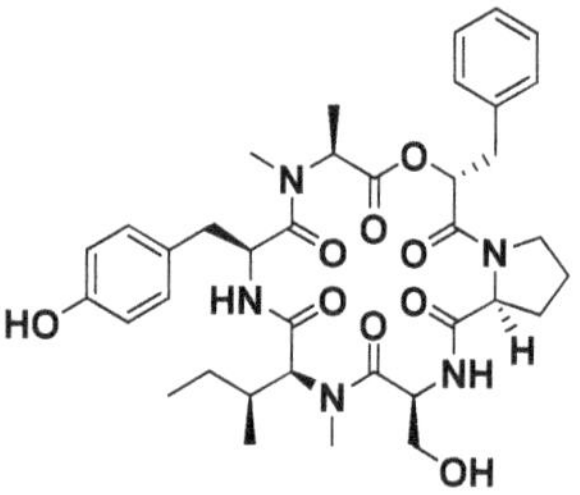

Deprenylpullularin A

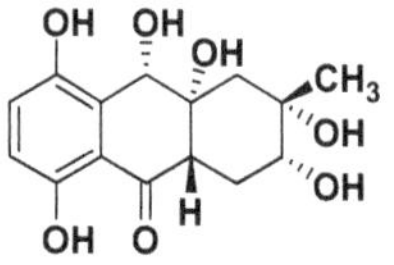

Nigrosporone A

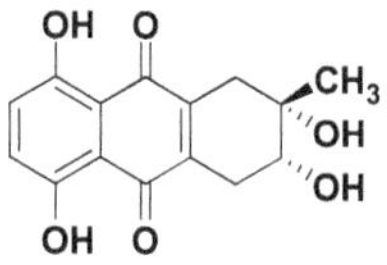

Nigrosporones B

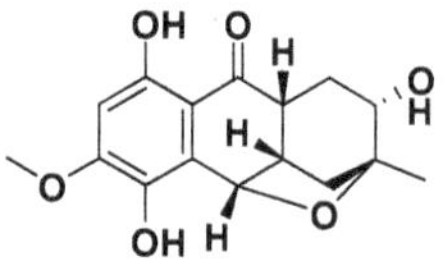

Fusaquinon A

Fig. 10.7 Structure of compounds isolated from fungal endophytes against *Plasmodium* sp.

with IC_{50} values of 3.6, 3.3, 9.8, and >20 μg/mL using the microculture radioisotope technique (Isaka et al. 2007).

Nigrospora sp. BCC 4778, an endophytic fungus, was isolated from the leaf of *Choerospondias axillaris* (Anacardiaceae) collected from the Nakhon Ratchasima Province, Thailand and identified. The fungus was grown on potato dextrose agar for extraction of compounds. A total of 13 compounds were isolated from the ethyl acetate extract of fungus. Among all compounds, Nigrosporone A, Nigrosporone B, and Fusaquinon A (Fig. 10.7) were tested for its activity against *P. falciparum*, *M. tuberculosis, Bacillus cereus*, and *E. faecium*. Nigrosporone A, nigrosporone B, and fusaquinon A exhibited activity against *P. falciparum* (K1, multidrug-resistant strain) with IC_{50} values of >33.97, 10.81, and >32.67 µg/Ml, respectively (Kornsakulkarn et al. 2018).

Nine compounds were isolated from the endophytic fungus *Penicillium* sp. BCC1605. The fungus was isolated from the grass belonging to the family Poaceae, collected from Northern Thailand. Seven compounds such as penicolinate A, penicolinate B, penicolinate C, phenopyrrozin, p-hydroxyphenopyrrozin, gliotoxin, and bisdethiobis(methylthio)gliotoxin were tested against *P. falciparum*, *M. tuberculosis, B. cereus*, and *C. albicans*. Penicolinate A, Penicolinate B, Penicolinate C, Phenopyrrozin (Fig. 10.8), and Gliotoxin (Fig. 10.3) exhibited antimalarial activity against *P. falciparum* K-1, the multidrug-resistant strain with IC_{50} values of 3.25, 1.40, 3.07, 3.95, and 0.42 µg/Ml, respectively. While P-hydroxyphenopyrrozin and Bisdethiobis(methylthio)gliotoxin (Fig. 10.8) exhibited activity with IC_{50} value of >10 µg/mL (maximum concentration tested) (Intaraudom et al. 2013).

Exserohilum rostratum, an endophytic fungus, was isolated from the leaves of a *Stemona* sp. Monocerin, 11-hydroxymonocerin, and 12-hydroxymonocerin were isolated from the ethyl acetate fungal extract. Another two compounds were also prepared by acetylation of monocerin and 11-hydroxymonocerin. Monocerin and 11-hydroxymonocerin (Fig. 10.8) exhibited the antimalarial activity against *P. falciparum* (K1, multidrug-resistant strain) with IC_{50} values of 0.68 and 7.70 µM, respectively, while acetylated compounds of monocerin and 11-hydroxymonocerin exhibited antimalarial activity with IC_{50} value of 0.82 and 9.10 µM, respectively (Sappapan et al. 2008).

Xylaria sp. was isolated from the leaves of *Sandoricum koetjape* (Meliaceae). The ethyl acetate fungal extract was subjected to separation by silica gel column chromatography. With an elution mixture of dichloromethane/hexane (70:30), 2-chloro-5-methoxy-3-methylcyclohexa-2,5-diene-1,4-dione and xylariaquinone A were isolated. 2-Hydroxy-5-methoxy-3-methylcyclohexa-2,5-diene1,4-dione and 4-hydroxymellein were isolated from the elution with a mixture of dichloromethane/MeOH (90:10). All four compounds were tested for antimalarial activity. 2-Chloro-5-methoxy-3-methylcyclohexa-2,5-diene-1,4-dione (Fig. 10.2) and Xylariaquinone A (Fig. 10.8) exhibited antimalarial activity against *P. falciparum* (K1, multidrug-resistant strain) with IC_{50} values of 1.84 and 6.68 µM, respectively, while another two compounds, 2-hydroxy-5-methoxy-3-methylcyclohexa-2,5-diene1,4-dione and 4-hydroxymellein, were inactive against *P. falciparum* (K1, multidrug-resistant strain) (Tansuwan et al. 2007).

Preussia sp. strain CAD4, an endophytic fungus, is associated with the plant *Enantia chlorantha Oliv* (Annonaceae). Eight compounds were isolated from the endophytic fungal extract. Preussiafuran A, Preussiafuran B, Cissetin, and Asterric acid (Fig. 10.8) were tested for antimalarial activity, and they exhibited activity against erythrocytic stages of chloroquine-resistant *P. falciparum* (NF54) with IC_{50}

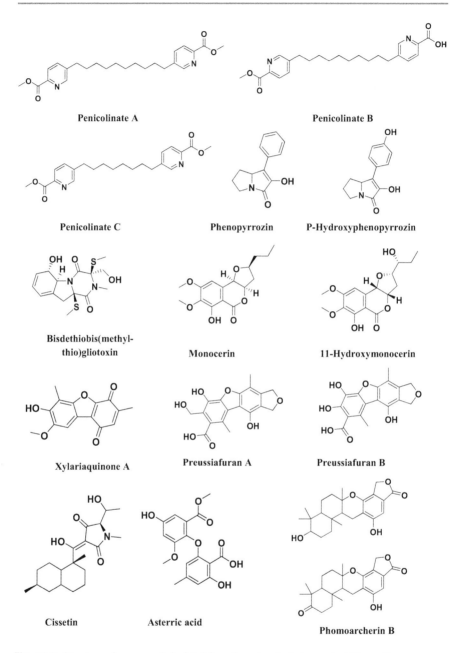

Fig. 10.8 Structure of compounds isolated from fungal endophytes against Plasmodium sp.

values of 8.76, 15.0, 10.3, and 8.67 μg/mL, respectively. These compounds also exhibited moderate cytotoxicity on L6 cell lines (Talontsi et al. 2014).

The endophytic fungus *Phomopsis archeri* was isolated from the stem of *Vanilla albida* (Orchidaceae), collected from the Thailand. Phomoarcherins A, phomoarcherins B, phomoarcherins C, and kampanol A were isolated from the fungus. Phomoarcherin A and Phomoarcherin B (Fig. 10.8) exhibited antimalarial activity against *P. falciparum* (K1, multidrug-resistant strain) with IC_{50} values of >20 and 0.79 μg/mL, respectively. The cytotoxicity of phomoarcherins A and B was also tested against the KB cell lines (Hemtasin et al. 2011).

Aspergillus versicolor, an endophytic fungus, was isolated from the roots of *Pulicaria crispa* belonging to the family Asteraceae. Aspernolide L, aspernolide M, butyrolactones I, and butyrolactones VI were isolated from the fungus. The compounds are tested for antimicrobial, antileishmanial, and antimalarial activity. The antimalarial activity was tested against *P. falciparum* (D6 clone), *P. falciparum* (D6 S1 clone), *P. falciparum* (W2 clone, chloroquine resistant), and *P. falciparum* (W2 S1 clone, chloroquine-sensitive clones). Aspernolide L, Aspernolide M, Butyrolactone I, and Butyrolactone VI exhibited (Fig. 10.9) activity against *P. falciparum* (W2 clone, chloroquine resistant) with IC_{50} values of 59.1 ± 2.11, 200, 119.5 ± 8.23, and 200 mM, respectively (Ibrahim and Asfour 2018).

Nemania sp. UM10M, an endophytic fungus, was isolated from the leaf of diseased *Torreya taxifolia* (Taxaceae). The fractionization of ethyl acetate extract from endophytic fungus afforded three compounds known as 19,20-epoxycytochalasins C, 19,20-epoxycytochalasins D, and 18-deoxy-19,20-epoxy-cytochalasin C. The antiplasmodial activity and cytotoxicity of isolated compounds were evaluated. The compounds named 19,20-Epoxycytochalasin C and 19,20-epoxycytochalasin D (Fig. 10.9) exhibited potent antimalarial activity against both chloroquine-sensitive (D6) and chloroquine-resistant (W2) strains of *P. falciparum* with IC_{50} value in the range of 0.04 to 0.07 μM. While 18-deoxy-19,20-epoxy-cytochalasin C (Fig. 10.9) exhibited moderate antimalarial activity with IC_{50} values of 0.56 and 0.19 μM against chloroquine-sensitive (D6) and chloroquine-resistant (W2) strains, respectively (Kumarihamy et al. 2019).

10.4   Fungal Endophytes Against Resistant Mycobacterium Sp.

Tuberculosis (TB) is caused by bacteria *M. tuberculosis*. The most commonly affected organs are the lungs, and it can easily spread to another person by cough or sneeze from infected person. It will become a life-threatening condition if the patient does not respond to treatment. According to the WHO, ten million people are affected, and 1.5 million people died in 2018 by TB. The first-line drugs for TB include rifampin, isoniazid, pyrazinamide, ethambutol, and streptomycin. The second-line treatment includes ofloxacin, levofloxacin, moxifloxacin, ciprofloxacin, kanamycin, amikacin, and capreomycin.

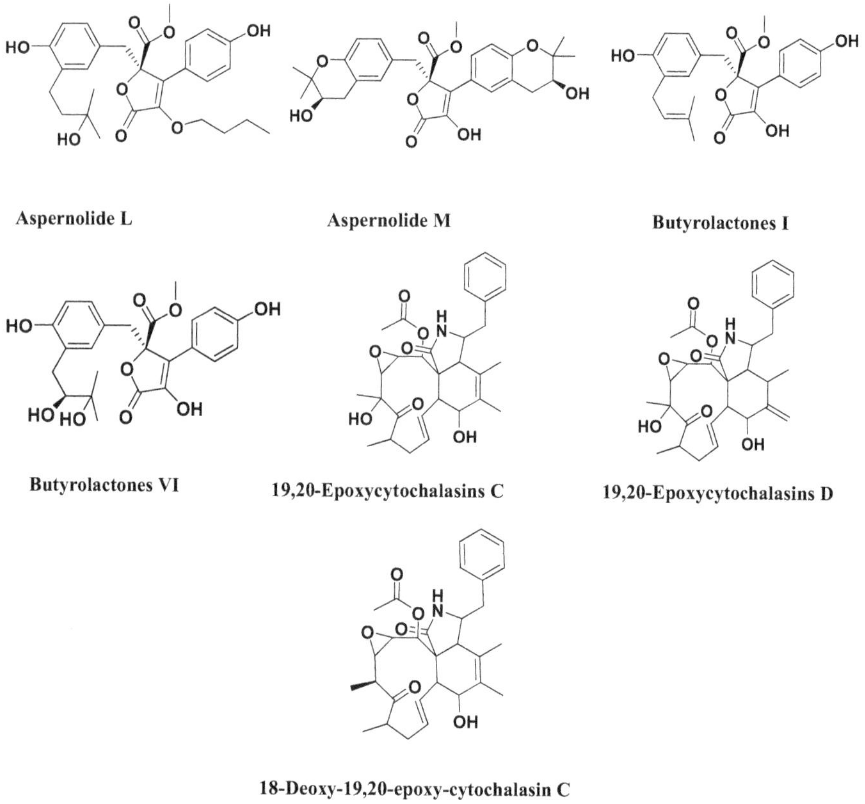

Aspernolide L   Aspernolide M   Butyrolactones I

Butyrolactones VI   19,20-Epoxycytochalasins C   19,20-Epoxycytochalasins D

18-Deoxy-19,20-epoxy-cytochalasin C

Fig. 10.9 Structure of compounds isolated from fungal endophytes against *Plasmodium* sp.

In 1948, resistant strains of tuberculosis were identified after starting treatment with streptomycin. Then combination therapy was started for the treatment of TB. Later so many drugs are introduced for the treatment of TB; subsequently, the resistant strains were identified. Then MDR-TB evolved in different countries on multiple occasions and continues to spread across the globe reducing the effectiveness of drugs against TB (Gygli et al. 2017). The WHO reported 558,000 MDR/RR-TB cases globally and 230,000 deaths in 2017. Deaths occurred mostly in China and India. Among all reported cases, 8.5% are reported to be XDR-TB.

Rifampicin is one of the most effective antitubercular drugs introduced in the year 1972, and it is the most prescribed first-line drug in combination with isoniazid. It inhibits the DNA-dependent RNA polymerase by binding to the β-subunit of the RNA polymerase which further suppresses the RNA synthesis. Resistance was developed due to the mutations in the *rpoB* gene that encodes for the β-subunit of the RNA polymerase that leads to reducing the affinity of drug toward the binding site due to conformational changes. Isoniazid interferes with the cell wall synthesis by inhibiting the mycolic acid through the NADH-dependent enoyl-acyl carrier protein (ACP)-reductase that is encoded by *inhA* gene. Mutations in *katG, inhA*, *ahpC*, *kasA*, and NDH genes are associated with resistance to isoniazid. In other first-line drugs like ethambutol, pyrazinamide resistance was developed due to mutations in the *embB and pncA* genes, respectively. The resistance in streptomycin is due the mutations in the rpsL and *rrs* genes. Recent studies also reported that mutations in *gid*B gene also confer low level of resistance to streptomycin (Palomino and Martin 2014). MDR is defined as resistance to at least both isoniazid and rifampicin. XDT-TB is defined as resistance to any fluoroquinolone and at least one of the three second-line drugs including capreomycin, kanamycin, and amikacin in addition to MDR (Seung et al. 2015). Compounds are isolated from the fungal endophytes that exhibited activity against drug-resistant strains against TB mentioned here.

Fusaric acid was isolated from the extract of endophytic fungi *Fusarium Sp.* DZ-27. The fungus was isolated from the bark of mangrove plant *Kandelia candel* (Rhizophoraceae) from the region of South China Sea. Antimycobacterial activity was performed for fusaric acid alone and metal complexes of fusaric acid. Copper, iron, zinc, manganese, lead, and cadmium were used for complexation with fusaric acid. Antimicrobial activity for fusaric acid and its metal chelates were tested on *M. bovis* BCG (ATCC 35734), *M. tuberculosis* H37Rv (ATCC 27294), clinical multidrug-resistant *M. tuberculosis* strains, and clinical extensively drug-resistant *M. avium-intracellulare* strains. The formation of metal chelates increased the antimycobacterial activity only in the case of *M. bovis* BCG (MIC values are in the range of 4–25 µg/mL with highest activity being exhibited by complex with candium, i.e., 4 µg/mL). For the remaining strains, the MIC values were in the range of 10–60 µg/mL (Pan et al. 2011).

A mangrove-derived endophytic fungus *Nigrospora* sp. was collected from the region of South China Sea. Two compounds named 4-deoxybostrycin and nigrosporin were isolated from the endophytic fungi. The antimycobacterial activity of these two compounds was tested against different strains including resistance strains. The antimycobacterial activity was tested against *M. bovis* BCG, *M. tuberculosis* H37Rv strain, clinical MDR *M. tuberculosis* strain (K2903531, resistant to ethambutol, isoniazid, rifampicin, and streptomycin), clinical MDR *M. tuberculosis* strains (0907961, resistant to ethambutol and streptomycin), clinical drug-resistant *M. tuberculosis* strain (K0903557, resistant to INH), clinical drug-sensitive *M. tuberculosis* strain (0907762), *M. avium* reference strain, *M. intracellulare* reference strain (ATCC 13950), and clinical extensively drug-resistant (XDR) *M. avium-intracellulare* strain (K0803182, resistant to SM, INH, RFP, levofloxacin [LVFX], protionamide, and isoniazid aminosalicylate). 4-Deoxybostrycin and nigrosporin exhibited antimycobacterial activity with MIC values in the range of 5 to >60 µg/mL and 15 to >60 µg/mL, respectively (Wang et al. 2013).

A total of 64 endophytes were isolated from the *Piper longum* (Piperaceae). Among all, 42 were identified as endophytic fungi, 17 were bacterial endophytes, and remaining 5 were endophytic actinomycetes. The isolated endophytes were tested for their antimycobacterial activity. Among all, endophytic fungi belong to *Periconia* sp. That exhibited the highest activity against *Mycobacterium* sp. This strain was isolated from the leaf tissue of *Piper longum*. The compounds that were responsible for exhibiting antimycobacterial activity was isolated and identified as well-known alkaloid piperine. Antimycobacterial bioassay was performed by disk diffusion method against clinical multidrug-resistant strain of *M. tuberculosis* and *M. smegmatis* with four different concentrations of piperine, i.e., at 2, 5, 10, and 15 mg/mL. The increase in bacterial growth inhibition was linear with increase in concentrations of piperine. The zone of inhibition was found to be 10, 13, 14, and 16 mm against multidrug-resistant strain of *M. tuberculosis* and 6, 8, 11, and 15 mm against *M. smegmatis* at 2, 5, 10, and 15 mg/mL concentration, respectively (Verma et al. 2011).

A natural tetramic acid named vermelhotin was isolated from endophytic fungus MEXU 26343, collected from the plant *Hintonia latiflora* (Rubiaceae) (Leyte-Lugo et al. 2012). Vermelhotin exhibited antimycobacterial activity with MIC value of 3.1 mg/mL against *M. tuberculosis* H37Ra strain. It also exhibited activity against clinical strains of MDR-TB with MIC 1.5–12.5 g/mL. The clinical strains are resistant to isoniazid (INH), rifampin (RMP), ofloxacin (OFX), streptomycin (SM), and ethambutol (EMB) (Ganihigama et al. 2015).

Muscodor crispans, an endophytic fungus, was isolated from the pineapple plant, *Ananas ananassoides* (Ananas ananassoides). The endophytic fungi *Muscodor crispans* was the major source for the production of volatile organic compounds. The activity of the mixture of volatile organic compounds produced by endophytic fungus was tested against various pathogenic microbes. The volatile organic compounds inhibited the growth of *Yersinia pestis*, *S. aureus*, *Salmonella* enteric, and *M. marinum* including drug resistance strain of *M. tuberculosis* (resistant to isoniazid, streptomycin, and ethambutol). These volatile organic compounds also exhibited activity against various plant pathogens (Mitchell et al. 2010).

10.5 Fungal Endophytes Against Resistant *Candida albicans*

Candidiasis is fungal infection caused by the pathogenic yeast *C. albicans*. It may cause a wide range of infections including oral, vaginal, systemic, and urinary tract infections in humans. It is also associated with HIV. The antifungal drugs used include amphotericin B, azoles, echinocandin, nystatin, and clotrimazole. Fluconazole is the most widely used drug as antifungal agent due to its broad activity against various fungal species. It acts by inhibiting to the enzyme lanosterol 14-α-demethylase that is responsible for converting lanosterol to ergosterol, thus interfering with the cell wall synthesis. The fluconazole resistance was due to the overexpression of *ERG11*gene (enzymes responsible for the biosynthesis of ergosterol). *CDR1* and *CDR2* (ATP-dependent efflux pumps) and MDR1 gene overexpression also causes efflux of drug which causes resistance in *C. albicans* (White et al. 2002). Compounds obtained from fungal endophytes with antifungal activity against resistant *C. albicans* are mentioned here.

Cytochalasans are a group of fungal metabolites isolated from the ethyl acetate extract of endophytic fungus *Xylaria longipes*. Three cytochalasans (curtachalasins C, curtachalasins D, and curtachalasins E) were isolated and identified by different spectral analytical techniques. Curtachalasin C and curtachalasin E were tested for resistance reversal activity against fluconazole-resistant *C. albicans*. Curtachalasins C exhibited resistance reversal activity when combined with 10 µg/mL of fluconazole. A significant improvement in inhibitory ratio was observed in combination, i.e., 50% increase in inhibitory ratio was observed compared to fluconazole alone, while curtachalasin E at a concentration of 128 µg/mL exhibited weak resistance reversal activity (Wang et al. 2019).

Compounds derived from *Tolypocladium cylindrosporum*, an endophytic fungus of *Lethariella zahlbruckner* (lichen), showed antifungal activity against various drug resistance strains of *C. albicans*. Pyridoxatin and N-deoxy-pyridoxatin were isolated from the endophytic fungus *Tolypocladium cylindrosporum*. The activity of pyridoxatin was tested against various strains of *C. albicans*, *C. krusei, C. glabrata* and *C. tropicalis*. Pyridoxatin exhibited antifungal activity against azole-resistant strain of *C. albicans* with MIC value of 1 µg/mL. It also showed antifungal activity against fluconazole-susceptible and fluconazole-resistant isolates with MIC values in the range of 1–2 µg/mL (Chang et al. 2015).

A diphenyl ether derivative, diorcinol D, was isolated from the endophytic fungus *Aspergillus versicolor* derived from the lichen *Lobaria quercizans*. Diorcinol D exhibited antifungal activity against five fluconazole-resistant isolates, five fluconazole-sensitive isolates, and six mutant strains with the MIC values being in the range of 8–32 µg/mL. Diorcinol D showed synergistic and drug-resistant reversing effects in combination with fluconazole. The combination of fluconazole along with diorcinol D resulted in reduction of MIC values of fluconazole against all isolates tested. A 2–16-fold decrease in MIC value was observed for fluconazole-sensitive isolates, while >250-fold decrease in MIC value was observed for fluconazole-resistant isolates (Li et al. 2015).

Ibrahim and Asfour (2018) also reported about the activity of aspernolide L, aspernolide M, butyrolactone I, and butyrolactone VI compounds isolated from the endophytic fungus *Aspergillus versicolor* against *Aspergillus fumigates*, *Cryptococcus neoformans*, *S. aureus*, *P. aeruginosa*, *E. coli*, and methicillin-resistant *C. albicans*. Aspernolide L, aspernolide M, butyrolactone I, and butyrolactone VI exhibited activity against methicillin-resistant *C. albicans* with IC_{50} values of 4.31 ± 0.17, 5.41 ± 0.25, >20, and > 20 mM, respectively.

10.6   Miscellaneous

Ola et al. (2013a, b) reported about the compounds extracted from the endophytic fungi *Fusarium tricinctum* when cocultured with *B. subtilis*. Among all isolated compounds Enniatin A1 exhibited good activity against clarithromycin-, erythromycin-, moxifloxacin-, telithromycin-, and vancomycin-resistant enterococci, with MIC of 4 µg/mL, while both enniatin B1 and lateropyrone showed activity with MIC value of 8 µg/mL. All remaining compounds exhibited activity with MIC value of >64 µg/mL.

Talaromyces sp. ZH-154, an endophytic fungus, was isolated from the stem bark of *Kandelia candel* belonging to the family Rhizophoraceae. A total of seven compounds were isolated from the endophytic fungus. All compounds were tested against various human pathogens including drug-resistant *P. aeruginosa*. 7-epiaustdiol, 8-O-methylepiaustdiol, stemphyperylenol, skyrin, secalonic acid, emodin, and norlichexanthone that exhibited antimicrobial activity against multidrug-resistant *P. aeruginosa* with MIC values of 6.25, 25.0, 12.5, 12.5, 12.5, 12.5, and 25.0 μg/mL, respectively (Liu et al. 2010).

Ola et al. (2013a, b) tested the compound neosartorin against various drug-resistant pathogens. The compound was derived from the endophytic fungi *Aspergillus fumigatiaffinis*. Neosartorin exhibited antimicrobial activity against *E. faecalis* (resistant to clarithromycin, erythromycin, moxifloxacin, and telithromycin) and *E. faecium* (resistant to clarithromycin, erythromycin, and telithromycin) with MIC values of 16 and 32 μg/mL, respectively. At >64 μg/mL, it exhibited activity against *P. aeruginosa* (resistant to ceftazidime, ciprofloxacin, cefepime, gentamycin, meropenem, piperacillin/tazobactam), *E. coli* (resistant to ciprofloxacin), and *K. pneumoniae* (resistant to doxycycline and kanamycin).

Five polyketides along with a new alkaloid named (±)- preisomide were isolated from the endophytic fungi *Preussia isomera*, resident of the stem of *Panax notoginseng* (Araliaceae). The isolated compounds are evaluated for antifungal and antibacterial activities against various pathogens. Among all tested compounds, setosol (a polyketide) exhibited antibacterial activity with MIC value of 25 μg/mL against multidrug-resistant *E. faecalis*, MRSA, and multidrug-resistant *E. faecium*. It also exhibited antifungal activity with MIC value of 50 μg/mL against *Gibberella saubinetii* (Chen et al. 2020).

Yang et al. (2019) isolated two new C13-polyketides along with five known compounds from the endophyte *Chaetomium globosum* XL-1198. The fungus was isolated from the aerial parts of *Salvia miltiorrhiza* (Lamiaceae). All compounds are tested for antibacterial activity against drug-resistant pathogens and plant pathogenic fungi. Among seven compounds, equisetin exhibited antibacterial activity against multidrug-resistant *S. epidermidis*, multidrug-resistant *E. faecalis*, MRSA, and multidrug-resistant *E. faecium* with MIC value in the range of 3.13–6.25 μg/mL.

Didymella sp. IEA-3B, an endophytic fungus was isolated from the leaves of *Terminalia catappa* (Combretaceae). The isolated compound Ascomylactam C showed antimicrobial activity against gram positive bacteria including drug-resistant strains. Ascomylactam C exhibited antimicrobial against MRSA, Vancomycin-resistant strain of *E. Faecium* ,Vancomycin-resistant strain of *E. faecalis*, with MIC values of 12.5, 12.5, 50 mM (Ariantari et al. 2020).

2'-acetyl-4',4-dimethoxybiphenyl-2- carbaldehyde was isolated from the endophyte *Pestalotiopsis zonata* resident of the plant *Cyrtotachys lakka* (Arecaceae). This compound exhibited weak antimicrobial activity against MRSA and vancomycin-resistant *E.faecium* with MIC values of 0.84 and 0.87 μm/mL respectively (Yang et al. 2011) (Fig. 10.10).

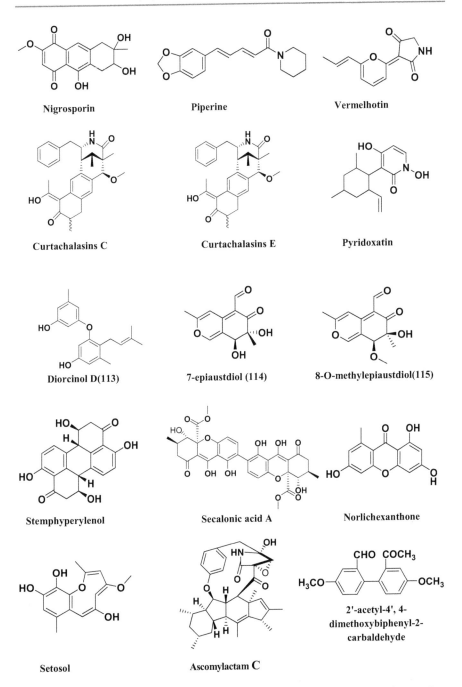

Fig. 10.10 Structure of compounds isolated from fungal endophytes against drug-resistant pathogenic microbes

10.7 Discussion and Conclusion

The pathogenic microorganisms which will cause major infectious diseases in humans include bacteria, fungi, protozoa, virus, etc. All these pathogens will cause a wide range of infections across various parts of human body. A number of antibiotics are available for the treatment of infectious diseases, e.g., penicillins, cephalosporins, tetracyclines, etc. AMR results in the reduction of activity of antibiotics in the biological tissues. As a result, the severity of the infection increasing gradually leads to failure of the treatment. Various factors influence the acquiring of resistance in microorganisms. By using different mechanisms, microorganisms are developing resistance to bioactive compounds. The resistance may be intrinsic to microorganisms or acquired by genetic transfer from other microbes, as we know that plant-endophyte in symbiotic relation will produce secondary metabolites that will show various therapeutic activities in the biological system. Fungal endophytes are very common and have high diversity within various plant tissues that will produce various secondary metabolites in mutualism with plants. The compounds isolated from fungal endophytes against various pathogenic microbes, most commonly MRSA, MDR-TB, resistant *P. falciparum*, and resistant *C. albicans*, were reported in this chapter. The estimated endophytic fungi species are ~1 million (Deshmukh et al. 2014). The vast variety of novel antibiotics are produced by fungal endophytes that have the ability to eradicate AMR. The reported endophytic fungi with anti-MRSA activity include *Alternaria alternata, Aspergillus fumigatiaffinis, Aspergillus terreus, Aspergillus* sp. TJ23, *Botryosphaeria mamane* PSU-M76, *Botryosphaeria rhodina* PSU-M35, *Botryosphaeria rhodina* PSU-M11, *Chaetomium globosum, Cytospora* sp. CR200, *Eurotium chevalieri* KUFA, *Fusarium oxysporum, Fusarium tricinctum, Fusarium* sp. TP-G1, *Guignardia* sp., *Hypocrea virens, Penicillium* sp. GD6, *Penicillium* sp., *Pestalotia* sp., *Phomopsis fukushii, Mycoleptodiscus* sp., *Nigrospora* sp. MA75, *Talaromyces wortmannii, Xylaria cubensis* BCRC 09F 0035, and *Xylaria cubensis* PSU-MA34.

Fungal endophytes having activity against drug-resistant *Plasmodium* sp. include *Aspergillus versicolor, Codinaeopsis gonytrichoides, Diaporthe miriciae, Exserohilum rostratum, Geotrichum* sp., *Nemania* sp. UM10M, *Nodulisporium* sp., *Nigrospora* sp. BCC 4778, *Penicillium* sp. BCC1605, *Pestalotiopsis* sp., *Phomopsis* sp. BCC 1323, *Phomopsis archeri, Preussia* sp. strain CAD4, *Pullularia* sp. BCC 8613, and *Xylaria* sp. Endophytic fungi having activity against drug-resistant *Mycobacterium* sp. include *Fusarium Sp., Nigrospora* sp., *Periconia* sp., and *Muscodor crispans*. Fungal endophytes against resistant *Candida albicans* include *Xylaria longipes, Tolypocladium cylindrosporum,* and *Aspergillus versicolor*. The endophytic fungi that are reported include *Fusarium tricinctum, Talaromyces* sp. ZH-154, *Aspergillus fumigatiaffinis, Preussia isomera, Chaetomium globosum* XL-1198, *Didymella* sp. IEA-3B, and *Pestalotiopsis zonata* that exhibited activity against various resistant pathogenic microorganisms like vancomycin-resistant enterococci, *P. aeruginosa*, multidrug-resistant *E. faecalis*, multidrug-resistant *E. faecium*, and multidrug-resistant *S. epidermidis*.

The activity against the resistance pathogens was determined with MIC and zone of inhibition values. Encapsulation of the active compounds isolated from the fungal endophytes in suitable delivery system will further enhance the activity against pathogenic microbes. Further research is needed for the identification of more active compounds against AMR and should bring in pharmaceutical level for producing in large scale. The fungal endophytes are considered as one of the promising approaches to eradicate various diseases including AMR.

Acknowledgments The authors acknowledge DST-SEED, New Delhi, for grant SP/YO/385/ 2018-G under the SYST scheme.

References

Ariantari NP et al (2020) Didymellanosine, a new decahydrofluorene analogue, and Ascolactone C from: Didymella Sp. IEA-3B.1, an endophyte of Terminalia Catappa. RSC Adv 10 (12):7232–7240

Arivudainambi U, Ezhil S et al (2011) Novel bioactive metabolites producing endophytic fungus Colletotrichum gloeosporioides against multidrug-resistant Staphylococcus aureus. FEMS Immunol Med Microbiol 61(3):340–345

Bara R et al (2013) Antibiotically active metabolites from Talaromyces wortmannii, an endophyte of Aloe vera. J Antibiot 66(8):491–493

Boucher HW, Ralph Corey G (2008) Epidemiology of methicillin-resistant Staphylococcus aureus. Clin Infect Dis 46(5):S344–S349

Chang W et al (2015) Lichen endophyte derived pyridoxatin inactivates Candida growth by interfering with ergosterol biosynthesis. Biochim Biophys Acta Gen Subj 1850(9):1762–1771

Chatterjee SS, Otto M (2013) Improved understanding of factors driving methicillin-resistant Staphylococcus Aureus epidemic waves. Clin Epidemiol 5(1):205–217

Chen HL et al (2020) (±)-Preisomide: a new alkaloid featuring a rare naturally occurring tetrahydro-2H-1,2-oxazin skeleton from an endophytic fungus Preussia isomera by using OSMAC strategy. Fitoterapia 141:104475

Chinworrungsee M et al (2001) Antimalarial halorosellinic acid from the marine fungus Halorosellinia oceanica. Bioorg Med Chem Lett 11(15):1965–1969

Cui HB et al (2008) Antibacterial constituents from the endophytic fungus Penicillium Sp. 0935030 of mangrove plant Acrostichum aureum. Chin J Antibiotics 33(7):407–410

Debbab A, Aly AH, Proksch P (2013) Mangrove derived fungal endophytes—a chemical and biological perception. Fungal Divers 61(1):1–27

Deshidi R et al (2017) Isolation and quantification of alternariols from endophytic fungus, Alternaria alternata: LC-ESI-MS/MS analysis. ChemistrySelect 2(1):364–368

Deshmukh SK, Verekar SA, Bhave SV (2014) Endophytic fungi: a reservoir of antibacterials. Front Microbiol 5:1–47

Dissanayake RK et al (2016a) Antimicrobial activities of endophytic fungi of the Sri Lankan aquatic plant nymphaea Nouchali and Chaetoglobosin A and C, produced by the endophytic fungus Chaetomium globosum. Mycology 7(1):1–8

Dissanayake RK et al (2016b) Antimicrobial activities of mycoleptodiscin B isolated from endophytic fungus Mycoleptodiscus Sp. of Calamus thwaitesii Becc. J Appl Pharmaceut Sci 6 (1):001–006

Elkhayat ES, Ibrahim SRM, Mohamed GA, Ross SA (2016) Terrenolide S, a new antileishmanial butenolide from the endophytic fungus Aspergillus terreus. Nat Prod Res 30(7):814–820

Ferreira MC, de Carvalho CR, Zani CL, Rosa LH (2019) Antimycobacterial and antiplasmodial compounds produced by endophytic fungi: an overview. In: Advances in endophytic fungal research. Springer, Cham, pp 17–33

Ganihigama DU et al (2015) Antimycobacterial activity of natural products and synthetic agents: Pyrrolodiquinolines and Vermelhotin as anti-tubercular leads against clinical multidrug resistant isolates of Mycobacterium Tuberculosis. Eur J Med Chem 89:1–12

Gygli SM, Borrell S, Trauner A, Gagneux S (2017) Antimicrobial resistance in Mycobacterium tuberculosis: mechanistic and evolutionary perspectives. FEMS Microbiol Rev 41(3):354–373

Hemtasin C et al (2011) Cytotoxic pentacyclic and tetracyclic aromatic Sesquiterpenes from Phomopsis archeri. J Nat Prod 74(4):609–613

Ibrahim SRM, Asfour HZ (2018) Bioactive γ-butyrolactones from endophytic fungus Aspergillus versicolor. Int J Pharm 14(3):437–443

Ibrahim SRM et al (2015) Aspernolides F and G, new butyrolactones from the endophytic fungus Aspergillus terreus. Phytochem Lett 14:84–90

Intaraudom C et al (2013) Penicolinates A-E from endophytic Penicillium Sp. BCC16054. Tetrahedron Lett 54(8):744–748

Isaka M et al (2001) Phomoxanthones A and B, novel Xanthone dimers from the endophytic fungus Phomopsis species. J Nat Prod 64(8):1015–1018

Isaka M et al (2007) Antiplasmodial and antiviral cyclohexadepsipeptides from the endophytic fungus Pullularia Sp. BCC 8613. Tetrahedron 63(29):6855–6860

Iswarya B, Vijaya Ramesh K (2019) Promising antimicrobial compounds from an endophytic fungus Penicillium daleae EF 4 isolated from the seaweed Enteromorpha Flexuosa Linn. Res Rev 9(1):26–39

Jiang CS, Zhen-Fang ZH, Xiao-Hong YA, Le-Fu LA, Yu-Cheng GU, Bo-Ping YE, Yue-Wei GU (2018) Antibacterial sorbicillin and diketopiperazines from the endogenous fungus Penicillium sp. GD6 associated Chinese mangrove Bruguiera gymnorrhiza. Chin J Nat Med 16:358–365

Jiménez-Romero C, Eduardo O-B, Elizabeth Arnold A, Cubilla-Rios L (2008) Activity against Plasmodium Falciparum of lactones isolated from the endophytic fungus Xylaria Sp. Pharm Biol 46(10–11):700–703

Kaur DC, Chate SS (2015) Study of antibiotic resistance pattern in methicillin resistant Staphylococcus aureus with special reference to newer antibiotic. J Glob Infect 7(2):78–84

Kjer J et al (2009) Xanalteric acids I and II and related phenolic compounds from an endophytic Alternaria Sp. isolated from the mangrove plant Sonneratia alba. J Nat Prod 72(11):2053–2057

Klaiklay S et al (2012) Metabolites from the mangrove-derived fungus Xylaria cubensis PSU-MA34. Arch Pharm Res 35(7):1127–1131

Kongsaeree P et al (2003) Antimalarial Dihydroisocoumarins produced by Geotrichum Sp., an endophytic fungus of Crassocephalum crepidioides. J Nat Prod 66(5):709–711

Kontnik R, Clardy J (2008) Codinaeopsin, an antimalarial fungal polyketide. Org Lett 10 (18):4149–4151

Kornsakulkarn J et al (2018) Bioactive hydroanthraquinones from endophytic fungus Nigrospora Sp. BCC 47789. Phytochem Lett 24:46–50

Kumarihamy M et al (2019) Antiplasmodial and cytotoxic cytochalasins from an endophytic fungus, Nemania Sp. UM10M, isolated from a diseased Torreya taxifolia leaf. Molecules 24 (4):777

Kuritzkes DR (2011) Drug resistance in HIV-1. Curr Opin Virol 1(6):582–589

Leyte-Lugo M et al (2012) (+)-Ascosalitoxin and Vermelhotin, a calmodulin inhibitor, from an endophytic fungus isolated from Hintonia Latiflora. J Nat Prod 75(9):1571–1577

Li Y et al (2015) Synergistic and drug-resistant reversing effects of Diorcinol D combined with fluconazole against Candida albicans. FEMS Yeast Res 15(2):1

Liu F et al (2010) The bioactive metabolites of the mangrove endophytic fungus Talaromyces Sp. ZH-154 isolated from Kandelia Candel (L.) Druce. Planta Med 76(2):185–189

Macías-Rubalcava ML, Sánchez-Fernández RE (2017) Secondary metabolites of endophytic Xylaria species with potential applications in medicine and agriculture. World J Microbiol Biotechnol 33(1):1–22

McKeegan KS, Ines Borges-Walmsley M, Walmsley AR (2002) Microbial and viral drug resistance mechanisms. Trends Microbiol 10(10):8–14

Mei WL et al (2012) Meroterpenes from endophytic fungus A1 of mangrove plant Scyphiphora hydrophyllacea. Mar Drugs 10(9):1993–2001

Mitchell AM et al (2010) Volatile antimicrobials from Muscodor crispans, a novel endophytic fungus. Microbiology 156(1):270–277

Nurunnabi TR et al (2018) Anti-MRSA activity of oxysporone and xylitol from the endophytic fungus Pestalotia Sp. growing on the Sundarbans mangrove plant Heritiera fomes. Phytother Res 32(2):348–354

Ola ARB et al (2013a) Inducing secondary metabolite production by the endophytic fungus fusarium Tricinctum through Coculture with Bacillus subtilis. J Nat Prod 76(11):2094–2099

Ola ARB et al (2013b) Absolute configuration and antibiotic activity of neosartorin from the endophytic fungus Aspergillus fumigatiaffinis. Tetrahedron Lett 55(5):1020–1023

Palomino JC, Martin A (2014) Drug resistance mechanisms in Mycobacterium tuberculosis. Antibiotics 3(3):317–340

Pan JH et al (2011) Antimycobacterial activity of fusaric acid from a mangrove endophyte and its metal complexes. Arch Pharm Res 34(7):1177–1181

Pandey PK et al (2014) Fungal endophytes: promising tools for pharmaceutical science. Int J Pharmaceut Sciences Rev Res 25(2):128–138

Pantosti A, Sanchini A, Monaco M (2007) Mechanisms of antibiotic resistance in Staphylococcus aureus. Future Microbiol 2(3):323–334

Parlet CP et al (2019) Apicidin attenuates MRSA virulence through quorum-sensing inhibition and enhanced host defense. Cell Rep 27(1):187–198

Peacock SJ, Paterson GK (2015) Mechanisms of methicillin resistance in Staphylococcus aureus. Annu Rev Biochem 84(1):577–601

Pongcharoen W, Rukachaisirikul V, Phongpaichit S, Sakayaroj J (2007) A new dihydrobenzofuran derivative from the endophytic fungus Botryosphaeria Mamane PSU-M76. Chem Pharm Bull 55(9):1404–1405

Prabpai S, Wiyakrutta S, Sriubolmas N, Kongsaeree P (2015) Antimycobacterial dihydronaphthalenone from the endophytic fungus Nodulisporium Sp. of Antidesma ghaesembilla. Phytochem Lett 13:375–378

Prestinaci F, Pezzotti P, Pantosti A (2015) Antimicrobial resistance: a global multifaceted phenomenon. Pathog Glob Heal 109(7):309–318

Qiao Y et al (2018) Aspermerodione, a novel fungal metabolite with an unusual 2,6-dioxabicyclo [2.2.1]heptane skeleton, as an inhibitor of penicillin-binding protein 2a. Sci Rep 8:1

Radić N, Štrukelj B (2012) Endophytic fungi - the treasure chest of antibacterial substances. Phytomedicine 19(14):1270–1284

Rajamanikyam M, Vadlapudi V, Amanchy R, Upadhyayula SM (2017) Endophytic fungi as novel resources of natural therapeutics. Braz Arch Biol Technol 2017:60

Ratnaweera PB, Dilip de Silva E, Wijesundera RLC, Andersen RJ (2016) Antimicrobial constituents of Hypocrea virens, an endophyte of the mangrove-associate plant Premna serratifolia L. J Natl Sci Found 44(1):43–51

Ratnaweera PB, Dilip de Silva E, Williams DE, Andersen RJ (2015) Antimicrobial activities of endophytic fungi obtained from the arid zone invasive plant Opuntia dillenii and the isolation of Equisetin, from endophytic Fusarium Sp. BMC Complement Altern Med 15:1

Reygaert WC (2013) Antimicrobial resistance mechanisms of Staphylococcus aureus. Microbial pathogens and strategies for combating them: science, technology and education, pp 297–305. https://www.researchgate.net/publication/267695121_Antimicrobial_resistance_mechanisms_ of_Staphylococcus_aureus. Accessed 17 May 2020

Rout S, Mahapatra RK (2019) Plasmodium falciparum: multidrug resistance. Chem Biol Drug Des 93(5):737–759

Rukachaisirikul V et al (2009) Metabolites from the endophytic fungi Botryosphaeria rhodina PSU-M35 and PSU-M114. Tetrahedron 65(51):10590–10595

Sappapan R et al (2008) 11-Hydroxymonocerin from the plant endophytic fungus Exserohilum rostratum. J Nat Prod 71(9):1657–1659

Seung KJ, Keshavjee S, Rich ML (2015) Multidrug-resistant tuberculosis and extensively drug-resistant tuberculosis. Cold Spring Harb Perspect Med 5(9):1–20

Shi S et al (2018) Biological activity and chemical composition of the endophytic fungus Fusarium Sp. TP-G1 obtained from the root of dendrobium Officinale Kimura et Migo. Rec Nat Prod 12 (6):549–556

Singh SR, Krishnamurthy NB, Mathew BB (2014) A review on recent diseases caused by microbes. J Appl Environ Microbiol 2(4):106–115

Sommart U et al (2008) Hydronaphthalenones and a dihydroramulosin from the endophytic fungus PSU-N24. Chem Pharm Bull 56(12):1687–1690

Talontsi FM et al (2014) Antiplasmodial and cytotoxic dibenzofurans from Preussia Sp. Harboured in Enantia Chlorantha Oliv. Fitoterapia 93:233–238

Tansuwan S et al (2007) Antimalarial benzoquinones from an endophytic fungus, Xylaria Sp. J Nat Prod 70(10):1620–1623

Ventola CL (2015) The antibiotic resistance crisis: causes and threats. P & T J 40(4):277–283

Verma VC et al (2011) Piperine production by endophytic fungus Periconia Sp. isolated from Piper Longum L. J Antibiot 64(6):427–431

Wang C et al (2013) Anti-mycobacterial activity of marine fungus-derived 4-deoxybostrycin and nigrosporin. Molecules 18(2):1728–1740

Wang QX et al (2011) Chemical constituents from endophytic fungus Fusarium oxysporum. Fitoterapia 82(5):777–781

Wang WX et al (2019) Cytochalasans from the endophytic fungus Xylaria Cf. Curta with resistance reversal activity against fluconazole-resistant Candida albicans. Org Lett 21(4):1108–1111

White TC et al (2002) Resistance mechanisms in clinical isolates of Candida albicans. Antimicrob Agents Chemother 46(6):1704–1713

Xu J (2015) Bioactive natural products derived from mangrove-associated microbes. RSC Adv 5 (2):841–892

Yang HY et al (2017) Three new naphthalene derivatives from the endophytic fungus Phomopsis Fukushii. Phytochem Lett 22:266–269

Yang SX et al (2019) Aureonitols A and B, two new C13-polyketides from Chaetomium globosum, an endophytic fungus in salvia Miltiorrhiza. Chem Biodivers 16:9

Yang XL et al (2011) A new biphenyl from the fermentation broth of plant endophytic fungus Pestalotiopsis zonata isolated from Cyrtostachys Lakka. Chin J Nat Med 9(2):101–104

Zeng YB et al (2012) A fatty acid glycoside from a marine-derived fungus isolated from mangrove plant Scyphiphora hydrophyllacea. Mar Drugs 10(3):598–603

Zin M, War W et al (2017) Antibacterial and antibiofilm activities of the metabolites isolated from the culture of the mangrove-derived endophytic fungus Eurotium chevalieri KUFA 0006. Phytochemistry 141:86–97

Antimicrobial Peptides as Effective Agents Against Drug-Resistant Pathogens

11

Pragya Tiwari, Yashdeep Srivastava, and Vinay Kumar

Contents

P. Tiwari (✉)
Molecular Metabolic Engineering Lab, Department of Biotechnology, Yeungnam University, Gyeongsan, Gyeongbuk, Republic of Korea

Y. Srivastava
Department of Biotechnology, Invertis University, Bareilly, Uttar Pradesh, India

V. Kumar (✉)
Department of Biotechnology, Modern College of Arts, Science and Commerce, Savitribai Phule Pune University, Pune, India

Department of Environmental Science, Savitribai Phule Pune University, Pune, India

© The Author(s), under exclusive license to Springer Nature Singapore Pte Ltd. 2022
V. Kumar et al. (eds.), *Antimicrobial Resistance*,
https://doi.org/10.1007/978-981-16-3120-7_11

289

Abstract

The rising incidence of antimicrobial resistance continues to project a global healthcare concern. The spread of drug-resistant pathogens and indiscriminate use of the existing antibiotics has a profound effect on the economy of developing and under-developing countries. However, the drying pipeline of antibiotic arsenals and little progress in this direction necessitate the discovery and characterization of novel antimicrobials from natural sources. Antimicrobial peptides (AMPs) are gaining momentum as antimicrobial therapeutics with potent efficacy to tackle rising drug-resistant bacterial strains. The natural and synthetic AMPs as novel antimicrobials highlight remarkable therapeutic potential via diverse mechanism of action. Recent advances in antimicrobial research have improved our knowledge on the structure, properties, and function of AMPs; however, there is still a long way ahead for complete exploitation of these therapeutic candidates. With an overview on the emerging popularity of AMPs in countering diverse infectious diseases and drug-resistant pathogens, the chapter provides a detailed insight on the history and development of AMPs and its production in plant-based expression systems. The contribution of combinational chemistry co-integrated with computational biology in AMP research and development, projected bottlenecks, and prospects of success are further discussed.

Keywords

Antimicrobial peptides · Clinical trials · Drug-resistant microbes · *Pseudomonas aeruginosa* · Peptide-based drugs · Transgenic plant systems

11.1   Introduction

The spread of multidrug-resistant (MDR) bacteria has been an alarming cause of concern, emerging as a global threat worldwide. According to the World Health Organization, the rising drug resistance due to indiscriminate antibiotic use would be a key cause of poverty and lead to ten million deaths (de Kraker et al. 2016). The increasing prevalence of antibiotic resistance and spread of drug-resistant pathogens will profoundly affect the economy of the developing countries (Aslam et al. 2018; Zaman et al. 2017). The global spread of antimicrobial resistance (AMR) has been declared a global risk by the World Economic Forum, owing to its adverse implications (WEF 2013).

The discovery of Penicillin by Alexander Fleming in 1928 was a landmark in medical revolution and greatly expanded the drug arsenals for the treatment of infections (Van Epps 2006a, b). The untreatable infections and its adverse effects in the earlier days can now be effectively treated with advances in medicine;

however, the emergence and rapid rise of drug-resistant microbes projects serious concerns. The discovery of penicillin by Alexander Fleming in 1928 was a landmark in the "antibiotic era" and was one of the best known antibiotics. The discovery of streptomycin (from *Streptomyces griseus*) by Selman Waksman (father of antibiotics) marked the golden age of antibiotics and progressed with the discovery of ß-lactams, sulfonamides, tetracyclines, aminoglycosides, cephalosporin, amphenicols, and oxazolidinones, among others (Roncevic et al. 2019). However, the indiscriminate use and drying pipeline of new antibiotics, together with rising incidence of drug-resistant strains, further contributes to the challenge (Mwangi et al. 2019a, b). In addition to public health, the spread of drug-resistant strains poses a concern in clinical applications like surgeries, organ transplantations, and treatment of cancer (Gudiol and Carratalà 2014; Lupei et al. 2010). The present trends in upsurge of antimicrobial resistance (AMR) may be attributed to multiple factors comprising of indiscriminate/overuse, absence of proper guidelines for marketing (Ayukekbong et al. 2017), and poor sanitation practices (Davies and Davies 2010). In the present context, the prevalent MDR bacteria include vancomycin-resistant MRSA, *Mycobacterium tuberculosis* (extensively drug-resistant XDR), methicillin-resistant *Staphylococcus aureus* (MRSA), vancomycin-resistant enterococci (VRE), and carbapenem-resistant *Acinetobacter baumannii*, *E. coli*, and *Klebsiella pneumonia* (Levin et al. 1999; Miller et al. 2005). Some bacterial strains showed resistance to all antibiotics; for example, the indiscriminate (excessive) use of colistin for the treatment of MDR strains (*K. pneumonia*, *A. baumannii*, *Pseudomonas aeruginosa*) led to MCR-1 gene-mediated antibiotic resistance (Liu et al. 2016; Macnair et al. 2018). The concerns associated with rising AMR worldwide necessitate the discovery and characterization of new antimicrobials from natural sources.

In the last few decades, multiple classes of compounds entered phases of development and clinical investigations (Draenert et al. 2015); however, very few exhibited promising results. The oxazolidinones and lipopeptides were commercialized for Gram-negative pathogens (Yu et al. 2020); however, the emergence of staphylococcal resistance for oxazolidinones followed soon after (Eliopoulos et al. 2004). The adverse effects associated with tigecycline and telithromycin led to its discontinuation by the Food and Drug Administration (FDA) (Gleason and Shaughnessy 2007; Dixit et al. 2014). Another example of MDR bacteria, *M. tuberculosis*, developed resistance to fluoroquinolones and injectable drugs (capreomycin, kanamycin, or amikacin) (Prestinaci et al. 2015). Daptomycin (lipopeptide) showed some success in the treatment; however, it was later removed by the WHO (WHO 2019). The WHO in 2017 published a list of critical MDR strains comprising of the third-generation cephalosporin-resistant *Pseudomonas aeruginosa*, *Acinetobacter baumannii*, and *Enterobacteriaceae* with a critical need to address the evolving MDR strains (WHO 2018). The antimicrobial peptides (AMPs) highlighted prospects as alternatives to conventional treatment for therapeutic use, owing to its low potential to elicit resistance (Mahlapuu et al. 2016a, b).

Considering the trends of evolving MDR bacteria, an urgent requirement of novel anti-infective agents displaying multiple mode of action, is needed. Antimicrobial peptides (AMPs) are gaining momentum as promising therapeutics against fungal and bacterial pathogens (Brogden and Brogden 2011). A diverse class of molecules,

AMPs, are components of innate immunity and produced as first line of defense by multicellular organisms (Zhang and Gallo 2016). To date, more than 3000 AMPs have been discovered and characterized Wang et al. (2016) and the US Food and Drug Administration (FDA) has approved seven peptides for commercial applications (Chen and Lu 2020). However, an understanding of the mechanism of action is essential for designing or discovering novel AMPs with potent biological efficacy. With rising importance of AMPs in combating AMR, the chapter provides a detailed insight into the prospects and challenges associated with AMP discovery, function as antimicrobial therapeutics and bottlenecks in maximum utilization. Recent advance in designing/redesigning of novel AMPs, its success stories and popularity as alternatives to conventional antibiotics and novel therapeutics to tackle AMR, is highlighted.

11.1.1 Natural Products as Prospective Source to Counter Antimicrobial Resistance

In the present era, natural products from plants, animals, and microbes are extensively explored for their antimicrobial activity (Deshmukh et al. 2015; da Silva et al. 2017; Gyawali and Ibrahim 2014). Antibiotic-producing microbes were known from diverse environmental niches, including extreme environments (hypersaline locations) (Jose and Jebakumar 2014), plants (endophytes) (Santos et al.), soil (Hassan et al. 2016), and marine sponges (Saurav et al. 2016), among others. The antimicrobial compounds produced from these microbes demonstrate significant potential to inhibit the growth of both bacteria (de Oliveira et al. 2016) and fungi (Kumar et al. 2014, 2018) and demonstrate important biotechnological applications. These microbes show potential antimicrobial activity, and several antibiotics of potent efficacy were produced from different microbial strains. Moreover, some bacterial strains were found to inhibit the toxin production and formation of biofilms (Papa et al. 2015; Saurav et al. 2016). *Lactococcus lactis* produces a lantibiotic (nisin) which inhibits biofilm formation in multiple species by removing biofilm or inhibiting its formation and does not adversely affect the human cells. Moreira et al. (2016) showed that a macrolide antibiotic boromycin (produced by *Streptomyces antibioticus*) targets ion gradient in transmembrane and acts against nongrowing and growing drug-resistant cells.

Plant-derived natural products are gaining popularity as antimicrobials attributed to their huge chemical diversity and potent efficacy in treating microbial infections. Plants are a source of high-value metabolites and have been used in traditional system of medicine. Various plant parts including essential oils (Yu et al. 2020), triterpenoids (Wu et al. 2015), plant extracts (Silva et al. 2016), and plant proteins (Patiño-Rodríguez et al. 2013) were effective in tackling microbial infection on multiple animal models. Plant secondary metabolites demonstrating antimicrobial activity comprise of phenolics, coumarins, saponins, alkaloids, terpenoids, and tannins, among others (Ciocan and Bara 2007; Gyawali and Ibrahim 2014). Table 11.1 provides a comprehensive account of different classes of antimicrobial

Table 11.1 Antimicrobial compounds from natural sources, bioactive constituents, and their mechanism

S. no.	Natural source	Antimicrobial compounds	Mechanism of action	References
1.	*Plant-derived compounds*			
	Phenolic compounds	Carvacrol, thymol	Hydroxyl group increase delocalization of electrons, disruption of membrane, cellular leakage	Lai and Roy (2004), Xue et al. (2013)
	Phenolic compounds	Eugenol, isoeugenol	Antimicrobial activity against *Campylobacter jejuni* and *Listeria*	Friedman et al. (2002)
	Essential oil from *Salvia fruticosa*	–	Efflux pump inhibition in *Staphylococcus epidermidis* (clinical isolates)	Chovanova et al. (2015)
	Essential oil from *Chenopodium ambrosioides*	–	Efflux pump Tet(K) inhibition in *Staphylococcus aureus* IS-58	Limaverde et al. (2017)
	Alkaloid	Capsaicin	Efflux pump NorA inhibition in *S. aureus* SA-1199B	Kalia et al. (2012)
	Alkaloid	Catharanthine	Efflux pump EtBr inhibition in *P. aeruginosa*	Dwivedi et al. (2017)
	Flavonoid	Baicalein	Efflux pump NorA inhibition in *S. aureus* SA-1199B	Chan et al. (2011)
	Triterpenoids	Ursolic acid and derivatives	Efflux pump AcrA/B, MacB, TolC, and YojI inhibition in MDR *E. coli* (KG4)	Dwivedi et al. (2014)
2.	*Plant by-products in food processing*			
	Grape pomace	Phenolic compounds	Growth inhibition of Enterobacteriaceae, *S. aureus*, *Salmonella*, yeasts, and molds	Sagdic et al. (2011)
	Pomegranate fruit peels extracts	Phenolics, flavonoids	Growth inhibition of foodborne pathogens including *L. monocytogenes*, *S. aureus*, *E. coli*, *Y. enterocolitica*, and *B. cereus*	Agourram et al. (2013)
	Coconut husk	Phytochemical including phenolics and tannins	Growth inhibition of *L. monocytogenes*, *S. aureus*, and *V. cholera*	Wonghirundecha and Sumpavapol (2012)
3.	*Animal-origin compounds*			
	Lactoferrin	Glycoprotein	Growth inhibition of foodborne pathogens *L. monocytogenes*, *E. coli*, *Carnobacterium*	Lonnerdal (2011)

(continued)

Table 11.1 (continued)

S. no.	Natural source	Antimicrobial compounds	Mechanism of action	References
	Chitosan	Polycationic biopolymer compound	Antibacterial activity against *S. aureus*, *L. monocytogenes*, *B. cereus*, *E. coli*, *Shigella dysenteriae*	Tiwari et al. (2009)
	Lysozyme	Bacteriolytic enzyme	Lysozyme hydrolyze the ß-1,4 linkage between N-acetylmuramic acid and N-acetylglucosamine in the peptidoglycan of the microbial cell wall	Tiwari et al. (2009)
4.	*Antimicrobials of bacterial origin*			
	Bacteriocin	Nisin	Growth inhibition of gram-positive and spore-producing bacteria in food	Lucera et al. (2012)
	Reuterin	ß-hydroxypropionaldehyde	Antimicrobial activity against foodborne pathogens	Arques et al. (2004)
5.	*Antimicrobials from algae and mushrooms*			
	Phlorotannins from marine brown algae	–	Antimicrobial activity against *S. aureus*, MRSA, *Salmonella* spp., *E. coli*	Eom et al. (2012)
	Grifolin, pleuromutilin from macrofungi	–	Antimicrobial activity against *S. aureus*, *B. cereus*, *L. monocytogenes*, *E. coli*	Bala et al. (2012)
	Fatty acids, ß-carotene-linoleic acid, flavonoids from *Agaricus* spp.	–	Antimicrobial activity against *Micrococcus luteus*, *M. flavus*, *B. subtilis*, *B. cereus*	Oztürk et al. (2011)

compounds from natural sources, bioactive constituents, and their mechanism. The enormous structural diversity among plant metabolites influences the antimicrobial function displaying different mechanisms of action. For example, the –OH group present in phenolic compounds results in inhibition of microbes (Lai and Roy 2004) by disrupting the bacterial membrane and causing cellular leakage (Xue et al. 2013) as exemplified by thymol and carvacrol. The presence of double bond in eugenol renders it 13-fold more effective than isoeugenol against *Listeria* and *Campylobacter jejuni* (Friedman et al. 2002). With respect to terpenes, even a little structural change in the position of –OH group may significantly affect antimicrobial activity, for example, K^+ leakage from *E. coli* cells was induced by terpinen-4-ol at low concentrations compared to α-terpineol. The antimicrobial action is due to the ability of oxygenated terpenes to disrupt membrane and cause K^+ leakage (Griffin et al. 2005). Plant essential oils demonstrate potent antimicrobial activity attributed to the presence of phenolic structure (the –OH groups and allylic chains improve efficacy of essential oils). A key example shows better efficacy of limonene (1-methyl-4-(1-methylethenyl)-cyclohexene) than p-cymene in countering microbial infections (Dorman and Deans 2000). Similarly, several plant by-products generated during food processing were also found to demonstrate antimicrobial activity, in addition to other functions (Balasundram et al. 2006; Tiwari et al. 2009). The large amount of by-products generated during food processing (seeds, pulps, husks, fruit pomace, peels) has high content of phenolic compounds. Therefore, attempts toward commercial utilization of these by-products as antimicrobial agents and their maximum recovery would define an important application. Antimicrobial compounds from animal origin include chitosan, lactoferrin, lactoperoxidase, lipids, and defensins, among others (Gyawali and Ibrahim 2014). Chitosan as food preservative has gained special interest but has limited use due to its insolubility at neutral and higher pH, which needs to be addressed (Du et al. 2009). The chitosan derivatives exhibited potent antibacterial activity against *B. cereus*, *E. coli*, *S. aureus*, *Shigella dysenteriae*, etc., as acid-soluble chitosan substitute (Chung et al. 2011). Bioactive compounds in milk (e.g., casein) were found to demonstrate multiple functions; antimicrobial is one of them (Phelan et al. 2009). Other food by-products with antimicrobial properties include coffee peels and husks (Esquivel and Jimenez 2012), bergamot peel (by-product of essential oil) (Mandalari et al. 2007), pomegranate juice by-products (Reddy et al. 2007), coconut husk (Wonghirundecha and Sumpavapol 2012), and some others. In recent times, honey is emerging as source of bioactive molecules, owing to the presence of bioactive compounds consisting of proteins, sugars, hydroxymethylfurfural, etc. The presence of glycoproteins in honey imparts antimicrobial activity against MDR isolates (Brudzynski et al. 2015). Several other bioactive compounds demonstrating antimicrobial activity have been isolated from different natural sources; a few key examples were enlisted (Table 11.1).

11.1.2 Antimicrobial Drugs: History and Developments

The antimicrobial resistance has been a matter of concern since the discovery and clinical use of first antibiotics in 1940. Although the discovery of antibiotics was regarded as one of the landmark advances in medicine, the rising statistics of AMR has threatened the development and use of antimicrobials, limiting the treatment options. The need to tackle emerging AMR has necessitated the discovery and development of new antimicrobial agents and the efficacy improvement of the existing ones (Powers 2014). The limited availability of the existing antibiotics and the similarities in their mode of action have prompted clinical researchers to investigate anti-infective therapies from natural sources.

The beginning of "antibiotic era" can be traced back to the discovery of penicillin by Alexander Fleming in 1928; since then, the antimicrobial arsenal to treat infectious diseases has greatly expanded (Van Epps 2006a, b). The emergence of "antibiotic era" between the 1940s and 1960s witnessed the discovery of majority of antibiotics, currently in use contributing immensely to decline in global mortality rate (Davies 2006). The sulfanilamide decreased the death rate in meningococcal meningitis drastically compared to pre-antibiotic era (Schwentker et al. 1937). Unfortunately, since then the emergence of drug-resistant microbes compromised the applications of the existing antimicrobial drugs. In a classical example, methicillin was discovered in 1959 and was used to treat *Staphylococcus aureus* (which demonstrated penicillin resistance). Similarly, streptomycin was used for the treatment of tuberculosis, but resistance was developed by *Mycobacterium tuberculosis* strains sooner due to the presence of rRNA mutations (Springer et al. 2001). Most of the known antibiotics face drug resistance, with the most recent example being lipopeptides which shows the emergence of resistant strains to daptomycin (Humphries et al. 2013). Moreover, safety issues over the use of antimicrobials raised critical concerns for public health. For instance, deaths caused due to the use of sulfanilamide raised concerns and led to the passing of the Food, Drug, and Cosmetic (FD&C) Act in 1938 (Temple 1995) which necessitated the sponsors to provide details about the drug efficacy in study. Several new classes of drugs with novel mechanism were discovered comprising of chloramphenicol, tetracyclines, sulfonamides, rifamycins, tetracyclines, quinolones with lipopeptides, and oxazolidinones being the recent ones. Since that time, majority of the introduced antimicrobials are chemical derivatives of the known classes of drugs (Powers 2014). The third generation of cephalosporins was effective against Gram-negative organisms and in bacterial meningitis compared to first-generation cephalosporins. Other antibiotics introduced showed better efficacy compared to previous generation of the same class of drugs. The FDA approved 29 new antibacterial drugs during 1980–1989 including ß-lactams, which belonged to cephalosporin class (Powers 2014). Furthermore, 22 new antimicrobial drugs classified in ß-lactams and quinolone class were approved by the FDA between 1990 and 1999 period. The lack of new drug classes led to a decline in drug arsenal with FDA approving only 2.9 new antibacterial drugs (per year) in the 1990s and 2.2 drugs (per year) in 2000. Since then, there has been a sharp decline in discovery and approval of new antimicrobial drugs which highlights an urgent requirement of developing new therapeutic agents against microbes. The high cost associated with drug development and its commercialization and its low efficacy are some factors which led to decline in approval of new drugs in all therapeutic classes. However, the development of antimicrobials has some advantages in terms of short mean and median time of clinical development (Reichert 2003), compared to other therapeutic classes, and attributed to animal models and in vitro testing methods (DiMasi 1995). Presently, many biotechnology firms are involved in research on antimicrobials, with new prospective drugs from natural sources being investigated for their antimicrobial potential and improving the efficacy of existing ones by adjuvant therapies. In this direction, FDA attempts to devise a methodical approach toward drug development, by streamlining design of clinical trials (including pharmacokinetic data on drugs) and small size of samples for clinical trials (Powers 2014). In 2013, the US Center for Disease Control and Prevention (CDC) reported two million people were infected by drug-resistant pathogen and 23,000 deaths resulted to infection projecting a statistics of ten million deaths due to MDR bacteria (O'Neil 2016). The ESKAPE pathogens (*Enterococcus faecium*, *S. aureus*, *Klebsiella pneumoniae*, *Acinetobacter baumannii*, *P. aeruginosa*, and *Enterobacter*) constitute a new paradigm by resisting antibacterial action of antibiotics (Boucher et al. 2009). Another report published by the European Center for Disease Control (2014) discussed the rising incidences of resistance in the ESKAPE pathogens, raising alarming concerns (EDCD, antimicrobial-resistance Europe 2014). The indiscriminate use of the existing antibiotics in human medicine, food industry, veterinary, and agriculture has contributed significantly to the spread of AMR. There is an urgent requirement of new antimicrobials to tackle the rising menace of drug-resistant microbes.

11.2 Antimicrobial Peptides: The Peptide-Based Drugs as Novel Class of Therapeutics to Tackle Antibiotic Resistance

Antimicrobial peptides (AMPs) are an emerging class of antimicrobials demonstrating advantages such as broad anti-biofilm activity, slower emergence of resistance, and modulation of immune response of the host (Magana et al. 2020). AMPs comprise of a group of small bioactive proteins and act as part of first line of defense for inactivation of pathogens. The mechanism of action of AMPs comprises of immune response modulation, disruption of cell membranes, and regulation of inflammation (Mookherjee and Hancock 2007). AMPs are isolated from diverse sources including marine organisms and soil employing several methods, leading to higher availability of AMPs. There is an urgent requirement of new antimicrobials to tackle the rising menace of drug-resistant microbes. Although more than 2800 AMPs have been discovered to date, only few have shown promising effects in clinical trials. Moreover, the physicochemical properties of AMPs define its antibacterial activity, bioavailability, and toxicity (Magana et al. 2020). Both natural and synthetic analogues of AMPs have been used as antimicrobials, with studies on the safe use of synthetic AMPs in livestock, aquaculture, and as food preservatives (Liu et al.

2017; Choi et al. 2013). Therefore, in the absence of effective antibiotics and rise of AMR, the demand for AMPs has seen an upsurge in the last decade.

11.2.1 Identification and Properties

Regarded as conserved molecules during evolution, AMPs are present in diverse organism, ranging from prokaryotes to humans (Hancock 2000a, b). The main features of AMPs include the following: (1) amphipathic molecules ($\geq$30% of hydrophobic residues), (2) positive charge molecules (+2 to +9 net positive charge), and (3) ability to undergo posttranslational modifications (Huerta-Cantillo and Navarro-García 2016). Furthermore, AMPs are coded in biological genomes as one or more copies and derived from larger precursor molecules. The posttranslational modifications comprise of proteolytic processing, halogenations and cyclization, amino-acid isomerization, and carboxyl-terminal amidation (Huerta-Cantillo and Navarro-García 2016).

The first AMPs were identified in *Drosophila melanogaster* infected with fungi or bacteria in the 1990s (Lemaitre et al. 1996). A common origin was suggested for defensins from *Drosophila* and human β-defensins, while other AMPs (in *Drosophila*) show homology with insect AMPs (Mylonakis et al. 2016). The coevolution of AMPs in insects with other species highlights the specificity for antimicrobial function, indicating synergistic action and variation in specificity toward specific microbes. AMP proteins demonstrate broad-spectrum activity to kill yeasts, bacteria, viruses, cancer cells, and fungi (Hancock 2000a, b; Zhang and Gallo 2016). AMPs are employed by plants and insects to fight against pathogenic microbes, while microbes employ AMPs for maintaining their niches. The AMPs exist in many secondary structures, namely, α-helices, β-strands with disulfide bridges, and loop/extended structures leading to broad-spectrum activity (Hancock 2001). The AMPs were classified into cationic (thanatin, penaeidins) and non-cationic peptides (lactoferricin, secretolytin) exhibiting different structural features. Additionally, other properties like charge, hydrophobicity, stereo-geometry, size, and self-association contribute to broad antimicrobial functions (Pushpanathan et al. 2013). Moreover, another property of AMPs, namely, modulation of immune response by host cell interaction, may have led to coevolution of AMPs (Niyonsaba et al. 2020) but may have effect on the health of living organisms. The effectiveness of AMPs as antimicrobials is governed by the application route, target tissue, duration, dose, and formulation, among others. For example, the intestinal microbiota-mediated expression of AMP maintains bacterial colonization and decreases resulting inflammation (Lhocine et al. 2008). AMP may initiate the action of innate immune cells at the infection site and induce chemokine through various cells (Hancock et al. 2016). The immunomodulatory properties of AMPs are diverse (specific for AMP type) and comprise of growth-factor-like effects and cytokines implemented in immune homeostasis (Zhang and Gallo 2016). Structural features of AMPs play a critical role in its antimicrobial function which is responsible for specificity of AMPs toward the target cells. AMPs consist of less than

100 amino acid residues with +2 to +9 net positive charge and hydrophobic residues. During the course of evolution, AMPs present in different biological species were found to be conserved, and the information about the conserved structural elements formed the basis of designing novel peptides (Yount and Yeaman 2004). For example, penaeidins comprise of chimeric cationic peptide with a conserved chitin binding domain with a cysteine-rich domain at the C-terminal and PRP domain at the N-terminal and demonstrate antimicrobial activity against fungi and Gram-positive bacteria (Tassanakajon et al. 2010). The AMP, LL-37, is an amphipathic peptide consisting of XBBXBX motifs for binding to heparin and a helical structure (Andersson et al. 2004). Another important aspect is the structural and physico-chemical properties on AMPs that play a crucial role in affecting toxicity against particular cells. For example, tachystatin is an AMP, isolated from *Tachypleus tridentatus* (horseshoe crab), and demonstrates antimicrobial activity against Gram-positive and Gram-negative bacteria and fungi. The cytotoxic activity of tachystatin is attributed to the presence of amphiphilic β-sheet at C-terminal end (Osaki et al. 1999). However, the presence of conserved domain has not been reported for antifungal peptides. In recent times, studies have attempted on improving the efficacy of AMPs by employing random mutagenesis, peptide structure alteration (by cyclization or increasing hydrophobicity), or computational approaches. Furthermore, AMPs display multifaceted properties as vector for drug delivery (Henriques et al. 2006), tumor-inducing/mitogenic agent owing to its ability to interact with cell membranes (Utsugi et al. 1991), and signaling molecules (Blomqvist et al. 1999), among other significant ones.

11.2.2 Classification and Structure of AMPs

AMPs display enormous structural and functional diversity and are emerging as novel antimicrobial candidates in the "post-antibiotic era." Being evolutionary conserved in genomes during evolution, AMPs comprise of important constituents of innate immunity, as first line of defense against microbial infections. In nature, AMPs are synthesized by ribosomal translation of mRNA or peptide synthesis (nonribosomal) (Hancock and Chapple 1999). The AMPs synthesized from ribosomal translation are produced by all biological organisms and encoded genetically, while nonribosomally synthesized peptides have bacterial origin (Hancock and Chapple 1999). The ribosomally synthesized AMPs have been gaining attention in immune responses and as therapeutic agents (Hancock and Chapple 1999; Hancock 2000a, b; Mahlapuu et al. 2016a, b). Sometimes, AMPs are produced as inactive precursors, and their regulation depends on proteases and their own expression (Lai and Gallo 2009). Commonly, AMPs are classified based on structure into α-helical, ß-sheet, or peptides with extended/random-coil structure (Takahashi et al. 2010; Lombardi et al. 2019), with majority of AMPs classified in α-helical, ß-sheet structures. The α-helical peptides do not have a defined structure but form an amphipathic structure in association with membrane (Pasupuleti et al. 2012). Human lactoferricin and LL-97 are the two most studied peptides from this group.

The ß-sheet peptides comprise of the largest group of AMPs and are found in many species of plants, amphibians, marine invertebrates, etc. (Oppenheim et al. 2003; Wang et al. 2016). The ß-sheet peptides display a rigid and ordered structure in aqueous state and do not undergo a conformational change as helical peptides. The ß-sheet peptides demonstrate diverse antimicrobial activities such as antibacterial, antiviral, antifungal, and anti-inflammatory (Kumar et al. 2018). Some classical members of the class include protegrins, defensins, and tachyplesins. Defensins are the best studied ß-sheet peptides and produced as inactive precursor molecule in macrophages, epithelial cells, and neutrophils (Lai and Gallo 2009; Pasupuleti et al. 2012). These AMPs are present in vertebrates, invertebrates, and plants and demonstrate strong antimicrobial activity (Mwangi et al. 2019a, b). The third class of AMPs includes loops/extended-coil structure. The extended-coil structure shows the absence of α-helices and ß-sheets and consists of amino-acid residues including proline, tryptophan, and arginine (Mahlapuu et al. 2016a, b). The broad-spectrum function of these AMPs consists of action against Gram-negative bacteria via disruption of membranes and antitumor activity (Falla et al. 1996). Important examples include indolicidin, a 13-amino-acid peptide with potent antimicrobial activity (Falla et al. 1996), besides histatin (human saliva) and tritrpticin being other key examples (Mwangi et al. 2019a, b).

11.2.3 Antimicrobial Peptides and Their Mechanism of Action

As compared to classical antibiotics, AMPs have advantage as they are less likely to generate microbial resistance. These peptides exert action at the membranes and disrupt pathogen-associated processes and activate host immune response (Sierra et al. 2017); however, several challenges need to be addressed for broader AMP application as antimicrobials. Some associated concerns highlight toxicity parameters, production costs, and optimization (stability of AMPs). The AMPs have been extensively studied for their mechanism of action which is important for the development of novel AMPs as therapeutic agents. AMPs are classified as membrane-acting and non-membrane-acting peptides. The membrane-acting AMPs are basically the cationic peptides resulting in disruption of membranes, while the non-membrane peptides move across the membrane without its disruption (Hancock and Patrzykat 2002). For example, antibacterial peptides such as LL-37 (Xhindoli et al. 2016), defensins (Shafee et al. 2017), and melittin (Sun et al. 2017) form transmembrane pores in the target membrane. Some examples of AMPs which translocate through membranes (without causing disruption) include dermaseptin (Belmadani et al. 2018), buforin (Perez et al. 2016), and pleurocidin (Zhang et al. 2016), among others. The mechanism of these AMPs is characterized by translocation across cell membrane and disruption of regular functions (Lohner 2017). The cationic peptide interacts with cell membrane (negatively charged due to the presence of lipopolysaccharides) causing cell permeability and disruption of membranes (Da Costa et al. 2015; Sani and Separovic 2016). Furthermore, these AMPs exhibit structural dynamics while interacting with microbial membranes (Mingeot-Leclercq and Décout 2016). The other mechanisms of action comprise of inhibiting processes, namely, nucleic acid synthesis, protein synthesis, and cell wall synthesis (Cudic and Otvos Jr 2002). Moreover, fluidity of the membranes affects AMP insertion into the cell membranes. Other factors which contribute in AMP action consist of outer membrane charge, membrane fluidity, and architecture which are necessary for translocation of AMPs across the membranes (Claro et al. 2018). Moreover, some AMPs target components intracellularly by moving through lipid bilayer, which blocks cellular functions (nucleic acid synthesis, protein synthesis) (Falanga and Galdiero 2017). Several models, namely, carpet model (Han et al. 2017), toroidal pore model (Phoenix et al. 2016), and barrel-stave mechanism (Shabir et al. 2018), have discussed the mechanism of action of AMPs on bacterial membrane, in detail.

11.2.4 AMPs in Clinical Trials as Antimicrobial Therapeutics

The AMPs highlight multiple advantages for the development and application as antimicrobials, namely, high efficacy, low toxicity, and little accumulation in tissues. The AMPs are emerging candidates in pharmaceutical sector, and the clinicians aim to develop its therapeutic usage by subjecting to clinical trials and validation (Bach 2018). Recent advances in peptide technologies consisting of peptide drug conjugates, multifunctional peptides, and cell-penetrating peptides will facilitate broader application of AMPs as therapeutics (Raucher and Ryu 2015). On a global level, the United States has the highest production and marketing of peptide-based drugs, with Europe coming second in production and the companies, namely, Vicuron Pharmaceuticals and Theravance primarily dedicated to research and development of peptide therapeutics (Ben Lagha et al. 2017). Presently, several novel peptides are undergoing clinical development and trials with around 60 peptide drugs being marketed for use (Lau and Dunn 2018). However, challenges associated with peptide therapeutics still need to be addressed with some peptides being not successful in clinical trials, for example, MSI-78 (pexiganan acetate from magainin) failed in phase III clinical trials and had efficacy against diabetic foot ulcer infections (Lee and Lee 2015); the use of magainin was discontinued owing to inadequate trial design by the FDA in 1999. In another trial, three antimicrobial peptides (related to indolicidin) were introduced into clinical trials (Sandreschi et al. 2016). The recently introduced one in clinical trial is MBI-226, progressing to phase III for preventing catheter-related bloodstream infections (Shabir et al. 2018). The clinical trials showed the efficacy of MBI-226 in animal models by successfully restricting bacterial colonization in catheter-related symptoms (Riool et al. 2017) and antifungal activity against *Candida albicans* in guinea pig (Żelechowska et al. 2016). In another trial, Micrologix carried clinical trials for indolicidin-like peptides for acute acne treatment (phase III trials) and inhibiting MRSA strains (phase Ib trials) (Tasiemski et al. 2014; Cal et al. 2017). To a considerable extent, AMPs and derivatives have been successful in the treatment of infections and have witnessed global market (Deslouches and Di 2017). The global initiatives on antimicrobials have provided in-depth knowledge on various aspects of AMPs, about their efficacy, mechanism, safety, and other parameters (Wang 2017). Furthermore, more investments in AMP research are required to develop peptide antibiotics as drug arsenals.

11.3 Recent Progress in the Development of Peptide-Based Drugs

Peptide-based medicines are now gaining more and more interest from the scientific community in therapeutic and biomedical applications, due to increase in antibiotic resistance and the intermittent appearance of new pathogens. However, the low level of cellular absorption and toxicity was the key issue for these forms of pharmacological compounds, which demote their application to extracellular targets. Many new methods have been created in recent years to bypass the inherent issue of pharmaceuticals of this kind (Fox 2013). The production of *stapled peptides* proved as a solution for these problems (Moiola et al. 2019). *Stapled peptides* contain peptide that forces the peptide structure into a α -helical one. The cross-link is obtained by linking the side chains of modified amino acids posed at the right distance inside the peptide chain (Moiola et al. 2019). Controlled drug delivery was first suggested by Paul Ehrlich who conceptualized "magic bullet" therapy for drug delivery to specific target cells. Due to their biochemical and biophysical properties, peptide-based materials are used as carriers for drug delivery and are preferred over synthetic materials (Varanko et al. 2020). Self-assembling peptide research is also gaining significant interest in the scientific community (Acar et al. 2017). A study accounts the use of polyelectrolyte complexes (PECs), as a carrier for drug delivery (Folchman-Wagner et al. 2017). PECs are structures that arise in a solution based on the combination of oppositely charged macromolecules. In contrast to conventional nanoparticle-based carriers, PECs have several advantages, including controllable size, biodegradability, biocompatibility, and lack of toxicity (Edwards-Gayle and Hamley 2017). Recent progress in immunology has allowed a greater understanding of the immunological effects of associated diseases by researchers. Subunit vaccines based on peptides are of particular interest for modern immunotherapy as they are safer, simple, and easy to produce. Thorough understanding of immunological stimulation events associated with the mode of action of an adjuvant and the discovery of unique mechanisms of signaling leading to an adaptive immune reaction gives a hope for development of an efficient peptide-based vaccine in the future (Beutler et al. 2003; Azmi et al. 2014). Despite continued advancements in therapies for HIV/AIDS, still there are some major limitations which are associated with these therapies. New classes of potent HIV-1 protease inhibitors with novel ligands and features have been developed (Ghosh et al. 2017). A number of protease inhibitors have novel structural features with favorable resistance profiles, showing excellent clinical potential. Hydrogelation of peptide is a technique in which a charged synthetic peptide-based noncytotoxic hydrogelator was employed in encapsulation, storage, and sustainable release of different kinds of drugs, including an anticancer drug, an antibiotic, and proteins. These peptide hydrogels are used as platforms for sustained release of antitumor and antimicrobial drugs (Roy et al. 2020). Recombinant silk-like peptides (SLPs) have also been used to create nanoparticles with reproducible sizes for drug and gene delivery.

11.3.1 Plant-Based Expression System for the Production of AMPs

Antimicrobial peptides (AMP) protect plants or animals against pathogenic attacks (Nawrot et al. 2014). More than 1700 natural AMPs have been found to date. Using natural AMPs as models, a number of derivatives and analogues have been designed by computational system or synthetically generated systems. Advancement in biotechnology permitted the use of plant as bioreactors for mass production of proteins, peptides, and pharmaceutical compounds (Holaskova et al. 2015). Antimicrobial peptide (AMP) production in plants is carried due to major reasons including molecular farming for large-scale production of these peptides in plants and to provide protection to the plant against pathogens. Plants are preferred for production of AMPs due to large-scale production, high quality of peptides (proper folding, disulfide bond formation and glycosylation), and low cost of purification of peptides after production (Vriens et al. 2014). Also plant-based molecular farming is safer in comparison to bacterial and yeast expression systems since there is almost no risk of product contamination with animal/human pathogens or endotoxins, as plants lack these contaminants. Production of AMPs in plants can be achieved by two major strategies, namely, stable integration of AMP genes to the plant genome or transient transformation strategy (Sinha 2019). Figure 11.1 provides a schematic representation of molecular farming of antimicrobial peptides in plants.

11.3.1.1 Plant Tissue Culture-Based Expression Systems

In vitro plants are maintained as either tissue culture, cell suspension cultures, callus culture, or hairy root culture. In vitro plant tissue culture technique is a powerful method for analyzing diverse expression procedures before production of a stable transgenic plant (Espinosa-Leal et al. 2018). This is because in vitro cultures are independent of climatic factors and plant cells or tissues are cultured by controlled application of phytohormones and culture medium. Hairy root cultures are used in plant biotechnology for production of phytochemical and therapeutic proteins via *Agrobacterium rhizogenes* (Hussain et al. 2012; Jiao et al. 2018). Figure 11.2 discusses the various platforms for molecular farming of antimicrobial peptides in plants.

Callus cultures can be obtained from every live plant tissue, especially young and meristematic tissues are used as starting material for callus cultures (Efferth 2019). Callus cultures containing transgene are obtained either by explant transformation or direct transformation of callus and successive culture after transformation (Zhang et al. 2000; Carciofi et al. 2012). Mostly, monocotyledonous crop plants lack a screening platform for expression, and they are recalcitrant to in vitro regeneration. Hence, callus cultures serve as alternative tools for screening stable transformants after expression. The transgenics were prepared to express barley protein via callus

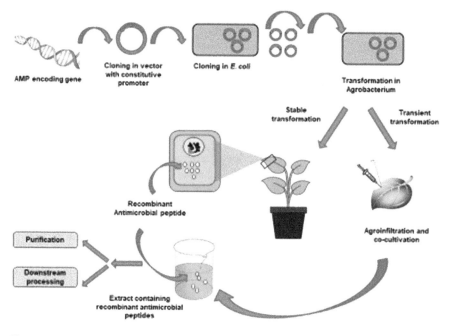

Fig. 11.1 Molecular farming of antimicrobial peptides in plants

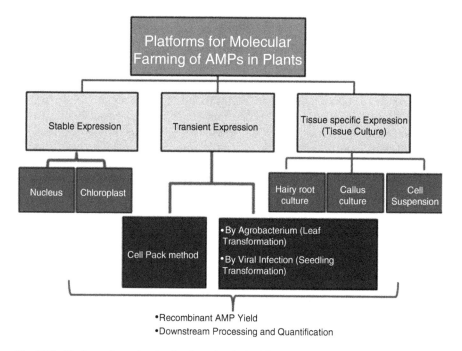

Fig. 11.2 Platforms for molecular farming of antimicrobial peptides in plants

cultures from barley (Imani et al. 2011). Hairy root cultures (HRCs) are produced from the infection of a plant by *Rhizobium rhizogenes* (earlier known as *A. rhizogenes*) bacteria. It can be produced from a wide range of crops and facilitates highly diverse molecules to be generated (Gutierrez-Valdes et al. 2020). HRCs accumulate recombinant proteins to levels equivalent to that of whole plants and with stable biosynthetic capacity. Root inducing plasmids from *A. rhizogenes* is accountable for stable integration of gene (including the gene of interest) into the genome of a host plant cell (Gutierrez-Valdes et al. 2020). Lactoferrampin and lactoferricin chimera antimicrobial peptides from bovine lactoferrin were expressed in tobacco hairy root cultures using *A. rhizogenes* (Chahardoli et al. 2018). Therapeutic proteins are expressed under standardized conditions in in vitro plant systems that is an attractive production approach in terms of product consistency and the downstream purification process (Madeira et al. 2016). Nowadays, "next-generation human therapeutic antibodies" are produced by using hairy root cultures of plants (Drake et al. 2009; Donini et al. 2015; Lallemand et al. 2015). Moreover, for the development of next-generation therapeutic human antibodies, hairy root cultures of *N. benthamiana* were exploited for expressing the tumor-targeting antibody -mAb H10 (Lonoce et al. 2016).

11.3.1.2 Genetically Engineered Plant Systems

In plant molecular farming, plant cells or tissues are exploited for the expression and processing of recombinant proteins or peptides of pharmaceutical value. This strategy is regarded as one of most effective techniques for AMP production (Holásková et al. 2018). In earlier researches, potato and tobacco were mostly used as model plants for the production of recombinant proteins, antigens, and pharmaceuticals (Ma et al. 2003). Alteration of host plant systems during production (subcellular targeting, translation, and posttranslational modifications) and control of transcription by promoter are the two strategies which affects the production of recombinant protein in a plant (Makhzoum et al. 2014; Obembe et al. 2011; Iyappan et al. 2019; da Cunha et al. 2019) (Fig. 11.2). Constitutive expression of cecropin P1, antimicrobial peptide in transgenic potato, produced resistance to the plant against causative fungal pathogen of white rot and potato blight (Zakharchenko et al. 2013). Majority of the studies uses constitutive expression of AMPs under *ubiquitin* and *CaMV35S* promoters (Abdallah et al. 2010; Jung and Kang 2014; Ribeiro et al. 2017; Ilyas et al. 2017; Tiwari et al. 2020). Citrus Huanglongbing (HLB) associated with citrus canker and *Candidatus* Liberibacter asiaticus' (Las) disease caused by *Xanthomonas citri* are the most injurious diseases to the citrus plants worldwide. Transgenic expression of D2A21 antimicrobial peptide was achieved to create resistant varieties against canker and HLB disease. Compared to control plants, transgenic tobacco plants expressing D2A21 displayed remarkable disease resistance (Hao et al. 2017). Transgenic barley was also developed for production of the LL-37 antimicrobial peptide (Holásková et al. 2018). In a study, potato plants were transformed with a gene producing antifungal peptide derived from alfalfa plant (Sathoff et al. 2019). Similarly, tachyplesin I was derived from horseshoe crab (Lipsky et al. 2016) which provides protection from bacterial and fungal pathogens. In another study, tachyplesin I was expressed in transgenic *Ornithogalum* plants (Lipsky et al. 2016). To study the environmental and ecological effects of plant-microbe interactiont, *Nicotiana attenuata* (wild tobacco) was transformed to constitutively express an antimicrobial peptide (Mc-AMP1) from common rice plant. As a result, the transgenic plants showed *in planta* activity against plant-beneficial bacteria (Weinhold et al. 2018). Transgenic tobacco hairy root (HR) culture was used for expression of a dermaseptin B1 recombinant antimicrobial peptide (C2-B1) by direct and indirect transformation strategies. Transgenic clones derived from both direct and indirect transformation methods and in liquid medium exhibited distinction in terms of total recombinant protein content. The study reveals that hairy root derived from indirect transformation method produced a significantly higher amount of recombinant peptide (Varasteh-Shams et al. 2020). Antimicrobial peptide, BTD-S, possesses very high antimicrobial activity against economically important plant pathogen *Verticillium dahliae*. This peptide was expressed into wild-type *Arabidopsis thaliana* (ecotype Columbia-0) plants. BTD-S-transgenic lines increased resistance against *V. dahliae* in both in vivo and in vitro assays (Li et al. 2016). The heterologous expression of antimicrobial peptide PaDef from Mexican avocado fruit by an N-terminal 6X-His tagged recombinant PaDef was achieved in *Pichia pastoris*. The secreted peptide completely inhibited the growth of *Escherichia coli* and *Staphylococcus aureus* (Meng et al. 2017). Table 11.2, discuss key examples of CRISPR-Cas-based genome editing, generation of antimicrobials, and applications against drug-resistant pathogens.

11.3.1.3 Strategies for Transient Expression

Transient transformation is a very well-suited technique for expression of a gene in plant and production of recombinant proteins, due to its high expression levels and comparatively shorter time with high yield (Fisher et al. 2019). Using a genetically engineered vector, normally *Agrobacterium* or a plant virus, the gene of interest is inserted into plant cells during the transient expression system, and output (recombinant protein or peptide) is achieved by extrachromosomal gene expression within plant cells. In this technique, expression starts within 24 h and lasts for up to a week (Streatfield 2007; Dirisala et al. 2016). Natural characteristics of causing infection of plant viruses and *Agrobacterium* are exploited for transient expression. Components used in transient transformation include hairy root culture or plant suspension cell culture or whole leaf of plant (Rage et al. 2020). Several peptides and proteins such as human interleukin-2, bovine lysozyme, human α-galactosidase A, bovine aprotinin, and others have been produced by transient expression based on virus infection (Pogue et al. 2010). Recombinant antimicrobial peptide aprotinin was produced using TMV-based vector in tobacco plants. In the next method, DNA delivery was carried out using *Agrobacterium*, combined with a plant virus with capacity of rapid replication and high level expression (Gleba et al. 2007; Diamos and Mason 2018). This approach was used to produce heterologous proteins such as bacterial and viral antigens, interferons, cytokines, and growth hormones (Gleba et al. 2007). Tobacco plants were transformed by vacuum infiltration by *Agrobacterium* cultures. Transient systems are even used for the development of

Table 11.2 CRISPR-Cas-based genome editing, generation of antimicrobials, and applications against drug-resistant pathogens

S. no.	Genome-editing tool	Host organism	Methodology	Key application	References
1.	CRISPR-Cas9	*E. coli*	CRISPR-Cas9-mediated creation of antimicrobials, RNA-guided nucleases (RGNs)	RGNs modulated of bacteria by selective knockdown of targeted strains	Citorik et al. (2014)
2.	CRISPR-Cas9	Mouse	Mouse skin infection with *S. aureus* and phagemids using the CRISPR system	The phagemids significantly reduced bacterial infections	Bikard et al. (2014)
3.	CRISPR-Cas9	*E. coli*	Extended-spectrum β-lactamase (ESBL)-producing *E. coli* mutants were targeted to restore antibiotic sensitivity in bacteria	Resensitization of antibiotic-resistant *E. coli*	Kim et al. (2016)
4.	CRISPR-Cas9	*E. coli*	Plasmid-mediated CRISPR-Cas9 system was used to efficiently resensitize *E. coli* to colistin	Reversal of *mcr-1*-mediated Colistin resistance	Wan et al. (2020)
5.	CRISPR-Cas9 and optical DNA mapping	*E. coli*	CRISPR-Cas9 plasmid digestion into linear forms for gene detection	Direct identification of antibiotic resistance genes on the plasmid	Muller et al. (2016)
6.	CRISPR-Cas9	*K. pneumoniae*	Cas9 from *Streptococcus pyogenes* and a single guide RNA (sgRNA) were used to modify a virulent *Klebsiella* bacteriophage, phiKpS2	CRISPR-Cas9-based phage genome editing as therapeutic against AMR	Shen et al. (2018)
7.	CRISPR-Cas9	*E. coli*	CRISPR/Cas9/sgRNA-mediated gyrA mutations in quinolone-susceptible *E. coli*	Reversed quinolone resistance	Qiu et al. (2018)
8.	CRISPR-Cas9	*S. aureus*	CRISPR-Cas9-mediated genetic engineering of	CRISPR/Cas9) system to create programmable	Park et al. (2017)

(continued)

Table 11.2 (continued)

S. no.	Genome-editing tool	Host organism	Methodology	Key application	References
			phage-based delivery system for CRISPR/Cas9 antimicrobials	gene-specific antimicrobials	

pharmaceutical proteins or peptides of industry-grade. In a study, recombinant aprotinin was produced using *Nicotiana sp.* via TMV-based vector. This study was the first example of production of antimicrobial peptides via transient transformation strategy for molecular farming. Avian H5N1 influenza vaccine was produced using transient expression in tobacco leaf tissues (Mbewana et al. 2015). A broad-spectrum antimicrobial peptide, antimicrobial peptide protegrin-1 (PG-1), was expressed in *Nicotiana tabacum*, using *Agrobacterium tumefaciens*-mediated transient expression system. This peptide was biologically active against strains of bacteria and fungus (Patiño-Rodríguez et al. 2013). Inducible expression of the de novo designed AMP SP1–1 targeted to the apoplast protected tomato fruits against bacterial spot disease which is the primary infection site for plant pathogens of *Raphanus sativus* (Herrera Diaz et al. 2016).

11.3.1.4 Cell Pack Method

Plant cell suspension cultures require constant and defined conditions, which gives a better quality control option with chemically well-defined media, specifically during continuous production processes (Fisher et al. 2019). But the major problems associated with plant cell culture are the scalability of plant cells, which is less in comparison to whole plants. Also, the methods optimized for plant cell cultures are also not always directly transferable to plants (Arya et al. 2020). Thus, further screening and translational research is a must. A plant cell pack (PCP), also known as "cookies," is generated to form a link between plant cells and the scalability of whole plants. This platform is based on plant cell cultures that lack liquid medium. Plant cell packs therefore provide a suitable platform for standard recombinant protein expression approaches, metabolic engineering, and synthetic biology that require the high-throughput screening of several of constructs for proficient product development (Moon et al. 2020). PCP technology was exploited for molecular farming human antibody chain proteins (Rademacher et al. 2019). Method for the generation and cultivation of a plant cell pack was created by Rademacher in 2020. Transiently expressed recombinant proteins via whole plants or plant cells are produced in a low quantity, which makes a limiting factor for industrial adaptation of plant-based expression systems as an effective alternative of production as compared with that of existing expression systems. PCPs provide a rapid and high-throughput screening system with increased expression of recombinant protein. PCPs worked successfully in cell suspension cultures of plants, including *A. thaliana, Catharanthus roseus, Daucus carota, N. tabacum,* and

N. benthamiana. The technique also makes it easier to purify the product relative to leaf-based expression systems, due to fewer secondary metabolites and host proteins. An expression system containing three-dimensional, porous plant cell aggregates deprived of cultivation medium containing the infusion of *Agrobacterium tumefaciens* (plant cell packs, PCPs) was created (Poborilova et al. 2020). These PCPs are well suited and more efficient as compared to the transient expression system in liquid plant cell culture for plant species such as *Nicotiana* spp. and *D. carota* (Rademacher et al. 2019) (Fig. 11.2). In a study, the PCP technology was used to study co-expression and transient expression of proteins in BY-2 cell packs of *Nicotiana* spp. (Rademacher et al. 2019). This uses both fast-growing cell suspension cultures (Xue et al. 2013) and transient protein expression assays to study the efficiency of replicating expression vectors (derived from the pRIC vector). The results suggested the ability of replicating vectors to transiently express proteins in BY-2 plant cell packs detected as red and green fluorescent proteins (Poborilova et al. 2020). The high-throughput screening of recombinant protein expression is now possible for plants due to the advent of plant cell packs (PCPs). A study integrated the plate casted with the PCPs with a fully automated laboratory liquid-handling station for high-throughput detection of expression in plant cells. These are termed as "robot cookies." The accumulation of fluorescent protein in cell organelles is easily detected using an integrated plate reader. The new automated method provides low-cost platform with increased number of sample detection (Gengenbach et al. 2020). A method for automated transformation of plant cell packs was invented by the same group of researchers (Gengenbach et al. 2020).

11.4 Computational Biology-Based Antimicrobial Research

The advances in computational biology has facilitated the development of vaccines/drugs with improved efficacy. The development of antibacterial focused on combinational chemistry approach co-integrated with computational modeling of 3D structure of antibacterial compound employing energy minimization and other techniques (Baker and Sali 2001) for the identification of a better docked compound to virulent gene, limiting the pathogen growth. With the deciphering of biological genomes and genome analysis through automated tools, the capacity to archive proteomic/genomic conclusions and retrieve biological information has addressed the challenges associated with tackling drug-resistant pathogens (Bansal 2008). In recent times, bioinformatics approaches have been employed to study potential virulent genes and their mechanisms in silico in a cost-effective way, compared to time-taking and costly experimental methods. The function of the microbes at systemic level may be understood through the integration of biochemical methods co-integrated with computational approaches (Franklin and Snow 1998; Bansal et al. 2018). Drug development employing bioinformatics approaches aims to gain insights on microbial function and dynamics in a cost-effective manner and to identify potential antimicrobial candidates with good efficacy for further development. Any prospective drug candidate should possess good efficacy in in vivo

Table 11.3 Computational tools and resources for antimicrobial peptides research in drug discovery

S. no.	Webtool/ software	Description	Link
1.	APD3 database	The AMP database	http://aps.unmc.edu/AP/main.php
2.	SATPdb database	Collection of structurally annotated therapeutic peptides	http://crdd.osdd.net/raghava/satpdb/
3.	LAMP2 database	Useful tool for studies on AMPs	https://databases.lovd.nl/shared/genes/LAMP2
4.	CPPpred	For prediction of cell-penetrating peptides	http://bioware.ucd.ie/~compass/biowareweb/Server_pages/cpppred.php
5.	ProtParam	Assessment of chemical and physical variables of a protein	https://web.expasy.org/protparam/
6.	GROMACS	For molecular dynamic studies	http://www.gromacs.org/
7.	DRAMP 2·0 database	Collection of information on patent, clinical, and general AMPs	http://dramp.cpu-bioinfor.org/
8.	ExPASy	Web resource which provides access to software and databases	https://www.expasy.org/
9.	CAMPR3 database	Information on antimicrobial peptides	http://www.camp3.bicnirrh.res.in/
10.	BaAMPs database	Database on AMPs tested against microbial biofilms	http://www.baamps.it
11.	YADAMP database	Collection of information on antimicrobial activity against common bacterial strains	http://www.yadamp.unisa.it
12.	AMPep database	Sequence-based prediction of AMPs	https://omictools.com/ampep-tool
13.	AntiBP2	Predict the presence of antibacterial peptides in protein	http://crdd.osdd.net/raghava/antibp2/
14	Joker	An algorithm for designing AMPs	https://github.com/williamfp7/Joker

conditions and should have minimum toxicity and side effects (Bansal 2008). Additionally, several bioinformatics tools and databases have been developed to provide information on antimicrobial peptides, information of microbes, and in silico tools to analyze human-pathogen interactions. Table 11.3 provides information about various computational tools and resources for antimicrobial peptide research in drug discovery. Most of the research on the development of antibacterial drugs till now focused on generation of compounds and derivatives highlighting potent activity in blocking pathogen growth either by inhibition of nucleic acid synthesis, restricting biosynthesis of peptidoglycan, or blocking the essential metabolic pathways, among other functions (Bansal 2008; Franklin and Snow 1998). Computational biology has significantly contributed in drug discovery initiatives by gene identification, whole-genome sequencing, comparative genomics-based metabolic pathway reconstruction, and identification of plasmid genes that block the pathway

(Bansal 2008). However, there is still a long way to go in complete integration of bioinformatics and biochemical analysis for the development of potential drug candidates. The ongoing progress in bioinformatics would improve knowledge on virulent genes and presence of surface antigens. A better understanding and information about microbial dynamics is necessary to develop effective antimicrobial candidate for drug discovery and development.

11.5 Challenges Associated with Development of AMPs as Antimicrobial Therapeutics

Antimicrobial peptides (AMPs), an inherent component of innate immune response, have been extensively studied as small molecule antimicrobial candidates. The saturation in antimicrobial arsenal discovery and drying pipelines has necessitated the exploration of compounds from natural sources. To date, more than 3000 AMPs have been discovered and characterized (Bansal 2008; Wang et al. 2016); however, a majority of them are not suitable for human treatment (in natural state) (Chen and Lu 2020), and a number of them were unsuccessful during clinical trials. Presently, six AMPs, namely, vancomycin, telavancin, gramicidin D, dalbavancin, daptomycin, and oritavancin, have been approved by FDA and commercially marketed (Chen and Lu 2020). Currently, most of the peptide-based therapeutics bind receptors by either blocking or activating to the receptors they bind and leading to a biological response, while others are peptides acting on membranes and pathway inhibitors. The stability of peptide-based therapeutics is essential for use as drug molecules – most of peptides approved by FDA show stability in vivo (Nguyen et al. 2010). For synthetic AMPs, it is necessary to design molecules possessing antimicrobial activity and clinical properties, namely, antimicrobial agents from Gram-positive bacteria and knowing their biochemical properties, adding non-canonical amino acids into peptide sequence, thereby increasing elimination half-lives, combinational use of AMPs synergistically with other compounds for improved efficacy, and development of in vitro models to analyze the antimicrobial effects (Chen and Lu 2020). A structural combination of a well-defined cationic region and the hydrophobic surface area is responsible for broad-spectrum activity of antimicrobial agents; however, rigidity in structure of AMPs hinder its functions (Nguyen et al. 2011). Efficacy and toxicity of AMPs are the major bottlenecks which limits its use. Several AMPs have shown side effects in vivo and therefore are used for tropical applications only (Hancock and Patrzykat 2002). Moreover, the presence of serum and high salt can adversely affect antimicrobial functions (Han et al. 2017). Another major drawback is the sensitivity of AMPs to digestion by host proteolytic enzymes which degrade AMPs, thereby affecting their pharmacokinetics and in vivo stability (Starr and Wimley 2017). Therefore, many AMPs are not suggested for oral administration and specified for tropical purpose only. In the present era, current researches have concluded a more diverse and complex mechanism for AMPs, highlighting broader and multi-prolonged strategy to counter microbial pathogens. In addition to their biological properties, other factors including in vivo stability (Nguyen et al. 2010), together with the associated side effects and cost of production, are to be considered. A comprehensive knowledge about structure-function studies of AMPs and its efficacy range will be important in designing AMPs with improved function.

11.6   Commercial Success and Prospects of AMPs as Antimicrobial Therapeutics

Despite many existing challenges associated with AMPs, these are emerging as next-generation antibiotics to counter diverse microbial infections, particularly MDR strains. The natural and synthetic AMPs as novel antimicrobials highlight remarkable therapeutic potential via diverse mechanism of action. Recent advances in antimicrobial research has improved our knowledge on the structure and properties of AMPs; however, there is still long way ahead for complete exploitation of these therapeutic candidates. Furthermore, modified AMPs such as hybrid peptides, peptide mimetics, immobilized peptides, and peptide conjugates have been developed from natural AMPs and demonstrate targeted or selective antimicrobial action (Brogden and Brogden 2011) with improved properties. These novel peptides project industrial and medical applications to tackle antibiotic-resistant bacterial strains and in food preservation. The potential of AMPs to tackle drug-resistant pathogens makes them ideal candidates in treating infectious diseases, highlighting the development of novel antimicrobial candidates for viral, fungal, and bacterial infections, and countering the rising AMR worldwide. In recent times, considerable progress has been made in designing and synthesis of short peptides with improved efficacy and less toxicity with short synthetic AMPs being widely employed to counter MDR microbial strains. The development of AMPs shows significant prospects to restore the drying antibiotic pipeline and to address the rising AMR across globe.

Acknowledgments   The authors acknowledge their respective institutions for encouragement and support.

Conflict of Interests   No conflict of interests was declared.

References

Abdallah NA, Shah D, Abbas D, Madkour M (2010) Stable integration and expression of a plant defensin in tomato confers resistance to fusarium wilt. GM Crops 1:344–350

Acar H, Srivastava S, Chung EJ, Schnorenberg MR, Barrett JC, LaBelle JL, Tirrell MJA (2017) Self-assembling peptide-based building blocks in medical applications. Adv Drug Deliv Rev 110:65–79

Agourram A, Ghirardello D, Rantsiou K, Zeppa G, Belviso S, Romane A et al (2013) Phenolic content, antioxidant potential and antimicrobial activities of fruit and vegetable by-product extracts. Int J Food Propert 16(5):1092–1104

Andersson E, Rydengard V, Sonesson A, Morgelin M, Bjorck L, Schmidtchen A (2004) Antimicrobial activities of heparin- binding peptides. Eur J Biochem 271(6):1219–1226

Arques JL, Fernandez J, Gaya P, Nunez M, Rodrıguez E, Medina M (2004) Antimicrobial activity of reuterin in combination with nisin against food-borne pathogens. Int J Food Microbiol 95 (2):225–229

Arya SS, Rookes JE, Cahill DM, Lenka SK (2020) Next-generation metabolic engineering approaches towards development of plant cell suspension cultures as specialized metabolite producing biofactories. Biotech Advances 45:107635

Aslam B, Wang W, Arshad MI, Khurshid M, Muzammil S, Rasool MH, Nisar MA, Alvi RF, Aslam MA, Qamar MU, Salamat MKF, Baloch Z (2018) Antibiotic resistance: a rundown of a global crisis. Infect Drug Resist 11:1645–1658

Ayukekbong JA, Ntemgwa M, Atabe AN (2017) The threat of antimicrobial resistance in developing countries: causes and control strategies. Antimicrob Resist Infect Control 6(1):47

Azmi F, Ahmad Fuaad AAH, Skwarczynski M, Toth I (2014) Recent progress in adjuvant discovery for peptide-based subunit vaccines. Hum Vaccin Immunother 10:778–796

Bach HA (2018) New Era without Antibiotics. Antibiotics 2018:1

Baker D, Sali A (2001) Protein structure prediction and structural genomics. Science 294:93–96

Bala N, Aitken EA, Cusack A, Steadman KJ (2012) Antimicrobial potential of Australian macrofungi extracts against foodborne and other pathogens. Phytother Res 26(3):465–469

Balasundram N, Sundram K, Samman S (2006) Phenolic compounds in plants and Agri-industrial by-products: antioxidant activity, occurrence, and potential uses. Food Chem 99(1):191–203

Bansal AK (2008) Role of bioinformatics in the development of new antibacterial therapy. Expert Rev Anti Infect Ther 6(1):51–65

Bansal A, Srivastava PA, Singh TR (2018) An integrative approach to develop computational pipeline for drug-target interaction network analysis. Sci Rep 8:10238. https://doi.org/10.1038/s41598-018-28577-6

Belmadani A, Semlali A, Rouabhia M (2018) Dermaseptin-S1 decreases *Candida albicans* growth, biofilm formation and the expression of hyphal wall protein 1 and aspartic protease genes. J App Microbial 125(1):72–83

Ben Lagha A, Haas B, Gottschalk M et al (2017) Antimicrobial potential of bacteriocins in poultry and swine production. Vet Res 48(22). https://doi.org/10.1186/s13567-017-0425-6

Beutler B, Hoebe K, Du X, Ulevitch RJ (2003) How we detect microbes and respond to them: the toll-like receptors and their transducers. J Leukoc Biol 74:479–485

Bikard D, Euler CW, Jiang W et al (2014) Exploiting CRISPR-Cas nucleases to produce sequence-specific antimicrobials. Nat Biotechnol 32(11):1146–1150

Blomqvist M, Bergquist J, Westman A (1999) Identification of defensins in human lymphocyte nuclei. Eur J Biochem 263(2):312–318

Boucher HW, Talbot GH, Bradley JS (2009) Bad bugs, no drugs: no ESKAPE! An update from the infectious diseases society of America. Clin Infect Dis 48:1–12

Brogden NK, Brogden KA (2011) Will new generations of modified antimicrobial peptides improve their potential as pharmaceuticals? Int J Antimicrob Agents 38:217–225

Brudzynski K, Sjaarda C, Lannigan R (2015) MRJP1-containing glycoproteins isolated from honey, a novel antibacterial drug candidate with broad spectrum activity against multi-drug resistant clinical isolates. Front Microbiol 6:711. https://doi.org/10.3389/fmicb.2015.00711

Cal PM, Matos MJ, Bernardes GJ (2017) Trends in therapeutic drug conjugates for bacterial diseases: a patent review. Expert Opin Ther Pat 27(2):179–189

Carciofi M, Blennow A, Nielsen MM, Holm PB, Hebelstrup KHJPM (2012) Barley callus: a model system for bioengineering of starch in cereals. Plant Methods 8:1–10

Chahardoli M, Fazeli A, Ghabooli M (2018) Recombinant production of bovine Lactoferrin-derived antimicrobial peptide in tobacco hairy roots expression system. Plant Physiol Biochem 123:414–421

Chan BC, Ip M, Lau CB, Lui SL, Jolivalt C, Ganem-Elbaz C et al (2011) Synergistic effects of baicalein with ciprofloxacin against NorA over-expressed methicillin-resistant *Staphylococcus aureus* (MRSA) and inhibition of MRSA pyruvate kinase. J Ethnopharmacol 137:767–773

Chen CH, Lu TK (2020) Development and challenges of antimicrobial peptides for therapeutic applications. Antibiotics 9:24. https://doi.org/10.3390/antibiotics9010024

Choi SC, Ingale SL, Kim JS, Park YK, Kwon IK, Chae BJ (2013) An antimicrobial peptide-A3: effects on growth performance, nutrient retention, intestinal and faecal microflora and intestinal morphology of broilers. Br Poultry Sci 54:738–746

Chovanova R, Mezovska J, Vaverkova S, Mikulasova M (2015) The inhibition the Tet(K) efflux pump of tetracycline resistant *Staphylococcus epidermidis* by essential oils from three *Salvia* species. Lett Appl Microbiol 61:58–62

Chung YC, Yeh JY, Tsai CF (2011) Antibacterial characteristics and activity of water-soluble chitosan derivatives prepared by the Maillard reaction. Molecules 16(10):8504–8514

Ciocan ID, Bara I (2007) Plant products as anti-microbial agents. Genetică și Biologie Molecul 8 (1):151–156

Citorik RJ, Mimee M, Lu TK (2014) Sequence-specific antimicrobials using efficiently delivered RNA-guided nucleases. Nat Biotechnol 32(11):1141–1145

Claro B, Bastos M, Garcia-Fandino R (2018) Design and applications of cyclic peptides. In: Koutsopoulos S (ed) Peptide applications in biomedicine, biotechnology and bioengineering. Woodhead Publishing, Sawston, pp 87–129

Cudic M, Otvos L Jr (2002) Intracellular targets of antibacterial peptides. Curr Drug Targets 3 (2):101–106

Da Costa JP, Cova M, Ferreira R, Vitorino R (2015) Antimicrobial peptides: an alternative for innovative medicines? Appl Microbiol Biotechnol 99(5):2023–2040

da Cunha NB, Leite ML, Dias SC, Vianna GR, Rech Filho EL (2019) Plant genetic engineering: basic concepts and strategies for boosting the accumulation of recombinant proteins in crops. Int J Latest Trans Eng Sci

da Silva LCN, da Silva MV, MTDS C (2017) Editorial: new frontiers in the search of antimicrobials agents from natural products. Front Microbiol 8:210. https://doi.org/10.3389/fmicb.2017.00210

Davies J (2006) Where have all the antibiotics gone? Can J Infect Dis Med Microbiol 17:287–290

Davies J, Davies D (2010) Origins and evolution of antibiotic resistance. Microbiol Mol Biol Rev 74(3):417–433

de Kraker MEA, Stewardson AJ, Harbarth S (2016) Will 10 million people die a year due to antimicrobial resistance by 2050? PLoS Med 13(11):e1002184

de Oliveira AG, Spago FR, Simionato AS, Navarro MO, da Silva CS, Barazetti AR, Cely MV, Tischer CA, San Martin JA, de Jesus Andrade CG, Novello CR, Mello JC, Andrade G (2016) Bioactive Organocopper compound from *Pseudomonas aeruginosa* inhibits the growth of *Xanthomonas citri* subsp. citri. Front Microbiol 7:113. https://doi.org/10.3389/fmicb.2016.00113

Deshmukh SK, Verekar SA, Bhave SV (2015) Endophytic fungi: a reservoir of antibacterials. Front Microbiol 5:715. https://doi.org/10.3389/fmicb.2014.00715

Deslouches B, Di YP (2017) Antimicrobial peptides with selective antitumor mechanisms: prospect for anticancer applications. Oncotarget 8(28):46635–46651

Diamos AG, Mason HS (2018) High-level expression and enrichment of norovirus virus-like particles in plants using modified geminiviral vectors. Protein Expr Purif 151:86–92

DiMasi JA (1995) Success rates for new drugs entering clinical testing in the United States. Clin Pharmacol Ther 58:1–14

Dirisala VR, Nair RR, Srirama K, Reddy PN, Rao KRSS, Satya Sampath Kumar N, Parvatam G (2016) Recombinant pharmaceutical protein production in plants: unraveling the therapeutic potential of molecular pharming. Acta Physiol Plant 39:18

Dixit D, Madduri RP, Sharma R (2014) The role of tigecycline in the treatment of infections in light of the new black box warning. Expert Rev Anti Infect Ther 12:397–400

Donini M, Lombardi R, Lonoce C, Di Carli M, Marusic C, Morea V, Di Micco PJB (2015) Antibody proteolysis: a common picture emerging from plants. Bioengineered 6:299–302

Dorman H, Deans S (2000) Antimicrobial agents from plants: antibacterial activity of plant volatile oils. J Appl Microbiol 88(2):308–316

Draenert R, Seybold U, Grützner E, Bogner JR (2015) Novel antibiotics: are we still in the pre-post-antibiotic era? Infection 43:145–151

Drake PM, Barbi T, Sexton A, McGowan E, Stadlmann J, Navarre C, Paul MJ, Ma JK (2009) Development of rhizosecretion as a production system for recombinant proteins from hydroponic cultivated tobacco. FASEB J 23:3581–3589

Du Y, Zhao Y, Dai S, Yang B (2009) Preparation of water-soluble chitosan from shrimp shell and its antibacterial activity. Innov Food Sci Emerg Technol 10(1):103–107

Dwivedi G, Maurya A, Yadav D, Khan F, Darokar M, Srivastava S (2014) Drug resistance reversal potential of ursolic acid derivatives against nalidixic acid- and multidrug-resistant *Escherichia coli*. Chem Biol Drug Des 86:272–283

Dwivedi GR, Tyagi R, Sanchita TS, Pati S, Srivastava SK et al (2017) Antibiotics potentiating potential of catharanthine against superbug *Pseudomonas aeruginosa*. J Biomol Struct Dyn 2017:1–15. https://doi.org/10.1080/07391102.2017.1413424

EDCD (2014) Antimicrobial-resistance-europe-2014. http://ecdc.europa.eu/en/publications/Publications/antimicrobial-resistance-europe-2014.pdf

Edwards-Gayle CJ, Hamley IW (2017) Self-assembly of bioactive peptides, peptide conjugates, and peptide mimetic materials. Org Biomol Chem 15:5867–5876

Efferth TJE (2019) Biotechnology applications of plant callus cultures. Engineering 5:50–59

Eliopoulos GM, Mek VG, Gold HS (2004) Antimicrobial Resistance to Linezolid. Clin Infect Dis 39:1010–1015

Eom SH, Kim YM, Kim SK (2012) Antimicrobial effect of phlorotannins from marine brown algae. Food Chem Toxicol 50(9):3251–3255

Espinosa-Leal CA, Puente-Garza CA, García-Lara SJP (2018) In vitro plant tissue culture: means for production of biological active compounds. Planta 248:1–18

Esquivel P, Jimenez VM (2012) Functional properties of coffee and coffee byproducts. Food Res Int 46(2):488–495

Falanga A, Galdiero S (2017) Emerging therapeutic agents on the basis of naturally occurring antimicrobial peptides. In: Ryadnov M, Hudecz F (eds) Amino acids, peptides proteins, vol 42. Royal Society of Chemistry, London, pp 190–227

Falla TJ, Karunaratne DN, Hancock REW (1996) Mode of action of the antimicrobial peptide indolicidin. J Biol Chem 271(32):19298–19303

Fisher AC, Kamga MH, Agarabi C, Brorson K, Lee SL, Yoon SJ (2019) The current scientific and regulatory landscape in advancing integrated continuous biopharmaceutical manufacturing. Trends Biotechnol 37:253–267

Folchman-Wagner Z, Zaro J, Shen WC (2017) Characterization of polyelectrolyte complex formation between anionic and cationic poly (amino acids) and their potential applications in pH-dependent drug delivery. Molecules 22:1089

Fox JL (2013) Antimicrobial peptides stage a comeback. Nat Biotechnol 31(5):379–382

Franklin TJ, Snow GA (1998) Biochemistry and molecular biology of antimicrobial drug action, 5th edn. Kluwer Academic Press, New York

Friedman M, Henika PR, Mandrell RE (2002) Bactericidal activities of plant essential oils and some of their isolated constituents against *Campylobacter jejuni*, *Escherichia coli*, *Listeria monocytogenes*, and *Salmonella enterica*. J Food Prot 65(10):1545–1560

Gengenbach BB, Opdensteinen P, Buyel JF (2020) Robot cookies-plant cell packs as an automated high-throughput screening platform based on transient expression. Front Bioeng Biotechnol 8:393

Ghosh AK, Rao KV, Nyalapatla PR, Osswald HL, Martyr CD, Aoki M, Hayashi H, Agniswamy J, Wang YF, Bulut HJ (2017) Design and development of highly potent HIV-1 protease inhibitors with a crown-like oxotricyclic core as the P2-ligand to combat multidrug-resistant HIV variants. J Med Chem 60(10):4267–4278

Gleason PP, Shaughnessy AF (2007) Telithromycin (Ketek) for treatment of community-acquired pneumonia. AFP 76:1857

Gleba Y, Klimyuk V, Marillonnet S (2007) Viral vectors for the expression of proteins in plants. Curr Opin Biotechnol 18:134–141

Griffin SG, Wyllie SG, Markham JL (2005) Antimicrobially active terpenes cause Kþ leakage in *E. coli* cells. J Essent Oil Res 17(6):686–690

Gudiol C, Carratalà J (2014) Antibiotic resistance in cancer patients. Expert Rev Anti Infect Ther 12 (8):1003–1016

Gutierrez-Valdes N, Häkkinen ST, Lemasson C, Guillet M, Oksman-Caldentey KM, Ritala A, Cardon FJ (2020) Hairy root cultures-a versatile tool with multiple applications. Front Plant Sci 11:5

Gyawali R, Ibrahim SA (2014) Natural products as antimicrobial agents. Food Contr 46:412–429

Han J, Zhao S, Ma Z, Gao L, Li H, Muhammad U (2017) The antibacterial activity and modes of LI–F type antimicrobial peptides against *Bacillus cereus* in vitro. J Appl Microbiol 123 (3):602–614

Hancock RE (2000a) Cationic antimicrobial peptides: towards clinical applications. Expert Opin Investig Drugs 9:1723–1729

Hancock RE (2000b) Cationic antimicrobial peptides: towards clinical applications. Expert Opin Investig Drugs 9:1723–1729

Hancock RE (2001) Cationic peptides: effectors in innate immunity and novel antimicrobials. Lancet Infect Dis 1(3):156–164

Hancock RE, Chapple DS (1999) Peptide antibiotics. Antimicrob Agents Chemother 43:1317–1323

Hancock RE, Haney EF, Gill EE (2016) The immunology of host defense peptides: beyond antimicrobial activity. Nat Rev Immunol 16:321–334

Hancock REW, Patrzykat A (2002) Clinical development of cationic antimicrobial peptides: from natural to novel antibiotics. Curr Drug Targets Infect Disord 2(1):79–83

Hao G, Zhang S, Stover EJ (2017) Transgenic expression of antimicrobial peptide D2A21 confers resistance to diseases incited by *Pseudomonas syringae* pv. tabaci and *Xanthomonas citri*, but not Candidatus Liberibacter asiaticus. PLoS One 12:e0186810

Hassan R, Shaaban MI, Abdel Bar FM, El-Mahdy AM, Shokralla S (2016) Quorum sensing inhibiting activity of *Streptomyces coelicoflavus* isolated from soil. Front Microbiol 7:659. https://doi.org/10.3389/fmicb.2016.00659

Henriques ST, Melo MN, Castanho MARB (2006) Cell penetrating peptides and antimicrobial peptides: how different are they? Biochem J 399(1):1–7

Herrera Diaz A, Kovacs I, Lindermayr C (2016) Inducible expression of the De-novo designed antimicrobial peptide SP1-1 in tomato confers resistance to Xanthomonas campestris pv. Vesicatoria. Plos One 11:e0164097

Holaskova E, Galuszka P, Frebort I, Oz MT (2015) Antimicrobial peptide production and plant-based expression systems for medical and agricultural biotechnology. Biotechnol Adv 33:1005–1023

Holásková E, Galuszka P, Mičúchová A, Šebela M, Öz MT, Frébort IJ (2018) Molecular farming in barley: development of a novel production platform to produce human antimicrobial peptide LL-37. Biotechnol J 13:1700628

Huerta-Cantillo J, Navarro-García F (2016) Properties and design of antimicrobial peptides as potential tools against pathogens and malignant cells. Investigación en Discapacidad 5:96–115

Humphries RM, Pollett S, Sakoulas G (2013) A current perspective on daptomycin for the clinical microbiologist. Clin Microbiol Rev 26:759–780

Hussain A, Qarshi IA, Nazir H, Ullah IJ (2012) Plant tissue culture: current status and opportunities. IntechOpen, Rijeka, pp 1–28

Ilyas H, Datta A, Bhunia AJ (2017) An approach towards structure based antimicrobial peptide design for use in development of transgenic plants: a strategy for plant disease management. Curr Med Chem 24:1350–1364

Imani J, Li L, Schaefer P, Kogel KH (2011) STARTS–A stable root transformation system for rapid functional analyses of proteins of the monocot model plant barley. Plant J 67:726–735

Iyappan G, Omosimua RO, Sathishkumar R (2019) Enhanced production of therapeutic proteins in plants: novel expression strategies. In: Advances in plant transgenics: methods and applications. Springer, Cham, pp 333–351

Jiao J, Gai QY, Yao LP, Niu LL, Zang YP, Fu Y (2018) Ultraviolet radiation for flavonoid augmentation in *Isatis tinctoria* L. hairy root cultures mediated by oxidative stress and biosynthetic gene expression. Ind Crop Prod 118:347–354

Jose PA, Jebakumar SR (2014) Unexplored hypersaline habitats are sources of novel actinomycetes. Front Microbiol 5:242. https://doi.org/10.3389/fmicb.2014.00242

Jung YJ, Kang KK (2014) Application of antimicrobial peptides for disease control in plants. Plant Breed Biotechnol 2:1–13

Kalia NP, Mahajan P, Mehra R, Nargotra A, Sharma JP, Koul S et al (2012) Capsaicin, a novel inhibitor of the NorA efflux pump, reduces the intracellular invasion of *Staphylococcus aureus*. J Antimicrob Chemother 67:2401–2408

Kim J, Cho D, Park M et al (2016) CRISPR/Cas9-mediated re-sensitization of antibiotic-resistant *Escherichia coli* harboring extended-spectrum β-lactamases. J Microbial Biotechnol 26(2):394

Kumar P, Kizhakkedathu JN, Straus SK (2018) Antimicrobial peptides: diversity, mechanism of action and strategies to improve the activity and biocompatibility in vivo. Biomol Ther 8(1):4

Kumar V, Naik B, Gusain O, Bisht GS (2014) An actinomycete isolate from solitary wasp mud nest having strong antibacterial activity and kills the *Candida* cells due to the shrinkage and the cytosolic loss. Front Microbiol 5:446. https://doi.org/10.3389/fmicb.2014.00446

Lai P, Roy J (2004) Antimicrobial and chemo-preventive properties of herbs and spices. Curr Med Chem 11(11):1451–1460

Lai Y, Gallo RL (2009) AMPed up immunity: how antimicrobial peptides have multiple roles in immune defense. Trends Immunol 30:131–141

Lallemand J, Bouché F, Desiron C, Stautemas J, de Lemos EF, Périlleux C, Tocquin PJ (2015) Extracellular peptidase hunting for improvement of protein production in plant cells and roots. Front Plant Sci 2015:6–37

Lau JL, Dunn MK (2018) Therapeutic peptides: historical perspectives, current development trends, and future directions. Bioorg Med Chem 26(10):2700–2707

Lee J, Lee DG (2015) Antimicrobial peptides (AMPs) with dual mechanisms: membrane disruption and apoptosis. J Microbiol Biotechnol 25(6):759–764

Lemaitre B, Nicolas E, Michaut L, Reichhart JM, Hoffmann JA (1996) The dorsoventral regulatory gene cassette spätzle/toll/cactus controls the potent antifungal response in *Drosophila* adults. Cell 86:973–983

Levin AS, Barone AA, Penço J, Santos MV, Marinho IS, Arruda EAG, Manrique EI, Costa SF (1999) Intravenous colistin as therapy for nosocomial infections caused by multidrug-resistant *Pseudomonas aeruginosa* and *Acinetobacter baumannii*. Clin Infect Dis 28(5):1008–1011

Lhocine N, Ribeiro PS, Buchon N (2008) PIMS modulates immune tolerance by negatively regulating *Drosophila* innate immune signaling. Cell Host Microbe 4:147–158

Li F, Shen H, Wang M, Fan K, Bibi N, Ni M, Yuan S, Wang XJ (2016) A synthetic antimicrobial peptide BTD-S expressed in Arabidopsis thaliana confers enhanced resistance to Verticillium dahliae. Front Plant Sci 291:1647–1661

Limaverde PW, Campina FF, da Cunha FA, Crispim FD, Figueredo FG, Lima LF et al (2017) Inhibition of the TetK efflux-pump by the essential oil of *Chenopodium ambrosioides* L. and a-terpinene against *Staphylococcus aureus* IS-58. Food Chem Toxicol 109:957–961

Lipsky A, Joshi JR, Carmi N, Yedidia IJ (2016) Expression levels of antimicrobial peptide tachyplesin I in transgenic *Ornithogalum* lines affect the resistance to *Pectobacterium* infection. J Biotechnol 238:22–29

Liu Q, Yao S, Chen Y (2017) Use of antimicrobial peptides as a feed additive for juvenile goats. Sci Rep 7:12254

Liu YY, Wang Y, Walsh TR, Yi LX, Zhang R, Spencer J, Doi Y, Tian G, Dong B, Huang XH, Yu LF, Gu DX, Ren HW, Chen XJ, Lv LC, He DD, Zhou HW, Liang Z, Liu J-H, Shen JZ (2016) Emergence of plasmid-mediated colistin resistance mechanism MCR-1 in animals and human beings in China: a microbiological and molecular biological study. Lancet Infect Dis 16 (2):161–168

Lohner K (2017) Membrane-active antimicrobial peptides as template structures for novel antibiotic agents. Curr Top Med Chem 17(5):508–519

Lombardi L, Falanga A, Del Genio V, Galdiero S (2019) A new hope: self assembling peptides with antimicrobial activity. Pharmaceutics 11(4):166

Lonnerdal B (2011) Biological effects of novel bovine milk fractions. Nestle nutrition workshop series. Paediatr Program 67:41–54

Lonoce C, Salem R, Marusic C, Jutras PV, Scaloni A, Salzano AM, Lucretti S, Steinkellner H, Benvenuto E, Donini MJ (2016) Production of a tumour-targeting antibody with a human-compatible glycosylation profile in N. benthamiana hairy root cultures. Biotechnol J 11:1209–1220

Lucera A, Costa C, Conte A, Del Nobile MA (2012) Food applications of natural antimicrobial compounds. Front Microbiol 3:287. https://doi.org/10.3389/fmicb.2012.00287

Lupei MI, Mann HJ, Beilman GJ, Oancea C, Chipman JG (2010) Inadequate antibiotic therapy in solid organ transplant recipients is associated with a higher mortality rate. Surg Infect (Larchmt) 11(1):33–39

Ma JK, Drake PM, Christou PJ (2003) The production of recombinant pharmaceutical proteins in plants. Biotechnol J 4:794–805

Macnair CR, Stokes JM, Carfrae LA, Fiebig-Comyn AA, Coombes BK, Mulvey MR, Brown ED (2018) Overcoming mcr-1 mediated colistin resistance with colistin in combination with other antibiotics. Nat Commun 9(1):458

Madeira LM, Szeto TH, Henquet M, Raven N, Runions J, Huddleston J, Garrard I, Drake PM, Ma JK (2016) High-yield production of a human monoclonal IgG by rhizosecretion in hydroponic tobacco cultures. Plant Biotechnol J 14:615–624

Magana M, Pushpanathan M, Santos AL, Leanse L, Fernandez M, Ioannidis A, Giulianotti MA, Apidianakis Y, Bradfute S, Ferguson AL, Cherkasov A, Seleem MN, Pinilla C, de la Fuente-Nunez C, Lazaridis T, Dai T, Houghten RA, Hancock REW, Tegos GP (2020) The value of antimicrobial peptides in the age of resistance. Lancet Infect Dis 20(9):e216–e230

Mahlapuu M, Håkansson J, Ringstad L, Björn C (2016a) Antimicrobial peptides: an emerging category of therapeutic agents. Front Cell Infect Microbiol 6:194. https://doi.org/10.3389/fcimb. 2016.00194

Mahlapuu M, Håkansson J, Ringstad L, Björn C (2016b) Antimicrobial peptides: an emerging category of therapeutic agents. Front Cell Infect Microbiol 6:194

Makhzoum A, Benyammi R, Moustafa K, Trémouillaux-Guiller JJ (2014) Recent advances on host plants and expression cassettes' structure and function in plant molecular pharming. BioDrugs 28:145–159

Mandalari G, Bennett R, Bisignano G, Trombetta D, Saija A, Faulds C et al (2007) Antimicrobial activity of flavonoids extracted from bergamot (Citrus bergamia Risso) peel, a byproduct of the essential oil industry. J Appl Microbiol 103(6):2056–2064

Mbewana S, Mortimer E, Pêra FFPG, Hitzeroth II, Rybicki EP (2015) Production of H5N1 influenza virus matrix protein 2 Ectodomain protein bodies in tobacco plants and in insect cells as a candidate universal influenza vaccine. Front Bioeng Biotechnol 6:5

Meng DM, Zhao JF, Ling X, Dai HX, Guo YJ, Gao XF, Dong B, Zhang ZQ, Meng X, Fan ZC (2017) Recombinant expression, purification and antimicrobial activity of a novel antimicrobial peptide PaDef in Pichia pastoris. Protein Expr Purif 130:90–99

Miller LG, Perdreau-Remington F, Rieg G, Mehdi S, Perlroth J, Bayer AS, Tang AW, Phung TO, Spellberg B (2005) Necrotizing fasciitis caused by community-associated methicillin-resistant Staphylococcus aureus in Los Angeles. N Engl J Med 352(14):1445–1453

Mingeot-Leclercq MP, Décout JL (2016) Bacterial lipid membranes as promising targets to fight antimicrobial resistance, molecular foundations and illustration through the renewal of aminoglycoside antibiotics and emergence of amphiphilic aminoglycosides. Med Chem Commun 7(4):586–611

Moiola M, Memeo MG, Quadrelli PJM (2019) Stapled peptides-a useful improvement for peptide-based drugs. Molecules 24:3654

Mookherjee N, Hancock RE (2007) Cationic host defence peptides: innate immune regulatory peptides as a novel approach for treating infections. Cell Mol Life Sci 64:922–933

Moon KB, Park JS, Park YI, Song IJ, Lee HJ, Cho HS, Jeon JH, Kim HS (2020) Development of systems for the production of plant-derived biopharmaceuticals. Plan Theory 9:30

Moreira W, Aziz DB, Dick T (2016) Boromycin kills mycobacterial persisters without detectable resistance. Front Microbiol 7:199. https://doi.org/10.3389/fmicb.2016.00199

Muller V, Rajer F, Frykholm K et al (2016) Direct identification of antibiotic resistance genes on single plasmid molecules using CRISPR/Cas9 in combination with optical DNA mapping. Sci Rep 6:37938

Mwangi J, Hao X, Lai R, Zhang ZY (2019b) Antimicrobial peptides: new hope in the war against multidrug resistance. Zool Res 40(6):488–505

Mwangi J, Hao X, Lai R, Zhang Z-Y (2019a) Antimicrobial peptides: new hope in the war against multidrug resistance. Zool Res 40(6):488–505

Mylonakis E, Podsiadlowski L, Muhammed M, Vilcinskas A (2016) Diversity, evolution and medical applications of insect antimicrobial peptides. Philos Trans R Soc Lond B Biol Sci 371:371

Nawrot R, Barylski J, Nowicki G, Broniarczyk J, Buchwald W, Goździcka-Józefiak AJFM (2014) Plant antimicrobial peptides. Folia Microbiol 59:181–196

Nguyen LT, Chau JK, Perry NA, de Boer L, Zaat SA, Vogel HJ (2010) Serum stabilities of short tryptophan and arginine-rich antimicrobial peptide analogs. PLoS One 5:e12684

Nguyen LT, Haney EF, Vogel HJ (2011) The expanding scope of antimicrobial peptide structures and their modes of action. Trends Biotechnol 29(9):464–472

Niyonsaba F, Song P, Yue H (2020) Antimicrobial peptide derived from insulin-like growth factor-binding protein 5 activates mast cells via mas-related G protein-coupled receptor X2. Allergy 75:203–207

O'Neil J (2016) Tackling drug-resistant infections globally: final report and recommendations. https://amr-review.org/sites/default/files/160525_Final paper_withcover.pdf Accessed 5 Dec 2016

Obembe OO, Popoola JO, Leelavathi S, Reddy SVJB (2011) Advances in plant molecular farming. Biotechnol Adv 29:210–222

Oppenheim JJ, Biragyn A, Kwak LW, Yang D (2003) Roles of antimicrobial peptides such as defensins in innate and adaptive immunity. Ann Rheum Dis 62(2):17–21

Osaki T, Omotezako M, Nagayama R (1999) Horseshoe crab hemocyte-derived antimicrobial polypeptides, tachystatins, with sequence similarity to spider neurotoxins. J Biol Chem 274 (37):26172–26178

Oztürk M, Duru ME, Kivrak S, Mercan Dogan N, Turkoglu A, Ozler MA (2011) In vitro antioxidant, anticholinesterase and antimicrobial activity studies on three *Agaricus* species with fatty acid compositions and iron contents: a comparative study on the three most edible mushrooms. Food Chem Toxicol 49(6):1353–1360

Papa R, Selan L, Parrilli E, Tilotta M, Sannino F, Feller G, Tutino ML, Artini M (2015) Anti-biofilm activities from marine cold adapted bacteria against *Staphylococci* and *Pseudomonas aeruginosa*. Front Microbiol 6:1333. https://doi.org/10.3389/fmicb.2015.01333

Park JY, Moon BY, Park JW, Thornton JA, Park YH, Seo KS (2017) Genetic engineering of a temperate phage-based delivery system for CRISPR/Cas9 antimicrobials against *Staphylococcus aureus*. Sci Rep 2017:44429. https://doi.org/10.1038/srep44929

Pasupuleti M, Schmidtchen A, Malmsten M (2012) Antimicrobial peptides: key components of the innate immune system. Crit Rev Biotechnol 32:143–171

Patiño-Rodríguez O, Ortega-Berlanga B, Llamas-González YY, Flores-Valdez MA, Herrera-Díaz A, Montes-de-Oca-Luna R, Korban SS, Alpuche-Solís ÁG (2013) Transient expression and characterization of the antimicrobial peptide protegrin-1 in Nicotiana tabacum for control of bacterial and fungal mammalian pathogens. Plant Cell Tissue Organ Cult 115:99–106

Perez RH, Ishibashi N, Inoue T, Himeno K, Masuda Y, Sawa N (2016) Functional analysis of genes involved in the biosynthesis of enterocin NKR-5-3B, a novel circular bacteriocin. J Bacteriol 198(2):291–300

Phelan M, Aherne A, FitzGerald RJ, O'Brien NM (2009) Casein-derived bioactive peptides: biological effects, industrial uses, safety aspects and regulatory status. Int Dairy J 19 (11):643–654

Phoenix DA, Dennison SR, Harris F (2016) Bacterial resistance to host defense peptides. In: Epand RM (ed) Host defense peptides and their potential as therapeutic agents. Springer, Cham, pp 161–204

Poborilova Z, Plchova H, Cerovska N, Gunter CJ, Hitzeroth II, Rybicki EP, Moravec T (2020) Transient protein expression in tobacco BY-2 plant cell packs using single and multi-cassette replicating vectors. Plant Cell Rep 39:1115–1127

Pogue GP, Vojdani F, Palmer KE, Hiatt E, Hume S, Phelps J, Long L, Bohorova N, Kim D, Pauly M et al (2010) Production of pharmaceutical-grade recombinant aprotinin and a monoclonal antibody product using plant-based transient expression systems. Plant Biotechnol J 8:638–654

Powers JH (2014) Antimicrobial drug development-the past, the present, and the future. Clin Microbiol Infect 10(4):23–31

Prestinaci F, Pezzotti P, Pantosti A (2015) Antimicrobial resistance: a global multifaceted phenomenon. Pathog Glob Health 109:309–318

Pushpanathan M, Gunasekaran P, Rajendhran J (2013) Antimicrobial peptides: versatile biological properties. Int J Pept 2013(675391):1–15

Qiu HX, Gong JS, Butaye P et al (2018) CRISPR/Cas9/sgRNA-mediated targeted gene modification confirms the cause-effect relationship between *gyrA* mutation and quinolone resistance in *Escherichia coli*. FEMS Microbiol Lett 365:13

Rademacher T, Sack M, Blessing D, Fischer R, Holland T, Buyel J (2019) Plant cell packs: a scalable platform for recombinant protein production and metabolic engineering. Pant Biotechnol J 17:1560–1566

Rage E, Marusic C, Lico C, Baschieri S, Donini M (2020) Current state-of-the-art in the use of plants for the production of recombinant vaccines against infectious bursal disease virus. Appl Microbiol Biotechnol 104:2287–2296

Raucher D, Ryu JS (2015) Cell-penetrating peptides: strategies for anticancer treatment. Trends Mol Med 21(9):560–570

Reddy MK, Gupta SK, Jacob MR, Khan SI, Ferreira D (2007) Antioxidant, antimalarial and antimicrobial activities of tannin-rich fractions, ellagitannins and phenolic acids from *Punica granatum* L. Planta Med 73(05):461–467

Reichert JM (2003) Trends in development and approval times for new therapeutics in the United States. Nat Rev Drug Discov 2:695–702

Ribeiro TP, Arraes FBM, Lourenço-Tessutti IT, Silva MS, Lisei-de-Sá ME, Lucena WA, Macedo LLP, Lima JN, Santos Amorim RM, Artico SJP (2017) Transgenic cotton expressing Cry10Aa toxin confers high resistance to the cotton boll weevil. Pant Biotechnol J 15:997–1009

Riool M, de Breij A, Drijfhout JW, Nibbering PH, Zaat SAJ (2017) Antimicrobial peptides in biomedical device manufacturing. Front Chem 5:63. https://doi.org/10.3389/fchem.2017.00063

Rončević T, Puizina J, Tossi A (2019) Antimicrobial peptides as anti-infective agents in pre-post-antibiotic era? Int J Mol Sci 20(22):5713. https://doi.org/10.3390/ijms20225713

Roy K, Pandit G, Chetia M, Sarkar AK, Chowdhuri S, Bidkar AP, Chatterjee SJAABM (2020) Peptide hydrogels as platforms for sustained release of antimicrobial and antitumor drugs and proteins. ACS Appl Bio Mater 3:6251–6262

Sagdic O, Ozturk I, Yilmaz MT, Yetim H (2011) Effect of grape pomace extracts obtained from different grape varieties on microbial quality of beef patty. J Food Sci 76(7):M515–M521

Sandreschi S, Piras AM, Batoni G, Chiellini F (2016) Perspectives on polymeric nanostructures for the therapeutic application of antimicrobial peptides. Nanomedicine (Lond) 11(13):1729–1744

Sani MA, Separovic F (2016) How membrane-active peptides get into lipid membranes. Acc Chem Res 49(6):1130–1138

Sathoff AE, Velivelli S, Shah DM, Samac DA (2019) Plant defensin peptides have antifungal and antibacterial activity against human and plant pathogens. Phytopathology 109:402–408

Saurav K, Bar-Shalom R, Haber M, Burgsdorf I, Oliviero G, Costantino V, Morgenstern D, Steindler L (2016) In search of alternative antibiotic drugs: quorum-quenching activity in sponges and their bacterial isolates. Front Microbiol 7:416. https://doi.org/10.3389/fmicb.2016.00416

Schwentker FF, Gelman S, Long PH (1937) The treatment of meningococcic meningitis. JAMA 108:1407–1408

Shabir U, Ali S, Magray AR, Ganai BA, Firdous P, Hassan T, Nazir R (2018) Fish antimicrobial peptides (AMP's) as essential and promising molecular therapeutic agents: a review. Microb Pathog 114:50–56

Shafee TM, Lay FT, Phan TK, Anderson MA, Hulett MD (2017) Convergent evolution of defensin sequence, structure and function. Cell Mol Life Sci 74(4):663–682

Shen J, Zhou J, Chen G-Q, Xiu Z-L (2018) Efficient genome engineering of a virulent *Klebsiella* bacteriophage using CRISPR-Cas9. J. Virology 92(17):e00534

Sierra JM, Fusté E, Rabanal F, Vinuesa T, Viñas M (2017) An overview of antimicrobial peptides and the latest advances in their development. Expert Opin Biol Ther 17(6):663–676

Silva APSA, Silva LCN, Fonseca CSM, Araújo JM, Santos Correia MT, Silva Cavalcanti M (2016) Antimicrobial activity and phytochemical analysis of organic extracts from *Cleome spinosa* Jaqc. Front Microbiol 7:963. https://doi.org/10.3389/fmicb.2016.00963

Sinha R, PJP S (2019) Antimicrobial peptides: Recent insights on biotechnological interventions and future perspectives. Protein Pept Lett 26:79–87

Springer B, Kidan YG, Prammananan T (2001) Mechanisms of streptomycin resistance: selection of mutations in the 16S rRNA gene conferring resistance. Antimicrob Agents Chemother 45:2877–2884

Starr CG, Wimley WC (2017) Antimicrobial peptides are degraded by the cytosolic proteases of human erythrocytes. Biochim Biophys Acta Biomembr 1859:2319–2326

Streatfield SJ (2007) Approaches to achieve high-level heterologous protein production in plants. Plant Biotechnol J 5:2–15

Sun D, Forsman J, Woodward CE (2017) Molecular simulations of melittin-induced membrane pores. J Phys Chem B 121(44):10209–10214

Takahashi D, Shukla SK, Prakash O, Zhang G (2010) Structural determinants of host defense peptides for antimicrobial activity and target cell selectivity. Biochimie 92:1236–1241

Tasiemski A, Salzet M, Gaill FUS (2014) Antimicrobial peptides. Patent No 8,652,514

Tassanakajon A, Amparyup P, Somboonwiwat K, Supungul P (2010) Cationic antimicrobial peptides in penaeid shrimp. Marine Biotechnol 12(5):487–505

Temple RJ (1995) Development of drug law, regulations, and guidance in the United States. In: Munson PL, Mueller RA, Breese G (eds) Principles of pharmacology basic concepts and clinical applications. Chapman & Hall, Boca Raton, pp 1643–1663

Tiwari BK, Valdramidis VP, O'Donnell CP, Muthukumarappan K, Bourke P, Cullen P (2009) Application of natural antimicrobials for food preservation. J Agric Food Chem 57 (14):5987–6000

Tiwari LD, Khungar L, Grover AJTPJ (2020) AtHsc70-1 negatively regulates the basal heat tolerance in *Arabidopsis thaliana* through affecting the activity of HsfAs and Hsp101. Plant J 103:2069–2083

Utsugi T, Schroit AJ, Connor J, Bucana CD, Fidler IJ (1991) Elevated expression of phosphatidylserine in the outer membrane leaflet of human tumor cells and recognition by activated human blood monocytes. Cancer Res 51(11):3062–3066

Van Epps HL (2006a) René Dubos: unearthing antibiotics. J Exp Med 203:259

Van Epps HL (2006b) René Dubos: unearthing antibiotics. J Exp Med 203:259

Varanko A, Saha S, Chilkoti A (2020) Recent trends in protein and peptide-based biomaterials for advanced drug delivery. Adv Drug Deliv Rev 156:133–187

Varasteh-Shams M, Nazarian-Firouzabadi F, Ismaili AJ (2020) The direct and indirect transformation methods on expressing a recombinant Dermaseptin peptide in tobacco transgenic hairy root clones. Curr Plant Biol 24:100177

Vriens K, Cammue B, KJM T (2014) Antifungal plant defensins: mechanisms of action and production. Molecules 19:12280–12303

Wan P, Cui S, Ma Z, Chen L, Li X, Zhao R, Xiong W, Zeng Z (2020) Reversal of *mcr-1*-mediated Colistin resistance in *Escherichia coli* by CRISPR-Cas9 system. Infect Drug Resist 13:1171–1178

Wang G (2017) Antimicrobial peptides: discovery, design and novel therapeutic strategies. CAB International, London

Wang G, Li X, Wang Z (2016) APD3: the antimicrobial peptide database as a tool for research and education. Nucleic Acids Res 44:D1087–D1093

WEF (World Economic Forum) (2013) Global risks 2013 eighth edition: an initiative of the risk response network. World Economic Forum, Geneva

Weinhold A, Dorcheh EK, Li R, Rameshkumar N, Baldwin ITJE (2018) Antimicrobial peptide expression in a wild tobacco plant reveals the limits of host-microbe-manipulations in the field. Elife 7:e28715

WHO (2018) Global priority list of antibiotic-resistant bacteria to guide research, discovery, and development of new antibiotics. http://www.who.int/medicines/publications/global-priority-listantibiotic-resistant-bacteria/en/

WHO (2019) Executive summary: the selection and use of essential medicines 2019. In: Proceedings of the report of the 22nd WHO expert committee on the selection and use of essential medicines: WHO, headquarters, Geneva, Switzerland, 1–5 April 2019

Wonghirundecha S, Sumpavapol P (2012) Antibacterial activity of selected plant by-products against food-borne pathogenic bacteria. Int Conf Nutr Food Sci 39:116–120

Wu D, Ding W, Zhang Y, Liu X, Yang L (2015) Oleanolic acid induces the type III secretion system of *Ralstonia solanacearum*. Front Microbiol 6:1466. https://doi.org/10.3389/fmicb.2015.01466

Xhindoli D, Pacor S, Benincasa M, Scocchi M, Gennaro R, Tossi A (2016) The human cathelicidin LL-37--a pore-forming antibacterial peptide and host-cell modulator. Biochim Biophys Acta 1858(3):546–566

Xue J, Davidson PM, Zhong Q (2013) Thymol nano-emulsified by whey protein-maltodextrin conjugates: the enhanced emulsifying capacity and anti-listerial properties in milk by propylene glycol. J Agric Food Chem 61:12720–12726

Yount NY, Yeaman MR (2004) Multidimensional signatures in antimicrobial peptides. Proc Natl Acad Sci U S A 101(19):7363–7368

Yu Z, Tang J, Khare T, Kumar V (2020) The alarming antimicrobial resistance in ESKAPEE pathogens: can essential oils come to the rescue? Fitoterapia 140:104433. https://doi.org/10.1016/j.fitote.2019.104433

Zakharchenko N, Buryanov YI, Lebedeva A, Pigoleva S, Vetoshkina D, Loktyushov E, Chepurnova M, Kreslavski V, Kosobryukhov AJ (2013) Physiological features of rapeseed plants expressing the gene for an antimicrobial peptide cecropin P1. Russian J Plant Physiol 60:411–419

Zaman SB, Hussain MA, Nye R, Mehta V, Mamun KT, Hossain N (2017) A review on antibiotic resistance: alarm bells are ringing. Cureus 9(6):e1403

Żelechowska P, Agier J, Brzezińska-Błaszczyk E (2016) Endogenous antimicrobial factors in the treatment of infectious diseases. Cent Eur J Immunol 41(4):419–425

Zhang L, Rybczynski J, Langenberg W, Mitra A, French RJ (2000) An efficient wheat transformation procedure: transformed calli with long-term morphogenic potential for plant regeneration. Plant Cell Rep 19:241–250

Zhang LJ, Gallo RL (2016) Antimicrobial peptides. Curr Biol 26:R1–R21

Zhang M, Wei W, Sun Y, Jiang X, Ying X, Tao R, Ni L (2016) Pleurocidin congeners demonstrate activity against *Streptococcus* and low toxicity on gingival fibroblasts. Arch Oral Biol 70:79–87

Essential Oils for Combating Antimicrobial Resistance: Mechanism Insights and Clinical Uses

12

Nasreddine El Omari, Saoulajan Charfi, Naoual Elmenyiy, Naoufal El Hachlafi, Abdelaali Balahbib, Imane Chamkhi, and Abdelhakim Bouyahya

Contents

N. El Omari
Laboratory of Histology, Embryology and Cytogenetic, Faculty of Medicine and Pharmacy, Mohammed V University in Rabat, Rabat, Morocco

S. Charfi
Laboratory of Biotechnology and Applied Microbiology, Team Biotechnology and Applied Microbiology, Department of Biology, Faculty of Sciences, Abdelmalek Essaâdi University, Tetouan, Morocco

N. Elmenyiy
Laboratory of Physiology, Pharmacology & Environmental Health, Faculty of Science, University Sidi Mohamed Ben Abdellah, Fez, Morocco

N. El Hachlafi
Microbial Biotechnology and Bioactive Molecules Laboratory, Sciences and Technologies Faculty, Sidi Mohmed Ben Abdellah University, Fez, Morocco

A. Balahbib
Laboratory of Biodiversity, Ecology, and Genome, Faculty of Sciences, Mohammed V University in Rabat, Rabat, Morocco

I. Chamkhi
Laboratory of Plant-Microbe Interactions, AgroBioSciences, Mohammed VI Polytechnic University, Ben Guerir, Morocco

A. Bouyahya (✉)
Laboratory of Human Pathologies Biology, Department of Biology, Faculty of Sciences, and Genomic Center of Human Pathologies, Faculty of Medicine and Pharmacy, Mohammed V University in Rabat, Rabat, Morocco

© The Author(s), under exclusive license to Springer Nature Singapore Pte Ltd. 2022
V. Kumar et al. (eds.), *Antimicrobial Resistance*,
https://doi.org/10.1007/978-981-16-3120-7_12

Abstract

With the emergence and the reemergence of bacterial resistance to different commercialized antibiotics, the identification of novel antibacterial drugs such as natural bioactive compounds can constitute a promising alternative to fight against infectious diseases. Essential oils (EO) or volatile compounds extracted from aromatic plants have been reported as remarkable antibacterial bioactive compounds with several mechanisms of action. Indeed, they affect membrane integrity and permeability, interfere with proteins, disturb cell membrane and respiratory processes, inhibit biofilm formation, and dysregulate quorum sensing communication. In their chapter, we report the antibacterial effect of EOs against multidrug strains and their different mechanism insights, as well as their potential use in clinical trials as alternative to commercialized antibiotics.

Keywords

Bacterial resistance · Essential oil · Antibacterial action · Clinical application

12.1 Introduction

Infectious diseases constitute a major cause of human death worldwide, representing around 50% of all deaths in tropical countries and more than 20% of deaths in the Americas (Panackal 2011). Indeed, the emergence of multidrug resistance (MDR) has led to an increase in the severity of diseases caused by pathogenic strains. According to the statistics of the World Health Organization (WHO), multidrug-resistant bacteria is responsible for almost 700,000 deaths each year (World Health Organization 2018). In fact, there is a need of new and effective alternative strategies to act facing this problem of resistance to antibiotics. Thanks to their valuable therapeutic compounds, essential oils represent an alternative solution fight against the prevalence of microbial resistance. Essential oils also called "volatile compounds" are complex mixture of natural compounds belonging to several chemical classes and produced by different plant parts of aromatic plants as secondary metabolites (Dhifi et al. 2016).

Essential oils have reported to exhibit a wide range of biological activities such as antibacterial, antifungal, antioxidant, anticancer, anti-inflammatory, insecticidal, antidiabetic, antileishmanial, hepatoprotection, and anti-acetylcholinesterase activities (Caputo et al. 2017; Poma et al. 2019; Bouyahya et al. 2017a, 2021). In recent years, several investigations have described the mechanism of antibacterial action of essential oils, underlying their effectiveness against multidrug-resistant strains (Bouyahya et al. 2017a; Gadisa et al. 2019). Indeed, essential oils and their major components can act against bacterial strains through various mechanisms of action including inhibition of bacterial quorum sensing targeting several pathways (N-acyl homoserine lactone production (AHL), AHL receptors, and biofilm formation and dissemination) (Bouyahya et al. 2017b), the lysis of bacterial cell, the leakage of cytoplasmic contents such as proteins and nucleic acids, and the disruption of cell membrane integrity (Dhifi et al. 2016; Ait-Sidi-Brahim et al. 2019; Bouyahya et al. 2019a). Moreover, essential oils rich in phenolic compounds such as carvacrol, thymol, and eugenol have shown an interesting antibacterial effect. Hence, these components are responsible for the inhibition of bacterial protein synthesis, the perturbation of cell membrane, and the proton driving force, and they affect the electron flow and active transport (Pauli 2001; Churklam et al. 2020). Overall, these studies provide clear evidence that essential oils and their components can be used in clinical applications to prevent and treat infectious diseases.

Most importantly, numerous in vitro studies have found that the combination of conventional antimicrobial agents and essential oils inhibits the growth of multidrug-resistant bacteria such as methicillin-resistant *Staphylococcus aureus* (MRSA), vancomycin-resistant *enterococci* (VRE), *Acinetobacter baumannii*, and *Enterococcus faecalis* (Rosato et al. 2007; Mahboubi and Ghazian Bidgoli 2010; Guerra et al. 2012; Yap et al. 2014). These findings lead to novel therapeutic strategies based on the use of essential oils with standard drugs to overcome the emergence of microbial resistance.

12.2 Antibacterial Effects of Essential Oils

Several studies showed the antibacterial activities of different essential oils from different parts of plants (Benali et al. 2020a, b; Bouyahya et al. 2017b, c, d, e, 2020b; Akarca 2019; Hu et al. 2019; Imane et al. 2020; Jugreet and Mahomoodally 2020; Larrazabal-Fuentes et al. 2019; Mohamed et al. 2019; Moussii et al. 2020; Risaliti et al. 2019; Sameh et al. 2019; Vasilijević et al. 2019; Walia et al. 2020). The table outlines the representative studies that evaluated the antibacterial activity of numerous plants, including essential oils origin, the used part, type of antibacterial assay, tested strains, and main results. A literature survey indicated that several researchers investigated the antibacterial potential of essential oils against the most critical pathogenic agents belonging to Gram-positive and Gram-negative bacteria. Benali et al. (2020a) have tested in vitro the antibacterial effect of *Cistus ladanifer* essential oils (CLEO) and *Mentha suaveolens* essential oils (MSEO) against *Pseudomonas savastanoi pv. savastanoi* 2636-40 and *Clavibacter michiganensis* subsp. *michiganensis* 1616-3 using agar disc diffusion and micro-dilution assay. The results showed that *C. michiganensis* (a Gram+) is more sensitive to plant EOs tested ($\Phi = 41.33 \pm 0.57$ mm, MIC = 0.78 mg/mL for CLEO, and $\Phi = 43 \pm 1$ mm, MIC = 0.78 mg/mL for MSEO) compared with the *P. savastanoi* (Gram–) ($\Phi = 19.66 \pm 0.57$ mm, MIC = 0.19 mg/mL for CLEO, and $\Phi = 24.66 \pm 1.52$ mm,

MIC = 0.78 mg/mL for MSEO). In another study, the authors reported the antibacterial activity of *Achillea odorata subsp. pectinata* (AOpEO) and *Ruta montana* (RMEO) essential oils using the same methods against Gram-positive bacteria (*Staphylococcus aureus* CECT 976, *Bacillus subtilis* DSM 6633, and *Listeria innocua* CECT 4030), Gram-negative bacteria (*Escherichia coli* K12, *Pseudomonas aeruginosa* CECT 118, and *Proteus mirabilis*)). Results showed that *Bacillus subtilis* DSM 6633 was the most sensitive strain to the AOpEO and RMEO, with an inhibition zone of 31 ± 1 mm and 21.33 ± 1.52 mm, respectively, and MIC values of 0.19 mg/mL and 0.39 mg/mL, respectively. Moreover, the MBC/MIC values mean that AOpEO and RMEO exhibit bacteriostatic effects against this bacterial strain. Benali et al. (2020b) and Bouyahya et al. (2017e) have investigated the antibacterial activity of *Lavandula stoechas* L. (*L. stoechas*) essential oil against pathogenic strains (*Escherichia coli* K12 and *Staphylococcus aureus* MBLA, *Staphylococcus aureus* CECT 976, *Staphylococcus aureus* CECT 994, *Listeria monocytogenes serovar* 4b CECT 4032, *Proteus mirabilis, Pseudomonas aeruginosa* IH, and *Bacillus subtilis* 6633) using the diffusion method, the minimum inhibitory concentration (MIC), and the minimum bactericidal concentration (MBC) by microtitration assay. This study revealed that *Listeria monocytogenes* was the most sensitive strain with inhibition zones of 23 ± 0.85 mm and MIC value equal to the MBC values of 0.25% (v/v) which indicated a bactericidal action. The same bacterial strains and the same methods are used in another work to study the antibacterial activity of *Mentha pulegium L* essential oil *(*MPEO) and *Rosmarinus officinalis* L. essential oil (ROEO). In this study, the MPEO was most active against *S. aureus* MBLA and presented bactericidal effect (MIC = MBC = 0.25% (v/v)), while ROEO has exhibited a bactericidal effect against *Listeria monocytogenes* (MIC = MBC = 0.5% (v/v)) (Bouyahya et al. 2017e). In another study, Bouyahya et al. (2017e) reported the antibacterial activity of *Origanum compactum* essential oils at three developmental stages. The essential oil of flowering stage showed important antibacterial activities against *Proteus mirabilis* and *S. aureus* with an inhibition diameter of 43 ± 2.12 and 35 ± 0.94 mm, respectively, and presented bactericidal effects (MIC = MBC = 0.125% (v/v) and MIC = MBC = 0.0312% (v/v), respectively). Moreover, *Mentha viridis* essential oil demonstrated an important activity against *Staphylococcus aureus* with an inhibition diameter of 32.00 ± 2.65 mm and MIC value of 0.25% (v/v). Bouyahya et al. (2019a, b) have assessed the antibacterial activity of *Centaurium erythraea* essential oil (CEEO) at three developmental stages by determining the diameters of inhibition, MIC, and MBC. The essential oils showed important antibacterial inhibition at three seasonal stages and particularly against *Staphylococcus aureus* (28–34 mm) and *Listeria monocytogenes* (26–31 mm). Moreover, CEEO at post-flowering stage was the most active against the strains tested and presented bactericidal effects especially against *S. aureus* (MIC = MBC = 0.125% (v/v)). Recently, Bouyahya et al. (2020b) have investigated the antibacterial activity of *Ajuga iva* essential oil (AIEO) at three phenological stages (post-flowering, flowering, and vegetative stages) and showed that the highest inhibitor zone was recorded against *S. aureus* by AIEO at flowering and vegetative stages (31.83 ± 0.55 and 30.33 ± 1.11 mm, respectively), while

AIEO post-flowering stage showed important bacterial inhibition effects with concentrations reaching 0.125% (v/v) against the same bacterial strain.

Akarca (2019) studied the antibacterial activity of *Hibiscus surattensis L.* essential oil (HSEO) using Agar disc diffusion and micro-dilution methods against six bacterial strains (*(ATCC) 25922, Listeria monocytogenes ATCC 51774, Staphylococcus aureus ATCC 6538, Enterococcus aerogenes ATCC 13048, Salmonella typhimurium ATCC 14028, and Shigella flexneri ATCC 12022*). Results showed that HSEO exhibits the highest antibacterial effect on *Listeria monocytogenes* with a diameter zone of 25.26 mm and with MIC and MBC values of 0.156 ± 0.05 and 0.083 ± 0.04 mg/L, respectively. Hu et al. (2019) used time-kill curve assay to evaluate the antibacterial activity of *Litsea cubeba* essential oil (LCEO) against methicillin-resistant *Staphylococcus aureus (MRSA)*. This study showed that LCEO proved a high inhibitory activity against MRSA (MIC = 0.5 mg/mL and MBC = 1.0 mg /mL). This EO is able to decline 99.99% in the population of MRSA at 0.25 mg/mL and 0.5 mg/mL. Imane et al. (2020) tested the antibacterial activity of six selected EOs (*R. officinalis, Z. officinale, M. alternifolia, C. winteriarus, S. sclarea, and S. aromaticum*) against six bacterial *Staphylococcus aureus* MRSA NCTC 12493, *Escherichia coli* ESBL, *Enterococcus faecalis* ERV, *Klebsiella pneumonia* CRK ATCC 700603, *Staphylococcus aureus*, and *Escherichia coli* using the disc diffusion method and micro-dilution methods. According this study, all extracts exhibited promising antibacterial potency with varying degrees of inhibition zone diameter and MIC values. The most sensitive bacteria to *Rosmarinus officinalis* L. and *Zingiber officinale* Roscoe was *Staphylococcus aureus* MRSA NCTC 12493 with an inhibition diameter of 22.0 ± 2.8 and 11.5 ± 0.7 mm, respectively, and MIC values of 0.67 and 0.30 mg/L, respectively. *Klebsiella pneumoniae* ATCC 700603 was most sensitive bacteria to *Melaleuca alternifolia* Cheel (Φ = 37.2 ± 0.3 mm, MIC = 8.84 mg/L), *Enterococcus faecalis* ERV was most sensitive to *Cymbopogon winterianus* (Φ = 18.5 ± 2.8 mm, MIC = 8.07 mg/L), *Escherichia coli* ESBL was most sensitive to *Salvia sclarea* L. (Φ = 31.2 ± 1.7 mm, MIC = 11.05 mg/L), and *Escherichia coli* ATCC 25922 was most sensitive to *Syzygium aromaticum* (Φ = 19.9 ± 0.4 mm, MIC = 0.21 mg/L, MBC = 0.21 mg/L).

Jugreet and Mahomoodally (2020) determined the in vitro antibacterial activity of ten EOs prepared from two endemic (*Pittosporum senacia putterl. Subsp. Senacia and Syzygium coriaceum J. Bosser and J. Gu_eho*) and seven exotic (*Cinnamomum camphora (L.) Nees and Eberm, Citrus aurantium L., Curcuma longa L, Morinda citrifolia L., Petroselinum crispum (Mill.) Fuss, Plectranthusamboinicus (Lour.) Sprengel, and Syzygium samarangense (Blume) Merr. and L. M. Perry*) aromatic medicinal plants against eight bacterial strains (*Escherichia coli (ATCC 29194), Staphylococcus epidermidis (ATCC 12228), Staphylococcus aureus (ATCC 25923), methicillin-resistant Staphylococcus aureus (MRSA), Pseudomonas aeruginosa (ATCC 27853), Bacillus spizizenii (ATCC 6633), Enterococcus faecalis, and Klebsiella pneumonia*). In this study, the micro-dilution method demonstrated that the EOs were found to possess varying degree of antibacterial potency. The most active EOs were found to have minimum inhibitory concentration (MIC) of 0.25 ± 4 mg/ mL and minimum bactericidal concentration (MBC) of 0.25 ± 16 mg/mL. In particular, at its MIC values, *P. amboinicus* EO showed bactericidal effects against four strains. However, all tested bacteria were insensitive to *P. senecia* and *C. aurantium* fruit peel EOs. *Bacillus spizizenii* was found to be the most susceptible strain to the active EOs.

Other studies of in vitro antibacterial activities involve inhibition zone diameter (expressed by mm) and MICs against 32 bacterial strains for two oil samples of *Acantholippia deserticola* (RREO) and *Artemisia copa* (CCEO). Results showed that CCEO exerted a significant inhibitory effect (> 20 mm) against nine bacteria, including *Listeria monocytogenes* and *Staphylococcus aureus*, and RREO significantly inhibited *Escherichia coli*. The CCEO MIC values were 0.39µg/mL for *Corynebacterium*. The RREO MIC value was 6.25µg/mL for *S. viridians* (Larrazabal-Fuentes et al. 2019). In addition, Moussii et al. (2020) have investigated the antibacterial effects of three essential oils (*Artemisia herba alba, Lavandula angustifolia,* and *Rosmarinus officinalis)* against three bacterial strains (*Staphylococcus aureus, Escherichia coli,* and *Pseudomonas aeruginosa*). This study showed that *Lavandula angustifolia, Rosmarinus officinalis,* and *Artemisia herba-alba* essential oils inhibited the bacterial growth at 1.33, 3.33, and 42.67µl/mL; 1.33 1.67, and 42.67µL/mL; and 4.00, 6.67, and 42.67µL/mL for *S. aureus, E. coli,* and *P. aeruginosa*, respectively.

Other studies of in vitro antibacterial activities involve inhibition zone diameter and MICs against five bacterial isolates (*Staphylococcus aureus (RCMB010010), methicillin-resistant Staphylococcus aureus* clinical isolate 'MRSA' *(RCMB 2658), Escherichia coli (RCMB 010052, ATCC 25955), Salmonella typhimurium (RCMB006 (1) ATCC 14028),* and *Helicobacter pylori (RCMB 031124, ATCC 43504))* for essential oils of fresh plant organs (leaves, flowers, and fruits) of *Spondias pinnata*. This study concluded that that essential oils of leaf, flower, and fruit exerted antimicrobial effect against most tested bacteria. *MRSA* was the most susceptible Gram-positive bacteria to essential oil of leaf with MIC of 156.25µg/mL and inhibition zone diameter of 16.07 mm. The same essential oil exerted the highest activity against the Gram-negative bacteria *S. typhimurium (*$\Phi = 9.97 \pm 0.35$ mm and MIC $= 125.0$µg/mL) (Sameh et al. 2019).

The mechanisms of action of EOs have been studied. The antibacterial activity of EOs of plants is linked to the presence of secondary metabolites, which were able to inhibit the growth of bacteria strains by increasing the bacterial membrane permeability, which causes the leakage of vital cell contents (Bouyahya et al. 2019b). Moreover, other mechanisms of action have been demonstrated such as enzymes inhibitory, DNA fragmentation, and quorum sensing inhibition (Bouyahya et al. 2017b).

Table 12.1 Antibacterial effects of essential oils

Essential oils (origin)	Methods	Bacterial strains	Key findings	References
Cistus ladaniferus subsp. *ladanifer* (CL) and *Mentha suaveolens* (MS)	Agar disc diffusion and micro-dilution methods	*Pseudomonas savastanoi pv. savastanoi* 2636–40	For CL: $\Phi = 19.66 \pm 0.57$ mm MIC $= 0.19$ mg/mL For MS: $\Phi = 24.66 \pm 1.52$ mm MIC $= 0.78$ mg/mL	Benali et al. (2020a)
Cistus ladaniferus subsp. *ladanifer* (CL) and *Mentha suaveolens* (MS)	Agar disc diffusion and micro-dilution methods	*Clavibacter michiganensis* subsp. *michiganensis* 1616–3	For CL: $\Phi = 41.33 \pm 0.57$ mm MIC $= 0.78$ mg/mL For MS: $\Phi = 43 \pm 1$ mm MIC $= 0.78$ mg/mL	Benali et al. (2020a)
Achillea odorata subsp. *pectinata* (AO$_p$EO) and *Ruta montana* (RMEO)	Agar disc diffusion and micro-dilution methods	*Bacillus subtilis DSM 6633*	For AO$_p$EO: $\Phi = 31 \pm 1$ mm MIC $= 0.19$ mg/mL MBC $= 3.12$ mg/mL For RMEO: $\Phi = 21.33 \pm 1.52$ mm MIC $= 0.39$ mg/mL MBC $= 6.25$ mg/mL	Benali et al. (2020b)
Lavandula stoechas L.	Agar-well diffusion and micro-dilution methods	*Listeria monocytogenes*	MIC = MBC $= 0.25\%$ v/v (bactericidal action) $\Phi = 23 \pm 0.85$ mm	Bouyahya et al. (2017c)
Mentha pulegium L.	Agar-well diffusion and micro-dilution methods	*Bacillus subtilis 6633*	$\Phi = 30 \pm 1.43$ mm MIC and MBC are not determined	Bouyahya et al. (2017d)

(continued)

Table 12.1 (continued)

Essential oils (origin)	Methods	Bacterial strains	Key findings	References
Mentha pulegium L.	Agar-well diffusion and micro-dilution methods	*Staphylococcus aureus* MBLA	MIC = MBC = 0.25% v/v	Bouyahya et al. (2017d)
Rosmarinus officinalis L.	Agar-well diffusion and micro-dilution methods	*Listeria monocytogenes*	MIC = MBC = 0.5% v/v	Bouyahya et al. (2017d)
Origanum compactum Benth	Agar-well diffusion and micro-dilution methods	*Proteus mirabilis*	$\Phi = 47 \pm 2.73$ mm (flowering stage) MIC = MBC = 0.125% (v/v) (flowering stage)	Bouyahya et al. (2017e)
Origanum compactum Benth	Agar-well diffusion and micro-dilution methods	*Staphylococcus aureus* MBLA	$\Phi = 35 \pm 0.94$ mm (flowering stage) MIC = MBC = 0.0312% (v/v) (flowering stage)	Bouyahya et al. (2017e)
Mentha viridis	Agar-well diffusion and micro-dilution methods	*Staphylococcus aureus* CECT 994	$\Phi = 32.00 \pm 2.65$ mm MIC = MBC = 0.25% (v/v)	Bouyahya et al. (2019b)
Centaurium erythraea	Agar-well diffusion and micro-dilution methods	*Staphylococcus aureus* CECT 994	$\Phi = 34 \pm 0.5$ mm (vegetative stage) MIC = MBC = 0.125% (v/v) (post-flowering stage)	Bouyahya et al. (2019b)
Ajuga Iva	Agar-well diffusion and micro-dilution methods	*Staphylococcus aureus* CECT 994	$\Phi = 31.83 \pm 0.55$ (flowering stage) MIC = MBC = 0.125% (v/v) (post-flowering stage)	Bouyahya et al. (2020b)

			References
Hibiscus surattensis L.	Agar disc diffusion and micro-dilution methods	Listeria monocytogenes	$\Phi = 25.26$ mm MIC $= 0.156 \pm 0.05$ mg/L MBC $= 0.083 \pm 0.04$ mg/L	Akarca (2019)
Litsea cubeba	Time-kill curve assay	Methicillin-resistant Staphylococcus aureus(MRSA)	MIC $= 0.5$ mg/L MBC $= 1.0$ mg/L 99.99% decline in the population of MRSA after 2 h of the 0.25 mg/mL treatment	Hu et al. (2019)
Rosmarinus officinalis L.	Agar disc diffusion and micro-dilution methods	Staphylococcus aureus MRSA NCTC12493	$\Phi = 22.0 \pm 2.8$ mm MIC $= 0.67$ mg/L MBC $= 2.70$ mg/L	Imane et al. (2020)
Zingiber officinale roscoe	Agar disc diffusion and micro-dilution methods	Staphylococcus aureus MRSA NCTC12493	$\Phi = 11.5 \pm 0.7$ mm MIC $= 0.30$ mg/L MBC $= 1.23$ mg/L	Imane et al. (2020)
Melaleuca alternifolia Cheel	Agar disc diffusion and micro-dilution methods	Klebsiella pneumoniae ATCC 700603	$\Phi = 37.2 \pm 0.3$ mm MIC $= 8.84$ mg/L	Imane et al. (2020)
Cymbopogon winterianus	Agar disc diffusion and micro-dilution methods	Enterococcus faecalis ERV	$\Phi = 18.5 \pm 2.8$ mm MIC $= 8.07$ mg/L	Imane et al. (2020)
Salvia sclarea L.	Agar disc diffusion and micro-dilution methods	Escherichia coli ESBL	$\Phi = 31.2 \pm 1.7$ mm MIC $= 11.05$ mg/L	Imane et al. (2020)

(continued)

Table 12.1 (continued)

Essential oils (origin)	Methods	Bacterial strains	Key findings	References
Syzygium aromaticum	Agar disc diffusion and micro-dilution methods	*Escherichia coli* ATCC 25922	Φ = 19.9 ± 0.4 mm MIC = 0.21 mg/L MBC = 0.21 mg/L	Imane et al. (2020)
Citrus aurantium (leaf)	Micro-dilution method	*Pseudomonas aeruginosa*	MIC/MBC: 8 mg/mL	Jugreet and Mahomoodally (2020)
Curcuma longa	Micro-dilution method	*Bacillus spizizenii* *Staphylococcus aureus*	MIC/MBC: 4 mg/mL	Jugreet and Mahomoodally (2020)
Morinda citrifolia	Micro-dilution method	*Bacillus spizizenii*, methicillin-resistant *Staphylococcus aureus* (MRSA), and *Staphylococcus aureus*	MIC/MBC: 4 mg/mL	Jugreet and Mahomoodally (2020)
Plectranthus amboinicus	Micro-dilution method	*Staphylococcus epidermidis*	MIC/MBC: 1 mg/mL	Jugreet and Mahomoodally (2020)
Petroselinum crispum	Micro-dilution method	*Staphylococcus epidermidis*	MIC/MBC: 4 mg/mL	Jugreet and Mahomoodally (2020)
Artemisia copa Phil. (Copa copa)	Agar disc diffusion and micro-dilution methods	*Corynebacterium*	Φ = 30 mm MIC = 0.39μg/mL	Larrazabal-Fuentes et al. (2019)
Acantholippia deserticola (Phil.) Moldenke (Rica rica)	Agar disc diffusion and micro-dilution methods	*Streptococcus viridians*	Φ = 25 mm MIC = 6.25μg/mL	Larrazabal-Fuentes et al. (2019)

Lantana camara (aerial parts) at 5000μg/mL	Disc diffusion method	Eight isolates of *Ralstonia solanacearum* phylotype II	Φ = 17.50 mm for Rs48 Φ = 17.33 mm for RsMo2 Φ = 16.33 mm for RsScl	Mohamed et al. (2019)
Corymbia citriodora (leaves) at 5000μg/mL	Disc diffusion method	Eight isolates of *Ralstonia solanacearum* phylotype II	Φ = 14.00 mm for RsBe2 Φ = 13.66 mm for RsFr4 Φ = 13.66 mm for RsNe1	Mohamed et al. (2019)
Cupressus sempervirens (aerial parts) at 5000μg/mL	Disc diffusion method	Eight isolates of *Ralstonia solanacearum* phylotype II	Φ = 13.00 mm for RsNe1 Φ = 13.00 mm forRsIs2 Φ = 11.33 mm forRsBe2	Mohamed et al. (2019)
Rosmarinus officinalis (100%)	Agar disc diffusion and micro-dilution methods	*Escherichia coli*	Φ > 35 mm MIC = 1.67 ± 0.44μl/mL MBC = 10.67 ± 3.56μl/mL	Moussii et al. (2020)
Lavandula angustifolia (100%)	Agar disc diffusion and micro-dilution methods	*Escherichia coli*	Φ > 40 mm MIC = 3.33 ± 0.89μl/mL MBC = 10.67 ± 3.56μl/mL	Moussii et al. (2020)
Artemisia herba alba (100%)	Agar disc diffusion and micro-dilution methods	*Escherichia coli*	Φ > 35 mm MIC = 6.67 ± 1.78μl/mL MBC = 16 ± 0.00μl/mL	Moussii et al. (2020)
Rosmarinus officinalis (100%)	Agar disc diffusion and micro-dilution methods	*Staphylococcus aureus*	Φ > 40 mm MIC = 1.33 ± 0.44μl/mL MBC = 13.33 ± 3.56μl/mL	Moussii et al. (2020)
Lavandula angustifolia (100%)	Agar disc diffusion and micro-dilution methods	*Staphylococcus aureus*	Φ > 40 mm MIC = 1.33 ± 0.44μl/mL MBC = 6.67 ± 1.78μl/mL	Moussii et al. (2020)

(continued)

Table 12.1 (continued)

Essential oils (origin)	Methods	Bacterial strains	Key findings	References
Artemisia herba alba (100%)	Agar disc diffusion and micro-dilution methods	*Staphylococcus aureus*	$\Phi > 40$ mm MIC $= 4 \pm 0.00 \mu l/mL$ MBC $= 16 \pm 0.00 \mu l/mL$	Moussii et al. (2020)
Rosmarinus officinalis (100%)	Agar disc diffusion and micro-dilution methods	*Pseudomonas aeruginosa*	$\Phi > 20$ mm MIC $= 42.67 \pm 14.22$ MBC $= 85.33 \pm 28.44$	Moussii et al. (2020)
Lavandula angustifolia (100%)	Agar disc diffusion and micro-dilution methods	*Pseudomonas aeruginosa*	$\Phi > 15$ mm MIC $= 42.67 \pm 14.22 \mu l/mL$ MBC $= 85.33 \pm 28.44 \mu l/mL$	Moussii et al. (2020)
Artemisia herba alba (100%)	Agar disc diffusion and micro-dilution methods	*Pseudomonas aeruginosa*	$\Phi > 15$ mm MIC $= 42.67 \pm 14.22 \mu l/mL$ MBC $= 85.33 \pm 28.44 \mu l/mL$	Moussii et al. (2020)
Salvia triloba L.	Agar disc diffusion method	*Klebsiella pneumoniae*	$\Phi = 19.0 \pm 1.0$ mm	Risaliti et al. (2019)
Rosmarinus officinalis L.	Agar disc diffusion method	*Klebsiella pneumoniae*	$\Phi = 12.7 \pm 0.6$ mm	Risaliti et al. (2019)
Spondias pinnata (Linn. F.) Kurz (fruits)	Agar-well diffusion and micro-dilution methods	*Staphylococcus aureus*	$\Phi = 11.93 \pm 0.40$ mm MIC $= 78.13 \mu g/mL$	Sameh et al. (2019)
Spondias pinnata (Linn. F.) Kurz (flowers)	Agar-well diffusion and micro-dilution methods	MRSA	$\Phi = 10.93 \pm 0.60$ mm MIC $= 156.25 \mu g/mL$	Sameh et al. (2019)

Spondias pinnata (Linn. F.) Kurz (flowers)	Agar-well diffusion and micro-dilution methods	Escherichia coli ATCC 25955	$\Phi = 9.90 \pm 0.56$ mm MIC $= 12.5 \mu$g/mL	Sameh et al. (2019)
Spondias pinnata (Linn. F.) Kurz (leaves)	Agar-well diffusion and micro-dilution methods	Salmonella typhimurium ATCC 14028	$\Phi = 9.97 \pm 0.35$ mm MIC $= 125.0 \mu$g/mL	Sameh et al. (2019)
Spondias pinnata (Linn. F.) Kurz (leaves)	Agar-well diffusion and micro-dilution methods	Helicobacter pylori	91% inhibition MIC $= 1.95 \mu$g/mL	Sameh et al. (2019)
Spondias pinnata (Linn. F.) Kurz (leaves)	Agar-well diffusion and micro-dilution methods	Mycobacterium tuberculosis	84% inhibition MIC $= 0.49 \mu$g/mL	Sameh et al. (2019)
Daucus carota subsp. drepanensis (Lojac.) Heywood (EOCD)	Micro-dilution method	Salmonella typhimurium ATCC 1408, Escherichia coli ATCC 35218, Staphylococcus aureus ATCC 25923, Enterococcus faecalis ATCC 29212	Absence of activity	Snene et al. (2020)
D. carota subsp. hispidus (ball) Heywood (EOCH)	Micro-dilution method	Enterococcus faecalis ATCC 29212	MIC $= 1.25$ mg/mL MBC $= 5$ mg/mL	Snene et al. (2020)
Juniperus communis	Micro-dilution method Time-kill assay in vitro	Listeria monocytogenes: Strain ATCC19111 and LMB (beef carcass isolate)	MIC $= 0.5\%$ MBC $= 2\%$ Synergism for isolate from LMB and its absence for ATCC19111 strain	Vasilijević et al. (2019)

(continued)

Table 12.1 (continued)

Essential oils (origin)	Methods	Bacterial strains	Key findings	References
Satureja montana	Micro-dilution method Time-kill assay in vitro	*Listeria monocytogenes*: Strain ATCC19111 and LMB	MIC = 0.5% MBC = 1% Synergism for isolate from LMB and its absence for ATCC19111 strain	Vasilijević et al. (2019)
Tagetes minuta L. from locations L3 (Kotgarh), L9 (Gogardhar), L11 (Jia), and L15 (Makhan)	Agar-well diffusion and micro-dilution methods	*Staphylococcus aureus* MTCC 96	Φ >9 mm MIC of 25–30% (v/v)	Walia et al. (2020)
Tagetes minuta L. from location L3 (Kotgarh)	Agar-well diffusion and micro-dilution methods	*Klebsiella pneumoniae* MTCC 109	Φ = 4 mm	Walia et al. (2020)
Tagetes minuta L. from location L11 (Jia)	Agar-well diffusion and micro-dilution methods	*Micrococcus luteus* MTCC 2470	Φ = 8 mm	Walia et al. (2020)

12.3 Antibacterial Mechanisms of EOs and Their Bioactive Compounds Against Bacteria

12.3.1 Antibacterial Mechanisms of EOs

The mechanism of action of some Eos is reported in Table 12.1. In *Salmonella enteritidis* ATCC 13076 cells, *Origanum vulgare* EO induced a stress response characterized by differential expressions of chaperones and cellular protein synthesis mediated by the bacterial signaling system. In addition, an interference in protein regulation and DNA synthesis was observed (Barbosa et al. 2020). Moreover, *Origanum vulgare* and *Salvia officinalis* Eos showed an antibiofilm activity against *Streptococcus pyogenes,* as they were able to prevent and eradicate the biofilm (Wijesundara and Rupasinghe 2018).

The mechanism of action of *Origanum compactum* EO was investigated against several bacteria. Against *Pseudomonas aeruginosa* ATCC 27853 and *Staphylococcus aureus* ATCC 29213, Bouhdid et al. (2009) found that the EO acted essentially on the cell membrane causing the loss of membrane potential and permeability and consequently the loss of cell viability. Additionally, structural alterations were observed by transmission electron microscopy; mesosome-like structures in *Staphylococcus aureus* cells, and cytoplasmic material coagulation, membrane vesicle liberation, and intracellular material leakage in *Pseudomonas aeruginosa* (Bouhdid et al. 2009). Against *Escherichia coli* and *Bacillus subtilis,* EO from *Origanum compactum* collected at three phenological stages acted also on the cell membrane, causing disturbance of its integrity, increase of its permeability, and consequently release of cell constituents (DNA and RNA). The EO also inhibited the formation of biofilm, known as a quorum sensing regulated phenotype (Bouyahya et al. 2019a). Accordingly, *Origanum compactum* EO targeted primary the cell membrane leading to the release of intracellular material and consequently to bacterial death.

The mechanism of action of *Cinnamomum verum* EO against *Pseudomonas aeruginosa* and *Staphylococcus aureus* was investigated (Bouhdid et al. 2010). The treatment of *Pseudomonas aeruginosa* with the EO altered its cell membrane leading to collapse of the membrane potential and loss of membrane permeability. In addition, the respiratory activity was inhibited leading to cell death. Observation by transmission electron microscopy showed coagulation and leakage of cytoplasmic material. Against *Staphylococcus aureus, Cinnamomum verum* EO caused the cells to enter a viable but noncultivable state. Initially, a decrease in the metabolic and respiratory activities and in the replication capacity was observed, followed by the loss of membrane integrity. Moreover, fibers extending from the cell surface were observed by transmission electron microscopy. Therefore, *Cinnamomum verum* EO action was characterized by the loss of membrane integrity and structural alteration. On another hand, Wang et al. (2018) investigated the mechanism of action of *Cinnamomum zeylanicum* EO against *Porphyromonas gingivalis* and observed the increase of nucleic acid and protein leakage with increasing EO concentration. Thus, the EO acted on the cell membrane and impaired its integrity by enhancing cell permeability. Also, it induced cell membrane destruction as revealed by scanning electron microscopy. Additionally, *Cinnamomum zeylanicum* EO had an antibiofilm activity and was able to inhibit its formation by 74.5% at sub-MIC levels.

The treatment of *Staphylococcus aureus* ATCC 25923 with *Syzygium aromaticum* EO caused the destruction of cell wall and membranes, leading to the loss of vital intracellular materials, the inhibition of the normal synthesis of DNA and proteins, and eventually the bacterial death. Therefore, the effect of *Syzygium aromaticum* EO on *Staphylococcus aureus* may be at the molecular level (Xu et al. 2016). Similarly, Yoo et al. (2021) also found that the treatment of *Staphylococcus aureus* and *Escherichia coli* O157:H7 with *Syzygium aromaticum* EO caused membrane disruption, including significant changes in cell structure, and increased membrane permeability and intracellular material release.

Li et al. (2019) investigated the mechanism of action of *Citrus medica* L. var. *sarcodactylis* EO against common foodborne bacteria: *Escherichia coli, Staphylococcus aureus, Bacillus subtilis,* and *Micrococcus luteus.* They have found that EO caused changes and damages in bacterial morphology that increased with the increase of EO concentration and exposure time, leading to lysis of the cell wall, leakage of intracellular ingredient, and bacterial death. Zhang et al. (2020) showed that *Litsea cubeba* EO caused damage and rupture of *Stenotrophomonas maltophilia* cell membrane, as determined by fluorescent dyes and field emission scanning electron microscopy. This activity was associated with citral, its key active component.

Consequently, the studies have shown that Eos primary target is the cell membrane. The damages caused in the membrane lead to leakage of intracellular material (DNA, RNA, protein) and eventually to bacterial death. Other cellular functions can be affected such as inhibition of protein and DNA synthesis, decrease of respiratory and metabolic activities, and decrease of replication capacity. The alteration of ultrastructure was also observed, along with antibiofilm activity.

12.3.2 Antibacterial Mechanisms of Volatile Bioactive Compounds

EOs contain several compounds belonging to terpenoid chemical class such as β-caryophyllene, 1,8-cineole, myrtenol, geraniol, carvacrol, thymol, L-carvone, phytol, limonene, and linalool (Fig. 12.1). The mechanism of action of some volatile bioactive compounds is reported in Table 12.2. The action of linalool, the active component of several EOs, was reported against *Acinetobacter baumannii* (Alves et al. 2016), *Pseudomonas aeruginosa* (Liu et al. 2020), and *Salmonella typhimurium* (Prakash et al. 2019). Against *Acinetobacter baumannii,* linalool changed the bacterial adhesion to surfaces and interfered with the quorum sensing system. It also had an antibiofilm activity by inhibiting biofilm formation and dispersing established one (Alves et al. 2016). Against *Pseudomonas aeruginosa,* linalool induced morphology disruption and acted on the cell membrane. It destroyed the membrane integrity as demonstrated by the decrease in the membrane potential and the release of nucleic acids, leading to cell death. Additionally, it caused damages in the respiratory chain (Liu et al. 2020). The encapsulation of

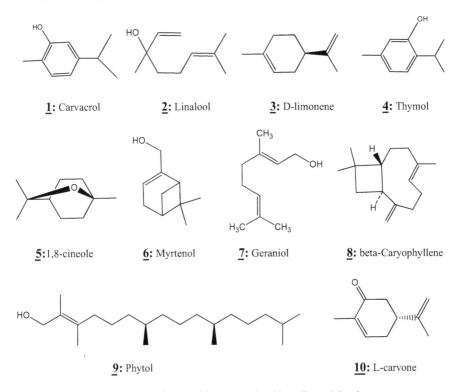

Fig. 12.1 Chemical structures of terpenoid compounds with antibacterial action

linalool into nanoemulsions using Tween 80 (ratio 1:3 (v/v)) increased its ability to disrupt *Salmonella typhimurium* cell membrane integrity and its antibiofilm activity (11.5% higher than the pure linalool) (Prakash et al. 2019).

The major component of citrus fruits Eos, (+)-limonene, was reported to act on bacterial cell membrane. Espina et al. (2013) observed permanent permeabilization of the cell membrane of *Escherichia coli* BJ4 and sublethal damages in *Escherichia coli* MC4100 *lptD4213* mutant at pH 4.0. Also, Han et al. (2020) confirmed the damages to the cell membrane of *Listeria monocytogenes* by the increase in conductivity and the loss of intracellular content (nucleic acids and proteins). Also, the destruction of the cell integrity and wall structure was observed, along with the inhibition of the respiratory chain complex function which affected respiration and energy metabolism. On another hand, Subramenium et al. (2015) reported limonene antibiofilm potential against *Streptococcus pyogenes* (SF370). Indeed, limonene reduced the biofilm formation in a concentration-dependent way, possibly by inhibiting bacterial adhesion to surfaces and thus preventing the biofilm formation cascade. Against *Streptococcus mutans,* limonene showed anticariogenic efficacy, with inhibition of acid production and downregulation of the *vicR* gene. Against both

Table 12.2 Mechanisms of antibacterial action of essential oils

Essential oils/ bioactive compounds	Methods	Bacterial strains	Mechanisms	References
Origanum vulgare	Bacterial protein profile, enzymatic activities, and DNA synthesis	*Salmonella Enteritidis* ATCC 13076	Stress response with differential expressions of chaperones. Cellular protein synthesis mediated by the bacterial signaling system. Affected the DNA synthesis + interference in protein regulation	Barbosa et al. (2020)
Origanum compactum	Plate count, potassium leakage, flow cytometry, and transmission electron microscopy(TEM) analysis	*Pseudomonas aeruginosa* ATCC 27853 and *Staphylococcus aureus* ATCC 29213	Membrane damage showed by the leakage of potassium and uptake of propidium iodide and bis-oxonol. Ultrastructural alterations and the loss of cell viability	Bouhdid et al. (2009)
Origanum compactum (at three phenological stages)	Antibacterial kinetics assay. Cell membrane integrity. Leakage of proteins through the bacterial membrane. Permeability of cell membrane. Anti-quorum sensing (QS) activity	*Escherichia coli* and *Bacillus subtilis*	Significant inhibition of kinetic growth. Increased the membrane permeability. Significant increase in cell constituent release (DNA and RNA) in an essential oil concentration-dependent manner. Inhibited the formation of the biofilms, a phenotype that has been known to be QS regulated	Bouhdid et al. (2009)
Origanum vulgare and *Salvia officinalis*	MTT assay. SEM analysis	*Streptococcus pyogenes*	Inhibited bacterial biofilm formation with dual actions, preventing and eradicating the biofilm	Wijesundara and Rupasinghe (2018)

Cinnamomum verum	Plate count, potassium leakage, flow cytometry, and TEM analysis	*Pseudomonas aeruginosa* ATCC 27853	Alterations of the bacterial membrane: Collapse of the membrane potential and loss of membrane-selective permeability + inhibited the respiratory activity Loss of membrane integrity + structural alteration: Coagulation of cytoplasmic material and leakage of intracellular material	Bouhdid et al. (2010)
Cinnamomum verum	Plate count, potassium leakage, flow cytometry, and TEM analysis	*Staphylococcus aureus* ATCC 29213	+ viable but noncultivable (VNC) state Inhibited the respiratory activity Decrease in metabolic activity and bacterial cell replication capacity Loss of membrane integrity + structural alteration: Fibers extending from the cell surface	Bouhdid et al. (2010)
Cinnamomum zeylanicum	Propidium iodide uptake assays SEM Crystal violet staining method	*Porphyromonas gingivalis*	Increased nucleic acid and protein leakage with increasing concentrations of essential oil Impaired bacterial membrane integrity by enhancing cell permeability Induced cell membrane destruction Inhibited bacterial biofilm formation by 74.5% at sub-MIC levels	Bouhdid et al. (2010)
Syzygium aromaticum L.	Kill-time analysis Cell permeabilization assay DNA-binding assays	*Staphylococcus aureus* ATCC 25923	Destruction of cell wall and membrane Loss of vital intracellular materials Inhibited the normal synthesis of DNA and proteins	Xu et al. (2016)

(continued)

Table 12.2 (continued)

Essential oils/ bioactive compounds	Methods	Bacterial strains	Mechanisms	References
Syzygium aromaticum L.	Cell leakage determination Confocal laser SEM TEM analysis	*Escherichia coli* O157:H7 and *Staphylococcus aureus*	Cell membrane disruption, including significant changes in cell structure, increased membrane permeability, and increased leakage of intracellular materials	Yoo et al. (2021)
Citrus medica L. var. *sarcodactylis*	Scanning electron microscopy (SEM), time-kill analysis, and permeability of cell and membrane integrity	*Escherichia coli, Staphylococcus aureus, Bacillus subtilis,* and *Micrococcus luteus*	Damaged the morphology of the tested bacteria with increased concentration and exposure time of essential oil Lysis of the cell wall and intracellular ingredient leakage	Li et al. (2019)
Litsea cubeba	Fluorescent dyes and field emission SEM	*Stenotrophomonas maltophilia*	Damage and rupture of cell membrane	Zhang et al. (2020)
Linalool	Assessment of biofilm formation on different surfaces Quantification of biofilm biomass—CV staining Anti-QS activity assay	*Acinetobacter baumannii*	Inhibited the biofilm formation and dispersed established biofilms Changed the bacterial adhesion to surfaces and interfered with the QS system	Alves et al. (2016)
Linalool	SEM, cell membrane permeability, membrane potential, and respiratory chain dehydrogenase determination	*Pseudomonas aeruginosa*	Disrupted the normal morphology of the cell Destroyed the integrity of the bacterial membrane Damaged the respiratory chain	Liu et al. (2020)
Linalool nanoemulsions	Integrity of cell membrane Biofilm inhibition studies	*Salmonella typhimurium*	Increased the ability to disrupt cell membrane integrity Higher antibiofilm activity (11.5%)	Prakash et al. (2019)

(+)-limonene	Cell permeabilization assay	Escherichia coli BJ4	Permanent permeabilization of the bacterial membrane	Espina et al. (2013)
(+)-limonene	Cell permeabilization assay	Escherichia coli lptD4213	Sublethal damages in the cytoplasmic membrane at pH 4.0	Espina et al. (2013)
Limonene	SEM analysis Conductivity measurement Determination of the effect of limonene on the respiratory chain complex I ~ V	Listeria monocytogenes	Increase in cell membrane permeability Destruction of cell integrity and bacterial wall structure Affected respiration and energy metabolism by inhibiting the function of the respiratory chain complex	Han et al. (2020)
Limonene	Analysis of antibiofilm potential, SEM, and cell viability assay	Streptococcus pyogenes (SF370)	Reduction in the biofilm formation in a concentration-dependent manner + targeted the surface-associated proteins reducing surface-mediated virulence factors	Subramenium et al. (2015)
Limonene	Analysis of antibiofilm potential, SEM, and cell viability assay	Streptococcus mutans	Anticariogenic efficacy, with inhibition of acid production and downregulation of the vicR gene + targeted the surface-associated proteins reducing surface-mediated virulence factors	Subramenium et al. (2015)
Phytol	Measurement of reactive oxygen species (ROS) The NAD$^+$ cycling assay Detection of DNA damage Membrane depolarization assay	Pseudomonas aeruginosa	Increased the level of intracellular ROS and transient NADH depletion Induced the DNA damage through oxidative stress + caused cell filamentation Triggered the inhibition of bacterial cell division Induced the membrane depolarization	Lee et al. (2016)

(continued)

Table 12.2 (continued)

Essential oils/ bioactive compounds	Methods	Bacterial strains	Mechanisms	References
L-carvone	QS interference approach In silico analysis and RT-qPCR studies	Hafnia alvei	Inhibited the QS activity by the reduction in swinging motility (61.49%), swarming motility (74.94%), biofilm formation (52.41%), and acyl-homoserine lactone (AHL) production (0.5μL/ mL)	Li et al. (2019)
Carvacrol	SEM Antibiofilm activity	Escherichia coli and Staphylococcus aureus	Reduced the biofilm formation on the EVA 14 polymeric films surface	Nostro et al. (2012)
Carvacrol	SEM analysis Crystal violet assay MTT assay	Salmonella enterica serotype typhimurium	+ shriveled and retracted appearance At $4 \times$ MIC Strong reduction in biofilm biomass (1.719 OD_{550}) Strong reduction in metabolic activity (0.089 OD_{550})	Trevisan et al. (2018)
Carvacrol and thymol	Flow cytometry Fluorescent dyes	Escherichia coli	Disruption of the cytoplasmic membrane by permeabilizing and depolarizing its potential	Xu et al. (2008)
Carvacrol	Flow cytometric analysis TEM analysis	Listeria monocytogenes	Increase in membrane permeability and depolarization Change in respiratory activity Degenerative changes of cell wall and cytoplasmic membrane Structural disruption of bacterial cells	Churklam et al. (2020)

Compound	Assay/Method	Organism	Effect	Reference
Geraniol	Antibiofilm activity Quantification of biofilm biomass	*Staphylococcus aureus*	Reduced the biofilm biomass up to 100% between 0.5 and 4 mg/mL concentrations Reduced the cell viability (1 mg/mL)	Pontes et al. (2019)
Myrtenol	Ring biofilm inhibition assay Autolysis assay Extraction of eDNA Autoaggregation assay Extraction of staphyloxanthin	Methicillin-resistant *Staphylococcus aureus*	Inhibited the biofilm formation Inhibited the synthesis of major virulence factors including slime, lipase, α-hemolysin, staphyloxanthin, and autolysin Decreased the amount of eDNA release Reduced the autoaggregation Inhibited the staphyloxanthin production in a concentration-dependent manner	Selvaraj et al. (2019)
Myrtenol	Antibiofilm effect	*Staphylococcus aureus*	Strong ability to inhibit the formation of the biofilm	Cordeiro et al. (2020)
1,8-cineole	TEM analysis Validation of proteomics analysis	*Salmonella* sp. D194–2	Damage the structure of cell walls and membranes Downregulated the carbohydrate metabolism and membrane protein-related genes at the mRNA level	Sun et al. (2018)
β-Caryophyllene	Confocal laser scanning microscope Real-time RT-PCR	*Streptococcus mutans*	Inhibited the biofilm formation (above 0.32%) Reduced the expression of *gtf* genes at a non-killing concentration	Yoo and Jwa (2018)
β-Caryophyllene	Zeta-potential measurement Measurement of UV-absorbing materials	*Bacillus cereus*	Altered the membrane permeability and integrity, leading to membrane damage and intracellular content leakage	Moo et al. (2020)

Streptococcus species, limonene had an antagonistic mechanism on the surface-associated proteins reducing surface-mediated virulence factors.

Phytol, a constituent of chlorophyll and a diterpene in Eos, induced an oxidative stress response in *Pseudomonas aeruginosa* treated cells, characterized by an increase of the level of intracellular reactive oxygen species (ROS) and transient NADH depletion. Additionally, the oxidative stress induced by phytol damages severely the DNA which caused cell filamentation and cell division inhibition. Also, phytol caused membrane depolarization and therefore cell death (Lee et al. 2016). On another hand, L-carvone showed a plausible molecular mechanism for quorum sensing inhibition against *Hafnia alvei,* characterized by the reduction in swinging motility (61.49%), swarming motility (74.94%), biofilm formation (52.41%), and acyl-homoserine lactone (AHL) production (0.5μL/mL) (Lee et al. 2016).

Carvacrol, a monoterpene present in several Eos, showed an antibiofilm activity against *Staphylococcus aureus* and *Escherichia coli* when it was incorporated to a polyethylene-co-vinylacetate (EVA) film (Nostro et al. 2012). Moreover, carvacrol and its isomer thymol were able to depolarize and permeabilize the cytoplasmic membrane of *Escherichia coli* (Xu et al. 2008). Against *Salmonella enterica* serotype Typhimurium, Trevisan et al. (2018) observed a shriveled and retracted appearance in cells treated by carvacrol, demonstrating alteration in structure and membrane. Moreover, strong metabolic and antibiofilm activities were reported, especially at 4 × MIC. Also, Churklam et al. (2020) found that carvacrol acted on the cell wall and cytoplasmic membrane of *Listeria monocytogenes*, causing degenerative changes and increasing membrane permeability and depolarization and thus leading to structural disruption. Furthermore, inhibition of respiratory activity was observed.

Geraniol, a monoterpene, was able to reduce *Staphylococcus aureus* viability at 1 mg/mL and showed an antibiofilm activity by reducing the biofilm biomass up to 100% between 0.5 and 4 mg/mL concentrations (Pontes et al. 2019). The antibiofilm activity was also reported in *Staphylococcus aureus* after treatment with myrtenol, a bicyclic alcohol monoterpene (Cordeiro et al. 2020; Selvaraj et al. 2019). In methicillin-resistant *Staphylococcus aureus,* myrtenol also had antivirulence potential and was able to inhibit the synthesis of major virulence factors including slime, lipase, α-hemolysin, staphyloxanthin, and autolysin. Additionally, it impaired eDNA releasing autolysis causing reduction in autoaggregation (Selvaraj et al. 2019).

1,8-Cineole, a monoterpene compound also known as eucalyptol or cajeputol, acted on the cell wall and membrane of *Salmonella* sp. D194–2 causing structural damages. Moreover, it was able to downregulate carbohydrate metabolism and membrane protein-related genes at the mRNA level (Sun et al. 2018). The treatment of *Streptococcus mutans* by β-caryophyllene, a bicyclic sesquiterpene, caused the inhibition of cell growth and biofilm formation. Moreover, it had an inhibitory effect on biofilm-related factor and was able to reduce the expression of *gtf* genes at a non-killing concentration (Yoo and Jwa 2018). In *Bacillus cereus cells,* β-caryophyllene caused an alteration of membrane permeability and integrity, leading to membrane damage, intracellular content release, and eventually cell death (MOO et al. 2020).

Therefore, the mechanism of action of these bioactive compounds involved cell membrane damage and structure alteration. In addition, inhibition of quorum sensing, antibiofilm activity, and respiratory and metabolism activities were also reported. This shows that these compounds can be used as potential antibiotics, anticariogenic agents, antimicrobial agent, and natural food preservative.

12.4 Clinical Investigation of Bioactive Compounds from Essential Oils Against Bacteria

To confirm and validate the efficacy, safety, and tolerance of a treatment (cause and effect relationship), clinical trials are carried out in human medical therapy after nonclinical experimental studies (in vitro and in vivo) (Table 12.3). In this sense, natural molecules have been the subject of numerous clinical studies in the evaluation of antibacterial properties. However, little work has been performed to investigate the relevance of terpenoid activity at the clinical level (Casetti et al. 2012; Sosto et al. 2011; Baygin et al. 2014; Shim et al. 2019).

Indeed, Sosto et al. (2011) investigated the effect of a vaginal douche at the base of two monoterpenes (thymol + eugenol) in 221 cases of bacterial vaginosis (BV) in a multicenter, parallel group, randomized trial (Sosto et al. 2011). Consequently, the use of a douche/day for 1 week resulted in a significant decrease in the severity of itching, inflammatory signs, and vaginal pH. In fact, BV is a type of vaginal inflammation caused by the proliferation of bacteria naturally present in the vagina. This shows that this application (thymol + eugenol) can be prescribed for minor vaginal infections. In the same year, Casetti and coworkers evaluated the antibacterial effect of essential coriander oil (ECO) on clinical strains of bacteria responsible for superficial skin infections using the micro-dilution method as well as the skin tolerance of this oil, at effective antibacterial concentrations, in 40 volunteers in a prospective, randomized, placebo-controlled double-blind study (Casetti et al. 2012). Therefore, none of the volunteers showed skin irritation. This indicates that the cream containing ECO may be used as a treatment for bacterial skin infections with *Streptococcus pyogenes* and methicillin-resistant *Staphylococcus aureus* (MRSA), which were the most sensitive to this preparation. Moreover, thymol has also been studied in combination with chlorhexidine in the prevention of caries in 90 disabled children randomly divided into three groups (Baygin et al. 2014). Effectively, the authors tested the antibacterial effect of this combination against two of oral cavity bacteria, namely, salivary mutans streptococci (MS) and lactobacilli (LB) at four stages. They found that this treatment was able to reduce bacteria values compared to the control group at 1 month and 6 months after treatment.

Furthermore, Shim and collaborators studied the antibacterial activity of the administration of β-caryophyllene (126 mg/day) for 8 weeks in 33 patients with *Helicobacter pylori* infection, in a randomized, double-blind, placebo-controlled

Table 12.3 Essential oil-based antimicrobials in clinical trial phases

Molecules/ essential oils	Experience	Bacterial strains	Treatment	Effects	References
Essential coriander oil	A prospective, randomized, placebo-controlled double-blind study Preparations applied on the skin of 40 healthy volunteers using the occlusive patch test Macro-dilution method	*Streptococcus pyogenes* (Lancefield group A)	Concentrations of 0.5% and 1%v/v	Excellent skin tolerance No adverse events MIC = 0.04% v/v	Casetti et al. (2012)
Essential coriander oil	A prospective, randomized, placebo-controlled double-blind study Preparations applied on the skin of 40 healthy volunteers using the occlusive patch test Macro-dilution method	Methicillin-resistant *Staphylococcus aureus* (MRSA)	Concentrations of 0.5% and 1%v/v	Excellent skin tolerance No adverse events MIC = 0.25% v/v	Casetti et al. (2012)
Thymol+ eugenol	A multicenter, parallel group, randomized study 221 bacterial vaginosis (BV) cases	Vaginal strain	1 douche/day for 1 week	Significant reduction of the severity of itching, burning, vulvovaginal erythema and edema, vaginal dryness, dyspareunia, and vaginal discharge versus the baseline values Reduced the vaginal pH	Sosto et al. (2011)
Thymol	90 patients (3–17 years) treated under general anesthesia were randomly assigned into three groups CRT (caries risk test) bacterial counts for each individual	Salivary mutans streptococci (MS) and lactobacilli (LB)	4 stages: T_0, before general anesthesia; T_1, one month after treatment; T_2, six months after treatment; T_3, 12 months after treatment	Antibacterial treatment regimens (groups 1 and 2) dramatically decreased MS and LB values compared to group 3 at T_1 and T_2	Baygin et al. (2014)

	patient at four stages (T_0, T_1, T_2, and T_3)			At T_0 and T_3: no significant differences	
β-Caryophyllene	Randomized double-blind, placebo-controlled trial 33 patients received β-caryophyllene 33 patients received a placebo preparation	*Helicobacter pylori*	126 mg/day for 8-week	No complete eradication No significant change in the urea breath test Improvement of nausea and epigastric pain Decrease in serum IL-1β levels	Shim et al. (2019)

trial in 33 patients who received a placebo preparation (Shim et al. 2019). In addition, the eradication rates and the inflammation level were determined with the urea breath test (UBT) and endoscopically before and after molecule administration in both groups. Interestingly, this research team recorded promising results such as relief of epigastric pain, nausea, and decrease in serum IL-1β levels with no significant change in UBT.

From these studies, it can be deduced that terpenoids may be used in the treatment of various bacterial infections of different locations. Nevertheless, further clinical studies are needed to fully demonstrate this potential at this level.

12.5 Conclusions and Perspectives

The emergence of multidrug-resistant bacteria makes it necessary to find a new and effective antibacterial agent to counteract this microbial resistance. Essential oils owing to the presence of several bioactive compounds possess an interesting antimicrobial activity against a broad spectrum of bacterial strains including multidrug-resistant. The present chapter provides a comprehensive information on the use of essential oils as promising antimicrobial agents, giving an insight of their effect, mode of action, and their potential clinical applications against infectious diseases caused by various multidrug-resistant bacteria. Indeed, several studies have well designed the antimicrobial action of essential oils obtained from various aromatic plants against various multidrug-resistant bacteria. However, the antibacterial action of numerous plant EOs is still not reported. Therefore, further studies investigating the mechanisms of antibacterial action of these essential oils and their individual bioactive components are required. Essential oils have demonstrated their effectiveness to prevent and inhibit the growth of numerous foodborne pathogens in vitro as well as in food system model, suggesting the fact that these essential oils can be used in foods as natural additives to conserve their microbiological safety. In addition, although there are several published works and various proposed practical applications, it is more interesting to conduct further in vivo studies in order to validate the results of the in vitro investigations and also to broaden the scientific understanding of the toxicological aspects. The pharmaceutical field has a need for ecofriendly alternative drug to treat infectious diseases and other pathology. Thus, essential oils could be a potential source of alternative antimicrobial agents and could play a pivotal role in the discovery of new drugs for the treatment of several diseases caused by pathogenic microbial strains in the near future.

References

Ait-Sidi-Brahim M, Markouk M, Larhsini M (2019) Chapter 5 - Moroccan medicinal plants as antiinfective and antioxidant agents. In: Ahmad Khan MS, Ahmad I, Chattopadhyay D (eds) New look to phytomedicine. Academic Press, Amsterdam, pp 91–142. https://doi.org/10.1016/B978-0-12-814619-4.00005-7

Akarca G (2019) Composition and antibacterial effect on food borne pathogens of Hibiscus surattensis L. calyces essential oil. Ind Crop Prod 137:285–289. https://doi.org/10.1016/j. indcrop.2019.05.043

Alves S, Duarte A, Sousa S, Domingues FC (2016) Study of the major essential oil compounds of Coriandrum sativum against Acinetobacter baumannii and the effect of linalool on adhesion, biofilms and quorum sensing. Biofouling 32:155–165. https://doi.org/10.1080/08927014.2015. 1133810

Barbosa LN, Alves FCB, Andrade BFMT, Albano M, Rall VLM, Fernandes AAH, Buzalaf MAR, de Leite AL, de Pontes LG, dos Santos LD, Fernandes Junior A (2020) Proteomic analysis and antibacterial resistance mechanisms of salmonella Enteritidis submitted to the inhibitory effect of Origanum vulgare essential oil, thymol and carvacrol. J Proteomics 214:103625. https://doi. org/10.1016/j.jprot.2019.103625

Baygin O et al (2014) Antibacterial effects of fluoride varnish compared with chlorhexidine plus fluoride in disabled children. Oral Health Prev Dent 12:373–382. https://doi.org/10.3290/j.ohpd. a32129

Benali T, Bouyahya A, Habbadi K, Zengin G, Khabbach A, Achbani EH, Hammani K (2020a) Chemical composition and antibacterial activity of the essential oil and extracts of Cistus ladaniferus subsp ladanifer and Mentha suaveolens against phytopathogenic bacteria and their ecofriendly management of phytopathogenic bacteria. Biocatal Agric Biotechnol 28:101696. https://doi.org/10.1016/j.bcab.2020.101696

Benali T, Habbadi K, Khabbach A, Marmouzi I, Zengin G, Bouyahya A, Chamkhi I, Chtibi H, Aanniz T, Achbani EH, Hammani K (2020b) GC–MS analysis, antioxidant and antimicrobial activities of Achillea odorata subsp. Pectinata and Ruta Montana essential oils and their potential use as food preservatives. Foods 9:668. https://doi.org/10.3390/foods9050668

Bouhdid S, Abrini J, Amensour M, Zhiri A, Espuny MJ, Manresa A (2010) Functional and ultrastructural changes in *Pseudomonas aeruginosa* and *Staphylococcus aureus* cells induced by *Cinnamomum verum* essential oil. J Appl Microbiol 4:1139–1149. https://doi.org/10.1111/j. 1365-2672.2010.04740.x

Bouhdid S, Abrini J, Zhiri A, Espuny MJ, Manresa A (2009) Investigation of functional and morphological changes in Pseudomonas aeruginosa and Staphylococcus aureus cells induced by Origanum compactum essential oil. J Appl Microbiol 106:1558. https://doi.org/10.1111/j. 1365-2672.2008.04124.x

Bouyahya A, Abrini J, Dakka N, Bakri Y (2019a) Essential oils of Origanum compactum increase membrane permeability, disturb cell membrane integrity, and suppress quorum-sensing phenotype in bacteria. J Pharmaceut Anal 9:301–311. https://doi.org/10.1016/j.jpha.2019.03.001

Bouyahya A, Bakri Y, Et-Touys A, Talbaoui A, Khouchlaa A, Charfi S, Abrini J, Dakka N (2017a) Résistance aux antibiotiques et mécanismes d'action des huiles essentielles contre les bactéries. Phytothérapie 2017:1–11. https://doi.org/10.1007/s10298-017-1118-z

Bouyahya A, Belmehdi O, El Jemli M, Marmouzi I, Bourais I, Abrini J, Faouzi MEA, Dakka N, Bakri Y (2019b) Chemical variability of Centaurium erythraea essential oils at three developmental stages and investigation of their in vitro antioxidant, antidiabetic, dermatoprotective and antibacterial activities. Ind Crop Prod 132:111–117. https://doi.org/10.1016/j.indcrop.2019.01. 042

Bouyahya A, Chamkhi I, Benali T, Guaouguaou F-E, Balahbib A, El Omari N, Taha D, Belmehdi O, Ghokhan Z, El Menyiy N (2021) Traditional use, phytochemistry, toxicology, and pharmacology of Origanum majorana L. J Ethnopharmacol 265:113318. https://doi.org/10. 1016/j.jep.2020.113318

Bouyahya A, Chamkhi I, Guaouguaou F-E, Benali T, Balahbib A, El Omari N, Taha D, El-Shazly M, El Menyiy N (2020a) Ethnomedicinal use, phytochemistry, pharmacology, and food benefits of Thymus capitatus. J Ethnopharmacol 259:112925. https://doi.org/10.1016/j.jep. 2020.112925

Bouyahya A, Dakka N, Et-Touys A, Abrini J, Bakri Y (2017b) Medicinal plant products targeting quorum sensing for combating bacterial infections. Asian Pac J Trop Med 10:729–743. https://doi.org/10.1016/j.apjtm.2017.07.021

Bouyahya A, Dakka N, Talbaoui A, Et-Touys A, El-Boury H, Abrini J, Bakri Y (2017c) Correlation between phenological changes, chemical composition and biological activities of the essential oil from Moroccan endemic oregano (Origanum compactum Benth). Ind Crop Prod 108:729–737. https://doi.org/10.1016/j.indcrop.2017.07.033

Bouyahya A, Et-Touys A, Abrini J, Talbaoui A, Fellah H, Bakri Y, Dakka N (2017d) Lavandula stoechas essential oil from Morocco as novel source of antileishmanial, antibacterial and antioxidant activities. Biocatal Agric Biotechnol 12:179–184. https://doi.org/10.1016/j.bcab.2017.10.003

Bouyahya A, Et-Touys A, Bakri Y, Talbaui A, Fellah H, Abrini J, Dakka N (2017e) Chemical composition of Mentha pulegium and Rosmarinus officinalis essential oils and their antileishmanial, antibacterial and antioxidant activities. Microb Pathog 111:41–49. https://doi.org/10.1016/j.micpath.2017.08.015

Bouyahya A, Omari NE, Belmehdi O, Lagrouh F, Jemli ME, Marmouzi I, Faouzi MEA, Taha D, Bourais I, Zengin G, Bakri Y, Dakka N (2020b) Pharmacological investigation of Ajuga Iva essential oils collected at three phenological stages. Flavour Fragr J. https://doi.org/10.1002/ffj.3618

Caputo L, Nazzaro F, Souza LF, Aliberti L, De Martino L, Fratianni F, Coppola R, De Feo V (2017) Laurus nobilis: composition of essential oil and its biological activities. Molecules 22:930. https://doi.org/10.3390/molecules22060930

Casetti F, Bartelke S, Biehler K, Augustin M, Schempp CM, Frank U (2012) Antimicrobial activity against bacteria with dermatological relevance and skin tolerance of the essential oil from Coriandrum sativum L. fruits. Phytother Res 26:420–424. https://doi.org/10.1002/ptr.3571

Churklam W, Chaturongakul S, Ngamwongsatit B, Aunpad R (2020) The mechanisms of action of carvacrol and its synergism with nisin against Listeria monocytogenes on sliced bologna sausage. Food Control 108:106864. https://doi.org/10.1016/j.foodcont.2019.106864

Cordeiro L, Figueiredo P, Souza H, Sousa A, Andrade-Júnior F, Barbosa-Filho J, Lima E (2020) Antibacterial and Antibiofilm activity of Myrtenol against Staphylococcus aureus. Pharmaceuticals 13:133. https://doi.org/10.3390/ph13060133

Dhifi W, Bellili S, Jazi S, Bahloul N, Mnif W (2016) Essential oils' chemical characterization and investigation of some biological activities: a critical review. Medicine 3:25. https://doi.org/10.3390/medicines3040025

Espina L, Gelaw TK, de Lamo-Castellví S, Pagán R, García-Gonzalo D (2013) Mechanism of bacterial inactivation by (+)-limonene and its potential use in food preservation combined processes. PLoS One 8:e56769. https://doi.org/10.1371/journal.pone.0056769

Gadisa E, Weldearegay G, Desta K, Tsegaye G, Hailu S, Jote K, Takele A (2019) Combined antibacterial effect of essential oils from three most commonly used Ethiopian traditional medicinal plants on multidrug resistant bacteria. BMC Complement Altern Med 19:24. https://doi.org/10.1186/s12906-019-2429-4

Guerra FQS, Mendes JM, de Sousa JP, Morais-Braga MFB, Santos BHC, Coutinho HDM, de Lima EO (2012) Increasing antibiotic activity against a multidrug-resistant Acinetobacter spp by essential oils of Citrus limon and Cinnamomum zeylanicum. Nat Prod Res 26:2235–2238. https://doi.org/10.1080/14786419.2011.647019

Han Y, Sun Z, Chen W (2020) Antimicrobial susceptibility and antibacterial mechanism of limonene against Listeria monocytogenes. Molecules 25:33. https://doi.org/10.3390/molecules25010033

Hu W, Li C, Dai J, Cui H, Lin L (2019) Antibacterial activity and mechanism of Litsea cubeba essential oil against methicillin-resistant Staphylococcus aureus (MRSA). Ind Crop Prod 130:34–41. https://doi.org/10.1016/j.indcrop.2018.12.078

Imane NI, Fouzia H, Azzahra LF, Ahmed E, Ismail G, Idrissa D, Mohamed K-H, Sirine F, L'Houcine O, Noureddine B (2020) Chemical composition, antibacterial and antioxidant activities of some essential oils against multidrug resistant bacteria. European J Integr Med 35:101074. https://doi.org/10.1016/j.eujim.2020.101074

Jugreet BS, Mahomoodally MF (2020) Essential oils from 9 exotic and endemic medicinal plants from Mauritius shows in vitro antibacterial and antibiotic potentiating activities. S Afr J Bot 132:355–362. https://doi.org/10.1016/j.sajb.2020.05.001

Larrazabal-Fuentes M, Palma J, Paredes A, Mercado A, Neira I, Lizama C, Sepulveda B, Bravo J (2019) Chemical composition, antioxidant capacity, toxicity and antibacterial activity of the essential oils from Acantholippia deserticola (Phil.) Moldenke (Rica rica) and Artemisia copa Phil. (Copa copa) extracted by microwave-assisted hydrodistillation. Ind Crop Prod 142:111830. https://doi.org/10.1016/j.indcrop.2019.111830

Lee W, Woo E-R, Lee DG (2016) Phytol has antibacterial property by inducing oxidative stress response in Pseudomonas aeruginosa. Free Radic Res 50:1309–1318. https://doi.org/10.1080/10715762.2016.1241395

Li T, Mei Y, He B, Sun X, Li J (2019) Reducing quorum sensing-mediated virulence factor expression and biofilm formation in hafnia alvei by using the potential quorum sensing inhibitor L-Carvone. Front Microbiol 9:3324. https://doi.org/10.3389/fmicb.2018.03324

Liu X, Cai J, Chen H, Zhong Q, Hou Y, Chen W, Chen W (2020) Antibacterial activity and mechanism of linalool against Pseudomonas aeruginosa. Microb Pathog 141:103980. https://doi.org/10.1016/j.micpath.2020.103980

Mahboubi M, Ghazian Bidgoli F (2010) Antistaphylococcal activity of Zataria multiflora essential oil and its synergy with vancomycin. Phytomedicine 17:548–550. https://doi.org/10.1016/j.phymed.2009.11.004

Mohamed AA, Behiry SI, Younes HA, Ashmawy NA, Salem MZM, Márquez-Molina O, Barbabosa-Pilego A (2019) Antibacterial activity of three essential oils and some monoterpenes against Ralstonia solanacearum phylotype II isolated from potato. Microb Pathog 135:103604. https://doi.org/10.1016/j.micpath.2019.103604

MOO C-L, YANG S-K, OSMAN M-A, YUSWAN MH, LOH J-Y, LIM W-M, LIM S-H-E, LAI K-S (2020) Antibacterial activity and mode of action of β-caryophyllene on Bacillus cereus. Pol J Microbiol 69:49–54. https://doi.org/10.33073/pjm-2020-007

Moussii I, Nayme K, Timinouni M, Jamaleddine J, Filali H, Hakkou F (2020) Synergistic antibacterial effects of Moroccan Artemisia herba alba, Lavandula angustifolia and Rosmarinus officinalis essential oils. Synergy 10:100057. https://doi.org/10.1016/j.synres.2019.100057

Nostro A, Scaffaro R, D'Arrigo M, Botta L, Filocamo A, Marino A, Bisignano G (2012) Study on carvacrol and cinnamaldehyde polymeric films: mechanical properties, release kinetics and antibacterial and antibiofilm activities. Appl Microbiol Biotechnol 96:1029–1038. https://doi.org/10.1007/s00253-012-4091-3

Panackal AA (2011) Global climate change and infectious diseases: invasive mycoses. J Earth Sci Climatic Change 2:1–5. https://doi.org/10.4172/2157-7617.1000108

Pauli A (2001) Antimicrobial properties of essential oil constituents. Int J Aromather 11:126–133. https://doi.org/10.1016/S0962-4562(01)80048-5

Poma P, Labbozzetta M, Zito P, Alduina R, Ramarosandratana AV, Bruno M, Rosselli S, Sajeva M, Notarbartolo M (2019) Essential oil composition of Alluaudia procera and in vitro biological activity on two drug-resistant models. Molecules 24:2871. https://doi.org/10.3390/molecules24162871

Pontes EKU, Melo HM, Nogueira JWA, Firmino NCS, de Carvalho MG, Catunda Júnior FEA, Cavalcante TTA (2019) Antibiofilm activity of the essential oil of citronella (Cymbopogon nardus) and its major component, geraniol, on the bacterial biofilms of Staphylococcus aureus. Food Sci Biotechnol 28:633–639. https://doi.org/10.1007/s10068-018-0502-2

Prakash A, Vadivel V, Rubini D, Nithyanand P (2019) Antibacterial and antibiofilm activities of linalool nanoemulsions against salmonella typhimurium. Food Biosci 28:57–65. https://doi.org/10.1016/j.fbio.2019.01.018

Risaliti L, Kehagia A, Daoultzi E, Lazari D, Bergonzi MC, Vergkizi-Nikolakaki S, Hadjipavlou-Litina D, Bilia AR (2019) Liposomes loaded with Salvia triloba and Rosmarinus officinalis essential oils: in vitro assessment of antioxidant, antiinflammatory and antibacterial activities. J Drug Deliv Sci Technol 51:493–498. https://doi.org/10.1016/j.jddst.2019.03.034

Rosato A, Vitali C, De Laurentis N, Armenise D, Antonietta Milillo M (2007) Antibacterial effect of some essential oils administered alone or in combination with Norfloxacin. Phytomedicine 14:727–732. https://doi.org/10.1016/j.phymed.2007.01.005

Sameh S, Al-Sayed E, Labib RM, Singab ANB (2019) Comparative metabolic profiling of essential oils from Spondias pinnata (Linn. F.) Kurz and characterization of their antibacterial activities. Ind Crop Prod 137:468–474. https://doi.org/10.1016/j.indcrop.2019.05.060

Selvaraj A, Jayasree T, Valliammai A, Pandian SK (2019) Myrtenol attenuates MRSA biofilm and virulence by suppressing Sar a expression dynamism. Front Microbiol 10:5. https://doi.org/10.3389/fmicb.2019.02027

Shim, Song HI, Shin DJ, Yoon CM, Park H, Kim YS, Lee N, Ho D (2019) Helicobacter pylori-negative gastric cancer in South Korea: incidence and clinicopathologic characteristics. Helicobacter pylori 74:199–204

Snene A, Mokni RE, Mahdhi A, Joshi RK, Hammami S (2020) Comparative study of essential oils composition and in vitro antibacterial effects of two subspecies of Daucus carota growing in Tunisia. S Afr J Bot 130:366–370. https://doi.org/10.1016/j.sajb.2020.01.028

Sosto F, Benvenuti C, Group CS (2011) Controlled study on thymol + eugenol vaginal douche versus econazole in vaginal candidiasis and metronidazole in bacterial vaginosis. Arzneimittelforschung 61:126–131. https://doi.org/10.1055/s-0031-1296178

Subramenium GA, Vijayakumar K, Pandian SK (2015) Limonene inhibits streptococcal biofilm formation by targeting surface-associated virulence factors. J Med Microbiol 64:879–890. https://doi.org/10.1099/jmm.0.000105

Sun Y, Cai X, Cao J, Wu Z, Pan D (2018) Effects of 1,8-cineole on carbohydrate metabolism related cell structure changes of Salmonella. Front Microbiol 9:1078. https://doi.org/10.3389/fmicb.2018.01078

Trevisan DAC, da Silva AF, Negri M, de Abreu Filho BA, Machinski Junior M, Patussi EV, Campanerut-Sá PAZ, Mikcha JMG, Trevisan DAC, da Silva AF, Negri M, de Abreu Filho BA, Machinski Junior M, Patussi EV, Campanerut-Sá PAZ, Mikcha JMG (2018) Antibacterial and antibiofilm activity of carvacrol against Salmonella enterica serotype typhimurium. Braz J Pharm Sci 54:5. https://doi.org/10.1590/s2175-97902018000117229

Vasilijević B, Mitić-Ćulafić D, Djekic I, Marković T, Knežević-Vukčević J, Tomasevic I, Velebit B, Nikolić B (2019) Antibacterial effect of Juniperus communis and Satureja montana essential oils against Listeria monocytogenes in vitro and in wine marinated beef. Food Control 100:247–256. https://doi.org/10.1016/j.foodcont.2019.01.025

Walia S, Mukhia S, Bhatt V, Kumar R, Kumar R (2020) Variability in chemical composition and antimicrobial activity of Tagetes minuta L essential oil collected from different locations of Himalaya. Ind Crop Prod 150:112449. https://doi.org/10.1016/j.indcrop.2020.112449

Wang Y, Zhang Y, Shi Y, Pan X, Lu Y, Cao P (2018) Antibacterial effects of cinnamon (Cinnamomum zeylanicum) bark essential oil on Porphyromonas gingivalis. Microb Pathog 116:26–32. https://doi.org/10.1016/j.micpath.2018.01.009

Wijesundara NM, Rupasinghe HPV (2018) Essential oils from Origanum vulgare and Salvia officinalis exhibit antibacterial and anti-biofilm activities against Streptococcus pyogenes. Microb Pathog 117:118–127. https://doi.org/10.1016/j.micpath.2018.02.026

World Health Organization (2018) Antimicrobial resistance. The global risks report 2018. https://wef.ch/2JvW5Wz. Accessed 23 Sept 20

Xu J, Zhou F, Ji B-P, Pei R-S, Xu N (2008) The antibacterial mechanism of carvacrol and thymol against Escherichia coli. Lett Appl Microbiol 47(3):174–179

Xu J-G, Liu T, Hu Q-P, Cao X-M (2016) Chemical composition, antibacterial properties and mechanism of action of essential oil from clove buds against Staphylococcus aureus. Molecules 21:1194. https://doi.org/10.3390/molecules21091194

Yap PSX, Yiap BC, Ping HC, Lim SHE (2014) Essential oils, a new horizon in combating bacterial antibiotic resistance. Open Microbiol J 8:6–14. https://doi.org/10.2174/1874285801408010006

Yoo H-J, Jwa S-K (2018) Inhibitory effects of β-caryophyllene on Streptococcus mutans biofilm. Arch Oral Biol 88:42–46. https://doi.org/10.1016/j.archoralbio.2018.01.009

Yoo JH, Baek KH, Heo YS, Yong HI, Jo C (2021) Synergistic bactericidal effect of clove oil and encapsulated atmospheric pressure plasma against Escherichia coli O157:H7 and Staphylococcus aureus and its mechanism of action. Food Microbiol 93:103611. https://doi.org/10.1016/j.fm.2020.103611

Zhang Y, Wei J, Chen H, Song Z, Guo H, Yuan Y, Yue T (2020) Antibacterial activity of essential oils against Stenotrophomonas maltophilia and the effect of citral on cell membrane. LWT 117:108667. https://doi.org/10.1016/j.lwt.2019.108667

Antimicrobial Resistance and Medicinal Plant Products as Potential Alternatives to Antibiotics in Animal Husbandry

13

Sagar Reddy, Pramod Barathe, Kawaljeet Kaur, Uttpal Anand, Varsha Shriram, and Vinay Kumar

Contents

Sagar Reddy, Pramod Barathe and Kawaljeet Kaur contributed equally.

S. Reddy · V. Shriram (✉)
Department of Botany, Prof. Ramkrishna More College, Savitribai Phule Pune University, Pune, Maharashtra, India

Department of Biotechnology, Modern College of Arts, Science and Commerce, Savitribai Phule Pune University, Pune, Maharashtra, India
e-mail: vds_botany@pdearmacs.edu.in

P. Barathe · K. Kaur · V. Kumar (✉)
Department of Biotechnology, Modern College of Arts, Science and Commerce, Savitribai Phule Pune University, Pune, Maharashtra, India
e-mail: vinay.kumar@moderncollegegk.org

U. Anand
Department of Life Sciences, The National Institute for Biotechnology in the Negev, Ben-Gurion University of the Negev, Beer-Sheva, Israel

© The Author(s), under exclusive license to Springer Nature Singapore Pte Ltd. 2022
V. Kumar et al. (eds.), *Antimicrobial Resistance*,
https://doi.org/10.1007/978-981-16-3120-7_13

Abstract

With increasing problems of antimicrobial resistance (AMR) in animal husbandry, quick solutions are needed in order to control the persistence and emergence of drug-resistant microbial pathogens. Medicinal plant resources such as probiotics, prebiotics, synbiotics, enzymes, phytogenics, and antimicrobial peptides have shown great antimicrobial potentials in recent years and are looked upon as potential alternatives to conventional veterinary antibiotics. The study of these medicinal plant products has confirmed their putative role in targeting the pathogenicity of microbes. In this chapter, we are discussing the use of antibiotics in animal husbandry and related sectors, the overall impact of heavy usage of antibiotics in animal husbandry and on public health, and the emergence of AMR and available alternative options to tackle it with special emphasis on plant resources. Major drug resistance determinants and targeting them with phytochemicals have also been discussed.

Keywords

AMR · Probiotics · Antimicrobial peptides · Quorum sensing · Efflux pumps · Medicinal plant products

13.1 Introduction

The increasing concern of antimicrobial resistance (AMR) globally has led to the induction of policies and various guidelines for the usage of antibiotics in poultry, livestock, and agricultural sectors, to tackle the incidences of AMR (Walia et al. 2019). Antibiotics are used in veterinary practices for the prevention, control, and treatment of various diseases that can pose a potentially serious threat to the production of livestock (Hosain et al. 2021). Various microbial strains showing AMR are often related and attributed to the administration of antibiotics into animal feeds (Tiwari et al. 2021). These antibiotics if used heavily in animal husbandry can lead to the production of drug-resistant bacteria, which causes pathogenic infections and results in poultry and animal diseases. Henceforth, alternatives to these antibiotics have been proposed and implemented in various countries for safety considerations. Medicinal plant resources are hailed for their potentials in supplementing or replacing the antibiotics used in animal husbandry and agricultural practices.

India, being one of the leading countries facing serious AMR challenge, recently is seen as a global hot spot with increased rates of AMR in several sectors including animal husbandry, which has the potential to impact the exports from this sector.

Mehndiratta and Bhalla (2014) in their review reported the production of MRSA ST 398 animal-associated clones from animal food products, posing a serious threat to public health. In Indian animal husbandry, major industries such as animal agriculture, livestock, and dairy farming consist of a large number of the work force that is directly exposed to these animals and thus drug-resistant microbial strains (Mehndiratta and Bhalla 2014). Studies showed the presence of *E. coli*, *K. pneumoniae*, and *S. lentus,* isolated from the animal litter used in farms as fertilizers, showing resistance to 16 drugs (Mahalmani et al. 2019). Colistin is considered as the last line of antibiotic defence; microbial strains resistant to colistin in veterinary practices have already been reported (Vidovic and Vidovic 2020). Thus, effective alternatives to antibiotics are required urgently, with increasing demands of poultries and animals for human consumption, leading to the emergence of the microbial strains resistant to multiple drugs (AMR) and causing drug-resistant infections.

This chapter primarily focuses on the usage of various antibiotics (often injudiciously) and the emergence of AMR in veterinary practices. To tackle the problem of antibiotic resistance and how medicinal plant resources can be explored for their actions as probiotics, prebiotics, synbiotics, enzymes, phytogenics, and antimicrobial peptides, different mechanisms of action are discussed. Further, medicinal plant resources for targeting major drug resistance determinants like quorum sensing inhibitors, efflux pump inhibitors, bacterial virulence inhibitors, and biofilm inhibitors have also been discussed.

13.2 Antibiotic Use in Animal Husbandry

Since the discovery of penicillin, antibiotics have become an important part of human life (Cheng et al. 2014; Van de Vijver et al. 2016). Globally, changes in living standards have driven the increased demands of animal protein for human consumption and have forced intensive farming (Manyi-Loh et al. 2018; Tilman et al. 2011; Bengtsson and Greko 2014). The use of antibiotics in agriculture began with applications of synthetic chemicals, such as sulphonamides and prontosil (sulphochrysoidine) (Wainwright 2007) and extensive use of gramicidin (McManus et al. 2002).

Animal farming includes different animal species such as avian, bee, bovine, caprine, camel, equine, rabbit, ovine, fish, and swine (Vaarten 2012). It was estimated that increase in animal protein in Asian countries by 7 grams per capita per day (PCPD) in 1960 to about 25 grams PCPD by 2013, whilst overall rice and wheat decrease in high-income adults (Van Boeckel et al. 2015). Since then increase in animal protein consumption has shifted animal farming towards highly concentrated and intense animal feeding activities with an eye on cost-efficient methods (Van Boeckel et al. 2015), leading to the invitation of a variety of contagious diseases causing illnesses to animals and economical loss in food production (Duff and Galyean 2007; Hogeveen et al. 2011; Vaarten 2012; Page and Gautier 2012). The growth in the animal food industry has prompted increased usage of veterinary antibiotics (VAs) to protect animals and increased feed efficacy and productivity (Boxall et al. 2002; Halling-Sørensen et al. 2002; Van Boeckel et al. 2015; Kirchhelle 2018). Currently, about 45 mg kg^{-1}, 148 mg kg^{-1}, and 172 mg kg^{-1} of VAs are used to produce cattle, chicken, and pigs, respectively (Van Boeckel et al. 2015). Manyi-Loh et al. (2018) and Van Boeckel et al. (2015) estimated that increased meat consumption in middle- and lower-income countries results in a 69% increase in global agricultural antibiotic usage between 2010 and 2030.

Antibiotics include a wide range of natural, semisynthetic, and synthetic chemicals that are mainly used for bacteriostatic and bactericidal effects (Martínez 2012; Mili'c et al. 2013; Gillings 2013). Apart from these effects, VAs have also been reported for nontherapeutic purposes such as growth promotion, feed efficiency, and weight gain (Kumar et al. 2012). The usage of VAs varies across the geopolitical regions of the world, depending on the 'food animal species', regional production patterns and types of a production system, intensive or extensive farming, the purpose of farming (commercial, industrial, or domestic), lack of clear legislative framework or policies on the use of antibiotics, and the size and socioeconomic status of the farmers (Manyi-Loh et al. 2018). An unprecedented increase in animal protein mainly in BRICS country, an acronym for Brazil, Russia, India, China, and South Africa, has resulted in higher VA consumption. The consumption of VA is expected to rise by 67% globally (Van Boeckel et al. 2015). The VAs are classified based on the mechanism of action or by the structure of the molecule (Cunningham 2008) and are used in four different ways, namely, therapeutic, metaphylactic, prophylactics, and growth promotion (Aarestrup 2005; Phares et al. 2020; Van et al. 2020). Scientists reported more than 250 different chemicals being used as VAs (Kümmerer and Henninger 2003) whose functionality within the same molecule may change under different pH conditions such as neutral, cationic, anionic, or zwitterionic (Cunningham 2008).

13.3   Antimicrobial Resistance in Animal Husbandry

Microbial capacity to survive the predefined concentration of the antimicrobial agents or antibiotics is known as AMR (Acar and Röstel 2001; Ronquillo and Hernandez 2017; Laloučková and Skřivanová 2019). This definition may differ depending on the scientific orientation and objectives to achieve, viz. clinical, pharmacological, molecular, and epidemiological. Along with the antibiotics, antimicrobial metal ions such as copper and zinc also contribute to resistance through co-selection, whereas the use of metals (chromium, nickel, lead, and iron) in feed can cause the acquisition of drug resistance genes. Scientists have reported copper resistance in bacteria that can show resistance to ampicillin as well as sulphanilamide, known as co-resistance (Cheng et al. 2019). The growth in animal protein demand and animal farming has significantly contributed to higher production and usage of the VAs for therapeutic and growth-promoting purposes (Laloučková and Skřivanová 2019; Acar and Röstel 2001). Frequent use of small doses of VAs in farm animals, apart from disease control, has contributed to the AMR (Van et al. 2020). The resistant microorganisms are becoming 'more prevalent, more virulent, and more diverse' (CDDEP 2016). It is estimated that unprecedented use of antimicrobials will cause ten million deaths annually and economic loss of about $100 trillion globally by the year 2050 (Cheng et al. 2019). Since the antimicrobials are not taken up 100% (Van de Vijver et al. 2016) by the animals, they are released into the environment through animal wastes (Kumar et al. 2012; Manyi-Loh et al. 2018). This is one of the major reasons for microbial resistance to veterinary antimicrobials and environmental transmission of AMR via food chains causing toxicity to human life, animals, and plants which has raised major concerns for public health over years (Kumar et al. 2012; Bengtsson and Greko 2014; Cheng et al. 2019). Resistance to antimicrobials varies from region to region, with a degree of VA consumption, and the regulatory structure of the particular country (Benbachir et al. 2001; Sahoo et al. 2012), with a large number of AMR, spread through horizontal transfer of R-genes amongst the species (Acar and Röstel 2001).

The AMR involves phenotypic and genotypic resistance mechanisms. The phenotypic resistance is established between the strains of bacterial species without any gene alterations (Acar and Röstel 2001). These different levels of resistance can be caused by multiple mechanisms, operating simultaneously or by mechanism integration. On the other hand, the genotypic resistance mechanisms involve chromosomal/plasmid mutations or the acquired (de novo acquisition) resistance (Acar and Röstel 2001; CDDEP 2016; Vidovic and Vidovic 2020). The location of the genetic mutations may be critical for spreading of the resistance, for example, the vertical transmission caused by the chromosomal mutation most commonly provide 'exclusive resistance to a specific antibiotic or class of antibiotics, whereas the de novo mutations generate resistance to a diverse group of functionally unrelated antibiotics. There are certain microorganisms such as methicillin-resistant *Staphylococcus aureus* (MRSA) (resistant to all ß-lactam antibiotics), *E. coli* (plasmid-mediated colistin resistance mechanism—MCR-1), *Pseudomonas aeruginosa*, *Acinetobacter baumannii*, *Salmonella typhimurium* DT104, *Klebsiella pneumoniae*, *Citrobacter freundii*, and fluoroquinolone-resistant *Campylobacter* which are of great concern globally because of their ability to transfer R-genes, hence posing a threat to human life (CDDEP 2016; Van de Vijver et al. 2016). These microorganisms have led to the increased incidences of antibiotic-resistant infections with severe implications on human healthcare systems. Therefore, continuous monitoring and controlled use of VAs is essential, besides search for newer, more effective, and safer options.

13.4   Medicinal Plant Resources as Classes of Alternatives

The usage of antimicrobials in animal husbandry was introduced before the Second World War with commercially available synthetic sulphonamides. In 1951, the subtherapeutic use of antibiotics as growth promoters (AGPs) in animal feed additives without any veterinary prescription was approved by the US Food and Drug Administration (FDA; Castanon 2007). The use of antibiotics as AGPs and crossover use of some antibiotics in agricultural/animal settings caused the inadvertent emergence of antibiotic-resistant bacteria. An escalating AMR in microbial pathogens was linked with the use of AGPs (Gyles 2008). This affected the nutritional and economic potential of animal production severely and also disturbed both animal and human ecosystems. As stated in the 'No time to wait' report published by the World Health Organization (WHO 2019), AMR is rightly identified as one of the grave global threats to public health, food security, and animal welfare. Therefore, the use of antibiotics as AGPs besides therapeutics in animal production is banned, and severe restrictions are enforced by multiple countries. However, the ban on antibiotic usage in animal production harmed animal health and productivity. Therefore, effective alternatives to antibiotics, such as AGPs and therapeutics, are urgently needed to help maintain current animal production levels without threatening public health. Researchers are in search of safe substitutes for antibiotics especially from natural sources like herbs and medicinal plants. To overcome the increased rate of mortality and morbidity caused by the ban of in-feed antibiotics, several substitutes such as probiotics, prebiotics, synbiotics, phytogenics, antimicrobial peptides (AMPs), and feed enzymes have been proposed (Seal et al. 2013). Some alternatives to the antibiotics are encapsulated in Table 13.1.

13.4.1 Probiotics

Probiotics are the 'microorganisms which when administrated in adequate amount confer benefits to health of the host' (Hill et al. 2014). Probiotics destroy harmful pathogenic bacteria by several potent mechanisms via the production of bacteriocins, surfactin, subtilin, organic acid, and hydrogen peroxide antimicrobial substances (Elshaghabee et al. 2017); reduced intestinal availability of oxygen leading to the changes in the pH causing an unfavourable environment for the growth of pathogenic microorganisms (Indira et al. 2019); direct competition with the pathogenic bacteria for the nutrient availability and binding sites; promoting gut homeostasis by enhancing gut epithelial barrier function to decrease the epithelial permeability of pathogen (Nair et al. 2017); and activating innate as well as an adaptive immune system against pathogenic microbes to stimulate IFN-gamma, NK cell, and IL-10 (Chang et al. 2020), thus reducing the pro-inflammatory gene expressions (Dargahi et al. 2019).

The synthesis of vitamins K and B_{12} by the probiotic bacteria helps in the adherence of beneficial bacteria to the intestinal cell membrane (Giri et al. 2018). Sometimes, they produce short-chain polysaccharides (butyric acid, acetic acid), amino acids (lysine, arginine), and enzymes (lipases, amylases, esterase) that have a beneficial role in the alleviation of diseases. Direct antagonistic activity of *Lactobacillus plantarum* against highly resistant and WHO priority pathogenic strains like *Acinetobacter baumannii* and *Pseudomonas aeruginosa* has been reported (Dallal et al. 2017). Further, probiotics are also known for total cholesterol level reduction, improved feed conversion ratio, body weight gainer, nutrient utilizer, and intestinal architecture of the host showing potential to use as an alternative to antibiotics in

Table 13.1 Different antibiotic alternatives based on their effects on growth and performance

Microorganisms	Effects on growth and performance	References
[A] *Probiotics*		
Lactobacillus salivarius strain CI1, CI2, CI3	Improved FCR and body weight gain, reduces a load of harmful pathogenic bacteria (*Escherichia coli*), bacteria enzymes such as beta-glucosidase, beta-glucuronidase, and total cholesterol levels	Shokryazdan et al. (2017)
Lactobacillus fermentum	Positive regulation of pro-inflammatory cytokine and reduced impact of *Campylobacter jejuni* on body weight	Šefcová et al. (2020)
Bacillus subtilis	Improve ability to cope with heat stress effectively via microbiota modulated immunity regulation	Wang et al. (2018)
Lactobacillus casei	Competitive reduction of *Campylobacter jejuni* and *Salmonella enterica* from the chicken gut by effectively colonizing various parts of chicken gut	Tabashsum et al. (2020)
Bifidobacterium infantis BL2416, B. breve JCM1192, Lactobacillus casei ATTC334	Improved weight gain, FCR, and prevented mortality, the deleterious effect caused by *Salmonella* infection via cytokine release and commutative exclusion mechanism	El-Sharkawy et al. (2020)
Lactobacillus fermentum Biocenol CCM 7514	The improved architecture of intestine with larger villi and deeper crypts in chickens infected with *Campylobacter jejuni*	Šefcová et al. (2021)
Bacillus subtilis, B. velezensis	Improve growth performance and intestinal properties such as duodenal villi height	Ramlucken et al. (2020)
Bacillus subtilis KKU213	Improve growth and health of broilers by reducing *Salmonella* infection	Khochamit et al. (2020)
Enterococcus faecium	Decreased Bacteroidetes and increased serum FSH level and egg weight	Wang et al. (2021)
Lactobacillus acidophilus	Improved beneficial microbes such as *Lactobacillus reuteri, Lactobacillus crispatus*, etc., and functional genes in broiler crop	De Cesare et al. (2020).
[B] *Prebiotics*		
Partially hydrolysed guar gum (PHGG)	Improve both feeding behaviour and food passage from the crop in growing chicks	Hajati and Rezaei (2010)
Wheat	Increase relative amounts of *bifidobacteria and lactobacilli*, which may affect Fe bioavailability in long-term use	Tako et al. (2014)

(continued)

Table 13.1 (continued)

Microorganisms	Effects on growth and performance	References
Beta-glucan from an edible mushroom	Act as an immunomodulator on the innate (*Pleurotus florida*) immune responses	Paul et al. (2012)
Non-starch polysaccharides (NSP)	Cause changes in gut micro-environment and from chicory gut morphology	Lindberg (2014)
Dietary resistant potato starch	Alters immunological status, limit *Salmonella* Increase butyrate production, role in maintaining colonic homeostasis	Trachsel et al. (2019)
Arabinoxylooligosachrides (AXOS) FOS and MOS	Decrease *Clostridia* and *E. coli* populations and increase in lactobacilli populations and diversity and total bacterial populations	Kim et al. (2011)
Essential oil + cobalt lactate	Synergistically enhance ruminal development, improve fibre digestion, improve body weight and frame growth rates, reduce the gut health challenges	Liu (2020)
Mannan-oligosaccharide and *Saccharomyces cerevisiae*	Significantly improves the gut health of broiler chickens	Padihari et al. (2014)
Inulin	Improve intestinal microflora and gut morphology. Increased *Bifidobacterium* counts and decreased *E. coli* counts in faecal contents	Nabizadeh (2012)
Fructans	Increase mineral absorption of calcium and phosphorous and improve the hardness of the eggshell and cholesterol diminishing of the yolk	Curbelo et al. (2012)
[C] *Phytogenics or phytobiotics*		
Olive pulp (*Olea europaea L var. chemlal*)	Anticoccidial effect against *Eimeria spp.*	Debbou-Iouknane et al. (2019)
Garlic rhizome meal and moringa leaf meal	Enhance the antioxidative status of boilers with increased serum catalase and glutathione peroxidase and lowering the lipid peroxidation improving the shelf life of meat	Gbore et al. (2021)
Macleaya cordata	Increase in strength and thickness of eggshell, increase in progesterone, estradiol, follicle-stimulating hormone, and luteinizing hormone with a significant decrease in tumour necrosis factor-alpha	Zhou et al. (2020)
Cinnamon bark oil	Improved antioxidant status and cholesterol level in the blood of broiler chickens as compared to antibiotic growth promoter	Chowdhury et al. (2018)

(continued)

Table 13.1 (continued)

Microorganisms	Effects on growth and performance	References
Horsetail (*Equisetum arvense*) and spirulina (*Spirulina platensis*)	Increase eggshell thickness and yolk colour also lowering yolk total cholesterol level	Tufarelli et al. (2021)
Cinnamomum verum	Increased body weight gain, improved breast weight, elevated serum albumin concentration	Qaid et al. (2021b)
Curcuma longa and *Nigella sativa*	Both act as an immune enhancer in boiler infected with *Pasteurella multocida*	Raheem et al. (2021)
Black cumin seeds	Improve nutrient utilization and immunity and decrease nitrogen excretion in the environment	Kumar et al. (2017)
Pomegranate (*Punica granatum L.*)	Enhanced performance in terms of carcass, digestibility, and organ indices as compared to birds fed with alpha-tocopherol also decreases the concentration of serum aspartate aminotransferase	Akuru et al. (2021)
Clove seeds (*Syzygium aromaticum*)	Increased internal organ weight	Suliman et al. (2021)
Cinnamomum verum	Anticoccidial effect of cinnamon in broiler chickens infected with *Eimeria tenella*	Qaid et al. (2021a)
Black cumin seeds	Improved slaughter body weight and beneficial fatty acid concentration	Singh and Kumar (2018)
Olea europaea	Anti-inflammatory response	Herrero-Encinas et al. (2020)
Green tea and pomegranate extract	Improved total blood antiradical activity with a positive effect on gut microbial ecology including the increased abundance of lactic acid bacteria	Perricone et al. (2020)
Lavender essential oil	Decreased *E. coli* populations in ileum	Barbarestani et al. (2020)
Persicaria odorata leaf and *Piper betle* leaf	Increased lactobacillus count in comparison to tetracycline, halquinol, and also reduction in populations of *E. coli*, *Salmonella*, and *Staphylococcus aureus*	Basit et al. (2020)

animal husbandry. Some of the commonly used probiotics are presented in Table 13.2.

The WHO defined probiotics into various criteria for allowing specific microbial strains to qualify as probiotics to be used in food and dietary supplements (Binda et al. 2020). Some of the criteria for the microorganisms to qualify as probiotics are as follows: they must be able to survive extreme conditions of the gastrointestinal

Table 13.2 Examples of probiotic microorganisms

Lactic acid bacteria (LAB)	Bifidobacteria	Propionibacteria	Enterobacteria	Sporulated bacteria
Genus: *Lactobacillus*	Genus: *Bifidobacterium*	Genus; *Propionibacterium*	Genus: *Enterococcus*	Genus: *Bacillus*
Species:	Species:	Species:	Species:	Species:
L. amylovorus	*B. bifidum*	*P. acidipropionici*	*E. faecium*	*B. alcalophilus*
L. acidophilus	*B. breve*	*P. jensenii*	*E. faecalis*	*B. clausii*
L. brevis	*B. animalis*	*P. freudenreichii*		*B. subtilis*
L. bulgaricus	*B. adolescentis*	*P. thoenii*		*B. coagulans*
L. casei	*B. longum*			*B. cereus*
L. fermentum	*B. thermophilum*			
L. rhamnosus	*B. infantis*			
L. curvatus	*B. essensis*			
L. reuteri				
L. plantarum				
L. crispatus				
L. lactis				
L. cellobiosus				
L. johnsonii				
L. gallinarum				
L. paracasei				
L. salivarius				

tract like low pH and bile salts; adhere to intestinal epithelial cells; stabilize intestinal microflora; have nonpathogenic nature; multiply fast with either temporary or permanent colonization of the gastrointestinal tract; be sufficiently characterized; be supported by at least one positive clinical trial according to conventional scientific standard or as per the recommendation of local and national authorities; be safe for the intended use; and be able to survive in the product throughout the shelf life at an efficacious dose (Tomasik and Tomasik 2003). However, there are some of the challenges that are needed to overcome for the introduction of probiotics in animal nutrition by determining the appropriate growth stage at which probiotics should be implemented; method of administration; the optimum dosage of probiotics; animal age; health and nutritional status (Al-Shawi et al. 2020); avoiding the inactivation of potent probiotic strain during the transport, processing, and storage processes; difficulty in reaching a high enough number of viable cell count that can colonize the intestine; and lack of adequate regulation and standards leading to the no proper labelling of probiotic products (Cheng et al. 2014).

13.4.2 Prebiotics

Prebiotics have been proved as one of the effective alternatives to the antibiotics used as AGPs as well as disease control and immune enhancement. The use of prebiotics as feed additives initiated in the late 1980s and started in China during the late 1990s. Prebiotics can be defined as the non-digestible (by the host) and non-nutritive compounds that help probiotic bacteria to grow and flourish. They induce the growth or activity of beneficial microorganisms by helping and outcompeting harmful bacteria. Roberfroid (2007) redefined prebiotics as 'a selectively fermented ingredient that allows specific changes, both in the composition and/or activity in the gastrointestinal microflora that confers benefits upon host well-being and health'.

Prebiotics are in general considered as colonic foods that provide energy; are the metabolic substrate and essential micronutrients to beneficial bacteria, thus showing similar effects to an antibiotic; and have no residues and not developing any resistance (Selaledi et al. 2020). Hence, prebiotics have an economical and medical justification and are safe for the ecosystem. Some examples of the common prebiotic substances include fructooligosaccharides (FOS), trans-galacto-oligosaccharide (TOS), insulin, galactooligosaccharides, a polymer of enzymatically altered lactose, polydextrose, lactosucrose, soy-oligosaccharides, xylooligosaccharides, pyrodextrins, isomaltooligosaccharides, polysaccharides, natural plant extracts, protein hydrolysates, polyols, and lactulose along with the new, emerging prebiotic compounds such as pectic oligosaccharides, lactosucrose, the sugar alcohols, glucooligosaccharides, levans, resistant starch, and polysaccharides. Similar to probiotics, prebiotics needs some criteria to qualify such as the following: should neither be hydrolysed nor absorbed in the upper part of the gastrointestinal tract, reduce gastric acidity and hydrolysis by mammalian enzymes, ferment intestinal microflora, increase gastrointestinal absorption, selectively stimulate the growth/activity of intestinal bacteria, able to alter the colonic flora in favour of a healthier composition, and induce luminal or systemic effects that are beneficial to the host's health (Gibson and Roberfroid 1995). Prebiotics have various mechanisms of action and beneficiary effects that prohibit the growth of detrimental intestinal microbes (*E. coli, Salmonella typhimurium, Clostridium botulinum, C. sporogenes*) and prevent pathogen colonization (by altering the pH, competition for substrates, mucosal attachment sites of intestinal content, production of toxic compounds); stimulate the growth of advantageous bacteria (e.g. *B. longum, L. casei, L. acidophilus, L. delbrückei*); stimulate the growth of intestinal absorptive cells; improve the enteric immune system; facilitate better growth performance and health status; increase the production of volatile fatty acids, biomass, and stool bulking; reduce inflammatory reactions; improve mineral absorption and metabolism; enhance animal performance; lower serum cholesterol; decrease ammonia, urea excretion, and carcass contamination; prevent cancer; and show antiviral activity (Davani-Davari et al. 2019).

However, there are some drawbacks of presently available prebiotic products as they often fail to inhibit/kill pathogens and thus cannot be considered as antibiotics. Further, excessive feeding of prebiotics may cause bloating, diarrhoea, and other adverse reactions caused by the fermentation in the gastrointestinal tract (De Vrese and Schrezenmeir 2008). The relationship between the structure and physiological function of prebiotics is yet not clear due to which the efficacy of prebiotics varies with different animal species, ages, and physical conditions. Additionally, the high cost of prebiotic production limits their application in the animal husbandry industry.

A detailed study of immunogenic, transcriptomic, and proteomic profiling of prebiotics is advocated for a better understanding of the functionality of prebiotics in numerous biological and immunological processes of the host. A comprehensive study of action mechanisms through which the immune system improves and information on structure to function, as well as individual metabolic profiles of target bacteria, can further help to tailor prebiotics for specific health attributes.

13.4.3 Synbiotics

The concept of 'synbiotic' was first introduced by Gibson and Roberfroid (1995) as 'a mixture of prebiotics and probiotics which beneficially affect the host by improving the implementation and survival of live, microbial, dietary supplements in the gastrointestinal tract by selectively stimulating the growth and/or activating metabolism of one or a limited number of health-promoting bacteria, and thus enhancing the welfare of host' (Peng et al. 2019). This definition of synbiotic was further updated by the International Scientific Association for Probiotics and Prebiotics (ISAPP) in May 2019. The primary purpose of using synbiotic is to extend the survival probiotics inside the digestive system of the host by the successful utilization of prebiotics as a food source which increases tolerance of probiotics to low pH, bile salt, oxygen, and temperature (Malik et al. 2019). Briefly, synbiotics are known to beneficially impact the host by improving survival and implementation of probiotic strain and enhancing survival and growth of beneficial microorganisms residing inside the gastrointestinal tract (Gyawali et al. 2019). Batista et al. (2020) in his review stated some of the major criteria for the selection of synbiotics such as the selection of suitable prebiotic and probiotic pairs that exert a beneficial effect in host alone and the combination should be able to stimulate the growth of selected microorganisms which positively influences host health with no or limited stimulation of growth of other microorganisms.

   Prebiotics provide nutrition to the probiotics which lead to the improvement in the growth of probiotics and decrease in the growth of pathogenic microorganism by producing organic acids such as lactic acid and acetic acid. Besides this, on the one hand, probiotics block the pathogenic microorganisms from colonizing gut epithelium by competitive exclusion, increasing mucus production, and forming a biofilm, whereas, on the other hand, prebiotics prevent pathogenic bacteria from colonizing the intestinal epithelium by utilizing the ability of some pathogenic bacteria to bind to the sugar moieties of prebiotics, leading to the failure of pathogenic bacteria in colonizing the gut epithelium, hence acting in a synergistic way to inhibit the growth of pathogenic bacteria (Peng et al. 2019). However, before the mass production of various combinations of probiotics and prebiotics, they are needed to be evaluated with empirical evidence in animal nutrition. The gut microbiome being exceptionally diverse, there is a possibility that the introduction of prebiotics or probiotics could also lead to the disturbance of bacterial homeostasis resulting in undesirable effects like diarrhoea. Sometimes, prebiotics can also act as a food source for other pathogenic bacteria which may lead to suppression of growth of health-promoting bacteria either by utilizing their nutrients or by making uninhabitable environment for the growth of health-promoting bacteria, making it very necessary to evaluate the predefined procedures. Due to the natural anaerobic bacterial fermentation in the rumen of ruminant species, such as sheep and cattle, administration of probiotics becomes challenging because compounds that are normally non-digestible easily get degraded before reaching the small intestine (Peng et al. 2019).

13.4.4 Enzymes

Animals utilize the endogenous enzymes produced either by the microorganism residing inside the animal gut or from the animal system itself for the digestion of animal feed. However, most of the animals are not able to digest the greater part of their feed because of non-digestible deleterious factors present that slow down the digestive processes. Most of the plant-based feed contains a broad range of anti-nutritional factors, such as non-starch polysaccharides, cell wall carbohydrates (pectins, beta-glucans, hemicellulose, xylans, and alpha-galactosides), phytic acids, and protease inhibitors that hinders nutrient utilization. Further, the lower digestibility of several feed ingredients due to unavailability of an enzyme (s) required for the breakdown of complex cell wall structure (Castillo and Gatlin 2015), and allowing the microbial population to assimilate a greater proportion of feed ingredients contained within the feed (Thacker 2013), reduces the nutritional value of feed ingredients. Therefore, to alleviate the anti-nutritional effects of non-starch polysaccharides, phytic acid, and nutrient utilization, supplementation of exogenous enzymes in the feed is extensively practised for the last many decades. Some of the commercially available formulations for animal feed are summarized in Table 13.3.

For decades, enzyme supplementation is known to reduce intestinal viscosity and increase nutrient digestibility and nutrient availability by efficient mixing of digesta and degradation of ingredients present in animal feed. Besides, enzyme supplementation has also been documented to improve intestinal mucosal integrity via reducing the inflammation response and modification of host gut microflora, which leaves beneficial effects on the health of the host (Suresh et al. 2020). The most commonly used enzymes in the animal diet include xylanases, beta-glucanases, amylase, phytases, mannanases, pectinases, hemicellulases, proteases, lipases, pentosanase, cellulose, and alpha-galactosidases. The degradation of anti-nutritional factors and feed ingredients that endogenous enzyme cannot digest, increased nutrient availability by degrading deleterious factors responsible for reducing the digestion process and increasing viscosity of the gut, modification of internal morphology, modulation in the microbial population residing in the gastrointestinal tract by a change in the composition of enzymes, and feed solubilisation are some of the mechanisms of action of enzymes (Ojha et al. 2018).

Table 13.3 Commercially available formulations for animal feed

1. Probiotics		
Targeted animal	*Commercial probiotic formulations*	*Microorganisms*
Poultry and pigs	Acid-Pak 4-way (Alltech)	*Enterococcus faecium, Lactobacillus acidophilus*
	Probion (Woogene B and G Co. Ltd.)	*Lactobacillus acidophilus, Clostridium butyricum, Bacillus subtilis*
	Calsporin (ORFFA)	*Bacillus subtilis*
Calves and pigs	Provita LE (Schaumann)	*Enterococcus faecium, Lactobacillus rhamnosus*
	Cernivet LBC (Cerbios)	*Enterococcus faecium*
Calves and poultry	UltraCruz (Santa Cruz Animal Health)	*Lactobacillus acidophilus, Lactobacillus plantarum, Enterococcus faecium*
	Probiomix	*Bifidobacterium bifidum, Enterococcus faecium, Lactobacillus amylovorus*
Poultry, pigs, and calves	Doctor Em® (Biotron)	*Saccharomyces cerevisiae, Lactococcus lactis, Lactobacillus plantarum*
	Bio Plus 2B® (Chr. Hansen)	*Bacillus licheniformis, Bacillus subtilis*
	Oralin® (Chevita GmbH)	*Enterococcus faecium*

2. Synbiotics		
Targeted animal	*Commercial probiotic formulations*	*Probiotic microorganisms and prebiotic substances*
Horses	DigestAid™	*Saccharomyces cerevisiae, Saccharomyces boulardii* and MOS, beta-glucan
Calves, pigs, and poultry	Biomin® IMBO	*Enterococcus faecium* and FOS
Poultry	Synbiotic poultry (Vetafarm)	*Lactobacillus acidophilus, Lactobacillus salivarius, Lactobacillus plantarum, Lactobacillus casei, Streptococcus thermophiles*, and inulin
Poultry	PoultryStar®	*Pediococcus acidilactici, Lactobacillus salivarius, Enterococcus faecium*, and inulin

3. Enzymes		
Targeted animal	*Commercial enzyme formulations*	*Enzyme*
Porcine and poultry	Phytase SP	3-Phytase
Porcine and poultry	Phyzyme XP	6-Phytase
Turkeys	Feedlyve AXC	Endo-1,4-beta-xylanase
Dairy cows	Ronozyme	Alpha-amylase
Piglets, poultry species	Hostazym C	Endo-1,4-beta-glucanase

13.4.5 Phytogenics

For decades, plants are known for their advantageous multifunctional properties like prevention and treatment of various diseases of farm animals and hence have been used as feed additives in animal husbandry. These feed additives can be obtained from a wide variety of plants including herbs, spices, and plant products, called phytogenics that can be categorized as essential oil, crude or processed plant parts, mixtures of powders or extracts, processed extracts, and phytochemicals. Phytogenics are naturally available, safe to consume, residue-free, environmentally friendly, and effective due to the presence of a varying degree of growth-promoting nutraceuticals.

13.4.6 Essential Oils

Essential oils are volatile or ethereal oils obtained from medicinal and aromatic plant materials, having their characteristic odour and flavour. As reported earlier, they increase the release of digestive enzymes, reduce the number of nutrients available for the growth of bacteria in the lumen of the gut, and have antimicrobial properties to check the growth of harmful bacteria. There are numerous examples of essential oils such as cineol and eucalyptol of *Eucalyptus,* garlic oil, cinnamaldehyde, eugenol, carvacrol, and thymol which can be used as medicinal plant source as an alternative to antibiotics (Swamy et al. 2016).

13.4.7 Phytochemicals

Phytochemicals are plant-derived bioactive chemicals or compounds that display positive effects in animal production. They show antimicrobial properties that reduce the levels of pathogenic infections through their antimicrobial actions by changing permeability or disrupting the cell membrane of microbes and interfering with virulence properties of the microbes by increasing the hydrophobicity, which may influence the surface characteristics of microbial cells. Further, augmentation of digestive secretions, stimulation of blood circulation, exerting antioxidant properties, immunostimulant action, stimulating the proliferation and growth of absorptive cells (villus and crypt) in the gastrointestinal tract leading to the reduced levels of coccidian parasites in broiler chicken, preventing diarrhoea in cattle, exhibiting anti-inflammatory and antiparasitic activities, enhancing dietary palatability, and improving the gut functions and immune status are some of the essential properties possessed by phytochemicals. Phytochemicals are comprised of terpenoids, phenolics, glycosides, alkaloids, alcohols, aldehydes, ketones, esters, ethers, saponins, tannins, and lactones (Huyghebaert et al. 2011). Ginger, pepper, coriander, bay, oregano, rosemary, sage, thyme, cloves, mustard, cinnamon, garlic, lemon, citrus peel (lime, lemon, orange), tobacco, and mint are some of the common plants used for their antibacterial properties (Chouhan 2017). Rahimi-Mianji et al.

(2005) reported increased feed intake, feed conversion ratio, and body weight improvement when plant extracts were supplemented in the animal food.

Phytogenics, including powder/extracts also called as called botanical supplements, are known to improve nutrient utilization, absorption, and stimulation of the immune system. However, phytogenics are still considered a very complex blend of bioactive components with a lot of variations in the composition because of their factors including plant species, growing location, and harvest conditions. Therefore, in-depth study of action mechanisms, quantitative evaluation of their effects, investigation of their interaction with the microbiota, analysis of their benefits at specific life stages, compatibility with diet, toxicity, and safety assessment is advocated.

13.4.8 Antimicrobial Peptides

Antimicrobial peptides (AMPs) are low-mass cationic oligopeptides with antibacterial, antiviral, antifungal, and antiparasitic effects and are produced from various microorganisms, plants, invertebrates, and vertebrates. At present, more than 3000 different AMPs with a vast and extremely diverse group of molecules are known. Over 74% of all AMPs are produced from animals such as bovine, caprine, ovine, porcine, chicken, turkey, and other animal species origin. These are considered as a potentially promising alternative for growth promotion in addition to disease prevention and treatment. They are also acknowledged for their immune-modulatory, health-protective, and anti-carcinogenic effect. In addition to the growth promotion, they exhibit irreversible damage to the microbial cell membrane, resulting in cell lysis of harmful bacteria, DNA and protein synthesis, and food storage improvement as they prevent food spoilage as well as contamination. They have been reported to show antibacterial effects against animal pathogens, viz. *Mannheimia haemolytica, E. coli. Klebsiella pneumoniae, Pseudomonas aeruginosa,* and other Gram-positive and Gram-negative bacteria and fungi (Chouhan et al. 2017). In vitro as well as in vivo studies have provided strong circumstantial evidence that AMPs have great potentials in being used for improving intestinal health, increasing beneficial bacteria in the guts, and suppressing harmful bacteria leading to increased daily weight gain when compared to weight gain in antibiotic control animals.

AMPs produced by both probiotic Gram-negative and Gram-positive bacteria are called bacteriocins and are ribosomally synthesized by digestive enzymes. They are highly effective as they can act at very low concentrations, i.e. pico- or nano-molar concentrations. They are categorized as per their chemical structure, mechanism of action, molecular weight, and action spectrum. However, pure bacteriocins being sensitive to pH and proteolytic enzymes limit their supply. There are several products available including nisin, lactoferrin, cecropin, and defensin which are used in animal husbandry. Nisin being used for sanitizing the udder before milking has demonstrated significant reductions in udder pathogens, whereas lactoferrin is used for the treatment of cows and suckling piglets against mastitis pathogens.

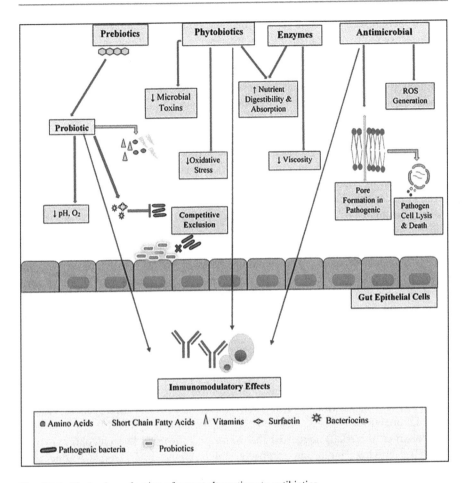

Fig. 13.1 Mechanism of action of potent alternatives to antibiotics

Cecropin, lactoferrin, and defensin are also being effectively used for treating piglets and were found to reduce the frequency of bacterial diarrhoea. A snapshot of the overall view of the mechanism of action of these potent alternatives to antibiotics is depicted in Fig. 13.1.

13.5 Medicinal Plant Products Targeting Pathogenicity

13.5.1 Quorum Sensing Inhibitors

The communication system in the bacterial population that allows the expression of new genes to respond against repellents and towards attractants is called quorum sensing (QS). Gene regulation in bacteria by quorum sensing leads to the enhancement of pathogenicity and antibiotic resistance in the bacterial population. This system participates to form a biofilm, bacteria's communal weapon against antibiotics which makes them highly antibiotic-resistant. Bouyahya et al. (2017) reviewed various QS signalling pathways such as AHL molecule-based QS, peptide secretion-based QS, and production of AI-2 system. In AHL molecule-based QS, autoinducer-like AHL molecules are produced by lactone synthase. Once cell density is increased, i.e. reached quorum, it diffuses into cells and interacts with transcription regulators further activating numerous transcription regulators such as *lasI, lasB*, and *toxA*, whereas in peptide secretion-based QS, peptide formed from oligopeptide called autoinducing oligopeptide (AIPs), are considered under the agrD gene family; after expressing the AIPs, membrane-bound molecule changes its conformation by the thiolactone addition, leading to the exportation of a molecule in the form of an oligopeptide. As the threshold extracellular concentration of AIPs is reached, it binds to the AgrC membrane receptor and activates AgrA, which ultimately activates a further cascade of protein expression in QS like agrBDCA. Besides AHL molecule-based QS and peptide secretion-based QS, the production of the AI-2 system is another QS signalling pathway where AI-2 (furanosyl borate diester) molecule, found in several pathogenic bacteria, such as *Vibrio harveyi, Streptococcus gordonii, and Salmonella typhimurium*, plays an important role. S-adenosyl homocysteine precursor for *luxS* (metalloenzyme) in *P aeruginosa luxS* performs the synthesis of AI-2. *luxS* further regulates the synthesis of AI-2. These mechanisms can be targeted with the help of plant-derived products as quorum sensing inhibitor (QSI). *Campylobacter jejuni* colonized in the broiler chicken (poultry) have shown to have QS which is treated by monoterpene (−)-α-pinene leading to a reduction in quorum sensing of poultry (Šimunović et al. 2020). Henceforth, the QSI study would help in the inhibition of various bacterial infections.

Sharchi et al. (2020) suggested that *Satureja sahendica* hydroalcoholic extract (SSHE) showing inhibition property inhibits bacterial isolate *Salmonella typhimurium* causing quorum sensing and virulence in the poultry. SSHE has antimicrobial as well as anti-QS property which is affected by decreasing the expression of the *sdiA* gene. Licochalcone A (LAA) and epigallocatechin-3 gallate (EGCG) as a natural plant product are found to be effective against *S. typhimurium* isolated from poultry to evade its pathogenicity. The impact of LAA and EGCG sub-inhibitory concentration was studied by targeting QS-associated genes, *sdiA and luxS,* resulting in decreased expression of QS-associated genes, showing its putative role in therapeutics against salmonellosis (Hosseinzadeh et al. 2020). Hence, researchers are using plant cells as future QS inhibitors, though it is still challenging to conduct extensive research to get pharmacokinetics and pharmacodynamics.

13.5.2 Efflux Pump Inhibitors

Besides QS, bacterial efflux pumps are also considered as powerful drug resistance determinants in bacteria and are thus being targeted with medicinal plant products (Yu et al. 2020). Bacteria pump/extrude antibiotics from their lumen to the outer environment to protect themselves using efflux pumps. As novel antibiotic availability is decreasing due to the increasing antibiotic resistance mechanisms, one of the alternatives used by the researchers is to target the bacterial efflux pumps. Inhibitory molecules that help to regain the activity of antibiotics no longer effective can be isolated from synthetic as well as natural sources. Bacteria extrude antibiotics, which make them antibiotic-resistant. The efflux pumps are classified into five major families, namely, small multidrug-resistant family, major facilitator superfamily, adenosine triphosphate-binding cassette superfamily, multidrug and toxic compound extrusion family, and resistance nodulation division superfamily. Efflux pumps of resistance nodulation division superfamily contribute to intrinsic and acquired resistance for a wide variety of antibiotics (Seukep et al. 2020). The mechanism of action of potent plant product targeting pathogenicity is shown in Fig. 13.2.

Bacteria use these mechanisms to avoid antibiotic intrusion, which helps them easily escape and grow. Medicinal plant products have been proven effective against efflux pumps due to their diverse nature in a pharmacological property (reviewed by Shriram et al. 2018). Therefore, EP contributes to bacterial survival and stress response. Investigation of characterizing efflux pump inhibitor to block drug extrusion and restoring antibacterial susceptibility has become the priority of investigation making it challenging to develop a new clinically proven efflux pump inhibitor. Mutations that cause the overexpression of genes encoding efflux pump proteins usually lead to the development of AMR in bacteria (Vidovic and Vidovic 2020). In silico approaches including molecular docking and simulation are being successfully utilized to screen and predict molecular interactions between efflux pump and its inhibitor (Shriram et al. 2018).

13.5.3 Bacterial Virulence Inhibitor

The bacterial virulence is promoted by the combined effect of bacterial virulence factors, which include QS, biofilms, bacterial motility, toxins, pigments, surfactants, and enzymes. The combined effect can be due to the host invasion, tissue colonization, tissue damage, and host defence invasion. Heading towards the post-antibiotic era, due to the longer effectiveness of antibiotics, we are tending to face problems in healthcare against various microbial threats. Therefore, anti-virulence therapy has come up as an alternative approach that targets bacterial virulence. These virulence factors not essential for microbial survival can lead us to generate antimicrobials targeting novel action mechanisms, check the range of pharmacological targets, and develop mild pressure on developing antimicrobial resistance (Silva et al. 2016). Erfan and Marouf (2019) found *Staphylococcus aureus sed* gene, *Escherichia coli stx1* gene, *Avibacterium paragallinarum HPG-2* gene, *Pasteurella multocida ptfA* gene, *Mycoplasma gallisepticum Mgc2* gene, and *Ornithobacterium rhinotracheale adk* gene from poultry bacterial agents, which were downregulated by the cinnamon oil isolated from respiratory bacterial agents in poultry. Though the adjuvant effect may not be effective alone to combat bacterial infection, a combination strategy can

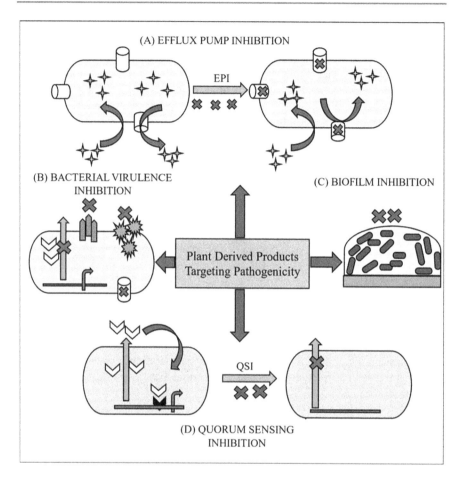

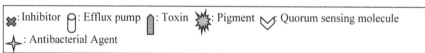

Fig. 13.2 Mechanism of action of potent plant products targeting pathogenicity. *EPI* efflux pump inhibitor, *QSI* quorum sensing molecule

be applied to treat with anti-virulence agents and antibiotics; however, still more work needs to be undertaken in this direction for its full potential exploration.

13.5.4 Biofilm Inhibitors

The protective layer formed by bacteria with polysaccharides, lipoproteins, and fibrinogen to evade many extreme conditions is called biofilm (Benzaid et al.

2021). To eradicate biofilm, antibiotics were used, such as for protein synthesis inhibition (oxazolidinones and tetracyclines), cell membrane and wall-active antibiotics (lipopeptides and glycolipids), and DNA and RNA synthesis inhibition (rifampin) are usually used (Cheng et al. 2014). Hall and Mah (2017) described resistance mechanisms such as nutrient gradient with less nutrient availability in the core of the biofilm, stress responses (oxidative stress response, stringent response, etc.), discrete genetic determinants that are specifically expressed in biofilms and whose gene products act to reduce biofilm susceptibility via diverse mechanisms (*ndvB, brlR*), intercellular interactions (horizontal gene transfer, QS, multispecies communication), and persister cells (dormant and non-growing) taken up by the bacterial strains.

Single-drug therapy is often ineffective against these biofilms because it involves intricate phenomena like bacterial cell adhesion, quorum sensing regulation, and biofilm maturation (Cheng et al. 2014). Therefore, a combinational approach is needed where combinations of different antibiotics or antibiotics with the biofilm inhibitors can be used (Cheng et al. 2014). The plant-derived compounds in *Citrus aurantium* essential oil showed decreased growth of *Streptococcus mutants* leading to biofilm degradation. Seyedtaghiya et al. (2021) reported *Satureja hortensis* essential oil showing anti-biofilm property, resulting in decreased salmonellosis and colibacillosis severity, caused by poultry isolates. The cost-effectiveness of antibiotics on *Staphylococcus aureus* biofilm, regained with polylactic acid (PLA) nanoparticles, showed improved retention of rifampicin-loaded PLA with better antibiotic efficacy (Drulis-kawa 2021). Lu et al. (2019) described some of the natural anti-biofilm agents such as garlic, *Cocculus trilobus,* and *Coptis chinensis* extracts against biofilm formation; cranberry polyphenols and components affecting glucan-binding proteins causing inhibition of the biofilm-forming ECM, proteolytic activities, and coaggregation; *Herba patriniae* extract showing interference between the formation of mature biofilm in *P. aeruginosa*; and phloretin inhibiting toxin genes (*hlyE* and *stx*(2)), autoinducer-2 importer genes (lsrACDBF), curli genes (*csgA* and *csgB*), and prophage genes in *E. coli* O157:H7 biofilm cell showing the anti-biofilm property. Quercetin is another plant polyphenol that showed anti-biofilm activity in *Streptococcus mutans and Enterococcus faecalis* (Yang et al. 2020).

Combinational treatment of biofilm is complicated and needs an effective study of bioactive compounds. It is challenging to treat biofilm-related infections, though essential oil extract showed to be effective against biofilm; there is a need for comprehensive investigations and clinical trials.

References

Aarestrup FM (2005) Veterinary drug usage and antimicrobial resistance in bacteria of animal origin. Basic Clin Pharmacol Toxicol 96(4):271–281. https://doi.org/10.1111/j.1742-7843. 2005.pto960401.x

Acar J, Röstel B (2001) Antimicrobial resistance: an overview. Rev Sci Tech 20(3):797–810. https://doi.org/10.20506/rst.20.3.1309

Akuru EA, Mpendulo CT, Oyeagu CE, Nantapo CWT (2021) Pomegranate (*Punica granatum* L.) peel powder meal supplementation in broilers: effect on growth performance, digestibility, carcase and organ weights, serum and some meat antioxidant enzyme biomarkers. Ital J Anim Sci 20:119–131. https://doi.org/10.1080/1828051X.2020.1870877

Al-Shawi SG, Dang DS, Yousif AY et al (2020) The potential use of probiotics to improve animal health, efficiency, and meat quality: a review. Agri 10(10):1–14. https://doi.org/10.3390/agriculture10100452

Barbarestani YS, Jazi V, Mohebodini H et al (2020) Effects of dietary lavender essential oil on growth performance, intestinal function, and antioxidant status of broiler chickens. Livest Sci 233:103958. https://doi.org/10.1016/j.livsci.2020.103958

Basit MA, Kadir AA, Loh TC et al (2020) Comparative efficacy of selected phytobiotics with halquinol and tetracycline on gut morphology, ileal digestibility, cecal microbiota composition and growth performance in broiler chickens. Animals (Basel) 10(11):2150. https://doi.org/10.3390/ani10112150

Batista VL, da Silva TF, de Jesus LCL et al (2020) Probiotics, prebiotics, synbiotics, and paraprobiotics as a therapeutic alternative for intestinal mucositis. Front Microbiol 11:544490. https://doi.org/10.3389/fmicb.2020.544490

Benbachir M, Benredjeb S, Boye CS, Dosso M, Belabbes H, Kamoun A, Kair O, Elmdaghri N (2001) Two-year surveillance of antibiotic resistance in *Streptococcus pneumoniae* in four African cities. Antimicrob Agents Chemother 45(2):627–629. https://doi.org/10.1128/AAC.45.2.627-629.2001

Bengtsson B, Greko C (2014) Antibiotic resistance-consequences for animal health, welfare, and food production. Ups J Med Sci 119(2):96–102. https://doi.org/10.3109/03009734.2014.901445

Benzaid C, Belmadani A, Tichati L et al (2021) Effect of *citrus aurantium* l. essential oil on streptococcus mutans growth, biofilm formation and virulent genes expression. Antibiotics (Basel) 10(1):54. https://doi.org/10.3390/antibiotics10010054

Binda S, Hill C, Johansen E et al (2020) Criteria to qualify microorganisms as "probiotic" in foods and dietary supplements. Front Microbiol 11:1662. https://doi.org/10.3389/fmicb.2020.01662

Bouyahya A, Dakka N, Et-Touys A et al (2017) Medicinal plant products targeting quorum sensing for combating bacterial infections. Asian Pac J Trop Med 10(8):729–743. https://doi.org/10.1016/j.apjtm.2017.07.021

Boxall ABA, Blackwell PA, Cavallo R, Kay P, Tolls J (2002) The sorption and transport of a sulphonamide antibiotics in soil systems. Toxicol Lett 131(1–2):19–28. https://doi.org/10.1016/s0378-4274(02)00063-2

Castanon JIR (2007) History of the use of antibiotic as growth promoters in European poultry feeds. Poult Sci 86(11):2466–2471. https://doi.org/10.3382/ps.2007-00249

Castillo S, Gatlin DM (2015) Dietary supplementation of exogenous carbohydrase enzymes in fish nutrition: a review. Aquaculture 435:286–292. https://doi.org/10.1016/j.aquaculture.2014.10.011

Center for Disease Dynamics, Economics & Policy (CDDEP) (2016) Antibiotic use and resistance in food animals current policy and recommendations. https://cddep.org/wp-content/uploads/2017/06/india_abx_report-2.pdf

Chang CH, Teng PY, Lee TT, Yu B (2020) Effects of multi-strain probiotic supplementation on intestinal microbiota, tight junctions, and inflammation in young broiler chickens challenged with *Salmonella enterica subsp. enterica*. Asian-Australasian J Anim Sci 33(11):1797–1808. https://doi.org/10.5713/ajas.19.0427

Cheng G, Ning J, Ahmed S, et al (2019) Selection and dissemination of antimicrobial resistance in Agri-food production. Antimicrob Resist Infect Control 8:158. https://doi.org/10.1186/s13756-019-0623-2

Cheng G, Hao H, Xie S et al (2014) Antibiotic alternatives: the substitution of antibiotics in animal husbandry? Front Microbiol 5:217. https://doi.org/10.3389/fmicb.2014.00217

Chouhan S, Sharma K, Guleria S (2017) Antimicrobial activity of some essential oils-present status and future perspectives. Medicines (Basel) 4(3):58. https://doi.org/10.3390/medicines4030058

Chowdhury S, Mandal GP, Patra AK et al (2018) Different essential oils in diets of broiler chickens: 2. Gut microbes and morphology, immune response, and some blood profile and antioxidant enzymes. Anim Feed Sci Technol 236:39–47. https://doi.org/10.1016/j.anifeedsci.2017.12.003

Curbelo YG, López MG, Bocourt R et al (2012) Prebiotics in the feeding of monogastric animals. Cuba J Agric Sci 46(3)

Cunningham VL (2008) Special characteristics of pharmaceuticals related to environmental fate. ChemInform 36:23–34. https://doi.org/10.1007/978-3-540-74664-5_2

Dallal SMM, Davoodabadi A, Abdi M et al (2017) Inhibitory effect of Lactobacillus plantarum and Lb. fermentum isolated from the faeces of healthy infants against nonfermentative bacteria causing nosocomial infections. New Microbes New Infect 15:9–13. https://doi.org/10.1016/j.nmni.2016.09.003

Dargahi N, Johnson J, Donkor O et al (2019) Immunomodulatory effects of probiotics: can they be used to treat allergies and autoimmune diseases? Maturitas 119:25–38. https://doi.org/10.1016/j.maturitas.2018.11.002

Davani-Davari D, Negahdaripour M, Karimzadeh I et al (2019) Prebiotics: definition, types, sources, mechanisms, and clinical applications. Foods 8(3):92. https://doi.org/10.3390/foods8030092

De Cesare A, Sala C, Castellani G et al (2020) Effect of Lactobacillus acidophilus D2/CSL (CECT 4529) supplementation in drinking water on chicken crop and caeca microbiome. PLoS One 15 (1):e0228338. https://doi.org/10.1371/journal.pone.0228338

Debbou-Iouknane N, Nerín C, Amrane M et al (2019) In vitro Anticoccidial activity of olive pulp (Olea europaea L. var. Chemlal) extract against Eimeria Oocysts in broiler chickens. Acta Parasitol 64:887–897. https://doi.org/10.2478/s11686-019-00113-0

De Vrese M, Schrezenmeir J (2008) Probiotics, prebiotics and synbiotics. In: Stahl U, Donalies UEB, Nevoigt E (eds) Food biotechnology, advances in biochemical engineering/biotechnology. Springer, Berlin, pp 1–66

Drulis-kawa Z (2021) Bacteriophages and biofilms. Viruses 13(2):257. https://doi.org/10.3390/v13020257

Duff GC, Galyean ML (2007) Board-invited review: recent advances in management of highly stressed, newly received feedlot cattle. J Anim Sci 85(3):823–840. https://doi.org/10.2527/jas.2006-501

Elshaghabee FMF, Rokana N, Gulhane RD et al (2017) Bacillus as potential probiotics: status, concerns, and future perspectives. Front Microbiol 8:1490. https://doi.org/10.3389/fmicb.2017.01490

El-Sharkawy H, Tahoun A, Rizk AM et al (2020) Evaluation of bifidobacteria and lactobacillus probiotics as alternative therapy for Salmonella typhimurium infection in broiler chickens. Animals 10(6):1023. https://doi.org/10.3390/ani10061023

Erfan AM, Marouf S (2019) Cinnamon oil downregulates virulence genes of poultry respiratory bacterial agents and revealed significant bacterial inhibition: an in vitro perspective. Vet World 12(11):1707–1715. https://doi.org/10.14202/vetworld.2019.1707-1715

Gbore FA, Oloruntola OD, Adu OA et al (2021) Serum and meat antioxidative status of broiler chickens fed diets supplemented with garlic rhizome meal, moringa leaf meal and their composite. Trop Anim Health Prod 53(1):26. https://doi.org/10.1007/s11250-020-02438-9

Gibson GR, Roberfroid MB (1995) Dietary modulation of the human colonic microbiota: introducing the concept of prebiotics. J Nutr 125(6):1401–1412. https://doi.org/10.1093/jn/125.6.1401

Gillings MR (2013) Evolutionary consequences of antibiotic use for the resistome, mobilome, and microbial pangenome. Front Microbiol 4:4. https://doi.org/10.3389/fmicb.2013.00004

Giri SS, Sukumaran V, Sen SS, Park SC (2018) Use of a potential probiotic, Lactobacillus casei L4, in the preparation of fermented coconut water beverage. Front Microbiol 9:1976. https://doi.org/10.3389/fmicb.2018.01976

Gyawali R, Nwamaioha N, Fiagbor R et al (2019) The role of prebiotics in disease prevention and health promotion. Elsevier, San Diego, pp 151–167. https://doi.org/10.1016/B978-0-12-814468-8.00012-0

Gyles CL (2008) Antimicrobial resistance in selected bacteria from poultry. Anim Health Res Rev 9 (2):149–158. https://doi.org/10.1017/S1466252308001552

Hall CW, Mah TF (2017) Molecular mechanisms of biofilm-based antibiotic resistance and tolerance in pathogenic bacteria. FEMS Microbiol Rev 41(3):276–301. https://doi.org/10.1093/femsre/fux010

Hajati H, Rezaei M (2010) The application of prebiotics in poultry production. Int J Poult Sci 9(3). https://doi.org/10.3923/ijps.2010.298.304

Halling-Sørenson B, Sengelov G, Tjornelund J (2002) Toxicity of tetracyclines and tetracycline degradation products to environmentally relevant bacteria, including selected tetracycline resistant bacteria. Arch Environ Contam Toxicol 42(3):236–271. https://doi.org/10.1007/s00244-001-0017-2

Herrero-Encinas J, Blanch M, Pastor JJ et al (2020) Effects of a bioactive olive pomace extract from Olea europaea on growth performance, gut function, and intestinal microbiota in broiler chickens. Poult Sci 99(1):2–10. https://doi.org/10.3382/ps/pez467

Hill C, Guarner F, Reid G et al (2014) Expert consensus document: the international scientific association for probiotics and prebiotics consensus statement on the scope and appropriate use of the term probiotic. Nat Rev Gastroenterol Hepatol 11:506–514. https://doi.org/10.1038/nrgastro.2014.66

Hogeveen H, Huijps K, Lam TJ (2011) Economic aspects of mastitis: new developments. New Zealand Vet J 59(1):16–23. https://doi.org/10.1080/00480169.2011.547165

Hosain MZ, Lutful Kabir SM, Kamal MM (2021) Antimicrobial uses for livestock production in developing countries. Vet World 14(1):210–221. https://doi.org/10.14202/VETWORLD.2021.210-221

Hosseinzadeh S, Saei HD, Ahmadi M, Zahraei-Salehi T (2020) Anti-quorum sensing effects of licochalcone a and epigallocatechin-3-gallate against Salmonella typhimurium isolates from poultry sources. Vet Res Forum 11(3):273–279. https://doi.org/10.30466/vrf.2019.95102.2289

Huyghebaert G, Ducatelle R, Van Immerseel F (2011) An update on alternatives to antimicrobial growth promoters for broilers. Vet J 187(2):182–188. https://doi.org/10.1016/j.tvjl.2010.03.003

Indira M, Venkateswarulu TC, Abraham Peele K et al (2019) Bioactive molecules of probiotic bacteria and their mechanism of action: a review. 3 Biotech 9:306. https://doi.org/10.1007/s13205-019-1841-2

Kim GB, Seo YM, Kim CH, Paik IK (2011) Effect of dietary prebiotic supplementation on the performance, intestinal microflora, and immune response of broilers. Poult Sci 90(1):75–82. https://doi.org/10.3382/ps.2010-00732

Khochamit N, Siripornadulsil S, Sukon P, Siripornadulsil W (2020) Bacillus subtilis and lactic acid bacteria improve the growth performance and blood parameters and reduce salmonella infection in broilers. Vet World 13(1):2663–2672. https://doi.org/10.14202/vetworld.2020.2663-2672

Kirchhelle C (2018) Pharming animals: a global history of antibiotics in food production (1935–2017). Palgrave Commun 4:96. https://doi.org/10.1057/s41599-018-0152-2

Kumar P, Patra AK, Mandal GP, Samanta I, Pradhan S (2017) Effect of black cumin seeds on growth performance, nutrient utilisation, immunity, gut health, and nitrogen excretion by broiler chickens. J Sci Food Agric 97(11):3742–3751. https://doi.org/10.1002/jsfa.8237

Kumar RR, Lee JT, Cho JY (2012) Fate, occurrence, and toxicity of veterinary antibiotics in environment. J Korean Soc Appl Biol Chem 55:701–709. https://doi.org/10.1007/s13765-012-2220-4

Kumar P, Patra AK, Mandal GP et al (2017) Effect of black cumin seeds on growth performance, nutrient utilization, immunity, gut health and nitrogen excretion in broiler chickens. J Sci Food Agric 97(11):3742–3751. https://doi.org/10.1002/jsfa.8237

Kümmerer K, Henninger A (2003) Promoting resistance by the emission of antibiotics from hospitals and households into effluents. Clin Microbiol Infect 9(12):1203–1214. https://doi.org/10.1111/j.1469-0691.2003.00739.x

Laloučková K, Skřivanová E (2019) Antibiotic resistance in livestock breeding: a review. Sci Agric Bohem 50(1):15–22. https://doi.org/10.2478/sab-2019-0003

Lindberg JE (2014) Fiber effects in nutrition and gut health in pigs. J Anim Sci Biotechnol 5(1):15. https://doi.org/10.1186/2049-1891-5-15

Lu L, Hu W, Tian Z et al (2019) Developing natural products as potential anti-biofilm agents. Chinese Med 14:11. https://doi.org/10.1186/s13020-019-0232-2

Liu T, Chen H, Bai Y et al (2020) Calf starter containing a blend of essential oils and prebiotics affects the growth performance of Holstein calves. J Dairy Sci 103(3). https://doi.org/10.3168/jds.2019-16647

Mahalmani V, Sarma P, Prakash A, Medhi B (2019) Positive list of antibiotics and food products: current perspective in India and across the globe. Indian J Pharm 51(4):231–235. https://doi.org/10.4103/ijp.IJP_548_19

Malik JK, Prakash A, Srivastava AK, Gupta RC (2019) Synbiotics in animal health and production. In: Gupta R, Srivastava A, Lall R (eds) Nutraceuticals in veterinary medicine. Springer, Cham. https://doi.org/10.1007/978-3-030-04624-8_20

Manyi-Loh C, Mamphweli S, Meyer E, Okoh A (2018) Antibiotic use in agriculture and its consequential resistance in environmental sources: potential public health implications. Molecules 23(4):795. https://doi.org/10.3390/molecules23040795

Martínez JL (2012) Natural antibiotic resistance and contamination by antibiotic resistance determinants: the two ages in the evolution of resistance to antimicrobials. Front Microbiol 3:1. https://doi.org/10.3389/fmicb.2012.00001

McManus PS, Stockwell VO, Sundin GW, Jones AL (2002) Antibiotic use in plant agriculture. Annu Rev Phytopathol 40:443–465. https://doi.org/10.1146/annurev.phyto.40.120301.093927

Mehndiratta PL, Bhalla P (2014) Use of antibiotics in animal agriculture & emergence of methicillin-resistant *Staphylococcus aureus* (MRSA) clones: need to assess the impact on public health. Indian J Med Res 140(3):339–344

Milic N, Milanovic M, Letic NG, Sekulic MT, Radonic J, Mihajlovic I, Miloradov V (2013) Occurrence of antibiotics as emerging contaminant substances in the aquatic environment. Int J Environ Health Res 23:296–310. https://doi.org/10.1080/09603123.2012.733934

Nabizadeh A (2012) The effect of inulin on broiler chicken intestinal microflora, gut morphology, and performance. J Anim Feed Sci 21(4):725–734. https://doi.org/10.22358/jafs/66144/2012

Nair SM, Amalaradjou MA, Venkitanarayanan K (2017) Antivirulence properties of probiotics in combating microbial pathogenesis. Adv Appl Microbiol 98:1–29. https://doi.org/10.1016/bs.aambs.2016.12.001

Ojha BK, Singh PK, Shrivastava N (2018) Enzymes in the animal feed industry. In: Kuddus M (ed) Enzymes in food biotechnology. Academic Press, Amsterdam, pp 93–109. https://doi.org/10.1016/B978-0-12-813280-7.00007-4

Padihari VP, Tiwari SP, Sahu T, Gendley MK (2014) Effects of Mannan Oligosaccharide and Saccharomyces cerevisiae on gut morphology of broiler chickens. J World's Poult Res 4(3):56–59

Paul I, Isore DP, Joardar SN et al (2012) Orally administered β-glucan of edible mushroom (Pleuratus florida) origin upregulates innate immune response in broiler. Indian J Anim Sci 82(07):745–748

Page SW, Gautier P (2012) Use of antimicrobial agents in livestock. Rev Sci Tech 31:145–188. https://doi.org/10.20506/rst.31.1.2106

Peng M, Patel P, Nagarajan V et al (2019) Feasible options to control colonization of enteric pathogens with designed synbiotics. In: Dietary interventions in gastrointestinal diseases. Elsevier, San Diego, pp 135–149. https://doi.org/10.1016/B978-0-12-814468-8.00011-9

Perricone V, Comi M, Giromini C et al (2020) Green tea and pomegranate extract administered during critical moments of the production cycle improves blood antiradical activity and alters cecal microbial ecology of broiler chickens. Animals 10(5):785. https://doi.org/10.3390/ani10050785

Phares CA, Danquah A, Atiah K, Agyei FK, Michael OT (2020) Antibiotics utilization and farmers' knowledge of its effects on soil ecosystem in the coastal drylands of Ghana. PLoS One 15:2. https://doi.org/10.1371/journal.pone.0228777

Qaid MM, Al-Mufarrej SI, Azzam MM, Al-Garadi MA (2021b) Anticoccidial effectivity of a traditional medicinal plant, *Cinnamomum verum*, in broiler chickens infected with Eimeria tenella. Poult Sci 100(3):100902. https://doi.org/10.1016/j.psj.2020.11.071

Qaid MM, Al-Mufarrej SI, Azzam MM et al (2021a) Growth performance, serum biochemical indices, duodenal histomorphology, and cecal microbiota of broiler chickens fed on diets supplemented with cinnamon bark powder at prestarter and starter phases. Animals 11(1):94. https://doi.org/10.3390/ani11010094

Raheem MA, Jiangang H, Yin D et al (2021) Response of lymphatic tissues to natural feed additives; curcumin (*Curcuma longa*) and black cumin seeds (*Nigella sativa*) in broilers against Pasteurella multocida. Poult Sci. https://doi.org/10.1016/j.psj.2021.01.028

Rahimi-Mianji G, Rezaei M, Hafezian H (2005) The effect of intermittent lighting schedule on broiler performance. Int J Poult Sci 4(6):396–398. https://doi.org/10.3923/ijps.2005.396.398

Ramlucken U, Ramchuran SO, Moonsamy G et al (2020) A novel *Bacillus* based multi-strain probiotic improves growth performance and intestinal properties of *Clostridium perfringens* challenged broilers. Poult Sci 99(1):331–341. https://doi.org/10.3382/ps/pez496

Roberfroid M (2007) Prebiotics: the concept revisited. J Nutr 137(3):830S–837S. https://doi.org/10.1093/jn/137.3.830s

Ronquillo MG, Hernandez JCA (2017) Antibiotic and synthetic growth promoters in animal diets: review of impact and analytical methods. Food Control 72:255–267. https://doi.org/10.1016/j.foodcont.2016.03.001

Sahoo KC, Tamhankar AJ, Sahoo S, Sahu PS, Klintz SR, Lindborg CS (2012) Geographical variation in antibiotic-resistant *Escherichia coli* isolates from stool, cow-dung and drinking water. Int J Environ Res Public Health 9(3):746–759. https://doi.org/10.3390/ijerph9030746

Seal BS, Lillehoj HS, Donovan DM, Gay CG (2013) Alternatives to antibiotics: a symposium on the challenges and solutions for animal production. Anim Health Res Rev 14(1):78–87. https://doi.org/10.1017/S1466252313000030

Šefcová M, Larrea-Álvarez M, Larrea-Álvarez C et al (2020) Effects of *Lactobacillus fermentum* supplementation on body weight and pro-inflammatory cytokine expression in *Campylobacter Jejuni*-challenged chickens. Vet Sci 7(3):121. https://doi.org/10.3390/vetsci7030121

Šefcová MA, Larrea-Álvarez M, Larrea-Álvarez CM et al (2021) The probiotic *Lactobacillus fermentum* biocenol CCM 7514 moderates campylobacter jejuni-induced body weight impairment by improving gut morphometry and regulating cecal cytokine abundance in broiler chickens. Animals 11(1):235. https://doi.org/10.3390/ani11010235

Selaledi LA, Hassan ZM, Manyelo TG, Mabelebele M (2020) The current status of the alternative use to antibiotics in poultry production: an African perspective. Antibiotics 9(9):594. https://doi.org/10.3390/antibiotics9090594

Seukep AJ, Kuete V, Nahar L et al (2020) Plant-derived secondary metabolites as the main source of efflux pump inhibitors and methods for identification. J Pharm Anal 10(2):277–290. https://doi.org/10.1016/j.jpha.2019.11.002

Seyedtaghiya MH, Fasaei BN, Peighambari SM (2021) Antimicrobial and antibiofilm effects of *Satureja hortensis* essential oil against *Escherichia coli* and *Salmonella* isolated from poultry. Iran J Microbiol 13(1):74–80. https://doi.org/10.18502/ijm.v13i1.5495

Sharchi R, Shayegh J, Somayyeh H (2020) Anti-quorum sensing and antibacterial activities of Satureja sahendica hydroalcoholic extract against Salmonella typhimurium isolated from poultry flocks. Iran J Vet Sci Technol 12:71. https://doi.org/10.22067/veterinary.v12i1.85433

Shokryazdan P, Jahromi MF, Liang JB et al (2017) Effects of a Lactobacillus salivarius mixture on performance, intestinal health and serum lipids of broiler chickens. PLoS One 12(5):e0175959. https://doi.org/10.1371/journal.pone.0176065

Shriram V, Khare T, Bhagwat R, Shukla R, Kumar V (2018) Inhibiting bacterial drug efflux pumps via phyto-therapeutics to combat threatening antimicrobial resistance. Front Microbiol 9:2990. https://doi.org/10.3389/fmicb.2018.02990

Silva LN, Zimmer KR, Macedo AJ, Trentin DS (2016) Plant natural products targeting bacterial virulence factors. Chem Rev 116(16):9162–9236. https://doi.org/10.1021/acs.chemrev. 6b00184

Šimunović K, Sahin O, Kovač J et al (2020) (−)-α-Pinene reduces quorum sensing and *Campylobacter jejuni* colonization in broiler chickens. PLoS One 15(4):1–16. https://doi.org/10.1371/journal.pone.0230423

Suliman GM, Alowaimer AN, Al-Mufarrej SI et al (2021) The effects of clove seed (*Syzygium aromaticum*) dietary administration on carcass characteristics, meat quality, and sensory attributes of broiler chickens. Poult Sci 100(3):100904. https://doi.org/10.1016/j.psj.2020.12. 009

Singh PK, Kumar A (2018) Effect of dietary black cumin (Nigella sativa) on the growth performance, nutrient utilization, blood biochemical profile and carcass traits in broiler chickens. Anim Nutr Feed Technol 18(3):409. https://doi.org/10.5958/0974-181X.2018.00038.0

Suresh G, Santos DU, Rouissi T et al (2020) In-field poultry tests to evaluate efficacy of bioformulation consisting of enzymes and yeast biomass. Anim Feed Sci Technol 262:114398. https://doi.org/10.1016/j.anifeedsci.2020.114398

Swamy MK, Akhtar MS, Sinniah UR (2016) Antimicrobial properties of plant essential oils against human pathogens and their mode of action: an updated review. Evid Based Complement Alternat Med 2016:3012462. https://doi.org/10.1155/2016/3012462

Tabashsum Z, Peng M, Alvarado-Martinez Z et al (2020) Competitive reduction of poultry-borne enteric bacterial pathogens in chicken gut with bioactive *Lactobacillus casei*. Sci Rep 10:16259. https://doi.org/10.1038/s41598-020-73316-5

Tako E, Glahn RP, Knez M, Stangoulis JC (2014) The effect of wheat prebiotics on the gut bacterial population and iron status of iron deficient broiler chickens. Nutr J 13:58. https://doi.org/10. 1186/1475-2891-13-58

Thacker PA (2013) Alternatives to antibiotics as growth promoters for use in swine production: a review. J Anim Sci Biotechnol 4:35. https://doi.org/10.1186/2049-1891-4-35

Tilman D, Balzer C, Hill J, Befort BL (2011) Global food demand and the sustainable intensification of agriculture. Proc Natl Acad Sci U S A 108(50):20260–20264. https://doi.org/10.1073/pnas.1116437108

Tiwari P, Khare T, Shriram V, Bae H, Kumar V (2021) Plant synthetic biology for producing potent phyto-antimicrobials to combat antimicrobial resistance. Biotechnol Adv. https://doi.org/10. 1016/j.biotechadv.2021.107729

Tomasik PJ, Tomasik P (2003) Probiotics and prebiotics. Cereal Chem 80(2):113–117. https://doi. org/10.1094/CCHEM.2003.80.2.113

Trachsel J, Briggs C, Gabler NK et al (2019) Dietary resistant potato starch alters intestinal microbial communities and their metabolites, and markers of immune regulation and barrier function in swine. Front Immunol 10:1381. https://doi.org/10.3389/fimmu.2019.01381

Tufarelli V, Baghban-kanani P, Azimi-youvalari S et al (2021) Effects of horsetail (*Equisetum arvense*) and spirulina (*Spirulina platensis*) dietary supplementation on laying hens productivity and oxidative status. Animals (Basel) 11(2):335. https://doi.org/10.3390/ani11020335

Vaarten J (2012) Clinical impact of antimicrobial resistance in animals. OIE Rev Sci Tech 31 (1):221–229. https://doi.org/10.20506/rst.31.1.2110

Van Boeckel TP, Brower C, Gilbert M et al (2015) Global trends in antimicrobial use in food animals. Proc Natl Acad Sci U S A 112(18):5649–5654. https://doi.org/10.1073/pnas. 1503141112

Van de Vijver L, Verwer C, Smolders G, Hospers-Brands M, Van Eekeren N (2016) The cycle of veterinary antibiotics in the ecosystem. https://www.louisbolk.org/downloads/3159.pdf. Accessed 15 Jan 21

Van TTH, Yidana Z, Smooker PM, Coloe PJ (2020) Antibiotic use in food animals worldwide, with a focus on Africa: pluses and minuses. J Global Antimicrob Res 20:170–177. https://doi.org/10.1016/j.jgar.2019.07.031

Vidovic N, Vidovic S (2020) Antimicrobial resistance and food animals: influence of livestock environment on the emergence and dissemination of antimicrobial resistance. Antibiotics 9 (2):52. https://doi.org/10.3390/antibiotics9020052

Wainwright M (2007) The first miracle drugs: how the sulfa drugs transformed medicine (review). Perspect Biol Med 50(4):639–642. https://doi.org/10.1353/pbm.2007.0057

Walia K, Sharma M, Vijay S, Shome BR (2019) Understanding policy dilemmas around antibiotic use in food animals & offering potential solutions. Indian J Med Res 149(2):107–118. https://doi.org/10.4103/ijmr.IJMR_2_18

Wang J, Wan C, Shuju Z et al (2021) Differential analysis of gut microbiota and the effect of dietary *Enterococcus faecium* supplementation in broiler breeders with high or low laying performance. Poult Sci 100(2):1109–1119. https://doi.org/10.1016/j.psj.2020.10.024

Wang WC, Yan FF, Hu JY et al (2018) Supplementation of *Bacillus subtilis*-based probiotic reduces heat stress-related behaviors and inflammatory response in broiler chickens. J Anim Sci 96(5):1654–1666. https://doi.org/10.1093/jas/sky092

WHO (2019) No time to wait: securing the future from drug-resistant infections. Artforum Int. Available at: https://www.who.int/docs/defaultsource/documents/no-time-to-wait-securing-the-future-from-drug-resistant-infections-en.pdf?sfvrsn=5b424d7_6

Yang D, Wang T, Long M, Li P (2020) Quercetin: its main pharmacological activity and potential application in clinical medicine. Oxid Med Cell Longev 2020:1–13. https://doi.org/10.1155/2020/8825387

Yu Z, Tang J, Khare T, Kumar V (2020) The alarming antimicrobial resistance in ESKAPEE pathogens: can essential oils come to the rescue? Fitoterapia 140:104433. https://doi.org/10.1016/j.fitote.2019.104433

Zhou H, Chen R, Wang J et al (2020) Effect of Macleaya cordata extract on laying performance, egg quality, and serum indices in Xuefeng black-bone chicken. Poult Sci 100(4):101031. https://doi.org/10.1016/j.psj.2021.101031

Recent Updates on Bacterial Secondary Metabolites to Overcome Antibiotic Resistance in Gram-Negative Superbugs: Encouragement or Discontinuation?

14

Manoj Jangra, Parminder Kaur, Rushikesh Tambat, Vrushali Raka, Nisha Mahey, Nishtha Chandal, Shobit Attery, Vikas Pathania, Vidhu Singh, and Hemraj Nandanwar

Contents

M. Jangra · P. Kaur · R. Tambat · V. Raka · N. Mahey · N. Chandal · S. Attery · V. Pathania · V. Singh · H. Nandanwar (✉)
Clinical Microbiology and Antimicrobial Research Laboratory, CSIR-Institute of Microbial Technology, Chandigarh, India
e-mail: hemraj@imtech.res.in

V. Kumar et al. (eds.), *Antimicrobial Resistance*,
https://doi.org/10.1007/978-981-16-3120-7_14

385

Abstract

The rapid spread of COVID-19 has dramatically changed our perspective about how we should be well prepared for upcoming health disasters in the future. Like COVID-19, the world does not seem prepared to fight the slow-moving pandemic, i.e., antimicrobial resistance (AMR). At present, more than 7,00,000 people per year across the globe succumb to drug-resistant infections. According to several reports, if we fail to respond, AMR could lead to the loss of ten million lives and trillions of money by 2050. Among the different pathogens affecting human health, the World Health Organization (WHO) has recently announced a priority list of drug-resistant bacteria to pave the way for the development of new antibiotics. Gram-negative bacteria such as *Escherichia coli, Klebsiella pneumoniae, Acinetobacter baumannii,* and *Pseudomonas aeruginosa* are the most notorious ones and are responsible for the majority of healthcare-associated infections. These pathogens come under the critical threat category because they express resistance to all of the current antibiotics. The modern combinatorial chemistry approaches and chemical genomics have been unsuccessful to provide enough new antibiotics. In stark contrast to this, natural products have been gifted with remarkable chemical diversity and biological activity. Our modern antibiotic armamentarium was built from microbes' natural products, especially *Streptomyces* spp. and *Bacillus* spp. isolated in the golden era. Today, the antibiotic discovery pipeline has almost dried up, in part due to the rediscovery of already known compounds from bacteria, and no new classes emerged from bacteria until recently. These novel natural antibacterial agents from bacteria resurged a spark in the exploitation of bacteria to find new chemical entities. This chapter mainly focuses on natural antimicrobials and adjuvants isolated from the bacterial domain in the last two decades, i.e., from 2001 to 2020, and their status to fight drug-resistant Gram-negative superbugs. We have also described briefly the discovery of synthetic compounds based on natural scaffolds. In conclusion, the bacterial natural products comprise a goldmine to fight superbugs, and future research should be focused on exploring new antimicrobials from bacterial diversity.

Keywords

Antimicrobial resistance · Gram-negative bacteria · Natural metabolites · Resistance-modifying agents

Abbreviations

AHLs	Acyl-homoserine lactones
AMR	Antimicrobial resistance
BAM	β-barrel assembly machinery
BLIs	β-lactamase inhibitors
CDC	Centers for Disease Control and Prevention
CRE	Carbapenem-resistant *Enterobacteriaceae*
Da	Dalton
EPIs	Efflux pump inhibitors
EPS	Extracellular polymeric substances
ESBL	Extended-spectrum β-lactamase
FDA	Food and Drug Administration
g	Grams
h	Hour
IC_{50}	Half-maximal inhibitory concentration
ICU	Intensive care unit
LD_{50}	Lethal dose, 50%
LPS	Lipopolysaccharides
M.S.	Mass spectrometry
MBC	Minimum bactericidal concentration
mcr	Mobilized colistin-resistant gene
MDR	Multiple drug resistant
MIC	Minimum inhibitory concentration
mM	Millimolar
MTT	3-(4,5-Dimethylthiazol-2-yl)-2,5-diphenyl tetrazolium bromide
MW	Molecular weight
NMR	Nuclear magnetic resonance
NRP(s)	Nonribosomal peptide(s)
NRPS	Nonribosomal peptide synthetases
OM	Outer membrane
PK	Pharmacokinetics
PABA	*Para*-aminobenzoic acid
PD	Pharmacodynamics
PDR	Pan drug resistant
PNBA	*Para*-nitrobenzoic acid
QS	Quorum sensing
QSIs	Quorum sensing inhibitors
RBCs	Red blood cells
Ripp(s)	Ribosomally synthesized and posttranslationally modified peptide(s)

RMAs   Resistance modifying agents
RND   Resistance nodulation and cell division
SAR   Structure-activity relationship
WHO   World Health Organization
XDR   Extensively drug resistant
µg   Microgram
µl   Microliter

14.1 Introduction

14.1.1 Antibiotic Resistance: A Perfect Storm

The development of antimicrobial resistance is a stochastic process that magnifies under the selection pressure of drugs. In other words, microorganisms change themselves, mainly through genetic manipulations, when they are exposed to antimicrobial agents (Fig. 14.1). Improper and excessive use of antimicrobials accelerates this process (WHO 2016). There are reports available that demonstrate the presence of infectious pathogens of animal origin in humans (Lebreton et al. 2013; Reich et al. 2013; van der Mee-Marquet et al. 2011; Wang et al. 2012). Antibiotic consumption in livestock and agriculture is more than humans. For instance, in the USA alone, approximately 51 tons is the daily consumption of total antibiotics, 80% of which is used for livestock (FDA 2015; Giorgi 2016). In 2010, antibiotic consumption was the highest in India. Between 2000 and 2010, global antibiotic consumption increased by 36% (of which 76% mainly attributed to BRICS countries, i.e., Brazil, Russia, India, China, and South Africa). It is proposed to go up to 67% by 2030 (Van Boeckel et al. 2014).

The World Health Organization (WHO) has categorized antibiotic-resistant pathogens into three different priority levels to pave the way for discovering and developing new antimicrobial compounds (Tacconelli et al. 2018). Figure 14.2 represents the global-priority bacterial pathogens. Among them, "ESKAPE" (*Enterococcus faecium, Staphylococcus aureus, Klebsiella pneumoniae, E. coli, Acinetobacter baumannii, Pseudomonas aeruginosa,* and *Enterobacter* species) pathogens defined by Rice et al. (2008) are responsible for the major nosocomial and community infections worldwide (Santajit and Indrawattana 2016; Tzouvelekis et al. 2012; Vasoo et al. 2015). These pathogens can escape out all the drug regimens available today.

14.1.2 Socioeconomic Impact of Antibiotic Resistance

Antibiotic resistance possesses a looming danger to the globe, associated with high morbidity and mortality. It also impacts the nation's healthcare systems' economic status because of prolonged treatments in ICUs (Kunz and Brook 2010). For example, in Europe, antibiotic-resistant infections cost around $1.5 billion per

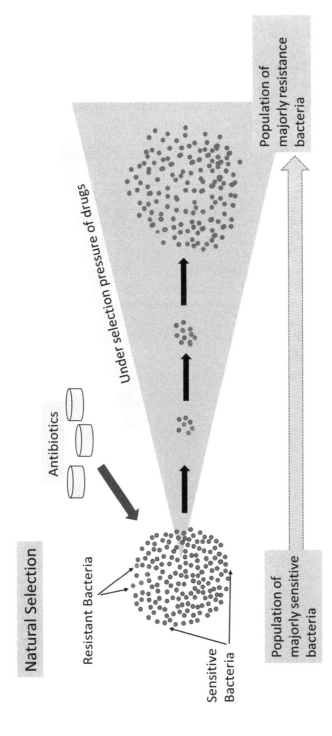

Fig. 14.1 Antibiotic resistance is part of a natural evolution, facilitated by the selection pressure of antibiotics. Source: https://www.reactgroup.org/toolbox/understand/antibiotic-resistance/mutation-and-selection/

Priority 1: CRITICAL

- *Acinetobacter baumannii*, carbapenem-resistant
- *Pseudomonas aeruginosa*, carbapenem-resistant
- *Enterobacteriaceae*, carbapenem-resistant, 3rd generation cephalosporin-resistant

Priority 2: HIGH

- *Enterococcus faecium*, vancomycin-resistant
- *Staphylococcus aureus*, methicillin-resistant, vancomycin intermediate and resistant
- *Helicobacter pylori*, clarithromycin-resistant
- *Campylobacter*, fluoroquinolone-resistant
- *Salmonella spp.*, fluoroquinolone-resistant
- *Neisseria gonorrhoeae*, 3rd generation cephalosporin-resistant, fluoroquinolone-resistant

Priority 3: MEDIUM

- *Streptococcus pneumoniae*, penicillin-non-susceptible
- *Haemophilus influenzae*, ampicillin-resistant
- *Shigella spp.*, fluoroquinolone-resistant

Fig. 14.2 WHO priority pathogen list for R&D of new antibiotics (Tacconelli et al. 2018)

annum, while in the USA, the estimated cost is $55 billion (Control and Prevention 2013).

Due to the new resistance mechanisms emerging, antimicrobial resistance is now an alarming threat to our society, making common infections complicated and untreatable. As reported by CDC, two million people in the USA acquire infections due to drug-resistant strains annually, of which 23,000 people die. In Europe, 25,000 deaths are attributed to antibiotic-resistant infections (Control and Prevention 2013). In India, the situation is even worse. At present, India's crude infection mortality rate is as high as 416.75 per 100,000 people (Laxminarayan et al. 2013). It is tough to assess the burden of resistance for the overall population, but neonates and older adults are more susceptible to infections and, therefore, associated with many more complications. According to one estimation, more than 58,000 neonates in India die each year due to drug-resistant infections (Laxminarayan and Chaudhury 2016).

14.2   The Need for New Antibiotics

14.2.1 The Golden Era of Antibiotics Versus the "Void" in the Discovery Pipeline

The history of antimicrobial agents dates back to 1910, when a Nobel laureate and German-Jewish physician and scientist Dr. Paul Ehrlich discovered Salvarsan, the first antimicrobial agent for syphilis (Saga and Yamaguchi 2009). In 1928, Alexander Fleming observed that contamination of blue mold (later identified as *Penicillium* sp.) on the agar plate inoculated with *S. aureus* did not allow the bacteria to grow around it (Fleming 1929). This finding led to the theory that microorganisms produce substances that inhibit the growth of other microbes. This observation originated the concept of "antibiotic" and the compound produced by Fleming mold was named penicillin. The discovery of penicillin transformed the world of medical science and saved millions of lives during World War II (Brown and Wright 2016; Saga and Yamaguchi 2009). In the early 1940s, Selman Waksman investigated soil microbial diversity, especially *Streptomyces* spp. for their potential to produce antimicrobial compounds. In 1943, he discovered streptomycin from the soil bacterium *Streptomyces griseus*, the first drug to treat tuberculosis (Lewis 2012). The antibiotic discovery approach used by Waksman was named as Waksman platform. Several researchers and pharmaceutical companies adopted this approach to uncover many new antibiotic classes from the soil microorganisms in the consecutive two decades. This period was termed as "golden age of antibiotics." Polymyxins, chloramphenicol, tetracycline, aminoglycoside, macrolides, glycopeptides (e.g., vancomycin), etc., were discovered during that time. Figure 14.3 shows the antibiotic discovery timeline.

In the late 1960s, it became harder and harder to get novel and effective antibiotics due to the overmining of terrestrial microorganisms (Brown and Wright 2016). Also, the poor pharmacological properties of several natural products limited their use in clinics. The replication of already discovered compounds caused a screeching halt, suggesting the decline in the antibiotic pipeline. In contrast, the emergence of resistant pathogens left its infancy and was causing a problem. Due to these concerns, medicinal chemistry appeared to play a role in the innovation of drug discovery. The improvements in the already existing class of antibiotics led to the development of next-generation antibiotics and new antimicrobial agents. The chemical modifications resulted in better safety profiling of drugs and desirable pharmacological properties (Wright 2017). Examples of synthetic antibiotics and semisynthetic analogues are trimethoprim (pyrimidine), levofloxacin (quinolone), amoxicillin (β-lactam), first- to fifth-generation cephalosporins (β-lactam), imipenem and meropenem (carbapenems), etc. Table 14.1 lists the main classes of developed antibiotics and their analogues. Apparently, some types showed the feasibility of analogue development over the others, and it became expensive to make effective analogues, compared to the emergence of new resistance mechanisms (Coates et al. 2011). Notably, no new class of antibiotics was approved for the next

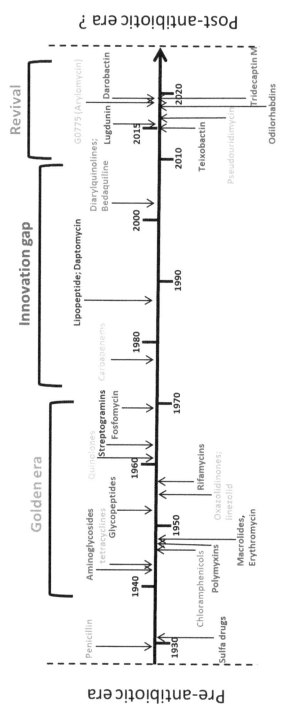

Fig. 14.3 Antibiotic discovery timeline and different phases of development of new antibiotics. Failing to develop enough new antibacterial therapies could lead us to a postantibiotic era where once-curable common infections can become deadly again

Table 14.1 Main antibiotic classes and their developed analogues (Coates et al. 2011)

Class	Subclass	Examples
β-Lactams	Penicillins	Penicillin G, penicillin V, methicillin, oxacillin, cloxacillin, dicloxacillin, nafcillin, ampicillin, amoxicillin, carbenicillin, ticarcillin, mezlocillin, piperacillin, azlocillin, temocillin
	Cephalosporins	*First generation*: Cephalothin, cephapirin, cephaloridine, cefazolin *Second generation*: Cefuroxime, cephalexin, cefprozil, cefaclor, cefoxitin *Third generation*: Cefotaxime, ceftizoxime, ceftriaxone, ceftazidime, cefixime, cefpodoxime, ceftibuten, cefdinir *Fourth generation*: Cefepime *Fifth generation*: Ceftaroline, ceftobiprole, ceftolozane
	Carbapenems	Imipenem, meropenem, doripenem, ertaoenem
	Monobactams	Aztreonam
	β-Lactamase inhibitors	Clavulanate, sulbactam, tazobactam
	Non-β-lactam lactamase inhibitors	Avibactam, vaborbactam
Aminoglycosides		Streptomycin, neomycin, kanamycin, paromomycin, gentamicin, tobramycin, amikacin, netilmicin, spectinomycin, sisomicin, dibekacin, isepamicin, plazomicin
Tetracyclines		Tetracycline, chlortetracycline, demeclocycline, minocycline, oxytetracycline, methacycline, doxycycline, tigecycline, omadacycline
Rifamycins		Rifampicin (or rifampin), rifapentine, rifabutin, bezoxazinorifamycin, rifaximin
Macrolides		Erythromycin, azithromycin, clarithromycin
Ketolides		Telithromycin
Lincosamides		Lincomycin, clindamycin
Glycopeptides		Vancomycin, teicoplanin, telavancin, dalbavancin, oritavancin
Lipopeptides/peptides	Daptomycin	Daptomycin
	Bacitracin	Bacitracin (A-E)
	Polymyxin	Polymyxin B, colistin
Streptogramins		Quinupristin, dalfopristin, pristinamycin
Oxazolidinones		Linezolid
Quinolones		Nalidixic acid, norfloxacin, pefloxacin, enoxacin, ofloxacin/levofloxacin, ciprofloxacin, temafloxacin, lomefloxacin, fleroxacin, sparfloxacin, trovafloxacin, clinafloxacin, gatifloxacin, moxifloxacin
Sulfonamides		Sulfanilamide, para-aminobenzoic acid, sulfamethoxazole
Others		Metronidazole, trimethoprim

40 years (from 1964 to the early 2000s), creating the innovation gap or the "void" in our discovery pipeline (Fig. 14.3).

Despite the high-throughput screening platforms and vast combinatorial libraries of millions of compounds, the success rate has been low (Fox et al. 2006). Pharmaceutical companies spent years of time and a lot of money to develop new antibiotics but ended up with nothing. It is estimated that one in ten million compounds will make its way to the market (Zhu et al. 2013). Therefore, synthetic scaffolds have not been quality drugs. To mention, only four new classes of antibiotics have been introduced to the market in the late 2000s and early 2010s, after the long void of four decades (Coates et al. 2011; Srivastava et al. 2011). These classes comprise of cyclic lipopeptides (such as daptomycin), glycylcyclines (tigecycline), oxazolidinones (linezolid), and lipiarmycins (fidaxomicin). The antibiotics which are currently in development or those discovered recently will be discussed later in this chapter.

The emergence of multiple drug-resistant (MDR), extensively drug-resistant (XDR), and pan drug-resistant (PDR) strains worldwide is a significant challenge due to limited, efficient antibiotics (Kollef et al. 2011; Magiorakos et al. 2012). Treating common infection with global capacity relies on the constant supply of effective antibiotics. In the past decade, 20 new antibiotics were approved, but only four were active against critical priority pathogens (Butler and Paterson 2020). There is a lack of effective antibiotics to treat Gram-negative pathogens. According to the Pew Charitable Trusts analysis of the antibiotic pipeline (Project 2019), only 11 antibiotics (no novel class) in clinical development could address infections caused by critical pathogens mentioned in the WHO priority list. We are in a standoff with these superbugs, and we are poised to lose because our antibiotic reservoir now appears empty. Suppose we fail to discover new treatments in the coming time, common infections will become deadly again, and modern medical procedures such as major surgery, organ transplantation, cancer chemotherapy, etc., would be a thing of the past. Given the lesser chances of approval of a newly discovered antibiotic for final use and the profound crisis in available therapies against difficult-to-treat-with pathogens, we must tackle this global challenge by continuously screening for new antibiotics or alternative solutions.

14.3 Can We Rely Upon Bacterial Hidden Treasure to Find New Antibiotics?

14.3.1 Natural Products Versus Synthetic Compounds

Natural products have been a foundation for modern medicines for decades, such as aspirin (Jeffreys 2008), morphine (Courtwright 2009), penicillin, and many other antibiotics (Wright 2017). One of the prominent reasons for the outstanding achievements of natural products as antibiotics is their structural and chemical complexity (Rossiter et al. 2017). The antibiotics we consume today were mined in the golden era and most of them were derived from natural products of bacteria. To note, only six classes of entirely synthetic compounds have been approved as antibiotics sulfonamides, diaminopyrimidine, nitroimidazole, fluoroquinolones, oxazolidinones, and diarylquinolines (Rossiter et al. 2017; Wright 2014). Moreover, the antimicrobial spectrum of synthetic compounds is comparatively narrow (Lewis 2017). Despite the advancements in high-throughput screening technology, synthetic chemical compounds' success rate has been negligible (Fox et al. 2006). This may primarily be because of the fact that synthetic approaches have mainly focused on quantity over quality. Compared to synthetic chemists, Mother Nature is a great architect designing diverse scaffolds with potential biological activity. Natural products are already undergoing primary screening during evolution (Lewis 2013). The low-quality chemical scaffolds will not survive, and therefore the producer microorganism has to synthesize something better, something unknown to its neighboring microbes. There is no doubt that nature-driven screening beats the chemical diversity in synthetic libraries. Another reason for the failure of synthetic compounds was that these readily available libraries were not synthesized for microbial biology purpose (Wright 2014); rather, these preferred human bioactivity. Moreover, the hits identified in combinatorial libraries are selected on the basis of Lipinski rules, which focus on the absorption of molecules after oral administration. The natural compounds to be used as antibiotics, on the other hand, simply don't follow these rules (Lewis 2013, 2017).

Undoubtedly, the synthetic libraries identified hits against various bacterial targets, but the penetration into whole bacterial cells remained a major bottleneck in antibiotic development. In contrast, natural products have been selected for their intracellular penetration through evolution (Wright 2014). Although natural products have been gifted remarkable chemical complexities and biological activities, they have their limitations. More than 25,000 natural antibiotic products have been discovered so far (Bérdy 2012), and only a few have been developed as drugs. Why? A straightforward explanation is that most natural products are not suitable as drugs (Wright 2014). In contrast, synthetic analogues of these compounds have proved to be better in terms of safety and druggability.

14.3.2 Untapped Bacterial Diversity as a Source of New and Effective Antibiotics

The ever-increasing burden of antimicrobial resistance is a global health crisis. We must find solutions to tackle this problem by using several alternatives. Having said that the resistance is an inevitable process, and it will eventually develop against any antibiotic that we discover; though very late in some cases, we must continuously search for new and effective antibiotics to fight difficult-to-treat pathogens. Natural products from microbial source hold the potential to provide new chemical scaffolds which can be developed further as antibiotic candidates (Lewis 2013; Stratton et al. 2015; Wright 2014, 2017). Natural products built our antibiotic arsenal in the golden era and continued to provide new drugs with the advancement of medicinal chemistry (Brown and Wright 2016). Despite being the revolutionary work in antibiotic history, the traditional Waksman platform to find natural antibiotics is challenged by replicating already known compounds. One prospect to reboot or revive this platform is the exploitation of microbes from untapped environments or to culture the unculturables. Table 14.2 demonstrates the recent discoveries of novel natural antibiotic products or their derivatives (isolated from the bacterial domain) against Gram-negative pathogens. The exemplary discovery of such antibiotics stimulated a resurgence in exploiting new natural products to fight difficult-to-address pathogens (Clardy et al. 2006). Microbes represent an essential resource on the earth, next to the plants, and a massive repertoire of new chemical compounds. This chapter mainly focuses on natural antimicrobials isolated from the bacterial domain in the last two decades, i.e., from 2001 to 2020. We have also described briefly the discovery of synthetic compounds based on natural scaffolds. The number of compounds mentioned in this chapter is not comprehensive; rather, it is illustrative. We have preferably included the compounds that (1) are active against Gram-negative bacteria or at least have been shown to have activity in one of the critical category pathogens, (2) are well characterized chemically, and (3) have been investigated for detailed in vitro studies or for which in vivo efficacy has been performed. Additionally, we have mentioned bacterial metabolites that do not possess direct antibacterial activity by themselves but can either potentiate the traditional antibiotics or modify the pathogens' resistance mechanisms.

14.3.2.1 Novel Tetracyclines

Tetracyclines are a large family of antibiotics that were initially discovered in 1945 from the *Streptomyces* sp. (Jukes 1985). Afterward, there have been continuous efforts to develop novel analogues (Nelson and Levy 2011). They comprise one of the most successful chemical scaffolds to provide new antibiotics against drug-resistant bacteria. Two novel tetracyclines that have been developed since the 2000s and possess activity against critical threat pathogens are eravacycline and omadacycline. Eravacycline is a fully synthetic fluorocycline antibiotic developed by Tetraphase Pharmaceuticals which was derived by modification on C7 and C9 on the tetracyclic core (Sutcliffe et al. 2013). It inhibited the multidrug-resistant strains of *A. baumannii, K. pneumoniae,* and *E. coli* at a MIC of 0.016–8 µg/mL, 0.03–16 µg/mL, and 0.016–4 µg/mL, respectively. The compound has moderate activity on *Pseudomonas sp.* (MIC, 1 to >32 µg/mL). The drug was approved in 2019 by USFDA for complicated intra-abdominal infections and the treatment of complicated urinary tract infections caused by Gram-negative bacteria (Lee and Burton 2019). Omadacycline is approved to treat Gram-positive bacteria, but it also inhibited *E. coli* strains (MIC of 0.5–2 µg/mL). In vivo studies with *E. coli* indicated the ED_{50} of omadacyline to be 2.02 mg/kg (Macone et al. 2014). However, more data is required to prove the clinical utility of this antibiotic against Gram-negative bacteria.

14.3.2.2 Cefiderocol

Cefiderocol is a recently approved antibiotic that belongs to the cephalosporin class of antibiotics (Aoki et al. 2018). It was derived by conjugating a siderophore (catechol) to the cephalosporin core. Cefiderocol retained potent activity in

Table 14.2 Promising natural anti-Gram-negative compounds (or derivatives thereof) discovered from bacteria since 2001

Sr. no.	Antibiotic name	Category	Year of discovery	Source	Activity spectrum[a]	Reference
1	Omadacycline[b] (PTK0796)	Tetracycline	2003	–	Ec, Kp	Macone et al. (2014), Macone et al. (2003)
2	Cefiderocol	B-lactam (cephalosporin derivative)	2008	–	Ab, pa, Kp, Ec	Aoki et al. (2018), Zhanel et al. (2019)
3	Enterocin E 760	Bacteriocin	2008	Enterococcus spp.	Kp, Ec	Line et al. (2008)
4	Macrolactin S	Macrolide	2008	Bacillus sp. AT28	Ec	Lu et al. (2008)
5	Bacteriocin B 602	Bacteriocin	2009	Paenibacillus polymyxa	Ab, pa, Kp, Ec	Svetoch et al. (2009)
6	Bacteriocin E50–52	Bacteriocin	2009	Enterococcus faecium	Ab, pa, Kp, Ec	Svetoch et al. (2009)
7	Plazomicin[b] (ACHN-490)	Aminoglycoside (derived from sisomicin)	2009		Ec, Kp	Aggen et al. (2009), Aggen et al. (2010)
8	CB-182 804[b]	Polymyxin derivative	2010		Ec, pa, ab, Kp	Leese (2013), Quale et al. (2012)
9	Eravacycline[b] (TP-434)	Tetracycline (fluorocycline)	2010	–	Ec, pa, ab, Kp	Hunt et al. (2010)
10	Pulvomycin	Polyketide	2010	Streptomyces flavopersicus	Ab, pa, Ec	McKenzie et al. (2010)
11	Battacin (Octapeptin 5)	NRP (lipopeptide)	2011	Paenibacillus tianmuensis	Ec, pa, ab, Kp	Qian et al. (2012)
12	Paenibacterin	NRP	2012	Paenibacillus thiaminolyticus	Ec, pa, ab, Kp	Guo et al. (2012)
13	Cystobactamids	NRP	2014	Cystobacter sp.	Ab, Ec	Baumann et al. (2014)
14	FADDI-002[b]	Polymyxin derivative	2014	–	Ec, pa, ab, Kp	Velkov et al. (2014)
15	Paenipeptins	NRP	2017	Paenibacillus sp.	Ec, pa	Huang et al. (2017)
16	Brevicidine	NRP (lipopeptide)	2018	Brevibacillus laterosporus DSM 25	Ec, pa, ab, Kp	Li et al. (2018)

(continued)

Table 14.2 (continued)

Sr. no.	Antibiotic name	Category	Year of discovery	Source	Activity spectrum[a]	Reference
17	Laterocidine	NRP (lipopeptide)	2018	*Brevibacillus laterosporus* ATCC 9141	*Ec, pa, ab, Kp*	Li et al. (2018)
18	Odilorhabdins	NRP	2018	*Xenorhabdus nematophila*	*Ec, Kp*	Pantel et al. (2018)
19	G0775[b]	Macrocycle (optimized arylomycin)	2018	–	*Ab, Ec, Kp, Pa*	Smith et al. (2018)
20	Tridecaptin M	NRP (lipopeptide)	2019	*Paenibacillus* sp.	*Ec, Kp*	Jangra et al. (2019)
21	Darobactin	RiPP	2019	*Photorhabdus khanii*	*Ab, Ec, Kp, Pa*	Imai et al. (2019)
22	Picolinamycin	Picolinamide	2020	*Streptomyces* sp.	*Ec, Kp, Pa*	Maiti et al. (2020)

[a]*Ab Acinetobacter baumannii, Ec Escherichia coli, Kp Klebsiella pneumoniae, Pa Pseudomonas aeruginosa.*
[b]Synthetic or semisynthetic derivative of natural products of bacteria. The source of original scaffold is mentioned in the text.

carbapenem- and cephalosporin-resistant *Enterobacteriaceae*, *P. aeruginosa*, and *A. baumannii* with MIC_{90} <4 µg/mL in sensitive strains (Zhanel et al. 2019). Cefiderocol is one example of how we can improve existing natural products' activity to design superior next-generation antibiotics.

14.3.2.3 Bacteriocins

Microorganisms produce many antimicrobial compounds, among which bacteriocins are the antimicrobial peptides produced by bacteria. Though several bacteriocins have been reported to inhibit drug-resistant pathogens (Rebuffat et al. 2020), none of them has been advanced to human clinical trials. Some promising candidates having activity specifically against Gram-negative bacteria are mentioned below.

Enterocin E 760

Bacteriocins produced by enterobacteria are termed as enterocins. Enterocin E 760 was isolated and purified from *Enterococcus* species (Line et al. 2008). It belongs to the class IIa bacteriocin category with 71% similarity to enterocin P. It exhibited broad-spectrum inhibitory activity on Gram-positive and Gram-negative bacteria, including *K. pneumoniae* (MIC 0.1–3.2 ug/mL). It significantly reduced the colonization of *C. jejuni* and other *Campylobacter* spp. in broiler chickens when mixed in feed.

14.3.2.4 Bacteriocin E 50–52 and B 602

Bacteriocin E 50–52 and bacteriocin B 602 were isolated from *Enterococcus faecium* and *Paenibacillus polymyxa,* respectively (Svetoch et al. 2009). They were classified as class IIa bacteriocin depending on their amino acid sequence analysis. Previously published bacteriocins of class IIa are reported for narrow-spectrum antibacterial activity, mainly against Gram-positive bacteria, and they act by forming pores in the bacterial cell membrane. On the contrary, bacteriocin E50–52 and B 602 showed broad-spectrum activity, including Gram-positive and Gram-negative pathogens. Their efficacy was tested for causative agents of nosocomial infections collected from Russian hospitals. Gram-positive bacterial strains like *S. aureus* were susceptible to MIC values of 0.025–0.2 µg/mL and for Gram-negative bacterial strains like *A. baumannii* and *Proteus* spp. MIC values ranged from 0.05 to 0.2 µg/mL, but for *P. aeruginosa* strains, MIC values were comparatively high (0.1 to 1.6 µg/mL). The bacteriocins displayed activity against a large panel of clinical isolates ($n = 64$) tested, including MDR strains.

14.3.2.5 Macrolactin S

Macrolactin S, a macrolide, was isolated from a soil bacterium *Bacillus sp.* AT28 (Lu et al. 2008). Structure elucidation by M.S. and NMR confirmed that it is a 24-membered lactone and is a derivative of macrolactin A. Fatty acid synthesis is an essential bacterial process that is carried out by multiple enzymes encoded by numerous genes, unlike mammalian fatty acid synthetases. Among those genes, FabG gene is well conserved in bacterial family and is thought to be a new target for pathogenic bacterial antibiotics. Compounds active against other fatty acid synthase have been reported, but macrolactin S is the first inhibitor of FabG synthase isolated from bacterial origin. Its IC_{50} against purified FabG of *S. aureus* was 1.3 mM, and against *E. coli* and *B. subtilis,* it showed weaker activities (MIC 64 µg/mL).

14.3.2.6 Pulvomycin

Genome analysis of *Streptomyces coelicolor* has shown that biosynthetic pathways for secondary metabolites are encoded by multiple-gene clusters which are influenced by regulators like absA (McKenzie et al. 2010). It encodes a kinase that regulates other response genes by phosphorylation. So the *absA1* gene from *S. coelicolor* was used to alter the expression of secondary metabolites in *Streptomyces'* heterologous system, which led to a finding that it could activate the genes for new antimicrobial compounds. Strains used for heterologous expression had minimal effects on both *P. aeruginosa* and *B. cenocepacia* MDR strains, but the introduction of *absA1* gene altered the expression and the activity of not all but a few *Streptomyces* against MDR strains by increasing its antimicrobial activity. Notably, the expression of metabolites in *S. flavopersicus* NRRL2820 was checked and extracted for their mass analysis and thin-layer chromatography to identify the molecule responsible for inhibition. NMR studies solved the structured to be similar to the known molecule pulvomycin that acts as an inhibitor of translation elongation factor EF-TU. MICs of pulvomycin were determined to be 8 µg/mL against *B. cenocepacia* and *B. vietnamiensis* and 16–32 µg/mL against *A. baumannii* and *P. aeruginosa* strains.

14.3.2.7 Plazomicin

Plazomicin was developed by Achaogen and initially reported in 2009 (Aggen et al. 2009). It is classified as semisynthetic aminoglycoside that was derived from sisomicin, a natural product of *Micromonospora* sp. (Aggen et al. 2010; Weinstein et al. 1970) Plazomicin has excellent activity against carbapenem-resistant *Enterobacteriaceae* (MIC <2 µg/mL) and is superior to other aminoglycosides such gentamicin or tobramycin. Most of the *P. aeruginosa* and *A. baumannii* strains are less susceptible to plazomicin (MIC >4 µg/mL) and generally considered as resistant because clinical breakpoint for plazomicin is 2 µg/mL (Eljaaly et al. 2019). In 2018, the USFDA approved plazomicin to treat complicated urinary tract infections caused by CRE and ESBL-producing *Enterobacteriaceae*.

14.3.2.8 Novel Polymyxin Derivatives

Polymyxins are cationic lipopeptides produced by the *Paenibacillus* genus. They were initially reported in the late 1940s and came into use a few years after their discovery. Though several members of this class have been reported to date (Brown and Dawson 2017; Jangra et al. 2018), only two of the natural polymyxins, i.e., polymyxin B and colistin (polymyxin E), are used in clinical settings. Unluckily, this class faced the exodus from the market very soon because (1) they were associated with severe nephrotoxicity and neurotoxicity and (2) many safer and better tolerable antibiotics were available in the market. It was not until the start of the twenty-first century that this group made a comeback, thanks to the evolving superbugs that expressed resistance to most of the available antibiotics, especially carbapenems. Even today, polymyxins are considered last resort of therapy to treat severe carbapenem-resistant and ESBL-producing Gram-negative pathogens. Therefore, several researchers across the world have tried to develop new analogues of polymyxins that are safe and are active in polymyxin-resistant bacteria. The readers can have details of polymyxin analogues in several other reports published on this class (Brown and Dawson 2017; Vaara 2019; Velkov and Roberts 2019). The present chapter highlights the discovery of the latest polymyxin analogues studied extensively for their in vitro and in vivo potential. CB-182804, a derivative of polymyxin B, was developed by Cubist Pharmaceuticals and taken for further studies. It contains a 2-chlorophenyl moiety at its N-terminus, which is linked to polymyxin B nonapeptide through urea. The compound had comparable antimicrobial activity in Gram-negative bacteria as polymyxin B. Interestingly, it showed less toxicity than polymyxin B in cynomolgus monkeys (Brown and Dawson 2017). CB-182804 was advanced into phase 1 clinical trial, but no further information is available. Another analogue of polymyxin class is FADDI-002, developed by Monash University (Velkov et al. 2014). This molecule is different from polymyxin B in two ways: L-octyl glycine at position 7 and octanoic acid at N-terminus. FADDI-002 showed better activity in polymyxin-resistant bacteria and exhibited in vivo efficacy in the lung infection model of *P. aeruginosa*. Additionally, the in vivo toxicity profile of this analogue was better than polymyxin B. The current status of this compound is not available. To mention, most of the polymyxin analogues have been synthesized based on polymyxin B core. Our group recently reported that polymyxin A was superior to polymyxin B or colistin in most of the clinical strains tested (Jangra et al. 2018). Moreover, polymyxin A displayed reduced in HEK293 and THP-1 cell lines compared to standard polymyxins. These results indicate that future analogues based on polymyxin A core may have superior activity and safety profile.

14.3.2.9 Octapeptins

Akin to polymyxins, octapeptins are nonribosomal lipopeptides produced by *Bacillus* sp. They have high structure similarity with polymyxins, consisting of cyclic heptapeptide core and an N-terminal acyl chain linked to the core by one free amino acid. One noticeable difference between polymyxins and octapeptins is the hydroxyl substituted acyl chain in the latter group. Octapeptins display broad-spectrum antimicrobial activity, which is in stark contrast to polymyxins. Also, octapeptins retain activity in colistin-resistant Gram-negative bacteria (Abou Fayad and Herrmann 2018; Velkov et al. 2018). Although they were discovered four decades ago, there is not enough information to prove the clinical utility of this class. Recently, several researchers from Monash University, the University of Melbourne, and the University of Queensland, Australia, started pioneering work on this class and studied their biosynthesis and detailed SAR (Velkov et al. 2018). Since then, several analogues have been synthesized, and they have been studied to act on colistin-resistant MDR Gram-negative bacteria. Octapeptin B5/battacin is the latest member of this class isolated from *Paenibacillus tianmuensis* in 2013. It showed high binding affinity toward lipid A as compared to colistin and polymyxin B. Battacin exhibited a MIC of 2–4 µg/mL in clinical isolates of *E. coli* and *P. aeruginosa,* while MIC in *K. pneumoniae* and *A. baumannii* is slightly higher, ranging from 2 to 8 µg/mL and 4 to 16 µg/mL, respectively (Qian et al. 2012). Battacin was nonhemolytic up to 750 µg/mL and had no cytotoxicity at a concentration of 128 µg/mL when tested in HEK293 cell line. The LD_{50} of battacin in mice was 15.46 mg/kg, significantly higher than polymyxin (6.52 mg/kg). The compound also displayed in vivo efficacy intraperitoneal infection model of MDR *E. coli* with an intravenous dose of 4 mg/kg. The other novel analogues of octapeptins such FADDI-117, FADDI-118 showed superior in vivo efficacy compared to natural octapeptins and polymyxins, but they were also associated with a high degree of nephrotoxicity than natural octapeptin C4. These results indicate that octapeptins have significant in vivo biological activity and can lead to new antibiotic candidates' development if toxicity can be managed. More research is needed to investigate the clinical potential of octapeptins.

14.3.2.10 Paenibacterin

Paenibacterin is 1604 Da cyclic lipopeptide produced by *Paenibacillus thiaminolyticus* (Guo et al. 2012). It is a nonribosomally synthesized peptide that acts mainly by membrane depolarization. The secondary mechanism of action also involves the formation of radical hydroxyl species that leads to cell death. MIC of paenibacterin against Gram-negative bacteria is 2–8 µg/mL (Huang 2013). The activity of paenibacterin is reasonably good, but the toxicity or in vivo data is not reported yet. Future investigations should focus on its in vivo and PK/PD studies.

14.3.2.11 Cystobactamids

Cystobactamids comprise novel antibacterial compounds reported from *Cystobacter* sp. (Baumann et al. 2014). They also belong to NRP category that contains unusual amino acids, i.e., *para*-nitrobenzoic acid (PNBA) and *para*-aminobenzoic acid units (PABA). These amino acids were not previously reported in any of the antibiotic classes. The compounds showed good antimicrobial activity in both Gram-positive and Gram-negative pathogens. Later, the researchers performed the total chemical synthesis and devised various analogues that possessed superior antibacterial activity, even in resistant strains. The analogues showed MIC of 0.25–2 µg/mL, 1–8 µg/ mL in carbapenem-resistant *E. coli*, and carbapenem-resistant *P. aeruginosa* strains, respectively (Hüttel et al. 2017). Their activity is attributed to bacterial topoisomerases' inhibition, similar to quinolones (Elgaher et al. 2020). Recently, a new analogue of this class, 22, was found to have the best in vivo activity in the *E. coli* infection model (Testolin et al. 2020). These findings suggest that cystobactamids are highly effective lead compounds against Gram-negative superbugs.

14.3.2.12 Paenipeptins

Paenipeptins A and B are linear counterparts of cyclic lipopeptides (pelgipeptin C and pelgipeptin B, respectively) reported from *Paenibacillus* sp. (Huang et al. 2017) in 2017. Paenipeptin C is the new member of this family and was first reported in its cyclic and linear form. Paenipeptins A and C were not produced in liquid media in enough quantity, and a strong resemblance in their structure made it difficult to separate them into pure forms. Paenipeptin was isolated, and its complete structure was solved. Later, chemical variants of paenipeptin B were prepared using chemical synthesis and designated as paenipeptins A', B', and C'. Paenipeptins A', B', and C' displayed potent inhibitory activity against *E. coli* (8–16 µg/mL, 1–4 µg/mL, 0.5–1 µg/mL, respectively) and *P. aeruginosa* (8–16 µg/mL, 4 µg/mL, 1–2 µg/mL, respectively).

14.3.2.13 Brevicidine and Laterocidin

The traditional discovery of antibiotics from microorganisms faced the rediscovery of already known compounds which led people to invent new and alternative approaches to drug discovery. Genome mining is one of them in which bioinformatic investigations lead to the identification of previously uncharacterized biosynthetic gene clusters that may yield new chemical entities. Brevicidine and laterocidine are such examples of new classes of cyclic depsipeptides produced by *B. laterosporus* that were discovered through the global genome-mining approach (Li et al. 2018). Their structure comprises a unique linear cationic segment with three positively charged ornithine residues and a small hydrophobic tetrapeptide/pentapeptide ring which are crucial for its antimicrobial activity. MIC of brevicidine in colistin-resistant *E. coli* and *K. pneumoniae* is 2 µg/mL; in *P. aeruginosa,* 1 µg/mL; and in *A. baumannii*, 16 µg/mL, while MIC of laterocidine is 2 µg/mL for *P. aeruginosa* and 4 µg/mL on *A. baumannii*. Both compound also exerted in vivo efficacy in a mouse thigh infection model of *E. coli* strain.

Moreover, the peptides were nontoxic against HeLa cell lines up to a concentration of 128 µg/mL. The mode of action studies suggested that these peptides bind with bacterial LPS and disrupt the outer membrane integrity. Still, no membrane disruption was seen at $2 \times$ MIC, indicating that they might have multiple targets as exhibited by other cationic peptides. This unique mode of action does not coincide with other antibiotic and makes this class an attractive drug candidate. Recently, Al-Ayed et al. posted a preprint on the total synthesis of brevicidine and laterocidine, which opens new ways to develop novel analogues (Al-Ayed et al. 2021).

14.3.2.14 Odilorhabdins

Odilorhabdins were isolated from nematode-symbiotic bacteria *Xenorhabdus nematophila*. They are also the products of nonribosomal peptide synthetases and show activity against both Gram-positive and Gram-negative pathogens by binding to smaller subunit of ribosomes (Pantel et al. 2018). Three variants of odilorhabdins were reported from fermentation broth of *X. nematophila*, namely, NOSO-95A (MW, 1296 Da), NOSO-95B (MW, 1280 Da), and NOSO-95C (MW, 1264 Da). All of them showed strong activity against carbapenem-resistant *Enterobacteriaceae*

(CRE). To develop a preclinical candidate, the researchers performed the chemical synthesis of this class and synthesized several analogues based on the structure of NOSO-95C (Sarciaux et al. 2018). One of the analogues, NOSO-95179, demonstrated best in vitro and in vivo activity in *E. coli* (MIC 2–4 µg/mL) and *K. pneumoniae* strains (MIC 1–2 µg/mL). Additionally, no cytotoxicity was seen till the concentration of 256 µg/mL in mammalian HepG2 and HK-2 cell lines. This compound was further optimized to design new drug candidates based on their pharmacokinetic (P.K.) properties, plasma protein binding, and efficacy in a murine sepsis infection model. Finally, a preclinical candidate, NOSO-502, was selected for further studies. This compound showed a MIC of 0.5–4 µg/mL in CRE strains (Racine et al. 2018). NOSO-502 demonstrated in vivo efficacy in several infection models such as the lung infection model, sepsis model, UTI model, etc. Moreover, the compound exhibited an excellent safety profile and good pharmacokinetic properties (Zhao et al. 2018). The exciting data on odilorhabdins indicate that this class makes an attractive drug candidate to develop antibiotics against one of the deadliest superbugs (Racine and Gualtieri 2019). This is another example of a bacterial natural product that may lead to the development of future therapies.

14.3.2.15 Optimized Arylomycins

Arylomycins are the biaryl-bridged lipopeptides produced by actinomycetes first reported in 2002 (Schimana et al. 2002). Arylomycins showed activity against Gram-positive bacteria only. Arylomycins act by inhibiting type I bacterial signal peptidase (SPase), so any mutation in SPase will render bacteria resistant to arylomycin. To increase the spectrum of arylomycin, several analogues were synthesized (Liu et al. 2013). Among them, G0775 showed the most promising results. It exhibited 500-fold more potency than arylomycin against MDR Gram-negative pathogens. G0775 acts by LepB inhibition, hence decreasing the rate of mutation as compared to parent arylomycin. G0775 inhibited 90% of the MDR strains of *A. baumannii* and *P. aeruginosa* within the MIC of 4 µg/mL and 16 µg/mL, respectively. The compound exhibited excellent activity in XDR strains of *E. coli* and *K. pneumoniae* (MIC_{90} ≤0.25 µg/mL). In murine neutropenic thigh infection model with *E. coli* and *K. pneumoniae*, there was >2-log reduction in cfu at a dose of 1 mg/kg and 5 mg/kg, respectively. A similar animal model with *P. aeruginosa* ATCC 27853 and *A. baumannii* ATCC 17978 showed a significant decrease in bacterial burden. Additionally, the compound demonstrated in vivo efficacy in lung infection and mucin peritonitis survival model with *K. pneumoniae* strain; a 100% survival rate was observed with two doses of 5 mg/kg (Smith et al. 2018).

14.3.2.16 Tridecaptins

Tridecaptins, which were initially discovered in 1978, are linear nonribosomally synthesized cationic lipopeptide antibiotics that retain activity against Gram-negative bacteria (Bann et al. 2021; Jangra et al. 2019). They possess a different mechanism of action than polymyxins and kill Gram-negative bacteria by binding to lipid II and causing membrane disruption. Recently, a new member of this class, i.e., tridecaptin M, was isolated from mud bacterium *Paenibacillus* sp. M152, and for the first time, the report demonstrates the antibacterial effects of this class in colistin-resistant *Enterobacteriaceae* (Jangra et al. 2019). Tridecaptin M showed potent antimicrobial activity against polymyxin-resistant XDR clinical isolates of *K. pneumoniae* and *mcr-1*-positive *E. coli* strains (MIC 2–8 μg/mL).

Moreover, the compound appeared safe to red blood cells (RBCs) and mammalian cell lines at a much higher concentration than its effective concentration in bacteria. Also, no acute toxicity was observed in the animal model when administered to mice at $12 \, \text{mg} \, \text{kg}^{-1}$ every 2 h until an accumulated dose of $72 \, \text{mg} \, \text{kg}^{-1}$. In contrast to this, colistin was lethal to mice at the same dose. Above all, the compound exhibited remarkable efficacy in a thigh infection model with a colistin-resistant strain of *K. pneumoniae*. Nevertheless, this compound requires detailed investigations regarding its PK and PD to advance this molecule further in clinical trials. Currently, tridecaptin M is in the preclinical stage and is being developed by CSIR-Institute of Microbial Technology, India.

14.3.2.17 Darobactin

Darobactin is a 965 Da molecule that is ribosomally synthesized and posttranslationally modified heptapeptide. It was isolated from *Photorhabdus khanii,* which is an inhabitant of the nematophilic gut. These nematophilic bacteria protect the host against pathogenic bacteria, especially Enterobacteriaceae family members, by producing antimicrobial compounds in the surroundings (Imai et al. 2019). Darobactin inhibits the BAM complex, which is involved in outer membrane protein folding. The MIC values of darobactin range from 2 μg/mL for *E. coli* and *P. aeruginosa* to 8 μg/mL for *A. baumannii*. In vivo studies on mice showed that darobactin has good exposure and pharmacokinetics at a dose of $50 \, \text{mg} \, \text{kg}^{-1}$ given intraperitoneally. Septicemia model with wild-type and polymyxin-resistant *P. aeruginosa*, carbapenemase-producing *K. pneumoniae*, and wild-type and polymyxin-resistant *E. coli* showed a 100% survival rate in mice treated with $25 \, \text{mg} \, \text{kg}^{-1}$ of darobactin. In another mouse thigh infection model with *E. coli mcr-1*, a significant pathogen burden was reduced. These results encourage future investigations on this class and highlight animal symbionts' potential to produce novel antibacterial compounds to fight superbugs.

14.3.2.18 Picolinamycin

Picolinamycin is the new antimicrobial class reported from *Streptomyces* sp. isolated from Kashmir Himalayan soil (Maiti et al. 2020). The authors selected this source for bacteria isolation because of its unique climatic conditions and having a diversity of actinomycetes (Maiti and Mandal 2019). Picolinamycin is a 1013 Da antimicrobial compound that consists of three substructures—A, B, and C—and has picolinamide moiety in the center. It showed MIC of 0.02 μg/mL in *K. pneumoniae*, 2.56 μg/mL in *E. coli*, and 5.12 μg/mL in *P. aeruginosa*. Moreover, it inhibited the MDR strains of *P. aeruginosa* at 2.56 μg/mL. Compared to the MIC values, the MBC of picolinamycin was relatively high, indicating that this compound is bacteriostatic. The authors also reported the cytotoxic studies on the human A549 cell line and found no cytotoxic effects up to a concentration of 31.25 μg/mL. These studies are very preliminary but showcase the antimicrobial potential of new secondary metabolites and provide an opportunity to find new classes of chemical compounds that may have antimicrobial potential.

14.4 Bacteria as a Potential Source of Antibiotic Adjuvants/ Resistance-Modifying Agents (RMAs)

The rapid decline in antibiotic usefulness has led some clinicians to estimate the future utility of available antibiotics to be limited to a few years against multidrug-resistant Gram-negative pathogens. The common mechanisms of multidrug resistance (MDR) exhibited by Gram-negative bacteria are drug inactivation by enzymes, reduced membrane permeability, increased efflux of antibiotics, and physical blockade to antibiotics through biofilm formation. Developing alternative strategies targeting the resistance mechanisms themselves is needed, thereby (1) forestalling the emergence of resistance and (2) retrieving antibiotic potency. Natural products from the bacterial domain have often been selected to penetrate both the outer and inner membranes of Gram-negative bacteria. A countermeasure by antibiotic-producing bacteria is to co-evolve inhibitors of their competitor's resistance mechanism to boost their own antibiotics' efficacy (Wright 2014). This section of the chapter summarizes antibiotic adjuvants/RMAs from bacteria as an alternative to traditional antibiotic monotherapy.

14.4.1 β-Lactamase Inhibitors (BLI)

β-Lactam antibiotics are used to treat several bacterial infections and are among the most prescribed drugs in the world. The BLIs are the drugs that are coadministered with β-lactam antibiotics to thwart antimicrobial resistance by inhibiting β-lactamases, which are enzymes that inactivate the β-lactam ring, which is a common chemical scaffold to all β-lactam antimicrobials. Therefore, BLIs are primarily prescribed for infections by Gram-negative bacteria, as they produce this enzyme. Clavulanic acid (the first microbe-derived β-lactamase inhibitor) was discovered around 1974–1975 by British scientists working at the drug company "Beecham" from the bacteria *Streptomyces clavuligerus* (Sutherland 1991). Clavulanic acid is a suicide inhibitor, covalently bonding to a serine residue in the active site of the β-lactamase. This restructures the clavulanic acid molecule, creating a much more reactive species that attacks another amino acid in the active site, thus permanently inactivating the enzyme. This inhibition restores the efficacy of β-lactam antibiotics against lactamase-producing resistant bacteria. To the best of our knowledge, no other BLI has been discovered from bacterial or *Actinomycetes* resources. So, there exists an opportunity to discover novel BLIs from untapped microbial reservoirs.

14.4.2 Efflux Pump Inhibitors (EPIs)

Efflux systems are universally present in all organisms and contribute to both intrinsic and acquired resistance to antimicrobials. These pumps can expel a variety of different antimicrobials. The resistance nodulation and cell division (RND) family of efflux pumps stand as major players in antibiotic resistance in Gram-negative bacteria (Amaral et al. 2014). Consequently, EPIs will simultaneously increase the intracellular accumulation of several antibiotics, making bacteria more susceptible to antimicrobials in a general way. This increased susceptibility has two major consequences: First, it will allow resistant bacteria's resensitization to current antibiotics in use. Second, it will reduce the chances of the emergence of resistant mutants. Even if these mutants are present in the population, the low level of resistance achieved by them in the presence of EPIs will impede their selection (Blanco et al. 2018). Most efforts in developing adjuvants for treating Gram-negative bacterial infections have concentrated on the development of anti-RND EPIs. Many plant-derived compounds and screening of synthetic libraries have already populated the search of EPIs. Considering the intense competition faced by bacteria, it is not difficult to think about coevolutionary processes in which bacterial secondary metabolites may act as EPIs. Some examples are enlisted in Table 14.3.

14.4.3 Quorum Sensing Inhibitors (QSIs)

Quorum sensing (Q.S.) is a process of intracellular communication, being one of the best-studied types of interactions among bacterial communities in a diversity of ecological niches. This signaling process is also involved in the expression of genes critical to the production of virulence factors, host colonization, biofilm formation, and antibiotic resistance in a number of Gram-negative pathogenic bacteria (Borges et al. 2016). The synthesis of signaling molecules called acyl-homoserine lactones (AHLs) among Gram-negative bacteria is under the control of the synthase gene *luxI* and its regulator *luxR* (Zhang et al. 2002). They coordinate the release of protease, elastase, exotoxin A, pyocyanin, hydrogen cyanide, rhamnolipids, and lectins through AHLs (Gupta et al. 2011). Inhibition of the Q.S. system could assist in the termination of the bacterial resistance without killing the bacteria. Various QSIs have been identified from either plant resources or chemical libraries. For decades, marine bacteria and *Streptomyces* have been considered as essential resources for antibiotics and other metabolites. This section highlights the importance of bacterial metabolites as QSIs and is enlisted in Table 14.4.

14.4.4 Biofilm Inhibitors

Biofilms are aggregates of microorganisms with distinct sessile cells followed by cell division to form small clusters, microcolonies, and more enormous sums (Flemming

Table 14.3 Microbe-derived EPIs

Compounds	Source	Target organism	Target efflux pumps	Antibiotics potentiated	References
3,4-Dibromopyrrole-2,5-dione	*Pseudoalteromonas* sp.	*Enterobacteriaceae* and *P. aeruginosa*	AcrAB-TolC, MexAB-OprM, and MexXY-OprM	CHL, CIP, ERY, KAN, LEV, OXA, PIP, TET	Whalen et al. (2015)
EA-371α	*Streptomyces vellosus*	*P. aeruginosa*	MexAB-OprM, MexCD-OprJ, and MexEF-OprN	LEV	Lee et al. (2001)
EA-371δ					

CHL chloramphenicol, *CIP* ciprofloxacin, *ERY* erythromycin, *KAN* kanamycin, *LEV* levofloxacin, *OXA* oxacillin, *PIP* piperacillin, *TET* tetracycline.

Table 14.4 Quorum sensing inhibition by bacteria and their metabolites

Source	Metabolite(s)	Indicator organism(s)	Q.S. inhibition related phenotype	References
Rhizobium sp.	AHL analogues (C4-AHL)	*C. violaceum*, *P. aeruginosa*	Inhibition of violacein, biofilm formation, and virulence factors	Chang et al. (2017)
H. salinus	N-(20-phenylethyl)-isobutyramide and 3-methyl-N-(20-phenylethyl)-butyramide	*C. violaceum* and *V. harveyi*	Inhibition of violacein and luminescence emission	Teasdale et al. (2009)
Bacillus, Halobacillus, Streptomyces, and *Micromonospora*	Phenethylamides and a cyclic dipeptide	*C. violaceum* and *V. harveyi*	Inhibition of violacein and luminescence emission	Teasdale et al. (2011)
Streptomyces sp.	Butenolides and 3-hydroxy-butyrolactones	–	Competition with AHL signaling molecules	Cho et al. (2001)
Streptomyces sp.	Cinnamic acid, linear dipeptides proline–glycine and N-amido-proline	*C. violaceum*, *P. aeruginosa*	Inhibition of violacein and virulence factors	Naik et al. (2013)
Streptomyces parvulus	Actinomycin D and cyclic (4-hydroxy-pro-Phe)	*C. violaceum*, *P. aeruginosa*	Inhibition of violacein and biofilm	Miao et al. (2017)
S. Saprophyticus	Cyclo(pro-Leu)	*C. violaceum*	Inhibition of violacein	Liu et al. (2013)
Proteobacteria, Firmicutes, Actinobacteria, and Bacteroidetes phylum	Licochalcone A, malyngamide J, isomitomycin A, ansamitocin P-3, pederin, nisamycin, and kanglemycin A	*C. violaceum*, *P. aeruginosa, V. fischeri* and *A. tumefaciens*	Inhibition of bioluminescence, violacein production, and virulence factors	Saurav et al. (2016)
B. amyloliquefaciens	Cyclo(L-leucyl-l-prolyl)	*S. marcescens*	Inhibition of prodigiosin and virulence factors	Gowrishankar et al. (2019)
S. hominis	DL-homocysteine thiolactone	*C. violaceum*, *P. aeruginosa*	Inhibition of violacein, biofilm formation	Saurav et al. (2017)
S. hydnoides and *L. majuscule*	Malyngolide	*C. violaceum*	Inhibition of violacein	Dobretsov et al. (2010)
Streptomyces coelicoflavus	Docosanoic acid, borrelidin, and 1H-pyrrole-2-carboxylic acid	*C. violaceum*, *P. aeruginosa*	Inhibition of violacein and virulence factors	Hassan et al. (2016)

Table 14.5 Bacterial biosurfactants with anti-biofilm activity against Gram-negative pathogens

Source	Class (dose)	Pathogen	Effect on biofilm	References
Acinetobacter junii	Lipopeptide biosurfactants (1250 μ g/mL)	*Proteus mirabilis* and *P. aeruginosa*	Biofilm disruption 10% and 32%	Ohadi et al. (2020)
Pontibacter korlensis	Pontifactin (2 mg/ mL)	*Salmonella typhi* and *Vibrio cholerae*	99% anti-adhesive activity	Balan et al. (2016)
Burkholderia thailandensis	Rhamnolipids (6.25 mg/mL)	*Neisseria mucosa*	Biofilm inhibition 70%	Elshikh et al. (2017)
Pandoraea pnomenusa	Exopolysaccharides (0.25 mg/mL)	*Burkholderia cepacia*	Inhibit biofilm formation	Sacco et al. (2019)

and Wuertz 2019). The film underneath the biofilm is only in direct contact with the substratum in a multilayered heterogeneous microbial mat. Cells in biofilm survive harsh growth conditions as biofilms are surrounded by high molecular weight extracellular polymeric substances (EPS) that attach cells. The EPS are composed of polysaccharides, lipids, proteins, and extracellular nucleic acids and play an essential part in the pathogenesis of numerous microbial infections (Ch'ng et al. 2019). The excess production of EPS limits the penetration and diffusion of antibiotics; thus, cells in the biofilm get more time to become tolerant. Therefore, therapeutic strategies that destabilize and inhibit EPS are the best anti-biofilm approaches to inhibit biofilms and reduce significant antimicrobial resistance problems. Lipopeptides or biosurfactants produced by microbes hinder biofilm formation by modifying the surface physicochemical property to reduce adhesion, destabilize cell membranes, change the outer membrane hydrophobicity, and interfere with the electron transport system (K Satputea et al. 2016). The failure of conventional antibiotic therapies indicates that biofilm treatments with bacterial biosurfactants selectively eradicate the persistent biofilms and allow the diffusion of antibiotics into the biofilm matrix. A few of such agents are summarized in Table 14.5.

14.4.5 Outer Membrane Permeabilizers

The outer membrane (O.M.) of Gram-negative bacteria performs the crucial role of providing an extra layer of protection to the bacteria without compromising the exchange of materials required for sustaining life. The O.M. acts as a selective barrier by combining a highly hydrophobic lipid bilayer with pore-forming porins of specific size-exclusion properties. The permeability properties of this barrier, therefore, have a significant impact on the susceptibility of the Gram-negative bacteria to antibiotics, which, to date, are essentially targeted at intracellular processes. Small hydrophilic drugs, such as β-lactams, use the pore-forming porins to

Table 14.6 Peptide antibiotic potentiators derived from bacterial sources against Gram-negative bacteria

Source	Peptide	Pathogen	Antibiotics potentiated	References
Bacillus and *Paenibacillus* sp.	Tridecaptin A$_1$	*E. coli, Salmonella enterica, K. pneumoniae, A. baumannii*	DAP, VAN, RIF, PEN, TET, STM, CIP, AMP	Cochrane and Vederas (2014)
Paenibacillus dendritiformis	Tridecaptin M	*A. baumannii*	RIF, VAN, CRY, IMI, CFT	Jangra et al. (2020)
Bacillus polymyxa	Colistin	*E. coli, E. cloacae, E. aerogenes, S. typhimurium, K. pneumoniae*	MIN, NOV, MUP, AZM, ERY, CRY, RIF	MacNair et al. (2018)
Bacillus polymyxa	Polymyxin B	*K. pneumoniae*	RIF, IMI	Elemam et al. (2010)
–	SPR741	*E. coli, A. baumannii, K. pneumoniae*	RIF, CRY, AZM	Corbett et al. (2017)
Lactococcus lactis	Nisin Z	*Pseudomonas fluorescens*	PEN, AMP, VAN, KAN, TET, STM, CHL, RIF	Naghmouchi et al. (2012)
Pediococcus acidilactici	Pediocin PA-1/AcH			

DAP daptomycin, *VAN* vancomycin, *RIF* rifampicin, *PEN* penicillin, *TET* tetracycline, *STM* streptomycin, *CIP* ciprofloxacin, *AMP* ampicillin, *CRY* clarithromycin, *IMI* imipenem, *CFT* ceftazidime, *MIN* minocycline, *NOV* novobiocin, *MUP* mupirocin, *AZM* azithromycin, *ERY* erythromycin, *KAN* kanamycin, *CHL* chloramphenicol.

gain access to the cell interior, while other hydrophobic antibiotics diffuse across the lipid bilayer. The existence of drug resistance in many bacterial strains due to modifications in the lipid or protein composition of the O.M. indeed highlights the importance of the O.M. barrier in antibiotic susceptibility (Delcour 2009). So, the agents causing membrane destabilization that leads to cell envelope permeability, leakage of cellular contents, and ultimately lytic cell death can rejuvenate antibiotic sensitivity toward Gram-negative bacteria. Bacteria have proven to be a rich source of membrane-active antimicrobial peptides. The following Table 14.6 enlists the peptide antibiotic potentiators derived from bacterial sources. Bacterial products and their semisynthetic analogues continue to be the monumental source of potential drug leads. One such example is SPR741, a cationic peptide derived from polymyxin B and an excellent potentiator molecule under phase 1 clinical development to treat serious Gram-negative bacterial infections. SPR741 exhibits minimal intrinsic antibacterial activity but retains the ability to permeabilize Gram-negative bacteria's outer membrane, thus sensitizing them to hydrophobic antibiotics. Importantly, these structural changes also significantly improve the safety profile of SPR741 compared to that of polymyxin B, which suffers severe, dose-limiting nephrotoxicity in humans (Corbett et al. 2017).

14.5 Concluding Remarks

The dwindling antibiotic pipeline and the growing nonresponsiveness of antibiotic-resistant Gram-negative pathogens to colistin have propelled us to the verge of a "postantibiotic era." This scenario requires our immediate actions to address these organisms by developing new drugs and finding appropriate solutions to control the spread of resistance. Natural products built our antibiotic arsenal in the golden era and continued to provide new drugs with the advancement of medicinal chemistry. Examples of such drugs include new-generation penicillins, cephalosporins, tetracyclines, etc. In the early years of the twenty-first century, there was a sharp decline in the discovery of new classes of antibiotics because the conventional approaches for isolating bacterial natural products faced the replication of already discovered compounds, and it became challenging to find new antibacterial agents. Recently, the discovery of new classes from the untapped microbes or unculturable bacteria such as teixobactin (Ling et al. 2015) and lugdunin (Zipperer et al. 2016) sparked a resurgence in the exploitation of new natural products of bacteria to fight difficult-to-treat pathogens. Though it may not be enough, a few new antibiotics have been discovered from bacteria in recent years that are active against Gram-negative superbugs, viz., tridecaptin M, darobactin, brevicidine, laterocidin, odilorhabdins, etc. It is noteworthy that most of these new antibiotics belong to NRP category. Genome mining of various bacterial species has shown that there are several uncharacterized biosynthetic gene clusters for NRPs that may lead to the discovery of new antibiotics (Li et al. 2018). Also, to mention, few natural product-based compounds that are in preclinical or early-stage clinical development, such as novel analogues of polymyxins (SPR741, CB-182804, FADDI-002, etc.), octapeptins, and optimized arylomycins (G0775), are also derived from NRPs of bacteria. These examples highlight the antimicrobial potential of bacterial secondary metabolites. We have not even explored the 1% of the total bacterial diversity for antimicrobial production, so there is no doubt that we can find potential new antibiotic candidates from bacteria in the future. The rich chemical diversity of bacterial metabolites also makes them an attractive source of RMAs. However, the development of these adjuvants is still in its infancy, and therefore more efforts are necessary to discover and take these bacterial-derived antibiotic adjuvants further. In conclusion, a more holistic approach is necessary to develop new antibacterial therapies against Gram-negative pathogens that include innovative techniques to isolate novel bacterial diversity, chemical genomics to identify uncharacterized biosynthetic cluster and expression of cryptic gene clusters, medicinal chemistry to improve the physicochemical and biological properties of the compounds, etc.

References

Abou Fayad A, Herrmann J, Müller R (2018) Octapeptins: lipopeptide antibiotics against multidrug-resistant superbugs. Cell Chem Biol 25:351–353

Aggen JB et al (2009) Synthesis, structure and in vitro activity of the neoglycoside ACHN-490. In: 49th annual interscience conference on antimicrobial agents and chemotherapy (ICAAC), California, pp 12–15

Aggen JB et al (2010) Synthesis and spectrum of the neoglycoside ACHN-490. Antimicrob Agents Chemother 54:4636–4642

Al-Ayed K, Ballantine RD, Zhong Z, Li Y, Cochrane S, Martin N (2021) Total synthesis of the Brevicidine and Laterocidine family of lipopeptide antibiotics

Amaral L, Martins A, Spengler G, Molnar J (2014) Efflux pumps of Gram-negative bacteria: what they do, how they do it, with what and how to deal with them. Front Pharmacol 4:168

Aoki T et al (2018) A new siderophore cephalosporin exhibiting potent activities against Pseudomonas aeruginosa and other gram-negative pathogens including multidrug resistant bacteria: structure activity relationship. Eur J Med Chem 155:847–868

Balan SS, Kumar CG, Jayalakshmi S (2016) Pontifactin, a new lipopeptide biosurfactant produced by a marine Pontibacter korlensis strain SBK-47: purification, characterization and its biological evaluation. Process Biochem 51:2198–2207

Bann SJ, Ballantine RD, Cochrane SA (2021) The tridecaptins: nonribosomal peptides that selectively target gram-negative bacteria. RSC Med Chem 12:538

Baumann S et al (2014) Cystobactamids: myxobacterial topoisomerase inhibitors exhibiting potent antibacterial activity. Angew Chem Int Ed 53:14605–14609

Bérdy J (2012) Thoughts and facts about antibiotics: where we are now and where we are heading. J Antibiot 65:385

Blanco P, Sanz-García F, Hernando-Amado S, Martínez JL, Alcalde-Rico M (2018) The development of efflux pump inhibitors to treat Gram-negative infections. Expert Opin Drug Discovery 13:919–931

Borges A, Abreu AC, Dias C, Saavedra MJ, Borges F, Simões M (2016) New perspectives on the use of phytochemicals as an emergent strategy to control bacterial infections including biofilms. Molecules 21:877

Brown ED, Wright GD (2016) Antibacterial drug discovery in the resistance era. Nature 529:336

Brown P, Dawson MJ (2017) Development of new polymyxin derivatives for multidrug resistant Gram-negative infections. J Antibiot 70:386–394

Butler MS, Paterson DL (2020) Antibiotics in the clinical pipeline in October 2019. J Antibiot 73:329–364

Ch'ng J-H, Chong KK, Lam LN, Wong JJ, Kline KA (2019) Biofilm-associated infection by enterococci. Nat Rev Microbiol 17:82–94

Chang H, Zhou J, Zhu X, Yu S, Chen L, Jin H, Cai Z (2017) Strain identification and quorum sensing inhibition characterization of marine-derived rhizobium sp. NAO1. R Soc Open Sci 4:170025

Cho KW, Lee H-S, Rho J-R, Kim TS, Mo SJ, Shin J (2001) New lactone-containing metabolites from a marine-derived bacterium of the genus Streptomyces. J Nat Prod 64:664–667

Clardy J, Fischbach MA, Walsh CT (2006) New antibiotics from bacterial natural products. Nat Biotechnol 24:1541

Coates AR, Halls G, Hu Y (2011) Novel classes of antibiotics or more of the same? Br J Pharmacol 163:184–194

Cochrane SA, Vederas JC (2014) Unacylated tridecaptin A1 acts as an effective sensitiser of gram-negative bacteria to other antibiotics. Int J Antimicrob Agents 44:493–499

Control CfD, Prevention (2013) Antibiotic resistance threats in the United States, 2013. Centres for Disease Control and Prevention, U.S. Department of Health and Human Services

Corbett D et al (2017) Potentiation of antibiotic activity by a novel cationic peptide: potency and spectrum of activity of SPR741. Antimicrob Agents Chemother 61:8

Courtwright DT (2009) Forces of habit. Harvard University Press, Cambridge

Delcour AH (2009) Outer membrane permeability and antibiotic resistance. Biochim et Biophys Acta 1794:808–816

Dobretsov S, Teplitski M, Alagely A, Gunasekera SP, Paul VJ (2010) Malyngolide from the cyanobacterium Lyngbya majuscula interferes with quorum sensing circuitry. Environ Microbiol Rep 2:739–744

Elemam A, Rahimian J, Doymaz M (2010) In vitro evaluation of antibiotic synergy for polymyxin B-resistant carbapenemase-producing Klebsiella pneumoniae. J Clin Microbiol 48:3558–3562

Elgaher WA et al (2020) Cystobactamid 507: concise synthesis, mode of action and optimization toward more potent antibiotics. Chemistry: European J 26:7219

Eljaaly K, Alharbi A, Alshehri S, Ortwine JK, Pogue JM (2019) Plazomicin: a novel aminoglycoside for the treatment of resistant Gram-negative bacterial infections. Drugs 79:243–269

Elshikh M, Funston S, Chebbi A, Ahmed S, Marchant R, Banat IM (2017) Rhamnolipids from non-pathogenic Burkholderia thailandensis E264: physicochemical characterization, antimicrobial and antibiofilm efficacy against oral hygiene related pathogens. N Biotechnol 36:26–36

FDA USFaDA (2015) Drugs@FDA data files

Fleming A (1929) On the antibacterial action of cultures of a *penicillium*, with special reference to their use in the isolation of B. influenzae. Br J Exp Pathol 10:226

Flemming H-C, Wuertz S (2019) Bacteria and archaea on earth and their abundance in biofilms. Nat Rev Microbiol 17:247–260

Fox S, Farr-Jones S, Sopchak L, Boggs A, Nicely HW, Khoury R, Biros M (2006) High-throughput screening: update on practices and success. J Biomol Screen 11:864–869

Giorgi EE (2016) The antibacterial resistance threat: are we heading toward a postantibiotic era? http://www.huffingtonpost.in/entry/the-antibacterial-resista_b_9579234. Accessed 3 Oct 2016

Gowrishankar S, Pandian SK, Balasubramaniam B, Balamurugan K (2019) Quorum quelling efficacy of marine cyclic dipeptide-cyclo (L-leucyl-L-prolyl) against the uropathogen Serratia marcescens. Food Chem Toxicol 123:326–336

Guo Y, Huang E, Yuan C, Zhang L, Yousef AE (2012) Isolation of a Paenibacillus sp. strain and structural elucidation of its broad-spectrum lipopeptide antibiotic. Appl Environ Microbiol 78:3156–3165

Gupta RK, Setia S, Harjai K (2011) Expression of quorum sensing and virulence factors are interlinked in Pseudomonas aeruginosa: an in vitro approach. Am J Biomed Sci 3:116–125

Hassan R, Shaaban MI, Abdel Bar FM, El-Mahdy AM, Shokralla S (2016) Quorum sensing inhibiting activity of Streptomyces coelicoflavus isolated from soil. Front Microbiol 7:659

Huang E (2013) A novel broad-spectrum lipopeptide antimicrobial agent, paenibacterin, against drug-resistant bacteria: structural elucidation, biosynthesis, and mechanisms of action

Huang E, Yang X, Zhang L, Moon SH, Yousef AE (2017) New Paenibacillus strain produces a family of linear and cyclic antimicrobial lipopeptides: cyclization is not essential for their antimicrobial activity. FEMS Microbiol Lett 364:8

Hunt D et al. (2010) TP-434 is a novel broad-spectrum fluorocycline. In: 50th interscience conference on antimicrobial agents and chemotherapy conference, Boston

Hüttel S et al (2017) Discovery and total synthesis of natural Cystobactamid derivatives with superior activity against gram-negative pathogens. Angew Chem Int Ed 56:12760–12764

Imai Y et al (2019) A new antibiotic selectively kills gram-negative pathogens. Nature 576:459–464

Jangra M, Raka V, Nandanwar H (2020) In vitro evaluation of antimicrobial peptide Tridecaptin M in combination with other antibiotics against multidrug resistant Acinetobacter baumannii. Molecules 25:3255

Jangra M et al (2018) Purification, characterization and in vitro evaluation of polymyxin a from Paenibacillus dendritiformis: an underexplored member of the polymyxin family. Front Microbiol 9:2864

Jangra M et al (2019) Tridecaptin M, a new variant discovered in mud bacterium, shows activity against colistin-and extremely drug-resistant Enterobacteriaceae. Antimicrob Agents Chemother 63:6

Jeffreys D (2008) Aspirin: the remarkable story of a wonder drug. Bloomsbury Publishing, New York

Jukes TH (1985) Some historical notes on chlortetracycline. Rev Infect Dis 7:702–707

Kollef MH, Golan Y, Micek ST, Shorr AF, Restrepo MI (2011) Appraising contemporary strategies to combat multidrug resistant gram-negative bacterial infections–proceedings and data from the gram-negative resistance summit. Clin Infect Dis 53:S33–S55

Kunz AN, Brook I (2010) Emerging resistant Gram-negative aerobic bacilli in hospital-acquired infections. Chemotherapy 56:492–500

Laxminarayan R, Chaudhury RR (2016) Antibiotic resistance in India: drivers and opportunities for action. PLoS Med 13:e1001974

Laxminarayan R et al (2013) Antibiotic resistance—the need for global solutions. Lancet Infect Dis 13:1057–1098

Lebreton F et al (2013) Emergence of epidemic multidrug-resistant enterococcus faecium from animal and commensal strains. MBio 4:e00534

Lee MD et al (2001) Microbial fermentation-derived inhibitors of efflux-pump-mediated drug resistance. Farmaco 56:81–85

Lee YR, Burton CE (2019) Eravacycline, a newly approved fluorocycline. Eur J Clin Microbiol Infect Dis 38:1787–1794

Leese RA (2013) Antibiotic compositions for the treatment of gram negative infections. Google Patents

Lewis K (2012) Antibiotics: recover the lost art of drug discovery. Nature 485:439

Lewis K (2013) Platforms for antibiotic discovery. Nat Rev Drug Discov 12:371

Lewis K (2017) New approaches to antimicrobial discovery. Biochem Pharmacol 134:87–98

Li Y-X, Zhong Z, Zhang W-P, Qian P-Y (2018) Discovery of cationic nonribosomal peptides as Gram-negative antibiotics through global genome mining. Nat Commun 9:1–9

Line J et al (2008) Isolation and purification of enterocin E-760 with broad antimicrobial activity against gram-positive and gram-negative bacteria. Antimicrob Agents Chemother 52:1094–1100

Ling LL et al (2015) A new antibiotic kills pathogens without detectable resistance. Nature 517:455–459

Liu J, Smith PA, Steed DB, Romesberg F (2013) Efforts toward broadening the spectrum of arylomycin antibiotic activity. Bioorg Med Chem Lett 23:5654–5659

Lu X, Xu QZ, Shen Y, Liu X, Jiao B, Zhang W, Ni K (2008) Macrolactin S, a novel macrolactin antibiotic from marine Bacillus sp. Nat Prod Res 22:342–347

MacNair CR, Stokes JM, Carfrae LA, Fiebig-Comyn AA, Coombes BK, Mulvey MR, Brown ED (2018) Overcoming mcr-1 mediated colistin resistance with colistin in combination with other antibiotics. Nat Commun 9:1–8

Macone A, Donatelli J, Dumont T, Levy S, Tanaka S, Levy S (2003) In-vitro activity of PTK0796 against gram-positive and gram-negative organisms. In: Proceedings of the 43rd interscience conference on antimicrobial agents and chemotherapy, American Society of Microbiology, Washington, DC

Macone A et al (2014) In vitro and in vivo antibacterial activities of omadacycline, a novel aminomethylcycline. Antimicrob Agents Chemother 58:1127–1135

Magiorakos AP et al (2012) Multidrug-resistant, extensively drug-resistant and pandrug-resistant bacteria: an international expert proposal for interim standard definitions for acquired resistance. Clin Microbiol Infect 18:268–281

Maiti PK, Das S, Sahoo P, Mandal S (2020) Streptomyces sp SM01 isolated from Indian soil produces a novel antibiotic picolinamycin effective against multi drug resistant bacterial strains. Sci Rep 10:1–12

Maiti PK, Mandal S (2019) Majority of actinobacterial strains isolated from Kashmir Himalaya soil are rich source of antimicrobials and industrially important biomolecules. Adv Microbiol 9:220

McKenzie NL, Thaker M, Koteva K, Hughes DW, Wright GD, Nodwell JR (2010) Induction of antimicrobial activities in heterologous streptomycetes using alleles of the Streptomyces coelicolor gene absA1. J Antibiot 63:177–182

Miao L, Xu J, Yao Z, Jiang Y, Zhou H, Jiang W, Dong K (2017) The anti-quorum sensing activity and bioactive substance of a marine derived Streptomyces. Biotechnol Biotechnol Equip 31:1007–1015

Naghmouchi K, Le Lay C, Baah J, Drider D (2012) Antibiotic and antimicrobial peptide combinations: synergistic inhibition of Pseudomonas fluorescens and antibiotic-resistant variants. Res Microbiol 163:101–108

Naik D, Wahidullah S, Meena R (2013) Attenuation of Pseudomonas aeruginosa virulence by marine invertebrate–derived Streptomyces sp. Lett Appl Microbiol 56:197–207

Nelson ML, Levy SB (2011) The history of the tetracyclines. Ann N Y Acad Sci 1241:17–32

Ohadi M, Forootanfar H, Dehghannoudeh G, Eslaminejad T, Ameri A, Shakibaie M, Adeli-Sardou M (2020) Antimicrobial, anti-biofilm, and anti-proliferative activities of lipopeptide biosurfactant produced by Acinetobacter junii B6. Microb Pathog 138:103806

Pantel L et al (2018) Odilorhabdins, antibacterial agents that cause miscoding by binding at a new ribosomal site. Mol Cell 70:83–94

Project AR (2019) Antibiotics currently in global clinical development. The PEW Charitable Trusts, Philadelphia

Qian C-D et al (2012) Battacin (Octapeptin B5), a new cyclic lipopeptide antibiotic from Paenibacillus tianmuensis active against multidrug-resistant gram-negative bacteria. Antimicrob Agents Chemother 56:1458–1465

Quale J et al (2012) Activity of polymyxin B and the novel polymyxin analogue CB-182,804 against contemporary gram-negative pathogens in New York City. Microb Drug Resist 18:132–136

Racine E, Gualtieri M (2019) From worms to drug candidate: the story of odilorhabdins, a new class of antimicrobial agents. Front Microbiol 10:2893

Racine E et al (2018) In vitro and in vivo characterization of NOSO-502, a novel inhibitor of bacterial translation. Antimicrob Agents Chemother 62:9

Rebuffat SF, Telhig S, Said LB, Zirah S, Ismail F (2020) Bacteriocins to thwart bacterial resistance in Gram-negative bacteria. Front Microbiol 11:2807

Reich F, Atanassova V, Klein G (2013) Extended-spectrum β-lactamase-and AmpC-producing enterobacteria in healthy broiler chickens. Germany Emerg Infect Dis 19:1253–1259

Rice LB (2008) Federal funding for the study of antimicrobial resistance in nosocomial pathogens: no ESKAPE. J Infect Dis 197:1079–1081

Rossiter SE, Fletcher MH, Wuest WM (2017) Natural products as platforms to overcome antibiotic resistance. Chem Rev 117:12415–12474

Sacco LP, Castellane TCL, Polachini TC, de Macedo Lemos EG, LMC A (2019) Exopolysaccharides produced by Pandoraea shows emulsifying and anti-biofilm activities. J Polymer Res 26:1–11

Saga T, Yamaguchi K (2009) History of antimicrobial agents and resistant bacteria. JMAJ 52:103

Santajit S, Indrawattana N (2016) Mechanisms of antimicrobial resistance in ESKAPE pathogens. Biomed Res Int 2016:1–8

Sarciaux M et al (2018) Total synthesis and structure–activity relationships study of Odilorhabdins, a new class of peptides showing potent antibacterial activity. J Med Chem 61:7814–7826

Satputea KS, Banpurkar GA, Banat MI, Sangshetti NJ, Patil HR, Gade NW (2016) Multiple roles of biosurfactants in biofilms. Curr Pharm Des 22:1429–1448

Saurav K, Costantino V, Venturi V, Steindler L (2017) Quorum sensing inhibitors from the sea discovered using bacterial N-acyl-homoserine lactone-based biosensors. Mar Drugs 15:53

Saurav K et al (2016) In search of alternative antibiotic drugs: Quorum-quenching activity in sponges and their bacterial isolates. Front Microbiol 7:416

Schimana J et al (2002) Arylomycins a and B, new Biaryl-bridged Lipopeptide antibiotics produced by Streptomyces sp. Tü 6075 I. J Antibiotics 55:565–570

Smith PA et al (2018) Optimized arylomycins are a new class of gram-negative antibiotics. Nature 561:189–194

Srivastava A et al (2011) New target for inhibition of bacterial RNA polymerase: 'switch region'. Curr Opin Microbiol 14:532–543

Stratton CF, Newman DJ, Tan DS (2015) Cheminformatic comparison of approved drugs from natural product versus synthetic origins. Bioorg Med Chem Lett 25:4802–4807

Sutcliffe J, O'Brien W, Fyfe C, Grossman T (2013) Antibacterial activity of eravacycline (TP-434), a novel fluorocycline, against hospital and community pathogens. Antimicrob Agents Chemother 57:5548–5558

Sutherland R (1991) β-lactamase inhibitors and reversal of antibiotic resistance. Trends Pharmacol Sci 12:227–232

Svetoch E et al (2009) Antimicrobial activities of bacteriocins E 50–52 and B 602 against antibiotic-resistant strains involved in nosocomial infections. Probiot Antimicrob Protein 1:136–142

Tacconelli E et al (2018) Discovery, research, and development of new antibiotics: the WHO priority list of antibiotic-resistant bacteria and tuberculosis. Lancet Infect Dis 18:318–327

Teasdale ME, Donovan KA, Forschner-Dancause SR, Rowley DC (2011) Gram-positive marine bacteria as a potential resource for the discovery of quorum sensing inhibitors. Marine Biotechnol 13:722–732

Teasdale ME, Liu J, Wallace J, Akhlaghi F, Rowley DC (2009) Secondary metabolites produced by the marine bacterium Halobacillus salinus that inhibit quorum sensing-controlled phenotypes in gram-negative bacteria. Appl Environ Microbiol 75:567–572

Testolin G et al (2020) Synthetic studies of cystobactamids as antibiotics and bacterial imaging carriers lead to compounds with high in vivo efficacy. Chem Sci 11:1316–1334

Tzouvelekis L, Markogiannakis A, Psichogiou M, Tassios P, Daikos G (2012) Carbapenemases in Klebsiella pneumoniae and other Enterobacteriaceae: an evolving crisis of global dimensions. Clin Microbiol Rev 25:682–707

Vaara M (2019) Polymyxins and their potential next generation as therapeutic antibiotics. Front Microbiol 10:1689

Van Boeckel TP, Gandra S, Ashok A, Caudron Q, Grenfell BT, Levin SA, Laxminarayan R (2014) Global antibiotic consumption 2000 to 2010: an analysis of national pharmaceutical sales data. Lancet Infect Dis 14:742–750

van der Mee-Marquet N et al (2011) Emergence of unusual bloodstream infections associated with pig-borne–like Staphylococcus aureus ST398 in France. Clin Infect Dis 52:152–153

Vasoo S, Barreto JN, Tosh PK (2015) Emerging issues in gram-negative bacterial resistance: an update for the practicing clinician. In: Mayo clinic proceedings, vol 3. Elsevier, Amsterdam, pp 395–403

Velkov T, Roberts KD (2019) Discovery of novel polymyxin-like antibiotics polymyxin antibiotics: from laboratory bench to bedside. Springer, Cham, pp 343–362

Velkov T, Roberts KD, Nation RL, Wang J, Thompson PE, Li J (2014) Teaching 'old' polymyxins new tricks: new-generation lipopeptides targeting gram-negative 'superbugs'. ACS Chem Biol 9:1172–1177

Velkov T et al (2018) Structure, function, and biosynthetic origin of octapeptin antibiotics active against extensively drug-resistant gram-negative bacteria. Cell Chem Biol 25:380–391

Wang Y et al (2012) Identification of New Delhi metallo-β-lactamase 1 in Acinetobacter lwoffii of food animal origin. PLoS One 7:e37152

Weinstein MJ, Marquez JA, Testa RT, Wagman GH, Oden EM, Waitz JA (1970) Antibiotic 6640, a new micromonospora-produced aminoglycoside antibiotic. Journal of Antibiotics 23:551–554

Whalen KE, Poulson-Ellestad KL, Deering RW, Rowley DC, Mincer TJ (2015) Enhancement of antibiotic activity against multidrug-resistant bacteria by the efflux pump inhibitor 3, 4-dibromopyrrole-2, 5-dione isolated from a Pseudoalteromonas sp. J Nat Prod 78:402–412

WHO (2016) Antimicrobial resistance: fact sheet. http://www.who.int/mediacentre/factsheets/fs194/en/. Accessed 2 Oct 2016

Wright GD (2014) Something old, something new: revisiting natural products in antibiotic drug discovery. Can J Microbiol 60:147–154

Wright GD (2017) Opportunities for natural products in 21st century antibiotic discovery. Nat Prod Rep 34:694–701

Zhanel GG et al (2019) Cefiderocol: a siderophore cephalosporin with activity against carbapenem-resistant and multidrug-resistant gram-negative bacilli. Drugs 79:271–289

Zhang RG et al (2002) Structure of a bacterial quorum-sensing transcription factor complexed with pheromone and DNA. Nature 417:971–974

Zhao M, Lepak AJ, Marchillo K, VanHecker J, Andes DR (2018) In vivo pharmacodynamic characterization of a novel odilorhabdin antibiotic, NOSO-502, against Escherichia coli and Klebsiella pneumoniae in a murine thigh infection model. Antimicrob Agents Chemother 62:5

Zhu T et al (2013) Hit identification and optimization in virtual screening: practical recommendations based on a critical literature analysis: miniperspective. J Med Chem 56:6560–6572

Zipperer A et al (2016) Human commensals producing a novel antibiotic impair pathogen colonization. Nature 535:511–516

Plant Essential Oils for Combating Antimicrobial Resistance via Re-potentiating the Fading Antibiotic Arsenal

15

Tuyelee Das, Samapika Nandy, Anuradha Mukherjee, Potshanghbam Nongdam, and Abhijit Dey

Contents

T. Das · S. Nandy · A. Dey (✉)
Department of Life Sciences, Presidency University, Kolkata, West Bengal, India
e-mail: abhijit.dbs@presiuniv.ac.in

A. Mukherjee
MMHS, Joynagar, West Bengal, India

P. Nongdam
Department of Biotechnology, Manipur University, Imphal, Manipur, India

© The Author(s), under exclusive license to Springer Nature Singapore Pte
Ltd. 2022
V. Kumar et al. (eds.), *Antimicrobial Resistance*,
https://doi.org/10.1007/978-981-16-3120-7_15

419

Abstract

The rattling severity and progression of multidrug resistance (MDR) throughout
the globe; continuous surge in new infection categories; highly evolved
mechanisms of resistance; emergence of newer strains almost every day; inadequate availability of broad-spectrum antibiotics; high susceptibility of immunocompromised patients; lacunae in the first-line diagnosis; absence of
infrastructure and well-thought policies to counter-attack the threat; prolonged
duration of treatment; lack of awareness; and mounting burden on healthcare
facilities and public health budget of both developed and underdeveloped
countries have widened both the scope and necessity to explore therapeutic
leads from natural compounds. Plant-derived essential oils, or EOs, are an
integral part of myriads of natural compounds, which are presently passing
through various stages of clinical and preclinical trials. A large number of EOs
are already reported against resistant strains of microbes, but comparatively very
few papers have elucidated the underlying mode of action. However, the versa-
tility of the mechanism is prominent in attenuation of MDR via the application of
EOs which includes primarily replication blockage; halt in cell cycle progression;
altered membrane permeability; and multiphase immunomodulation. The
disturbances in normal microbial growth kinetics are closely related with reduced
ATP production, morphed DNA and protein synthesis, altered enzymatic activity,
changes in pH, as well as hampered intracytoplasmic ion uptake which could get
triggered by EOs. In many instances, it regulates the resistance-modifying mechanism via the activation of synergistic pathways. EOs, in combination with
promising drugs, restore the chemo-sensitization; act in sync with the
bacteriophages; modulate the protein-protein interaction; and potentiate target-
oriented drug transport and action. The effectiveness of EOs, mostly the
cinnamaldehydes and derivative acids, carvacrol, eugenol, and thymol, was
already established against hospital-acquired MRSA, VRE, ESBL *E. coli*, and

many other Gram-negative resistant strains, drug-resistant fungal strains, and drug-resistant protozoal strains and viruses. The present review also described plant family-wise antimicrobial activity against drug-resistant microbes. The present review is a specific mechanism-oriented approach that aims to highlight the physic-chemical changes forced by EOs against MDR strains and how the application of bioinformatics tools and transgenic experiments could aid in emphasizing the efficacy.

Keywords

Essential oils · Antimicrobial resistance · Natural products · Drug resistance · Medicinal plants

15.1 Introduction

Globally antimicrobial resistance (AMR) trends are highly shocking that showed many complex threats but only a few solutions are present. AMR is one of the main concerns for economic loss. Worldwide, many strains of human pathogenic microbes are already resistant to antibiotics with developing new resistance strategies that threatened our ability to cope up with diseases. Antimicrobial resistance to antibiotics gains risks of "Millennium Development and Sustainable Development Goals." Antimicrobial resistance means pathogens, like bacteria, fungi, viruses, and protozoa, show resistance to drugs naturally or develop defense over time gradually; as a result, drugs become ineffective and the pathogen survives and continues its multiplication (Tanwar et al. 2014). Each year over millions of people get infected with antimicrobial-resistant microbes and the number of affected people will increase each passing day that would hamper global health and the economy. Antimicrobial resistance is causing 700,000 deaths annually, and by 2050, it will result in around ten million deaths on a global scale (Hermsen et al. 2020).

The treatment failures and constant use of antimicrobial agents lead to the development of microbial multidrug-resistant forms. Some multidrug-resistant microbial strain examples are mainly *Staphylococcus aureus*, *Enterococcus faecalis*, *Streptococcus pneumonia*, *Pseudomonas aeruginosa*, *Mycoplasma genitalium*, *Clostridium difficile*, *Acinetobacter baumannii*, *Bacillus cereus*, *Escherichia coli*, *Campylobacter*, herpesviruses, influenza viruses, and several species of *Trypanosoma*, *Leishmania*, *Plasmodium*, *Aspergillus* sp., and *Candida* sp. (Barnes and Sampson 2011; Hansra and Shinkai 2011). MDR strains of bacteria resistant to beta-lactam antibiotics (carbapenem, methicillin, tazobactam, penicillin), macrolide antibiotics (erythromycin), lincomycin antibiotics (clindamycin), glycopeptide antibiotics (vancomycin), tetracycline antibiotics (tetracycline, minocycline), rifamycins (rifampin), quinolone, fluoroquinolone, kanamycin, and capreomycin (Chambers and DeLeo 2009; Garbati and Al Godhair 2013; Pachori et al. 2019; Partridge 2015; Theresa et al. 1999; Yayan et al. 2015). MDR fungi are resistant to mainly azole drugs (Ksiezopolska and Gabaldón 2018; Martinez-Rossi et al. 2018; Méndez-Tovar et al. 2007; Morace et al. 2014; Sarma and Upadhyay 2017);

protozoal pathogens are resistant to miltefosine, paromomycin, and amphotericin B (Pramanik et al. 2019); and viruses are mainly resistant to acyclovir, lamivudine, amantadine, amantadine, and emtricitabine drugs, as represented in Table 15.1 (Hussain et al. 2017; Zoulim 2011). According to the 2013 and 2019 reports, the Centers for Disease Control and Prevention (CDC) included 15 antibiotic-resistant bacteria and 3 antifungal-resistant fungi into three categories: urgent, serious, and concerning (https://www.cdc.gov/drugresistance/biggest-threats.html). Apart from this, many drug-resistant microbes are also present; however, CDC did not include drug-resistant protozoa and virus species. In 2017, WHO (World Health Organization) categorized some neglected tropical diseases caused by mainly protozoa or viruses (https://www.cdc.gov/globalhealth/ntd/index.html). The major neglected tropical diseases are Chagas disease, human African trypanosomiasis, leishmaniasis, leprosy, onchocerciasis, onchocerciasis, onchocerciasis, etc., caused by drug-resistant microbes (Akinsolu et al. 2019). These drug-resistant microbes cause a serious health-related threat that becomes a worldwide concern that has heightened the search for alternative therapeutic agents (Igwaran et al. 2017). The Centers for Disease Control and Prevention estimated and published in 2019 that 2.8 million infections occurred in the United States due to antimicrobial resistance each year and resulted in more than 35,000 deaths per year (Brockhouse and Scott 2020). The European Antimicrobial Resistance Surveillance Network (EARS-Net) estimated through databases from 2015 to 2018 showed that 670,000 infections happened in Europe due to bacterial resistance to antibiotics. Bacteria developed acquired multiple resistances that can spread horizontally (ECDC/EMEA 2009). Viruses become resistant to antiviral drugs due to resistance mutation in the virus gene that is constantly changing (Strasfeld and Chou 2010).

A synthetic chemical as an antimicrobial drug used for the management of pathogenic microorganisms is not significantly effective and limited due to its toxicity and environmental hazard potential. Therefore, the utilization of plant essential oils against pandemic multidrug-resistant pathogenic microorganisms can be helpful to reverse resistance and stop infectious diseases. Plants produce natural compounds referred to as essential oils that have bioactive properties and may block the growth of various microbes. Essential oils accumulate in the glandular trichomes, secretary canals, or cavities (Arumugam et al. 2016). Chemical constituents of essential oils show more bioactivity in the oxygenated form (Degenhardt et al. 2009; Sinniah 2015). Essential oils are mainly composed of low molecular weights terpenes, terpenoids, and other aromatic and aliphatic constituents (Nazzaro et al. 2013). The concentration and nature of chemical constituents in essential oils may differ as per species, genus, and even external factors like temperature and geographical locations that may influence the difference in antimicrobial activities against microbes (Swamy et al. 2015). The essential oils presented in this review were chosen for their proven antimicrobial activity against drug-resistant microbes. The present article aims to review published studies on the activity of essential oils and their constituents against multidrug-resistant microbes and to ensure perspectives for the future that limit the prolonged hospital stays, higher medical costs, and decreased mortality. Greater innovation and investment are required in the

Table 15.1 List of selected drug-resistant species and genus of microbes

Resistant microbes (pathogen)	Drug-resistant species family	Drug(s) resistant to	Human diseases/infections or associated diseases	Reference
Bacteria				
Acinetobacter baumannii	Moraxellaceae	Carbapenem	Pneumonia, wound infections, bloodstream infections, urinary tract infections, meningitis, bacteremia), endocarditis	(Almaghrabi et al. 2018; Manchanda et al. 2010)
Bacillus cereus	Bacillaceae	Ciprofloxacin, cloxacillin, erythromycin, tetracycline, streptomycin	Gastrointestinal illness	(Citron and Appleman 2006)
Bordetella pertussis	Alcaligenaceae	Erythromycin, clarithromycin, or azithromycin (macrolide)	Whooping cough	(Guillot et al. 2012)
Campylobacter	Campylobacteraceae	Ciprofloxacin and azithromycin	Diarrhea, fever, abdominal cramps	(Schiaffino et al. 2019)
Clostridium difficile	Clostridiaceae	Clindamycin, moxifloxacin, rifampin	Diarrhea and colitis	(Peng et al. 2017)
Enterococci	Enterococcaceae	Vancomycin, penicillin and ampicillin	Bloodstream, surgical site, and urinary tract infections	(Miller et al. 2014)
Escherichia coli	Enterobacteriaceae	Cephalosporins, fluoroquinolone	Cholecystitis, bloodstream infections, bacteremia, cholangitis, urinary tract infection, traveler's diarrhea	(Partridge 2015, Unemo and Shafer 2014)
Klebsiella pneumoniae	Enterobacteriaceae	Cephalosporins, carbapenems	Pneumonia, meningitis, bloodstream infections, wound, and urinary tract infections	(Garbati and Al Godhair 2013; Partridge 2015)
Mycobacterium tuberculosis	Mycobacteriaceae	Isoniazid, rifampin, fluoroquinolone, kanamycin, capreomycin or amikacin	Tuberculosis	(Palomino and Martin 2014)

(continued)

Table 15.1 (continued)

Resistant microbes (pathogen)	Drug-resistant species family	Drug(s) resistant to	Human diseases/infections or associated diseases	Reference
Mycoplasma genitalium	*Mycoplasmataceae*	Azithromycin, doxycycline, moxifloxacin, rifampicin, isoniazid, fluoroquinolone	Tuberculosis, cervicitis, pelvic inflammatory disease	(Jensen et al. 2016; Yew et al. 2011)
Neisseria gonorrhoeae	*Neisseriaceae*	Cephalosporins, macrolides	Gonorrhea, urethra, cervix, pharynx, or rectum inflammation	(Tanwar et al. 2014; Unemo and Shafer 2014)
Pseudomonas aeruginosa	*Pseudomonadaceae*	Carbapenem, piperacillin, tazobactam, cefepime, ceftazidime	Urinary tract infections, soft tissue infections, bacteremia, bone infections, joint infections, respiratory system infections, dermatitis	(Pachori et al. 2019; Yayan et al. 2015)
Salmonella serotype Typhi	*Enterobacteriaceae*	Ampicillin, chloramphenicol, trimethoprim-sulfamethoxazole	Typhoid fever, diarrhea, fever, and abdominal cramps, bloodstream infections	(Partridge 2015)
Staphylococcus aureus	*Staphylococcaceae*	Penicillin, methicillin, tetracycline, erythromycin, vancomycin	Septic arthritis, gastroenteritis, toxic shock Syndrome, urinary tract infections, food poisoning, wound, and bloodstream infections	(Theresa et al. 1999) (Chambers and DeLeo 2009)
Streptococcus pneumoniae	*Streptococcaceae*	Penicillin, clindamycin, methicillin, tetracycline, erythromycin, vancomycin	Meningitis, acute otitis media, bronchitis, pneumonia, sepsis, bacteriemia, sinusitis	(Mamishi et al. 2014)
Shigella spp.	*Enterobacteriaceae*	Fluoroquinolones	Diarrhea, fever, stomach cramps	(Partridge 2015)
Fungi				
Aspergillus fumigatus	*Trichocomaceae*	Azoles	Aspergillosis	(Howard and Arendrup 2011; Szalewski et al. 2018)
Candida sp.	*Saccharomycetaceae*	Fluconazole, echinocandins	Candidiasis	(Ksiezopolska and Gabaldón 2018; Morace et al. 2014; Sarma and Upadhyay 2017)

Trichophyton spp.	Arthrodermataceae	Azoles	Athlete's foot, jock itch, infections of the nail, beard, skin, and scalp	(Martínez-Rossi et al. 2018; Méndez-Tovar et al. 2007)
Plasmodium				
Entamoeba	Entamoebidae	Metronidazole, trifluoromethionine, emetine	Amoebiasis	(Bansal et al. 2006; Pramanik et al. 2019)
Leishmania spp.	Trypanosomatidae	Pentavalent antimonials, pentamidine, miltefosine, paromomycin, amphotericin B	Leishmaniasis	(Pramanik et al. 2019)
Plasmodium spp.	Plasmodiidae	Chloroquine, artemisinin, sulfadoxine/pyrimethamine, piperaquine, mefloquine, amodiaquine, atovaquone	Malaria	(Bloland and Bloland 2001; Pramanik et al. 2019)
Toxoplasma gondii	Sarcocystidae	Artemisinin, atovaquone, sulfadiazine	Toxoplasmosis	(McFadden et al. 2000; Nagamune et al. 2007; Pramanik et al. 2019)
Trichomonas vaginalis	Trichomonadidae	Nitroimidazoles	Trichomoniasis	(Muzny and Schwebke 2013)
Trypanosoma	Trypanosomatidae	Melarsoprol, suramin, nifurtimox, nitrofuran, benznidazole	Trypanosomiasis	(Pramanik et al. 2019)
Viruses				
Human immunodeficiency virus (HIV)	Retroviridae	Antiretroviral drugs	AIDS	(Cortez and Maldarelli 2011)
Influenza viruses	Orthomyxoviridae	Amantadine, amantadine, neuraminidase inhibitors	Flu, asthma, influenza	(Hurt 2014; Hussain et al. 2017)
Hepatitis B viruses	Hepeviridae	Lamivudine	Hepatitis B	(Suppiah et al. 2014; Zoulim 2011)

(continued)

Table 15.1 (continued)

Resistant microbes (pathogen)	Drug-resistant species family	Drug(s) resistant to	Human diseases/infections or associated diseases	Reference
Herpes simplex viruses (HSV)	*Herpesviridae*	Acyclovir, ganciclovir, famciclovir, valacyclovir	Herpes labialis, gingivostomatitis, herpes simplex	(Pellet and Roizman 2007; Wutzler 1997)
Varicella zoster virus (VZV)	*Herpesviridae*	Acyclovir and valacyclovir	Chicken pox	(Wutzler 1997)

research and development of new antimicrobial medicines, vaccines, and diagnostic tools.

15.2 Methodology

Relevant scientific papers were retrieved from the databases like Google Scholar, PubMed, ScienceDirect, SpringerLink, and Research Gate. The search strings contain specific species and family names. Other keywords like "antimicrobial plant essential oil," "medicinal plant essential oil, antimicrobial activity," "plant extract essential oil antibacterial effect," "synergy of essential oil," "essential oil mechanism against drug-resistant bacteria or protozoa or fungi or viruses," essential oils' role in microbial cell membrane disruption, anti-quorum activity, efflux pump inhibition, replication inhibition, and mitochondrial disruption were also used. A total of 17 families and 48 species (21 for antibacterial activity; 16 for antifungal activity; 15 for anti-plasmodial activity, 16 for antiviral activity) is presented in textual and tabulated format for in vitro antimicrobial activity, and 21 species of plant-wise description is presented in Table 15.7 for its synergistic activity. Table 15.1 described the drug-resistant human pathogenic microbes and their resistance to particular drugs. Table 15.2 represents the mode of action of essential oils from plants in bacteria, fungi, protozoa, and viruses. Tables 15.3, 15.4, 15.5, and 15.6 represent the plant-wise description of antibacterial, antifungal, antiprotozoal, and antiviral activities, respectively. Figure 15.1 demonstrates the approaches of essential oil combating antimicrobial resistance. Figure 15.2 demonstrates the chemical structures of the major bioactive components in plant essential oils that showed antimicrobial activity. All chemical structures of the major components were retrieved from https://pubchem.ncbi.nlm.nih.gov/ and http://www.chemspider.com/.

15.3 Possible Mechanism of Action of Essential Oil from Plants in Drug-Resistant Microbes

Miscellaneous mechanisms have been described to explain the activity of an EO on microbial cells. In drug-resistant bacteria and fungi, essential oils act on the cell wall, alter membrane permeability, and suppress efflux pumps on the cell membrane. The permeability barrier provided by cell membranes is indispensable to many cellular functions, including maintaining the energy status of the cell, membrane-coupled energy-transducing processes, solute transport, and metabolic regulation. The cell membranes act as an integral part for controlling the turgor pressure, whereas in drug-resistant viruses, the mechanism of essential oil is to interfere with viral replication. In fungus and protozoal cells, essential oils mainly stop mitochondrial function and reduce membrane potential that may alter the resistance of microbes towards antimicrobial drugs. The mechanism of essential oils in drug-resistant microbes compactly describes with the experimental details in Table 15.2 and Fig. 15.1.

Table 15.2 Mode of action of essential oils against drug-resistant microbes

Plant from which essential oil is derived	Mechanism of action	Microorganism targeted	Assays/method/techniques	Reference
Anethum graveolens L.	Mitochondrial hyperpolarization	*Candida albicans*	Flow cytometry, colorimetric method by XTT assay	(Chen et al. 2013)
Artemisia arborescens	Inhibiting cell-to-cell virus diffusion	HSV-1, HSV-2	MTT assay	(Saddi et al. 2007)
Chenopodium ambrosioides L.	Efflux pump inhibition	*Staphylococcus aureus*	Efflux pump inhibition assay	(de Morais Oliveira-Tintino et al. 2018)
Chenopodium ambrosioides L.	Efflux pump inhibition (TetK)	*Staphylococcus aureus* IS-58	Ethidium bromide assays	(Figueredo et al. 2017)
Coriandrum sativum L.	Cell membrane integrity disrupts	*Candida albicans*, *Candida tropicalis*	Germ tube formation assays, flow cytometry	(Silva et al. 2011)
Citrus sinensis (L.) Osbeck	Cytoplasm loss in fungal hyphae	*Aspergillus niger*	Determination of mycelial weight, spore germination assay, scanning electron microscopy	(Sharma and Tripathi 2008)
Curtisia dentata (Burm.f.) C.A. Sm.	Increased permeability of the membrane	*Escherichia coli*, *Acinetobacter* spp.	Na + and K+ leakage assay	(Doughari et al. 2012)
Cudrania tricuspidata (Carrière) Bureau ex Lavalle	Membrane integrity	*Bacillus cereus*, *Escherichia coli*	Scanning electron microscopy observation, potassium ions efflux assay	(Bajpai et al. 2013)
Cymbopogon citrates (DC.) Stapf.	Elicit morphological changes, inhibit septum and spheroplast formation, production of mesosomes, developed abnormal cell shape	*Escherichia coli* K-12	Electron microscope observation	(Ogunlana et al. 1987)
Cymbopogon citrates (DC) Stapf.	Changes in cell shape	*Leishmania chagasi*	Flow cytometry, ultrastructural assay	(Oliveira et al. 2009)

Plant/compound	Mechanism	Organism	Methods	Reference
Eugenol* (Syzygium aromaticum. (L.) Merr. & L. M.Perry)	Hyperpermeability and leakage of ions in the cell membrane, extensive loss of other cellular contents (proteins), cell death	Salmonella typhi	Crystal violet assay, measurement of the release of 260 nm absorbing material by UV–VIS spectrophotometer, Fourier transform infrared spectroscopy, atomic force microscopy, and scanning electron microscopic observation	(Devi et al. 2010)
Eugenol* (Syzygium aromaticum. (L.) Merr. & L. M.Perry)	Cell leakage, causing the release of cellular material, and increasing cell permeability	Candida albicans	Crystal Violet assay, scanning electron microscopy, and atomic force microscopy observation	(Latifah-Munirah et al. 2015)
Forsythia koreana Nakai	Loss of membrane integrity, increased membrane permeability	Foodborne and other pathogenic bacteria	Potassium ion flux assay, cellular material release assay, scanning electron microscopic observation	(Yang et al. 2015)
Foeniculum vulgare Mill.	Cell membrane disruption	Shigella dysenteriae	Cell membrane permeability assay, time-kill analysis assay, electron microscopic observation, SEM observation	(Diao et al. 2014)
Lavandula luisieri L.	Mitochondrial membrane potential loss, cell cycle arrest G(0)/G1	Leishmania infantum	Viability assays, transmission and scanning electron microscopy observation, cell cycle analysis, phosphatidylserine externalization analysis double, assessment of mitochondrial membrane potential, cathepsin D activity assay	(Machado et al. 2019)
Lavandula multifida L.	Cytoplasmic membrane disruption and cell death	Candida albicans	Flow cytometry	(Zuzarte et al. 2012)
Lippia sidoides Cham.	Changes in cell shape	Leishmania chagasi	Flow cytometry, ultrastructural assay	(Oliveira et al. 2009)

(continued)

Table 15.2 (continued)

Plant from which essential oil is derived	Mechanism of action	Microorganism targeted	Assays/method/techniques	Reference
Melaleuca alternifolia Cheel	Increased membrane permeability	*Escherichia coli, Staphylococcus aureus*	Cell viability assays, potassium ion-selective electrode	(Cox et al. 2000)
Melaleuca alternifolia Cheel	Mesosomes formation and cytoplasmic contents loss increased	*Staphylococcus aureus*	Bacterial killing assays, bacteriolysis, electron microscopic observation of terpinen-4-ol-treated bacteria	(Lazcano et al. 2002)
Eucalyptus camaldulensis Dehnh.	β-Lactamase inhibitory activity	*Staphylococcus aureus*	–	(Chaves et al. 2018)
Origanum vulgare L.	β-Lactamase- inhibitor	*Escherichia coli*	Checkerboard microtiter test	(Si et al. 2008)
Origanum compactum Benth.	Cell membrane integrity disruption, membrane permeability increases, inhibits biofilm formation	*Escherichia coli, Bacillus subtilis*	Antibacterial kinetics assay, anti-quorum sensing activity	(Bouyahya et al. 2019)
Ocotea odorifera (Vell.) Rohwer	Efflux pump inhibition (NorA and MepA)	*Staphylococcus aureus*	Ethidium bromide assay, antibiotic resistance modulation assay	(Almeida et al. 2020)
Ocimum canum Sims.	Discontinuity of the nucleus membrane and exocytic activity by the flagellar pocket	*Leishmania amazonensis*	Cytotoxicity assay, transmission electron microscopy	(da Silva et al. 2018)
Ocimum gratissimum L.,	Changes in cell shape	*Leishmania chagasi*	Flow cytometry, ultrastructural assay	(Oliveira et al. 2009)
Origanum vulgare L., *Thymus vulgaris* L.	Morphological alterations and cytoplasmic swelling	*Trypanosoma cruzi*	Transmission electron microscopy, SEM, flow cytometry	(Santoro et al. 2007)
Clove (*Syzygium aromaticum*	Deterioration of yeast cell membrane integrity	*Saccharomyces cerevisiae*	Extracellular conductivity measurement, extracellular pH measurement	(Konuk and Erguden 2018)

Plant/compound	Mechanism	Microorganism	Method	Reference
Rosmarinus officinalis L.	efflux pump inhibition	*Acinetobacter baumannii, Pseudomonas aeruginosa*	Flow cytometry analysis	(Saviuc et al. 2016)
Syzygium aromaticum (L.) Merr. & L.M. Perry	Quorum sensing inhibition	*Pseudomonas aeruginosa*	Scanning electron microscopy observation, cellular constituents	(Khan et al. 2009).
Thymus broussonetii Boiss., *Thymus maroccanus* Ball, *Thymus pallidus* Coss.	Efflux pump inhibition	*Escherichia coli, Enterobacter aerogenes, Klebsiella pneumoniae, Salmonella enterica*	–	(Fadii et al. 2011)
Punica granatum L.	Inhibits efflux pump (NorA)	*Staphylococcus aureus*	Time-kill assay, ethidium bromide efflux assay	(Braga et al. 2005)
Salvia fruticosa Mill.	Inhibits efflux pump (TetK)	*Staphylococcus epidermidis*	Spectro-fluorometry	(Chovanova et al. 2015)
Zingiber officinale	β-Lactamase- inhibitor	*Pseudomonas aeruginosa*	Scanning electron microscopy observation, transmission electron microscopy	(El-Shouny et al. 2018)
Thymol, carvacrol	Inhibits efflux pumps (MDR1, CDR1)	*Candida albicans*	Drug efflux activity was determined using two dyes, rhodamine 6G (R6G) and fluorescent Hoechst 33342	(Ahmad et al. 2013)

Table 15.3 Antibacterial activity of essential oil isolated from different plants

Family	Plant name	Plant part	Major chemicals present in essential oil	Assayed against bacteria	MIC/ concentration used in the studies	Experiment details	Allied studies; analytical study conducted [name of the study (if any)]	Reference
Apiaceae	*Foeniculum vulgare* Mill.	Seeds, Yaodu County of Sichuan Province	Trans-anethole (68.53%), estragole (10.42%), limonene (6.24%), fenchone (5.45%)	*Shigella dysenteriae*	MIC: 0.125 mg/mL	Antimicrobial activity test by Oxford cup method; 107 CFU/mL	GC-MS, time-kill curve assay	(Diao et al. 2014)
Aristolochiaceae	*Aristolochia mollissima* Hance	Rhizome and aerial part, Wuhan University, China	2,7,7-Tetramethyltricyclo [6.2.1.0(1,6)]undec-4-en-3-one (15.9%), (E)-santalol acetate (10.3%), camphene (6.7%), bornyl formate (5.3%), cedrenol acetate (5.2%)	*Staphylococcus aureus* ATCC 25923, *Escherichia coli* ATCC 25922, *Pseudomonas aeruginosa* ATCC 27853, *Candida albicans* ATCC 14053	0.39–25.00 mg/ml	Disk diffusion and broth microdilution method 2 × 10⁸ cfu/ml	GC-MS	(Yu et al. 2007)
Brassicaceae	*Sinapis arvensis* var. orientalis (L.)	Stem and flower	Benzyl isothiocyanate (15.15%), cubenol (15.12%), dimethyl trisulfide (6.12%), 6,10,14-trimethylpentadecane-2-one (3.85%), indole (1.91%), dimethyl tetrasulfide (2.22%), 1-butenyl isothiocyanate (18.4%), Thymol	*Bacillus subtilis* MTCC 441, *Staphylococcus aureus* NCTC7428, and three Gram-negative including *Escherichia coli* MTCC 739, *Klebsiella pneumonia* MGH 78578, and *Pseudomonas*	–	Agar disk diffusion; 1 × 108 CFU/ml	GC-MS, DPPH method	(Rad et al. 2013)

Family	Plant	Source	Composition	Microorganism	Concentration	Method	Analysis	Reference
	Tagetes minuta L.	Flowers, Cala community, South Africa	(3.44%), octadecane (4.14%)	aeruginosa MTCC 2453		Antimicrobial actiity test	GC-MS analyses, determination of antioxidant property, DPPH assay, ABTS scavenging assay, thiobarbituric acid reactive species (TBARS) assay	(Igwaran et al. 2017)
			–	Staphylococcus aureus ATCC 29213	0.125 mg/mL			
				Mycobacterium smegmatis ATCC 19420	0.125 mg/mL			
				Listeria ivanovii ATCC 19119	0.06 mg/mL			
				Enterobacter cloacae ATCC 13047	0.06 mg/mL			
				Escherichia coli	0.06 mg/mL			
				Vibrio spp.	0.06 mg/mL			
Lamiaceae	Ocimum basilicum L.	Rhodopes mountain (Bulgaria)	Linalool (54.95%), methyl chavikol (11.98%), methyl cinnamate (7.24%), linolen (0.14%)	Pseudomonas aeruginosa ATCC 9027, Staphylococcus aureus ATCC 6538, Staphylococcus epidermidis, Enterococcus faecalis	0.0030% to 0.0007% (v/v)	Saline test solution, susceptibility test, 10^5 cells/ml	Gas chromatography	(Opalchenova and Obreshkova 2003)
	Origanum vulgare L.	Leaves, Murcia, Spain	Carvacrol (39.08–49.43%), trans-sabinene hydrate (19.81–25.11%), cis-piperitol (5.51–7.77%), borneol (2.82–5.14%), terpinen-4-ol (3.29–	Staphylococcus aureus ATCC 25923, Bacillus subtilis ATCC 6633, Pseudomonas aeruginosa ATCC 10145, Escherichia coli ATCC 11775	Inhibition zone: 17.5–25.3 mm	Antimicrobial activity by the disk diffusion method, broth dilution method; 10^7 CFU/ml	GC-MS	(Santoyo et al. 2006)

(continued)

Table 15.3 (continued)

Family	Plant name	Plant part	Major chemicals present in essential oil	Assayed against bacteria	MIC/concentration used in the studies	Experiment details	Allied studies; analytical study conducted [name of the study (if any)]	Reference
	Origanum vulgare L.	Aerial parts, Mersin (Büyükceceli-Gülnar), Turkey	3.62%), linalool (2.50–3.25%)	Methicillin-resistant Staphylococcus aureus (MRSA)	250 µg/ml	Antimicrobial test by broth microdilution technique; 5×10^5 CFU/ml	–	(Özkalp et al. 2010)
				Escherichia coli RSKK 340	250 µg/ml			
				Klebsiella pneumoniae RSKK 06017	250 µg/ml			
				Pseudomonas aeruginosa RSKK 06021	64 µg/ml			
				Salmonella enteritidis RSKK 96046	128 µg/ml			
				Streptococcus pyogenes RSKK 413/214	128 µg/ml			
				Staphylococcus aureus RSKK 96090	64 µg/ml			
	Origanum vulgare subsp. glandulosum	Leaves, Tunisia	Carvacrol (61.08–83.37%), p-cymene (3.02–9.87%), c-terpinene (4.13–6.34%)	Escherichia coli ATCC 8739, Salmonella typhimurium ATCC 14028, Pseudomonas aeruginosa NCTC	125–600 µg/mL	Antimicrobial activity by agar disc diffusion, ZOI: 9–36 mm; 10^6 CFU/mL	Gas chromatography	(Béjaoui et al. 2013)

Origanum vulgare L.	Nova Lubovna, Slovakia	–	10418, *Staphylococcus aureus* ATCC 6538		Disk diffusion technique	DPPH assay	(Kačániová et al. 2012)
			Bacillus cereus CCM 2010	0.75, 0.375, 0.188, 0.094 ml ml^{-1}		
			Escherichia coli CCM 3988	0.75, 0.375 ml ml^{-1}		
Origanum vulgare L.	Leaves, Latakia in Syria	Terpinen-4-ol (24.90%), gamma-terpinene (10.57%), o-cymene (8.90%), cis-beta-terpineol (8.73%), alpha-terpinene (6.67%), beta-phellandrene (4.84%), alpha-terpineol (4.18%), carvacrol (3.90%)	*P. aeruginosa, B. cereus*	1.56–50 µl/ml	Minimum inhibitory concentration assay	DPPH assay, GC-MS	(Jnaid et al. 2016)
Origanum vulgare L.	Leaves, Norte de Santander, Colombia	β-Myrcene 1.6%, ∝ – terpinene 15.7%, 1.8-cineol 3.8%, γ-terpineol 2.6%, terpine-4-ol 1.1%, thymol methyl ether 17.4%, thymol 30.6%, carvacrol 8.1%, trans-β-caryophyllene 6.3%, ∝ – humulene 1%, caryophyllene oxide 3.1%	*Escherichia coli* ATCC 25922	15.62 mg/mL	10^8 CFU/ mL	Chromatographic analysis	(Arámbula et al. 2019)
			Staphylococcus aureus ATCC 25923	15.62 mg/mL		
			Pseudomonas aeruginosa ATCC 27853	125 mg/mL		
Rosmarinus officinalis L.	Leaves, Ankober	α-Pinene (50.83%), camphene (5.211%),	*Salmonella typhi, Salmonella*	15.75– 36.33 mg/mL		–	(Mekonnen et al. 2016)

(continued)

Table 15.3 (continued)

Family	Plant name	Plant part	Major chemicals present in essential oil	Assayed against bacteria	MIC/ concentration used in the studies	Experiment details	Allied studies; analytical study conducted [name of the study (if any)]	Reference
		District, Kundi	β-Pinene (2.068%), camphor (3.84%), 1,8-cineole (24.42%)	paratyphi, Salmonella typhimurium, Shigella species, Pseudomonas aeruginosa, Staphylococcus aureus, and Escherichia coli		Antibacterial activity by agar diffusion method		
	Thymus vulgaris L.	Nova Lubovna, Slovakia	–	Escherichia coli CCM 3988	0.375 ml ml^{-1}	Disk diffusion technique; 10^5 CFU ml^{-1}	DPPH activity	(Kačániová et al. 2012)
	Thymus schimperi Ronniger	Leaves, Kundi, Ankober District Ethiopia	Carvacrol (71.024%), p-Myrcene (15.250%), α-pinene (50.830%), 1,8-cineole (24.425%)	Salmonella typhi, Salmonella paratyphi, Salmonella typhimurium, Shigella species, Pseudomonas aeruginosa, Staphylococcus aureus, Escherichia coli	Lower than 15.75 mg/ml	Antimicrobial assays antimicrobial sensitivity testing by agar well diffusion method	GC-MS	(Mekonnen et al. 2016)
Moraceae	Cudrania tricuspidata	Fruit, Korea	Diethyl phthalate (36.24%), scyllitol (23.94%), (4S)1.1-	Bacillus cereus ATCC 13061	250 mg/mL	Antimicrobial test by disk diffusion assay and viable	GC-MS	(Bajpai et al. 2013)

Family	Plant	Part, Location	Chemical composition	Microorganism	Concentration	Result	Method	Reference
	(Carrière) Bureau ex Lavalle		difluoro-4 vinylspiropentane (8.76%), sulfamide (5.54%), ethyl-N-methylcarbamate (4.64%), boric acid (4.61%), 8-chloro-6-(2-fluorophenyl) imidazole (1,2-a)-[1,4]-benzodiazepine (3.26%)			count assay; ZOI: 25.5 mm; 10^7 CFU/mL		
	Artocarpus heterophyllus Lam.	Fruit, Delhi, India	–	*Escherichia coli*	1.50 mg/mL	Disk diffusion assay ZOI: 29.78 mm	–	(Jha and Srivastava 2012)
				Pseudomonas aeruginosa	2.93 mg/mL	ZOI: 17.44 mm		
				Staphylococcus aureus	5.20 mg/mL	ZOI: 5.23 mm		
Myrtaceae	*Eucalyptus citriodora* Hook.	Leaves, Algiers, Algeria	Citronellal (69.77%), citronellol (10.63), isopulegol (4.66%)	*Escherichia coli* ATCC 25922, *Pseudomonas aeruginosa* ATCC 27853, *Salmonella enteritidis*, *Salmonella typhimurium*, *Salmonella gallinarum* and *pullorum*, *Salmonella senftenberg*, *Enterococcus faecalis*,	20 µl/ml, 0.3 µl/ml	Antimicrobial tests by disk diffusion method and vapor diffusion assay; 108 CFU/ml); ZOI: 10–16 mm.	Gas chromatographic, DPPH assay	(Tolba et al. 2015)

(continued)

Table 15.3 (continued)

Family	Plant name	Plant part	Major chemicals present in essential oil	Assayed against bacteria	MIC/ concentration used in the studies	Experiment details	Allied studies; analytical study conducted [name of the study (if any)]	Reference
				Corynebacterium striatum, Bacillus subtilis, Staphylococcus aureus ATCC 25923, Staphylococcus epidermidis, Staphylococcus saprophyticus, and two yeasts *Candida albicans ATCC 10231* and *Saccharomyces cerevisiae*				
	Eucalyptus citriodora Hook.	Twigs and fruits, Taiwan	1,8-Cineole (17.7%), p-cymene (17.1%), caryophyllene oxide (13.8%), and isopulegol (12.2%); the fruit oil's main constituents were p-cymene (23.2%), α-pinene (16.1%), α-cadinol (11.5%), and τ-cadinol (8.4%)	*Bacillus cereus, Staphylococcus aureus, Staphylococcus epidermidis, Enterobacter aerogenes, Klebsiella pneumonia, Pseudomonas*	125–250 μg/mL	Antibacterial activity ZOI: 32–39 mm	GC-FID, GC-MS	(Su et al. 2017)

		Bacteria	Concentration	Activity	Technique	Reference
Eucalyptus camaldulensis Dehn.	Leaves, North West of Algeria	—	aeruginosa, Vibrio parahaemolyticus Staphylococcus aureus, Escherichia coli		Antibacterial activity by aromatogram test, microatmosphere test, broth dilution method	—	(Ghalem and Mohamed 2008)
Eucalyptus globules Labill.	Leaves, North West of Algeria	—	Escherichia coli, Staphylococcus aureus	13 µg mL^{-1}	Antibacterial activity agar disk diffusion and dilution broth methods, ZOI: 8–26 mm	—	(Bachir and Benali 2012)
Eucalyptus globules Labill.	Leaves, Ankober District, Kundi	1,8-Cineole (63.00%), α-pinene (16.101%), β-pinene (0.461%)	Salmonella typhi, Salmonella paratyphi, Salmonella typhimurium, Shigella species, Pseudomonas aeruginosa, Staphylococcus aureus, and Escherichia coli	15.75–31.25 mg/mL	Antibacterial activity by agar diffusion method	GC-MS	(Mekonnen et al. 2016)
Eucalyptus camaldulensis Dehn.	Leaves, Bom Jesus county, Piauí state, Brazil	1,8-Cineole (76.93%), β-pinene (11.49%), and α-pinene (7.15%), eugenol, b-caryophyllene, iso-caryophyllene, naphthalene, 8a-hexahydro-4,	Staphylococcus aureus, Escherichia coli	1000 µg mL^{-1}	Modulation activity; 10^6 CFU mL^{-1}	GC-MS	(Chaves et al. 2018)

(continued)

Table 15.3 (continued)

Family	Plant name	Plant part	Major chemicals present in essential oil	Assayed against bacteria	MIC/concentration used in the studies	Experiment details	Allied studies; analytical study conducted [name of the study (if any)]	Reference
Oleaceae	Forsythia koreana Nakai	Leaves, Kyungbuk, South Korea	7-dimethyl-1-(1-methyl ethyl), 1,6-octadiene-ol-,3,7-dimethyl acetate, caryophyllene, caryophyllene oxide; Trans-phytol (42.73%), cis-3-hexenol (12.95%), b-linalool (10.68%), trans-2-hexenal (8.86%), trans-2-hexenol (8.86%), myrcenol (3.86%), 4-vinylphenyl acetate (3.86%), (4Z)-4,6-heptadien-1-ol (3.18%)	Salmonella enteritidis KCTC 12243	0.4% (v/v)	Antibacterial activity with zone diameters of 12.3 mm; 10^7 CFU/mL	Gas chromatography coupled with mass spectrometry, DPPH assay, nitric oxide scavenging activity, superoxide anion radical scavenging assay	(Yang et al. 2015)
				Escherichia coli ATCC 8739	0.3% (v/v)	Antibacterial activity with zone diameters of 8.0 mm; 10^7 CFU/mL		
				Staphylococcus aureus ATCC 6538	0.2% (v/v)	Antibacterial activity with zone diameters of 9.3 mm; 10^7 CFU/mL		
Poaceae	Cymbopogon schoenanthus (L.) Spreng	Leaves, Houintopka	Piperitone (68.4%), δ-2-carene (11.5%), α-eudesmol (4.6%)	Staphylococcus aureus ATCC 25923	2.63 ± 0.16 mg/mL	Antibacterial activity; 10^6 cfu/mL	Gas chromatography-flame ionization detection (GC-FID) and gas chromatography-	(Alitonou et al. 2012)
				Escherichia coli ATCC 25922	2.63 ± 0.16 mg/mL			

	Cymbopogon giganteus Chiov.	Leaves, Houintopka	Trans-p-mentha-2,8-dien-1-ol (16.4%), 3,9-epoxymentha-1,8(10)-diene (6.4%), cis-p-mentha-1(7),8-dien-2-ol (19.4%), trans-isopiperitenol (7.1%)	Staphylococcus aureus ATCC 25923	0.32 ± 0.02 mg/mL	Antibacterial activity; 10^6 cfu/mL	Gas chromatography-flame ionization detection (GC-FID) and gas chromatography-mass spectrometry (GC-MS)	(Alitonou et al. 2012)
				Escherichia coli ATCC 25922	0.64 ± 0.34 mg/mL			
Rutaceae	Ruta chalepensis L.	Leaves, bark, and wood	2-Undecanone (39.4%), nonanone (37.1%), decanone (2.8%), nonyl acetate (2.2%)	Escherichia coli ATCC 25922	500 mg/mL	10^8 CFU/mL	Chromatographic analysis, steam distillation	(Arámbula et al. 2019)
				Staphylococcus aureus ATCC 25923	1000 mg/mL			
				Pseudomonas aeruginosa ATCC 27853	500 mg/mL			
Schisandraceae	Kadsura longipedunculata	Stem and bark, China	d-Cadinene (21.79%), camphene (7.27%), borneol (6.05%), cubenol (5.12%), and d-cadinol (5.11%)	Enterococcus faecalis, Streptococcus pyogenes	60 mg/ml	Antimicrobial diffusion test; 1¥ 10^6 CFU/ml	Capillary gas chromatography (GLC-FID) and gas chromatography-mass spectrometry (GLC-MS, DPPH	(Mulyaningsih et al. 2010)

(continued)

Table 15.3 (continued)

Family	Plant name	Plant part	Major chemicals present in essential oil	Assayed against bacteria	MIC/ concentration used in the studies	Experiment details	Allied studies; analytical study conducted [name of the study (if any)]	Reference
Zingiberaceae	*Aframomum sceptrum* K. Schum	Rhizomes	b-Pinene (12.7%), caryophyllene oxide (10.03%), cyperene (5.99%), 3,4-dimethylcyclohex-3-ene-1-carbaldehyde (3.40%), a-pinene (2.78%), b-caryophyllene (2.33%), a-terpineol (2.12%), d-limonene (1.89%), 1,8-cineole (1.84%), and a-caryophyllene (1.1%)	*Staphylococcus aureus*	inhibition zone diameter: $16.0 + -0.7$; 10^6 CFU/ml	Antibacterial screening by disk diffusion method	GC-MS	(Cheikh-Ali et al. 2011)
				Staphylococcus epidermidis	inhibition zone diameter: 20.52.1; 10^6 CFU/ml			
				Bacillus subtilis	inhibition zone diameter: 23.52.1; 10^6 CFU/ml			

Table 15.4 Antifungal activity of essential oil isolated from different plants

Family	Plant name	Collected plant part, location	Major essential oils present in plant (%)	Assayed against drug-resistant fungal species	Minimum inhibitory concentrations (MIC)	Antimicrobial assays	Other assays	References
Asteraceae	Artemisia sieberi Besser	Aerial parts, Iran	–	Candida glabrata	37.4–4781.3 mg/ml	Antimicrobial test by agar well diffusion method; 11–40 mm zone of inhibition, minimal inhibitory concentration (MIC)	GC-MS, TEM	(Khosravi et al. 2011)
Lauraceae	Cinnamomum camphora Meisn.	Leaves, Tezpur Assam	Camphor (18–20%), safrole (30–35%), cineol (12–15%), borneol (10–12%), camphene (6–10%), dipentene (10–15%), terpineol (5–8%)	Candida albicans	3180 μg/mL	Antimicrobial activity with disc diffusion method, zone of inhibition 10.67 mm; 4.8×10^3 cells ml^{-1}	GC-MS	(Dutta et al. 2007)
Lamiaceae	Origanum vulgare L.	Aerial parts anhydrous sodium sulfate (100 g)	Linalool (42%), thymol (25.1%), a-terpineol (10%)	Candida glabrata	0.5 to 1100 mg/ml	Antimicrobial test agar well diffusion method and broth macrodilution method, zone of inhibition 18–40 mm; 1.5×10^8 CFU/ml	GC-MS, TEM	(Khosravi et al. 2011)
	Origanum vulgare L.	Aerial parts, Mersin (Büyükeceli-Gülnar), Turkey	Carvacrol (39.08 to 49.43%), trans-sabinene hydrate (19.81 to 25.11%), cis-piperitol (5.51–7.77%), borneol (2.82–5.14%), terpinen-4-ol (3.29–3.62%), linalool (2.50–3.25%)	Candida albicans	128 μg/ml	Antimicrobial test by broth microdilution technique; 5×10^5 cfu/m	–	(Özkalp et al. 2010)

(continued)

Table 15.4 (continued)

Family	Plant name	Collected plant part, location	Major essential oils present in plant (%)	Assayed against drug-resistant fungal species	Minimum inhibitory concentrations (MIC)	Antimicrobial assays	Other assays	References
	Origanum vulgare L.	Leaves, Murcia, Spain	Carvacrol (39.08–49.43%), trans-sabinene hydrate (19.81–25.11%), cis-piperitol (5.51–7.77%), borneol (2.82–5.14%), terpinen-4-ol (3.29–3.62%), linalool (2.50–3.25%)	*Candida albicans* ATCC 60193	1.75 to 1.48 mg/ml	Antimicrobial activity by the disk diffusion method, broth dilution method; maximal inhibition zones 22.2 to 25.3 mm; 10⁶ CFU/ml	GC-MS	(Santoyo et al. 2006)
	Rosmarinus officinalis L.	Leaves, Ankober District, Kundi	α-Pinene (50.83%), camphene (5.211%), β-pinene (2.068%), camphor (3.84%), 1,8-cineole (24.42%)	*Aspergillus* spp.	15.75–36.33 mg/mL	Antibacterial activity by agar diffusion method; ZOI 17 mm	GC-MS	(Mekonnen et al. 2016)
	Lavandula multifida L.	Aerial part, South of Portugal	Carvacrol, (42.8%), cis-β-ocimene (27.4%)	*Candida albicans* ATCC 10231	1.25 µL/mL	Antifungal activity Broth macrodilution methods; 10⁶ CFU/mL	GC-MS	(Zuzarte et al. 2012)
Myrtaceae	*Eucalyptus citriodora* Hook.	Leaves, Tezpur Assam	–	*Candida albicans*	318 µg/mL	Disk diffusion method, ZOI: 8.50 m; 4.8 × 10³ cells ml⁻¹	GC-MS	(Dutta et al. 2007)
	Eucalyptus globules Labill.	Leaves, Ankober District, Kundi	1,8-Cineole (63.00%), α-pinene (16.101%), β-pinene (0.461%)	*Trichophyton* spp.	–	Antibacterial activity by agar diffusion method; ZOI 27.3 mm	GC-MS	(Mekonnen et al. 2016)
Poaceae	*Cymbopogon citratus* (DC.) Stapf.	Leaves, Tezpur Assam	–	*Candida albicans*	322 µg/mL	Antimicrobial activity with disk diffusion method, zone of inhibition 21.17 mm; 4.8 × 10³ cells ml⁻¹	Gas chromatography-mass spectrometry (GC-MS)	(Dutta et al. 2007)

							GC-MS analyses	
	Cymbopogon citratus (DC.) Stapf.	Leaves, Blida city, Algeria	Geranial (42.2%), neral (31.5%), and b-myrcene (7.5%), geranyl acetate (4.3%) and isopulegol (1.4%)	Candida albicans	20–60 μg/mL	Antimicrobial activity with disk diffusion method and vapor diffusion method. zone of inhibition 35–90 mm; 10^8 CFU/ml	GC-MS analyses	(Boukhatem et al. 2014)
	Cymbopogon nardus	Malaysia	Citronellal (11.35%), z-citral (11.34%), β-myrcene (6.70%), β-trans-ocimene (6.03%)	Candida albicans (ATCC 10231)	–	Agar well diffusion assay; IC_{50}: 26.70 mm	GC-MS	(Ahmad Kamal et al. 2020)
Rutaceae	Citrus limon (L.) Osbeck	Peel, Tezpur Assam	–	Candida albicans	3775 μg/mL	Disk diffusion method, ZOI: 9.67 mm; 10^3 cells ml^{-1}	GC-MS	(Dutta et al. 2007)
Umbelliferae	Anethum graveolens L.	Seeds, Xinjiang, China	–	Candida albicans ATCC64550, Candida albicans 09–5304, Candida albicans 09–1502	1.56 μg ml^{-1} to 100 μg ml^{-1}	Macrodilution broth method; 1×10^3 CFU ml^{-1}	–	(Chen et al. 2013)
Zingiberaceae	Aframomum sceptrum K. Schum	Rhizomes	b-Pinene (12.7%), caryophyllene oxide (10.03%), cyperene (5.99%), 3,4-dimethylcyclohex-3-ene-1-carbaldehyde (3.40%), a-pinene (2.78%), b-caryophyllene (2.33%), a-terpineol (2.12%), d-limonene (1.89%), 1,8-cineole (1.84%), and a-caryophyllene (1.1%)	Aspergillus fumigates Candida albicans	5.000 mg/ml 1.250 mg/ml	Antifungal activity test by broth dilution method, CFU: 10^3–5×10^3 CFU/ml	GC-MS	(Cheikh-Ali et al. 2011)

Table 15.5 Antiprotozoal activity of plant essential oils

Plant	Family	Part used	Active chemicals	Assayed against drug-resistant protozoal species	Half maximal inhibitory concentration (IC_{50})/minimum lethal concentration (MLC)	Assays/techniques	Reference
Annonaceae	*Annona coriacea* Mart.	Leaves	Bicyclogermacrene (39.8%)	*Leishmania chagasi*	IC_{50}: 39.93 µg/mL	GC-MS, GC-FID, MTT assay	(Siqueira et al. 2011)
Apiaceae	*Foeniculum vulgare* Mill.	Seeds, Iran	E-Anetholeb (88%), phenylpropene (88%)	*Trichomonas vaginalis*	MLC: 360 µg/ml	GC-MS, GC-FID	(Karami et al. 2019)
Asteraceae	*Baccharis dracunculifolia* DC.	Leaves, Franca, São Paulo State, Brazil	Nerolidol (33.51%), spathulenol (16.24%)	*Leishmania donovani*	IC_{50}: 42 mg/ml	GC, GC-MS	(Parreira et al. 2010)
Myrtaceae	*Eugenia gracillima* Kiaersk.	Leaves, Brazil	germacrene D (16.10%), cG muurolene (15.60%), bicyclogermacrene (8.53%), germacrene B (7.43%), D-elemene (6.06%)	*Leishmania braziliensis*	IC_{50}: 74.64 mg/mL	GC, GC-MS	(Gomes Vidal Sampaio et al. 2019)
				Leishmania infantum	IC_{50}: 80.4 mg/mL		
Lamiaceae	*Origanum onites* L.	Aerial parts, Turkey	Carvacrol (70.6%), linalool (9.7%), p-cymene (7%), γ-terpinene (2.1%), thymol (1.8%)	*Trypanosoma brucei rhodesiense*	IC_{50} 0.18 µg/mL	GC-FID, GC-MS, antiprotozoal activity	(Tasdemir et al. 2019).

Ocimum canum Sims.	Leaves	Thymol (42.15%), p-cymene (21.17%), γ-terpinene (19.81%)	*Leishmania amazonensis*; *Trypanosoma cruzi*	IC_{50} of 17.4 µg/mL; IC_{50}: 60.0 µg/m	Gas chromatography with ZB-5 ms capillary column TEM, MTT assay	(da Silva et al. 2018)
Ocimum canum Sims.	Leaves, Cameroon	Hydrocarbon monoterpenes, oxygenated monoterpenes, hydrocarbon sesquiterpenes	*Plasmodium falciparum*	IC_{50}: 20.6 lg/mL	Radioisotopic method, gas chromatography and GC-MS, anti-plasmodial test	(Akono Ntonga et al. 2014)
Ocimum gratissimum L.	Leaves, Piauí, Brazil	1,8-Cineole (24%), eugenol (40%)	*Leishmania chagasi*	IC_{50}: 0 75 µg/mL	GC-MS, nuclear magnetic resonance	(Oliveira et al. 2009)
Ocimum basilicum L.	Leaves, Cameroon	Hydrocarbon monoterpenes, oxygenated monoterpenes, hydrocarbon sesquiterpenes	*Plasmodium falciparum*	IC50: 21.0 lg/mL	Radioisotopic method, GC, GC-MS, larvicidal tests, anti-plasmodial test	(Akono Ntonga et al. 2014)
Lavandula luisieri L	Inflorescences	α-Pinene (2.3%), linalool (3.1%), 1,8-cineol (18.9%)	*Leishmania infantum*; *Leishmania major*; *Leishmania tropica*	IC_{50}/24 h: 63 µg/mL; IC_{50}/48 h: 31 µg/mL; IC_{50}/24 h: 38 µg/mL)	Gas chromatography, MTT assay	(Machado et al. 2019)
Lavandula viridis L'Hér.	Aerial parts	α-Pinene (9.2%), camphene (2.7%), 1,8-cineol (29.7%), linalool (9.0%)	*Leishmania infantum*	IC_{50}/24 h: 263 µg/mL	Gas chromatography, tetrazolium-dye, MTT assay SEM, TEM	(Machado et al. 2019)

(continued)

Table 15.5 (continued)

Plant	Family	Part used	Active chemicals	Assayed against drug-resistant protozoal species	Half maximal inhibitory concentration (IC$_{50}$)/minimum lethal concentration (MLC)	Assays/techniques	Reference
			camphor (10.0%), borneol (2.7%)				
	Origanum vulgare L.	Aerial parts	3-Ciclohen-1-ol (26.2%), p-Cymene (2.3%), γ-terpinene (16.0%), α-terpineol (12.3%)	*Trypanosoma cruzi*	IC$_{50}$:115 µg/ml	GC-MS, GC-FID, TEM, SEM, flow cytometry	(Santoro et al. 2007)
	Thymus vulgaris L.	Aerial parts	Thymol (80.4%), limonene (5.2%), carvacrol (6.0%)	*Trypanosoma cruzi*	IC$_{50}$: 38 µg/ml	GC-MS, GC-FID, TEM, SEM, Flow cytometry	(Santoro et al. 2007)
Poaceae	*Cymbopogon citrates* (DC.) Stapf.	Leaves	Geranial, neral	*Plasmodium falciparum*	IC$_{50}$: 4.2 lg/mL	Anti-plasmodial activity by radioisotopic method	(Akono Ntonga et al. 2014)
Poaceae	*Cymbopogon citratus* (DC) Stapf.	Leaves	Myrcene (10.8%), neral (40.4%), geranial (41.1%)	*Leishmania chagasi*	IC$_{50}$: 45 µg/mL	GC-MS, nuclear magnetic resonance, Flow cytometry	(Oliveira et al. 2009)

Zingiberaceae	*Aframomum sceptrum* K. Schum	Rhizomes, Ivory Coast	b-Pinene (12.7%), caryophyllene oxide (10.03%), cyperene (5.99%), 3,4-dimethylcyclohex-3-ene-1-carbaldehyde (3.40%), a-pinene (2.78%), b-caryophyllene (2.33%), a-terpineol (2.12%), d-limonene (1.89%), 1,8-cineole (1.84%), and a-caryophyllene (1.1%)	*Trypanosoma brucei*	MLC: 1.51 mg/ml	GC-MS, antiprotozoal activity assay	(Cheikh-Ali et al. 2011)
				Trichomonas vaginalis	IC$_{50}$: 0.120 mg/ml MLC: 1.72 mg/ml		
Verbenaceae	*Lippia sidoides* Cham.	Leaves	p-Cymene (14.4%), thymol (68.3%)	*Leishmania chagasi*	89 µg/mL	GC-MS, mass spectrometry, nuclear magnetic resonance, flow cytometry	(Oliveira et al. 2009)

Table 15.6 Antiviral activities of essential oils in resistant viruses

Family	Essential oil	Major components	Assayed against drug-resistant virus species	IC_{50} (μg/mL)	Cell lines	Assays/techniques	References
Apocynaceae	*Cynanchum stauntonii* (Decne.) Schltr. ex H. Lév.	(2E,4E)-decadienal (23.0%), γ-nonalactone (4.2%), 5-pentyl-2(3H)-furanone (3.8%), 3-isopropyl-1-pentanol (3.5%)	Influenza type A (H1N1)	64	–	–	(Setzer 2016)
Asteraceae	*Artemisia arborescens* L.	β-Thujone (45.0%), camphor (6.8%), chamazulene (22.7%)	HSV-1 and HSV-2	2.4 and 4.1	Vero cells	Plaque reduction assay, Penetration assay, MTT assay	(Saddi et al. 2007)
	Artemisia mendozana DC.	Camphor (22.4%), artemiscole (11.7%), artemisia alcohol (10.8%), borneol (7.2%)	DENV-2, and HSV-1	129.3 and 153.7	Vero cells	Plaque reduction assay	(Duschatzky et al. 2005)
	Matricaria recutita L.	α-Bisabolol oxide A (13.4–55.9%), α-bisabolol oxide B (8.4–25.1%), bisabolone oxide A (2.9–11.4%), cis-bicycloether (spiroether) (3.6–17.7%).	HSV-2	1.5	RC-37	Plaque reduction assay	(Koch et al. 2008)

Family	Plant species	Composition	Virus	Value	Cell line	Assay	Reference
Lauraceae	Cinnamomum zeylanicum Blume (syn. Cinnamomum verum J. Presl)	Eugenol (75–85%), linalool (1.6–8.5%), (E)-cinnamaldehyde (0.6–1.5%), cinnamyl acetate (0.7–2.6%), β-caryophyllene (0.5–6.7%), eugenyl acetate (0.1–2.9%), benzyl benzoate (0.1–8.3%) (E)-β-farnesene (1.9–10.4%)	Influenza type A (H1N1)	–	–	–	(Setzer 2016)
	Cinnamomum zeylanicum Blume (syn. Cinnamomum verum J. Presl)	(E)-Cinnamaldehyde (63.9%), eugenol (7.0%), (E)-cinnamyl acetate (5.1%)	HSV-2	82	HeLa cells	Plaque reduction assay	(Bourne et al. 1999)
Lamiaceae	Mentha suaveolens Ehrh.	Piperitenone oxide	HSV	5.1 ± 0.48	Vero cells	Plaque reduction assay	(Civitelli et al. 2014)
	Mentha piperita L.[a]	Menthol (42.8%), menthone (14.6%), isomenthone (5.9%), menthylacetate (4.4%), cineole (3.8%), limonene (1.2%) carvone (0.6%)	HSV	2.4	RC-37	Plaque reduction assay	(Schuhmacher et al. 2003)
	Salvia fruticosa Mill.	1,8-Cineole (47.5%), camphor (9.0%), β-thujone (7.6%), α-thujone (4.3%)	HSV-1, HSV-2	1300	Vero cells	Plaque reduction assay	(Sivropoulou et al. 1997)
	Salvia desoleana Atzei & V. Picci		HSV-2	–	Vero cells		(Cagno et al. 2017)

(continued)

Table 15.6 (continued)

Family	Essential oil	Major components	Assayed against drug-resistant virus species	IC$_{50}$ (µg/mL)	Cell lines	Assays/techniques	References
		Linalyl acetate, alpha terpinyl acetate, germacrene				Plaque reduction assay	
	Satureja hortensis L.	Carvacrol (32.4%), γ-terpinene (32.0%), thymol (10.0%), p-cymene (6.6%)	HSV-1	80	Vero cells	Plaque reduction assay	(Gavanji et al. 2015)
	Thymus vulgaris L.	Thymol (43.9%), carvacrol (14.4%), p-cymene (10.5%), β-caryophyllene (7.0%), and γ-terpinene (5.1%)	HSV-2	7	RC-37	Plaque reduction assay	(Koch et al. 2008)
	Zataria multiflora Boiss.	Thymol (33.1%), carvacrol (25.9%), p-cymene (11.3%)	HSV-1	30	Vero cells	Plaque reduction assay	(Gavanji et al. 2015)
Myrtaceae	Eucalyptus camaldulensis Dehnh.	α-Terpinene (26.3%), α-terpineol (9.1%), camphene (8.5%)	HSV-1	90% at 10% EO concentration	–	Plaque reduction assay,	(El-baz et al. 2015)
Santalaceae	Santalum album L.	α – Santalol (45.2%), β – santalol (25.4%), trans – α – bergamotol (7.8%)	HSV-2	5	RC-37	Plaque reduction assay	(Koch et al. 2008)
Schisandraceae	Illicium verum Hook. f.	Trans-anethol (89.1%), estragole (3.6%), linalool	HSV-2	30	RC-37		(Koch et al. 2008)

						Plaque reduction assay	
		(1.1%), a-terpineol (0.2%), cis-anethol (0.2%)					
Verbenaceae	Lippia graveolens Kunth	Carvacrol (56.8%), o-cymene (32.1%), γ-terpinene (3.7%)	HSV 1	55.9	Madin-Darby bovine kidney (MDBK) cells, MA104 cells, HEp-2 cells	HPLC, (MTT) assay, plaque reduction assay	(Pilau et al. 2011)
Zingiberaceae	Zingiber officinale Roscoe[a]	Zingiberene (18.9%), limonene/cineol (15.5%), b-sesquiphellandrene (6.8%), Camphene (6.2%), pinocamphone (6.8%)	HSV-2	1	RC-37	Plaque reduction assay	(Koch et al. 2008)

[a]Essential oil composition not reported; essential oil composition of commercial

Table 15.7 List of essential oils/conventional antimicrobial combinations showing combinatory effects against a panel of drug-resistant microorganisms (FIC < 1 = synergy)

Essential oils	Antibiotics	Microorganisms	FIC index μg ml^{-1}	Methods	Reference
Chenopodium ambrosioides	Pentamidine	*Leishmania amazonensis*	0.453	Isobologram analysis	(Monzote et al. 2007)
Pelargonium graveolens	Norfloxacin	*Staphylococcus aureus, Bacillus cereus*	–	Agar dilution assay, checkerboard assay	(Rosato et al. 2007)
Oregano vulgare L.	Fluoroquinolones/ doxycycline/ lincomycin/ maquindox	*Escherichia coli*	0.375– 0.500	Broth microdilution, checkerboard assay	(Si et al. 2008)
Eucalyptus oil	Chlorhexidine digluconate	*Staphylococcus epidermidis*	–	Broth microdilution, checkerboard assay	(Karpanen et al. 2008)
Melaleuca alternifolia	Ciprofloxacin	*Staphylococcus aureus*	1.58–7.7	Checkerboard assay	(Van Vuuren et al. 2009)
		Klebsiella pneumonia	0.73–1.85		
Eugenol	Vancomycin/b-lactam antibiotics	*Escherichia coli, Enterobacter aerogenes, Proteus vulgaris, Pseudomonas aeruginosa, Salmonella typhimurium*	–	Broth microdilution, checkerboard assay	(Hemaiswarya and Doble 2009)
Croton zehntneri Pax et Hoffm.	Gentamicin	*Staphylococcus aureus, Pseudomonas aeruginosa*	–	Disk diffusion assay	(Rodrigues et al. 2009)
Rosmarinus officinalis	Ciprofloxacin	*Klebsiella pneumoniae*	0.97–0.55	Broth microdilution, checkerboard assay	(Van Vuuren et al. 2009)
Thymus vulgaris	Ciprofloxacin	*Staphylococcus aureus*	0.80–2.59	Checkerboard assay	(Van Vuuren et al. 2009)
		Klebsiella pneumoniae	0.71–1.40		
Myrtus communis	Amphotericin B	*Candida albicans*	0.25	Broth microdilution, checkerboard assay	(Mahboubi and Bidgoli 2010)

Zataria multiflora Shiraz	Vancomycin	*Staphylococcus aureus*	–	Broth microdilution, checkerboard assay	(Mahboubi and Ghazian Bidgoli 2010)
Cinnamaldehyde	Fluconazole	*Aspergillus fumigates, Trichophyton rubrum*	0.031	Disk diffusion assay, broth macrodilution, time-kill assay, checkerboard assay, microtiter test	(Khan and Ahmad 2011)
Citrus limon	Amikacin	*Acinetobacter baumannii*	0.04	Checkerboard assay	(Guerra et al. 2012)
Lippia sidoides	Amikacin	*Staphylococcus aureus, Pseudomonous aeruginosa*	–	Change in inhibition zone in the presence of EO vapor	(Veras et al. 2012)
Cinnamomum zeylanicum	Amikacin	*Acinetobacter* spp.	0.037	Broth microdilution, checkerboard assay	(Guerra et al. 2012)
Coriandrum sativum L.	Chloramphenicol/ciprofloxacin/gentamicin/tetracycline	*Acinetobacter baumannii*	0.047–0.375	Broth microdilution, checkerboard assay	(Duarte et al. 2012)
Lantana montevidensis	Amikacin	*Escherichia coli*	–	Broth microdilution, checkerboard assay	(De Sousa et al. 2013)
Mentha piperita L.	Amphotericin B	*Candida albicans*	0.128 g/ml	Broth microdilution, checkerboard assay	(Salleh et al. 2014)
Melissa officinalis L.	Oseltamivir	H9N2	–	Hemagglutination assay, real-time PCR	(Pourghanbari et al. 2016)
Carvacrol	Endolysin LysSA97	*Staphylococcus aureus*	0.28–0.29	Checkerboard assay	(Chang et al. 2016)
Eucalyptus camaldulensis	Ciprofloxacin/gentamicin	*Acinetobacter baumannii*	Less than 0.5	Time-kill curves	(Knezevic et al. 2016)
Boswellia serrata Roxb. ex Colebr.	Ketoconazole, fluconazole, posaconazole, or voriconazole	*Candida albicans*	–	Etest method	(Sadhasivam et al. 2016)

(continued)

Table 15.7 (continued)

Essential oils	Antibiotics	Microorganisms	FIC index µg ml^{-1}	Methods	Reference
Origanum onites L., *Cymbopogon citratus* (DC.) Stapf.	Acyclovir	HSV 1	100 TCID 50: 5.0 µg/ml	Cytotoxicity tests, checkerboard assay	(Duran and Kaya 2018)
Moringa oleifera Lam.	Amphotericin B	*Leishmania major*	0.375	Anti-amastigote assay, checkboard assay	(Hammi et al. 2020)
Momordica charantia L.	Aminoglycosides	*Staphylococcus aureus*	–	–	(Coutinho et al. 2010)

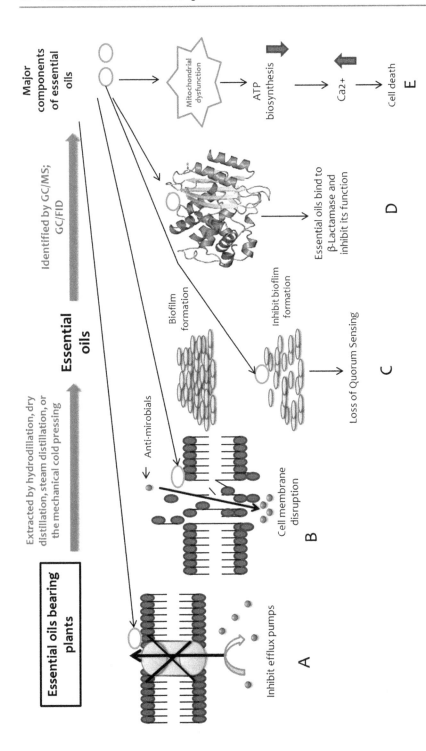

Fig. 15.1 Mode of action of essential oil for combating antimicrobial resistance in human pathogenic microbes. (**a**) Essential oil inhibits efflux pumps thus antimicrobials accumulated into the cell. (**b**) Essential oil disrupts the cell membrane and alters cell morphology; thus, antimicrobials easily pass through the cell membrane (**c**) Essential oil acts as an anti-quorum sensing. (**d**) Essential oil acts as a β-lactamase inhibitor; thus, it is unable to break β-lactam antibiotics. (**e**) Essential oil initiates mitochondrial dysfunction

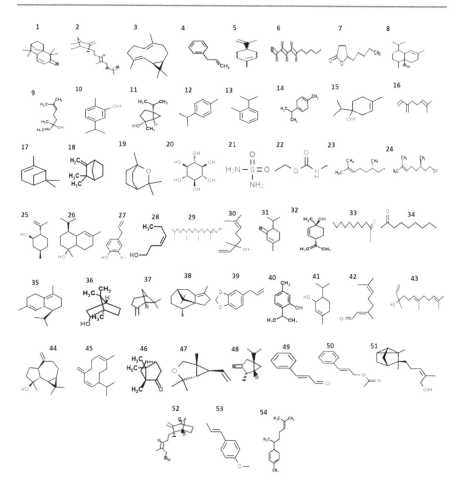

Fig. 15.2 Structures of major components present in essential oil. **1.** 2,7,7-Tetramethyltricyclo [6.2.1.0 (1,6)]undec-4-en-3-one (C$_{15}$H$_{22}$O) (Aristolochiaceae), **2.** (E)-Santalol acetate C$_{17}$H$_{26}$O$_2$ (Aristolochiaceae), **3.** Bicyclogermacrene C$_{15}$H$_{24}$ (Annonaceae, Myrtaceae), **4.** Phenylpropene C$_9$H$_{10}$ (Apiaceae), **5.** Limonene C$_{10}$H$_{16}$ (Apiaceae, Lamiaceae, Zingiberaceae), **6.** (2E,4E)-decadienal C$_{10}$H$_{16}$O (Apocynaceae), **7.** γ-Nonalactone C$_9$H$_{16}$O$_2$ (Apocynaceae), **8.** Cubenol C$_{15}$H$_{26}$O (Brassicaceae, Schisandraceae), **9.** Linalol C$_{10}$H$_{18}$O (Lamiaceae), **10.** Carvacrol (Lamiaceae, Verbenaceae)C$_{10}$H$_{14}$O, **11.** Trans-sabinene hydrate (Lamiaceae) C$_{10}$H$_{18}$O, **12.** p-cymeneC$_{10}$H$_{14}$ (Myrtaceae, Lamiaceae, Verbenaceae),**13.** o-cymene C$_{10}$H$_{14}$ (Lamiaceae), **14.** Terpinene C$_{10}$H$_{16}$ (Lamiaceae, Mytraceae), **15.** Terpinen-4-ol C$_{10}$H$_{18}$O (Lamiaceae), **16.** β-mircene C$_{10}$H$_{16}$ (Lamiaceae), **17.** α-Pinene C$_{10}$H$_{16}$O (Lamiaceae), **18.** Camphene C$_{10}$H$_{16}$, **19.** 1,8-cineole C$_{12}$H$_{20}$O$_3$ (Lamiaceae), **20.** scyllitol C$_6$H$_{12}$O$_6$ (Moraceae), **21.** Sulfamide H$_4$N$_2$O$_2$S (Moraceae), **22.** Ethyl-N-methylcarbamate C$_4$H$_9$NO$_2$ (Moraceae), **23.** Citronellal C$_{10}$H$_{18}$O (Oleaceae, Myrtaceae), **24.** Citronellol C$_{10}$H$_{20}$O (Myrtaceae), **25.** Isopulegol C$_{10}$H$_{18}$O (Myrtaceae), **26.** Cadinol C$_{15}$H$_{26}$O (Schisandraceae, Myrtaceae), **27.** Eugenol C$_{10}$H$_{12}$O$_2$ (Lauraceae, Myrtaceae, Lamiaceae), **28.** cis-3-hexenol C$_6$H$_{12}$O (Oleaceae), **29.** Trans-phytol C$_{20}$H$_{40}$O (Oleaceae), **30.** b-Linalool C$_{10}$H$_{18}$O (Oleaceae), **31.** Piperitone C$_{10}$H$_{16}$O (Oleaceae), **32.** Trans-p-mentha-2,8-dien-1-ol C$_{10}$H$_{16}$O (Oleaceae), **33.** 2-Undecanone C$_{11}$H$_{22}$O (Rutaceae), **34.** Nonanone C$_9$H$_{18}$O (Rutaceae), **35.** d-cadinene C$_{15}$H$_{24}$ (Schisandraceae), **36.** Borneol (Asteraceae, Lamiaceae, Lauraceae, Schisandraceae), **37:** b-pinene C$_{10}$H$_{16}$ (Zingiberaceae), **38.** Cyperene C$_{15}$H$_{24}$ (Zingiberaceae), **39.** Safrole C$_{10}$H$_{10}$O$_2$ (Lauraceae), **40.** Thymol C$_{10}$H$_{14}$O (Brassicaceae, Lamiaceae, Verbenaceae), **41.** Cis-piperitol C$_{10}$H$_{18}$O (Lamiaceae), **42.** Geranial C$_{10}$H$_{16}$O(Poaceae), **43.** Nerolidol C$_{15}$H$_{26}$O

15.3.1 Mode of Action of Essential Oils in Bacteria

15.3.1.1 Altered Membrane Permeability of Bacterial Cell

Antibiotics can pass through the bacterial cell membrane by either lipid-mediated pathway or by general diffusions. Bacterial cell membrane lipids and proteins are linked with its resistance to antibiotics. The Gram-negative bacterial intimidating outer membrane provides an extra layer of a protective barrier that must be overcome (Delcour 2009). The outer membrane acts as a selective barrier that increased the permeability associated with increased levels of antibiotic efflux. Essential oils have hydrophobic nature and destabilize the bacterial cell membrane and increased bacterial cell membrane permeability that disturbs cell architecture and membrane transport. The antimicrobial activity of essential oils of various plants was against human pathogenic drug-resistant bacteria reported to disrupt proton pumps and depletion of ATP and loss of ions (Turina et al. 2006).

For instance, the destruction of *Staphylococcus aureus* and *Escherichia coli* outer membrane might be due to the penetration of the *Melaleuca alternifolia* essential oil at 0·25% (v/v) that resulted in increased bacterial susceptibility to antibiotics (Cox et al. 2000). Yang et al. studied the effects of *Forsythia koreana* Nakai leaf essential oil which acted on the cytoplasmic membrane that increased the permeability of cell and passage of H^+, K^+, Na^+, and Ca^{2+}; a cytoplasmic membrane is a permeability barrier of the cell. The authors attributed that *F. koreana* contains mainly trans-phytol (42.73%), cis-3-hexenol (12.95%), and b-linalool (10.68%) that have antimicrobial activity against foodborne and pathogenic bacteria (Yang et al. 2015). The essential oil of *Curtisia dentata* can damage membrane structure and increased leakage of Na^+ and K^+ when treated with 30 mg/ml of the crude ethanol extract (Doughari et al. 2012). *Staphylococcus aureus* cells treated with terpinen-4-ol of tea tree oil showed cascade effects and cytoplasmic material loss (Lazcano et al. 2002). Ogunlana et al. reported that *Escherichia coli* cytoplasmic losses and bleb formation occur after treatment with lemongrass oil (Ogunlana et al. 1987). *Foeniculum vulgare* Mill. essential oil treated against *Shigella dysenteriae* resulted in the disruption of the membrane integrity causing K^+, Ca_2^+, and Na^+ electrolyte loss and intercellular ingredient leakage (Diao et al. 2014). Eugenol, a phenol compound in the essential oil of clove, interacts with the bacterial cell membrane and has activity against *Salmonella typhi* within 60 min of exposure. Eugenol can increase membrane permeability, disruption of the cytosolic membrane, and release of intracellular proteins, observed and verified by the Crystal Violet assay, SDS-PAGE, SEM, AFM analysis, and FT-IR spectroscopy (Devi et al. 2010). Eugenol, a hydrophobic

---

Fig. 15.2 (continued) (Asteraceae), **44.** Spathulenol $C_{15}H_{24}O$, **45.** germacrene D $C_{15}H_{24}$ (Myrtaceae), **46.** Camphor $C_{10}H_{16}O$ (Asteraceae, Lamiaceae), **47.** Artemiseole $C_{10}H_{16}O$ (Asteraceae), **48.** β-Thujone $C_{10}H_{16}O$ (Asteraceae, Lamiaceae), **49.** Cinnamaldehyde C_9H_8O (Lauraceae), **50.** cinnamyl acetate $C_{11}H_{12}O_2$ (Lauraceae), **51.** α-Santalol $C_{15}H_{24}O$ (Santalaceae), **52.** β-Santalol $C_{15}H_{24}O$ (Santalaceae), **53.** Trans-anethol $C_{10}H_{12}O$, **54.** Zingiberene $C_{15}H_{24}$ (Zingiberaceae)

compound, comes into the Gram-negative bacteria cell through lipopolysaccharide and disrupts the cellular structure, which results in intercellular leakage. Another possibility of the eugenol mechanism is that the hydroxyl group of eugenol binds to proteins and prevents enzyme action in *Enterobacter aerogenes* (Burt 2004). *Cudrania tricuspidata* (Carrière) Bureau ex Lavalle fruit essential oil has an antimicrobial mode of action including morphological alterations against *Bacillus cereus* and *Escherichia coli* with minimum inhibitory concentration (MIC) and minimum bacterial concentration (MBC) of 250–1000 mg/mL, respectively. These fruit EOs have a significant inhibitory effect on the cell viability that gets detected by the release of extracellular ATP, loss of 260 nm absorbing materials, and leakage of potassium ions with scanning electron microscope (SEM) analysis. The addition of MIC showed instantly potassium ions released from the cells (Bajpai et al. 2013).

15.3.1.2 Bacterial Efflux Pump Inhibition

To kill the bacterial cells, antibiotics must be transported through the bacterial cell membrane, accumulate into the bacterial cells, and inhibit protein synthesis. Resistant bacteria overproduce membrane proteins that efflux out antibiotics from the cell that stops accumulations of antibiotics into cells. Drug efflux pumps present in several bacteria are responsible for drug resistance. Bacterial drug efflux pumps are plasma membrane-integrated protein, classified into six types such as major facilitator superfamily (MFS), multidrug and toxic compound extrusion (MATE); small multidrug resistance (SMR) family, resistance-nodulation-division (RND) superfamily; drug/metabolite transporter (DMT) superfamily, and ATP-binding cassette (ABC) superfamily (George and Gordon 1991; Walsh 2000). Plant essential oil, as an efflux pump inhibitor, has the potential to overcome the resistance of bacteria to antibiotics with limited side effects. Essential oils from 23 different plants have been reported and evaluated for their potential as efflux pump inhibitors based on efflux (Prasch and Bucar 2015). *Staphylococcus aureus* possesses NorA efflux protein pumps, major facilitator superfamily pumps, that exclude out hydrophilic fluoroquinolones and DNA intercalating dyes when treated with *Chenopodium ambrosioides* L. leaf crude extract to combine with antibiotics, reduced to norfloxacin and ethidium bromide efflux, and increase the susceptibility of *Staphylococcus aureus* bacteria with MIC $\geq$1024 µg/mL concentration (de Morais Oliveira-Tintino et al. 2018). Another study in the essential oil of *C. ambrosioides* and α-terpinene against *Staphylococcus aureus* showed significant effects for TetK efflux pump inhibition with similar MIC concentration $\geq$ 1024 µg/mL, the same as that of (de Morais Oliveira-Tintino et al. 2018) study. TetK is a tetracycline efflux pump, belonging to a major facilitator superfamily of *Staphylococcus aureus* (Figueredo et al. 2017). TetK efflux pump is also inhibited by essential oil of *Salvia fruticosa* Mill. that obtained by hydrodistillation against tetracycline-resistant *Staphylococcus epidermidis* strain. This study showed a significant result that reverses resistance and improves the clinical efficacy of antibiotics (Chovanova et al. 2015; Papadopoulos et al. 2008).

Fadli et al. demonstrate that essential oils of *Thymus broussonetii*, *Thymus maroccanus*, *Thymus pallidus*, and *Rosmarinus officinalis* can change efflux pump activity and improve chloramphenicol susceptibility in multidrug-resistant *Escherichia coli, Enterobacter aerogenes, Klebsiella pneumoniae*, and *Salmonella enterica* bacteria (Fadli et al. 2011). The essential oil, obtained by hydrodistillation of *Rosmarinus officinalis* and eucalyptol, was also evaluated against multidrug-resistant *Acinetobacter baumannii* and *Pseudomonas aeruginosa* with promising results. A flow cytometry analysis revealed that the essential oils, tested in 13.07–52.26 mg/mL and 1.86–41.86 mg/mL, respectively, have efflux pump inhibition activity (Saviuc et al. 2016). One recent study by Almeida et al. (2020) found a major constituent essential oil named safrole in Ocotea odorifera. It showed reverssal of antibiotic resistance in *S. aureus* by inhibition of the NorA and MepA (MATE) efflux pumps (Almeida et al. 2020). Ferruginol is an efflux inhibitor isolated from cones of *Chamaecyparis lawsoniana* that inhibited quinolone (NorA), tetracycline (TetK), and erythromycin (MsrA) resistance pumps in *S. aureus*. They also showed that because of ferruginol the activity of oxacillin antibiotics increased by 80-fold (Smith et al. 2007).

15.3.1.3 Essential Oil as a Beta-Lactamase Inhibitor

One of the mechanisms of clinically isolated multiple drug-resistant bacterial strains is that they can produce ESBLs (extended-spectrum b-lactamases). Beta-lactamase breaks the antibiotic molecular structure that contains a ring through hydrolysis which shutdowns the molecule's antibacterial properties. Some of the beta-lactam antibiotics are penicillin derivatives, cephalosporins, carbapenems, monobactams, and carbacephems. The combination of essential oils and antibiotics against beta-lactamase producers is currently limited; there are three reports on the synergistic effects of oregano essential oil in combination with fluoroquinolones, doxycycline, lincomycin, maquindox, and florfenicol against ESBL-producing *Escherichia coli* suggesting the possibility that essential oils might function as the ESBL inhibitor (Si et al. 2008). El-Shouny and colleagues reported that ginger essential oil has significant activity against drug resistance in ESβL-producing *Pseudomonas aeruginosa* (El-Shouny et al. 2018).

15.3.1.4 Anti-quorum Sensing

Quorum sensing is a cell-cell communication process that maintains cell density and gene expression in bacteria. Quorum sensing is responsible for regulating the genes involved in biofilm development (Solano and Echeverz 2014). In the human body, approximately 80% of recurrent microbial infections are because of bacterial biofilm (Sharma et al. 2019). Essential oils can inhibit the biofilm formation of bacteria by inhibiting quorum sensing. Clove oil showed significant quorum sensing inhibition activity in *Pseudomonas aeruginosa* (Khan et al. 2009). Besides that cinnamon, oregano, lavender and peppermint oils also have anti-quorum sensing activity *against E. coli, B. subtilis, P. aeruginosa* (Bouyahya et al. 2019).

15.3.2 Mode of Action of Essential Oils in Fungus

An antifungal resistance mechanism is not as broadly described as antibacterial resistance due to lack of reorganization of fungal diseases (Ghannoum and Rice 1999).

15.3.2.1 Cell Membrane Disruption, Alteration, and Inhibition of Cell Wall Formation

Essential oils act on the fungal plasma membrane and alternate cell morphology. Ergosterol is the main component of the fungal membrane and maintains cell function and integrity; applying an essential oil on fungus resulted in membrane integrity and decreased amount of ergosterol that leads to inhibition in wall formation. Moreover, membrane ATPase inhibition, cytokine interaction, and alternation of few gene expressions also take place by essential oil action. Carvacrol and cis-β-ocimene-rich compounds of *Lavandula multifida* L. responsible for cytoplasmic membrane disruption and cell death by inhibiting filamentation of *Candida albicans* fungus (Zuzarte et al. 2012). Limonene is the major essential oil *Citrus sinensis* (L.) Osbeck epicarp, that has growth-inhibiting and morphogenesis alternating activity in *Aspergillus niger*. The mycelial growth of fungus inhibited by fungal hyphae cytoplasm loss and hyphal tip loss resulted in cell wall disruption (Sharma and Tripathi 2008). Silva et al. also showed coriander oil can disrupt cell membrane of *Candida albicans* and *Candida tropicalis* without interfering with DNA (Silva et al. 2011). However, another study indicates that coriander oil interferes with DNA and membrane permeability (2007).

15.3.2.2 Dysfunction of the Fungal Mitochondria

Anethum graveolens L., *Origanum compactum*, *Artemisia herba alba*, and *Cinnamomum camphora* essential oils have a role in mitochondrial dysfunction by inhibiting mitochondrial dehydrogenases and ATP biosynthesis (Li et al. 2020). Mitochondrial dysfunction can initiate calcium signaling and Ca^{2+} in cytosol increased apoptotic death. *Anethum graveolens* L. seed essential oil disrupts the cell membrane that leads to cell death by inhibiting mitochondrial dehydrogenases (MDH). Mitochondrial dehydrogenase catalyzes ATP biosynthesis. Inhibition of the MDH enzyme may disturb the citric acid cycle (Chen et al. 2013). Bakkali et al. showed this study in *C. albicans* different strains where dill seed oil caused mitochondrial hyperpolarization. EOs of *O. compactum*, *A. herba alba*, and *C. camphora* showed cytotoxicity and an increase in mitochondrial damage by cytoplasmic petite mutations (Bakkali et al. 2006).

15.3.2.3 Inhibition of Efflux Pump in Fungal Cell Membrane

Efflux pump overexpression in the fungal cell may exclude out the accumulated drug that is being used to treat fungus. Several studies are presented that essential oil treatment against fungus causing human diseases can reverse drug resistance by inhibiting efflux pumps. H ± ATPase maintains a proton gradient in fungal cells and regulates intracellular pH and cell growth. Inhibition of H ± ATPase leads to intracellular acidification and cell death. Ahmad et al. showed that eugenol and thymol, principal compounds present in thyme oil, inhibit H $\pm$ ATPase activity and showed 70–90% of CDR1 (cerebellar degeneration-related protein 1) and MDR1 (multidrug resistance protein 1) efflux pump gene inhibition. These two monoterpenes (eugenol, thymol) have a synergistic antifungal effect with the azolic compounds, in *Candida* spp. (Ahmad et al. 2013).

15.3.3 Mode of Action of Essential Oils in Protozoa

Protozoa are free-living, microscopic unicellular eukaryotic organisms that have relatively complex internal structures and carry out complex metabolic activities. Protozoan-mediated diseases include mainly malaria, sleeping sickness, Chagas disease, amoebic dysentery, and toxoplasmosis caused by several species of *Plasmodium, Leishmania, Trypanosoma, Entamoeba, Toxoplasma*. Several drugs show limitations such as drug resistance and toxicity. Plant essential oils have been used traditionally since very long due to their analgesic, sedative, anti-inflammatory, spasmolytic, and local anesthetic properties. Essential oils involved in membrane disruption and induced antimicrobial activity. *Lavandula luisieri, Lavandula viridis, Cymbopogon citratus, Lippia sidoides*, and *Ocimum gratissimum* essential oils have antiparasitic activity through mitochondrial membrane potential loss, cell cycle arrest, and morphological alteration of the plasma membrane. Scanning electron microscopy (SEM) and transmission electron microscopy (TEM) study of treated cells showed morphological alterations and cytoplasmic swelling of *Trypanosoma cruzi* plasma membrane treated with extracted essential oil containing mainly p-cymene, terpinene, α-terpineol, 3-ciclohen-1-ol, limonene, thymol, and carvacrol components (Santoro et al. 2007). SEM study on *Leishmania chagasi* treated with essential oil of *Cymbopogon citratus* revealed changes in cell shape (Oliveira et al. 2009). *Lavandula* essential oil containing components such as 1,8-cineole, linalool, and borneol also has antiprotozoal activity against different species of *Leishmania* that initiates loss of mitochondrial membrane potential, cell cycle arrest, and then apoptosis (Machado et al. 2019). A study in BALB/c mice showed activity against promastigote and amastigote forms of *Leishmania* treated with *Chenopodium* oil that helps in the reduction of cytochrome c leading to loss of mitochondrial membrane potential. This essential oil also active against *Plasmodium falciparum* (Monzote et al. 2014). The ultrastructural alterations by basil essential oil observed in *Leishmania amazonensis* resulted in nucleus membrane discontinuity (da Silva et al. 2018).

15.3.4 Mode of Action of Essential Oils in Viruses

Antiviral drugs work on the virus by inhibiting viral replication and give time to the human body's immune system to quickly resume itself and neutralize the virus. The virus becomes resistant to drugs due to changes in its gene and continues replication even in the presence of an antiviral drug in the host. Some viruses reduce its susceptibility towards drugs by alternating its active binding site. The antiviral activity of essential oils was tested against many enveloped RNA and DNA viruses, such as herpes simplex virus type 1 and type 2, influenza virus, dengue virus type 2, adenovirus type 3, poliovirus, and coxsackievirus B1 (Wagstaff et al. 1994). Acyclovir, penciclovir, valacyclovir, and famciclovir are some antiviral agents that target DNA polymerase and inhibit HSV-1 virus replication (Gilbert et al. 2007). With the increasing application of antiviral therapy, the virus becomes resistant, that necessitates the discovery newer agents with novel targets and better side effect profiles. HSV-1 belonging to *Herpesviridae* family induced pro-oxidative state and reduced intercellular glutathione; treatment with piperitenone oxide an active component of *Mentha suaveolens* essential oil affected viral replication by interfering with redox-sensitive cellular pathways (Civitelli et al. 2014). The anti-herpetic activity of essential oil may be due to direct interaction with HSP-1 virons. One study indicates that essential oil treated to the virus after infection did not block cell receptor or virus biding site to the cell and not reduced virus plaque formation. However, viruses pretreated before cell infection found a reduction in plaque formation. This study also showed HSV-1's replication ability suppressed by application of 0.1% lemongrass essential oil (Minami et al. 2003). *Artemisia arborescens* oil reported a virucidal activity against herpesviruses, i.e., HSV-1 and HSV-2, by inhibiting cell-to-cell virus diffusion (Saddi et al. 2007).

15.4 Plant-Based Essential Oil Chemistry and Family-Wise Description of In Vitro Antimicrobial Activity of Essential Oil Against Drug-Resistant Bacteria, Fungi, Protozoa, and Viruses

Essential oils are known for the sense that they contain, the fragrance of the plant from where they are derived. Essential oils consist of very complex structures and contain up to 5000–7000 various chemical compounds of primary terpenoids, specially monoterpenes, sesquiterpenes, diterpenes (C20), phenylpropanoid derivatives, alcohol derivatives, and phenols. Carvacrol, eugenol, camphorene, thymol, geraniol, ketones, esters, and catechol are some major components in many essential oils where the chemical and structural differences among the compounds are minimal (Dhifi et al. 2016). The components of essential oils may be broadly classified as volatile and nonvolatile fractions that are mainly obtained by non-heated techniques, i.e., hydrodistillation or dry-distillation, or by a suitable mechanical process without heating, i.e., solvent extraction, cold pressing, sfumatura processes (Dhifi et al. 2016). The use of different essential oils is gaining global interest in aromatherapy or in alternative medicine (Lee et al. 2012). Gas chromatography-mass spectrometry (GC-MS) and gas chromatography-flame ionization detector (GC-FID) methods are used for screening chemical constituents present in essential oils. Currently, the main essential oils studied for antifungal activity are thyme oil, rich in thymol and carvacrol, tea tree oil rich in terpenes, and peppermint or clove oil, although many others have also been shown to be effective against bacteria, fungi, protozoa, and viruses. Essential oils have several other properties; besides being antimicrobial, they could also be used in complementary therapy.

15.4.1 Annonaceae

The essential oil of *Annona coriacea* Mart. leaf extraction was extracted by hydrodistillation and screened by GC-MS and GC-FID, and 60 compounds were found including sesquiterpenes and monoterpenes. This obtainable volatile oil has antileishmanial activity against *Leishmania chagasi* (Siqueira et al. 2011).

15.4.2 Apiaceae

Trichomonas vaginalis is a flagellated protozoan causing sexually transmitted infections worldwide, and it gradually becomes drug-resistant. In this study the antitrichomonal activity of methanolic extract of *Foeniculum vulgare* was investigated on clinically isolates of *Trichomonas vaginalis*, and this extract inhibited the growth of this clinical isolate at concentrations of MLC 360 μg/m, where the action was comparable with metronidazole (Karami et al. 2019). *Foeniculum vulgare* Mill., native to central Europe and the Mediterranean region, was tested using the Oxford cup method. Hydrodistillation of crude extract of seed has shown a Gram-negative microorganism with the highest antimicrobial activity. The diameter of the inhibition zone against *Shigella dysenteriae* showed 17.8 ± 1.5 (Diao et al. 2014).

15.4.3 Aristolochiaceae

Aristolochia mollissima essential oil contains mainly 2,2,7,7-tetramethyltricyclo [6.2.1.0(1,6)]undec4-en-3-one in the rhizome and the aerial parts that has significant bactericidal activity against 20 bacterial strains including methicillin-resistant *Staphylococcus aureus* and *Staphylococcus saprophyticus* using disk diffusion and broth microdilution method (Yu et al. 2007).

15.4.4 Asteraceae

Tagetes minuta L. is native to Mexico, commonly known as wild marigold, traditionally used as an insect repellent and for treatment of stomach and intestinal diseases. Essential oils mainly β-ocimene and m-tert-butyl-phenol obtained by hydrodistillation from the flower of the plant were applied for antibacterial test against seven different bacterial isolates, both Gram-positive and Gram-negative.

The MIC and MBC values range from 0.125 µg/mL to 0.05 µg/ mL (Igwaran et al. 2017). *Artemisia sieberi* Besser. essential oil exhibited notable antifungal activity against *Candida glabrata* (Khosravi et al. 2011).

Baccharis dracunculifolia DC., a native plant of Brazil, commonly known as "Vassoura" contains essential oils mainly spathulenol and nerolidol extracted by hydrodistillation technique which have in vitro antiprotozoal activity against *Leishmania donovani*, with IC50 values of 42 mg/ml; however, the study also observed that *B. dracunculifolia* oil is not active against anti-plasmodial assays (Parreira et al. 2010).

Another study also indicates that monoterpenes of *Pulicaria vulgaris* Gaertn. affected HSV-1 mostly before adsorption, but *Pulicaria vulgaris* showed the lowest selectivity index among all tasted oils of *Sinapis arvensis* L. and *Lallemantia royleana* Benth. (Sharifi-Rad et al. 2017).

Essential oils of *Sinapis arvensis* L., *Lallemantia royleana* Benth., and *Pulicaria vulgaris* Gaertn. affected HSV-1 mostly before adsorption, in which thymol showed a high selectivity index of 7 and *Pulicaria vulgaris* showed the lowest selectivity index, 1, among all tasted oils (Sharifi-Rad et al. 2017). Koch et al. showed chamomile oil significantly reduced plaque formation in HSV-2 strain. The main components in chamomile oils are α – bisabolol oxide A (13.4–55.9%), *cis* – bicycloether (=(Z) – spiroether) (3.6–17.7%), α – bisabolol oxide B (8.4–25.1%), bisabolone oxide A (2.9–11.4%), and (E) – β – farnesene (1.9–10.4%) (Koch et al. 2008). A similar inhibitory effect of *Artemisia arborescens* L. activity has been shown against DENV−2, HSV-1, and HSV-2 (Duschatzky et al. 2005; Saddi et al. 2007).

15.4.5 Brassicaceae

Rad et al. revealed *Sinapis arvensis* L. essential oil has antibacterial activity against *Bacillus subtilis*, *Staphylococcus aureus*, *Escherichia coli*, *Klebsiella pneumonia*, and *Pseudomonas aeruginosa* (Rad et al. 2013). *S. arvensis* L. also has significant antiviral activity against herpes simplex virus type 1 (Sharifi-Rad et al. 2017).

15.4.6 Lauraceae

A study by Dutta et al. showed that *Candida albicans* causing clinical mycoses could be completely inhibited with cinnamon oil main constituent camphor isolated from wood chips, showing 10.67 mm zone of inhibition (Dutta et al. 2007). Masood et al. studied *Cinnamomum cassia* (L.) J. Presl bark oil against 178 bacterial strains, and it showed high antibacterial activity against *Streptococcus* strains but did not show any positive activity against *Salmonella paratyphi* (Masood et al. 2006). Cinnamon as a natural product has significant activity against diabetes. Cinnamon essential oil also has antiviral properties. *Cinnamomum zeylanicum* Blume essential oils' main components are (E) – cinnamaldehyde, linalool, and eugenol that have antiviral activity against influenza type A (H1N1) and HSV-2 by plaque reduction assay (Bourne et al. 1999; Setzer 2016).

15.4.7 Lamiaceae

Ozkalp et al. demonstrated the inhibition of *Candida albicans* and eight bacterial strains with *Origanum vulgare* L. essential oil (Özkalp et al. 2010). Santoyo et al. also investigated *Origanum vulgare* L. antimicrobial activity by the disk diffusion and broth dilution methods against *Staphylococcus aureus*, *Bacillus subtilis*, *Pseudomonas aeruginosa*, *Escherichia coli*, *Candida albicans*, and *Aspergillus niger*. Carvacrol, sabinene hydrate, borneol, and linalool are the main components of *Origanum vulgare* L. essential oil extracted by supercritical fluid extraction method, where carvacrol is the most effective (Santoyo et al. 2006). (Arámbula et al. 2019; Béjaoui et al. 2013; Khosravi et al. 2011) also observed significant antibacterial and antifungal activity in vitro with *Origanum vulgare* L. essential oil. Kačániová et al. showed very strong antibacterial activity in *Thymus vulgaris* and *Origanum vulgare* L. against *Bacillus cereus* (Kačániová et al. 2012). Another species of the Lamiaceae family, *Origanum onites* L. Turkish oregano, has essential oil which has antiprotozoal activity against African trypanosome and American trypanosome such as *Trypanosoma brucei rhodesiense* and *Trypanosoma cruzi*, respectively, and *Leishmania donovani* and *Plasmodium falciparum* (Tasdemir et al. 2019).

This Ethiopian plants *Thymus schimperi* Ronniger and *Rosmarinus officinalis* are used to treat blood pressure, general pain syndrome, influenza, abdominal pain, ascariasis, intestinal parasites, menstrual problems, and hypertension (Benarba 2016; Damtie and Mekonnen 2016). Essential oils of Ethiopian thyme and rosemary were applied against five bacterial species (*Salmonella typhi*, *Salmonella paratyphi*, *Salmonella typhimurium*, *Shigella species*, *Pseudomonas aeruginosa*, *Staphylococcus aureus*, *Escherichia coli*) and three fungal (*Trichophyton* spp. and *Aspergillus* spp.) pathogens to determine the degree of antibacterial activities of these plants using the agar well diffusion method with lower than 15.75 mg/mL concentration of MIC (Mekonnen et al. 2016).

The *Ocimum* genus or basil leaves consist of 70 species and contain essential oil at a 0.2–1% concentration, with major components such as linalool, estragole, o-cymene, camphene, geraniol, and geranial. In vitro studies of *Ocimum canum* Sims., *Ocimum gratissimum* L., and *Ocimum basilicum* L. species showed significant antimicrobial activity. *Ocimum canum* has been used for the treatment of asthma, fevers, coughs, flu, bronchitis, and indigestion and contains major constituents as hydrocarbon monoterpenes, oxygenated monoterpenes, and hydrocarbon sesquiterpenes. da Silva et al. observed it has antimicrobial and antiprotozoal activity against *Staphylococcus aureus*, *Escherichia coli*, *Pseudomonas aeruginosa*, *Trypanosoma cruzi*, and *Leishmania amazonensis* (da Silva et al. 2018), whereas, Akono Ntonga et al. showed *Ocimum canum* essential oil activity against *Plasmodium falciparum with* IC50 concentration 20.6 lg/mL (Akono Ntonga et al. 2014). Other species of basil like sweet basil and clove basil (*Ocimum basilicum* and

Ocimum gratissimum) have been investigated for their antibacterial and antiprotozoal activity (Oliveira et al. 2009; Opalchenova and Obreshkova 2003).

Except for MRSA *S. aureus* (250 µg/ml), ampicillin had more effect on *K. pneumoniae* (128 µg/ml) and *C. albicans* (128 µg/ml). Another in vitro study against *Trypanosoma cruzi* treated with *Origanum onites* L. also showed no activity up to IC50 90 µg/mL concentration, but the same plant oil showed antileishmanial and anti-plasmodial activity with IC50 17.8 µg/mL and IC50 7.9 µg/mL, respectively (Tasdemir et al. 2019). Amoebiasis and giardiasis are protozoan diseases caused by *Entamoeba histolytica* and several *Giardia* species, respectively. Their drug resistance tendency can be cured by applying *Ocimum basilicum* chloroform extract and showed strong activity against *Entamoeba histolytica* and *Giardia duodenalis* with IC50 value 68.62 µg/ml and 53.31 µg/ml, respectively (El-badry et al. 2010). *Ocimum basilicum* and *Ocimum canum* leaves have anti-plasmodial properties. Geranial, 1,8-cineole, and linalool showed effectiveness in inhibiting the growth of *Plasmodium falciparum* in vitro (Akono Ntonga et al. 2014). HSV-1-ACV^{res} is an acyclovir-resistant HSV-1 strain inhibited by peppermint essential oil when used in a nontoxic concentration of IC50) 0.01%, with a reduced viral plaque formation by 99%. Peppermint oil inactivated HSV-1 before it enters the cell (Schuhmacher et al. 2003).

Another study also indicates that monoterpenes, i.e., thymol, carvacrol, and p-cymene, and *Sinapis arvensis* L., *Lallemantia royleana* Benth., and *Pulicaria vulgaris* Gaertn. essential oils affected HSV-1 mostly before adsorption, in which thymol showed a high selectivity index of 7 and *Pulicaria vulgaris* Gaertn. showed the lowest selectivity index, 1, among all tasted oils (Sharifi-Rad et al. 2017). In vitro activity of *Mentha suaveolens* oil and its active component piperitenone oxide that constitutes from 80 to 90% of *Mentha suaveolens* oil was tested against HSV-1 during viral adsorption or pretreated before viral infection (Civitelli et al. 2014).

Thymus vulgaris L. (Koch et al. 2008), *Mentha suaveolens* Ehrh. (Civitelli et al. 2014), *Salvia fruticosa* Mill. (Sivropoulou et al. 1997), *Salvia desoleana* Atzei & V. Picci (Cagno et al. 2017), *Satureja hortensis* L. (Gavanji et al. 2015), and *Zataria multiflora* Boiss. (Gavanji et al. 2015) are some plants of the Lamiaceae family that showed antiviral activities against HSV-2 and HSV-1, in vitro, and RC-37 cells. Thymol, carvacrol, piperitenone oxide, and 1,8-cineole are major constituents present in the essential oils of these plants.

However, the postinfection addition of *Salvia desoleana* Atzei and V. Picci essential oil has a significant suppression of HSV-2 replication with an EC50 value of 33.01 µg/ml (Cagno et al. 2017). Another genus from Lamiaceae, *Lavandula*, has been used in food, perfume, and cosmetic industries. Essential oils from *Lavandula* also have been reported for antibacterial, antispasmodic, antifungal, and antioxidant effects. *Lavandula luisieri* L. and *Lavandula viridis* L'Hér. have been reported for their antiprotozoal activity against *Leishmania infantum*, *Leishmania major*, and *Leishmania tropica* (Machado et al. 2019).

15.4.8 Moraceae

Art*ocarpus heterophyllus* Lam. is known for its uses in malarial fever, stomachache, ulcers, dysentery, diarrhea, and desecration. Jha and Srivastava showed effects against three bacteria (*Escherichia coli, Pseudomonas aeruginosa, Staphylococcus aureus*) with MIC ranging between 1.50 and 5.20 mg/mL (Jha and Srivastava 2012). Moreover, the study of Bajpai et al. demonstrated the effectiveness of silkworm thorn (*Cudrania tricuspidata*) essential oil extracted by microwave-assisted extraction technique against *Bacillus cereus* which showed a MIC of 250 mg/mL (Bajpai et al. 2013).

15.4.9 Myrtaceae

Myrtaceae family plants are known for their several therapeutic properties, mainly antioxidant, antimicrobial, and anti-inflammatory. *Eucalyptus* is a native plant of Australia and studied widely for its source of important and significant essential oils used for antibacterial, antifungal, and acaricidal activities. *Eucalyptus camaldulensis* Dehn., *Eucalyptus tereticornis, Eucalyptus globules*, and *Eucalyptus citriodora* are some species of *Eucalyptus*, which we included in our review, as an antimicrobial alternative against drug resistance microbes (Bachir and Benali 2012). Lemon-scented eucalyptus or *Eucalyptus citriodora* leaves or twig and fruit were isolated using hydrodistillation and characterized by GC-FID and GC-MS and showed significant antibacterial activity against various human pathogenic bacteria, fungi, and protozoa (Su et al. 2017; Tolba et al. 2015). Essential oils from *E. citriodora* showed higher activity than two synthetic antibiotics miconazole and clotrimazole against *Candida albicans* (Dutta et al. 2007). Chaves et al. studied the antimicrobial activity of *Eucalyptus camaldulensis* Dehn. essential oil activity against β-lactam-resistant *Staphylococcus aureus* and *Escherichia coli* with MIC of 1000 μg mL^{-1} (Chaves et al. 2018). *Eugenia gracillima* Kiaersk. essential oil which most abundantly contains germacrene D, c-muurolene, bicyclogermacrene, germacrene B, and D-elemene showed antiprotozoal activity against *Leishmania braziliensis* and *Leishmania infantum* promastigotes but no activity against *Trypanosoma cruzi* (Gomes Vidal Sampaio et al. 2019).

15.4.10 Oleaceae

Forsythia koreana Nakai fruits have been used for its anti-inflammatory, diuretic, and damp heat-clearing effects. The essential oil from the leaves of Korean *F. koreana* was tested against *Salmonella enteritidis, E. coli*, and *S. aureus* and showed great antibacterial potency (Yang et al. 2015).

15.4.11 Poaceae

Cymbopogon belongs to the Poaceae family which represents almost 120 species, among them 27 species are present in India that have been used for their perfumery characteristics. Besides this it has antifungal and antiprotozoal activity. Oliveira et al. and Akono Ntonga et al. observed lemongrass essential oil (*Cymbopogon citratus* (DC) Stapf.) antiprotozoal activity against *Leishmania chagasi* and *Plasmodium falciparum* due to the presence of major compounds, i.e., myrcene, neral, and geranial compounds, detected with the radioisotopic method (Akono Ntonga et al. 2014; Oliveira et al. 2009). Dutta et al. (2007) and Boukhatem et al. (2014) revealed *Cymbopogon citrates* antifungal activity against *Candida albicans*; its antimicrobial activity is detected with disk diffusion method and vapor diffusion method (Dutta et al. 2007; Boukhatem et al. 2014). In addition, citronella essential oil from *Cymbopogon nardus* showed active antifungal activity against *Candida albicans* (Ahmad Kamal et al. 2020). Oxygenated sesquiterpenes, monoterpene hydrocarbons, sesquiterpene hydrocarbons, and oxygenated monoterpenes are the main components of *Cymbopogon schoenanthus* (L.) Spreng and *Cymbopogon schoenanthus* Chiov. plant essential oil, endemic species native to West Africa, and have antibacterial activity against two human pathogenic bacteria with minimum inhibitory concentration 0.32–0.64 mg/mL and 2.63 mg/mL (Alitonou et al. 2012). *Cymbopogon giganteus* also showed a strong effect against chloroquine-resistant *Plasmodium* (Kimbi and Fagbenro-Beyioku 1996).

15.4.12 Rutaceae

Previous literature findings have indicated that the essential oils of the Rutaceae family plants constitute efficient antibacterial activity. Dutta et al. and Arámbula et al. studied two Rutaceae family plants *Citrus limon* (L.) Osbeck and *Ruta chalepensis* L., against one fungus (*Candida albicans*) and three bacteria (*Escherichia coli*, *Staphylococcus aureus*, *Pseudomonas aeruginosa*), respectively. *Citrus limon* peel contains important essential oil that blocked the growth of *Candida albicans* with MIC of 3775 μg/mL (Dutta et al. 2007). *Ruta chalepensis* L. is native to the Mediterranean region and has been used as a phytotoxic, abortifacient, anthelmintic, and antispasmodic. *R. chalepensis* essential oil contains main components like 2-undecanone, nonanone, decanone, and nonyl acetate that showed inhibitory effects against human pathogenic bacteria with MIC of 500 mg/mL (Arámbula et al. 2019).

15.4.13 Santalaceae

Three isolated essential oils α-santalol, (Z)-β-santalol, and *trans*-α-bergamotol from *Santalum album* L. demonstrated antiviral activity against HSV-2 using an antiviral assay. However, sandalwood oil exhibited a low selectivity index, with an IC50 (%) value of 5 (Koch et al. 2008).

15.4.14 Schisandraceae

Kadsura longipedunculata Finet and Gagnep. and *Illicium verum* Hook. f. from the *Schisandraceae family showed antibacterial and antiviral activity,* *respectively. K. longipedunculata* is an evergreen plant also known as Chinese Kadsura vine which contains D-cadinene, camphene, borneol, and cubenol as major constituents of its essential oil that are screened by capillary gas chromatography (GLC-FID) and gas chromatography-mass spectrometry (GLC-MS). This oil showed positive antimicrobial activity against *Enterococcus faecalis*, *Streptococcus pyogenes*, and *Streptococcus agalactiae* with MIC of 60 mg/ml (Mulyaningsih et al. 2010). *I. verum* is a Chinese aromatic spice that showed HSV-2 inhibition activity. Its major component is trans-anethol (Koch et al. 2008).

15.4.15 Verbenaceae

The Mexican oregano *Lippia graveolens* is effective in inhibiting properties against DNA virus acyclovir-resistant herpes simplex virus type 1 with a 50% effective concentration (EC50) of 48.6 μg/ml (Pilau et al. 2011).

15.4.16 Umbelliferae

Anethum graveolens L. and *Coriandrum sativum* L. are two herbs from the Umbelliferae family which showed antifungal activity against *Candida albicans*. Dill seed essential oil and coriander oil are used as a flavoring agents and can also reverse drug-resistant human pathogenic fungus (Chen et al. 2013; Silva et al. 2011).

15.4.17 Zingiberaceae

Aframomum sceptrum (Oliv. and T. Hanb.) K. Schum. essential oil, with components such as b-pinene, caryophyllene oxide, and cyrene, was analyzed by GC-MS, which has highly significant antiprotozoal activity against *Trichomonas vaginalis* and *Trypanosoma brucei*. Other than that, *A. sceptrum* essential oil also showed activity against three the Gram-positive bacteria and two fungi; however, it did not exhibit any bactericidal activity against Gram-negative bacteria *Pseudomonas aeruginosa*, *P. vulgaris*, and *Escherichia coli* (Cheikh-Ali et al. 2011).

15.5 Synergistic Formulations by Combination of Antimicrobials and Essential Oils to Reverse Resistance

A study over antimicrobial activity of selected plant essential oils already reveals their significant growth-inhibitory properties against multidrug-resistant microbes. However, it is required to evaluate the combined activity of standard drugs with extracted essential oil from plants with different chemical compositions. The antimicrobial properties of plant essential oils are already being studied, and their association with antibiotics represents a cure to the crisis of antimicrobial resistance. Synergy in combination with therapy happens when the combination effects of two or more compounds are greater than the individual effects. An additive effect will occur when the interaction between antimicrobial combinations is equal to the sum of the individual substances (Bhat and Ahangar 2007). In vitro synergy is measured by a checkerboard assay by an isobologram plot or fractional inhibitory concentration (FIC) index calculation (Bhat and Ahangar 2007). Combination therapy of essential oils with antibiotics or nanoparticles is a new concept to overcome drug resistance microbial disease (Gibbons et al. 2003).

Clove oil's major constituent eugenol was tested in combination with antibiotics against *E. coli*, *E. aerogenes*, *P. vulgaris*, *P. aeruginosa*, and *S. typhimurium*, and a synergistic relation was found between all tested antibiotics, among them are norfloxacin, ampicillin, vancomycin, polymyxin B, tetracycline, and rifampicin (Hemaiswarya and Doble 2009). Cinnamaldehyde, from cinnamon essential oil, in combination with fluconazole was shown to give an effect against multiple drug-resistant *Aspergillus fumigates* and *Trichophyton rubrum* (Khan and Ahmad 2011). Cis-cinnamic acid also found in cinnamon essential oil showed synergy with rifampicin and significant effects against clinically isolated bacterial species *Mycobacterium tuberculosis* (Chen et al. 2011). The chemical analysis of the *Eucalyptus* essential oil such as 1,8 cineole (76.93%), β-pinene (11.49%), and α-pinene (7.15%) showed synergistic effects, mainly with amoxicillin and ampicillin in β-lactamase-producing *S. aureus* and MRSA strains and cephalexin and cefuroxime for *Escherichia coli* (Chaves et al. 2018).

Several studies have described the synergistic activity for existing antibiotics and whole essential oils (Mahboubi and Bidgoli 2010; Monzote et al. 2007; Rosato et al. 2007; Si et al. 2008). Possibly one example of multitarget synergy was demonstrated in *Melaleuca alternifolia* or Australian tea tree essential oil that was reported to show synergistic interactions with aminoglycoside antibiotics (ciprofloxacin) in concentration-dependent interactions against *Streptococcus aureus* and *Klebsiella pneumonia*. The composition of this oil includes mainly 1,8 cineole (76.93%), β-pinene (11.49%), and α-pinene (7.15%) (Van Vuuren et al. 2009). Aminoglycoside is the inhibitor of protein synthesis, and tea tree oil damages the cytoplasmic membrane of bacteria (Van Vuuren et al. 2009). *Zataria multiflora* Boiss. (Shiraz oregano) essential oil was combinedly tested with vancomycin and showed significant activity against 12 clinical isolates of MRSA (Mahboubi and Bidgoli 2010). *Lantana montevidensis* L., native to South America, essential oil of this plant has antibiotic (amikacin)-modifying capacity against *E. coli* by broth microdilution assay (De Sousa et al. 2013). The synergistic combination between Calli oil and lawsone (hennotannic acid) showed synergistic effects in MDR microbes (*Staphylococcus aureus* (MRSA), Gram-negative bacteria) (Soliman et al. 2017). The essential oil of *Croton zehntneri* Pax et Hoffm. leaves is able to enhance the gentamicin activity by 42.8% against *Pseudomonas aeruginosa* with a bacterial culture of 10^5 CFU/mL by gaseous contact demonstrated by (Rodrigues et al. 2009). The essential oil of lemongrass and acyclovir showed significant synergism against HSV-1 (Duran and Kaya 2018). Lemon balm and oseltamivir combined therapy is effective against avian influenza A virus (H9N2) (Pourghanbari et al. 2016).

Extracts from *Moringa oleifera* plant have shown significant drug-modifying ability against *Leishmania major* in combination with amphotericin B with FIC indices of less than 0.5 (FIC: 0.375) (Hammi et al. 2020). *Boswellia serrata* oil was found to have synergistic activity with azole drugs (fluconazole, ketoconazole, posaconazole, or voriconazole) against azole-resistant *Candida albicans* strain (Sadhasivam et al. 2016). Amikacin, an antibiotic, combined with *Citrus limon* (Guerra et al. 2012) or *Lippia sidoides* (Veras et al. 2012) or *Cinnamomum zeylanicum* (Guerra et al. 2012) showed drug activity enhancement through gaseous contact against *Acinetobacter baumannii* (Guerra et al. 2012) or *Staphylococcus aureus, Pseudomonas aeruginosa* (Veras et al. 2012), or *Acinetobacter* spp., respectively (Guerra et al. 2012). Spatulenol, cryptone, p-cymene, 1,8-cineole, terpinen-4-ol, and β-pinene components from the essential oils of *Eucalyptus camaldulensis* and combination with standard lipopeptide and fluoroquinolone antibiotics (ciprofloxacin, gentamicin, and polymyxin B) against 23 *A. baumannii* strains were tested. Among them *E. camaldulensis* and ciprofloxacin combination, resulted synergy in two strains (Aba-4914 and Aba-5055) and one strain (ba-6673), showed an additive effect. A combination study with gentamicin expressed synergy in only one MDR strain Aba-4914 (Knezevic et al. 2016). A study investigating thyme oil reported synergistic effect in combination with ciprofloxacin against *S. aureus* and *K. pneumoniae* (Van Vuuren et al. 2009) Table 15.7 broadly describes the list of essential oil and antibiotic combination and their effects against drug-resistant microbes.

15.6 Concluding Remarks

Antimicrobial treatment failure is a serious threat to today's world and will increase with upcoming times further. As the evolution of microbial genes and improvement of their acquired resistance towards antimicrobial drugs increases day by day, it leaves an adverse impact on human life. Microbes such as bacteria, fungi, protozoa, and viruses gathered resistance to conventional drugs and are used as therapeutic agents. However, many antibiotics are used improperly and generate unwanted toxic molecules that can hamper the quality of human life. Microbes show resistance to drugs due to overexpression of efflux pumps in the cell membranes, cell outer membrane stiffness, beta-lactamase production, biofilm production, genetic material changes, etc. Antimicrobial agents are derived from natural resources, and researchers are currently investigating plant extracts, their synergistic effect, and the medicinal importance of bioactive compounds through various screening programs. A total of 300 essential oils are commercially available and possess high antimicrobial potential. Components present in essential oils are also in various types of plant secondary metabolites such as phenolics, terpenes, etc. In this review, we discussed several relevant reports that represent essential oils as effective against drug-resistant antimicrobial strains. According to available papers, essential oils are aromatic oily liquids that qualify on the parameters like bioavailability, cost-effectiveness, and safety and could be looked at as a "potential alternative" treatment against drug-resistant microbes.

Essential oils such as star anise, clove, cajuput, cinnamon, sandalwood, camphor, citronella, eucalyptus, geranium, mint, lavender, lemon, lemongrass, peppermint, lime, sage, rosemary, basil, etc. have been traditionally used by people for various purposes in different parts of the world for their antibacterial, antifungal, antidiabetic, anti-inflammatory, antioxidant, immunomodulatory, and anticancer activity. In this review, we discussed in vitro antimicrobial activity of essential oils asper plant family against drug-resistant microbes, most of them potentially result in positive activity against resistant strains. Drug-resistant superbugs and their resistance to antimicrobial conventional drugs are listed in Table 15.1 according to https://www.cdc.gov/drugresistance/biggest-threats.html and other scientific research papers. Table 15.3 summarizes the latest research studies about active essential oils isolated from *Apiaceae*, Aristolochiaceae, Brassicaceae, Moraceae, Myrtaceae, Oleaceae, *Poaceae*, Rutaceae, Schisandraceae, and Zingiberaceae that are known for antibacterial properties. Antifungal, antiprotozoal, and antiviral activities are also summarized in Tables 15.4, 15.5, and 15.6, respectively. Aerial parts, leaves, and seeds were used for the isolation of essential oils that were active against several strains of *Candida* spp., *Trichophyton* spp., and *Aspergillus* spp. Several protozoal species *L. chagasi*, *L. braziliensis*, *L. amazonensis*, *L. tropica*, *L. infantum*, *Trypanosoma cruzi*, and *Plasmodium falciparum* are inhibited by essential oil treatment. Finally, it is worth mentioning that the antimicrobial activities of essential oils are not best for all drug-resistant strains. 1,8-Cineole, carvacrol, eugenol, α-terpinene, γ-terpinene, thymol, *p*-cymene, *o*-cymene, α − santalol, α-pinene, β-pinene, b-caryophyllene, iso-caryophyllene, b-linalool, phytol, piperitone, trans-p-mentha-2, 8-dien-1-ol, and 2-undecanone are some major chemical compounds screened in this review. These compounds are synthesized through mevalonic acid, methyl-d-erythritol-4-phosphate, and malonic acid pathways.

Additionally, screening of essential oils will give an idea about potent essential oils that can further be studied in clinical studies or turned into major particles in drug development. Essential oils have been screened against several bacteria, fungi, protozoa, and viruses. In recent years, coronavirus SARS-CoV-2 strain outbreaks, and one study by da Silva et al. hypothesized that components present in essential oils may synergistically act as antiviral agents or may give relief to COVID-19. They reported a molecular docking study of 171 essential oil components with the SARS-CoV-2 main protease. No conventional drug is present for COVID-19 strains, but remdesivir received permission for treatment as an "emergency use authorization" (da Silva et al. 2018).

For the production of plant-based medicines, the most important challenge is to produce pure and cost-effective commercial drugs. Metagenomics, molecular biotechnology, nanotechnology, and in silico studies are some recent fields for the development of new drugs. We need to focus in in vivo studies too by applying significant plant essential oils against MDR microbes in the future. Hence, superior strategies must be needed without any delay to reverse AMR among microbe organisms and also to reinforce the drug discovery pipeline leading to worldwide disease control or prevention and management.

References

Ahmad Kamal HZ, Tuan Ismail TNN, Arief EM, Ponnuraj KT (2020) Antimicrobial activities of citronella (Cymbopogon nardus) essential oil against several oral pathogens and its volatile compounds. Padjadjaran J Dent 32:24966. https://doi.org/10.24198/pjd.vol32no1.24966

Ahmad A, Khan A, Manzoor N (2013) Reversal of efflux mediated antifungal resistance underlies synergistic activity of two monoterpenes with fluconazole. Eur J Pharm Sci 48:80–86. https://doi.org/10.1016/j.ejps.2012.09.016

Akinsolu FT, Nemieboka PO, Njuguna DW, Ahadji MN, Dezso D, Varga O (2019) Emerging resistance of neglected tropical diseases: a scoping review of the literature. Int J Environ Res Public Health. https://doi.org/10.3390/ijerph16111925

Akono Ntonga P, Baldovini N, Mouray E, Mambu L, Belong P, Grellier P (2014) Activity of Ocimum basilicum, Ocimum canum, and Cymbopogon citratus essential oils against Plasmodium falciparum and mature-stage larvae of Anopheles funestus. Parasite 21:–33. https://doi.org/10.1051/parasite/2014033

Alitonou GA, Avlessi F, Tchobo F, Noudogbessi J, Tonouhewa A, Yehouenou B, Menut C, Sohounhloue DK (2012) Chemical composition and biological activities of essential oils from the leaves of Cymbopogon giganteus Chiov. and Cymbopogon schoenanthus (L.) Spreng (Poaceae) from Benin. Int J Biol Chem Sci 6:1819–1827

Almaghrabi MK, Joseph MRP, Assiry MM, Hamid ME (2018) Multidrug-resistant Acinetobacter baumannii: an emerging health threat in Aseer Region, Kingdom of Saudi Arabia. Can J Infect Dis Med Microbiol 2018. https://doi.org/10.1155/2018/9182747

Almeida RS, Freitas PR, Araújo ACJ, Alencar Menezes IR, Santos EL, Tintino SR et al (2020) GC-MS profile and enhancement of antibiotic activity by the essential oil of Ocotea odorífera and safrole: inhibition of Staphylococcus aureus efflux pumps. Antibiotics 9:247. https://doi.org/10.3390/antibiotics9050247

Arámbula CI, Diaz CE, Garcia MI (2019) Performance, chemical composition and antibacterial activity of the essential oil of Ruta chalepensis and Origanum vulgare. J Phys Conf Ser. https://doi.org/10.1088/1742-6596/1386/1/012059

Arumugam G, Swamy MK, Sinniah UR (2016) Plectranthus amboinicus (Lour.) Spreng: botanical, phytochemical, pharmacological and nutritional significance. Molecules 21:369. https://doi.org/10.3390/molecules21040369

Bachir RG, Benali M (2012) Antibacterial activity of the essential oils from the leaves of Eucalyptus globulus against Escherichia coli and Staphylococcus aureus. Asian Pac J Trop Biomed 2:739–742. https://doi.org/10.1016/S2221-1691(12)60220-2

Bajpai VK, Sharma A, Baek KH (2013) Antibacterial mode of action of Cudrania tricuspidata fruit essential oil, affecting membrane permeability and surface characteristics of food-borne pathogens. Food Control 32:582–590. https://doi.org/10.1016/j.foodcont.2013.01.032

Bakkali F, Averbeck S, Averbeck D, Zhiri A, Baudoux D, Idaomar M (2006) Antigenotoxic effects of three essential oils in diploid yeast (Saccharomyces cerevisiae) after treatments with UVC radiation, 8-MOP plus UVA and MMS. Mutat Res 606:27–38. https://doi.org/10.1016/j. mrgentox.2006.02.005

Bansal D, Malla N, Mahajan RC (2006) Drug resistance in amoebiasis. Indian J Med Res 123:115–118

Barnes BE, Sampson DA (2011) A literature review on community-acquired methicillin-resistant Staphylococcus aureus in the United States: clinical information for primary care nurse practitioners. J Am Acad Nurse Pract 23:23–32. https://doi.org/10.1111/j.1745-7599.2010. 00571.x

Béjaoui A, Chaabane H, Jemli M, Boulila A, Boussaid M (2013) Essential oil composition and antibacterial activity of Origanum vulgare subsp. glandulosum Desf. at different phenological stages. J Med Food 16:1115–1120. https://doi.org/10.1089/jmf.2013.0079

Benarba B (2016) Medicinal plants used by traditional healers from South-West Algeria: an ethnobotanical study. J Intercult Ethnopharmacol 5:320–330. https://doi.org/10.5455/jice. 20160814115725

Bhat AS, Ahangar AA (2007) Methods for detecting chemical-chemical interaction in toxicology. Toxicol Mech Methods 17:441–450. https://doi.org/10.1080/15376510601177654

Bloland PB, Bloland PB (2001) Drug resistance in malaria. Malar. Epidemology Branch

Boukhatem MN, Ferhat MA, Kameli A, Saidi F, Kebir HT (2014) Lemon grass (cymbopogon citratus) essential oil as a potent anti-inflammatory and antifungal drugs. Libyan J Med 9:1–10. https://doi.org/10.3402/ljm.v9.25431

Bourne KZ, Bourne N, Reising SF, Stanberry LR (1999) Plant products as topical microbicide candidates: assessment of in vitro and in vivo activity against herpes simplex virus type 2. Antivir Res 42:219–226. https://doi.org/10.1016/S0166-3542(99)00020-0

Bouyahya A, Abrini J, Dakka N, Bakri Y (2019) Essential oils of Origanum compactum increase membrane permeability, disturb cell membrane integrity, and suppress quorum-sensing pheno-type in bacteria. J Pharm Anal 9:301–311. https://doi.org/10.1016/j.jpha.2019.03.001

Braga LC, Leite AAM, Xavier KGS, Takahashi JA, Bemquerer MP, Chartone-Souza E, Nascimento AMA (2005) Synergic interaction between pomegranate extract and antibiotics against Staphylococcus aureus. Can J Microbiol 51:541–547. https://doi.org/10.1139/w05-022

Brockhouse M, Scott L (2020) Identification of antibiotic producing soil bacteria against Bacillus subtilis Morgan Brockhouse and Dr. Lori Scott

Burt SA (2004) Antibacterial activity of essential oils : potential applications in food. Int J Food Microbiol 94:223–253

Cagno V, Sgorbini B, Sanna C, Cagliero C, Ballero M, Civra A, Donalisio M, Bicchi C, Lembo D, Rubiolo P (2017) In vitro anti-herpes simplex virus-2 activity of Salvia desoleana Atzei & V. Picci essential oil. PLoS One 12:1–12. https://doi.org/10.1371/journal.pone.0172322

Chambers HF, DeLeo FR (2009) Waves of resistance: Staphylococcus aureus in the antibiotic era. Nat Rev Microbiol 7:629–641. https://doi.org/10.1038/nrmicro2200

Chang Y, Yoon H, Kang D, Chang P, Ryu S (2016) Endolysin LysSA97 is synergistic with carvacrol in controlling Staphylococcus aureus in foods. Int J Food Microbiol 224:19–26. https://doi.org/10.1016/j.ijfoodmicro.2016.12.007

Chaves TP, Pinheiro REE, Melo ES (2018) Essential oil of Eucalyptus camaldulensis Dehn potentiates β-lactam activity against Staphylococcus aureus and Escherichia coli resistant strains. Ind Crop Prod 112:70–74. https://doi.org/10.1016/j.indcrop.2017.10.048

Cheikh-Ali Z, Adiko M, Bouttier S, Bories C, Okpekon T, Poupon E, Champy P (2011) Composi-tion, and antimicrobial and remarkable antiprotozoal activities of the essential oil of rhizomes of Aframomum sceptrum K. Schum. (Zingiberaceae). Chem Biodivers 8:658–667. https://doi.org/ 10.1002/cbdv.201000216

Chen Y, Huang S, Sun F, Chiang Y, Chiang C, Tsai C, Weng C (2011) Transformation of cinnamic acid from trans- to cis-form raises a notable bactericidal and synergistic activity against multiple-drug resistant Mycobacterium tuberculosis. Eur J Pharm Sci 43:188–194. https://doi.org/10.1016/j.ejps.2011.04.012

Chen Y, Zeng H, Tian J, Ban X, Ma B, Wang Y (2013) Antifungal mechanism of essential oil from Anethum graveolens seeds against Candida albicans. J Med Microbiol 62:1175–1183. https://doi.org/10.1099/jmm.0.055467-0

Chovanova R, Mezovska J, Vaverková Š, Mikulášová M (2015) The inhibition the Tet (K) efflux pump of tetracycline resistant S taphylococcus epidermidis by essential oils from three S alvia species. Lett Appl Microbiol 61:58–62. https://doi.org/10.1111/lam.12424

Citron DM, Appleman MD (2006) In vitro activities of daptomycin, ciprofloxacin, and other antimicrobial agents against the cells and spores of clinical isolates of Bacillus species. J Clin Microbiol 44:3814–3818. https://doi.org/10.1128/JCM.00881-06

Civitelli L, Panella S, Marcocci ME, De Petris A, Garzoli S, Pepi F, Vavala E, Ragno R, Nencioni L, Palamara AT, Angiolella L (2014) In vitro inhibition of herpes simplex virus type 1 replication by Mentha suaveolens essential oil and its main component piperitenone oxide. Phytomedicine 21:857–865. https://doi.org/10.1016/j.phymed.2014.01.013

Cortez KJ, Maldarelli F (2011) Clinical management of HIV drug resistance. Viruses. https://doi.org/10.3390/v3040347

Coutinho HD, Costa JG, Falcão-Silva VS, Siqueira-Júnior JP, Lima EO (2010) Effect of Momordica charantia L. in the resistance to aminoglycosides in methicilin-resistant Staphylococcus aureus. Comp Immunol Microbiol Infect Dis 33:467–471. https://doi.org/10.1016/j.cimid.2009.08.001

Cox SD, Mann CM, Markham JL, Bell HC, Gustafson JE, Warmington JR, Wyllie SG (2000) The mode of antimicrobial action of the essential oil of Melaleuca alternifolia (Tea tree oil). J Appl Microbiol 88:170–175. https://doi.org/10.1046/j.1365-2672.2000.00943.x

da Silva VD, Almeida-Souza F, Teles AM, Neto PA, Mondego-Oliveira R, Mendes Filho NE, Taniwaki NN, Abreu-Silva AL (2018) Chemical composition of Ocimum canum Sims. essential oil and the antimicrobial, antiprotozoal and ultrastructural alterations it induces in Leishmania amazonensis promastigotes. Ind Crop Prod 119:201–208. https://doi.org/10.1016/j.indcrop.2018.04.005

Damtie D, Mekonnen Y (2016) Thymus species in Ethiopia: Distribution, medicinal value, economic benefit, current status and threatening factors. Ethiop J Sci Technol 8:81. https://doi.org/10.4314/ejst.v8i2.3

de Morais Oliveira-Tintino CD, Tintino SR, Limaverde PW, Figueredo FG, Campina FF, da Cunha FAB, da Costa RHS, Pereira PS, Lima LF, de Matos YMLS, Coutinho HDM, Siqueira-Júnior JP, Balbino VQ, da Silva TG (2018) Inhibition of the essential oil from Chenopodium ambrosioides L. and α-terpinene on the NorA efflux-pump of Staphylococcus aureus. Food Chem 262:72–77. https://doi.org/10.1016/j.foodchem.2018.04.040

De Sousa EO, Rodrigues FFG, Campos AR, Lima SG, Da Costa JGM (2013) Chemical composition and synergistic interaction between aminoglycosides antibiotics and essential oil of Lantana montevidensis Briq. Nat Prod Res 27:942–945. https://doi.org/10.1080/14786419.2012.678351

Degenhardt J, Köllner TG, Gershenzon J (2009) Monoterpene and sesquiterpene synthases and the origin of terpene skeletal diversity in plants. Phytochemistry 70:1621–1637. https://doi.org/10.1016/j.phytochem.2009.07.030

Delcour AH (2009) Outer membrane permeability and antibiotic resistance. Biochim Biophys Acta 1794:808–816. https://doi.org/10.1016/j.bbapap.2008.11.005

Devi KP, Nisha SA, Sakthivel R, Pandian SK (2010) Eugenol (an essential oil of clove) acts as an antibacterial agent against Salmonella typhi by disrupting the cellular membrane. J Ethnopharmacol 130:107–115. https://doi.org/10.1016/j.jep.2010.04.025

Dhifi W, Bellili S, Jazi S, Bahloul N, Mnif W (2016) Essential oils' chemical characterization and investigation of some biological activities: a critical review. Medicine 3:25. https://doi.org/10.3390/medicines3040025

Diao WR, Hu QP, Zhang H, Xu JG (2014) Chemical composition, antibacterial activity and mechanism of action of essential oil from seeds of fennel (Foeniculum vulgare Mill.). Food Control 35:109–116. https://doi.org/10.1016/j.foodcont.2013.06.056

Doughari JH, Ndakidemi PA, Human IS, Benade S (2012) Antioxidant, antimicrobial and antiverotoxic potentials of extracts of Curtisia dentata. J Ethnopharmacol 141:1041–1050. https://doi.org/10.1016/j.jep.2012.03.051

Duarte A, Ferreira S, Silva F, Domingues FC (2012) Synergistic activity of coriander oil and conventional antibiotics against Acinetobacter baumannii. Phytomedicine 19:236–238. https://doi.org/10.1016/j.phymed.2011.11.010

Duran N, Kaya DA (2018) Chemical composition of essential oils from Origanum onites L. and Cymbopogon citratus, and their synergistic effects with Acyclovir against HSV-1, pp 243–248. https://doi.org/10.24264/icams-2018.iv.1

Duschatzky CB, Possetto ML, Talarico LB, García CC, Michis F, Almeida NV, De Lampasona MP, Schuff C, Damonte EB (2005) Evaluation of chemical and antiviral properties of essential oils from South American plants. Antivir Chem Chemother 16:247–251. https://doi.org/10.1177/095632020501600404

Dutta BK, Karmakar S, Naglot A, Aich JC, Begam M (2007) Anticandidial activity of some essential oils of a mega biodiversity hotspot in India. Mycoses 50:121–124. https://doi.org/10.1111/j.0933-7407.2006.01332.x

ECDC/EMEA (2009) ECDC/EMEA joint technical report: the bacterial challenge: time to react. Stockholm. http://www.ecdc.europa.eu/en/publications/Publications/Forms/ECDC_DispForm.aspx?ID=444

El-badry AA, Al-ali KH, El-badry YA (2010) Activity of Mentha Longifolia and Ocimum Basilicum against Entamoeba Histolytica and Giardia Duodenalis. Sci Parasitol 11:109–117

El-baz FK, Mahmoud K, El-senousy WM, Darwesh OM, Elgohary AE (2015) Antiviral – antimicrobial and schistosomicidal activities of Eucalyptus camaldulensis essential oils. Int J Pharm Sci 31:262–268

El-Shouny WA, Ali SS, Sun J, Samy SM, Ali A (2018) Drug resistance profile and molecular characterization of extended spectrum beta-lactamase (ESβL)-producing Pseudomonas aeruginosa isolated from burn wound infections. Essential oils and their potential for utilization. Microb Pathog 116:301–312. https://doi.org/10.1016/j.micpath.2018.02.005

Fadli M, Chevalier J, Saad A, Mezrioui NE, Hassani L, Pages JM (2011) Essential oils from Moroccan plants as potential chemosensitisers restoring antibiotic activity in resistant Gram-negative bacteria. Int J Antimicrob Agents 38:325–330. https://doi.org/10.1016/j.ijantimicag.2011.05.005

Figueredo FG, Lima LF, Oliveira CDDM, De Matos YMLS, Morais-braga MFB, Irwin RA, Balbino VQ, Coutinho HDM, Siqueira-júnior JP (2017) Inhibition of the TetK efflux-pump by the essential oil of Chenopodium ambrosioides L. and α-terpinene against Staphylococcus aureus IS-58. Food Chem Toxicol 109:957–961. https://doi.org/10.1016/j.fct.2017.02.031

Garbati MA, Al Godhair AI (2013) The growing resistance of klebsiella pneumoniae; the need to expand our antibiogram: case report and review of the literature section of infectious diseases. Afr J Infect Dis 7:8–10

Gavanji S, Sayedipour SS, Larki B, Bakhtari A (2015) Antiviral activity of some plant oils against herpes simplex virus type 1 in Vero cell culture. J Acute Med 5:62–68. https://doi.org/10.1016/j.jacme.2015.07.001

George AJ, Gordon LA (1991) New mechanisms of bacterial resistance to antimicrobial agents. N Engl J Med 329:977–986. https://doi.org/10.1056/NEJM199309303291401

Ghalem BR, Mohamed B (2008) Antibacterial activity of leaf essential oils of Eucalyptus globulus and Eucalyptus camaldulensis. Afr J Pharm Pharmacol 2:211–215

Ghannoum MA, Rice LB (1999) Antifungal agents: mode of action, mechanisms of resistance, and correlation of these mechanisms with bacterial resistance. Clin Microbiol Rev 12:501–517. https://doi.org/10.1128/cmr.12.4.501

Gibbons S, Oluwatuyi M, Veitch NC, Gray AI (2003) Bacterial resistance modifying agents from Lycopus europaeus. Phytochemistry 62:83–87

Gilbert S, Corey L, Cunningham A et al (2007) An update on short-course intermittent and prevention therapies for herpes labialis. Herpes 14(suppl 1):13A–18A

Gomes Vidal Sampaio M, Bezerra Dos Santos CR, Cortez Sombra Vandesmet L, Souza Dos Santos B, Bianca Da Silva Santos I (2019) Chemical composition, antioxidant and antiprotozoal activity of Eugenia gracillima Kiaersk. leaves essential oil. Nat Prod Res 2019:1–5. https://doi.org/10.1080/14786419.2019.1644506

Guerra FQS, Mendes JM, De Sousa JP, Morais-Braga MFB, Santos BHC, Melo Coutinho HD, Lima EDO (2012) Increasing antibiotic activity against a multidrug-resistant Acinetobacter spp by essential oils of Citrus limon and Cinnamomum zeylanicum. Nat Prod Res 26:2235–2238. https://doi.org/10.1080/14786419.2011.647019

Guillot S, Descours G, Gillet Y, Etienne J, Floret D, Guiso N (2012) Macrolide-resistant bordetella pertussis infection in newborn girl. France Emerg Infect Dis 18:966–968. https://doi.org/10.3201/eid1806.120091

Hammi KM, Essid R, Tabbene O, Elkahoui S, Majdoub H, Ksouri R (2020) Antileishmanial activity of Moringa oleifera leaf extracts and potential synergy with amphotericin B. S Afr J Bot 129:67–73. https://doi.org/10.1016/j.sajb.2019.01.008

Hansra NK, Shinkai K (2011) Cutaneous community-acquired and hospital-acquired methicillin-resistant Staphylococcus aureus. Dermatol Ther 24:263–272. https://doi.org/10.1111/j.1529-8019.2011.01402.x

Hemaiswarya S, Doble M (2009) Synergistic interaction of eugenol with antibiotics against Gram negative bacteria. Phytomedicine 16:997–1005. https://doi.org/10.1016/j.phymed.2009.04.006

Hermsen ED, MacGeorge EL, Andresen ML, Myers LM, Lillis CJ, Rosof BM (2020) Decreasing the peril of antimicrobial resistance through enhanced health literacy in outpatient settings: an underrecognized approach to advance antimicrobial stewardship. Adv Ther 37:918–932. https://doi.org/10.1007/s12325-019-01203-1

Howard SJ, Arendrup MC (2011) Acquired antifungal drug resistance in Aspergillus fumigatus: epidemiology and detection. Med Mycol 49:90–95. https://doi.org/10.3109/13693786.2010.508469

Hurt AC (2014) The epidemiology and spread of drug resistant human influenza viruses. Curr Opin Virol 8:22–29. https://doi.org/10.1016/j.coviro.2014.04.009

Hussain M, Galvin HD, Haw TY, Nutsford AN, Husain M (2017) Drug resistance in influenza a virus: the epidemiology and management. Infect Drug Resist 10:121–134. https://doi.org/10.2147/IDR.S105473

Igwaran A, Iweriebor BC, Ofuzim Okoh S, Nwodo UU, Obi LC, Okoh AI (2017) Chemical constituents, antibacterial and antioxidant properties of the essential oil flower of Tagetes minuta grown in Cala community Eastern Cape, South Africa. BMC Complement Altern Med 17. https://doi.org/10.1186/s12906-017-1861-6

Jensen JS, Cusini M, Gomberg M, Moi H (2016) 2016 European guideline on Mycoplasma genitalium infections. J Eur Acad Dermatol Venereol 30:1650–1656. https://doi.org/10.1111/jdv.13849

Jha S, Srivastava AK (2012) Screening of antibacterial activity of the essential oil from seed of artocarpus heterophyllus. Int J Educ Res 2:92–96

Jnaid Y, Yacoub R, Al-Biski F (2016) Antioxidant and antimicrobial activities of Origanum vulgare essential oil. Int Food Res J 23:1706–1710

Kačániová M, Vukovič N, Hleba L, Bobková A, Rovná K, Arpášová H (2012) Antimicrobial and antiradicals activity of origanum vulgare l. and thymus vulgaris essential oils. J Sci Technol 2:263–271. https://doi.org/10.4314/star.v4i3.27

Karami F, Dastan D, Fallah M, Matini M (2019) In vitro activity of foeniculum vulgare and its main essential oil component trans-anethole on trichomonas vaginalis. Iran J Parasitol 14:631–638. https://doi.org/10.18502/ijpa.v14i4.2106

Karpanen TJ, Worthington T, Hendry ER, Conway BR, Lambert PA (2008) Antimicrobial efficacy of chlorhexidine digluconate alone and in combination with eucalyptus oil, tea tree oil and thymol against planktonic and biofilm cultures of Staphylococcus epidermidis. J Antimicrob Chemother 62:1031–1036. https://doi.org/10.1093/jac/dkn325

Khan MSA, Ahmad I (2011) Antifungal activity of essential oils and their synergy with fluconazole against drug-resistant strains of Aspergillus fumigatus and Trichophyton rubrum. Appl Microbiol Biotechnol 90:1083–1094. https://doi.org/10.1007/s00253-011-3152-3

Khan MSA, Zahin M, Hasan S, Husain FM, Ahmad I (2009) Inhibition of quorum sensing regulated bacterial functions by plant essential oils with special reference to clove oil. Lett Appl Microbiol 49:354–360. https://doi.org/10.1111/j.1472-765X.2009.02666.x

Khosravi AR, Shokri H, Kermani S, Dakhili M, Madani M, Parsa S (2011) Antifungal properties of Artemisia sieberi and Origanum vulgare essential oils against Candida glabrata isolates obtained from patients with vulvovaginal candidiasis. J Mycol Med 21:93–99. https://doi.org/10.1016/j.mycmed.2011.01.006

Kimbi HK, Fagbenro-Beyioku AF (1996) Efficacy of Cymbopogon giganteus and Enantia chrantha against chloroquine resistant Plasmodium yoelii nigeriensis. East Afr Med J 73(10):636–637

Knezevic P, Aleksic V, Simin N, Svircev E, Petrovic A, Mimica-Dukic N (2016) Antimicrobial activity of Eucalyptus camaldulensis essential oils and their interactions with conventional antimicrobial agents against multi-drug resistant Acinetobacter baumannii. J Ethnopharmacol 178:125–136. https://doi.org/10.1016/j.jep.2015.12.008

Koch C, Reichling J, Schneele J, Schnitzler P (2008) Inhibitory effect of essential oils against herpes simplex virus type 2. Phytomedicine 15:71–78. https://doi.org/10.1016/j.phymed.2007.09.003

Konuk HB, Erguden B (2018) Antifungal activity of various essential oils against Saccharomyces cerevisiae depends on disruption of cell membrane integrity. Biocell 41:13–18. https://doi.org/10.32604/biocell.2017.00013

Ksiezopolska E, Gabaldón T (2018) Evolutionary emergence of drug resistance in candida opportunistic pathogens. Gene 9:90461. https://doi.org/10.3390/genes9090461

Latifah-Munirah B, Himratul-Aznita WH, Mohd Zain N (2015) Eugenol, an essential oil of clove, causes disruption to the cell wall of Candida albicans (ATCC 14053). Front Life Sci 8:231–240. https://doi.org/10.1080/21553769.2015.1045628

Lazcano A, Carson CF, Mee BJ, Riley TV (2002) Mechanism of action of Melaleuca alternifolia (Tea Tree) oil on Staphylococcus aureus determined by time-kill, lysis, leakage, and salt tolerance assays and electron microscopy. Antimicrob Agents Chemother 46:1914–1920. https://doi.org/10.1128/AAC.46.6.1914

Lee MS, Choi J, Posadzki P, Ernst E (2012) Aromatherapy for health care: an overview of systematic reviews. Maturitas 71:257–260. https://doi.org/10.1016/j.maturitas.2011.12.018

Li Y, Zhang Y, Zhang C, Wang H, Wei X, Chen P, Lu L (2020) Mitochondrial dysfunctions trigger the calcium signaling-dependent fungal multidrug resistance. Proc Natl Acad Sci U S A 117:1711–1721. https://doi.org/10.1073/pnas.1911560116

Machado M, Martins N, Salgueiro L, Cavaleiro C, Sousa MC (2019) Lavandula luisieri and Lavandula viridis essential oils as upcoming anti-protozoal agents: a key focus on leishmaniasis. Appl Sci 9. https://doi.org/10.3390/app9153056

Mahboubi M, Bidgoli FG (2010) In vitro synergistic efficacy of combination of amphotericin B with Myrtus communis essential oil against clinical isolates of Candida albicans. Phytomedicine 17:771–774. https://doi.org/10.1016/j.phymed.2010.01.016

Mahboubi M, Ghazian Bidgoli F (2010) Antistaphylococcal activity of Zataria multiflora essential oil and its synergy with vancomycin. Phytomedicine 17:548–550. https://doi.org/10.1016/j.phymed.2009.11.004

Mamishi S, Moradkhani S, Mahmoudi S, Sadeghi RH (2014) Penicillin-resistant trend of Streptococcus pneumoniae in Asia: a systematic review. Microbiology 6:198–210

Manchanda V, Sinha S, Singh N (2010) Multidrug resistant acinetobacter. J Global Infect Dis 2:291. https://doi.org/10.4103/0974-777x.68538

Martinez-Rossi NM, Bitencourt TA, Peres NTA, Lang EAS, Gomes EV, Quaresemin NR, Martins MP, Lopes L, Rossi A (2018) Dermatophyte resistance to antifungal drugs: mechanisms and prospectus. Front Microbiol 9:1–18. https://doi.org/10.3389/fmicb.2018.01108

Masood N, Chaudhry A, Tariq P (2006) Anti-microbial activity of cinnamomum cassia against diverse microbial flora with its nutritional and medicinal impacts. Pak J Bot 38:169–174

McFadden DC, Tomavo S, Berry EA, Boothroyd JC (2000) Characterization of cytochrome b from Toxoplasma gondii and Q(o) domain mutations as a mechanism of atovaquone-resistance. Mol Biochem Parasitol 108:1–12. https://doi.org/10.1016/S0166-6851(00)00184-5

Mekonnen A, Yitayew B, Tesema A, Taddese S (2016) In vitro antimicrobial activity of essential oil of Thymus schimperi, Matricaria chamomilla, Eucalyptus globulus, and Rosmarinus officinalis. Int J Microbiol 2016:9545693. https://doi.org/10.1155/2016/9545693

Méndez-Tovar LJ, Manzano-Gayosso P, Velásquez-Hernández V, Millan-Chiu B, Hernández-Hernández F, Mondragón-González R, López-Martínez R (2007) Resistencia a compuestos azólicos de aislamientos clínicos de Trichophyton spp. Rev Iberoam Micol 24:320–322. https://doi.org/10.1016/S1130-1406(07)70065-7

Miller WR, Munita JM, Arias CA (2014) Mechanisms of antibiotic resistance in enterococci. Expert Rev Anti-Infect Ther 12:1221–1236. https://doi.org/10.1586/14787210.2014.956092

Minami M, Kita M, Nakaya T, Yamamoto T, Kuriyama H, Imanishi J (2003) The inhibitory effect of essential oils on herpes simplex virus type-1 replication in vitro. Microbiol Immunol 47:681–684. https://doi.org/10.1111/j.1348-0421.2003.tb03431.x

Monzote L, Montalvo AM, Scull R, Miranda M, Abreu J (2007) Combined effect of the essential oil from Chenopodium ambrosioides and antileishmanial drugs on promastigotes of Leishmania amazonensis. Rev Inst Med Trop 49:257–260. https://doi.org/10.1590/S0036-46652007000400012

Monzote L, García M, Pastor J, Gil L, Scull R, Maes L, Cos P, Gille L (2014) Essential oil from Chenopodium ambrosioides and main components: activity against Leishmania, their mitochondria and other microorganisms. Exp Parasitol 136:20–26. https://doi.org/10.1016/j.exppara.2013.10.007

Morace G, Perdoni F, Borghi E (2014) Antifungal drug resistance in Candida species. J Glob Antimicrob Resist 2:254–259. https://doi.org/10.1016/j.jgar.2014.09.002

Mulyaningsih S, Youns M, El-Readi MZ, Ashour ML, Nibret E, Sporer F, Herrmann F, Reichling J, Wink M (2010) Biological activity of the essential oil of Kadsura longipedunculata (Schisandraceae) and its major components. J Pharm Pharmacol 62:1037–1044. https://doi.org/10.1111/j.2042-7158.2010.01119.x

Muzny CA, Schwebke JR (2013) The clinical spectrum of Trichomonas vaginalis infection and challenges to management. Sex Transm Infect 89:423–425. https://doi.org/10.1136/sextrans-2012-050893

Nagamune K, Moreno SNJ, Sibley LD (2007) Artemisinin-resistant mutants of Toxoplasma gondii have altered calcium homeostasis. Antimicrob Agents Chemother 51:3816–3823. https://doi.org/10.1128/AAC.00582-07

Nazzaro F, Fratianni F, De Martino L, Coppola R, De Feo V (2013) Effect of essential oils on pathogenic bacteria. Pharmaceuticals 6:1451–1474. https://doi.org/10.3390/ph6121451

Ogunlana EO, Hoglund S, Onawunmi G, Skold O (1987) Effects of lemongrass oil on the morphological characteristics and peptidoglycan synthesis of Escherichia coli cells. Microbios 50:43

Oliveira VCS, Moura DMS, Lopes JAD, De Andrade PP, Da Silva NH, Figueiredo RCBQ (2009) Effects of essential oils from Cymbopogon citratus (DC) Stapf., Lippia sidoides Cham., and Ocimum gratissimum L. on growth and ultrastructure of Leishmania chagasi promastigotes. Parasitol Res 104:1053–1059. https://doi.org/10.1007/s00436-008-1288-6

Opalchenova G, Obreshkova D (2003) Comparative studies on the activity of basil - an essential oil from Ocimum basilicum L. - Against multidrug resistant clinical isolates of the genera Staphylococcus, Enterococcus and Pseudomonas by using different test methods. J Microbiol Methods 54:105–110. https://doi.org/10.1016/S0167-7012(03)00012-5

Özkalp B, Sevgi F, Özcan M, Özcan MM (2010) The antibacterial activity of essential oil of oregano (Origanum vulgare L.). Agric Environ 8:272–274

Pachori P, Gothalwal R, Gandhi P (2019) Emergence of antibiotic resistance Pseudomonas aeruginosa in intensive care unit; a critical review. Genes Dis 6:109–119. https://doi.org/10.1016/j.gendis.2019.04.001

Palomino JC, Martin A (2014) Drug resistance mechanisms in Mycobacterium tuberculosis. Antibiotics 3:317–340. https://doi.org/10.3390/antibiotics3030317

Papadopoulos CJ, Carson CF, Chang BJ (2008) Role of the MexAB-OprM efflux pump of Pseudomonas aeruginosa in tolerance to tea tree (Melaleuca alternifolia) oil and its monoterpene components terpinen-4-ol, 1,8-cineole, and -terpineol. Appl Environ Microbiol 74:1932–1935. https://doi.org/10.1128/AEM.02334-07

Parreira NA, Magalhães LG, Morais DR, Caixeta SC, De Sousa JPB, Bastos JK, Cunha WR, Silva MLA, Nanayakkara NPD, Rodrigues V, Da Silva Filho AA (2010) Antiprotozoal, schistosomicidal, and antimicrobial activities of the essential oil from the leaves of baccharis dracunculifolia. Chem Biodivers 7:993–1001. https://doi.org/10.1002/cbdv.200900292

Partridge SR (2015) Resistance mechanisms in Enterobacteriaceae. Pathology 47:276–284. https://doi.org/10.1097/PAT.0000000000000237

Pellet PE, Roizman B (2007) The family Herpesviridae: a brief introduction. In: Knipe DMPH, Griffin DE, Lamb RA, Martin MA, Roizman B, Straus SE (eds) Field' virology, 5th edn. Lippincott-Williams and Wilkins, New York, NY, pp 2479–2499

Peng Z, Jin D, Kim HB, Stratton CW, Wu B, Tang YW, Suna X (2017) Update on antimicrobial resistance in Clostridium difficile: resistance mechanisms and antimicrobial susceptibility testing. J Clin Microbiol 55:1998–2008. https://doi.org/10.1128/JCM.02250-16

Pilau MR, Alves SH, Weiblen R, Arenhart S, Cueto AP, Lovato LT (2011) Antiviral activity of the Lippia graveolens (Mexican oregano) essential oil and its main compound carvacrol against human and animal viruses. Braz J Microbiol 42:1616–1624. https://doi.org/10.1590/S1517-83822011000400049

Pourghanbari G, Nili H, Moattari A, Mohammadi A, Iraji A (2016) Antiviral activity of the oseltamivir and Melissa officinalis L. essential oil against avian influenza A virus (H9N2). Virus Dis 27:170–178. https://doi.org/10.1007/s13337-016-0321-0

Pramanik PK, Alam MN, Roy Chowdhury D, Chakraborti T (2019) Drug resistance in protozoan parasites: an incessant wrestle for survival. J Glob Antimicrob Resist 18:1–11. https://doi.org/10.1016/j.jgar.2019.01.023

Prasch S, Bucar F (2015) Plant derived inhibitors of bacterial efflux pumps: an update. Phytochem Rev 14:961–974. https://doi.org/10.1007/s11101-015-9436-y

Rad JS, Alfatemi MH, Rad MS, Jyoti D (2013) Phytochemical and antimicrobial evaluation of the essential oils and antioxidant activity of aqueous extracts from flower and stem of Sinapis arvensis L. Am J Adv Drug Deliv 1:001–100

Rodrigues FFG, Costa JGM, Coutinho HDM (2009) Synergy effects of the antibiotics gentamicin and the essential oil of Croton zehntneri. Phytomedicine 16:1052–1055. https://doi.org/10.1016/j.phymed.2009.04.004

Rosato A, Vitali C, De Laurentis N, Armenise D, Antonietta Milillo M (2007) Antibacterial effect of some essential oils administered alone or in combination with Norfloxacin. Phytomedicine 14:727–732. https://doi.org/10.1016/j.phymed.2007.01.005

Saddi M, Sanna A, Cottiglia F, Chisu L, Casu L, Bonsignore L, De Logu A (2007) Antimicrobials Antiherpevirus activity of Artemisia arborescens essential oil and inhibition of lateral diffusion in Vero cells. Ann Clin Microbiol Antimicrob 6. https://doi.org/10.1186/1476-0711-6-10

Sadhasivam S, Palanivel S, Ghosh S (2016) Synergistic antimicrobial activity of Boswellia serrata Roxb. ex Colebr. (Burseraceae) essential oil with various azoles against pathogens associated with skin, scalp and nail infections. Lett Appl Microbiol 63:495–501. https://doi.org/10.1111/lam.12683

Salleh WMNHW, Ahmad F, Yen KH (2014) Chemical composition and antimicrobial activity of essential oil of piper muricatum blume (Piperaceae). J Essent Oil-Bearing Plants 17:1329–1334. https://doi.org/10.1080/0972060X.2014.960271

Santoro GF, Das Graças Cardoso M, Guimarães LGL, Salgado APSP, Menna-Barreto RFS, Soares MJ (2007) Effect of oregano (Origanum vulgare L.) and thyme (Thymus vulgaris L.) essential oils on Trypanosoma cruzi (Protozoa: Kinetoplastida) growth and ultrastructure. Parasitol Res 100:783–790. https://doi.org/10.1007/s00436-006-0326-5

Santoyo S, Cavero S, Jaime L, Ibañez E, Señoráns FJ, Reglero G (2006) Supercritical carbon dioxide extraction of compounds with antimicrobial activity from Origanum vulgare L.: determination of optimal extraction parameters. J. Food Prot 69:369–375. https://doi.org/10.4315/0362-028X-69.2.369

Sarma S, Upadhyay S (2017) Current perspective on emergence, diagnosis and drug resistance in Candida auris. Infect Drug Resist 10:155–165. https://doi.org/10.2147/IDR.S116229

Saviuc C, Gheorghe I, Coban S, Drumea V, Chifiriuc MC, Banu O, Bezirtzoglou E, Lazăr V (2016) Rosmarinus Officinalis essential oil and eucalyptol act as efflux pumps inhibitors and increase ciprofloxacin efficiency against Pseudomonas Aeruginosa and Acinetobacter Baumannii MDR strains. Rom Biotechnol Lett 21:11796–11804

Schiaffino F, Colston JM, Paredes-Olortegui M, François R, Pisanic N, Burga R, Peñataro-Yori P, Kosek MN (2019) Antibiotic resistance of Campylobacter species in a pediatric cohort study. Antimicrob Agents Chemother 63:01911. https://doi.org/10.1128/AAC.01911-18

Schuhmacher A, Reichling J, Schnitzler P (2003) Virucidal effect of peppermint oil on the enveloped viruses herpes simplex virus type 1 and type 2 in vitro. Phytomedicine 10:504–510. https://doi.org/10.1078/094471103322331467

Setzer WN (2016) Essential oils as complementary and alternative medicines for the treatment of influenza. Am J Essent Oils Nat Prod 4:16–22

Sharifi-Rad JB, Salehi PS, Ayatollahi SA, Kobarfard F, Eisazadeh MF, Sharifi-Rad M (2017) Susceptibility of herpes simplex virus type 1 to monoterpenes thymol, carvacrol, p-cymene and essential oils of Sinapis arvensis L., Lallemantia royleana Benth. and Pulicaria vulgaris Gaertn. Cell Mol Biol 63:42–47

Sharma N, Tripathi A (2008) Effects of Citrus sinensis (L.) Osbeck epicarp essential oil on growth and morphogenesis of Aspergillus niger (L.) Van Tieghem. Microbiol Res 163:337–344. https://doi.org/10.1016/j.micres.2006.06.009

Sharma D, Misba L, Khan AU (2019) Antibiotics versus biofilm: an emerging battleground in microbial communities. Antimicrob Resist Infect Control 3:1–10

Si H, Hu J, Liu Z, Zeng ZL (2008) Antibacterial effect of oregano essential oil alone and in combination with antibiotics against extended-spectrum β-lactamase-producing Escherichia coli. FEMS Immunol Med Microbiol 53:190–194. https://doi.org/10.1111/j.1574-695X.2008.00414.x

Silva F, Ferreira S, Duarte A, Mendona DI, Domingues FC (2011) Antifungal activity of Coriandrum sativum essential oil, its mode of action against Candida species and potential synergism with amphotericin B. Phytomedicine 19:42–47. https://doi.org/10.1016/j.phymed.2011.06.033

Sinniah UR (2015) A comprehensive review on the phytochemical constituents and pharmacological activities of Pogostemon cablin Benth.: an aromatic medicinal plant of industrial importance. Molecules 20:8521–8547. https://doi.org/10.3390/molecules20058521

Siqueira CAT, Oliani J, Sartoratto A, Queiroga CL, Moreno PRH, Reimão JQ, Tempone AG, Fischer DCH (2011) Chemical constituents of the volatile oil from leaves of Annona coriacea and in vitro antiprotozoal activity. Braz J Pharmacogn 21:33–40. https://doi.org/10.1590/S0102-695X2011005000004

Sivropoulou A, Nikolaou C, Papanikolaou E, Kokkini S, Lanaras T, Arsenakis M (1997) Antimicrobial, cytotoxic, and antiviral activities of salvia fructicosa essential oil. J Agric Food Chem 45:3197–3201. https://doi.org/10.1021/jf970031m

Smith ECJ, Williamson EM, Wareham N, Kaatz GW, Gibbons S (2007) Antibacterials and
    modulators of bacterial resistance from the immature cones of Chamaecyparis lawsoniana.
    Phytochemistry 68:210–217. https://doi.org/10.1016/j.phytochem.2006.10.001
Solano C, Echeverz M (2014) Biofilm dispersion and quorum sensing. Curr Opin Microbiol
    18:96–104. https://doi.org/10.1016/j.mib.2014.02.008
Soliman SSM, Alsaadi AI, Youssef EG, Khitrov G, Noreddin AM, Husseiny MI, Ibrahim AS
    (2017) Calli essential oils synergize with lawsone against multidrug resistant pathogens.
    Molecules 22:1–13. https://doi.org/10.3390/molecules22122223
Strasfeld L, Chou S (2010) Antiviral drug resistance: Mechanisms and clinical implications. Infect
    Dis Clin N Am 24:809–833. https://doi.org/10.1016/j.idc.2010.07.001
Su YC, Hsu KP, Ho CL (2017) Composition, in vitro antibacterial and anti-mildew fungal activities
    of essential oils from twig and fruit parts of eucalyptus citriodora. Nat Prod Commun
    12:1647–1650. https://doi.org/10.1177/1934578x1701201031
Suppiah J, Zain RM, Nawi SH, Bahari N, Saat Z (2014) Drug-resistance associated mutations in
    polymerase (P) gene of hepatitis B virus isolated from Malaysian HBV carriers. Hepat Mon 14.
    https://doi.org/10.5812/hepatmon.13173
Swamy MK, Sinniah UR, Akhtar MS (2015) In vitro pharmacological activities and GC-ms
    analysis of different solvent extracts of Lantana camara leaves collected from tropical region
    of Malaysia. Altern Med. https://doi.org/10.1155/2015/506413
Szalewski D, Hinrichs VS, Zinniel DK, Barletta RG (2018) The pathogenicity of Aspergillus
    fumigatus, drug resistance and 2 nanoparticle delivery. Can J Microbiol 64:3
Tanwar J, Das S, Fatima Z, Hameed S (2014) Multidrug resistance: an emerging crisis. Interdiscip
    Perspect Infect Dis 2014. https://doi.org/10.1155/2014/541340
Tasdemir D, Kaiser M, Demirci B, Demirci F, Baser K (2019) Antiprotozoal activity of Turkish
    Origanum onites essential oil and its components. Molecules 24:4421
Theresa LS, Pearson ML, Wilcox KR, Cruz C, Lancaster MV, Robinson-Dunn B, Tenover FC,
    Zervos MJ, Band JD, White E, Jarvis WR (1999) Emergence of vancomycin resistance in
    Staphylococcus aureus. N Engl J Med 340:493–501
Tolba H, Moghrani H, Benelmouffok A, Kellou D, Maachi R (2015) Essential oil of Algerian
    Eucalyptus citriodora: chemical composition, antifungal activity. J Mycol Med 25:128–133.
    https://doi.org/10.1016/j.mycmed.2015.10.009
Turina AV, Nolan MV, Zygadlo JA, Perillo MA (2006) Natural terpenes: self-assembly and
    membrane partitioning. Biophys Chem 122:101–113. https://doi.org/10.1016/j.bpc.2006.02.
    007
Unemo M, Shafer WM (2014) Antimicrobial resistance in Neisseria gonorrhoeae in the 21st
    Century: past, evolution, and future. Clin Microbiol Rev 27:587–613. https://doi.org/10.1128/
    CMR.00010-14
Van Vuuren SF, Suliman S, Viljoen AM (2009) The antimicrobial activity of four commercial
    essential oils in combination with conventional antimicrobials. Lett Appl Microbiol
    48:440–446. https://doi.org/10.1111/j.1472-765X.2008.02548.x
Veras HNH, Rodrigues FFG, Colares AV, Menezes IRA, Coutinho HDM, Botelho MA, Costa
    JGM (2012) Synergistic antibiotic activity of volatile compounds from the essential oil of Lippia
    sidoides and thymol. Fitoterapia 83:508–512. https://doi.org/10.1016/j.fitote.2011.12.024
Wagstaff A, Faulds D, Gona KL (1994) Acyclovir. A reappraisal of its antiviral activity,
    pharmocokinetic properties and therapeutic efficacy. Drugs 47(1):153–205
Walsh C (2000) Molecular mechanisms that confer antibacterial drug resistance. Nature
    406:775–781. https://doi.org/10.1038/35021219
Wutzler P (1997) Antiviral therapy of herpes simplex and varicella-zoster virus infections.
    Intervirology 40(5–6):343–356
Yang XN, Khan I, Kang SC (2015) Chemical composition, mechanism of antibacterial action and
    antioxidant activity of leaf essential oil of Forsythia koreana deciduous shrub. Asian Pac J Trop
    Med 8:694–700. https://doi.org/10.1016/j.apjtm.2015.07.031

Yayan J, Ghebremedhin B, Rasche K (2015) Antibiotic resistance of pseudomonas aeruginosa in pneumonia at a single university hospital center in Germany over a 10-year period. PLoS One 10:1–20. https://doi.org/10.1371/journal.pone.0139836

Yew HS, Anderson T, Coughlan E (2011) Letters to the editor induced macrolide resistance in mycoplasma genitalium isolates from patients with recurrent nongonococcal urethritis. J Clin Microbiol 49:1695–1696. https://doi.org/10.1128/JCM.02475-10

Yu JQ, Liao ZX, Cai XQ, Lei JC, Zou GL (2007) Composition, antimicrobial activity and cytotoxicity of essential oils from Aristolochia mollissima. Environ Toxicol Pharmacol 23:162–167. https://doi.org/10.1016/j.etap.2006.08.004

Zoulim F (2011) Hepatitis B virus resistance to antiviral drugs: where are we going? Liver Int 31:111–116. https://doi.org/10.1111/j.1478-3231.2010.02399.x

Zuzarte M, Vale-Silva L, Gonçalves MJ, Cavaleiro C, Vaz S, Canhoto J, Pinto E, Salgueiro L (2012) Antifungal activity of phenolic-rich Lavandula multifida L. essential oil. Eur J Clin Microbiol Infect Dis 31:1359–1366. https://doi.org/10.1007/s10096-011-1450-4

Bacterial Drug Efflux Pump Inhibitors from Plants

16

Armel Jackson Seukep, Christophe Dongmo Fokoua-Maxime,
Hélène Gueaba Mbuntcha, Guilin Chen,
Jules Clément Nguedia Assob, Martin Tenniswood,
Satyajit Dey Sarker, Victor Kuete, and Guo Ming-Quan

Contents

A. J. Seukep (✉)
Department of Biomedical Sciences, Faculty of Health Sciences, University of Buea, Buea, Cameroon

CAS Key Laboratory of Plant Germplasm Enhancement and Specialty Agriculture, Wuhan Botanical Garden, Chinese Academy of Sciences, Wuhan, China

Sino-Africa Joint Research Center, Chinese Academy of Sciences, Wuhan, China

Innovation Academy for Drug Discovery and Development, Chinese Academy of Sciences, Shanghai, China
e-mail: seukep.armel@ubuea.cm

C. D. Fokoua-Maxime
Department of Epidemiology and Biostatistics, University of New York State - University at Albany School of Public Health, Albany, NY, USA

New York State Department of Health, Albany, NY, USA

H. G. Mbuntcha · V. Kuete
Department of Biochemistry, Faculty of Science, University of Dschang, Dschang, Cameroon

G. Chen · G. Ming-Quan
CAS Key Laboratory of Plant Germplasm Enhancement and Specialty Agriculture, Wuhan Botanical Garden, Chinese Academy of Sciences, Wuhan, China

Sino-Africa Joint Research Center, Chinese Academy of Sciences, Wuhan, China

Innovation Academy for Drug Discovery and Development, Chinese Academy of Sciences, Shanghai, China

V. Kumar et al. (eds.), *Antimicrobial Resistance*,
https://doi.org/10.1007/978-981-16-3120-7_16

Abstract

The current global antimicrobial resistance (AMR) crisis is a serious hazard to public health, food security, and development. Drug-resistant pathogens diminish the effectiveness of conventional antibiotics, thereby leading to frequent therapeutic failures and reducing the available antibiotic armory. There is an increasing risk of therapeutic impasse if no action is taken now. Novel resistance modalities are appearing within bacterial species and disseminating worldwide, undermining our ability to cure common infectious diseases. Among the identified resistance strategies developed by bacteria, efflux pump systems (EPs) are widely

J. C. N. Assob
Department of Biomedical Sciences, Faculty of Health Sciences, University of Buea, Buea,
Cameroon

M. Tenniswood
Cancer Research Center, University of New York State - University at Albany, Rensselaer, NY,
USA

S. D. Sarker
Centre for Natural Products Discovery, School of Pharmacy and Biomolecular Sciences, Liverpool
John Moores University, Liverpool, UK

acknowledged as the main mechanism leading to multidrug resistance (MDR). MDR is concomitant with the overexpression of the transporters (EPs) that identify and pump out from the cell a large array of structurally dissimilar compounds. The antibiotic pipeline is narrow, and the paucity of novel antimicrobial therapeutics in development is a growing concern. The inhibition of EPs in bacteria appears as a successful alternative in the fight against AMR. Indeed, efflux pump inhibitors (EPIs) offer considerable promise as therapeutic agents, under the hypothesis that they would increase the intracellular concentration of standard antibiotics and hence restore their antimicrobial activity. Considering the cellular toxicity of known synthetic EPIs, investigations are increasingly directed towards the discovery of naturally occurring agents. Plant-derived compounds are found as one of the most dependable sources of safe EPIs. In recent years, a plethora of plant-derived EPIs have been described, although so far, no compound has yet passed all the stages of drug development. The difficulties in developing plant-derived EPI drugs are ascribed to the complexity of EPs, the structural and functional requirements of EPIs, and their cellular toxicity and side effects. In this chapter, we emphasize the roles of EPs in multidrug resistance, describe the major plant-based EPIs, and lay out the challenges impeding the development of plant-derived EPI drugs.

Keywords

Antimicrobial resistance · Efflux pump systems · Efflux pump inhibitors · Plants · Phytochemicals

Abbreviations

ABC   Adenosine triphosphate (ATP)-binding cassette
AMR   Antimicrobial resistance
EPIs   Efflux pump inhibitors
EPs   Efflux pump systems
EtBr   Ethidium bromide
MATE   Multidrug and toxic extrusion
MDR   Multidrug resistance
MFS   Major facilitator superfamily
PACE   Proteobacterial antimicrobial compound efflux
PDR   Pandrug resistant
RND   Resistance-nodulation-cell division
SMR   Small multidrug resistance
XDR   Extensively drug resistant

16.1   Introduction

The current coronavirus pandemic (COVID-19) harshly reminds us about the vulnerability of the human population to infectious agents. In recent years, the world has witnessed an upsurge in deadly infectious diseases of various origins: bacterial, fungal, and viral. This ominous trend is in major part due to the multiple drug resistance (MDR) developed by infectious agents to standard therapeutic compounds. The advent of antibiotic therapy once sparked renewed hope in modern medicine for the control of many bacterial diseases. This hope soon faded due to the capacity developed by bacteria to resist one or more classes of antibiotics, thus reducing the available therapeutic arsenal. Antimicrobial resistance (AMR) is now present worldwide, with the emergence and spread of bacteria strains that are increasingly insensitive to antibacterial drugs. The global AMR crisis has developed into a global health hazard, with drastic consequences that include lengthier hospital stays, rising costs of care, and increased mortality (Dadgostar 2019). Novel resistance modalities are appearing and disseminating around the world, compromising therapeutic interventions to treat common infectious diseases. For an increasing toll of illnesses, such as pneumonia, tuberculosis, sepsis, gonorrhea, and foodborne illnesses, treatment becomes more challenging if not sometimes unfeasible, because of a dissipation of antibiotic potency (WHO 2020). AMR is a significant concern for both human and animal health, causing substantial socioeconomic damages. By 2030, AMR could push as many as 24 million individuals into extreme poverty, and by 2050 the projected number of AMR-related casualties may attain up to ten million per year (WHO 2019). The World Health Organization (WHO) has made the fight against AMR a priority. Particular attention is paid to bacteria of the ESKAPEE group (*Enterococcus faecium*, *Staphylococcus aureus*, *Klebsiella pneumoniae*, *Acinetobacter baumannii*, *Pseudomonas aeruginosa*, *Enterobacter* spp., and *Escherichia coli*) which are often involved in healthcare-associated infections and moreover express a marked resistance to several classes of antibiotics (WHO 2017). The Centers for Disease Control and Prevention (CDC) approximates that antibiotic-resistant ESKAPEE pathogens are responsible for more than two million infections and roughly 23,000 deaths a year (Najafi 2016). Besides, by comparison with non-MDR infections, Zhen et al. (2019) demonstrated that the hospital cost of MDR ESKAPEE-induced infections can be up to 7.21 times higher.

The facility with which bacterial strains adapt to unfavorable environments, in addition to their increased ability to exchange genetic material, highlights the inevitability of a more widespread AMR. The resistance crisis is being accelerated by several factors, which include overuse of antibiotics, nonstandard infection prevention and control operations, unadapted patient education, unavailability of diagnostic facilities, drug black market, the absence of state-of-the-art drug control systems, and the large-scale use of antimicrobials in animals (Aslam et al. 2018). So, if no urgent action is taken now, soon enough, the antibiotic era will end, and fatal small injuries and infections will be the new norm. The prerequisite to be effective against AMR is to understand the mechanisms and determinants of antimicrobial resistance. There are several patterns of antibiotic resistance: natural resistance

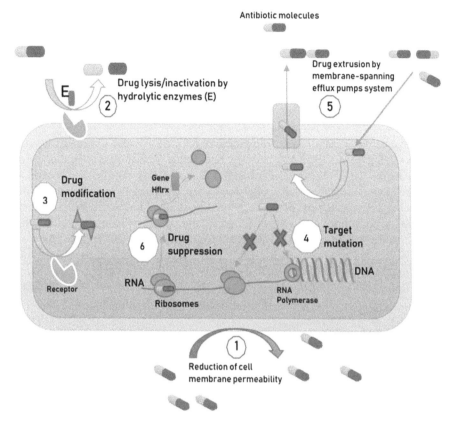

Fig. 16.1 Illustration of the major bacterial strategies of antibiotic resistance. Source: Seukep et al. (2020a)

(systematic), usual or current resistance, multi-resistance (MDR: bacteria multiresistant to antibiotics, carrying several resistance genes for different antibiotics), high resistance (HDR: highly resistant bacteria), ultra-resistance or extensively drug resistance (UDR or XDR), and pan-resistance (PDR) (Magiorakos et al. 2012). As illustrated in Fig. 16.1, the main mechanisms of antibiotic resistance consist of (1) decreasing absorption of drugs; (2) drug inactivation by hydrolytic enzymes; (3) drug modification; (4) target alteration; (5) drug extrusion by active efflux pumps; and (6) antibiotic suppression (Davies and Davies 2010; Duval et al. 2018). These resistance strategies are usually classified as intrinsic (natural) or acquired. The occurrence of resistance promoters on mobile genetic structures like plasmids and transposons added to the free mobility of human carriers has produced the dissemination of drug resistance to many different bacterial genera and geographical locations (Kumar and Varela 2013).

The efflux pump-mediated resistance is one of the major modes of multidrug resistance (MDR). Efflux pumps (EPs) are transmembrane proteins existing in all bacterial plasma membranes which recognize, transport, and expel antibiotics well before they reach their bacterial targets (Spengler et al. 2017). Overexpression of EPs is among the earmarks of recurrent antibiotic treatment failure (Sun et al. 2014). EPs (in prokaryotic cells) are grouped into six superfamilies; details on their structure and physiology as well as the bacterial species in which they are found are important points discussed in this chapter. Some efflux proteins impart resistance to a limited number of antibiotics (a single drug or a single class of drugs), while others expel a wide array (multidrug efflux pumps) of antimicrobials (Nikaido 1994). Numerous studies display the promiscuity of EPs in terms of their substrate specificity (Sun et al. 2014). EPs play key roles in the detoxification of endogenous intermediates, metabolic stress responses, and removal of detergents, antiseptics, solvents, dyes, heavy metals, bile salts, and importantly, EPs reduce the intrabacterial concentration of several families of antibiotics (fluoroquinolones, β-lactams, tetracyclines, aminoglycosides, etc.) that are the recommended therapies for serious bacterial infections (Sun et al. 2014; Karam et al. 2016; Spengler et al. 2017). The reduced efficacy of available drugs, added to the scarcity of new and effective antibacterial agents, is a rising concern. Therefore, the enquiry of the role of EPs in drug resistance and the development of EPIs as adjuvant agents are grabbing significant attention (Sun et al. 2014; Seukep et al. 2020a). The wide range of structurally diverse secondary metabolites (alkaloids, terpenes, and phenols) in medicinal plants makes them one of the most valuable natural resources for EPI screening. EPIs from plant sources are of growing interest, and numerous herbal products have been described as putative EPIs against MDR bacteria overexpressing EPs (Stavri et al. 2007; Kumar and Patial 2016; Spengler et al. 2017; Sharma et al. 2019; Seukep et al. 2020a). This chapter provides a summary of the relevant literature describing the diverse roles of efflux pump-mediated resistance, the methods for identification of plant-derived EPIs, as well as the compounds identified, and the challenges impeding the development and clinical applications of plant-derived EPI drugs.

16.2 Bacterial Efflux Pump Systems: An Overview

16.2.1 Background

Efflux phenomenon is a membrane transport process by which a single transporter (an efflux pump) is able to recognize, transport, and expel a substance from a cell (Dwivedi et al. 2017). Drug efflux was first discovered in 1976 when Juliano and Ling demonstrated that the P-gp protein was the culprit for the resistance of cancer cells to multiple chemotherapeutic regimens (Juliano and Ling 1976). The presence of drug efflux in bacteria was put in evidence through the pioneering research of Levy and McMurry (1978), who documented that tetracycline resistance in *E. coli* was affected by a Tet protein. Since then, many bacterial multidrug efflux pump systems have been identified, and their physiology has been better understood, thanks to the recent innovations in genome sequencing techniques and the vulgarization of such techniques in laboratory facilities across the globe (Li and Nikaido 2009). It has been ascertained that traditional efflux pumps (EPs) like the Tet protein each expels only one single drug or one single family of drugs (Nikaido 1994). Conversely, a multidrug EP can pump out a wide range of compounds (Nikaido 1994), and each bacteria can harbor different types of EPs (Paulsen et al. 1998).

Efflux systems were not developed solely for the purpose of multidrug resistance (MDR). In effect, they are part of a larger group of transporters that existed in the bacteria long before the advent of antibiotics, having functional roles relevant to bacterial physiology (Alvarez-Ortega et al. 2013). Of note, these transporters participate in cell-to-cell communication (also called quorum sensing), perform the uptake of vital nutrients and ions, and proceed to the elimination of metabolic end products and harmful compounds (Li and Nikaido 2009). Therefore, bacterial MDR through the expulsion of drugs by EPs is an evolutionary adaptation whereby the bacteria took advantage of an existing efficient physiologic mechanism to protect itself against the toxic effects of antibiotics (Saier et al. 1998). MDR is concomitant with the overexpression of the transporters that identify and pump out from the cell a large array of structurally dissimilar compounds (Zgurskaya and Nikaido 2000). Furthermore, acquired chromosomal mutations can further potentiate the drug efflux capacity of these transporters. An increasing body of evidence point that the activity of multidrug EPs could render the bacteria resistant to virtually all the different families of antibiotics, detergents, dyes, and organic solvents (Feng et al. 2020). Besides, multiple studies have provided grounds for the existence of a link between the physiological functions of EPs and the formation of biofilms (Matsumura et al. 2011; Baugh et al. 2014) which potentiate the virulence of the bacteria. Decades of research have consistently confirmed that the bacteria that develop biofilms are much more difficult to clear by bactericidal antimicrobials than are planktonic cells (Lewis 2001). In summary, the evolution has gifted the bacteria with an important arsenal of efficient abilities to survive in noxious environments of different types, hence the serious threat posed by efflux pump-mediated multidrug antibacterial resistance and the obligation to better understand the functioning of drug EPs.

16.2.2 Classification and Physiology of Efflux Pump Systems

16.2.2.1 Families of Drug Efflux Pumps

Efflux pumps (EPs) can be classified based on their structure, their energy source, the number of their folds across the bacterial membrane, their gene sequence, and the substrates they expulse. Using these features, literature has commonly grouped bacterial EPs into five families (Li and Nikaido 2009) (Fig. 16.2): (1) the adenosine triphosphate (ATP)-binding cassette (ABC) superfamily, (2) the major facilitator superfamily (MFS), (3) the small multidrug resistance (SMR) family, (4) the multidrug and toxic extrusion (MATE) family, and (5) the resistance-nodulation-cell division (RND) family. Research newly unearthed a sixth family of EPs called

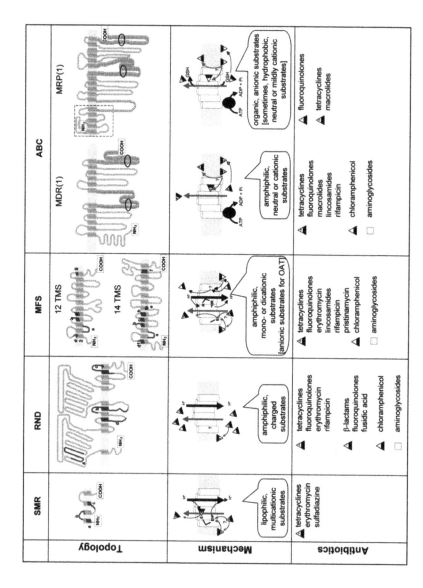

Fig. 16.2 Topology, modes of action, and characteristic antibiotic substrates of the major efflux pump systems. Source: Van Bambeke et al. (2000)

proteobacterial antimicrobial compound efflux (PACE) superfamily (Hassan et al. 2015). The five main families of EPs are mostly encoded by chromosomal genes (Nikaido 2003). RND, SMR, and MATE are exclusively found in prokaryotes, whereas MFS and ABC are shared by prokaryotes and eukaryotes (Dwivedi et al. 2017). MFS and RND pumps are the most frequent in the bacterial kingdom (Molnár et al. 2010), and RND pumps are the most common pumps in Gram-negative bacteria (Handzlik et al. 2013).

16.2.2.2 Structure of Drug Efflux Pumps

Bacterial EPs are structured in one of two fashions (Nikaido 1996) (Fig. 16.3). First, they can appear as a single-component (Li and Nikaido 2009) (Fig. 16.3b). This conformation is the architecture of all EPs of the Gram-positive bacteria, whereas it is the shape of only some of the EPs of Gram-negative bacteria (Nikaido 1998). Once this type of pumps expels drugs into the periplasmic space, the effect is counterbalanced by the spontaneous passive diffusion of lipophilic molecules through the lipid bilayer domains of the cytoplasmic membrane and back into the cytoplasm. Therefore, to be efficient, these pumps must have a throughput sufficiently high to surpass the spontaneous influx of drugs. The extremely wide specificity of drug molecules may not favor this essential requirement for a high turnover rate. This may in part explains the narrowness of the scope of antibacterial substrates to which Gram-positive bacteria are found to be resistant to (Nikaido 1996).

Bacterial EPs can appear secondarily as multiple component systems (Fig. 16.3a) that traverse both the cytoplasmic and outer membranes by utilizing three protein compounds: (1) the cytoplasmic membrane (CM) transporters (Li and Nikaido

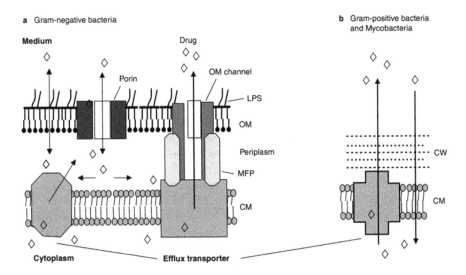

Fig. 16.3 Illustrative models of bacterial drug efflux pumps. Arrangement of characteristic drug efflux mechanisms in Gram-negative bacteria (**a**) and Gram-positive bacteria (**b**). Source: Li and Nikaido (2004)

2009), (2) the outer membrane (OM) channel proteins (Jacoby 2009), and (3) the periplasmic membrane fusion proteins (MFP) (Li 2005). These three protein systems are exclusively found in Gram-negative bacteria where they represent the most common type of drug EPs (Nikaido 1996). Proteins of the RND family are always part of these tripartite complexes. Well-known examples are the AcrAB system of *E. coli* and the MexAB-OprM system of *P. aeruginosa* (Nikaido 2011). The direct efflux of drugs through these three-proteins pumps confers a unique advantage to Gram-negative bacteria because; to reenter the cell, the expulsed drug molecules must overcome an OM which has a very low permeability to both hydrophobic and hydrophilic compounds (Koronakis et al. 2004). Hence, drug efflux and the low permeability OM work in tandem, a synergy that explains why Gram-negative bacteria are resistant to such a large variety of antibiotic families (Tseng et al. 1999).

16.2.2.3 Energy Sources of Drug Efflux Pumps

Bacterial EPs are active transporters since they utilize chemical energy to perform their duties (Fig. 16.3). The ABC superfamily EPs are classified as primary transporters because they use ATP hydrolysis as a source of energy, whereas the other families are named secondary active transporters (uniporters, symporters, or antiporters) as they function thanks to the energy of an electrochemical potential difference (or proton motive force) created by the excretion of hydrogen (H^+) or sodium (Na^+) ions outside the cell (Saier 2000).

16.2.2.4 Substrate Recognition

An important feature of multidrug transporters is their ability to carry numerous types of molecules. Two arguments have been advanced as the theory behind such observation. First, from a pharmacological standpoint, substrates need to share only a small number of structurally similar components in order to be detected by the same transporter (Van Bambeke et al. 2000). The unifying postulate is that almost all the transporters can identify any molecule carrying a polar and slightly charged head linked to a hydrophobic domain (Van Bambeke et al. 2000). Research has confirmed that the substrate specificity is very wide and overlaps many families of EPs. Consequently, a specific antibiotic can be a substrate for various types of pumps, a particular pump may expel not just several antibiotics but also several families of antibiotics, and finally a single bacteria may harbor a large and composite battery of EPs that make possible the expulsion of a vast panoply of drugs (Nikaido 1996; Paulsen et al. 1998).

The second argument explores the chemical and energetic requirements of transport by proteins and is soundly discussed by Neyfakh (2002). A summary of his statements is presented herein. It is generally conceived and accepted that the ligand-binding pocket of a protein should have a very tight "fit" in order to counteract the high energy required to bring compounds to a lipophilic protein interior. However, the lipophilic ligands that multidrug efflux pumps usually deal with only require weak van der Waals and staking interactions and neutralization charges, such that a loose-fitting binding pocket is enough (Neyfakh 2002). This low energy requirement explicates the high efficiency of drug EPs.

16.2.2.5 Mechanisms of Drug Efflux

Concerning the mechanisms of drug expulsion by drug EPs, the exact sequence of events from substrate recognition to expulsion is yet to be fully described (Van Bambeke et al. 2000). The current concepts advanced are generalizations of the well-known functioning mechanisms of physiological transporters, under the assumption that proteins originating from a common ancestor have remained somewhat similar in structure and function. Among the different concepts of drug efflux transport mechanisms that have been advanced, two stand out: one that considers the EPs as hydrophobic "vacuum cleaners" and the other which conceptualizes them as similar to flippases. The first concept hypothesizes that drug molecules circulate freely into the lipid portion of the membrane and then they attain the central channel of the efflux transporter protein which proceeds to actively pump them out. The second concept conceives that once the drug encounters the inner membrane protein, it is flipped to the outer layer (as described for the flippase-induced phosphatidylcholine flip-flop) (Van Bambeke et al. 2000). In summary, EPs are a group of complex structures that have an elaborate physiology. Each efflux pump family has specificities that are also important to discuss.

16.2.2.6 Specificities of Efflux Pump Families

ATP-Binding Cassette (ABC) Superfamily

ABC transporters are ubiquitous in the living world, and they use the energy from the hydrolysis of ATP to transport substrates across cell membranes (Fig. 16.3). In bacteria, gene sequencing techniques have discovered that this family hosts more than 300 member proteins regrouped in ca. 30 families (Saier et al. 1998). Each family targets only one of a vast collection of substrates including sugars, amino acids, ions, drugs, iron complexes, polysaccharides, and proteins (Saier et al. 1998; Li and Nikaido 2004). However, research noted that there exist a couple of these ABC-transporter families which have developed a broad specificity for hydrophobic molecules (Juliano and Ling 1976).

The minimal functional unit of all ABC-type efflux transporters consists of four domains: two transmembrane domains (TMDs) each consisting of six trans-membrane α-helical segments (TMSs) and two nucleotide ATP-binding domains (NBDs) located on the cytoplasmic side of the membrane (Pigeon and Silver 1994). When a drug molecule crosses the bacterial membrane, the two TMDs create a transmembrane chamber lined by hydrophobic and aromatic amino acids provided by the α-helices. The two NBDs form head-to-tail "sandwich" dimers in the intact protein, positioned in a way that allows each NBD contacts both TMDs. Two ATP-binding pockets are formed at the NBD dimer interface, with amino acids from each monomer contributing to each ATP-binding pocket that subsequently expels the drug. The drugs most commonly targeted by ABC-type EPs include tetracyclines, fluoroquinolones, aminoglycosides, macrolides, rifampicin, chloramphenicol, and lincosamides (Fig. 16.3) (Li and Nikaido 2009; Van Bambeke et al. 2000). ABC-type efflux transporters so far identified in bacteria include the LmrA of

Lactococcus lactis, the drug-specific pump MacAB of *E. coli* (Kobayashi et al. 2001), and Sav1866 from *S. aureus* (Dawson and Locher 2006).

Major Facilitator Superfamily (MFS)

The MFS is a historic superfamily of transporters with origins that date back as far as 3 billion years (Saier et al. 1998). This superfamily comprises over 300 sequenced proteins that belong to 17 families each specific for a single class of substrate (Pao et al. 1998); the sugar- and drug-specific MFS transporters are by far the most numerous (Dwivedi et al. 2017). MFS drug transporters have been categorized into two subfamilies: the 12 membrane-spanning helix transporters which are the most common, and the 14 membrane-spanning helix transporters (Paulsen et al. 1996a). Recently, research unearthed a third subfamily of MFS named the sugar porter family. Yet, a fourth subfamily of MFS transporters has been hypothesized to handle drugs; however, no member of this family has of date been physiologically portrayed (Pao et al. 1998). MFS drug EPs are secondary transporters energized by an electrochemical proton gradient (Saier 2000). The presumptive process of drug expulsion encompasses the following steps (Van Bambeke et al. 2000): (1) switch between the drug and a proton; (2) transport of the drug by a succession of conformational changes; (3) substitution of the drug by a proton on the extracellular face of the membrane; and (4) finally return to the initial conformation. A conserved arginine may be the molecule inducing the proton exchange in MFS EPs. The antibiotics most commonly expelled by MFS-type pumps are tetracyclines, fluoroquinolones, aminoglycosides, macrolides, rifampicin, chloramphenicol, lincosamides, and pristinamycin (Fig. 16.3) (Van Bambeke et al. 2000; Li and Nikaido 2009).

The MFS family is the most important EP group in Gram-positive bacteria, with the most researched members of this family being NorA from *S. aureus* and PmrA from *Streptococcus pneumonia* (Lorca et al. 2007). MFS-type transporters commonly operate as single-component pumps like the NorA of *S. aureus* (Yoshida et al. 1990), yet some MFS-type transporters in Gram-negative bacteria act in concert with MFP and OM components to form multiprotein pumps, like in the EmrAB-TolC multidrug efflux pump of *E. coli* (Lomovskaya and Lewis 1992).

Small Multidrug Resistance (SMR) Family

SMR proteins are exclusively found in prokaryotic cells (Chung and Saier 2001). They are the smallest in size among the multidrug transporters (Higgins 2007). SMR proteins contain only about 110 amino acids, and they comprise only four transmembrane α-helices, with no distinguished extra membrane domain. As opposed to most membrane proteins, SMR proteins are very hydrophobic, which makes them soluble in organic solvents. Research has isolated two subfamilies of SMR proteins: one subfamily that grants MDR and uses a drug-H^+ antiport mechanism and a second subfamily which neither participates in multidrug resistance nor catalyzes drug-H^+ antiport (Paulsen et al. 1996b).

The theoretical process of drug transport by SMR-type efflux pumps is somewhat similar to that of the MFS-type described above (Van Bambeke et al. 2000). The only difference between the drug transport mechanisms of these two types of pumps is that the compound responsible for proton exchange for SMR is a conserved glutamate instead of arginine. The antimicrobial medications most commonly targeted by SMR-type EPs are tetracyclines, erythromycin, and sulfadiazine (Fig. 16.3) (Van Bambeke et al. 2000; Li and Nikaido 2009). The most described examples of SMR efflux pumps include the Smr protein of *S. aureus* (Grinius et al. 1992) and the EmrE protein of *E. coli* (Schuldiner et al. 1997).

Multidrug and Toxic Extrusion (MATE) Family

The MATE family harbors a membrane that has an architecture similar to that of the MFS family, yet the sequence of proteins in the respective membranes of the MATE and MFS families is different. MATE transporters typically appear as 450-amino acid proteins arranged into 12 helices (Jack et al. 2001). The MATE transporters are among the latest of the EP families that have been brought to light, hence the dearth of evidence about their structure, function, and regulation (Van Bambeke et al. 2000). The members of the MATE family that have been identified so far are NorM of *Vibrio parahaemolyticus* and YdhE of *E. coli*; both are Na$^+$-antiport transporters that contribute to the resistance to cationic dyes, aminoglycosides, and fluoroquinolones(Yang et al. 2003). A recent line of evidence has suggested that bacterial MATE EPs may be responsible for the resistance to tigecycline, a new glycylcycline class antibiotic developed to overcome methicillin-resistant and vancomycin-resistant *S. aureus* (He et al. 2010).

Resistance-Nodulation-Cell Division (RND) Superfamily

Originally attributed only to the bacterial kingdom, the RND superfamily does also exist in eukaryotes (Tseng et al. 1999). Phylogenetic analyses have identified seven families within the RND superfamily, namely, (1) the heavy metal efflux (HME) family, (2) the hydrophobe/amphiphile efflux-1 (HAE1) family, (3) the nodulation factor exporter (NFE) family, (4) the SecDF protein-secretion accessory protein (SecDF) family, (5) the hydrophobe/amphiphile efflux-2 (HAE2) family, (6) the eukaryotic sterol homeostasis (ESH) family, and (7) the hydrophobe/amphiphile efflux-3 (HAE3) family (Tseng et al. 1999). The RND type EPs are cytoplasmic membrane proteins made up of 12 membrane-spanning segments and 2 large external (periplasmic) loops between MSS1 and MSS2 and between MSS7 and MSS8 (Ma et al. 1993). These transporters tend to be massive, comprising more than 1000 amino acids (Nakashima et al. 2011). RND EPs like AcrB of *E. coli* and MexB of *P. aeruginosa* are identified as the major drivers of the antibiotic multidrug tolerance observed in Gram-negative bacteria (Li and Nikaido 2004; Nakashima et al. 2011). RND EPs expel drugs after they have formed three-protein systems by associating themselves with two other elements: an OM component like the TolC of *E. coli* (Ma et al. 1993; Fralick 1996) or the OprM of *P. aeruginosa* (Li et al. 1995) and a periplasmic "adapter" protein as the AcrA of *E. coli* (Ma et al. 1993; Fralick 1996) or the MexA of *P. aeruginosa* (Li et al. 1995). Naturally occurring RND three-protein systems include the AcrB-AcrA-TolC system of *E. coli* (Ma et al. 1993; Fralick

1996) and the MexB-MexA-OprM of *P. aeruginosa* (Li et al. 1995); these appear as the most well-described tripartite transporters so far (Li and Nikaido 2004).

In Gram-negative bacteria, RND EPs extrude drugs following the interplay between the components of the three-protein system abovementioned. The two large extracellular loops of RND pumps interact with the OM and MFP components to form a "tripartite complex" that ejects the substrate directly out of the bacteria, thereby potentiating the efficacy of the transporter (Van Bambeke et al. 2000). Of all the EPs, RND-type multidrug transporters expulse the largest variety of substrates, including antibiotics, dyes, detergents, disinfectants, organic solvents, toxic lipids, and metabolic inhibitors (Dwivedi et al. 2015a). This wide diversity of expellable substances confers a polyvalent resistance capacity to RND efflux pump-harboring bacteria, allowing them to survive to most of the clinically relevant antimicrobials (Paulsen et al. 1998). In Gram-negative bacteria, RND EPs work synergistically with the low permeability OM to further reduce the intake of drugs. Consequently, the EPs of the RND superfamily contribute largely to the broad spectrum-multidrug resistance of Gram-negative bacteria (Tseng et al. 1999).

16.2.3 Public Health Significance of Multidrug Efflux Pumps

Since the introduction of the wonder drug penicillin in 1928, antibiotics have tremendously reshaped human lives, enabling considerable advances in modern medicine, agriculture, and industry (Pendleton et al. 2013). The high efficiency and numerous applications of antibiotics favored their extensive and uncontrolled use, soon provoking the emergence and rapid dissemination of antimicrobial resistance (AMR). The appealing results obtained with antibiotics further fostered their unlimited use and upscaled the spread of AMR. For example, it is now well documented that there is an increased resistance to carbapenems which are considered the last resort to cure life-threatening healthcare-associated infections (WHO 2014) and are the indicated treatment for infections caused by extended-spectrum beta-lactamase (ESBL)-producing bacteria (Okoche et al. 2015). WHO ranked AMR in the top 10 urgent threats for the year 2019 (WHO 2019), and experts have designated AMR as one of the greatest human health threats of the twenty-first century (Shriram et al. 2018). Despite these multiple warnings, antibiotic misuse has never waned down, and AMR soon led to the emergence of multidrug-resistant bacteria; the later have rendered entire classes of antibiotics irrelevant (Boucher et al. 2009) and have caused deadly hospital-acquired infections (HAIs). HAIs are a threat to worldwide health. The reported incidence of HAIs is approximately 5% in the USA, 7.1% in Europe (WHO 2014), and 3.8–8.7% across the UK, and it reaches a staggering 15.5% pooled incidence in developing countries where emergent and re-emergent infectious diseases remain the leading causes of morbidity and mortality (Vos et al. 2020).

The term ESKAPE was coined by Rice (2008) to describe the group of bacteria responsible for the majority of life-threatening HAIs (Rice 2010). The members of the ESKAPE group are *Enterococcus faecium*, *Staphylococcus aureus*, *Klebsiella*

(K) pneumoniae, Acinetobacter (A) baumannii, Pseudomonas (P) aeruginosa, and *Enterobacteriaceae. E. coli* is sometimes is considered separately from other *Enterobacteriaceae* because of its clinical significance, so that the ESKAPE group is at times referenced as ESKAPEE (Yu et al. 2020). The WHO lists the ESKAPE pathogens among the 12 bacteria for which new antibiotics are urgently needed (Tacconelli et al. 2018). In that list, the ESKAPE bacteria are given a "priority" status (Tacconelli et al. 2018); *A. baumannii, P. aeruginosa, K. pneumoniae*, and *Enterobacteriaceae* are listed in the critical priority list, whereas *E. faecium* and *S. aureus* are in the high-priority group. ESKAPE pathogens pose serious issues in clinical practice. Research has discovered that members of this group are MDR, XDR, and even PDR (Mulani et al. 2019), leaving clinicians with few to no options to treat them. Consequently, MDRESKAPE-induced infections are associated with higher morbidity and mortality rates, as well as increased hospital costs (Zhen et al., 2019). In a recent systematic review on the economic burden of antibiotic resistance in these organisms, Zhen et al. (2019) found that compared to non-MDR infections, MDR ESKAPE-induced infections can cost up to 7.21 times higher and the median cost difference can reach up to $180,948. Thus, MDR ESKAPE-induced infections do substantially contribute to the considerable societal and economic cost of AMR (Solomon and Oliver 2014) which deprives nations from much needed resources. It is therefore urgent to develop efficient strategies to tackle MDR ESKAPE. The pre-requisite for such task is a better understanding of the mechanisms of MDR in ESKAPE pathogens. It is now recognized that the intrinsic drug resistance of these pathogens is largely the faith of broad-specificity multidrug efflux pumps (EPs).

16.3 Efflux Pump Inhibitors (EPIs)

16.3.1 EPIs as Promising Therapeutic Agents to Reverse Bacterial MDR

Active efflux appears as the major mechanism leading to MDR. A single pump can expel panoply of antibiotic molecules; therefore, any inhibition of EPs may restore the efficacy of several families of antibiotics (Seukep et al. 2020a). The inactivation of EPs could be achieved by different mechanisms. Since most EPs involve the energy from H^+ and Na^+ electrochemical gradients for their functioning, the primary mechanism of action of EPIs includes targeting of H^+/Na^+ motive force (MF) of EPs through competitive or noncompetitive inhibition with the binding substrate. Other mechanisms comprise (1) downregulation of the expression of EPs genes by impeding the regulation of genes, (2) redesign antibiotics that are not anymore accepted as substrates, (3) inhibition of the assemblage of functional EPs, (4) blockage of EPs to refrain the substrate to effectively bind to the active site, and (5) collapse of the energy source (disruption of the hydrolysis of ATP thereby compromising activation of EPs) (Bhardwaj and Mohanty 2012). EPIs are molecules that impede EP functions by one or more of these strategies, resulting in the inactivation of drug extrusion. The overall effect is the effective accumulation of an antibiotic inside the bacterial cell.

Therefore, EPIs can be used as adjuvants in combination with standard antibiotics to improve their efficiency towards bacteria overexpressing EPs. The activities of standard antibiotics could thus be restored.

16.3.2 Properties of an Effective Efflux Pump Inhibitor

To set a clear difference between an EPI and a substrate (which has no inhibitory action), the characterization of effective EPIs involves the definition of kinetic parameters of both the inhibitor and substrate along with the interrelationships with the organization of EP components (Schweizer 2012). Besides, mastering how the EPs interfere with the drug extrusion outside the bacterial cell is one of the significant challenges in the screening for new EPIs (Krishnan et al. 2016). The development of EPIs is a significant challenge; it necessitates riding out many obstacles among which the selection of antibiotics to potentiate and the suitability of the bioactivity properties of the combination EPI antibiotic (Zgurskaya et al. 2015). In spite of the hope founded on the implementation of efficient EPIs, the main preoccupation leftover is their toxicity on cells that hampers their implementation in clinical settings. The development of EPIs acting on several EPs (broad-spectrum EPIs) is difficult because they will constitute a considerable danger (toxicity) to human cells, whereas selective inhibitors will be prone to give rise to mutants in which a secondary pump will undertake the role of those inactivated (Wang et al. 2016).

To be considered as an effective or successful EPI, a chemical compound would have to pass through a strict checklist (Sharma et al. 2019). Relevant conditions are hereby described. (1) No antibacterial activity: the chemical entity must not display inhibitory effects against tested bacteria. A molecule with antibacterial potential would definitely give rise to the selection of mutants resistant to its effect that will seriously affect its value as an EPI. (2) Selective toxicity: the chemical entity should be selective and not target any eukaryotic EPs. Since EPs are present either in prokaryotic and eukaryotic cells and their basic functional features tend to be comparable through the life forms, selective inhibition of bacterial EPs becomes a tricky problem. (3) Ideal pharmacological features: the chemical entity should have ideal pharmacological properties such as nontoxicity, high therapeutic and safety indices, good pharmacokinetic profile (ADMET: absorption, distribution, metabolism, excretion, and toxicity), and serum stability. (4) Cost-effective: to be commercially beneficial, the development of EPIs must be economically conceivable (Bhardwaj and Mohanty 2012; Sharma et al. 2019; Seukep et al. 2020a).

16.3.3 Classification of Efflux Pump Inhibitors

Efflux pump inhibitors (EPIs) are structurally diverse molecules of various origins, which act in different ways on the target EPs. Therefore, EP blockers are usually classified based on their mechanisms of action and origin.

16.3.3.1 Classes of Efflux Pump Inhibitors Based on their Mechanisms of Action

Considering the mechanism of action, EPIs can be grouped into two categories: (1) energy dissipation and (2) inhibition by direct binding.

Energy Dissipation

Since EPs are relying on energy from cells, the dissociation of the energy and efflux activity offers an attractive oncoming to the inhibition of efflux activity. The H^+ and Na^+ electrochemical gradients or the ATPase that provide energy to EPs have been experienced as targets of several EPIs. Such an inhibition scheme does not involve any direct interconnection of the inhibitor with the EP itself. This approach emerges to be beneficial as diverse EPs are reliant on the proton gradient, making this a universal scheme for their inhibition (Sharma et al. 2019). Some examples include synthetic EPIs known as carbonylcyanide-m-chlorophenylhydrazone (CCCP) and IITR08027 (Bhattacharyya et al. 2017). The CCCP has been reported to revive the activity of tetracycline in *Helicobacter pylori* (Anoushiravani et al. 2009) and *Klebsiella* spp. (Fenosa et al. 2009). IITR08027 was found to be very effective at reversing the resistance against fluoroquinolones in both recombinant *E. coli* and clinical strains of *A. baumannii* overexpressing the MATE EPAbeM (Bhattacharyya et al. 2017). Contrary to CCCP which depicted toxicity towards mammalian cells, IITR08027 did not have any antibacterial action of its own and exhibited low toxicity on animal cells. For that reason, IITR08027 has been considered to be very close to an ideal EPI. IITR08027 is currently evaluated for its preclinical potential (Bhattacharyya et al. 2017).

Direct Binding Inhibition

An alternative strategy of EP blockage is the binding of the EPIs to functional EPs, leading to a decreased capacity of EPs to interrelate with their substrates. This binding could be competitive (EPI competes with the substrates for the same binding site) or noncompetitive (the binding of EPI to the pump leads to the reduced affinity of the pump for its substrates). However, bacteria can still mutate their EPs to change the target sites of these EPIs, rendering them needless. Some examples include PAβN (phenylalanine-arginine β-naphthylamide or MC-207,110), verapamil, and 1-(1-napthylmethyl)-piperazine (NMP) (Bhattacharyya et al. 2017). PAβN (or MC-207,110) was the first inhibitor of the RND family pumps; it enhances the activity of levofloxacin (against *P. aeruginosa* cells expressing MexAB, MexCD, and MexEF pumps), erythromycin, and chloramphenicol (Lomovskaya et al. 2001). Verapamil inhibits the activity of MATE pumps; it improves the efficacy of bedaquiline and ofloxacin (Singh et al. 2011; Gupta et al. 2014). NMP potentiates the activity of levofloxacin in *E. coli* cells overexpressing AcrAB and AcrEF EPs (Bohnert and Kern 2005).

16.3.3.2 Classes of EPIs Based on their Origin

EPIs originate from various sources including plants, microorganisms, and chemical synthesis.

EPIs Originated from Plants

Plants are a rich source of a large array of structurally diverse chemical entities that could act as adjuvants, thereby contributing to the enhancement of standard antibacterial agent efficacy. Classes of plant-derived EPIs involve alkaloids, phenolic compounds (flavonoids, polyphenols), and terpenes. This section is further discussed later in this chapter.

EPIs Originated from Microorganisms

While most of the EPIs have their origin from phytochemicals or chemical synthesis, a few number of EPIs have been pointed out to derive from microorganisms. Two compounds derived from *Streptomyces* spp., namely, EA-371α and EA-371d, have been demonstrated to have an inhibitory effect against a *P. aeruginosa* efflux pump (MexAB-OprM) (Lee et al. 2001). The new structure of these chemical entities suggests an interesting opportunity to develop new by-products with improved efficacy and absorption and lessen harmful secondary effects. Computational studies could help to pick out the molecular interactivity between EPIs and MDR pumps. These studies are facilitated by the 3D crystal arrangement of familiar EPs. Investigations carried out by Coutinho et al. (2009) and Chaves et al. (2015) display the potential of ethyl alcohol extract of *Nasutitermes corniger* (termites) to synergistically act against *S. aureus* and *E. coli* expressing EPs once combined to aminoglycosides. This indicates a future key role of some organisms to inverse bacterial MDR.

EPIs Originated from Chemical Synthesis

Aside from the naturally occurring herbal products and those derived from microbes or other organisms, the discovery of novel compounds from semisynthetic or synthetic chemistry appears as a worthy method to develop putative EPIs. Considerable works have produced findings at different levels of achievement. The major classes of EPIs generated from semisynthetic or synthetic chemistry comprise (1) peptidomimetic compounds (e.g., PAβN) (Lomovskaya et al. 2001); (2) quinoline derivatives (e.g., pyridoquinolones, 2-phenyl-4(1H)-quinolone, and 2-phenyl-4-hydroxyquinoline) (Sabatini et al. 2011); (3) arylpiperidines and aryl piperazine derivatives (e.g., 3-arylpiperidine, phenylpiperidines, and NMP) (Thorarensen et al. 2001); and (4) pyridopyrimidine and pyranopyridine derivatives (e.g., pyridopyrimidine analogues D2 and D13-9001, MBX2319) (Mahmood et al. 2016).

16.4   Methods of Screening Efflux Pump Inhibitors

The direct measurement and accumulation assay are two methods for investigating the activity of EPs. The direct measurement consists of determining the EP substrate expelled, while the accumulation assay measured the quantity of EP substrate heaped inside the bacteria cell (Blair and Piddock 2016). The enhancement of the efficacy of efflux substrates combined with EPIs substantiates the expression of EPs in the bacteria. Evenly, any augmentation of the amount of the substrate, occurring solely once combined to an EPI, stipulates that the inhibitor acts by altering the efflux machinery (Dreier and Ruggerone 2015).

16.4.1 Direct Measurement of Efflux Activity

Several substances known as fluorescent dyes are used for the direct measurement of efflux activity. Alanine-b-napthylamide dye and rhodamine 160 6G (Ivnitski-Steele et al. 2009; Rajendran et al. 2011) are specifically used to detect MFS and ABC EPs. Other fluorescent dyes include 1,20-dinaphthylamine (mostly used for RND EPs detection) (Bohnert et al. 2011), Nile red (Bohnert et al. 2010), ethidium bromide (EtBr), and doxorubicin (Viveiros et al. 2008). MC-207110, CCCP (Nelson 2002; Askoura et al. 2011), and a plant alkaloid reserpine (Ahmed et al. 1993; Garvey and Piddock 2008) have been proven to be useful to evaluate the efflux activity in bacteria. MC-207,110 is usually applied for the screening of active EPs in Gram-negative bacterial profiling, whereas reserpine is used for Gram-positive bacteria (Van Bambeke et al. 2006).

16.4.2 Accumulation Assay

The checkerboard synergy assay has been exploited to detect putative inhibitors of EPs. The assay is applied to verify the interactivity and efficacy of two drug entities when used at the same time (Orhan et al. 2005). Besides, the berberine uptake assay is a useful tool for the rapid screening of many phytochemicals (Stermitz et al. 2000). Ethidium bromide (EtBr) is a substrate for many MDR EPs. The synergistic potential of a chemical entity could also be assessed through a method known as EtBr accumulation assay (Paixão et al. 2009). Moreover, fluorimetry is an essential tool to measure the action of EPIs; indeed there is a retention of fluorescence over time, proportionally to the decrease of the efflux activity. Acriflavine (or pyronin Y) and bisbenzimide H33342 assays can be executed similarly to that of the EtBr efflux inhibition assay (Paixão et al. 2009). Other most powerful screening methods comprise liquid chromatography/mass spectrometry (LC/MS)-based assay (Cai et al. 2009), fluorescein-di-β-D-galactopyranoside (FDG)-based assay (Matsumoto et al. 2011), in silico high-throughput virtual screening (HTVS) (Aparna et al. 2014; Rao et al. 2018), real-time (RT) PCR (Ramaswamy et al. 2017), and quantitative mass spectrometry (Brown et al. 2015).

The HTVS (Aparna et al. 2014; Rao et al. 2018) is a very important tool since it helps to select compounds with efflux substrate-like features. Indeed, the use of HTVS improves the selection process, thereby ameliorating the detection of inhibitors. The method necessitates a pre-established database of plant products with proven effects towards determined EPs. The matching with the standard pharmacophore models generated involving familiar efflux substrates is helpful to filter the process. The XP ligand docking of possible hits against defined transporters (e.g., AcrB of *P. aeruginosa* and MexB of *E. coli*) could be integrated with HTVS.

The in silico docking depicts several advantages such as the identification of the molecular interrelationship between the inhibitors and their targets. Mastering the molecular interactions will facilitate the modification of EPIs, thereby increasing the specificity of the inhibitor with the target, resulting in successful blockage of EPs (Rao et al. 2018). Usually, HTVS involves additional experiments known as in vitro validation assay; the tests include the determination of minimal inhibitory concentration (MIC), the checkerboard synergy assay of phytocompounds and standard antibacterial agents, and the fluorescence-based EtBr efflux inhibition testing (Aparna et al. 2014).

Molecular docking is an effective tool to master the interactivity of phytochemicals with their EP targets and to pick out the real target locations of EPIs (Ramaswamy et al. 2017). The determination of the spectrum of action of EPIs (action on a single or more EPs) as well as specific targets is an important challenge. The real-time PCR (RT-PCR) could be useful to detect if EPs are underexpressed owing to the effects of EPIs (Rao et al. 2018). Furthermore, several uncharacterized or even unknown EPs could be found in bacteria species. In that context, a microarray test is a substitute method that can facilitate the identification of unidentified EP genes regulated by the effects of EPIs (Ramaswamy et al. 2017).

The detection of natural compounds for efflux inhibition assay appears to be a tricky task, as some chemical entities may induce optical interaction when assessed with fluorescence-based techniques. Interestingly, quantitative mass spectrometry (MS) was found as an important tool to solve the issue. The MS quantifies the concentration to which a substrate has built up within bacteria cells by determining the reduction of the substrate from an exhausted liquid medium. For instance, high-performance liquid chromatography-electrospray ionization mass spectrometry (HPLC-ESI/MS) was performed to quantify ethidium bromide uptake to evaluate efflux inhibition activity in *S. aureus* by a crude plant extract and pure flavonoids (Brown et al. 2015).

The development of new and potent EPIs demands a determination of kinetic parameters and the structural correlation with the EP components which could be concerned. The quantitative structure-activity relationship (QSAR) analysis is a useful tool, which establishes the link between the pharmacological actions of a substance with the diverse physicochemical parameters. QSAR could be used to anticipate the EPI efficacy and quantify significant regions in chemical entities (Nargotra et al. 2008).

16.5   Efflux Pump Inhibitors (EPIs) from Plants

Plants have been used since antique times for their health beneficial effects on humans. The transmission of knowledge on the healing properties of medicinal plants has been made over the centuries within human communities (Kumar et al. 2009). Plants are one of the most valuable sources of novel drug entities (Seukep et al. 2020b). Many molecules called secondary metabolites are synthesized by plants, with the primary function of protecting themselves against the harmful effects of their environment as well as possible infections. These properties have been shown to be transferable to humans for the benefit of their health (Kumar et al. 2009; Seukep et al. 2020b). Three main groups of secondary metabolites can be distinguished: terpenoids, phenols, and alkaloids. These plant-derived compounds are widely acknowledged to possess a plethora of pharmacological properties (Kuete 2010; Seukep et al. 2020c). Numerous studies have demonstrated the key role of plant secondary metabolites in the development of antibacterials and especially their ability to reverse bacterial resistance to antibiotics towards bacteria overexpressing EPs; their roles are thereby highlighted as potent EPIs (Touani et al. 2014; Seukep et al. 2016; Spengler et al. 2017; Sharma et al. 2019; Seukep et al. 2020a). Naturally occurring EPIs could be better tolerated by human cells, unlike the synthetic molecules. In the present chapter, we summarize from the literature the reported EPIs from plant sources, grouped into different classes of secondary metabolites, their plant source, target EPs, bacteria, and identified substrates.

Plant-based EPIs are different plants pertaining to various families. It is clear that most EPIs act on the NorA, TetK, and MsrA pumps of *S. aureus* and other Gram-positive bacteria (Tables 16.1, 16.2, 16.3, and 16.4). Most investigations have been carried out against bacterial species including *S. aureus* (MRSA), *E. faecalis*, *B. subtilis*, and *S. pneumoniae* in Gram-positive bacteria, whereas *E. coli*, *A. baumannii*, *K. pneumonia*, and *P. aeruginosa* are the most investigated in Gram-negative. Several studies also documented EPIs towards *Mycobacterium* spp. EPs. Most of the EPIs are targeting the AcrAB-TolC and MexAB-OprM pumps in Gram-negative bacteria, whereas NorA and TetK are the main targets in Gram-positive bacteria. Common substrates of these naturally occurring EPIs are fluoroquinolones (norfloxacin, ciprofloxacin), tetracycline, erythromycin, berberine, isoniazid, and EtBr.

Although most of plant-EPIs are terpenoids, phenols, or alkaloids, some other classes of phytochemicals appear to be excellent EPIs (Table 16.4). These include resin glycosides (e.g., murucoidins; orizabins XIX, IX, and XV; stoloniferin) (Pereda-Miranda et al. 2006; Chérigo et al. 2008), polyyne (e.g., falcarindiol) (Garvey et al. 2011), arylbenzofuran aldehyde (e.g., spinosan A) (Belofsky et al. 2006), diarylheptanoids (e.g., trans,trans-1,7-diphenylhepta-4,6-dien-3-one) (Groblacher et al. 2012), fatty acids (e.g., oleic and linoleic acids) (Chan et al. 2015), and essential oils (Chovanová et al. 2015).

16.5.1 Terpenoids

Terpenoids are the most abundant natural products with a set of structurally diverse secondary metabolites in plants. Terpenoids play an increasingly promising role in the field of medicine and have various pharmacological activities including antimicrobial and other effects (Yang et al. 2020). Table 16.1 summarizes the putative terpenoid EPIs identified so far from the literature, and Fig. 16.4 displays the structures of common terpenoids with EPI activities.

Table 16.1 Terpenoids (and terpenes) class compound EPIs

EPIs	Plant source	Target efflux pump(s)	Bacterial strain (s)	Substrate(s)	References
Uvaol	*Carpobrotus edulis*	NorA	MRSA	Norfloxacin, oxacillin	Martins et al. (2011a)
Oleanolic acid	*Carpobrotus edulis*	/	*E. coli*	/	Martins et al. (2011b)
Karavilagenin C	*Momordica balsamina L.*	/	*E. faecalis, S. aureus, E. coli*	Fluoroquinolones	Ramalhete et al. (2011)
Balsaminagenin B	*Momordica balsamina L.*	/	*E. faecalis*	/	Ramalhete et al. (2011)
Balsaminol A	*Momordica balsamina L.*	NorA, AcrAB-TolC	*S. aureus E. faecalis*	/	Ramalhete et al. (2011)
Totarol	*Chamaecyparis nootkatensis*	NorA, MsrA, TetK	*S. aureus, Mycobacterium* spp.	Erythromycin, isoniazid	Smith et al. (2007a)
Jatropholone A and B	*Jatropha gossypiifolia*	NorA	S. aureus	/	Marquez et al. (2005)
Carnosic acid	*Rosmarinus officinalis*	NorA, MsrA	*S. aureus*	Tetracycline, erythromycin	Oluwatuyi et al. (2004)
Carnosol	*Rosmarinus officinalis*	MsrA, TetK	*S. aureus*	Tetracycline	Oluwatuyi et al. (2004)
Isopimarane diterpenes	*Lycopus europaeus*	TetK, MsrA	*S. aureus*	/	Gibbons et al. (2003)
Geranylgeranyl diterpenes	*Lycopus europaeus*	TetK, MsrA	*S. aureus*	/	Gibbons et al. (2003)

Abietane diterpenes	*Lycopus europaeus*	TetK, MsrA	*S. aureus*	/	Gibbons et al. (2003)
3-*O*-(β-Xylopyranosyl-(1→4)-β-galactopyranosyl)-oleanolic acid	*Acacia polyacantha* Willd.	/	Gram-negative	/	Mambe et al. (2019)
Ursolic acid and derivatives	*Eucalyptus tereticornis*	AcrAB-TolC, MacB, YojI	*E. coli*	/	Dwivedi et al. (2015b)
Isopimaric acid	*Pinus nigra*	NorA	MRSA	/	Smith et al. (2005)
Farnesol	Dietary and aromatic plants	Rv3065(mmr)	*Mycobacterium* spp.	EtBr	Jin et al. (2010)
Geraniol	Rose oil, palmarosa oil, and citronella oil	Rv3065(mmr)	*Mycobacterium* spp.	/	Jin et al. (2010)
Artemisinin (artesunate)	*Artemisia annua*	AcrAB-TolC	*E. coli*	Cefuroxime, cefoperazone, penicillin G, cefazolin, ampicillin	Li et al. (2011)
α-Terpinene	Cardamom and marjoram oils	TetK	*S. aureus*	Tetracycline	Limaverde et al. (2017)
Galbanic acid	*Ferula* spp.	NorA	*S. aureus*	Norfloxacin, ciprofloxacin	Fazly et al. (2010)
Clerodane diterpene	*Polyalthia longifolia*	NorA, NorB, NorC, MepA, MdeA	*S. aureus*	Norfloxacin, ciprofloxacin	Gupta et al. (2016)
Cuminaldehyde	Cumin seed oils	LmrS	*S. aureus*	EtBr	Kakarla et al. (2017)

EPIs efflux pump inhibitors; *EPs* efflux pumps; *MRSA* methicillin-resistant *S. aureus*; *EtBr* ethidium bromide

Table 16.2 Phenolic compound EPIs

EPIs	Plant source	Target efflux pump(s)	Bacterial strain(s)	Substrate(s)	References
Osthol	*Cnidii monnieri*	NorA, MdeA, TetK, MsrA	*P. aeruginosa*, *S. aureus*	/	Joshi et al. (2014)
Cumin	*Cuminum cyminum*	LmrS	*S. aureus*	/	Kakarla et al. (2017)
Silybin	*Silybum marianum*	/	*S. aureus*	/	Stermitz et al. (2001)
Chrysoplenetin	*Artemisia annua*	NorA	*S. aureus*	Berberine	Liu et al. (1992)
Chrysosplenol-D	*Artemisia annua*	NorA	*S. aureus*	Berberine	Liu et al. (1992)
Caffeoylquinic acids	*Artemisia absinthium*	/	*S. aureus*, *E. faecalis*	/	Fiamegos et al. (2011)
Crysoplenol	*Artemissia annua*	NorA	*S. aureus*	Berberine, norfloxacin	Stermitz et al. (2002)
Crysoplenetin	*Artemissia annua*	NorA	*S. aureus*	Berberine, norfloxacin	Stermitz et al. (2002)
Sarothrin	*Alkanna orientalis* (L.) Boiss.	NorA	*S. aureus*	/	Bame et al. (2013)
Gallotannin	*Terminalia chebula*	/	*E. coli*	/	Bag and Chattopadhyay (2014)
Ferruginol	*Chamaecyparis lawsoniana*	NorA, MsrA, TetK	*S. aureus*, *Mycobacterium*spp.	Tetracycline, erythromycin, norfloxacin isoniazid	Smith et al. (2007b)
Diospyrin	*Diospyros montana*	/	*Mycobacterium aurum*	/	Mukanganyama et al. (2012)
Liquiritin	*Glycyrrhiza uralensis*	/	*E. coli*	Fluoroquinolones	Junwei et al. (2013)
Pterocarpan	*Dalea spinosa*	NorA	*S. aureus*	/	Belofsky et al. (2006)
Genistein	*Lupinus argenteus*	NorA	*S. aureus*	Berberine	Morel et al. (2003)

Polyacylatedneohesperidosides	*Geranium caespitosum*	NorA	*S. aureus*	Berberine, rhein, ciprofloxacin, norfloxacin	Stermitz et al. (2003)
Coumarins	*Mesua ferrea*	NorA	*S. aureus*	Norfloxacin, ciprofloxacin	Roy et al. (2013)
Olympicin A	*Hypericum olympicum*	NorA	*S. aureus*	/	Shiu et al. (2013)
4-Hydroxy-α-tetralone	*Hypericum olympicum*; *Ammannia* spp.	NorA, yojI	*S. aureus, E. coli*	/	Shiu et al. (2013)
Baicalein	*Thymus vulgaris*	NorA, TetK	MRSA, *Salmonella enteritidis, E. coli*	Ciprofloxacin, tetracycline	Fujita et al. (2005), Chan et al. (2011)
Kaempferol rhamnoside	*Persea lingue*	NorA	*S. aureus*	Norfloxacin, ciprofloxacin	Holler et al. (2012)
Tiliroside	*Herissantiatiubae*	NorA	*S. aureus*	Norfloxacin, ciprofloxacin	Falcão-silva et al. (2009)
Gallic acid	*Punica granatum*	NorA	*S. aureus*	/	Dey et al. (2012)
N-trans-feruloyl 4'-*O*-methyldopamine	*Mirabilis jalapa* *Pisonia aculeata*	NorA	*S. aureus*	Norfloxacin, ciprofloxacin	Michalet et al. (2007)
Resveratrol	*Nauclea pobeguinii*	/	Gram-negative	/	Seukep et al. (2016)
Citropten	*Citrus paradisi*	NorA, ermA, ermB	MRSA	Norfloxacin, ciprofloxacin	Abulrob et al. (2004)
Furocoumarins	*Citrus paradisi*	NorA, ermA, ermB	MRSA	Norfloxacin, ciprofloxacin	Abulrob et al. (2004)
Phenylpropanoid (+) ailanthoidiol	*Zanthoxylum capense*	/	*S. aureus*	/	Cabral et al. (2015)
Phenylpropanoids (1'-*S*-1'-acetoxyeugenol acetate)	*Alpinia galanga*	Rv1145, Rv1146 Rv1877, Rv2846c Rv3065 (mmr)	*M. Smegmatis*	EtBr	Roy et al. (2012)

(continued)

Table 16.2 (continued)

EPIs	Plant source	Target efflux pump(s)	Bacterial strain(s)	Substrate(s)	References
Salicylic acid	*Salix alba*	/	*S. aureus*	/	Price et al. (2002)
Epigallocatechin gallate	*Camellia sinensis*	NorA, TetB, TetK	*S. aureus*	Tetracycline	Sudeno et al. (2004)
Curcumin	*Curcuma longa*	NorA, MdeA, TetK	*P. aeruginosa* *S. aureus*	Norfloxacin, ciprofloxacin	Joshi et al. (2014), Negi et al. (2014)
Caffeic and gallic acids	Variety of plants	NorA, MsrA	*S. aureus*	/	Dos Santos et al. (2018)
Tannic acids	Variety of plants	TetK, NorA	*S. aureus* *A. baumannii*	Tetracycline, norfloxacin	Chusri et al. (2009), Tintino et al. (2017)
Ellagic acid	Fruits and vegetables	/	*A. baumannii*		Chusri et al. (2009)
Thymol and carvacrol	Aromatic plants	/	Food-borne pathogens		Miladi et al. (2009)
p-Coumaric acid and derivatives	Variety of edible plants	MexAB-OprM	*P. aeruginosa*	/	Choudhury et al. (2016)
4',5'-*O*-dicaffeoylquinic acid	Variety plants	NorA	*S. aureus*	Berberine, norfloxacin	Fiamegos et al. (2011)
Silibinin	*Silybum marianum*	NorA	*S. aureus*	Norfloxacin	Mahmood et al. (2016)
Furocoumarins	*Heracleum* sp. and a variety of plants	NorA, ErmA, ErmB	*S. aureus*	Norfloxacin, ciprofloxacin	Rana et al. (2014)
Citropten	Variety of edible plants (citrus, lemons, and limes)	NorA, ErmA, ErmB	*S. aureus*	Norfloxacin, ciprofloxacin	Rana et al. (2014)
Diosmetin	Citrus fruits	MsrA, NorA	*S. aureus*	Erythromycin, rifloxacin	Chan et al. (2013)
Boeravinone B	*Boerhaavia diffusa*	NorA	*S. aureus*	Norfloxacin, ciprofloxacin	(Singh et al. 2017)

Biochanin A	*Trifolium pratense* (red clover)	NorA	*S. aureus*	Berberine, norfloxacin	(Morel et al. 2003)
Caffeoylquinic acid	Variety of plants	NorA	*Enterococcus faecalis, S. aureus*	Berberine	Fiamegos et al. (2011)
Chalcone	Variety of edible plants	NorA	*S. aureus*	Berberine, norfloxacin	Stavri et al. (2007)
Quercetin	Variety of plants and foods	Rv3065 (mmr)	*Mycobacterium* spp.	/	Song and Wu (2016)
Orobol	*Lupinus argenteus*	NorA	*S. aureus*	Berberine	Morel et al. (2003)

EPIs efflux pump inhibitors, *EPs* efflux pumps, *MRSA* methicillin-resistant *S. aureus*, *EtBr* ethidium bromide.

Table 16.3 Alkaloid class compound EPIs

EPIs	Plant source	Target efflux pump(s)	Bacterial strain(s)	Substrate(s)	References
Reserpine	*Rauwolfia* sp.	Bmr, NorA, TetK, LmrA, PmrA, MepA	*B. subtilis, S. aureus,* MRSA, *S. pneumoniae Lactococcus lactis*	Norfloxacin, ciprofloxacin, Tetracycline	Stavri et al. (2007)
Conessine	*Holarrhena antidysenterica*	MexAB-OprM, AdeIJK	*P. aeruginosa, A. baumannii*	Cefotaxime, levofloxacin, tetracycline, novobiocin, rifampicin	Siriyong et al. (2016), Siriyong et al. (2017)
Indirubin	*Wrightia tinctoria*	NorA	*S. aureus, S. epidermidis*	/	Ponnusamy et al. (2010)
Catharanthine	*Catharanthus roseus*	/	*P. aeruginosa*	/	Dwivedi et al. (2018)
Berberine	*Berberis* spp.	NorA, MexAB-OprM	*S. aureus P. aeruginosa*	/	Aghayan et al. (2017)
Palmatine	*Phellodendron amurense*	MexAB-OprM	*P. aeruginosa*	/	Aghayan et al. (2017)
Pheophorbide A	*Berberis* ssp.	NorA, MexAB-OprM	*Streptococcus aureus, P. aeruginosa*	Ciprofloxacin, berberine	Stermitz et al. (2000)
5′-MHC	*Berberis* ssp.	NorA	*S. aureus*	Berberine	Stermitz et al. (2000)
Lysergol and 17-O-3″,4″,5″-trimethoxybenzoyllysergol	*Ipomoea muricata*	YojI	*E. coli*	/	Maurya et al. (2013)
Propacine	*Jatropha elliptica*	/	*S. aureus*	/	Marquez et al. (2005)
2,6-Dimethyl-4-phenylpyridine-3,5-dicarboxylic acid diethyl ester	*Jatropha elliptica*	/	*S. aureus*	Ciprofloxacin, norfloxacin	Marquez et al. (2005)

Juliflorine	*Prosopis juliflora*	NorA	*S. aureus*	/	Morel et al. (2003)
Theobromine	*Theobroma cacao*	AcrAB-TolC, MexAB-OprM	*S. typhimurium*, *K. pneumoniae*	Ciprofloxacin, tetracycline	Piddock et al. (2010)
Tetrandrine	*Stephania tetrandra*	Rv2459 (jefA), Rv3728, Rv3065 (mmr)	*Mycobacterium* spp.	Isoniazid, ethambutol	Song and Wu (2016)
Piperine	*Piper nigrum*	NorA, MdeA, Rv1258c	*S. aureus*, *Mycobacterium* spp.	Norfloxacin, ciprofloxacin	Kumar et al. (2008), Sharma et al. (2010)
Quinine	Cinchona tree	/	MRSA	/	Mohtar et al. (2009)
Capsaicin	*Capsicum annuum* L.	NorA	*S. aureus*	Norfloxacin, ciprofloxacin	Kalia et al. (2012)
Harmaline	*Peganum harmala*	/	MRSA	/	Mohtar et al. (2009)
Cathinone	*Catha edulis*	AcrAB-TolC	*S. Typhimurium*	Ciprofloxacin	Piddock et al. (2010)
Ergotamine	*Claviceps purpurea*	/	/	Norfloxacin	Fujita et al. (2005)

EPIs efflux pump inhibitors, *MRSA* methicillin-resistant *S. aureus*.

Table 16.4 Other class compound EPIs from plants

EPIs (chemical class)	Plant source	Target efflux pump (s)	Bacterial strain(s)	Substrate(s)	References
Falcarindiol (polyyne)	*Levisticum officinale*	/	Gram-negative	/	Garvey et al. (2011)
Murucoidins (resin glycosides)	*Ipomoea murucoides*	NorA	*S. aureus*	Norfloxacin	Chérigo et al. (2008)
Orizabins XIX, IX, and XV (resin glycosides)	Mexican morning glory species	NorA	*S. aureus*	Norfloxacin, berberine	Pereda-Miranda et al. (2006)
Stoloniferin (resin glycosides)	*Ipomoea stolonifera*	NorA	*S. aureus*	Norfloxacin, ciprofloxacin	Chérigo et al. (2008)
Spinosan A (arylbenzofuran aldehyde)	*Dalea spinosa*	NorA	*S. aureus*	/	Belofsky et al. (2006)
Trans,trans-1,7-diphenylhepta-4,6-dien-3-one (diarylheptanoids)	*Alpinia katsumadai*	/	*M. Smegmatis*	/	Groblacher et al. (2012)
Essential oils	*Salvia fruticosa*	TetK	*S. aureus, S. epidermitis*	Tetracycline	Chovanová et al. (2015)
Oleic and linoleic acids (fatty acids)	*Portulaca oleracea*	MsrA	MRSA	Erythromycin	Chan et al. (2015)

EPIs efflux pump inhibitors, *MRSA* methicillin-resistant *S. aureus*.

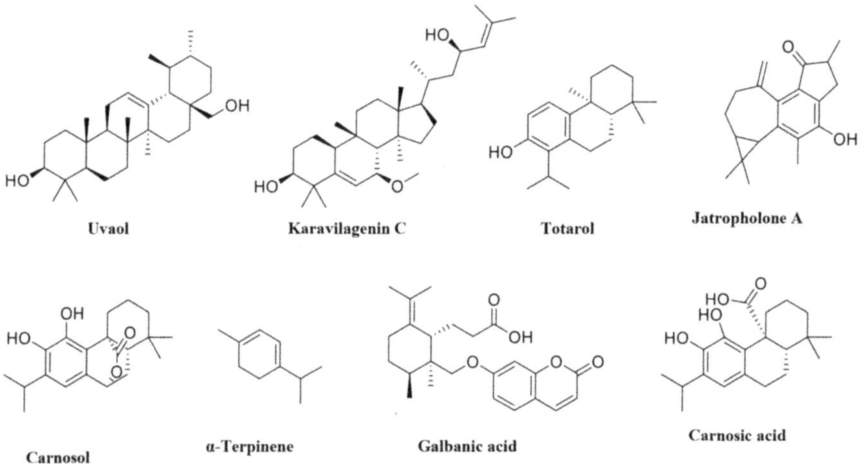

Uvaol   Karavilagenin C   Totarol   Jatropholone A

Carnosol   α-Terpinene   Galbanic acid   Carnosic acid

Fig. 16.4 Structures of common terpenoids with EPI activities

16.5.2 Phenolic Compounds

Phenolic compounds include a large class of naturally occurring products, mainly originated from plants. Nowadays, there is a growing interest in phenolic compounds owing to their diverse chemical structure and wide bioactivity important in the prevention of some chronic or degenerative disorders. These phytoconstituents are mostly found in functional foods or dietary plants. Phenolic compounds comprise flavonoids, phenolic acids, and tannins, among others. Flavonoids constitute the largest group of plant phenolics, accounting for over half of the eight thousand naturally occurring phenolic compounds (Harborne et al. 1999; Martins et al. 2011a; Nahar and Sarker 2019). The phenolic compounds with EPI properties are shown in Table 16.2. Some examples of the structure of phenolic compounds are given in Fig. 16.5. Among the plant-based EPIs, phenolic compounds are the most representative.

16.5.3 Alkaloids

Alkaloids are well-known nitrogen-containing natural phytochemicals. Literature unfolds that alkaloids possess numerous pharmacological actions. Alkaloids show a broad spectrum of activities including anti-infective potential (Adamski et al. 2020).

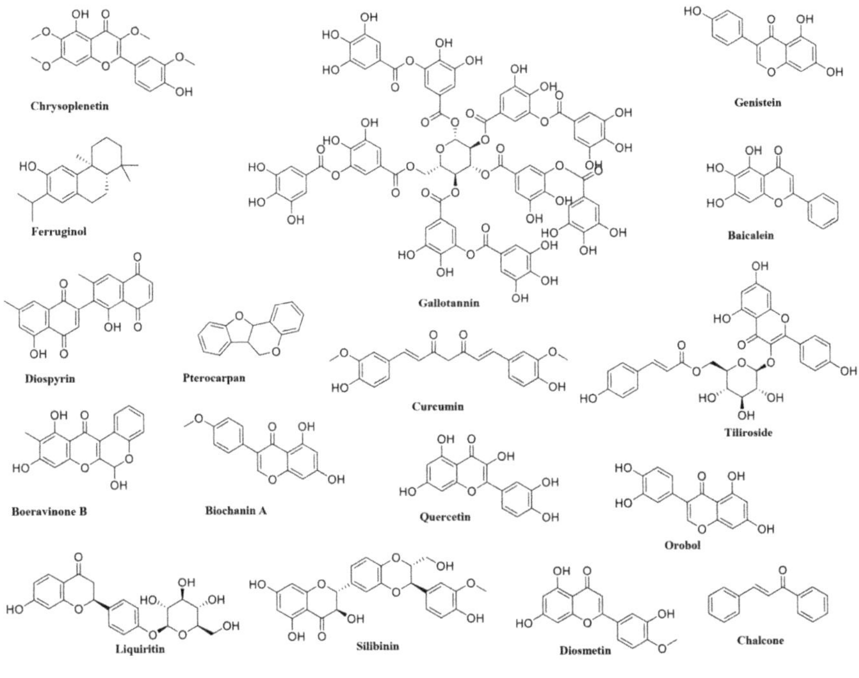

Fig. 16.5 Structures of some examples of phenolic compounds with EPI activities

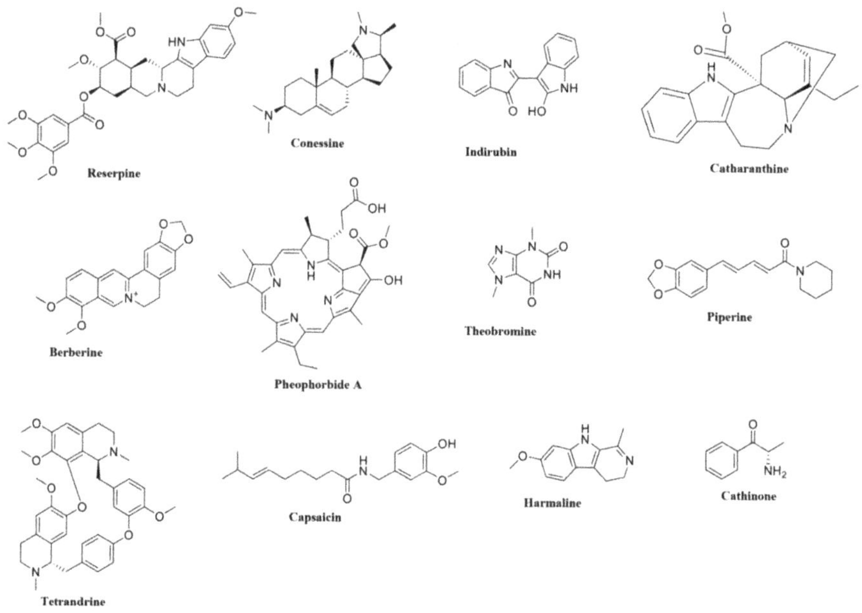

Fig. 16.6 Structures of some putative alkaloid EPIs

Alkaloids are the second most representative plant secondary metabolites with EPI activities after phenolic compounds. The well-known alkaloids EPIs comprise reserpine isolated from *Rauwolfia* sp. (Stavri et al. 2007), berberine, and 5'-MHC from *Berberis* sp. (Stermitz et al. 2000; Aghayan et al. 2017). Some of the alkaloids EPIs originated from dietary plants or spices include theobromine from *Theobroma cacao* (Piddock et al. 2010) and piperine from *Piper nigrum* (Kumar et al. 2008; Sharma et al. 2010). Most of these alkaloid EPIs have action on MexAB-OprM EPs of some highly resistant bacteria, namely, *P. aeruginosa*, *K. pneumoniae*, and *A. baumannii* (Table 16.3). The structures of some alkaloid EPIs are shown in Fig. 16.6.

16.6 Drugs from Plant-Derived EPIs: Current Stage and Challenges in Drug Development and Clinical Use

Infectious diseases are still a major cause of morbidity and mortality worldwide (Vos et al. 2020), and clinicians are urging for new tools to combat them efficiently. Based on the development stage, the world of antibiotic drug molecules can be partitioned into three categories: (1) approved, (2) in clinical trials, (3) in preclinical development (Lomovskaya and Bostian 2006). As of 2015 when the last batch of antibiotics was approved by the Food and Drug Administration (FDA) of the United States of America (USA), most of the existing molecules in these three categories did not have an adequate activity against any Gram-negative bacteria at all (Meyer 2005; Yu et al.

2020). In fact, recent research pointed that of the 160 antibiotics currently approved, there are very few that are still active against Gram-negative bacteria (Mahmood et al. 2016). EPs are identified as the main culprit of this observed reduced efficacy. Likewise, EPIs have appeared as the obvious solution, bearing the enticing prospect that all the abandoned and/or barely used antibiotics could regain clinical significance. The forecast of a market overload by new drugs did not come to fruition, owing to several challenges arising both inside and outside the laboratories.

16.6.1 Current Stage of Development of Plant-Derived EPI Drugs

Drug development is a five-step process: (1) discovery and development in the laboratory, (2) preclinical research with laboratory and animal testing to address questions about basic safety, (3) clinical trials on human subjects to assess the safety and efficacy, (4) review and approval by drug regulatory institutions, (5) post-market safety monitoring (FDA 2020). Microcide and Daiichi Pharmaceuticals were among the first companies in the late 1990s to engage into this five-step process to develop EPI drugs (Lomovskaya and Bostian 2006). Despite the promising results obtained in vitro with the molecule MC-207,110 (Georgopapadakou 2000), drug development was soon halted, owing to the unfavorable pharmacokinetic and toxicological profiles of the two basic moieties that had shown to be essential for activity (Lomovskaya and Bostian 2006).

Toxicity is a major hindrance to the development of EPIs drugs. Plant-derived EPIs then appeared as the panacea for toxicity, because naturally occurring compounds are usually more tolerated. Seukep et al. (2020a) have recently published an extensive review which reveals that there is a large number of available plant-derived EPIs. Despite the appealing number and the hypothetical "safety" nature of these compounds, many of them have failed to match such expectations. The plant alkaloid reserpine has not been applied in clinical settings because of its nephrotoxic nature (Pfeifer et al. 1976). For polyphenols, their effect as EPIs is observed only at high concentrations, which foresees a high risk of toxicity (Gibbons et al. 2004); hence, further in vivo and preclinical studies with these compounds have not been undertaken (Sudeno et al. 2004). Concerning the other plant-derived EPIs, to the best of our knowledge, there is no existing published studies or evidence which indicates that they are currently in drug development. The Infectious Diseases Society of America has vehemently critiqued the lack of new antibiotics in pharmaceutical pipelines to combat ESKAPE pathogens (Boucher et al. 2009). This scarcity might be due to the lack of EPI-focused financial investments, which in turn is most probably rooted in the multiple roadblocks that exist on the path leading to successful commercial plant-derived EPI drugs. We present herein some of these challenges.

16.6.2 Current Challenges in the Development and Clinical Use of Plant-Derived EPI Drugs

These challenges are of different origins: economic, scientific, and administrative (Sharma et al. 2019). Economic factors have substantially contributed to the lack of available plant-derived EPI drugs by seriously reducing the global workforce and infrastructure devoted to the development of these drugs. Most major pharmaceutical companies have stopped in-house natural product research and have outsourced this duty to smaller biotech organizations (Gibbons 2008). Unfortunately, the latter do not have the equivalent amount of experience and the level of infrastructure required to be successful in such a complex task. Pharmaceutical companies undertook this management U-turn probably because of the unattractive low market share values of antibiotics targeting drug-resistant germs. A closer look at the top 200 bestselling brand-name drugs in 2006 reveals that the first drug targeting drug-resistant bacteria ranked 139th and was the MRSA specific agent Zyvox (linezolid) of Pfizer, with a market sale value of 200 million US dollars, very far behind the 6.58 billion US dollars of the top-ranked drug Lipitor still from Pfizer (Gibbons 2008). Pharmaceutical companies have largely reoriented their investments towards drugs targeting lifestyle-related diseases such as cardiovascular diseases, diabetes, mental health, and cancers, because of the enticing economic dividends. When some investments are nevertheless made in the development of plant-derived EPIs, major scientific and administrative challenges step in.

One of the major scientific challenges is the toxicity of plant-derived EPIs described earlier. EPs, similar to those existing in bacteria, are also present in human cells (Spengler et al. 2017). Given the vital physiological functions bestowed into EPs, their inhibition could severely compromise the integrity of human cells, leading to adverse health outcomes. This joins the stream of the multiple difficulties inherent to the development of any new chemical entity (NCE), added to the substantial amount of time and capital initially required for the discovery of such NCE (Sharma et al. 2019). An extra impediment to the development of plant-derived EPI drugs is their complex and bulky structure which makes them difficult to synthesize (Sharma et al. 2019). The development of NCE from plant-derived EPIs is further hindered by the limited available knowledge about the functional assemblies of various efflux pumps, added to the scarcity of reliable biochemical, computational, and structural models of efflux pump functions (Opperman and Nguyen 2015). Another major conundrum is the choice of EPI targets. In effect, EPs are one of the many mechanisms used by bacteria to evade antibiotic toxicity. Furthermore, multiple EPs can be co-expressed at the same time in the same bacteria. Finally, there is an important substrate redundancy among various EPs. Together, all these factors could compromise the efficacy of EPI drugs at the community level (Sharma et al. 2019). Even if the abovementioned hurdles were crossed, another important challenge would loom: the compatibility between the EPI and the associated antibiotic (Sharma et al. 2019). In effect, EPIs are to be used in combination with the antibiotic that they intend to potentiate. In this regard, there must be a complementarity between the pharmacokinetics of both drug entities (Lomovskaya and Bostian 2006) to achieve the desired effect in a safe manner. The later pre-requisite is often ignored in laboratory experiments, such that many of the EPIs which are found to be successful in vitro usually fail in vivo (Sharma et al. 2019). The best example of such is the combination verapamil-clarithromycin which was the theme of an FDA warning because clarithromycin could cause the accumulation of verapamil to toxic levels, since clarithromycin targets a cytochrome responsible for verapamil metabolism (Gandhi et al. 2013).

If these scientific challenges were overcome, another major obstacle to cross will be the strict regulatory measures that control clinical trials (Lomovskaya and Bostian 2006). Even though certain EPI-antibiotic combinations can be proven efficient in vitro and in animal models (Drusano et al. 2001; Firsov et al. 2003; Jumbe et al. 2003; Zinner et al. 2003; Drusano et al. 2004), it is only the clinical endpoints that are relevant to the drug approval process. Given the dearth of evidence on the clinical endpoints of plant-derived EPI-antibiotic combinations, scientists are usually short of arguments to defend that these drugs have a similar activity against strains that are susceptible or resistant to the antibiotic of the combination (Lomovskaya and Bostian 2006). Hence, it is hard to advance from step 2 to step 3 of the drug development process. Drug regulatory bodies like the FDA are considering certain innovations in that regard (Lomovskaya and Bostian 2006), but they are yet to produce visible results.

16.7 Future Perspectives

Even though the use of EPIs as pharmacological agents encounters numerous challenges, their usefulness remains of key significance in the fight against the efflux pump-mediated multidrug resistance. In effect, the scarcity of new and effective antibacterial agents urges for the development of alternative therapeutic means. As such, EPIs provide a ray of hope since they can help to restore the efficacy of the standard antibiotics. Therefore, the utilization of EPIs removes the necessity to discover novel antibiotics, a scheme that spares a considerable amount of time, effort, and capital. Furthermore, it enables the clinicians to regain the advantage of the well-settled pharmacological properties of known antibiotics. Here lies a crucial implication of EPIs, if we consider the fact that the technology is available for a massive fabrication of pre-optimized and stockpiled antibiotics. Another noticeable benefit of EPIs is the very low probability to generate resistant mutants. The combination of antibiotics and EPI is, therefore, effective in not just eradicating the already resistant bacteria but also by offering a reprieve from the foreseen challenges of AMR. The use of EPIs is an enticing strategy, yet the current status of drug development is still far from clinical application. As discussed in the present chapter, there are many challenges that need to be addressed urgently. More studies are necessary to shed light on the scientific and economic advantages of EPIs. This would eventually revamp the interest of pharmaceutical industries and bring in more human and economic capital. A considerable amount of work has been accomplished in the laboratory; more consideration and efforts are required before the whole world can benefit from the advantages of EPIs.

16.8   Conclusion

EPs are the major mechanism responsible for the resistance to one and/or more classes of antibiotics in bacteria; consequently, the development of effective EPIs is of significant importance in the fight against AMR. EPIs display several advantages including the restoration of the activity of standard antibiotics by reversing the resistance strategy and the extremely low frequency of resistant mutants. Plants are a rich source of EPIs, and they are hypothetically considered less toxic than synthetic products. The low stability, selectivity, and high toxicity for mammalian cells significantly impede the clinical application of EPIs for the management of drug-resistant infections. These factors which restrict the use of EPIs should be considered as a priority in future investigations.

References

Abulrob AN, Suller MTE, Gumbleton M et al (2004) Identification and biological evaluation of grapefruit oil components as potential novel efflux pump modulators in methicillin-resistant *Staphylococcus aureus* bacterial strains. Phytochemistry 65:3021–3027

Adamski Z, Blythe LL, Milella L, Bufo SA (2020) Biological activities of alkaloids: from toxicologyto pharmacology. Toxins 12:210

Aghayan SS, Mogadam KH, Fazli M et al (2017) The effects of berberine and palmatine on efflux pumps inhibition with different gene patterns in Pseudomonas aeruginosa isolated from burn infections. Avicenna J Med Biotechnol 9:2–7

Ahmed M, Borsch CM, Neyfakh AA et al (1993) Mutants of the *Bacillus subtilis* multidrug transporter Bmr with altered sensitivity to the antihypertensive alkaloid reserpine. J Biol Chem 268:11086–11089

Alvarez-Ortega C, Olivares J, Martinez JL (2013) RND multidrug efflux pumps: what are they good for? Front Microbiol 4:7. https://doi.org/10.3389/fmicb.2013.00007

Anoushiravani M, Falsafi T, Niknam V (2009) Proton motive force-dependent efflux of tetracycline in clinical isolates of *Helicobacter pylori*. J Med Microbiol 58:1309–1313

Aparna V, Dineshkumar K, Mohanalakshmi N et al (2014) Identification of natural compound inhibitors for multidrug efflux pumps of *Escherichia coli* and *Pseudomonas aeruginosa* using in silico high-throughput virtual screening and in vitro validation. PLoS One 7:e101840

Askoura M, Mottawea W, Abujamel T et al (2011) Efflux pump inhibitors (EPIs) as new antimicrobial agents against *Pseudomonas aeruginosa*. Libyan J Med 6:86238

Aslam B, Wang W, Arshad IM et al (2018) Antibiotic resistance: a rundown of a global crisis. Infect Drug Resist 11:1645–1658

Bag A, Chattopadhyay RR (2014) Efflux-pump inhibitory activity of a gallotannin from *Terminalia chebula* fruit against multidrug-resistant uropathogenic *Escherichia coli*. Nat Prod Res 28:1280–1283

Bame JR, Graf TN, Junio HA et al (2013) Sarothrin from *Alkannaorientalis* is an antimicrobial agent and efflux pump inhibitor. Planta Med 79:327–329

Baugh S, Phillips CR, Ekanayaka AS et al (2014) Inhibition of multidrug efflux as a strategy to prevent biofilm formation. J Antimicrob Chemother 69:673–681. https://doi.org/10.1093/jac/dkt420

Belofsky G, Carreno R, Lewis K et al (2006) Metabolites of the 'smoke tree', *Dalea spinosa*, potentiate antibiotic activity against multidrug-resistant Staphylococcus aureus. J Nat Prod 69:261–264

Bhardwaj AK, Mohanty P (2012) Bacterial efflux pumps involved in multidrug resistance and their inhibitors: Rejuvinating the antimicrobial chemotherapy. Recent Pat Antiinfect Drug Discov 7:73–89

Bhattacharyya T, Sharma A, Akhter J, Pathania R (2017) The small molecule IITR08027 restores the antibacterial activity of fluoroquinolones against multidrug-resistant *Acinetobacter baumannii* by efflux inhibition. Int J Antimicrob Agents 50:219–226

Blair JM, Piddock LJ (2016) How to measure export via bacterial multidrug resistance efflux pumps. MBioi 7:e00840–e00816

Bohnert JA, Karamian B, Nikaido H et al (2010) Optimized Nile red efflux assay of AcrAB- TolC multidrug efflux system shows competition between substrates. Antimicrob Agents Chemother 54:3770–3775

Bohnert JA, Kern WV (2005) Selected arylpiperazines are capable of reversing multidrug resistance in *Escherichia coli* overexpressing RND efflux pumps. Antimicrob Agents Chemother 49:849–852

Bohnert JA, Schuster S, Szymaniak-Vits M et al (2011) Determination of real-time efflux phenotypes in *Escherichia coli*AcrB binding pocket phenylalanine mutants using a 1,20-dinaphthylamine efflux assay. PLoS One 6:e21196

Boucher HW, Talbot GH, Bradley JS et al (2009) Bad bugs, no drugs: no ESKAPE! an update from the infectious diseases society of America. Clin Infect Dis 48:1–12. https://doi.org/10.1086/595011

Brown AR, Ettefagh KA, Todd D et al (2015) A mass spectrometry-based assay for improved quantitative measurements of efflux pump inhibition. PLoS One 10:e0124814

Cabral V, Luo X, Junqueira E et al (2015) Enhancing activity of antibiotics against *Staphylococcus aureus*: *Zanthoxylumcapense* constituents and derivatives. Phytomedicine 22:469–476

Cai H, Rose K, Liang LH et al (2009) Development of a liquid chromatography/mass spectrometry-based drug accumulation assay in *Pseudomonas aeruginosa*. Anal Biochem 385:321–325

Chan BC, Han X, Lui S et al (2015) Combating against methicillin-resistant *Staphylococcus aureus*—two fatty acids from purslane (*Portulaca oleracea* L.) exhibit synergistic effects with erythromycin. J Pharm Pharmacol 67:107–116

Chan BC, Ip M, Gong H et al (2013) Synergistic effects of diosmetin with erythromycin against ABC transporter over-expressed methicillin-resistant *Staphylococcus aureus* (MRSA) RN4220/pUL5054 and inhibition of MRSA pyruvate kinase. Phytomedicine 20:611–614

Chan BC, Ip M, Lau CB, Lui et al (2011) Synergistic effects of baicalein with ciprofloxacin against NorA over-expressed methicillin-resistant *Staphylococcus aureus* (MRSA) and inhibition of MRSA pyruvate kinase. J Ethnopharmacol 137:767–773

Chaves TP, Clementino ELC, Felismino DC et al (2015) Antibiotic resistance modulation by natural products obtained from *Nasutitermescorniger* (Motschulsky, 1855) and its nest. Saudi J Biol Sci 22:404–408

Chérigo L, Pereda MR, Fragoso SM et al (2008) Inhibitors of bacterial multidrug efflux pumps from the resin glycosides of *Ipomoea murucoides*. J Nat Prod 71:1037–1045

Choudhury D, Talukdar AD, Chetia P et al (2016) Screening of natural products and derivatives for the identification of RND efflux pump inhibitors. Comb Chem High Throughput Screen 19:705–713

Chovanová R, Mezovská J, Vavebrková Š et al (2015) The inhibition of TetK efflux pump of tetracycline resistant Staphylococcus epidermidis by essential oils from three salvia species. Lett Appl Microbiol 61:58–62

Chung YJ, Saier MH (2001) SMR-type multidrug resistance pumps. Curr Opin Drug Discov Dev 4:237–245

Chusri S, Villanueva I, Voravuthikunchai SP et al (2009) Enhancing antibiotic activity: a strategy to control acinetobacter infections. J Antimicrob Chemother 64:1203–1211

Coutinho HDM, Vasconcellos A, Lima MA et al (2009) Termite usage associated with antibiotic therapy: enhancement of aminoglycoside antibiotic activity by natural products of *Nasutitermescorniger* (Motschulsky 1855). BMC Complement Altern Med 9:35

Dadgostar P (2019) Antimicrobial resistance: implications and costs. Infect Drug Resist 12:3903–3910

Davies J, Davies D (2010) Origins and evolution of antibiotic resistance. Microbiol Mol Biol Rev 74:417–433

Dawson RJP, Locher KP (2006) Structure of a bacterial multidrug ABC transporter. Nature 443:180–185. https://doi.org/10.1038/nature05155

Dey D, Debnath S, Hazra S et al (2012) Pomegranate pericarp extract enhances the antibacterial activity of ciprofloxacin against extended-spectrum ß- lactamase (ESBL) and metallo-ß-lactamase (MBL) producing gram-negative bacilli. Food Chem Toxicol 50:4302–4309

Dos Santos JF, Tintino SR, de Freitas TS et al (2018) In vitro and in silico evaluation of the inhibition of *Staphylococcus aureus* efflux pumps by caffeic and gallic acid. Comp Immunol Microbiol Infect Dis 57:22–28

Dreier J, Ruggerone P (2015) Interaction of antibacterial compounds with RND efflux pumps in *Pseudomonas aeruginosa*. Front Microbiol 6:660

Drusano GL, Preston SL, Fowler C et al (2004) Relationship between fluoroquinolone area under the curve: minimum inhibitory concentration ratio and the probability of eradication of the infecting pathogen, in patients with nosocomial pneumonia. J Infect Dis 189:1590–1597. https://doi.org/10.1086/383320

Drusano GL, Preston SL, Owens RC et al (2001) Fluoroquinolone pharmacodynamics. Clin Infect Dis 33:2091–2096. https://doi.org/10.1086/323748

Duval M, Dar D, Carvalho F et al (2018) HflXr, a homolog of a ribosome-splitting factor, mediates antibiotic resistance. PNAS 52:13359–13364

Dwivedi DRG, Singh DP, Sanchita et al (2017) Efflux pumps: warheads of gram-negative bacteria and efflux pump inhibitors. In: New approaches in biological research. Nova Science Publishers, New York, pp 35–71

Dwivedi GR, Gupta S, Maurya A et al (2015a) Synergy potential of indole alkaloids and its derivative against drug-resistant *Escherichia coli*. Chem Biol Drug Des 86:1471–1481. https://doi.org/10.1111/cbdd.12613

Dwivedi GR, Maurya A, Yadav DK et al (2015b) Drug resistance reversal potential of ursolic acid derivatives against nalidixic acid- and multidrug-resistant Escherichia coli. Chem Biol Drug Des 86:272–283

Dwivedi GR, Tyagi R, Sanchita et al (2018) Antibiotics potentiating potential of catharanthine against superbug *Pseudomonas aeruginosa*. J Biomol Struct Dyn 36:4270–4284

Falcão-silva V, Silva DA, de Souza MF et al (2009) Modulation of drug resistance in *Staphylococcus aureus* by a kaempferol glycoside from *Herissantiatiubae* (Malvaceae). Phytother Resm 10:1367–1370

Fazly BS, Iranshahi M, Naderinasab M et al (2010) Evaluation of the effects of galbanic acid from Ferulaszowitsiana and conferol from *F. badrakema*, as modulators of multi-drug resistance in clinical isolates of *Escherichia coli* and *Staphylococcus aureus*. Res Pharm Sci 5:21–28

FDA (2020) The drug development process. https://www.fda.gov/patients/learn-about-drug-and-device-approvals/drug-development-process Accessed 31 Oct 2020

Feng Z, Liu D, Wang L et al (2020) A putative efflux transporter of the ABC family, YbhFSR, in *Escherichia coli*functions in tetracycline efflux and Na+(Li+)/H+ transport. Front Microbiol 11:556. https://doi.org/10.3389/fmicb.2020.00556

Fenosa A, Fusté E, Ruiz L et al (2009) Role of tolC in *Klebsiella oxytoca* resistance to antibiotics. J Antimicrob Chemother 63:668–674

Fiamegos YC, Kastritis PL, Exarchou V et al (2011) Antimicrobial and efflux pump inhibitory activity of caffeoylquinic acids from *Artemisia absinthium* against gram-positive pathogenic bacteria. PLoS One 4:812–817

Firsov AA, Vostrov SN, Lubenko IY et al (2003) In vitro pharmacodynamic evaluation of the mutant selection window hypothesis using four fluoroquinolones against *Staphylococcus aureus*. Antimicrob Agents Chemother 47:1604–1613. https://doi.org/10.1128/AAC.47.5. 1604-1613.2003

Fralick JA (1996) Evidence that TolC is required for functioning of the mar/AcrAB efflux pump of *Escherichia coli*. J Bacteriol 178:5803–5805

Fujita M, Shiota S, Kuroda T et al (2005) Remarkable synergies between baicalein and tetracycline, and baicalein and ß-lactams against methicillin-resistant *Staphylococcus aureus*. Microbiol Immunol 49:391–396

Gandhi S, Fleet JL, Bailey DG et al (2013) Calcium-channel blocker-clarithromycin drug interactions and acute kidney injury. JAMA 310:2544–2553. https://doi.org/10.1001/jama. 2013.282426

Garvey M, Rahman M, Gibbons S et al (2011) Medicinal plant extracts with efflux inhibitory activity against gram-negative bacteria. Int J Antimicrob Agents 37:145–151

Garvey MI, Piddock LJV (2008) The efflux pump inhibitor reserpine selects multidrug-resistant *Streptococcus pneumoniae* strains that overexpress the ABC transporters PatA and PatB. Antimicrob Agents Chemother 52:1677–1685

Georgopapadakou NH (2000) Infectious disease 2000: drug resistance and new drugs. Drug Resistance Updates 3:265–269. https://doi.org/10.1054/drup.2000.0168

Gibbons S (2008) Phytochemicals for bacterial resistance - strengths, weaknesses and opportunities. Planta Med 74:594–602. https://doi.org/10.1055/s-2008-1074518

Gibbons S, Moser E, Kaatz GW (2004) Catechin gallates inhibit multidrug resistance (MDR) in *Staphylococcus aureus*. Planta Med 70:1240–1242. https://doi.org/10.1055/s-2004-835860

Gibbons S, Oluwatuyi M, Veitch N et al (2003) Bacterial resistance modifying agents from *Lycopuseuropaeus*. Phytochemistry 62:83–87

Grinius L, Dreguniene G, Goldberg EB et al (1992) A staphylococcal multidrug resistance gene product is a member of a new protein family. Plasmid 27:119–129. https://doi.org/10.1016/ 0147-619x(92)90012-y

Groblacher B, Kunert O, Bucar F (2012) Compounds of *Alpinia katsumadai* as potential efflux inhibitors in *Mycobacterium smegmatis*. Bioorg Med Chem 20:2701–2706

Gupta S, Cohen KA, Winglee K et al (2014) Efflux inhibition with verapamil potentiates bedaquiline in *Mycobacterium tuberculosis*. Antimicrob Agents Chemother 58:574–576

Gupta VK, Tiwari N, Gupta P et al (2016) A clerodane diterpene from *Polyalthialongifolia*as a modifying agent of the resistance of methicillin resistant *Staphylococcus aureus*. Phytomedicine 23:654–661

Handzlik J, Matys A, Kieć-Kononowicz K (2013) Recent advances in multi-drug resistance (MDR) efflux pump inhibitors of gram-positive bacteria S. *aureus*. Antibiotics 2:28–45. https://doi.org/ 10.3390/antibiotics2010028

Harborne JB, Baxter H, Moss GP (1999) Phytochemical dictionary: handbook of bioactive compounds from plants, 2nd edn. Taylor & Francis, London

Hassan KA, Liu Q, Henderson PJF et al (2015) Homologs of the *Acinetobacter baumannii*Aceltransporter represent a new family of bacterial multidrug efflux systems. Mbio 6(1):e01982. https://doi.org/10.1128/mBio.01982-14

He X, Szewczyk P, Karyakin A et al (2010) Structure of a cation-bound multidrug and toxic compound extrusion transporter. Nature 467:991–994. https://doi.org/10.1038/nature09408

Higgins CF (2007) Multiple molecular mechanisms for multidrug resistance transporters. Nature 446:749–757. https://doi.org/10.1038/nature05630

Holler JG, Christensen SB, Slotved HC et al (2012) Novel inhibitory activity of the *Staphylococcus aureus* NorA efflux pump by a kaempferol rhamnoside isolated from *Persealingue*nees. J AntimicrobChemother 67:1138–1144

Ivnitski-Steele I, Holmes AR, Lamping E et al (2009) Identification of nile red as a fluorescent substrate of the *Candida albicans* ATP-binding cassette transporters Cdr1p and Cdr2p and the major facilitator superfamily transporter Mdr1p. Anal Biochem 394:87–91

Jack DL, Yang NM, Saier MH (2001) The drug/metabolite transporter superfamily. Eur J Biochem 268:3620–3639. https://doi.org/10.1046/j.1432-1327.2001.02265.x

Jacoby GA (2009) AmpC beta-lactamases. Clin Microbiol Rev 22:161–182. https://doi.org/10.1128/CMR.00036-08

Jin J, Zhang J, Guo N et al (2010) Farnesol, a potential efflux pump inhibitor in *Mycobacterium smegmatis*. Molecules 15:7750–7762

Joshi P, Singh S, Wani A et al (2014) Osthol and curcumin as inhibitors of human Pgp and multidrug efflux pumps of *Staphylococcus aureus*: reversing the resistance against frontline antibacterial drugs. Med Chem Commun 5:1540–1547

Juliano RL, Ling V (1976) A surface glycoprotein modulating drug permeability in Chinese hamster ovary cell mutants. Biochim Biophys Acta 455:152–162. https://doi.org/10.1016/0005-2736(76)90160-7

Jumbe N, Louie A, Leary R et al (2003) Application of a mathematical model to prevent in vivo amplification of antibiotic-resistant bacterial populations during therapy. J Clin Invest 112:275–285. https://doi.org/10.1172/JCI200316814

Junwei W, Jing Z, Sanxia L et al (2013) Application of liquiritin in preparing *Escherichia coli* fluoroquinolone efflux pump inhibitor. Chinese Patent CN 102988400 (2013), 27 March 2013

Kakarla P, Floyd J, Mukherjee M et al (2017) Inhibition of the multidrug efflux pump LmrS from *Staphylococcus aureus* by cumin spice *Cuminumcyminum*. Arch Microbiol 199:465–474

Kalia NP, Mahajan P, Mehra R et al (2012) Capsaicin, a novel inhibitor of the NorA efflux pump, reduces the intracellular invasion of *Staphylococcus aureus*. J Antimicrob Chemother 67:2401–2408

Karam G, Chastre J, Wilcox MH, Vincent JL (2016) Antibiotic strategies in the era of multidrug resistance. Crit Care 20:136

Kobayashi N, Nishino K, Yamaguchi A (2001) Novel macrolide-specific ABC-type efflux transporter in *Escherichia coli*. J Bacteriol 183:5639–5644. https://doi.org/10.1128/JB.183.19.5639-5644.2001

Koronakis V, Eswaran J, Hughes C (2004) Structure and function of TolC: the bacterial exit duct for proteins and drugs. Annu Rev Biochem 73:467–489. https://doi.org/10.1146/annurev.biochem.73.011303.074104

Krishnan VR, Cacciotto P, Malloci G et al (2016) Multidrug efflux pumps and their inhibitors characterized by computational modeling. In: Li XZ, Elkins CA, Zgurskaya HI (eds) Efflux mediated antimicrobial resistance in Bacteria. Springer, Cham, Switzerland, pp 797–831

Kuete V (2010) Potential of Cameroonian plants and derived products against microbial infections: a review. Planta Med 76:1479–1491

Kumar A, Khan IA, Koul S et al (2008) Novel structural analogues of piperine as inhibitors of the NorA efflux pump of *Staphylococcus aureus*. J Antimicrob Chemother 61:1270–1276

Kumar M, Yash P, Anand VK (2009) An ethnobotanical study of medicinal plants used by the locals in Kishtwar, Jammu and Kashmir, India. Ethnobot Leaf 113:1240–1256

Kumar R, Patial SJP (2016) A review on efflux pump inhibitors of gram-positive and gram-negative bacteria from plant sources. Int J CurrMicrobiol App Sci 5(6):837–855

Kumar S, Varela MF (2013) Molecular mechanisms of bacterial resistance to antimicrobial agents. In: Méndez-Vilas A (ed) Microbial pathogens and strategies for combating them, science, technology, and education. Formatex Research Centre, Badajoz, Spain, pp 522–534

Lee MD, Galazzo JL, Staley AL et al (2001) Microbial fermentation-derived inhibitors of efflux-pump-mediated drug resistance. Farmaco 56:81–85

Levy SB, McMurry L (1978) Plasmid-determined tetracycline resistance involves new transport systems for tetracycline. Nature 276:90–92. https://doi.org/10.1038/276090a0

Lewis K (2001) Riddle of biofilm resistance. Antimicrob Agents Chemother 45:999–1007. https://doi.org/10.1128/AAC.45.4.999-1007.2001

Li B, Yao Q, Pan XC et al (2011) Artesunate enhances the antibacterial effect of {beta}-lactam antibiotics against *Escherichia coli* by increasing antibiotic accumulation via inhibition of the multidrug efflux pump system AcrAB-TolC. J Antimicrob Chemother 66:769–777

Li X-Z (2005) Quinolone resistance in bacteria: emphasis on plasmid-mediated mechanisms. Int J Antimicrob Agents 25:453–463. https://doi.org/10.1016/j.ijantimicag.2005.04.002

Li X-Z, Nikaido H (2004) Efflux-mediated drug resistance in bacteria. Drugs 64:159–204. https://doi.org/10.2165/00003495-200464020-00004

Li X-Z, Nikaido H (2009) Efflux-mediated drug resistance in bacteria: an update. Drugs 69:1555–1623. https://doi.org/10.2165/11317030-000000000-00000

Li XZ, Nikaido H, Poole K (1995) Role of mexA-mexB-oprM in antibiotic efflux in *Pseudomonas aeruginosa*. Antimicrob Agents Chemother 39:1948–1953

Limaverde PW, Campina FF, da Cunha FAB et al (2017) Inhibition of the TetK efflux-pump by the essential oil of *Chenopodium ambrosioides* L. and α-terpinene against *Staphylococcus aureus* IS-58. Food Chem Toxicol 109:957–961

Liu KCS, Yang SL, Roberts MF et al (1992) Antimalarial activity of *Artemisia annua* flavonoids from whole plants and cell cultures. Plant Cell Rep 11:637–640

Lomovskaya O, Bostian KA (2006) Practical applications and feasibility of efflux pump inhibitors in the clinic--a vision for applied use. Biochem Pharmacol 71:910–918. https://doi.org/10.1016/j.bcp.2005.12.008

Lomovskaya O, Lewis K (1992) Emr, an *Escherichia coli* locus for multidrug resistance. Proc Natl Acad Sci USA 89:8938–8942. https://doi.org/10.1073/pnas.89.19.8938

Lomovskaya O, Warren MS, Lee A et al (2001) Identification and characterization of inhibitors of multidrug resistance efflux pumps in *Pseudomonas aeruginosa*: novel agents for combination therapy. Antimicrob Agents Chemother 45:105–116

Lorca GL, Barabote RD, Zlotopolski V et al (2007) Transport capabilities of eleven gram-positive bacteria: comparative genomic analyses. Biochim Biophys Acta 1768:1342–1366. https://doi.org/10.1016/j.bbamem.2007.02.007

Ma D, Cook DN, Alberti M et al (1993) Molecular cloning and characterization of acrA and acrE genes of *Escherichia coli*. J Bacteriol 175:6299–6313. https://doi.org/10.1128/jb.175.19.6299-6313.1993

Magiorakos AP, Srinivasan A, Carey RB et al (2012) Multidrug-resistant, extensively drug-resistant and pan-drug-resistant bacteria: an international expert proposal for interim standard definitions for acquired resistance. Clin Microbiol Infect 18:268–281

Mahmood HY, Jamshidi S, Sutton JM et al (2016) Current advances in developing inhibitors of bacterial multidrug efflux pumps. Curr Med Chem 23:1062–1081. https://doi.org/10.2174/0929867323666160304150522

Mambe TF, Na-Iya J, Fotso WG et al (2019) Antibacterial and antibiotic modifying potential of crude extracts, fractions, and compounds from acacia polyacantha willd. Against MDR gram-negative bacteria. Evid Based Complement Alternat Med 2019:13

Marquez B, Neuville L, Moreau NJ et al (2005) Multidrug resistance reversal agent from *Jatropha elliptica*. Phytochemistry 66:1804–1811

Martins A, Vasas A, Viveiros M et al (2011a) Antibacterial properties of compounds isolated from *Carpobrotus edulis*. Int J Antimicrob Agents 37:438–444

Martins S, Mussatto SI, Martínez-Avila G et al (2011b) Bioactive phenolic compounds: production and extraction by solid-state fermentation. A review. Biotechnol Adv 29:365–373

Matsumoto Y, Hayama K, Sakakihara S et al (2011) Evaluation of multidrug efflux pump inhibitors by a new method using microfluidic channels. PLoS One 6:e18547

Matsumura K, Furukawa S, Ogihara H et al (2011) Roles of multidrug efflux pumps on the biofilm formation of *Escherichia coli* K-12. Biocontrol Sci 16:69–72. https://doi.org/10.4265/bio.16.69

Maurya A, Dwivedi G, Darokar M et al (2013) Antibacterial and synergy of clavine alkaloid lysergol and its derivatives against nalidixic acid-resistant *Escherichia coli*. Chem Biol Drug Des 81:484–490

Meyer AL (2005) Prospects and challenges of developing new agents for tough gram-negatives. Curr Opin Pharmacol 5:490–494. https://doi.org/10.1016/j.coph.2005.04.012

Michalet S, Cartier G, David B et al (2007) N-Caffeoylphenalkylamide derivates as bacterial efflux pump inhibitors. Bioorg Med Chem Lett 17:1755–1758

Miladi H, Zmantar T, Chaabouni Y et al (2009) Antibacterial and efflux pump inhibitors of thymol and carvacrol against food-borne pathogens. Microb Pathog 99:95–100

Mohtar M, Johari SA, Li AR et al (2009) Inhibitory and resistance-modifying potential of plant-based alkaloids against methicillin-resistant *Staphylococcus aureus* (MRSA). Curr Microbiol 59:181–186

Molnár J, Engi H, Hohmann J et al (2010) Reversal of multidrug resitance by natural substances from plants. Curr Top Med Chem 10:1757–1768. https://doi.org/10.2174/156802610792928103

Morel C, Stermitz FR, Tegos G et al (2003) Isoflavones as potentiators of antibacterial activity. J Agric Food Chem 51:5677–5679

Mukanganyama S, Chirisa E, Hazra B et al (2012) Antimycobacterial activity of diospyrin and its derivatives against *Mycobacterium aurum*. Res Pharm 2:1–13

Mulani MS, Kamble EE, Kumkar SN et al (2019) Emerging strategies to combat ESKAPE pathogens in the era of antimicrobial resistance: a review. Front Microbiol 10:539. https://doi.org/10.3389/fmicb.2019.00539

Nahar L, Sarker SD (2019) Chemistry for pharmacy students –general, organic and natural product chemistry, 2nd edn. Elsevier, Amsterdam

Najafi A (2016) There is no escape from the ESKAPE Pathogens. https://emerypharma.com/blog/author/anajafi/ Accessed 10 Feb 2019

Nakashima R, Sakurai K, Yamasaki S et al (2011) Structures of the multidrug exporter AcrB reveal a proximal multisite drug-binding pocket. Nature 480:565–569. https://doi.org/10.1038/nature10641

Nargotra A, Koul S, Sharma S et al (2008) Quantitative-structure-activity relationship (QSAR) of aryl alkenyl amides/imines for bacterial efflux pump inhibitors. Eur J Med Chem 44:229–238

Negi N, Prakash P, Gupta ML et al (2014) Possible role of curcumin as an efflux pump inhibitor in multi drug resistant clinical isolates of *Pseudomonas aeruginosa*. J Clin Diagn Res 10:04–07

Nelson ML (2002) Modulation of antibiotic efflux in bacteria. Antiinfect Agents Med Chem 1:35–54

Neyfakh AA (2002) Mystery of multidrug transporters: the answer can be simple. Mol Microbiol 44:1123–1130. https://doi.org/10.1046/j.1365-2958.2002.02965.x

Nikaido H (1994) Prevention of drug access to bacterial targets: permeability barriers and active efflux. Science 264:382–388. https://doi.org/10.1126/science.8153625

Nikaido H (1996) Multidrug efflux pumps of gram-negative bacteria. J Bacteriol 178:5853–5859

Nikaido H (1998) Multiple antibiotic resistance and efflux. Curr Opin Microbiol 1:516–523. https://doi.org/10.1016/S1369-5274(98)80083-0

Nikaido H (2003) Molecular basis of bacterial outer membrane permeability revisited. Microbiol Mol Biol Rev 67:593–656. https://doi.org/10.1128/MMBR.67.4.593-656.2003

Nikaido H (2011) Structure and mechanism of RND-type multi-drug efflux pumps. Adv Enzymol Relat Areas Mol Biol 77:1–60

Okoche D, Asiimwe BB, Katabazi FA et al (2015) Prevalence and characterization of carbapenem-resistant Enterobacteriaceae isolated from Mulago National Referral Hospital, Uganda. PLos One 10:e0135745. https://doi.org/10.1371/journal.pone.0135745

Oluwatuyi M, Kaatz GW, Gibbons S (2004) Antibacterial and resistance modifying activity of *Rosmarinusofficinalis*. Phytochemistry 65:3249–3254

Opperman TJ, Nguyen ST (2015) Recent advances toward a molecular mechanism of efflux pump inhibition. Front Microbiol 6:421. https://doi.org/10.3389/fmicb.2015.00421

Orhan G, Bayram A, Zer Y et al (2005) Synergy tests by E-test and checkerboard methods of antimicrobial combinations against *Brucella melitensis*. J Clin Microbiol Infect 43:140–143

Paixão L, Rodrigues L, Couto I et al (2009) Fluorometric determination of ethidium bromide efflux kinetics in *Escherichia coli*. J Biol Eng 3:18

Pao SS, Paulsen IT, Saier MH (1998) Major facilitator superfamily. Microbiol Mol Biol Rev 62:1–34

Paulsen IT, Brown MH, Littlejohn TG et al (1996a) Multidrug resistance proteins QacA and QacB from *Staphylococcus aureus*: membrane topology and identification of residues involved in substrate specificity. Proc Natl Acad Sci USA 93:3630–3635. https://doi.org/10.1073/pnas.93.8. 3630

Paulsen IT, Skurray RA, Tam R et al (1996b) The SMR family: a novel family of multidrug efflux proteins involved with the efflux of lipophilic drugs. Mol Microbiol 19:1167–1175. https://doi.org/10.1111/j.1365-2958.1996.tb02462.x

Paulsen IT, Sliwinski MK, Saier MH (1998) Microbial genome analyses: global comparisons of transport capabilities based on phylogenies, bioenergetics and substrate specificities. J Mol Biol 277:573–592. https://doi.org/10.1006/jmbi.1998.1609

Pendleton JN, Gorman SP, Gilmore BF (2013) Clinical relevance of the ESKAPE pathogens. Expert Rev Anti Infect Ther 11:297–308. https://doi.org/10.1586/eri.13.12

Pereda-Miranda R, Kaatz GW, Gibbons S (2006) Polyacylated oligosaccharides from medicinal Mexican morning glory species as antibacterials and inhibitors of multidrug resistance in *Staphylococcus aureus*. J Nat Prod 69:406–409

Pfeifer HJ, Greenblatt DK, Koch-Wester J (1976) Clinical toxicity of reserpine in hospitalized patients: a report from the Boston collaborative drug surveillance program. Am J Med Sci 271:269–276. https://doi.org/10.1097/00000441-197605000-00002

Piddock LJ, Garvey MI, Rahman MM, Gibbons S (2010) Natural and synthetic compounds such as trimethoprim behave as inhibitors of efflux in gram-negative bacteria. J Antimicrob Chemother 65:1215–1223

Pigeon RP, Silver RP (1994) Topological and mutational analysis of KpsM, the hydrophobic component of the ABC-transporter involved in the export of polysialic acid in *Escherichia coli* K1. Mol Microbiol 14:871–881. https://doi.org/10.1111/j.1365-2958.1994.tb01323.x

Ponnusamy K, Ramasamy M, Savarimuthu I et al (2010) Indirubin potentiates ciprofloxacin activity in the NorA efflux pump of *Staphylococcus aureus*. Scand J Infect Dis 42:500–505

Price CTD, Kaatz GW, Gustafson JE (2002) The multidrug efflux pump NorA is not required for salicylate-induced reduction in drug accumulation by *Staphylococcus aureus*. Int J Antimicrob Agents 20:206–213

Rajendran R, Mowat E, McCulloch E et al (2011) Azole resistance of *Aspergillus fumigatus* biofilms is partly associated with efflux pump activity. Antimicrob Agents Chemother 55:2092–2097

Ramalhete C, Spengler G, Martins A et al (2011) Inhibition of efflux pumps in methicillin-resistant *Staphylococcus aureus* and *Enterococcus faecalis* resistant strains by triterpenoids from *Momordica balsamina*. Int J Antimicrob Agents 37:70–74

Ramaswamy VK, Cacciotto P, Malloci G et al (2017) Computational modelling of efflux pumps and their inhibitors. Essays Biochem 61:141–156

Rana T, Singh S, Kaur N et al (2014) A review on efflux pump inhibitors of medically important bacteria from plant sources. Int J Pharm Sci Rev Res 26:101–111

Rao M, Padyana S, Dipin KM et al (2018) Antimicrobial compounds of plant origin as efflux pump inhibitors: new avenues for controlling multidrug resistant pathogens. J Antimicrob Agents 4:1–6

Rice LB (2008) Federal funding for the study of antimicrobial resistance in nosocomial pathogens: no ESKAPE. J Infect Dis 197:1079–1081. https://doi.org/10.1086/533452

Rice LB (2010) Progress and challenges in implementing the research on ESKAPE pathogens. Infect Control Hosp Epidemiol 31:7–10. https://doi.org/10.1086/655995

Roy SK, Kumari N, Pahwa S et al (2013) NorA efflux pump inhibitory activity of coumarins from *Mesuaferrea*. Fitoterapia 90:140–150

Roy SK, Pahwa S, Nandanwar H, Jachak SM (2012) Phenylpropanoids of *Alpinia galangal* as efflux pump inhibitors in *Mycobacterium smegmatis* mc2 155. Fitoterapia 83:1248–1255

Sabatini S, Gosetto F, Manfroni G et al (2011) Evolution from a natural flavones nucleus to obtain 2-(4-propoxyphenyl)quinoline derivatives as potent inhibitors of the *S. aureus*NorA efflux pump. J Med Chem 54:5722–5736

Saier MH (2000) A functional-phylogenetic classification system for transmembrane solute transporters. Microbiol Mol Biol Rev 64:354–411. https://doi.org/10.1128/mmbr.64.2.354-411.2000

Saier MH, Paulsen IT, Marek KS et al (1998) Evolutionary origins of multidrug and drug-specific efflux pumps in bacteria. FASEB J 12:265–274. https://doi.org/10.1096/fasebj.12.03.265

Schuldiner S, Lebendiker M, Yerushalmi H (1997) EmrE, the smallest ion-coupled transporter, provides a unique paradigm for structure-function studies. J Exp Biol 200:335–341

Schweizer HP (2012) Understanding efflux in gram-negative bacteria: opportunities for drug discovery. Expert Opin Drug Discov 7:633–642

Seukep AJ, Fan M, Sarker SD, Kuete V, Guo MQ (2020c) *Plukenetiahuayllabambana* fruits: analysis of bioactive compounds, antibacterial activity and relative action mechanisms. Plants 9:1111

Seukep AJ, Kuete V, Nahar L et al (2020a) Plant-derived secondary metabolites as the main source of efflux pump inhibitors and methods for identification. J Pharm Anal 10:277–290. https://doi.org/10.1016/j.jpha.2019.11.002

Seukep AJ, Sandjo LP, Ngadjui BT, Kuete V (2016) Antibacterial and antibiotic-resistance modifying activity of the extracts and compounds from *Naucleapobeguinii* against gram-negative multi-drug resistant phenotypes. BMC Compl Altern Med 16:193

Seukep AJ, Zhang Y-L, Xu Y-B, Guo MQ (2020b) In vitro antibacterial and antiproliferative potential of *Echinopslanceolatus*Mattf. (Asteraceae) and identification of potential bioactive compounds. Pharmaceuticals 13:59

Sharma A, Gupta VK, Pathania R (2019) Efflux pump inhibitors for bacterial pathogens: from bench to bedside. Indian J Med Res 149:129–145

Sharma S, Kumar M, Sharma S, Nargotra A, Koul S, Khan IA (2010) Piperine as an inhibitor of Rv1258c, a putative multidrug efflux pump of *Mycobacterium tuberculosis*. J Antimicrob Chemother 65:1694–1701

Shiu WK, Malkinson JP, Rahman MM et al (2013) A new plant-derived antibacterial is an inhibitor of efflux pumps in *Staphylococcus aureus*. Int J Antimicrob Agents 42:513–518

Shriram V, Khare T, Bhagwat R et al (2018) Inhibiting bacterial drug efflux pumps via phytotherapeutics to combat threatening antimicrobial resistance. Front Microbiol 9:2990. https://doi.org/10.3389/fmicb.2018.02990

Singh M, Jadaun GP, Ramdas et al (2011) Effect of efflux pump inhibitors on drug susceptibility of ofloxacin resistant *Mycobacterium tuberculosis* isolates. Indian J Med Res 133:535–540

Singh S, Kalia NP, Joshi P et al (2017) Boeravinone B, a novel dual inhibitor of NorA bacterial efflux pump of *Staphylococcus aureus* and human P-glycoprotein, reduces the biofilm formation and intracellular invasion of bacteria. Front Microbiol 8:1868

Siriyong T, Chusri S, Srimanote P, Tipmanee V, Voravuthikunchai SP (2016) *Holarrhenaantidysenterica* extract and its steroidal alkaloid, conessine, as resistance-modifying agents against extensively drug-resistant *Acinetobacter baumannii*. Microb Drug Resist 22:273–282

Siriyong T, Srimanote P, Chusri S et al (2017) Conessine as a novel inhibitor of multidrug efflux pump systems in *Pseudomonas aeruginosa*. BMC Compl Altern Med 17:405

Smith E, Williamson E, Zloh M et al (2005) Isopimaric acid from *Pinus nigra* shows activity against multidrug-resistant and EMRSA strains of *Staphylococcus aureus*. Phytother Res 19:538–542

Smith EC, Kaatz GW, Seo SM et al (2007a) The phenolic diterpene totarol inhibits multidrug efflux pump activity in *Staphylococcus aureus*. Antimicrob Agents Chemother 51:4480–4483

Smith EC, Williamson EM, Wareham N, Kaatz GW, Gibbons S (2007b) Antibacterials and modulators of bacterial resistance from the immature cones of *Chamaecyparislawsoniana*. Phytochemistry 68:210–217

Solomon SL, Oliver KB (2014) Antibiotic resistance threats in the United States: stepping back from the brink. AFP 89:938–941

Song L, Wu X (2016) Development of efflux pump inhibitors in antituberculosis therapy. Int J Antimicrob Agents 47:421–429

Spengler G, Kincses A, Gajdacs M et al (2017) New roads leading to old destinations: efflux pumps as targets to reverse multidrug resistance in bacteria. Molecules 22:1–25

Stavri M, Piddock LJV, Gibbons S (2007) Bacterial efflux pump inhibitors from natural sources. J Antimicrob Chemother 59:1247–1260

Stermitz F, Scriven LN, Tegos G et al (2002) Two flavonols from *Artemisia annua*, which potentiate the activity of berberine and norfloxacin against a resistant strain of *Staphylococcus aureus*. Planta Med 68:1140–1141

Stermitz FR, Beeson TD, Mueller PJ et al (2001) *Staphylococcus aureus* MDR efflux pump inhibitors from a Berberis and a Mahonia (*sensustrictu*) species. Biochem Syst Ecol 29:793–798

Stermitz FR, Cashman KK, Halligan KM et al (2003) Polyacylatedneohesperidosides from *Geranium caespitosum*: bacterial multidrug resistance pump inhibitors. Bioorg Med Chem Lett 13:1915–1918

Stermitz FR, Tawara-Matsuda J, Lorenz P et al (2000) 5'- Methoxyhydnocarpin-D and pheophorbide a: Berberis species components that potentiate berberine growth inhibition of resistant *Staphylococcus aureus*. J Nat Prod 63:1146–1149

Sudeno RA, Blanco AR, Giuliano F et al (2004) Epigallocatechin-gallate enhances the activity of tetracyclines in staphylococci by inhibiting its efflux from bacterial cells. Antimicrob Agents Chemother 48:1968–1973

Sun JJ, Deng ZQ, Yan AX (2014) Bacterial multidrug efflux pumps: mechanisms, physiology and pharmacological exploitations. Biochem Bioph Res Commun 453:254–267

Tacconelli E, Carrara E, Savoldi A et al (2018) Discovery, research, and development of new antibiotics: the WHO priority list of antibiotic-resistant bacteria and tuberculosis. Lancet Infect Dis 18:318–327. https://doi.org/10.1016/S1473-3099(17)30753-3

Thorarensen A, Presley-Bodnar AL, Marotti KR et al (2001) 3-arylpiperidines as potentiators of existing antibacterial agents. Bioorg Med Chem Lett 11:1903–1906

Tintino SR, Morais-Tintino CD, Campina FF et al (2017) Tannic acid affects the phenotype of *Staphylococcusaureus* resistant to tetracycline and erythromycin by inhibition of efflux pumps. Bioorg Chem 74:197–200

Touani FK, Seukep AJ, Djeussi DE, Fankam AG, Noumedem JA, Kuete V (2014) Antibiotic-potentiation activities of four Cameroonian dietary plants against multidrug-resistant gram-negative bacteria expressing efflux pumps. BMC Compl Altern Med 14:258

Tseng TT, Gratwick KS, Kollman J et al (1999) The RND permease superfamily: an ancient, ubiquitous and diverse family that includes human disease and development proteins. J Mol Microbiol Biotechnol 1:107–125

Van Bambeke F, Balzi E, Tulkens PM (2000) Antibiotic efflux pumps. Biochem Pharmacol 60:457–470. https://doi.org/10.1016/s0006-2952(00)00291-4

Van Bambeke F, Pages JM, Lee VJ (2006) Inhibitors of bacterial efflux pumps as adjuvants in antibiotic treatments and diagnostic tools for detection of resistance by efflux. Recent Pat Antiinfect Drug Discov 1:157–175. https://doi.org/10.2174/157489106777452692

Viveiros M, Martins A, Paixão L et al (2008) Demonstration of intrinsic efflux activity of *Escherichia coli* K-12 AG100 by an automated ethidium bromide method. Int J Antimicrob Agents 31:458–462

Vos T, Lim SS, Abbafati C et al (2020) Global burden of 369 diseases and injuries in 204 countries and territories, 1990–2019: a systematic analysis for the global burden of disease study 2019. Lancet 396:1204–1222. https://doi.org/10.1016/S0140-6736(20)30925-9

Wang Y, Venter H, Ma S (2016) Efflux pump inhibitors: a novel approach to combat efflux-mediated drug resistance in bacteria. Curr Drug Targets 17:702–719

WHO (2014) Antimicrobial resistance: global report on surveillance 2014. WHO. http://www.who.int/drugresistance/documents/surveillancereport/en/ Accessed 4 Nov 2020

WHO (2019) Ten threats to global health in 2019. Medium 2020. https://medium.com/who/ten-threats-to-global-health-in-2019-fbe019ca7edf Accessed 2 Nov 2020

World Health Organization (WHO) (2017) Prioritization of pathogens to guide discovery, research and developmentof new antibiotics for drug-resistant bacterial infections, including tuberculosis. World Health Organization, Geneva, Switzerland

World Health Organization (WHO) (2019) New report calls for urgent action to avert antimicrobial resistance crisis. https://www.who.int/news-room/detail/29-04-2019-new-report-calls-for-urgent-action-to-avert-antimicrobial-resistance-crisis

World Health Organization (WHO) (2020) Antibiotic resistance. https://www.who.int/news-room/fact-sheets/detail/antibiotic-resistance. Accessed 08 Nov 2020

Yang S, Clayton SR, Zechiedrich EL (2003) Relative contributions of the AcrAB, MdfA and NorE efflux pumps to quinolone resistance in *Escherichia coli*. J Antimicrob Chemother 51:545–556. https://doi.org/10.1093/jac/dkg126

Yang W, Chen X, Li Y, Guo S, Wang Z, Yu X (2020) Advances in pharmacological activities of terpenoids. Nat Prod Commun 15:1–13. https://doi.org/10.1177/1934578X20903555

Yoshida H, Bogaki M, Nakamura S et al (1990) Nucleotide sequence and characterization of the *Staphylococcus aureus*norA gene, which confers resistance to quinolones. J Bacteriol 172:6942–6949

Yu Z, Tang J, Khare T et al (2020) The alarming antimicrobial resistance in ESKAPEE pathogens: can essential oils come to the rescue? Fitoterapia 140:104433. https://doi.org/10.1016/j.fitote.2019.104433

Zgurskaya HI, Lopez CA, Gnanakaran S (2015) Permeability barrier of gram-negative cell envelopes and approaches to bypass it. ACS Infect Dis 1:512–522

Zgurskaya HI, Nikaido H (2000) Multidrug resistance mechanisms: drug efflux across two membranes. Mol Microbiol 37:219–225. https://doi.org/10.1046/j.1365-2958.2000.01926.x

Zhen X, Lundborg CS, Sun X et al (2019) Economic burden of antibiotic resistance in ESKAPE organisms: a systematic review. Antimicrob Resist Infect Control 8:137. https://doi.org/10.1186/s13756-019-0590-7

Zinner SH, Lubenko IY, Gilbert D et al (2003) Emergence of resistant *Streptococcus pneumoniae* in an in vitro dynamic model that simulates moxifloxacin concentrations inside and outside the mutant selection window: related changes in susceptibility, resistance frequency and bacterial killing. J Antimicrob Chemother 52:616–622. https://doi.org/10.1093/jac/dkg401

Anti-Quorum Sensing Agents from Natural Sources

17

Abdelhakim Bouyahya, Nasreddine El Omari, Naoual El Menyiy,
Fatima-Ezzahrae Guaouguaou, Abdelaali Balahbib, and
Imane Chamkhi

Contents

A. Bouyahya (✉)
Laboratory of Human Pathologies Biology, Department of Biology, Faculty of Sciences, and Genomic Center of Human Pathologies, Faculty of Medicine and Pharmacy, Mohammed V University in Rabat, Rabat, Morocco

N. El Omari
Laboratory of Histology, Embryology and Cytogenetic, Faculty of Medicine and Pharmacy, Mohammed V University in Rabat, Rabat, Morocco

N. El Menyiy
Laboratory of Natural Substances, Pharmacology, Environment, Modeling, Health and Quality of Life (SNAMOPEQ), Faculty of Sciences, University Sidi Mohamed Ben Abdellah, Fez, Morocco

F.-E. Guaouguaou
Mohammed V University in Rabat, LPCMIO, Materials Science Center (MSC), Ecole Normale Supérieure, Rabat, Morocco

A. Balahbib
Laboratory of Zoology and General Biology, Faculty of Sciences, Mohammed V Universityin Rabat, Rabat, Morocco

I. Chamkhi
Laboratory of Plant-Microbe Interactions, AgroBioSciences, Mohammed VI Polytechnic University, Ben Guerir, Morocco

© The Author(s), under exclusive license to Springer Nature Singapore Pte Ltd. 2022
V. Kumar et al. (eds.), *Antimicrobial Resistance*,
https://doi.org/10.1007/978-981-16-3120-7_17

Abstract

Quorum sensing, as intercellular signaling control of numerous multicellular behaviors, is implicated in microbial cell-cell coordination and involved in the development of multidrug resistance. Targeting quorum sensing or quorum quenching is a promising strategy not only to decrease microbial resistance but also to treat microbial infections. Various bioactive molecules, notably those isolated from natural resources, can deactivate quorum sensing effects via various mechanisms such as the inhibition of autoinducer releases, sequestration of quorum sensing-mediated molecules, and deregulation of quorum sensing gene expression.

Keywords

Quorum sensing · Bacterial resistance · Medicinal plants · Natural product

17.1   Introduction

Microorganisms, especially bacteria, coordinate the interactions between them, on the one hand, and develop resistance, on the other hand, by establishing a powerful communication system called bacterial pheromones or quorum sensing. Indeed, this QS system organizes the transcriptional regulation of several genes which are involved in several life functions of these microbes (Lee et al. 2013).

The QS is based essentially on a mode of regulation called upon molecules signaled according to the type and the nature of bacterial strain (Lee et al. 2013); (Trosko 2016); (Wang et al. 2020a); (Wang et al. 2020b). This system regulates a number of activities in bacteria such as pathogenicity, biofilm formation, and resistance to antibiotics. Indeed, QS sensing regulates several phenotypes in bacteria; in particular it is involved in antibiotic resistance mechanisms via stimulating the biofilm formation (Wang et al. 2020a); (Wang et al. 2020b).

Exploiting this mode of regulation, screening for anti-QS molecules, or quorum quenching could limit and inhibit the development of bacterial resistance to antibiotics and therefore the emergence of infectious diseases (Cai et al. 2010); (Frederick et al. 2011). In recent years, several research works have focused on the specific screening of antibacterial molecules with anti-QS mechanisms of action.

Indeed, reported studies showed that resources such as medicinal plants, full of a panoply of bioactive molecules belonging to secondary metabolites, highly specifically target QS mediators with different modes of action (Cai et al. 2010); (Frederick et al. 2011). These mechanisms include the transcriptional inhibition of QS signal molecules, interference with their action, degradation, and transport system between the intra- and extracellular medium (Vattem et al. 2007). Indeed, the results demonstrated with natural substances against QS come from the phenogenetic links existing between different organisms and molecular mimetic which has recently been demonstrated between QS intermediates and secondary metabolites of medicinal plants (Musk Jr. and Hergenrother 2006); (Chevrot et al. 2006); (Asif 2020). Moreover, the exploration of these metabolites could offer evidence of antibacterial drugs with anti-QS actions. In this chapter, the function of the QS system in Gram (−) and Gram (+) bacteria and their involvement in pathogenicity and resistance are explained. In addition, recent investigations on the action of terpenoids, flavonoids, and phenolic acids were also discussed.

17.2 Overview on Quorum Sensing

QS is the set of signaling molecules synthesized when a microbial community reaches a high concentration threshold. Indeed, the expression of these mediators depends first of all on the cell density which is significant for the expression of these molecules to take place. These QS molecules then activate sets of genes for the synthesis of proteins which are responsible for pathogenicity phenotypes but also for antibiotic resistance (Bassler 2002); (Xu 2016). From a biochemical point of view, the expression of QS molecules is different between Gram (−) and Gram (+) cells (Reading and Sperandio 2006).

17.2.1 Quorum Sensing in Gram-Positive Bacteria

In Gram (+) bacteria, the QS system functions the secretion of intracellular molecules called self-inducing peptides (AIP) (Fig. 17.1). These peptides are

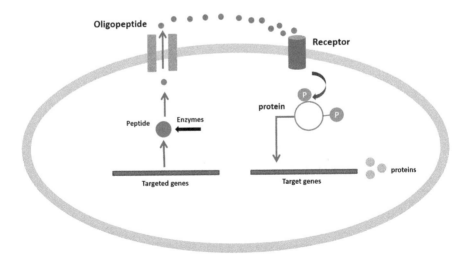

Fig. 17.1 Quorum sensing in gram-positive bacteria

exported to the extracellular medium in the form of oligopeptides which subsequently bind (depending on cell density) to external membrane receptors rich in histidine. Signal transduction activates signaling pathways that enable the activation of gene expression in specific ways. Indeed, each transcribed gene is regulated specifically by a given signal peptide (Reading and Sperandio 2006).

17.2.2 Quorum Sensing in Gram-Negative Bacteria

In Gram (−) bacteria, self-inducing molecules are produced from a basic molecule called N-acylhomoserine lactone (AHL). At high cell density, Gram (−) bacteria activate transcription of the lux operon which encodes the transcription of enzymes of the signal synthase family (LuxI) which catalyze the synthesis of AHL. These molecules, depending on bacterial density, can leave and rejoin the intracellular environment to regulate gene expression in a manner dependent on the extracellular environment (Reading and Sperandio 2006) (Fig. 17.2). In fact, with the increase in cell density, the AHL molecule diffuses inside cells and interacts with transcription regulators (intracellular receptors) to selectively activate the transcription of target genes.

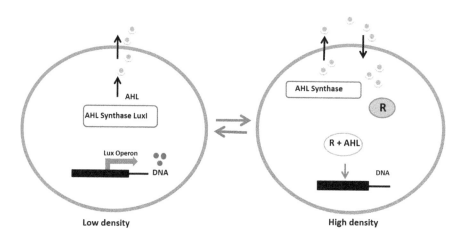

Fig. 17.2 Quorum sensing in Gram-negative bacteria

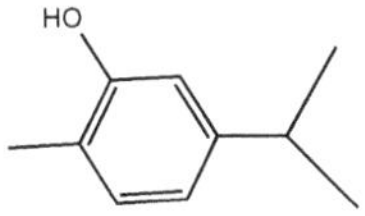

1: Carvacrol

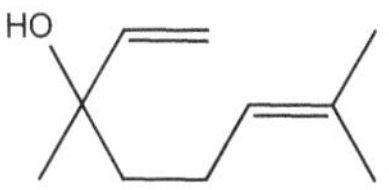

2: Linalool

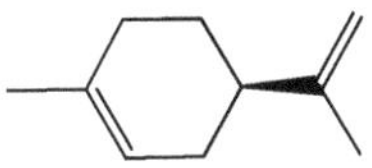

3: D-limonene

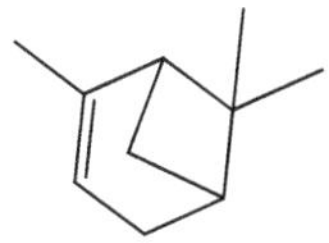

4: alpha-Pinene

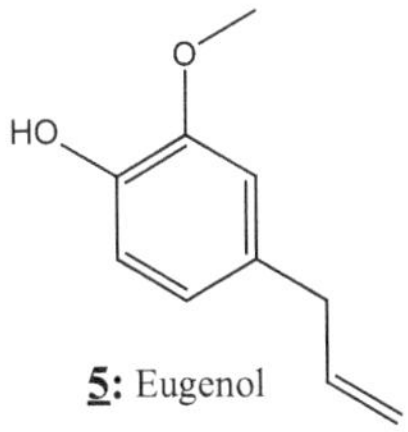

5: Eugenol

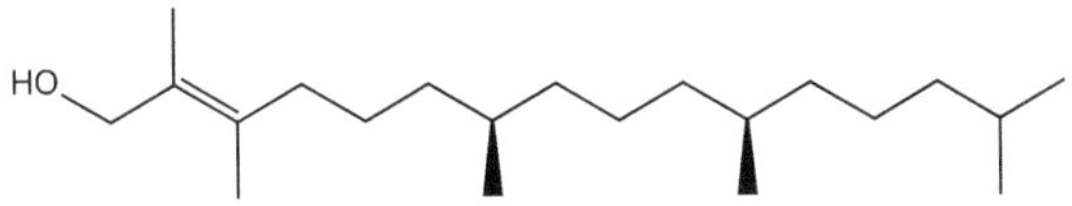

6: Phytol

Fig. 17.3 Structures of terpenoids with anti-quorum sensing effects

17.3 Quorum Quenching of Bioactive Compounds from Medicinal Plants

17.3.1 Quorum Quenching of Terpenes

Different types of terpenoids such as sesquiterpene (sesquiterpene lactone), monoterpene (carvacrol, linalool, d-limonene, and α-pinene), diterpene (phytol), and phenylpropene (eugenol) (Fig. 17.3) have also show an inhibitory activity of quorum sensing (Table 17.1).

Sesquiterpenoids such as sesquiterpene lactone were showed to have a potential quorum sensing inhibitory activity in *Pseudomonas aeruginosa* ATCC 27853 and *Chromobacterium violaceum* (Amaya et al. 2012; Aliyu et al. 2016, 2020). Indeed, Aliyu et al. (2020) showed that this compound induced a QS inhibitory activity (QSI $\geq$ 80%) at 1.31 mg/Ml. In another study, Amaya et al. (2012) reported that six sesquiterpene lactones (SLs) of the goyazensolide and isogoyazensolide type isolated from the Argentine herb *Centratherum punctatum* inhibited the production of N-acyl-homoserine lactones (AHLs) at 100μg/mL and the elastase activity at 0.5μg/ml by more than 50%.

Eugenol was reported as potent inhibitory activity in quorum sensing (Packiavathy et al. 2012; Zhou et al. 2013; Al-Shabib et al. 2017; Rathinam et al. 2017; Lou et al. 2019). This phenylpropene at subinhibitory concentrations inhibited the production of virulence factors, including violacein, elastase, and pyocyanin in *Pseudomonas aeruginosa* inhibited the las and pqs QS systems, and reduced biofilm formation by 43% at 400μM, (Zhou et al. 2013). At the same concentration, eugenol revealed a maximum reduction in biofilm formation on PAO1 (65.6%) and exhibited a marked effect on the production of QS signals (AIs) ($p < 0.001$) without affecting their chemical integrity (Rathinam et al. 2017). In another study, Lou et al. (2019)

Table 17.1 Anti-quorum sensing effects of terpenoids

Compounds	Methods used	Organisms tested	Key findings	References
Sesquiterpene lactone	Crystal violet staining assay Quorum sensing inhibition assay	*Pseudomonas aeruginosa* ATCC 27853	Altered biofilm formation, elastase activity, and production of N-acyl-homoserine lactones (AHLs)	Amaya et al. (2012)
	Violacein inhibition assays Agar diffusion double ring assays Molecular docking analysis	*Chromobacterium violaceum*	Affected cell-cell communication in silico and an antagonist effect against CviR protein to its receptor LuxR	Aliyu et al. (2016)
	Violacein inhibition assay	*Chromobacterium violaceum* ATCC 12472	Induced a QS inhibitory activity (QSI ≥ 80% at 1.31 mg/mL)	Aliyu et al. (2020)
Eugenol	Violacein inhibition assay Biofilm formation in 24-well MTP In situ visualization of biofilm	*Pseudomonas aeruginosa, Proteus mirabilis,* and *Serratia marcescens*	Reduced the AHL-dependent production of violacein, bioluminescence, and biofilm formation	Packiavathy et al. (2012)
	Elastin/Congo red assays Analysis of *lasB* and *pqsA* transcriptional Swarming assays	*Escherichia coli* *Pseudomonas aeruginosa*	Inhibited the production of virulence factors, including violacein, elastase, pyocyanin, and biofilm formation Inhibited the *las* and *pqs* QS systems	Zhou et al. (2013)
	Microtiter plate assay Quantitative estimation of violacein	Methicillin-resistant *Staphylococcus aureus*	Induced a significant anti-QS activity in CVO26 and reduced the QS-regulated production of elastase, protease, chitinase, pyocyanin, and EPS in PAO1 Reduced the biofilm biomass	Al-Shabib et al. (2017)
	Microtiter plate assay RT-qPCR	*Pseudomonas aeruginosa*	Reduced (at 400µM) the biofilm formation on virulence factors	Rathinam et al. (2017)

Compound	Method	Organism	Effect	Reference
	Qualitative and quantitative QS inhibition assays Real-time PCR CSLM	Pseudomonas aeruginosa	Induced a marked effect on the production of QS signals (AIs) Inhibited the expression of QS synthase genes with an expression level of 65% and 61% for *lasI* and *rhlI*, respectively Inhibited the expression of the *rhlA* gene (65%) responsible for the production of rhamnolipid Inhibited the biofilm formation (36%)	Lou et al. (2019)
Carvacrol	Polystyrene microtiter dishes Quantitative analysis of *cvil* gene expression QS inhibition assay	Chromobacterium violaceum	Inhibited the formation of biofilms at sub-lethal concentrations (< 0.5 mM) Reduced the expression of *cvil* (a gene coding for the AHL synthase), production of violacein, and chitinase activity (both regulated by QS)	Burt et al. (2014)
	Crystal violet staining assay	Pseudomonas aeruginosa	Inhibited the biofilm formation at 0.9–7.9 mM Reduced the pyocyanin production to 60% at 3.9 mM Reduced the violacein production to 50% at 0.7 mM	Tapia-Rodriguez et al. (2017)
	Quantification of AHLs Biofilm assays	Pseudomonas aeruginosa ATCC 10154	Reduced the AHL production to 60% at 1.9 mM Reduced the *LasI* synthase activity Reduced the expression of *lasR*, without affecting *lasI* gen Reduced the biofilm formation and swarming motility	Tapia-Rodriguez et al. (2019)
Phytol	Polystyrene microtiter plates	Pseudomonas aeruginosa PAO1	Reduced the biofilm formation (74.00–84.33%) Reduced bacterial twitching and flagella motility Inhibited the pyocyanin production (51.94%)	Pejin et al. (2014)
	SEM analysis Prodigiosin quantification assay	Serratia marcescens	Inhibited (at 10μg/mL) the prodigios in production to 92% Inhibited (at 10μg/mL) the QS-mediated protease production (68%) and biofilm formation (64%)	Srinivasan et al. (2016)

(continued)

Table 17.1 (continued)

Compounds	Methods used	Organisms tested	Key findings	References
Oleanolic aldehyde coumarate	XTT reduction assay CSLM Lipase quantification assay qPCR analysis	Serratia marcescens	Decreased (5 and 10μg/mL) the level of biofilm formation, lipase, and hemolysin production Inhibited the swarming motility and EPS productions Down-regulated the *fimA*, *fimC*, *flhC*, *flhD*, *bsmB*, *pigP*, and *shlA* gene expressions Reduced level of virulence enzymes (lipase and protease productions)	Ruckmani and Ravi (2017)
	Fluorescence microscopy Motility assays Homoserine lactone quantification	Pseudomonas aeruginosa PAO1	Inhibited the biofilm formation as well as the expression of the *las* and *rhl* QS systems Reduced the production of QS-controlled virulence factors including Inhibited the AHL production	Rasamiravaka et al. (2015)
Linalool	Polystyrene microtiter plates Crystal violet staining method	Acinetobacter baumannii	Inhibited the biofilm formation, changed the bacterial adhesion to surfaces, and interfered with the QS system	Alves et al. (2016)
D-limonene	Crystal violet staining EPS production assays Swimming and swarming assays Bioluminescence assay qRT-PCR assay	Escherichia coli	Inhibited the biofilm formation through the suppression of curli and EPS production Decreased swimming and swarming ability	Wang et al. (2018)
(−)-α-Pinene	AI-2 bioassays	Campylobacter jejuni	Reduced the QS signaling by >80%	(Šimunović et al. 2020)

reported that eugenol inhibited about 50% of the QS-mediated violacein production in *Chromobacterium violaceum* at a sub-MIC of 0.2 mg/mL, as well as the production of N-(3-oxododecanoyl)-L-homoserine lactone (3-oxo-C12-HSL) and C4-HSL N-acyl homoserine lactone signal molecules, pyocyanin, and swarming motility in *P. aeruginosa*. This compound inhibited also the expression of QS synthase genes with an expression level of 65% and 61% for *lasI* and *rhlI*, respectively, and 65% for *rhlA* gene as well as the biofilm formation (36%).

The eugenol was also effective against biofilms of *Pseudomonas aeruginosa*, *Proteus mirabilis*, *Serratia marcescens* clinical isolates (Packiavathy et al. 2012), and methicillin-resistant *Staphylococcus aureus* isolated from food handlers (Al-Shabib et al. 2017).

Carvacrol (2-methyl-5-[1-methylethyl]-phenol) was shown to be effective against quorum sensing and biofilm growth. In a study by Tapia-Rodriguez et al. (2017), carvacrol inhibited the formation of *Pseudomonas aeruginosa* biofilms at 0.9–7.9 mM and reduced the pyocyanin production to 60% at 3.9 mM and violacein production to 50% at 0.7 mM. Using the same model organism, the authors showed that carvacrol reduced the virulence of *Pseudomonas aeruginosa* by inhibiting LasI activity with the concomitant reduction on the expression of lasR, biofilm, and swarming motility without affecting *lasI* gen (Tapia-Rodriguez et al. 2019). This monoterpene also downregulated biofilm-associated genes in *Chromobacterium violaceum* ATCC 12472, *Salmonella enterica* subsp. *typhimurium* DT104, and *Staphylococcus aureus* 0074 at sublethal concentrations (< 0.5 mM) and reduced the expression of *cviI*, production of violacein, and chitinase activity (Burt et al. 2014).

Diterpene such as phytol showed the ability to inhibit biofilm growth and quorum sensing in *Pseudomonas aeruginosa* PAO1 and *Serratia marcescens* (Pejin et al. 2014); (Srinivasan et al. 2016); (Ruckmani and Ravi 2017). At a concentration of 10µg/mL, phytol inhibited the prodigiosin production (92%), the QS-mediated protease production (68%), and biofilm formation (64%) of *Serratia marcescens* (Srinivasan et al. 2016). Using the same bacterial strain and concentration, the authors reported that phytol decreased the level of biofilm formation, lipase, and hemolysin production and inhibited the swarming motility and EPS production. This compound also downregulated the *fimA*, *fimC*, *flhC*, *flhD*, *bsmB*, *pigP*, and *shlA* gene expressions and reduced the level of virulence enzymes (lipase and protease productions) (Ruckmani and Ravi 2017). In another study, phytol reduced the formation of *Pseudomonas aeruginosa* biofilm in the range of 74.00–84.33%. It also effectively reduced *P. aeruginosa* twitching and flagella motility and inhibited pyocyanin production (51.94%) (Pejin et al. 2014).

Oleanolic aldehyde coumarate, isolated from *D. trichocarpa* bark extract, showed inhibitory activities against *P. aeruginosa* biofilms as well as the expression of the *las* and *rhl* QS systems and AHL production and reduced the expression of lasI/R, rhlI/R and the global virulence factor activator gacA (Rasamiravaka et al. 2015). Furthermore, linalool isolated from *Coriandrum sativum* has been reported by (Alves et al. 2016) to inhibit biofilm formation and dispersed established biofilms of *A. baumannii*, changed the adhesion in *A. baumannii* to surfaces, and interfered with the quorum sensing system. Using *E. coli* bacterial strain, Wang et al. (2018) have shown that D-limonene nanoemulsion inhibited *E. coli* biofilm formation through the suppression of curli and extracellular polymeric substance (EPS) production without inhibiting cell growth and decreased swimming and swarming ability. On the other hand, the subinhibitory concentration of (−)-α-pinene (250 mg/L) has reduced quorum sensing signaling of *Campylobacter jejuni* by >80% (Šimunović et al. 2020).

17.3.2 Quorum Quenching of Flavonoids

Flavonoids are another type of phenolic which possesses several pharmacological properties. This group contains numerous bioactive (Fig. 17.4) that showed anti-QS and anti-biofilm effects (Table 17.2).

In a study by Hosseinzadeh et al. (2020), epigallocatechin showed an anti-biofilm activity against *Salmonella typhimurium* with downregulation of both *s di A* and *luxS* genes. Wu et al. (2018) reported that this compound decreased the biomass and acid production of *Streptococcus mutans* biofilms at a concentration of 250μg/mL and inhibited *Streptococcus mutans* (Sm) + probiotic *Lactobacillus casei* in Yakult (LcY) biofilm formation and acid production at a concentration of 500μg/mL. Using bioluminescence, motility assays and biofilm assays, Castillo et al. (2015) revealed that epigallocatechin disturbed the QS activity and reduced the motility and biofilm formation and decreased the AI-2 activity. Epigallocatechin also showed QS inhibitory activities and biofilm formation against *Burkholderia cepacia* and *Staphylococcus aureus* (Huber et al. 2003), *Listeria monocytogenes* (Nyila et al. 2012), and *Eikenella corrodens* (Matsunaga et al. 2010). On the other hand, at a concentration of 100 and 200μg/mL, naringenin inhibited *S. mutans* growth and biofilm formation, increased *S. mutans* surface hydrophobicity, reduced bacterial aggregation, and downregulated the mRNA expression of gtfB, gtfC, comD, comE, and luxS (Yue et al. 2018). This compound was also found to inhibit swimming motility in *Chromobacterium violaceum* and to be associated with the induction of the transcription levels of yenR, flhDC, and fliA (Truchado et al. 2012).

Similarly, quercetin has documented biofilm and quorum sensing inhibitory activities (Vikram et al. 2010; Gopu et al. 2015; Pejin et al. 2015; Ouyang et al. 2016; Al-Yousef et al. 2017; Grabski et al. 2017; Erdönmez et al. 2018; Ouyang et al. 2020). Vikram et al. (2010) reported that quercetin was an effective antagonist of cell-cell signaling. Furthermore, this flavonoid suppressed the biofilm formation in *Escherichia coli* O157:H7 and *Vibrio harveyi* at concentration of 80μg/mL quercetin and showed a significant reduction in QS-dependent phenotypes like violacein production, biofilm formation, and exopolysaccharide (EPS) production in *Chromobacterium violaceum* CV026, as well as motility and alginate production in a concentration-dependent manner (Gopu et al. 2015). Ouyang et al. (2016) reported anti-biofilm activities of this compound against *Pseudomonas aeruginosa* strain PAO1 as well as inhibition of production of virulence factors including pyocyanin, protease, and elastase at a lower concentration. The author revealed

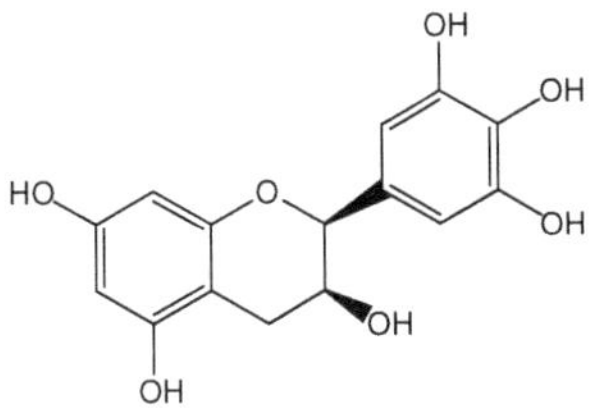

1: Epigallocatechin

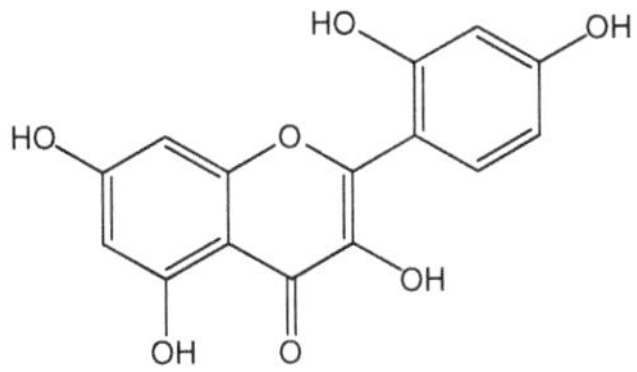

2: Morin

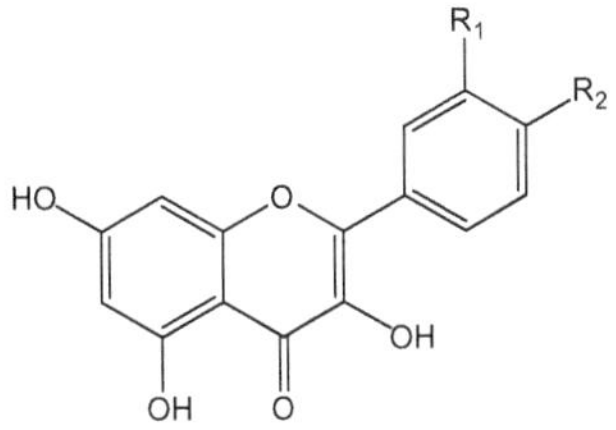

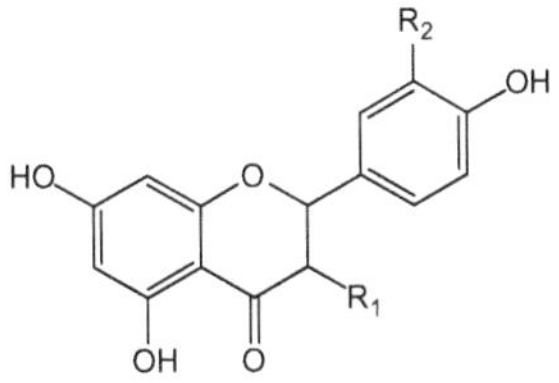

6: R_1=R_2=OH:Taxifolin
7: R_1=R_2=H:Naringenin

3: R_1=R_2=OH: Quercetin
4: R_1= OH, R_2=OGlu: Quercetin 4'-O-beta-D glucopyranoside
5: R_1= H, R_2=OH: Kaempferol

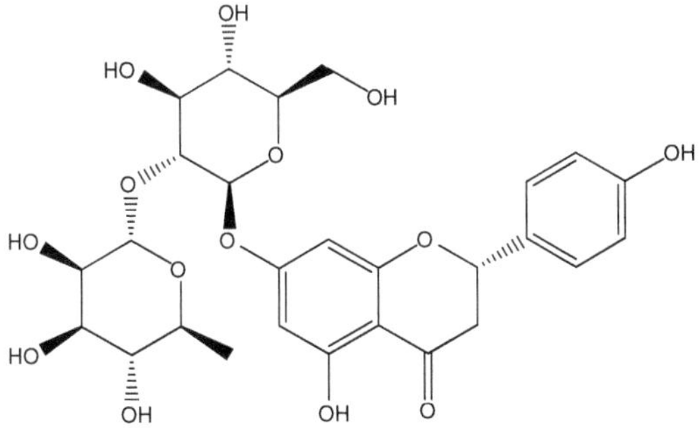

8: Naringin

Fig. 17.4 Structures of flavonoids with anti-quorum sensing effects

also that found that the expression levels of lasI, lasR, rhlI, and rhlR were reduced at a concentration of 16μg/mL. This compound also inhibited the QS circuitry by interacting with transcriptional regulator LasR in *Pseudomonas aeruginosa* (Grabski et al. 2017). In addition quercetin inhibited the QS-controlled virulence factors such as violacein, elastase, and pyocyanin in *Chromobacterium violaceum* CV12472 and *Pseudomonas aeruginosa* PAO1 (Al-Yousef et al. 2017). Using biofilm formation assay, Ouyang et al. (2020) reported that quercetin decreased adhesion and biofilm formation in *Pseudomonas aeruginosa*, as well as swarming motility and expression of biofilm-associated genes.

Table 17.2 Anti-quorum sensing effects of flavonoids

Compounds	Methods used	Organisms tested	Key findings	References
Epigallocatechin	Biofilm formation in polystyrene microtiter dishes	*Burkholderia cepacia* and *Staphylococcus aureus*	Reduced the biofilm by interference with AHL production	Huber et al. (2003)
	Adherence assay for the quantitation of biofilm production	*Eikenella corrodens*	Affected the QS mediated by autoinducer 2 (AI-2) and thus inhibited biofilm formation	Matsunaga et al. (2010)
	Confocal scanning laser microscopy (CSLM)	*Listeria monocytogenes*	Reduction in cell numbers and inhibition of biofilm formation	Nyila et al. (2012)
	Bioluminescence assay Motility assays Biofilm assays	*Campylobacter jejuni*	Disturbed the QS activity and reduced the motility and biofilm formation Decreased the AI-2 activity	Castillo et al. (2015)
	Crystal violet assay Scanning electron microscopy (SEM) Quantitative PCR assay	*Streptococcus mutans* (Sm) and probiotic *Lactobacillus casei* in Yakult (LcY)	Decreased the biomass and acid production of Sm biofilms (250µg/mL) Inhibited Sm + LcY biofilm formation and acid production (500µg/mL)	Wu et al. (2018)
	Quantitative real-time PCR assay	*Salmonella typhimurium*	Reduced the expression of QS-related genes (*sdtA* and *luxS*)	Hosseinzadeh et al. (2020)
Naringin	Analysis of the AHLs using LC-MS/MS Crystal violet assay Gene expression analyses	*Chromobacterium violaceum*	Inhibition of biofilm formation, swimming, and swarming motility Induction of some gene transcription such as yenR, flhDC, and fliA	Truchado et al. (2012)
	Analysis of the AHLs using LC-MS/MS Crystal violet assay Gene expression analyses	*Yersinia enterocolitica*	Decreased biofilm formation by decreasing AHL production	Truchado et al. (2012)
	CSLM crystal violet staining real-time PCR assay	*Streptococcus mutans*	Suppressed (at 100 and 200µg/mL) the second (bacterial adhesion) and third stages (biofilm maturation) of bacterial biofilm formation	Yue et al. (2018)

Compound	Assay/method	Organism	Effect	Reference
Quercetin	Assay for inhibition of intercellular signaling Biofilm formation Semiquantitative RT-PCR	Escherichia coli O157:H7 and Vibrio harveyi	Suppressed the biofilm formation Acted as an effective antagonist of cell-cell signaling	Vikram et al. (2010)
	QSI bioassay Microtiter plate assay Motility assay	Chromobacterium violaceum CV026	Reduced (at 80µg/mL) the QS-dependent phenotypes like violacein production, biofilm formation, EPS production, motility, and alginate production in a concentration-dependent manner	Gopu et al. (2015)
	Anti-biofilm activity Anti-QS effects	Pseudomonas aeruginosa PAO1	Reduced the bacterial biofilm formation (95%) and the twitching motility	Pejin et al. (2015)
	Crystal violet biofilm assay CSLM Fluorescence real-time quantitative PCR assay	Pseudomonas aeruginosa strain PAO1	Inhibited the biofilm formation and production of virulence factors including pyocyanin, protease, and elastase Reduced (at 16µg/mL) the expression levels of $lasI$, $lasR$, $rhlI$, and $rhlR$	Ouyang et al. (2016)
Quercetin 4'-O-β-D glucopyranoside	Microtiter plate assay Quantitative estimation of violacein	Chromobacterium violaceum CV12472 and Pseudomonas aeruginosa PAO1	Inhibited the QS-controlled virulence factors such as violacein, elastase, pyocyanin, and biofilm formation	Al-Yousef et al. (2017)
	Combination of molecular docking, molecular dynamic simulations, and machine learning techniques	Pseudomonas aeruginosa	Inhibited the QS circuitry by interacting with transcriptional regulator LasR	Grabski et al. (2017)
	Anti-QS disk diffusion and agar diffusion test Violacein pigment isolation	Chromobacterium violaceum ATCC 12472 and Chromobacterium violaceum CV026	Inhibited the bacterial communication system (QS) via: Inhibition of violacein pigment production Inhibition of the communication molecule (C6-AHL)	Erdönmez et al. (2018)

(continued)

Table 17.2 (continued)

Compounds	Methods used	Organisms tested	Key findings	References
	Biofilm formation assay Assay of bacterial adhesion ability Virulence factor assay Fluorescence real-time quantitative PCR	*Pseudomonas aeruginosa*	Decreased adhesion, biofilm formation, swarming motility, and expression of biofilm-associated genes Reduced the production of pyocyanin and protease activity Inhibited bacterial biofilm formation via the *vfr*-mediated *lasIR* system	Ouyang et al. (2020)
Taxifolin	Quantitative analysis of violacein production Homoserine lactone quantification RT-PCR assay	*Pseudomonas aeruginosa* PAO1	Reduced the production of pyocyanin and elastase Reduced the expression of several QS-controlled genes	Vandeputte et al. (2011)
Kaempferol	Crystal violet staining Fluorescence microscopy Fibrinogen-binding assay qRT-PCR assay	*Staphylococcus aureus*	Inhibited the biofilm formation Inhibited the primary attachment phase of biofilm formation Reduced the activity of *Staphylococcus aureus* ortaseA (SrtA) and the expression of adhesion-related genes	Ming et al. (2017)
Morin	In situ visualization biofilms SEM Anti-motility assay EPS production assay	*Staphylococcus aureus*	Inhibited the biofilm formation Reduced motility and spreading and EPS production	Chemmugil et al. (2019)
Naringenin	Real-time PCR assay Elastase activity Pyocyanin production	*Pseudomonas aeruginosa*	Inhibited the expression of QS-regulated genes, as well as the production of the QS-regulated virulence factors, pyocyanin, and elastase	Hernando-Amado et al. (2020)

Other flavonoids such as kaempferol, taxifolin, morin, and naringenin also showed similar effects. In a study by Hernando-Amado et al. (2020), naringenin was shown to inhibit the expression of QS-regulated genes, as well as the production of the QS-regulated virulence factors, pyocyanin, and elastase in *Pseudomonas aeruginosa* strains. Morin was reported by Chemmugil et al. (2019) to exhibit significant biofilm inhibition and reduce the motility and spreading and EPS production of *Staphylococcus aureus*. Using the same model organism, kaempferol inhibited biofilm formation by 80% at a concentration of 64μg/mL and reduced the activity of *Staphylococcus aureus* sortase A (SrtA) and the expression of adhesion-related genes (Ming et al. 2017). However, taxifolin revealed a significant decrease in the production of pyocyanin and elastase in *P. aeruginosa* without affecting bacterial growth. This compound also reduced the expression of several QS-controlled genes (i.e., lasI, lasR, rhlI, rhlR, lasA, lasB, phzA1, and rhlA) in *P. aeruginosa* PAO1 (Vandeputte et al. 2011).

17.3.3 Quorum Quenching of Phenolic Acids

Phenolic acids are important phenolic compounds (Fig. 17.5) that are present in several medicinal plants with anti-QS and anti-biofilm effects (Table 17.3).

Salicylic acid was reported by Joshi et al. (2015) to interfere with the quorum sensing (QS) system of two *Pectobacterium* species, *P. aroidearum* and *P. carotovorum* spp. *brasiliense* and affected QS machinery, consequently altering the expression of bacterial virulence factors. It also inhibited the expression of QS genes, including *expI*, *expR*, *PC1_1442* (*luxR* transcriptional regulator), and *luxS* (a component of the AI-2 system) and reduced the level of the AHL signal. Using motility and AHL production tests, treatment with salicylic acid reduced the biofilm formation by a significant decrease in twitching and swarming motility and AHL production in *Pseudomonas aeruginosa* (Chow et al., 2011). This activity was also confirmed by Lagonenko et al. (2013). In another study, the AHL production and biofilm formation in *Agrobacterium tumefaciens* were reduced by modulation of 103 genes family involved in virulence (Yuan et al. 2007).

Similarly, rosmarinic acid at 750μg/mL inhibited biofilm formation in *A. hydrophila* strains. At this concentration, RA reduced the QS-mediated hemolysin, lipase, and elastase production in *A. hydrophila* and downregulated the virulence genes such as *ahh1*, *aerA*, *lip*, and *ahyB* (Devi et al. 2016). Using molecular docking, Corral-Lugo et al. (2016) showed that this compound bound to the QS regulator RhlR of the *Pseudomonas aeruginosa* PAO1 and competed with the bacterial ligand N-butanoyl-homoserine lactone (C4-HSL) and stimulated a greater increase in RhlR-mediated transcription than that of C4-HSL. In *P. aeruginosa*, rosmarinic acid induced the QS-dependent gene expression and increased biofilm formation and the production of the virulence factors pyocyanin and elastase. In another study, rosmarinic acid induces the expression of 128 genes, among which many virulence factor genes triggered a broad QS response in *Pseudomonas aeruginosa* PAO1. This compound induced also seven sRNAs that were all encoded in regions close to

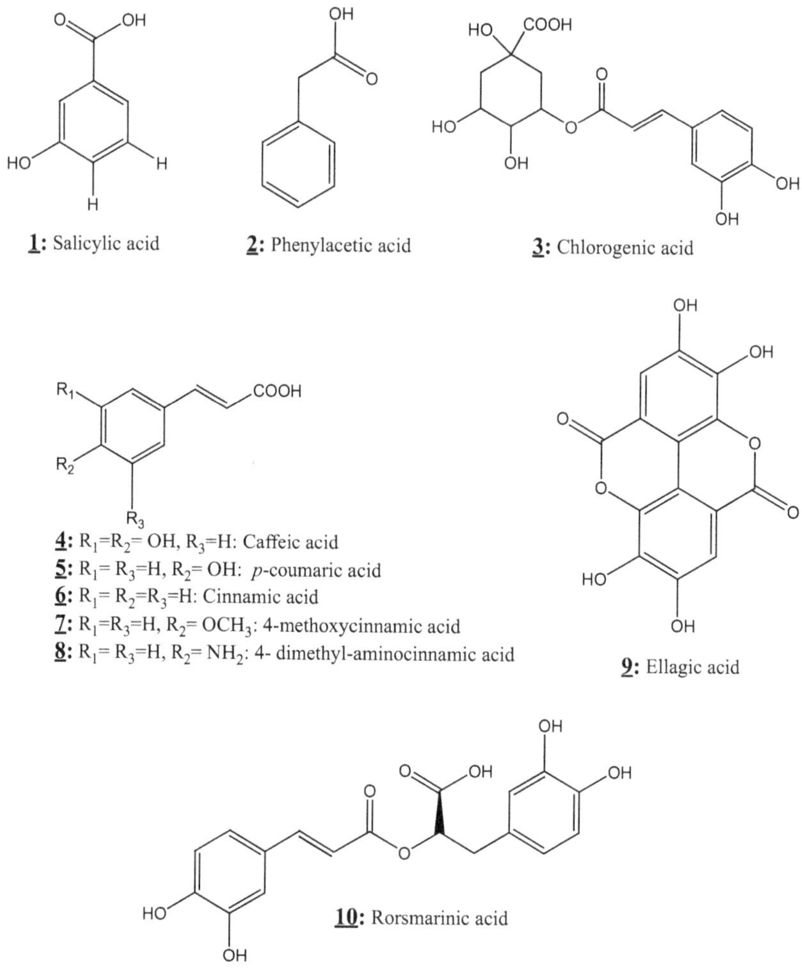

1: Salicylic acid   **2**: Phenylacetic acid   **3**: Chlorogenic acid

R_1 ... COOH

R_2

R_3

4: R_1=R_2= OH, R_3=H: Caffeic acid
5: R_1= R_3=H, R_2= OH: p-coumaric acid
6: R_1= R_2=R_3=H: Cinnamic acid
7: R_1=R_3=H, R_2= OCH$_3$: 4-methoxycinnamic acid
8: R_1=R_3=H, R_2= NH$_2$: 4- dimethyl-aminocinnamic acid

9: Ellagic acid

10: Rorsmarinic acid

Fig. 17.5 Structures of phenolic acids with anti-quorum sensing effects

QS-induced genes (Fernández Rodríguez et al. 2018). Using the same model organism, Walker et al. (2004) confirmed also this activity.

Cinnamic acid is another type of phenolic acid that has documented biofilm and quorum sensing inhibitory activities. At sublethal concentration, cinnamic acid effectively inhibited both the production of the QS-dependent virulence factors and biofilm formation in *P. aeruginosa* without affecting the viability of the bacterium (Rajkumari et al. 2018). In a study by (Joshi et al. 2015). Cinnamic acid affected the QS machinery of the two species (*Pectobacterium aroidearum* and *Pectobacterium carotovorum* spp. *brasiliense*) consequently altering the expression of bacterial virulence factors. This compound inhibited also the expression of QS genes, including *expl*, *expR*, *PC1_1442* (*luxR* transcriptional regulator), and *luxS*

Table 17.3 Anti-quorum sensing effects of phenolic acids

Compounds	Methods used	Organisms tested	Key findings	References
Salicylic acid	Assaying gene fusions Microarray Root infection assays	Agrobacterium tumefaciens	Diminution of biofilm and AHL production by modulation of 103 genes family involved in virulence	Yuan et al. (2007)
	Abiotic solid surface assay Swarming, swimming, and twitching assays	Pseudomonas aeruginosa	Decreased swimming, twitching, and swarming motility resulting in decreased biofilm formation	Chow et al. (2011)
	Motility test AHL production test Crystal violet staining assay	Pectobacterium carotovorum and Pseudomonas syringae pv. syringae	Inhibited the biofilm formation, motility, and AHL production	Lagonenko et al. (2013)
	qRT-PCR Qualitative assays for the detection of AHL molecules Bioluminescence-based assay	Pectobacterium aroidearum and Pectobacterium carotovorum spp. brasiliense	Affected the QS machinery of the two species, consequently altering the expression of bacterial virulence factors Inhibited the expression of QS genes, including expI, expR, PCI_1442 (luxR transcriptional regulator), and luxS (a component of the AI-2 system) Reduced the level of the AHL signal	Joshi et al. (2015)
Rosmarinic acid	CSLM Biofilm formation assay	Pseudomonas aeruginosa PAO1	Decreased the biofilm formation	Walker et al. (2004)
	Molecular docking, homology modeling, and structural alignment Gene expression studies Biofilm formation	Pseudomonas aeruginosa PAO1	Bound to the QS regulator RhlR of the bacteria and competed with the bacterial ligand N-butanoyl-homoserine lactone (C4-HSL) Stimulated a greater increase in	Corral-Lugo et al. (2016)

(continued)

Table 17.3 (continued)

Compounds	Methods used	Organisms tested	Key findings	References
			RhlR-mediated transcription than that of C4-HSL Induced the QS-dependent gene expression and increased biofilm formation and the production of the virulence factors: Pyocyanin and elastase	
	Gene expression analysis Microtiter plate assay Virulence factor inhibition assay	Aeromonas hydrophila	Biofilm inhibitory concentration was 750μg/mL Reduced the QS-mediated hemolysin, lipase, and elastase production Downregulated the virulence genes such as ahh1, aerA, lip, and ahyB	Devi et al. (2016)
	RNA-seq analysis RT-PCR	Pseudomonas aeruginosa PAO1	Induced the expression of 128 genes, among which many virulence factor genes Triggered a broad QS response Induced seven sRNAs that were all encoded in regions close to QS-induced genes	Fernández Rodríguez et al. (2018)
Cinnamic acid	Biosensor bioassay for detection of anti-QS activity Well microtiter plate	Pseudomonas aeruginosa PAO1	Inhibited both the production of the QS-dependent virulence factors and biofilm formation	Rajkumari et al. (2018)
	qRT-PCR qualitative assays for the detection of AHL molecules	Pectobacterium aroidearum and Pectobacterium carotovorum spp. brasiliense	Affected the QS machinery of the two species, consequently altering the expression of bacterial virulence factors	Joshi et al. (2015)

Compound	Assay	Organism	Effects	References
	Bioluminescence-based assay		Inhibited the expression of QS genes, including *expI*, *expR*, *PC1_1442* (*luxR* transcriptional regulator), and *luxS* (a component of the AI-2 system) Reduced the level of the AHL signal	
Two cinnamic acid derivatives: 4-dimethyl-aminocinnamic acid and 4-methoxycinnamic acid	qRT-PCR SEM CLSM AHL analysis Determination of virulence factors	*Chromobacterium violaceum* ATCC12472	Inhibited the levels of N-decanoyl-homoserine lactone (C10-HSL) Reduced the production of certain virulence factors, including violacein, hemolysin, and chitinase Downregulated the QS-related metabolites, such as ethanolamine and L-methionine Suppressed the expression of two QS-related genes (*cvil* and *cviR*) Inhibited the biofilm formation	Cheng et al. (2020)
Chlorogenic acid	Crystal violet method Pyocyanin assay Violacein inhibition assay Quantitative real-time PCR Molecular docking	*Pseudomonas aeruginosa*	Inhibited the formation of biofilm, the ability of swarming, and virulence factors Downregulated the expression of QS-related genes Formed hydrogen bonds with the three QS receptors	Wang et al. (2019)
	Crystal violet method Chitinolytic activity assay Violacein inhibition assay	*Chromobacterium violaceum*	Inhibited the biofilm formation, swarming motility, chitinolytic activity, and violacein production	Wang et al. (2019)

(continued)

Table 17.3 (continued)

Compounds	Methods used	Organisms tested	Key findings	References
p-Coumaric acid	QS inhibition assays	*Agrobacterium tumefaciens* NTL4, *Chromobacterium violaceum* 5999, and *Pseudomonas chlororaphis*	Fully inhibited the bacterial QS responses, with no influence on cell viability	Bodini et al. (2009)
p-Coumaric acid	Qualitative QS inhibition assay	*Chromobacterium violaceum* (CECT 494)	Inhibited the QS by inhibiting the violacein (1.10 ± 0.11 cm at 0.2 mg/mL)	Chen et al. (2020)
Caffeic acid	Flow cytometry Crystal violet staining assay SEM	*Staphylococcus aureus*	Influenced the bacterial adhesion properties Inhibited the production of α-hemolysin	Luís et al. (2014)
Ellagic acid	Well microtiter plate	*Burkholderia cepacia*	Reduced the biofilm formation	Huber et al. (2003)
Phenylacetic acid	Pyocyanin quantification assay Quantification of exopolysaccharide (EPS) Swimming inhibition assay	*Pseudomonas aeruginosa*	Competitive action with AHL signaling Diminution of pyocyanin production, EPS secretion, protease and elastase activities, and swimming motility	Musthafa et al. (2012)

(a component of the AI-2 system) and reduced the level of the AHL signal. In another study, anti-quorum sensing and anti-biofilm activities of two cinnamic acid derivatives, 4- dimethylaminocinnamic acid (DCA) and 4-methoxycinnamic acid (MCA), were shown against *Chromobacterium violaceum* ATCC12472 (Cheng et al. 2020). The author reported that both DCA (100μg/mL) and MCA (200μg/mL) inhibited the levels of N-decanoyl-homoserine lactone (C10-HSL) and reduced the production of certain virulence factors in *C. violaceum*, including violacein, hemolysin, and chitinase. In addition, DCA and MCA also downregulated the QS-related metabolites, such as ethanolamine and L-methionine, suppressed the expression of two QS-related genes (*cvil* and *cviR*), and inhibited the biofilm formation.

For chlorogenic acid, Wang et al. (2019) showed that CA inhibited the formation of biofilm in *Pseudomonas aeruginosa*, the ability of swarming, and virulence factors including protease and elastase activities and rhamnolipid and pyocyanin production and also showed similar inhibitory effects in *Chromobacterium violaceum* on its biofilm formation, swarming motility, chitinolytic activity, and violacein production. In a study by Bodini et al. (2009), *p*-coumaric acid inhibited QS responses of *Agrobacterium tumefaciens* NTL4, *Chromobacterium violaceum* 5999, and *Pseudomonas chlororaphis* with no influence on cell viability. Using qualitative QS inhibition assay by Chen et al. (2020) showed that at 0.2 mg/mL, this compound inhibited the QS in *Chromobacterium violaceum* (CECT 494) by inhibiting the violacein. Caffeic acid was reported by Luís et al. (2014) to have anti-QS and anti-biofilm effects in *Staphylococcus aureus* by inhibition of the production of α-hemolysin by this microorganism. In terms of biofilm formation, ellagic acid and phenylacetic acid were shown to be effective against *Burkholderia cepacia* (Huber et al. 2003) and *Pseudomonas aeruginosa* (Musthafa et al. 2012).

17.4 Conclusion

In this chapter, the QS system and its exploration as molecular targets to fight against infectious diseases were discussed focusing on the use of natural bioactive compounds. It has been shown that these molecules could be considered as alternative drug candidates to conventional antibiotics. Indeed, these drugs have quorum quenching effects with several mechanisms including the inhibition of the production, the action, and the transport of QS mediators. However, most of the discussed investigation were carried out using in vitro approaches and therefore further in vivo and clinical investigations should be carried out to validate the application of these drugs.

References

Aliyu AB, Koorbanally NA, Moodley B, Chenia HY (2020) Sesquiterpene lactones from *Polydora serratuloides* and their quorum sensing inhibitory activity. Nat Prod Res 2020:1–7. https://doi.org/10.1080/14786419.2020.1739037

Aliyu AB, Koorbanally NA, Moodley B, Singh P, Chenia HY (2016) Quorum sensing inhibitory potential and molecular docking studies of sesquiterpene lactones from Vernonia blumeoides. Phytochemistry 126:23–33. https://doi.org/10.1016/j.phytochem.2016.02.012

Al-Shabib NA, Husain FM, Ahmad I, Baig MH (2017) Eugenol inhibits quorum sensing and biofilm of toxigenic MRSA strains isolated from food handlers employed in Saudi Arabia. Biotechnol Biotechnol Equip 31(2):387–396

Alves S, Duarte A, Sousa S, Domingues FC (2016) Study of the major essential oil compounds of Coriandrum sativum against Acinetobacter baumannii and the effect of linalool on adhesion, biofilms and quorum sensing. Biofouling 32:155–165. https://doi.org/10.1080/08927014.2015.1133810

Al-Yousef HM, Ahmed AF, Al-Shabib NA, Laeeq S, Khan RA, Rehman MT, Alsalme A, Al-Ajmi MF, Khan MS, Husain FM (2017) Onion Peel Ethylacetate fraction and its derived constituent quercetin 4′-O-β-D Glucopyranoside attenuates quorum sensing regulated virulence and biofilm formation. Front Microbiol 8:1675. https://doi.org/10.3389/fmicb.2017.01675

Amaya S, Pereira JA, Borkosky SA, Valdez JC, Bardón A, Arena ME (2012) Inhibition of quorum sensing in Pseudomonas aeruginosa by sesquiterpene lactones. Phytomedicine 19:1173–1177. https://doi.org/10.1016/j.phymed.2012.07.003

Asif M (2020) Natural Anti-Quorum Sensing agents againstPseudomonas aeruginosa. J Chem Rev 2:57–69. https://doi.org/10.33945/SAMI/JCR.2020.1.4

Bassler BL (2002) Small talk: cell-to-cell communication in Bacteria. Cell 109:421–424. https://doi.org/10.1016/S0092-8674(02)00749-3

Bodini SF, Manfredini S, Epp M, Valentini S, Santori F (2009) Quorum sensing inhibition activity of garlic extract and p-coumaric acid. Lett Appl Microbiol 49:551–555. https://doi.org/10.1111/j.1472-765X.2009.02704.x

Burt SA, Ojo-Fakunle VTA, Woertman J, Veldhuizen EJA (2014) The natural antimicrobial Carvacrol inhibits quorum sensing in Chromobacterium violaceum and reduces bacterial biofilm formation at sub-lethal concentrations. PLoS One 9:e93414. https://doi.org/10.1371/journal.pone.0093414

Cai Y, Wang R, An M-M, Bei-Bei L (2010) Iron-depletion prevents biofilm formation in Pseudomonas aeruginosa through twitching motility and quorum sensing. Braz J Microbiol 41:37–41. https://doi.org/10.1590/S1517-83822010000100008

Castillo S, Heredia N, García S (2015) 2(5H)-Furanone, epigallocatechin gallate, and a citric-based disinfectant disturb quorum-sensing activity and reduce motility and biofilm formation of campylobacter jejuni. Folia Microbiol 60:89–95. https://doi.org/10.1007/s12223-014-0344-0

Chemmugil P, Lakshmi PTV, Annamalai A (2019) Exploring Morin as an anti-quorum sensing agent (anti-QSA) against resistant strains of Staphylococcus aureus. Microb Pathog 127:304–315. https://doi.org/10.1016/j.micpath.2018.12.007

Chen X, Yu F, Li Y, Lou Z, Toure SL, Wang H (2020) The inhibitory activity of p-coumaric acid on quorum sensing and its enhancement effect on meat preservation. CyTA J Food 18:61–67. https://doi.org/10.1080/19476337.2019.1701558

Cheng W-J, Zhou J-W, Zhang P-P, Luo H-Z, Tang S, Li J-J, Deng S-M, Jia A-Q (2020) Quorum sensing inhibition and tobramycin acceleration in Chromobacterium violaceum by two natural cinnamic acid derivatives. Appl Microbiol Biotechnol 104:5025–5037. https://doi.org/10.1007/s00253-020-10593-0

Chevrot R, Rosen R, Haudecoeur E, Cirou A, Shelp BJ, Ron E, Faure D (2006) GABA controls the level of quorum-sensing signal in agrobacterium tumefaciens. PNAS 103:7460–7464. https://doi.org/10.1073/pnas.0600313103

Chow S, Gu K, Jiang L, Nassour A (2011) Salicylic acid affects swimming, twitching and swarming motility in pseudomonas aeruginosa, resulting in decreased biofilm formation. J Exp Microbiol Immunol 15:8

Corral-Lugo A, Daddaoua A, Ortega A, Espinosa-Urgel M, Krell T (2016) Rosmarinic acid is a homoserine lactone mimic produced by plants that activates a bacterial quorum-sensing regulator. Sci Signal 9:ra1. https://doi.org/10.1126/scisignal.aaa8271

Devi KR, Srinivasan R, Kannappan A, Santhakumari S, Bhuvaneswari M, Rajasekar P, Prabhu NM, Ravi AV (2016) In vitro and in vivo efficacy of rosmarinic acid on quorum sensing mediated biofilm formation and virulence factor production in Aeromonas hydrophila. Biofouling 32:1171–1183. https://doi.org/10.1080/08927014.2016.1237220

Erdönmez D, Rad AY, Aksöz N, Erdönmez D, Rad AY, Aksöz N (2018) Anti-quorum sensing potential of antioxidant quercetin and resveratrol. Braz Arch Biol Technol 2018:61. https://doi.org/10.1590/1678-4324-2017160756

Fernández Rodríguez M, Corral-Lugo A, Krell T (2018) The plant compound rosmarinic acid induces a broad quorum sensing response in Pseudomonas aeruginosa PAO1. Environ Microbiol 20(12):4230–4244. https://doi.org/10.13039/501100000780

Frederick MR, Kuttler C, Hense BA, Eberl HJ (2011) A mathematical model of quorum sensing regulated EPS production in biofilm communities. Theor Biol Med Model 8:8. https://doi.org/10.1186/1742-4682-8-8

Gopu V, Meena CK, Shetty PH (2015) Quercetin influences quorum sensing in food borne Bacteria: in-vitro and in-silico evidence. PLoS One 10:e0134684. https://doi.org/10.1371/journal.pone.0134684

Grabski, H., Hunanyan, L., Tiratsuyan, S., Vardapetyan, H., 2017. Interaction of quercetin with transcriptional regulator LasR of Pseudomonas aeruginosa: Mechanistic insights of the inhibition of virulence through quorum sensing bioRxiv 239996. https://doi.org/10.1101/239996

Hernando-Amado S, Alcalde-Rico M, Gil-Gil T, Valverde JR, Martínez JL (2020) Naringenin inhibition of the Pseudomonas aeruginosa quorum sensing response is based on its time-dependent competition with N-(3-Oxo-dodecanoyl)-L-homoserine lactone for LasR binding. Front Mol Biosci 7:25. https://doi.org/10.3389/fmolb.2020.00025

Hosseinzadeh S, Dastmalchi Saei H, Ahmadi M, Zahraei-Salehi T (2020) Anti-quorum sensing effects of licochalcone a and epigallocatechin-3-gallate against Salmonella typhimurium isolates from poultry sources. Vet Res Forum 11:273–279. https://doi.org/10.30466/vrf.2019.95102.2289

Huber B, Eberl L, Feucht W, Polster J (2003) Influence of polyphenols on bacterial biofilm formation and quorum-sensing. Zeitschrift für Naturforschung C 58:879–884. https://doi.org/10.1515/znc-2003-11-1224

Joshi JR, Burdman S, Lipsky A, Yariv S, Yedidia I (2015) Plant phenolic acids affect the virulence of Pectobacterium aroidearum and P. carotovorum ssp. brasiliense via quorum sensing regulation. Mol Plant Pathol 4:487–500. https://doi.org/10.1111/mpp.12295

Lagonenko L, Lagonenko A, Evtushenkov A (2013) Impact of salicylic acid on biofilm formation by plant pathogenic bacteria. J Biol Earth Sci 3(2):176–181

Lee J, Wu J, Deng Y, Wang J, Wang C, Wang J, Chang C, Dong Y, Williams P, Zhang L-H (2013) A cell-cell communication signal integrates quorum sensing and stress response. Nat Chem Biol 9:339–343. https://doi.org/10.1038/nchembio.1225

Lou Z, Letsididi KS, Yu F, Pei Z, Wang H, Letsididi R (2019) Inhibitive effect of eugenol and its Nanoemulsion on quorum sensing–mediated virulence factors and biofilm formation by Pseudomonas aeruginosa. J Food Prot 82:379–389. https://doi.org/10.4315/0362-028X.JFP-18-196

Luís Â, Silva F, Sousa S, Duarte AP, Domingues F (2014) Antistaphylococcal and biofilm inhibitory activities of gallic, caffeic, and chlorogenic acids. Biofouling 30:69–79. https://doi.org/10.1080/08927014.2013.845878

Matsunaga T, Nakahara A, Minnatul KM, Noiri Y, Ebisu S, Kato A, Azakami H (2010) The inhibitory effects of Catechins on biofilm formation by the Periodontopathogenic bacterium,

Eikenella corrodens. Biosci Biotechnol Biochem 2010:1011022259. https://doi.org/10.1271/bbb.100499

Ming D, Wang D, Cao F, Xiang H, Mu D, Cao J, Li B, Zhong L, Dong X, Zhong X, Wang L, Wang T (2017) Kaempferol inhibits the primary attachment phase of biofilm formation in Staphylococcus aureus. Front Microbiol 8:2263. https://doi.org/10.3389/fmicb.2017.02263

Musk DJ Jr, Hergenrother PJ (2006) Chemical countermeasures for the control of bacterial biofilms: effective compounds and promising targets. Curr Med Chem 13:2163–2177. https://doi.org/10.2174/092986706777935212

Musthafa KS, Sivamaruthi BS, Pandian SK, Ravi AV (2012) Quorum sensing inhibition in Pseudomonas aeruginosa PAO1 by antagonistic compound Phenylacetic acid. Curr Microbiol 65:475–480. https://doi.org/10.1007/s00284-012-0181-9

Nyila MA, Leonard CM, Hussein AA, Lall N (2012) Activity of south African medicinal plants against listeria monocytogenes biofilms, and isolation of active compounds from Acacia karroo. S Afr J Bot 78:220–227

Ouyang J, Feng W, Lai X, Chen Y, Zhang X, Rong L, Sun F, Chen Y (2020) Quercetin inhibits Pseudomonas aeruginosa biofilm formation via the vfr-mediated lasIR system. Microb Pathog 149:104291

Ouyang J, Sun F, Feng W, Sun Y, Qiu X, Xiong L, Liu Y, Chen Y (2016) Quercetin is an effective inhibitor of quorum sensing, biofilm formation and virulence factors in Pseudomonas aeruginosa. J Appl Microbiol 120:966–974. https://doi.org/10.1111/jam.13073

Packiavathy V, Agilandeswari P, Musthafa KS, Pandian SK, Ravi AV (2012) Antibiofilm and quorum sensing inhibitory potential of Cuminum cyminum and its secondary metabolite methyl eugenol against Gram negative bacterial pathogens. Food Res Int 45:85–92

Pejin B, Ciric A, Glamoclija J, Nikolic M, Sokovic M (2014) In vitro anti-quorum sensing activity of phytol. Nat Prod Res 29(4):374–377

Pejin B, Ciric A, Markovic JD, Glamoclija J, Nikolic M, Stanimirovic B, Sokovic M (2015) Quercetin potently reduces biofilm formation of the strain Pseudomonas aeruginosa PAO1 in vitro. Curr Pharm Biotechnol 16:733–737

Rajkumari J, Borkotoky S, Murali A, Suchiang K, Mohanty SK, Busi S (2018) Cinnamic acid attenuates quorum sensing associated virulence factors and biofilm formation in Pseudomonas aeruginosa PAO1. Biotechnol Lett 40:1087–1100

Rasamiravaka T, Vandeputte OM, Pottier L, Huet J, Rabemanantsoa C, Kiendrebeogo M, Andriantsimahavandy A, Rasamindrakotroka A, Stévigny C, Duez P, Jaziri ME (2015) Pseudomonas aeruginosa biofilm formation and persistence, along with the production of quorum sensing-dependent virulence factors, are disrupted by a triterpenoid Coumarate Ester isolated from Dalbergia trichocarpa, a tropical legume. PLoS One 10:e0132791. https://doi.org/10.1371/journal.pone.0132791

Rathinam P, Vijay Kumar HS, Viswanathan P (2017) Eugenol exhibits anti-virulence properties by competitively binding to quorum sensing receptors. Biofouling 33:624–639. https://doi.org/10.1080/08927014.2017.1350655

Reading NC, Sperandio V (2006) Quorum sensing: the many languages of bacteria. FEMS Microbiol Lett 254:1–11. https://doi.org/10.1111/j.1574-6968.2005.00001.x

Ruckmani K, Ravi AV (2017) Exploring the anti-quorum sensing and Antibiofilm efficacy of phytol against Serratia marcescens associated acute pyelonephritis infection in Wistar rats. Front Cell Infect Microbiol 7:18

Šimunović K, Sahin O, Kovač J, Shen Z, Klančnik A, Zhang Q, Možina SS (2020) (−)-α-Pinene reduces quorum sensing and campylobacter jejuni colonization in broiler chickens. PLOS ONE 15:e0230423. https://doi.org/10.1371/journal.pone.0230423

Srinivasan R, Devi KR, Kannappan A, Pandian SK, Ravi AV (2016) Piper betle and its bioactive metabolite phytol mitigates quorum sensing mediated virulence factors and biofilm of nosocomial pathogen Serratia marcescens in vitro. J Ethnopharmacol 193:592–603

Tapia-Rodriguez MR, Bernal-Mercado AT, Gutierrez-Pacheco MM, Vazquez-Armenta FJ, Hernandez-Mendoza A, Gonzalez-Aguilar GA, Martinez-Tellez MA, Nazzaro F, Ayala-Zavala

JF (2019) Virulence of Pseudomonas aeruginosa exposed to carvacrol: alterations of the quorum sensing at enzymatic and gene levels. J Cell Commun Signal 13:531–537

Tapia-Rodriguez MR, Hernandez-Mendoza A, Gonzalez-Aguilar GA, Martinez-Tellez MA, Martins CM, Ayala-Zavala JF (2017) Carvacrol as potential quorum sensing inhibitor of Pseudomonas aeruginosa and biofilm production on stainless steel surfaces. Food Control 75:255–261

Trosko JE (2016) Evolution of microbial quorum sensing to human global quorum sensing: An insight into how gap junctional intercellular communication might be linked to the global metabolic disease crisis. Biology 5:29

Truchado P, Giménez-Bastida J-A, Larrosa M, Castro-Ibáñez I, Espin JC, Tomás-Barberán FA, Garcia-Conesa MT, Allende A (2012) Inhibition of quorum sensing (QS) in Yersinia enterocolitica by an Orange extract rich in glycosylated flavanones. J Agric Food Chem 60:8885–8894. https://doi.org/10.1021/jf301365a

Vandeputte OM, Kiendrebeogo M, Rasamiravaka T, Stévigny C, Duez P, Rajaonson S, Diallo B, Mol A, Baucher M, El Jaziri M (2011) The flavanone naringenin reduces the production of quorum sensing-controlled virulence factors in Pseudomonas aeruginosa PAO1. Microbiology 157:2120–2132

Vattem DA, Mihalik K, Crixell SH, McLean RJC (2007) Dietary phytochemicals as quorum sensing inhibitors. Fitoterapia 78:302–310

Vikram A, Jayaprakasha GK, Jesudhasan PR, Pillai SD, Patil BS (2010) Suppression of bacterial cell–cell signalling, biofilm formation and type III secretion system by citrus flavonoids. J Appl Microbiol 109:515–527

Walker TS, Bais HP, Déziel E, Schweizer HP, Rahme LG, Fall R, Vivanco JM (2004) Pseudomonas aeruginosa-plant root interactions. Pathogenicity, biofilm formation, and root exudation. Plant Physiol 134:320–331. https://doi.org/10.1104/pp.103.027888

Wang H, Chu W, Ye C, Gaeta B, Tao H, Wang M, Qiu Z (2019) Chlorogenic acid attenuates virulence factors and pathogenicity of Pseudomonas aeruginosa by regulating quorum sensing. Appl Microbiol Biotechnol 103:903–915. https://doi.org/10.1007/s00253-018-9482-7

Wang M, Zhu P, Jiang J, Zhu H, Tan S, Li R (2020a) Signaling molecules of quorum sensing in Bacteria. Rev Biotechnol Biochem 1:002

Wang R, Vega P, Xu Y, Chen C-Y, Irudayaraj J (2018) Exploring the anti-quorum sensing activity of a d-limonene nanoemulsion for Escherichia coli O157:H7. J Biomed Mater Res A 106:1979–1986. https://doi.org/10.1002/jbm.a.36404

Wang S, Payne GF, Bentley WE (2020b) Quorum sensing communication: molecularly connecting cells, their neighbors, and even devices. Ann Rev Chem Biomol Eng 11:447–468. https://doi.org/10.1146/annurev-chembioeng-101519-124728

Wu C-Y, Su T-Y, Wang M-Y, Yang S-F, Mar K, Hung S-L (2018) Inhibitory effects of tea catechin epigallocatechin-3-gallate against biofilms formed from Streptococcus mutans and a probiotic lactobacillus strain. Arch Oral Biol 94:69–77. https://doi.org/10.1016/j.archoralbio.2018.06.019

Xu G-M (2016) Relationships between the regulatory systems of quorum sensing and multidrug resistance. Front Microbiol 7:958. https://doi.org/10.3389/fmicb.2016.00958

Yuan Z-C, Edlind MP, Liu P, Saenkham P, Banta LM, Wise AA, Ronzone E, Binns AN, Kerr K, Nester EW (2007) The plant signal salicylic acid shuts down expression of the vir regulon and activates quormone-quenching genes in agrobacterium. PNAS 104:11790–11795. https://doi.org/10.1073/pnas.0704866104

Yue J, Yang H, Liu S, Song F, Guo J, Huang C (2018) Influence of naringenin on the biofilm formation of Streptococcus mutans. J Dent 76:24–31. https://doi.org/10.1016/j.jdent.2018.04.013

Zhou L, Zheng H, Tang Y, Yu W, Gong Q (2013) Eugenol inhibits quorum sensing at sub-inhibitory concentrations. Biotechnol Lett 35:631–637. https://doi.org/10.1007/s10529-012-1126-x

Plant-Assisted Plasmid Curing Strategies for Reversal of Antibiotic Resistance

Geetanjali M. Litake

Contents

Abstract

Antibiotic resistance attributed to mobile genetic elements (MGEs) has a universal concern in the management of treatment of infectious diseases. One of the modern-day strategies to have a check on these transmissible antibiotic resistance genes (ARGs) can be plasmid curing and anti-plasmid approaches. The in vitro tools implemented for plasmid curing make use of nontherapeutic chemicals, therapeutic drugs, and plant-derived compounds. The recent approach to deal with the reversal of multidrug resistance includes conjugation inhibition,

G. M. Litake (✉)
Department of Biotechnology, Modern College of Arts, Science and Commerce (Savitribai Phule Pune University), Ganeshkhind, Pune, India

V. Kumar et al. (eds.), *Antimicrobial Resistance*,
https://doi.org/10.1007/978-981-16-3120-7_18

biological strategies that employ plasmid incompatibility, the use of bacteriophages, chemically synthesized nanoparticles, and designed systems based on CRISPR-Cas9 and stress-free approach. In the notion to limit transmission of antibiotic resistance among diverse groups of bacterial species, an in vivo remedy is the need of time. An extensive research has shown that medicinal plants have a safe and effective therapeutic value to treat MDR infections. The phytochemicals from plant extracts and plant-derived compounds have the capacity to eliminate plasmids carrying ARGs and conjugal inhibition. Recent reports suggest that nanoparticles of plant origin can hold the promise for the reversal of antibiotic resistance. To explore the potential of medicinal plants as the promising supportive therapy to target the problem of antibiotic resistance can be the future option. This may help to reduce the prevalence of ARGs and reversal of multidrug resistance among infectious bacterial pathogens.

Keywords

Plasmid curing · Multidrug resistance · Plant-derived curing agents · Medicinal plant extracts · Plant-assisted nanoparticles

18.1 Introduction

Bacterial species are known to harbor a range of plasmids. The simple way to categorize plasmid DNA can be the genetic markers that they carry. Plasmids may confer distinctive characteristics to the host cell where an expression of an atypical phenotype in well-characterized bacterial species is an indication of plasmid conferred property (Grinsted and Bennett 1988).

The significance of plasmids in nosocomial infections is for the reason that they carry transferable genes for antibiotic resistance. Plasmids that bestow antibiotic resistance properties are referred to as R-plasmids. Plasmids participate in the dissemination of these features among different bacterial species with effective elimination of barriers. Plasmids play a crucial role where genes can collectively be able to reach distant bacterial species transforming them from being susceptible to resistant phenotype. The extent of resistance to therapeutic antibiotics can be directly associated with the copy number of the plasmid carrying antibiotic resistance genes (ARGs).

In the process of acquisition and transfer of ARGs, especially in Gram-negative bacteria, the plasmids play a significant role. In the case of bacterial pathogens, plasmids may contribute to a pathogenic potential, an unusual metabolic strength, or an ability to sustain in adverse environmental conditions. They also contribute to acquire resistance to the majority of the classes of antibiotics, particularly to β-lactams, aminoglycosides, polymyxins, and quinolones, which are current treatment options in the hospitals. This fact is applicable particularly to ESKAPE group of bacteria that cause most of the hospital-acquired infections (Stanisich 1988; Carattoli 2013; Ramirez et al. 2014; Vrancianu et al. 2020).

The emergence of resistant bacteria can be attributed to the widespread and irrational use of antibiotics in medicine, agriculture, and aquaculture. In recent years with the help of metagenomic studies, many novel ARGs have been demonstrated in various environments such as soil, activated sludge, and microbiota of animals and humans. These environments may act as reservoirs of ARGs (Cabello 2006; Economou and Gousia 2015; von Wintersdorff et al. 2016).

Natural consortia that represent a diverse group of organisms are considered as the set of connections of gene exchange that serves as a reservoir of antibiotic resistance called as "hot spots" for horizontal gene transfer. This leads to the dissemination of antibiotic resistance genes among pathogens and commensal microbes. This is an example of indigenous conjugation-assisted antibiotic resistance in diverse species.

Plasmids in extended-spectrum β-lactamases (ESBLs) producing *Escherichia coli* from wastewater treatment plants have shown to have an important role in the multiple antibiotic resistance linked transfer for β-lactam antibiotics in vitro (Li et al. 2019). Nonetheless plasmid-borne extended-spectrum β-lactamases have also been reported in the clinical isolates of *Acinetobacter baumannii* (Joshi et al. 2003).

ESBLs can effectively hydrolyze β-lactam antibiotics and help the strains turn into resistant phenotype to most of the β-lactam antibiotics, which is one of the prime mechanisms of β-lactam resistance among Gram-negative bacilli. Bacterial genes that encode for ESBLs are frequently located on the same plasmid carrying resistance genes for other antibiotics that leads to the prevalence of multiple drug resistance in bacteria (Li et al. 2019).

The unforeseen overconsumption of antibiotics in hospital practices imposes very high selective pressure that impels the evolution and spread of antibiotic resistance in nosocomial pathogens. These pathogens reside in the hospital environment and acquire different antibiotic resistance mechanisms through two principal routes, viz., chromosomal mutations and the acquisition of mobile genetic elements such as plasmids by horizontal gene transfer that ultimately leads to the emergence of untreatable superbugs (Millan 2018). The gut microbiome of hospitalized patients is one of the most important places for the transfer of antibiotic resistance genes. Conjugative plasmids play a significant role in the dissemination of resistance genes in the gut of patients especially true for the members of an *Enterobacteriaceae* family (Penders et al. 2013; Rozwandowicz et al. 2018; Singh et al. 2019).

The plasmids of varying size and number can be detected in diverse species of nosocomial pathogens. An example of single species multiple plasmids is *Acinetobacter*, one of the leading nosocomial pathogens that are known to be a natural reservoir of multiple plasmids (Bergogne-Berezin and Towner 1996).

Transfer of the antibiotic resistance markers from one bacterial species to another is an evidence of plasmid-borne or transposon-mediated dissemination. Also loss of the resistant phenotype corresponding to the loss of such genetic element from the cell is an evidence for plasmid involvement (Stanisich 1988). One of the mechanisms for the persistence of plasmids is horizontal gene transfer through conjugation. The persistence of plasmid is attributed to the fast or slow conjugation rate in the absence of selection. High plasmid transfer rates in an associated environment may favor a pool of local bacteria to share the beneficial phenotype. Disruption in such an environment may lead to the restoration of susceptible strains.

18.2   Why Target Plasmids?

Most of the pathogens have the ability to cope up with the new therapeutic options and adapt to this antibiotic stress/pressure. This leads to the acute need for the development of new effective therapeutic options. But there is always a limitation for new drug development. Therefore, there is a scope to develop new strategies to combat with AMR. Mere reduced antibiotic use is insufficient for the reversal of antibiotic resistance. One of the potential strategies to control AMR is to target plasmids/MGEs by their effective elimination from the resistant phenotype (Fig. 18.1). Another strategy is to inhibit/limit conjugation-mediated HGT. Reversal or suppression of conjugation mediated spread of antibiotic resistance by targeting parameters that are essential for plasmid maintenance in the bacterial community.

Thus, reversal of antibiotic resistance can be achieved by eliminating antibiotic usage, inhibiting conjugation, or promoting the rate of plasmid loss.

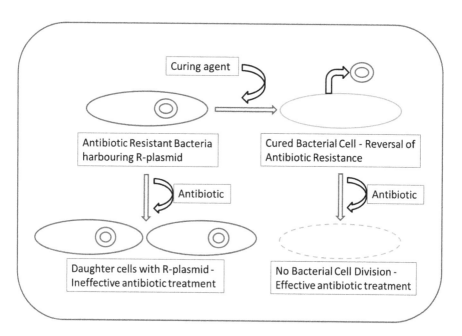

Fig. 18.1 Plasmid curing: an effective strategy for the reversal of antibiotic resistance

18.3   Strategies for Removing MGEs

To revert to a susceptible phenotype of a bacterium, it is mandatory to limit the exchange of genetic information contributing to multidrug resistance. Various strategies can be applied for the reversal of plasmid-encoded antibiotic resistance by targeting plasmids (Fig. 18.2).

18.3.1 Elimination of Plasmid

An elimination of plasmid from the bacterial host mostly occurs by two mechanisms, viz., inhibition of plasmid replication and interfering with plasmid segregation. Elimination of plasmid by using various chemical plasmid curing agents such as sodium dodecyl sulfate (SDS), ethidium bromide, acridine-orange, acriflavine,

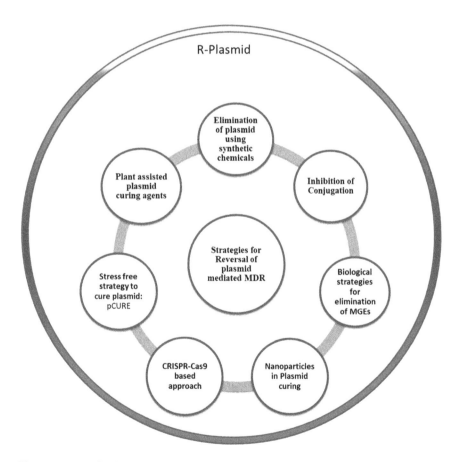

Fig. 18.2 Strategies for plasmid curing

triclosan (irgasan), fusidic acid, chitosan, and antibiotics has been effectively used to revert antibiotic resistance from various Gram-positive and Gram-negative bacteria.

Other chemical agents that are reported to be used for effective elimination of bacterial plasmids are ascorbic acid; bile; chlorpromazine; phenothiazine derivatives such as promethazine, thioridazine, and trifluoperazine; and drugs used in human medicine such as coumermycin A, novobiocin, rifampicin, and fluoroquinolone antibiotics. Lawsone, plumbagin, $1'$-acetoxychavicol acetate, and 8-epidiosbulbin E acetate are the well-explored chemical compounds that act as a plasmid curing agent which originated from plants (Spengler et al. 2006; Buckner et al. 2018; Vrancianu et al. 2020).

One should make a note that not all chemicals are recommended for reversal of all R-plasmids. Elimination of plasmids carrying MDR genes from different bacterial hosts can be achieved by different curing agents.

18.3.2 Inhibition of Conjugation

Conjugation plays a significant role in the dissemination of plasmids in a natural consortium. Therefore, the use of chemical inhibitors of conjugation is one of the strategies to prevent plasmid dissemination. It is understood that most of the mobile genetic elements use the common proteins for their transfer; the horizontal gene transfer can be blocked with the help of inhibitors of the natural conjugation process. The chemical agents such as intercalating agents, nitrofuran derivatives, unsaturated fatty acids, *TraE* inhibitors, bisphosphonates, antibodies, and chemical inhibitors of transposons recombination can be used as the conjugative inhibitor (Buckner et al. 2018; Vrancianu et al. 2020).

Unsaturated fatty acids have been shown to be effective conjugation inhibitors in laboratory studies on a variety of plasmids. The conjugation inhibition strategy was successfully applied to an engineered conjugation system by using synergistic action of linoleic acid (Lin) and phenothiazine (Pheno) to destabilize a plasmid (Lopatkin et al. 2017).

18.4 Biological Strategies for Elimination of MGEs

Biological strategies for the removal of mobile genetic elements include the use of bacteriophages for the insertion through the process of transduction. Phage exerts the genetic effects on bacteria that impose an impact on plasmid stability.

Incompatibility-based plasmid curing and toxin/antitoxin systems are based on the introduction of a second plasmid that destabilizes the transmission of the first plasmid indicating the two plasmids are incompatible. The reason for this is the fact that the two plasmids share the same replication and partitioning mechanisms. Therefore, in the presence of selective pressure, the inhabitant plasmid can be eliminated which results in the reversal of antibiotic resistance (Novick 1987; Vrancianu et al. 2020).

Various mechanisms of action are reported for plasmid curing (based on Trevors 1986):

1. Intercalating dyes such as acriflavine, acridine orange, ethidium bromide, and quinacrine have shown preferential inhibition of plasmid replication.
2. Coumermycin and novobiocin are known to inhibit DNA gyrase-dependent supercoiling of plasmid.
3. Mitomycin C is responsible for metabolic activation followed by nucleophilic attack on purine bases.
4. Rifampicin acts by inhibiting RNA polymerase.
5. The plasmid curing action of sodium dodecyl sulfate (SDS) is attributed to the presence of plasmid-specified pili on the cell surface of plasmid-containing cells.
6. Irgasan (Triclosan) acts by reducing plasmid carriage and overexpression of bacterial efflux pumps.
7. Elevated growth temperature may lead to either complete or partial deletions.
8. Thymine starvation can be the mechanism in the case of thymine-requiring auxotrophs.
9. Loss of certain plasmids may be due to protoplast formation and regeneration.
10. Plasmid incompatibility in the same cell is often associated with incompatibility curing.
11. The spontaneous loss of plasmid may be due to plasmidless segregants which arise during replication or portioning to daughter cells.

18.5   Nanoparticles in Plasmid Curing

There are reports of the use of various nanoparticles synthesized using silver, gold, zinc, copper, iron, titanium, platinum, nickel, etc. in control of MDR bacterial isolates (Baptista et al. 2018). Subinhibitory concentrations of platinum nanoparticles selectively cured plasmid from carbapenem-resistant *E.coli* strains in vitro with a remarkable decrease in MIC of the meropenem. The plasmid elimination effect is also demonstrated in vivo in an infected zebra fish (Bharathan et al. 2019).

18.6   CRISPR-Cas9-Based Approach to Plasmid Curing

It is demonstrated that by using replicon abundance and sequence conservation analysis, most of the bacterial vectors used in cloning and expression share sequence similarities. This property permits the CRISPR-Cas9 targeting. Reversal of colistin resistance mediated by plasmid borne mcr-1gene in *E. coli* by CRISPR-Cas9 system was reported as the potential strategy to resensitize the present MDR bacteria. It also helps to reduce the horizontal transfer of the conjugative plasmid (Wan et al. 2020).

A universal plasmid-curing system pFREE was constructed which can identify plasmid-free clones. This strategy was applied for the curing of plasmids commonly used for experimental purposes such as cloning and expression. It may be the future solution to target multiple plasmids. Several researchers are exploring the CRISPR-Cas system to target plasmids that harbor antibiotic resistance genes (Lauritsen et al. 2017).

18.7 Stress-Free Strategy to Cure Plasmid

The pCURE system for plasmid curing was developed in the Thomas laboratory. This system makes use of the combination of functions to block replication of a target plasmid and to neutralize any toxin-antitoxin system carried by plasmid (Lazdins et al. 2016). This system was based on the fact that the stability of the plasmids is because they contain multiple replicons and they encode post-segregational killing systems that result in reduced viability of plasmid-free segregants. This system was developed by combining important regions of the replicons, and the post-segregational killing loci into an unstable cloning vector that causes displacement of the plasmid leads to the construction of a plasmid-free strain. Different types of plasmids can be targeted with custom-made pCURE system. The limitation of this system is that it cannot be used in more complex environment (Hale et al. 2010).

18.8 Rationale for the Use of Plant Resources in Drug Resistance Reversal

The experimental use of chemical agents has been proven as an effective method to remove plasmids carrying antibiotic resistance markers. The use of these compounds to keep antibiotic resistance in check in humans is uncertain. This may be attributed to the fact that these compounds are harmful to the human body beyond a certain concentration. Earlier reports suggested the use of SDS is banned in humans and animals due to the adverse effects that it shows at the effective concentration required for the complete elimination of antibiotic resistance plasmids. Other chemicals such as intercalating agents used in the reversal of plasmid-mediated antibiotic resistance are associated with mutagenic effects. Chemically synthesized nanoparticles need to be evaluated for their safety and toxicological aspects as it is known that long-term exposure of nanoparticles leads to bioaccumulation in the body. In contrast to the use of chemical agents, the following aspects can support the use of plants in therapeutics to combat with MDR infections.

Plants are used widely as a dietary source and effective in traditional medical practices. The most important of it is associated with minimal side effects. The phyto-compounds are known to act as a source of medication to limit various types of infections. The economical considerations are also in favor of the use of plants.

Fig. 18.3 Plant-assisted plasmid curing strategies

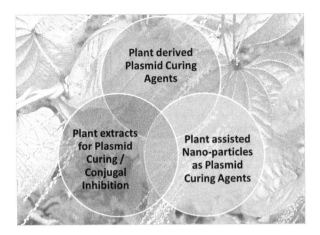

Supportive therapy using plant-derived natural products is widely reported for MDR infections. *Turnera ulmifolia* has been reported to be a source of plant-derived products with the capacity to modify antibiotic resistance that can be used effectively against multidrug-resistant bacteria such as methicillin-resistant *Staphylococcus aureus* (MRSA). The ethanol extract of *Turnera ulmifolia* has been used as a resistance modifying agent against the strain of *S. aureus* that is resistant to aminoglycoside (Coutinho et al. 2009).

Recent reports proposed amalgamation of plant-derived antimicrobials with plant-derived synergistic compounds as the future cost-effective approach to tackle the global antimicrobial resistance issue (Srivastava et al. 2014). The inclusion of herbal drug and antibiotic combination therapy will be a new promising approach for the treatment of infections involving multidrug-resistant bacteria (Bhardwaj et al. 2016). Medicinal plants are a rich source of bioactive compounds. Recently, it is reported that essential oils can be an effective antimicrobial option against wide-spectrum pathogens including the most significant ESKAPEE members (Yu et al. 2020). Plant resources can be explored for their potential to switch resistant bacterial phenotype into original susceptible one by using different approaches (Fig. 18.3).

18.9 Plant-Derived Plasmid Curing Agents

An effective in vitro plasmid curing potential of plant-derived compounds has been previously documented. These compounds included 8-epidiosbulbin E acetate from the bulbs of *Dioscorea bulbifera* (Shriram et al. 2008), 1′-acetoxychavicol acetate from *Alpinia galanga* (L.) Swartz, (Latha et al. 2009), plumbagin (5-hydroxy-2-methyl-1,4- naphthoquinone) from the root of the tropical/subtropical *Plumbago* species (Patwardhan et al. 2015a), and lawsone from *Plumbago zeylanica* (Patwardhan et al. 2018) (Table 18.1a).

8-Epidiosbulbin E acetate a bioactive compound that is isolated from the bulbs of an Ayurvedic medicinal plant *Dioscorea bulbifera* which belongs to the class of

Table 18.1 Plant-assisted plasmid curing/conjugal inhibition

(a) Plant-derived compounds as plasmid curing agents

Plant source	Part of the plant used	Bioactive compound	References
Dioscorea bulbifera	Bulbs	8-Epidiosbulbin E acetate	Shriram et al. (2008)
Alpinia galanga	Rhizome/root	1'-Acetoxychavicol acetate	Latha et al. (2009)
Tropical/subtropical *Plumbago* species	Root	Plumbagin (5-hydroxy-2-methyl-1,4- naphthoquinone)	Lakshmi et al. (1987), Patwardhan et al. (2015b), Patwardhan (2020)
Plumbago zeylanica	Root	Lawsone	Patwardhan et al. (2018)

(b) Plant extracts in plasmid curing/conjugal inhibition

Name of plant	Part of the plant used	Name of microorganisms	References
Allium sativum, Myrtus communis	Dry plant powder	*Proteus mirabilis*	Khder (2008)
Dioscorea bulbifera	Bulb	*Escherichia coli, Pseudomonas aeruginosa, Bacillus subtilis, Shigella sonnei,* and *Enterococcus faecalis*	Shriram et al. (2008), Osuntokun et al. (2019)
Quercus infectoria, Linusmus itatissium, Cinnamonum zeylanicum	Seeds, plant bark	*Escherichia coli*	Khider and Muhammed (2010)
Helicteres isora	Fruits	*Escherichia coli* (RP4), *Escherichia coli, Bacillus cereus,* and *Enterococcus faecalis*	Shriram et al. (2010)
Eugenia jambolana and *Elephantopus scaber*	Seed, whole plant	VR enterococci	Jasmine and Selvakumar (2011)
Cardamom (*Elettaria cardamom*)	Seed	*Proteus mirabilis, Staphylococcus aureus*	Akrayi (2012)
Rhus coriaria	Dried plant	*S. aureus*	Akrayi and Abdullrahman (2013)
Piper longum	Fruits (Aq)	*Enterococcus faecalis, Staphylococcus aureus, Salmonella typhi, Shigella sonnei*	Kumar et al. (2013)
Alpinia galanga	Rhizome (Aq)	*Enterococcus faecalis, Staphylococcus aureus, Salmonella typhi, Shigella sonnei*	Shriram et al. (2013)
Cuminum cyminum, Coriandrum sativum, Myristica fragrans	Seeds (methanolic extract)	*Acinetobacter* spp., *E. coli* and *Proteus* spp., *Klebsiella pneumonia,* and *Pseudomonas* spp.	Soman et al. (2015)

(continued)

Table 18.1 (continued)

(b) Plant extracts in plasmid curing/conjugal inhibition

Name of plant	Part of the plant used	Name of microorganisms	References
Terminalia chebula	Fruits	*Bacillus subtilis* (pUB110) and *Shigella sonnei* (pARI-815)	Srivastava et al. (2015)
Rosa canina	Fruit	*Escherichia coli* (conjugal inhibition)	Oyedemi et al. (2016)
Zingiber officinale, Ocimum gratissimum, Xylopia aethiopica	Rhizome, leaves, seeds	Citrate-negative motile *Salmonella species*	Iheukwumere et al. (2020)

diterpenes which is known as a plasmid curing agent. This compound has shown curing of plasmids from the reference strains of *Escherichia coli, Pseudomonas aeruginosa, Bacillus subtilis*, and clinical isolates of *Escherichia coli, Shigella sonnei*, and *Enterococcus faecalis* (Shriram et al. 2008).

An indigenous medicinal plant from Southeast Asian countries, *Alpinia galanga* (L.), has bioactive compound 1′-acetoxychavicol acetate. This compound when tested against nine bacterial reference strains harbors antibiotic resistance plasmids, and it was demonstrated that purified 1′-acetoxychavicol acetate has the potential to cure plasmids carrying MDR markers from *Escherichia coli, Enterococcus faecalis, Pseudomonas aeruginosa, Salmonella typhi*, and *Bacillus cereus* (Latha et al. 2009).

The root of the tropical/subtropical *Plumbago* species is a source of plumbagin (5-hydroxy-2-methyl-1,4-naphthoquinone). Plumbagin has shown its value as a plasmid curing agent by eliminating a conjugative MDR plasmid and the RP4 plasmid from *E.coli*. The reported mechanism is that plumbagin could eliminate the plasmids from *E.coli* by reducing plasmid copy number and the toxic effect of plasmid loss (Lakshmi et al. 1987; Lakshmi and Thomas 1996; Patwardhan et al. 2015b; Patwardhan 2020).

Another bioactive compound derived from the organic extract of *Plumbago zeylanica* roots is lawsone. It has a dual capacity to reverse antibiotic resistance by curing plasmids from reference strains and clinical isolates of *Acinetobacter baumannii* that are resistant to multiple antibiotics and effective in inhibition of interspecies plasmid transfer. It has also been shown to be nontoxic at lower concentrations (Patwardhan et al. 2018).

18.10 Plant Extracts in Plasmid Curing/Conjugal Inhibition

Various research workers are evaluating medicinal plant extracts and phyto-compounds as a plasmid curing option. Plant-based remedy in cure of various types of infections is well-known. Phytochemicals have potential antibacterial effects such as inhibition of efflux pump and elimination of R plasmids, which are the essential mechanisms involved in MDR.

The chemical plasmid curing agents may not be useful in therapeutic applications to combat with MDR. Extracts of medicinal plants can be safely used as therapeutic choice because they are safe, nontoxic, and non-mutagenic.

To combat with antimicrobial resistance, the use of plant extracts can be suggestive of a good approach to target plasmids. Many researchers are engaged in the evaluation of different plant extracts particularly the medicinal plants used in traditional medicine as a potential plasmid curing source against many bacterial species that harbor multiple plasmids. *Coptis chinensis* is a species of flowering plant used in traditional Chinese medicine and has also been evaluated for the elimination of R-plasmid from *E. coli* (Chen et al. 1996). A majority of the studies compared the plasmid curing capacities of plant extracts against the clinical isolates and the reference strains that harbor R-plasmids (Table 18.1b).

The extracts of *Allium sativum* and *Myrtus communis* dry plant powder obtained by steam distillation have proven to be effective to cure MDR plasmids from clinical isolates of multidrug-resistant *Proteus mirabilis* (Khder 2008).

Ethanolic extract of bulbs of *Dioscorea bulbifera* can cause the elimination of resistance plasmid from Gram-negative and Gram-positive bacteria and helps revert the resistant strain back into their original susceptible form (Shriram et al. 2008; Osuntokun et al. 2019). A subinhibitory concentration of crude acetone extract of *Alpinia galanga* rhizome is known to cure plasmids from *Escherichia coli, Salmonella typhi, Pseudomonas aeruginosa, Bacillus cereus*, and *Enterococcus faecalis* (Latha et al. 2009). Also according to another report, aqueous and methanol extracts of *Alpinia galanga* rhizomes have shown plasmid curing potential from multiple drug-resistant (MDR) clinical isolates of *Enterococcus faecalis, Staphylococcus aureus, Salmonella typhi*, and *Shigella sonnei*, along with reference plasmid-harboring strains of *Escherichia coli* (RP4) and *Bacillus subtilis* (pUB110) (Shriram et al. 2013).

Aqueous and alcohol extracts of *Quercus infectoria, Linum usitatissimum*, and *Cinnamonum zeylanicium* have shown potential to cure plasmid from MDR *Escherichia coli*. A comparative analysis carried out for plasmid elimination from *E. coli* E62 clinical isolate using subinhibitory concentrations of plant extracts showed that these extracts could cure plasmids with maximum plasmid curing efficiency by ethanol extract of *C. zeylanicum* that cured two plasmids. Ethanol extract of *L. usitatissimum* and aqueous and ethanol extract of *Q. infectoria* could cure single plasmid from E62 isolate (Khider and Muhammed 2010).

The active compound from acetone extract of *Helicteres isora* fruits cured R-plasmids from reference strain *Escherichia coli* (RP4) as well as clinical isolates of *Escherichia coli, Bacillus cereus*, and *Enterococcus faecalis* which resulted in the reversal of antibiotic resistance owing to the elimination of the plasmids encoding antibiotic resistance of cured strains (Shriram et al. 2010).

In the study of synergistic action of two Indian medicinal plants, it was suggested that the alcoholic extract of *Eugenia jambolana* and acetone extract of *Elephantopus scaber* has the potential to cure plasmids from vancomycin-resistant enterococci.

Reversal of resistant strains into sensitive phenotype due to loss of plasmid is attributed to the possibility that the extracts may have a capacity to inhibit plasmid replication at a subinhibitory concentration (Jasmine and Selvakumar 2011).

Alcoholic and aqueous seed extracts of cardamom (*Elettaria cardamomum*) have shown plasmid elimination from *Staphylococcus aureus* and *Proteus mirabilis*. The methanolic extract of cardamom (*Elettaria cardamom*) has a better capacity of curing than the ethanolic and aqueous extracts in *Staphylococcus aureus*, whereas in *Proteus mirabilis* plasmid curing was very weak (Akrayi 2012). The aqueous extract of dried plant *Rhus coriaria* has anti-plasmid potential to cure plasmid encoded antimicrobial resistant genes from *Staphylococcus aureus* (Akrayi and Abdullrahman 2013).

The aqueous and methanol extracts of *Piper longum* fruits could eliminate plasmids from multiple drug-resistant (MDR) clinical isolates of *Enterococcus faecalis*, *Staphylococcus aureus*, *Salmonella typhi*, and *Shigella sonnei*, along with reference plasmid-harboring strains of *Escherichia coli* (RP4) and *Bacillus subtilis* (pUB110) (Kumar et al. 2013). The methanolic extracts of seeds of *Cuminum cyminum*, *Coriandrum sativum*, and *Myristica fragrans* Houtt. could cure all three plasmid DNA present in the *Acinetobacter* spp., *Escherichia coli*, and *Proteus* spp. and only one plasmid DNA from *Klebsiella pneumoniae* and *Pseudomonas* spp. As methanolic extract can't be a choice for direct treatment, the active biomolecules from methanolic extracts of these seeds can be explored for future medicinal use against MDR bacteria (Soman et al. 2015).

The aqueous extract of *Terminalia chebula* fruit in citrate phosphate buffer exhibited plasmid curing activity against plasmids of *Bacillus subtilis* (pUB110) and *Shigella sonnei* (pARI-815) carrying resistance markers for kanamycin and gentamicin, respectively (Srivastava et al. 2015). Moderate conjugation inhibition was demonstrated against the transfer of TP114 and PKM101 from *Escherichia coli* by *Rosa canina* fruit extract (Oyedemi et al. 2016).

The aqueous suspensions of dried ethanolic extracts of seeds of *Xylopia aethiopica*, rhizome of *Zingiber officinale*, and leaves of *Ocimum gratissimum* were reported to show a reduction in the percentage of multiple antibiotics resistant strains. *Zingiber officinale* extract when compared exhibited the maximum plasmid curing potential which was similar to acridine orange (Iheukwumere et al. 2020).

In most of the reports of plasmid curing, the sublethal concentration of the extracts has been used as an effective concentration to eliminate R-plasmids from MDR bacteria. It is also documented that the plasmid curing frequency/efficacy determined in terms of percentage of curing can be varied for different plant extracts against different types of bacterial isolates. It is noteworthy that along with plasmid curing, these plant extracts may inhibit the essential virulence factors from the pathogen that are major contributors to the pathogenicity of the microorganism (Khder 2008).

There are different methods employed for the preparation of plant extracts using different solvent systems, viz., aqueous, ethanol, methanol, acetone, etc. All these extracts have shown their ability to reverse antimicrobial resistance by the elimination of R-plasmids. The organic plant extracts can be an essential step towards the isolation of an active biomolecule from these valued plants which may be a future medicine and can be used as therapeutic option in the treatment of MDR bacterial infections, whereas aqueous extracts can be an option for supportive therapy along with modern medicine.

18.11 Plant-Assisted Nanoparticles as Plasmid Curing Agents

An unrevealed option of green synthesis of nanoparticles as plasmid curing agent can be a potential choice of treatment to combat with MDR pathogens. It is an ecofriendly approach that may help avoid the use of undesirable toxic chemicals as a plasmid curing agent in vitro. It will be a valuable approach to limit environmental pollution attributed to experimental level use of chemical plasmid curing agents. Also it can be a future recommendation as a therapeutic choice.

The assumption of the action of nanoparticles is based on the formation of an electromagnetic force between the microorganism and the surface of nanoparticles due to the presence of opposite charges at the surface level, and their communication may force bacteria to bear oxidation effects. Also metal ions in nanoparticles may interact and get intercalated between DNA strands. Inside the microbial cell, nanoparticles can interrupt biochemical processes. In general the nanoparticles can target the outer and inner microbial cell organelles. Nanoparticles can be coupled with natural-based antimicrobials or other therapeutic compounds to inhibit the bacterial mechanisms attributing MDR properties and can be an approach to combat with multidrug-resistant bacteria (Baptista et al. 2018).

It is revealed that biosynthesized silver nanoparticles from aqueous extract of *Zingiber officinale* have the ability to cure plasmids of different bacterial isolates such as *E.coli*, *K. pneumoniae*, and *S. aureus* (Hashim 2017). The nanoemulsion prepared from *Alhagi maurorum* essential oil by the use of ionotropic gelation method and chitosan as a nano-carrier when tested against *P. aeruginosa*, *E. coli*, *S. aureus*, *K. pneumoniae*, *A. baumannii*, and *B. cereus* had shown noteworthy effect by curing R-plasmid from three antibiotic-resistant bacteria (Hassanshahian et al. 2020).

18.12 Future of Plant-Assisted Curing Agents

Elimination of plasmid will lead to the reversal of antibiotic resistance among bacterial population converting the resistance phenotype into sensitive one which will in turn bring once ineffective antibiotic therapeutic option in action again. The combination therapy using antibiotic along with curing agent can be possible as an effective therapeutic approach instead of antibiotic alone to tackle with plasmid-encoded multidrug-resistant bacterial infections particularly in hospital setup. An administration of a plasmid curing agent to an individual may help to restore drug-susceptible bacteria to the gut microbiome, and possible dissemination of MDR strains can be replaced by sensitive phenotype.

There is a scope to evaluate different underlying plasmid curing mechanisms of plant extracts or plant-derived/plant-synthesized compounds, and these approaches need to be evaluated by extensive studies to reveal safe and effective plant-based therapeutic option to cure plasmids responsible for multidrug resistance among bacterial species which in turn will help conquer global antibiotic resistance crisis.

References

Akrayi H (2012) Antibacterial effect of seed extracts of cardamom (*Elettaria cardamomum*) against *Staphylococcus aureus* and *Proteus mirabilis*. Tikrit J Pure Sci 17(2):14–17

Akrayi H, Abdullrahman Z (2013) Screening *in vitro* and *in vivo* the antibacerial activity of *Rhus Coriaria* extract against *S. aureus*. IJRRAS 15(3):390–397

Baptista PV, McCusker MP, Carvalho A, Ferreira DA, Mohan NM, Martins M, Fernandes AR (2018) Nano-strategies to fight multidrug resistant Bacteria—"a Battle of the titans". Frontiers in. Front Microbiol 9:1–26. https://doi.org/10.3389/fmicb.2018.01441

Bergogne-Berezin E, Towner KJ (1996) *Acinetobacter* spp. as nosocomial pathogens: microbiological, clinical and epidemiological features. Clin Microbiol Rev 9(2):148–165

Bharathan S, Sri Sundaramoorthy N, Chandrasekaran H, Rangappa G, Arun Kumar GP, Subramaniyan SB, Veerappan A, Nagarajan S (2019) Sub lethal levels of platinum nanoparticle cures plasmid and in combination with carbapenem, curtails carbapenem resistant *Escherichia coli*. Sci Rep 9:5305. https://doi.org/10.1038/s41598-019-41489-3

Bhardwaj M, Singh BR, Sinha DK, Kumar V, Prasanna Vadhana OR, Varan Singh S, Nirupama KR, Pruthvishree, Archana Saraf BS. (2016) Potential of herbal drug and antibiotic combination therapy: a new approach to treat multidrug resistant Bacteria. Pharm Anal Acta 7:11. https://doi.org/10.4172/2153-2435.1000523

Buckner MMC, Ciusa ML, Piddock LJV (2018) Strategies to combat antimicrobial resistance: anti-plasmid and plasmid curing. FEMS Microbiol Rev 42:781–804. https://doi.org/10.1093/femsre/fuy031

Cabello FC (2006) Heavy use of prophylactic antibiotics in aquaculture: a growing problem for human and animal health and for the environment. Environ Microbiol 8(7):1137–1144. https://doi.org/10.1111/j.1462-2920.2006.01054.x

Carattoli A (2013) Plasmids and the spread of resistance. Int J Med Microbiol 303(6–7):298–304. https://doi.org/10.1016/j.ijmm.2013.02.001

Chen Q, Chen NJ, Wang SC (1996) Experimental study of R plasmid eliminating action of Coptis chinensis on *E. coli*. Zhongguo Zhong Xi Yi Jie He Za Zhi 16(1):37–38

Coutinho HDM, Costa JGM, Lima EO, Falcão-Silva VS, Siqueira JP (2009) Herbal therapy associated with antibiotic therapy: potentiation of the antibiotic activity against methicillin – resistant *Staphylococcus aureus* by *Turnera ulmifolia* L. BMC Compl Altern Med 9:13–16

Economou V, Gousia P (2015) Agriculture and food animals as a source of antimicrobial-resistant bacteria. Infect Drug Resist 8:49–61. https://doi.org/10.2147/IDR.S55778

Grinsted J, Bennett PM (1988) Introduction: in methods in microbiology. In: Grinsted J, Bennett PM (eds) Plasmid technology, vol 21, 2nd edn, pp 1–10

Hale L, Lazos O, Haines AS, Thomas CM (2010) An efficient stress-free strategy to displace stable bacterial plasmids. Reports 48(3):223–228. www.BioTechniques.com

Hashim FJ (2017) Characterization and biological effect of silver nanoparticles synthesized by *Zingiber officinale* aqueous extract. RJPBCS 8(3):2490–2498

Hassanshahian M, Saadatfar A, Masoumipour F (2020) Formulation and characterization of nanoemulsion from *Alhagi maurorum* essential oil and study of its antimicrobial, antibiofilm, and plasmid curing activity against antibiotic-resistant pathogenic bacteria. J Environ Health Sci Eng 18:1015–1027

Iheukwumere IH, Dimejesi SA, Iheukwumere CM, Chude CO, Egbe PA, Nwaolisa CN, Amutaigwe EU, Nwakoby NE, Egbuna C, Olisah MC, Ifemeje JC (2020) Plasmid curing potentials of some medicinal plants against citrate negative motile *Salmonella* species. Eur J Biomed Pharm Sci 7(5):40–47

Jasmine R, Selvakumar B (2011) Synergistic action of two Indian medicinal plants on clinical isolates of vancomycin resistant *Enterococci species*. Pharmacolgyonline 2:898–904

Joshi SG, Litake GM, Ghole VS, Niphadkar KB (2003) Plasmid-borne extended-spectrum β-lactamase in a clinical isolate of *Acinetobacter baumannii*. J Med Microbiol 52 (12):1125–1127. https://doi.org/10.1099/jmm.0.05226-0

Khder AK (2008) Effect of *Allium sativum* and *Myrtus communis* on the elimination of antibiotic resistance and swarming of *Proteus mirabilis*. Jordan J Biol Sci 1(3):124–128

Khider A, Muhammed A (2010) Potential of aqueous and alcohol extracts of *Quercus infectoria, Linusm usitatissium* and *Cinnamomum zelyanicum as* antimicrobials and curing of antibiotic resistance in *E.coli*. Curr Res J Biol Sci 2(5):333–337

Kumar V, Shriram V, Mulla J (2013) Antibiotic resistance reversal of multiple drug resistant bacteria using *Piper longum* fruit extract. J Appl Pharm Sci 3(3):112–116

Lakshmi VV, Padma S, Polasa H (1987) Elimination of multidrug-resistant plasmid in bacteria by plumbagin, a compound derived from a plant. Curr Microbiol 16:159–161

Lakshmi VV, Thomas CM (1996) Curing of F-like plasmid TP181 by plumbagin is due to interference with both replication and maintenance functions. Microbiology 142:2399–2406

Latha C, Shriram VD, Jahagirdar SS, Dhakephalkar PK, Rojatkar SR (2009) Antiplasmid activity of 1′-acetoxychavicol acetate from Alpinia galanga against multi-drug resistant bacteria. J Ethnopharmacol 123(3):522–525

Lauritsen I, Porse A, Sommer MOA, Nørholm MHH (2017) Versatile one-step CRISPR-Cas9 based approach to plasmid-curing. Microb Cell Factories 16:135–144

Lazdins AM, Miller CE, Webber MA, Thomas CM (2016) Pcure: targeting plasmids to reduce the burden of antibiotic resistance. R & D Innovation AMR Control 2016:92–96

Li Q, Chang W, Zhang H, Hu D, Wang X (2019) The role of plasmids in the multiple antibiotic resistance transfer in ESBLs-producing *Escherichia coli* isolated from wastewater treatment plants. Front Microbiol 10:1–8. https://doi.org/10.3389/fmicb.2019.00633

Lopatkin AJ, Meredith HR, Srimani JK, Pfeiffer C, Durrett R, You L (2017) Persistence and reversal of plasmid-mediated antibiotic resistance. Nat Commun 8:1689–1698. https://doi.org/10.1038/s41467-017-01532-1

Millan AS (2018) Evolution of plasmid-mediated antibiotic resistance in the clinical context. Trends Microbiol 26(12):978–985. https://doi.org/10.1016/j.tim.2018.06.007

Novick RP (1987) Plasmid incompatibility. Microbiol Rev 51(4):381–395

Osuntokun OT, Mayowa A, Thonda OA, Aladejana OM (2019) Pre/post plasmid curing and killing kinetic reactivity of Dioscorea Bulbifera Linn against multiple antibiotics resistant clinical isolates, using Escherichia coli as a case study. Int J Cell Sci Mol Biol 6(2):46–56. https://doi.org/10.19080/IJCSMB.2019.06.555685

Oyedemi SO, Oyedemi BO, Prieto JM, Coopoosamy RM, Stapleton P, Gibbons S (2016) In vitro assessment of antibiotic-resistance reversal of a methanol extract from Rosa canina L. S Afr J Bot 105:337–342

Patwardhan RB (2020) Challenges and perspectives on plasmid curing, medicinal and pharmacological traits of Plumbago Zeylanica (Chitraka): a review. Int J Life Sci Pharma Res 10 (5):139–153. https://doi.org/10.22376/ijpbs/ijlpr.2020.10.5.P139-153

Patwardhan RB, Dhakephalkar PK, Chopade BA (2015a) Antibacterial and plasmid curing activities of root extracts of *Plumbago zeylanica*. Int J Herbo Med 2(1):13–25

Patwardhan RB, Dhakephalkar PK, Chopade BA, Dhavale DD, Bhonde RR (2018) Purification and characterization of an active principle, Lawsone, responsible for the plasmid curing activity of *Plumbago zeylanica* root extracts. Front Microbiol 9:1–10. https://doi.org/10.3389/fmicb.2018.02618

Patwardhan RB, Shinde PS, Chavan KR, Devale A (2015b) Reversal of plasmid encoded antibiotic resistance from nosocomial pathogens by using *Plumbago auriculata* root extracts. Int J Curr Microbiol App Sci 2:187–198

Penders J, Stobberingh EE, Savelkoul PHM, Wolffs PFG (2013) The human microbiome as a reservoir of antimicrobial resistance. Front Microbiol 4:1–7. https://doi.org/10.3389/fmicb.2013.00087

Ramirez MS, Traglia GM, Lin DL, Tran T, Tolmasky ME (2014) Plasmid-mediated antibiotic resistance and virulence in gram-negatives: the *Klebsiella pneumoniae* paradigm. Microbiol Spectr 2(5):1–15

Rozwandowicz M, Brouwer MSM, Fischer J, Wagenaar JA, Gonzalez-Zorn B, Guerra B, Mevius DJ, Hordijk J (2018) Plasmids carrying antimicrobial resistance genes in Enterobacteriaceae. J Antimicrob Chemother 73:1121–1137. https://doi.org/10.1093/jac/dkx488

Shriram V, Jahagirdar S, Latha C, Kumar V, Dhakephalkar P, Rojatkar S, Shitole MG (2010) Antibacterial & antiplasmid activities of *Helicteres isora* L. Indian J Med Res 132:94–99

Shriram V, Jahagirdar S, Latha C, Kumar V, Puranik V, Rojatkard S, Dhakephalkar PK, Shitole MG (2008) A potential plasmid-curing agent, 8-epidiosbulbin E acetate, from *Dioscorea bulbifera* L. against multidrug-resistant bacteria. Int J Antimicrob Agents 32:405–410

Shriram V, Kumar V, Mulla J, Latha C (2013) Curing of plasmid-mediated antibiotic resistance in multi-drug resistant pathogens using *Alpinia galanga* rhizome extract. Adv Bio Tech 13(1):1–5

Singh S, Verma N, Taneja N (2019) The human gut resistome: current concepts & future prospects. Indian J Med Res 150(4):345–358. https://doi.org/10.4103/ijmr.IJMR_1979_17

Soman YP, Mohite JA, Thakre SM, Raokhande SR, Mujumdar SS (2015) Plasmid curing activity by seed extracts of *Cuminum cyminum, Coriandrum sativum* and *Myristica fragrans* Houtt. And fruit peel extracts of orange, banana and pineapple against gram negative bacteria. Int J Curr Microbiol App Sci 2:302–316

Spengler G, Molnár A, Schelz Z, Amaral L, Sharples D, Molnár J (2006) The mechanism of plasmid curing in bacteria. Curr Drug Targets 7(7):823–841

Srivastava J, Chandra H, Nautiyal AR, Kalra SJS (2014) Antimicrobial resistance (AMR) and plant-derived antimicrobials (PDAms) as an alternative drug line to control infections. 3 Biotech 4:451–460. https://doi.org/10.1007/s13205-013-0180-y

Srivastava P, Wagh RS, Puranik NV, Puntambekar HM, Jahagirdar SS, Dhakephalkar PK (2015) *In vitro* plasmid curing activity of aqueous extract of *Terminalia Chebula* fruit against plasmids of *Bacillus Subtilis* and *Shigella Sonnei*. Int J Pharm Pharm Sci 7(4):298–301

Stanisich VA (1988) Identification and analysis of plasmids at the genetic level. In: Grinsted J, Bennett PM (eds) Methods in microbiology, Plasmid technology, vol 21, 2nd edn, pp 11–47

Trevors J (1986) Plasmid curing in bacteria. FEMS Microbiol Lett 32:149–157

von Wintersdorff CJH, Penders J, van Niekerk JM, Mills ND, Majumder S, van Alphen LB, Savelkoul Paul HM, Wolffs PFG (2016) Dissemination of antimicrobial resistance in microbial ecosystems through horizontal gene transfer. Front Microbiol 7:1–10. https://doi.org/10.3389/fmicb.2016.00173

Vrancianu CO, Popa LI, Bleotu C, Chifiriuc MC (2020) Targeting plasmids to limit acquisition and transmission of antimicrobial resistance. Front Microbiol 11:1–20. https://doi.org/10.3389/fmicb.2020.00761

Wan P, Cui S, Ma Z, Chen L, Li X, Zhao R, Xiong W, Zeng Z (2020) Reversal of mcr-1-mediated Colistin resistance in *Escherichia coli* by CRISPR-Cas9 system. Infect Drug Resist 13:1171–1178

Yu Z, Tang J, Khare T, Kumar V (2020) The alarming antimicrobial resistance in ESKAPEE pathogens: can essential oils come to the rescue? Fitoterapia 140:104433. https://doi.org/10.1016/j.fitote.2019.104433

Natural Product as Efflux Pump Inhibitors Against MRSA Efflux Pumps: An Update

19

Pallavi Ahirrao, Ritu Kalia, and Sanjay M. Jachak

Contents

P. Ahirrao (✉)
Department of Pharmaceutical Chemistry, Chandigarh College of Pharmacy, CGC, Landran-Mohali, Punjab, India
e-mail: pallavi.ccp@cgc.edu.in

R. Kalia · S. M. Jachak
Department of Natural Products, National Institute of Pharmaceutical Education and Research (NIPER), Sector-67, SAS Nagar, Punjab, India

Abstract

Among different nosocomial infections, *Staphylococcus aureus*, a Gram-positive bacterium, is a highly adaptive human pathogen. Over the years it had acquired resistance to multiple classes of antibiotics including methicillin. The multidrug resistance towards multiple antibiotics and poor pipeline of safe and effective drugs has rendered bacterial infections a life-threatening problem. Multidrug efflux pumps play an essential role in antibiotic resistance by extrusion of drugs via different mechanisms. Natural products especially derived from plants have emerged as an important source of effective efflux pump inhibitors. In this book chapter, different classes of plant- and microbe-derived natural products have been described as efflux pump inhibitors of MRSA that act synergistically in combination with antibiotics to modulate efflux pump-mediated extrusion of antibiotics and thereby help in combating the multidrug resistance.

Keywords

Natural products · Antimicrobial resistance · Methicillin-resistant Staphylococcus aureus (MRSA) · Efflux pumps · Efflux pump inhibitor · Multidrug resistance

19.1 Introduction

Staphylococcus aureus is a major cause of nosocomial- and community-acquired infections. Due to its ability to acquire mobile genetic elements encoding virulence and resistance determinants, the emergence of methicillin-resistant *Staphylococcus aureus* (MRSA) has occurred over the years which contributes to the development of new strains resistant against multiple classes of antibiotics (Alibayov et al. 2014; Lindsay and Holden 2004). Around 90–95% of S. *aureus* strains worldwide are found to be resistant to penicillin, and methicillin-resistant strains account for 70–80% of its total count in most Asian countries (Li et al. 2019). At least 50, 000 deaths are recorded due to S. *aureus* infection in Europe every year, and it is forecasted that infections due to drug resistance would be the reason for the deaths of nearly ten million people worldwide by 2050 (de Kraker et al. 2016). MRSA is a leading cause of endocarditis, bacteremia, soft tissue skin infections, and hospital-acquired infections (Alekshun and Levy 2007). Penicillin-binding protein 2a (PBP2a) has been reported to be encoded by the mecA gene, showing a low affinity to β-lactams that leads to resistance to this class of antibiotics. Since 1990, there has been a rapid spread of MRSA infections in the human community which poses a great challenge for their treatment. It is already known that staphylococci are able to act against each new antimicrobial agent by adopting one or more resistance mechanisms. Several mechanisms associated with the resistance to antibiotics include target protein mutation, antibiotic inactivation by enzymes, or antibiotic accumulation inhibition due to overexpression of the efflux system in bacterial cells. Among these, drug efflux is the most widespread reason for antimicrobial resistance (Alekshun and Levy 2007).

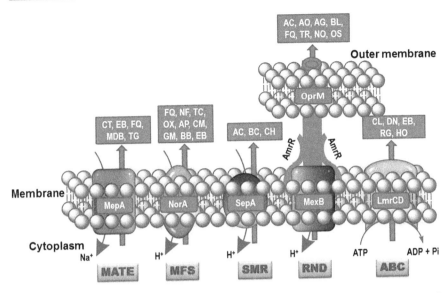

Fig. 19.1 EPs and the substrates that are effluxed out of the bacterial cell (Jachak et al. 2012) *ABC* ATP-binding cassettes, *RND* resistance nodulation division, *SMR* small multidrug resistance, *MFS* major facilitator superfamily, *MATE* multidrug and toxic compound extrusion, *AC* acriflavine, *AG* aminoglycoside, *AO* acridine orange, *AP* ampicillin, *BB* berberine, *BL* β-lactam, *BC* benzalkonium chloride, *CH* chlorhexidine, *CL* cholate, *CM* chloramphenicol, *CT* ceftazidime, *DN* daunomycin, *EB* ethidium bromide, *FQ* fluoroquinolones, *GM* gentamicin, *HO* Hoechst 33342, *MDB* monovalent and divalent biocides, *NF* norfloxacin, *NO* novobiocin, *OS* organic solvents, *OX* oxacillin, *RG* rhodamine, *TC* tetracyclines, *TG* tigecycline, *TR* triclosan

Gram-positive bacteria such as *S. aureus* lacks an outer membrane. Efflux pumps are helpful in limiting the accumulation of toxic compounds within the cell. So far, various efflux pumps have been discovered in microorganisms. These EPs are mainly classified in five different superfamilies, such as (1) adenosine triphosphate-binding cassette transporters (ABC), (2) multidrug and toxic compound extrusion (MATE), (3) major facilitator superfamily (MFS), (4) small multidrug resistance (SMR), and (5) resistance nodulation division (RND) family (Fig. 19.1) (Bohn and Bouloc 1998; Jachak et al. 2012; Zechini and Versace 2009). The above efflux pump families have been classified on the basis of amino acid sequence similarity, single or multiple components, specificity, energy source, and the number of transmembrane spanning regions. The ABC, MATE, SMR, and MFS superfamilies have been widely distributed in Gram (+)ve and Gram (−)ve bacteria whereas the RND family is distributed in Gram (−)ve bacteria only. For energy sources in the export of substrate, the efflux pump of the ABC family uses ATP hydrolysis whereas the rest other families utilize proton motive force. In Gram (+)ve bacteria, the most predominant efflux pump group is the MFS family, and the most studied member of the MFS family is NorA from *S. aureus* and PmrA from *S. pneumonia* (Jachak et al. 2012). A few of the bacterial efflux pumps of these superfamilies are summarized in Table 19.1 (Jachak et al. 2012).

Table 19.1 List of bacterial efflux pumps

Efflux pump(s)	Substrates	Organism(s)	References
Major facilitator superfamily (MFS)			
MdfA	Chloramphenicol, doxorubicin, fluoroquinolones, norfloxacin, tetracyclines	*Salmonella typhimurium, Escherichia coli*	Bohn and Bouloc (1998), Nishino et al. (2006)
MdeA	Benzalkonium chloride, dequalinium chloride, ethidium bromide, fusidic acid, Hoechst 33342, lincosamides, macrolides, mupirocin, virginiamycin, quaternary ammonium compounds, novobiocin, tetraphenylphosphonium	*Staphylococcus aureus, Staphylococcus haemolyticus, Bacillus cereus, Bacillus subtilis*	Huang et al. (2004), Yamada et al. (2006)
EmeA	Acriflavine, cholate, ethidium bromide, erythromycin, fluoroquinolones, novobiocin	*Enterococcus faecalis*	Jonas et al. (2001)
QacAa	Acriflavine, quaternary ammonium compounds, chlorhexidine, diamidines, ethidium bromide, crystal violet	*S. aureus*	Littlejohn et al. (1992)
Resistance nodulation division family (RND)			
MexB	Acriflavine, aminoglycosides, acridine orange, berberine, β-lactams, benzalkonium chloride, chloramphenicol, crystal violet, daunorubicin, ethidium bromide, erythromycin, fusidic acid, macrolides, novobiocin, organic solvents, rhodamine 6G, sodium dodecyl sulfate, sulfonamides, trimethoprim, triclosan	*Pseudomonas aeruginosa, Pseudomonas syringae*	Cao et al. (2004), Daigle et al. (2007), Li et al. (1995), Poole et al. (1996), Stoitsova et al. (2008)
CmeE	Acridine orange, ampicillin, cetyltrimethylammonium bromide, ethidium bromide, sodium dodecyl sulfate, polymyxin B, triclosan	*Campylobacter jejuni*	Akiba et al. (2006), Pumbwe et al. (2005)

(continued)

Table 19.1 (continued)

Efflux pump(s)	Substrates	Organism(s)	References
Major facilitator superfamily (MFS)			
SdeY	Acriflavine, erythromycin, norfloxacin, rhodamine 6G, tetracyclines	*Serratia marcescens*	Chen et al. (2003)
VexF	Benzalkonium chloride, bile salts, deoxycholate, ethidium bromide, erythromycin, norfloxacin, novobiocin, sodium dodecyl sulfate, tetracyclines, trimethoprim	*Vibrio cholera*	Rahman et al. (2007)
ATP-binding cassette (ABC) superfamily			
DrrAB	Daunorubicin, doxorubicin, ethidium bromide, tetracyclines, novobiocin	*Mycobacterium tuberculosis*	Choudhuri et al. (2002), Rossi et al. (2006)
Bcg0231	Ampicillin chloramphenicol, streptomycin, vancomycin	*Mycobacterium bovis* BCG	Danilchanka et al. (2008)
Rv0194	Ampicillin, erythromycin, novobiocin, vancomycin	*M. Tuberculosis*	Danilchanka et al. (2008)
SmdAB	Norfloxacin, 4',6-diamidino-2-phenylindole, tetracyclines, Hoechst 33342	*Serratia marcescens*	Matsuo et al. (2008)
VcaM	Daunomycin, 4',6-diamidino-2-phenylindole, doxorubicin, fluoroquinolones, Hoechst 33342, tetracyclines	*Vibrio cholerae*	Huda et al. (2003)
Multidrug and toxic compound extrusion family			
AbeM	Daunomycin, Hoechst 33342, acriflavine, fluoroquinolones, aminoglycosides, rhodamine 6G	*Acinetobacter baumannii*	Su et al. (2005)
NorMI	Acriflavine, berberine, fluoroquinolones, gentamicin, tetraphenylphosphonium	*Brucella melitensis*	Braibant et al. (2002)
HmrM	Acriflavine, berberine, deoxycholate, doxorubicin, ethidium bromide, tetraphenylphosphonium, fluoroquinolones	*Haemophilus influenzae*	Piddock (2006), Xu et al. (2003)
PmpM	Acriflavine, benzalkonium chloride, tetraphenylphosphonium, fluoroquinolones	*Pseudomonas aeruginosa*	He et al. (2004)

(continued)

Table 19.1 (continued)

Efflux pump(s)	Substrates	Organism(s)	References
Major facilitator superfamily (MFS)			
DinF	Berberine, acriflavine, ethidium bromide, ampicillin, tetraphenylphosphonium	*Ralstonia solanacearum*	Brown et al. (2007)
MepA	Fluoroquinolones, cetrimide, monovalent and divalent biocides, tigecycline ethidium bromide	*S. aureus*	Kaatz et al. (2006), Kaatz et al. (2005), McAleese et al. (2005)
Small multidrug resistance (SMR) family			
EmrE	Ethidium bromide, acriflavine, quaternary ammonium compounds, aminoglycosides	*E. coli, P. aeruginosa*	Li et al. (2003), Yerushalmi et al. (1995)
TehAB	Dequalinium, ethidium bromide, potassium tellurite, methyl viologen, proflavine	*E. coli*	Turner et al. (1997)
Mmr	Acriflavine, cetrimide, ethidium bromide, erythromycin, norfloxacin, tetraphenylphosphonium	*M. tuberculosis, M. smegmatis*	Turner et al. (1997)
MdtJI	Deoxycholate, sodium dodecyl sulfate, spermidine	*E.coli*	Higashi et al. (2008)
SsmE	Acriflavine, ethidium bromide, norfloxacin	*Serratia marcescens*	Minato et al. (2008)
SepA	Acriflavine, benzalkonium chloride, chlorhexidine	*S. aureus*	Narui et al. (2002)

Among all MDR microorganisms, MRSA strain is of main concern, liable for both hospital- and community-acquired infections (Craft et al. 2019). Among MRSA, overexpressed NorA strains are the most common ones (Costa et al. 2019). NorB and NorC are the other newly identified EPs which are responsible for resistance to quinolone in *S. aureus*. MgrA is a global regulator that regulates the NorA pump positively and NorB and NorC negatively (Belofsky et al. 2004; Truong-Bolduc et al. 2006). MsrA efflux pump of the ABC transporter family causes resistance to streptogramins and macrolides. MsrA is a transmembrane protein made up of 488 amino acids with two ATP-binding motifs. It functions independently in RN4220 strain (Costa et al. 2013).

Efflux pumps are an important and major antibacterial drug target. Hence, it is necessary to identify and develop potent efflux pump inhibitors (Edelsberg et al.

2014; Y Mahmood et al. 2016). Efflux pumps can be inhibited by the following strategies: (1) interference in genetic regulation by deregulating the EP expression, (2) antibiotic redesigning that are considered as substrates earlier, (3) suppressing the functional EPs assembly, (4) avoiding the substrate binding by blocking the active site, and (5) disintegrating the energy mechanism liable for reinvigorating the EPs (Sharma et al. 2019).

To tackle antibiotic resistance, drug resistance reversal agents especially efflux pump modulators/inhibitors would be promising leads. These compounds may possess either antimicrobial activity of their own or possess the ability to enhance the activity of ineffective antibiotics by inhibiting/modulating efflux pumps. Thus, the susceptibility or sensitivity of resistant strains to antibacterial agents can be reinstated with the aid of EPIs (Costa et al. 2013; Pagès and Amaral 2009; Sharma et al. 2019). In the development of EPIs, a few strategies can be employed. (a) In the case of the precise Tet protein EPs, the chemical modification on the antibiotic can be done to prevent its binding to the EP and transportation. (b) In multi-drug resistance EPs, it may be beneficial to test already known inhibitors of other efflux systems, e.g., eukaryotic systems. Since a large amount of data related to the safety and efficacy of an antibiotic is already available, testing of drugs apart from known antibiotics would be time-saving and cost-effective. (c) Screening of natural product libraries may allow identification of lead molecules that could be developed further by SAR studies (Schindler and Kaatz 2016).

The actions of EPI mechanisms are not precisely known yet. But, it has been suggested that the inhibitor binds directly to the pump and thus blocks it competitively or noncompetitively with the substrates. By inhibiting ATP binding or by disrupting the proton gradient, EPIs can cause depletion of energy. A complex of EPI with an antibiotic enhances the entry of an antibiotic into the bacterial cell and further inhibits the efflux of an antibiotic due to the larger complex size (Opperman and Nguyen 2015; Zloh et al. 2004).

To combat antibiotic resistance, presently the researchers are working on the development of synergistic antibiotic combinations to reduce the dose of antibiotics many times. In this regard, drug resistance reversal agents especially efflux pump inhibitors (EPIs)/modulators would be the promising agents.

19.2   Screening of Efflux Pump Inhibitors

Several pathogenic Gram (+)ve and Gram (−)ve bacteria remove antibiotics out of bacterial cells using efflux pumps, e.g., *S. aureus* and *Mycobacteria* use this mechanism to develop resistance. For bacterial infections, a combination of EPIs with a new or clinically used antibiotic may provide a new model of treatment. The bioassay methods to evaluate EPIs are described as follows.

19.2.1 Accumulation Assay (EtBr or Berberine)

Substrates like ethidium bromide (EtBr) or berberine are effluxed out by the resistant bacterial cells. When given along with EPI, EtBr/berberine is accumulated in the cells. The EPI activity of the potential inhibitors is measured fluorometrically (Paixão et al. 2009).

19.2.2 Susceptibility Testing

In susceptibility testing, growth inhibition of a compound showing EPI activity is determined with a subinhibitory concentration of berberine against Gram (+)ve bacteria and erythromycin for Gram (−)ve bacteria. This results in the elimination of synergy due to antibacterial activity. The National Centre for clinical Laboratory Standards (NCCLS) defined MDR inhibitor as the one that arreted cell growth completely at subinhibitory concentration at 37 °C for 18 h. Growth is assayed using a microtiter plate by measuring absorption at 600 nm (Bayot and Bragg 2020; Paixão et al. 2009).

19.2.3 MIC Synergy Testing in the Presence of EPI

Inactive antimicrobials are compounds possessing MICs >125 µg/mL. MIC for synergy is tested further with different concentrations of an antibiotic against a fixed concentration of test inhibitors in active efflux pumps in *S. aureus* strains. If the test inhibitor is found to possess efflux pump inhibitory property, then the antibiotic potency is enhanced at a subinhibitory concentration known as the modulation factor (MF) of that antibiotic (Abreu et al. 2017; Li et al. 2017).

$$MF = \frac{MIC(Antibiotic)}{MIC(Antibiotic + Modulator)}$$

19.2.4 Fractional Inhibitory Testing (FIC)

A fractional inhibitory concentration (FIC) method is used for compounds which exhibit $MF \geq 8$ to differentiate if the two compounds together demonstrate additive, synergistic, or antagonistic property against bacteria. FICI is determined and explained as ≤ 0.5 represented synergy, ≥ 4.0 for antagonism, and 4.0–2.0 for no interaction. The results between synergy and antagonistic tendency ($2.0 > FICI > 0.5$) is defined as additive or indifferent (Li et al. 2017; Odds 2003).

$$FICI = FIC(A) + FIC(B)$$

$$FIC(A) = \frac{MIC(A\,in\,presence\,of\,B)}{MIC(A\,alone)}$$

$$FIC(B) = \frac{MIC(B\,in\,presence\,of\,A)}{MIC(B\,alone)}$$

19.2.5 Time Kill Studies

The last step in testing a successful EPI is the time-kill test. It is considered as a basic microbiological method to determine the antimicrobial activity of a compound. This test is performed to evaluate the effect of antimicrobials on the reduction of microbial load in bacteria like *P. aeruginosa*, *A. niger*, *S. aureus*, and *E. coli*. The test compound is added to a known population (approximately 10^5 CFU/mL) of microorganisms for a specified temperature and period. Aliquots are removed at selected time intervals including zero time and kept in a neutralizer blank. The neutralizer dilutions are prepared and plated onto the agar medium, and the grown colonies of bacteria are counted. The log reduction and percent from an initial microbial population or test blank are calculated (Ahmed et al. 1993; Tsuji et al. 2008).

19.2.6 Natural Product Inhibitors of Efflux Pumps

Historically NPs have been a major source of biologically active molecules exhibiting numerous scaffolds and displaying various activities against both noninfectious and infectious diseases. The NPs are formed biogenetically by the processes catalyzed using enzymes that are highly regio-, enantio-, and diastereospecific. A few efflux pumps selectively efflux out a particular class of antibiotics, while other EPs extrude a diverse class of antibiotics; these are termed as MDR. EPIs may be useful in restoring the clinical efficacy of some earlier antibiotics, by enhancing the potency of antibiotics or by decreasing their resistance development. Numerous natural products are known to act as EPIs on different EPs located on the bacterial cell membrane (Ahmed et al. 1993; Shriram et al. 2018; Stavri et al. 2007; Tsuji et al. 2008; Zloh et al. 2004).

19.2.6.1 *S. aureus* NorA Multidrug Efflux Pump Inhibitors
In MF superfamily, the most studied example is NorA multidrug transporter that contributes to the resistance of *S. aureus*. Berberine and fluoroquinolones like norfloxacin and ciprofloxacin are effluxed out of the cell by NorA EP (Kumar et al. 2013). NPs that demonstrated EP inhibitory activity in *S. aureus* NorA EP are shown in Fig. 19.2.

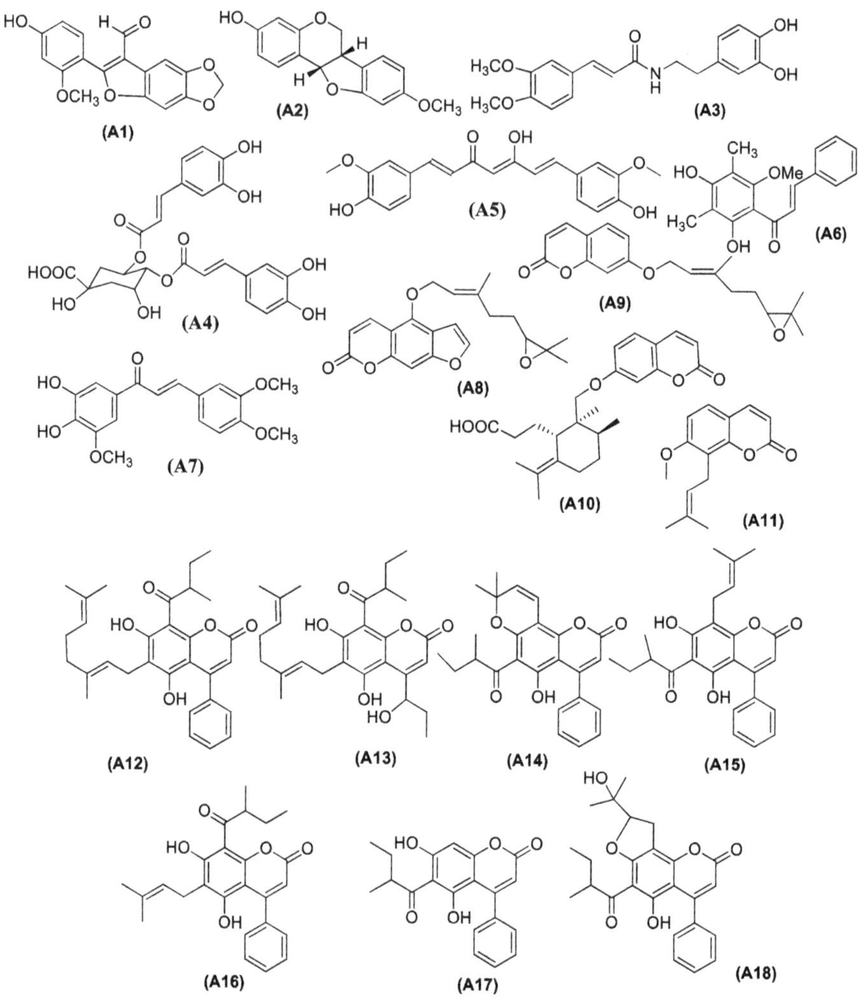

Fig. 19.2 NorA efflux pump inhibitors of natural product origin

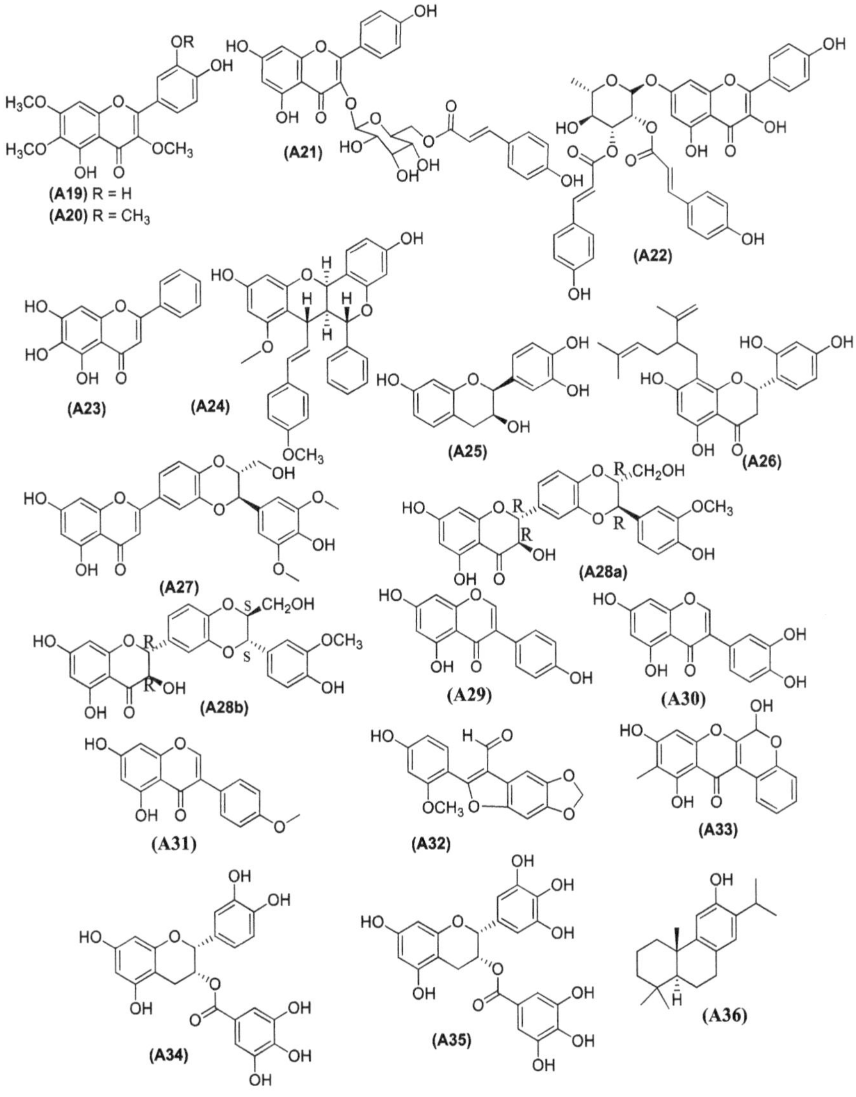

Fig. 19.2 (continued)

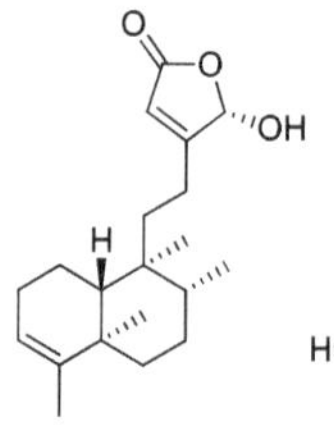

(A37) 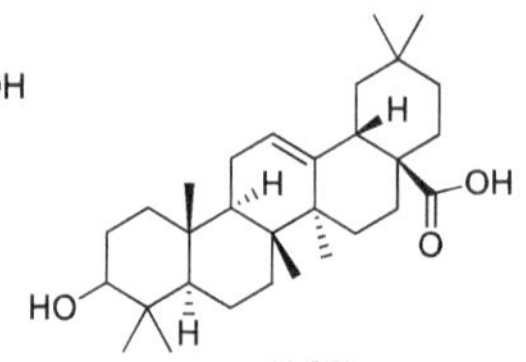 (A38) 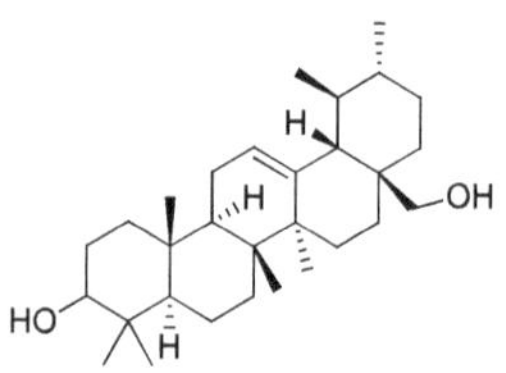 (A39)

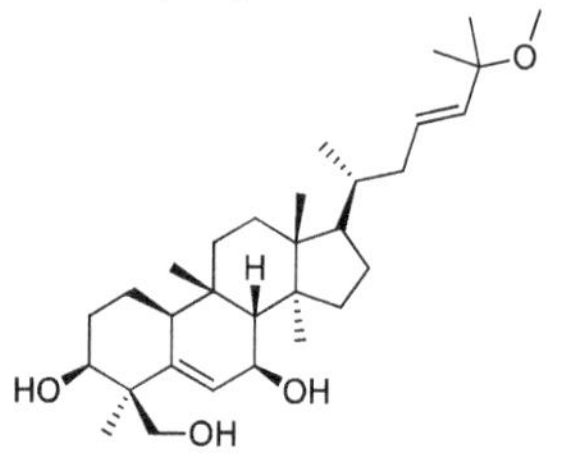

(A40)

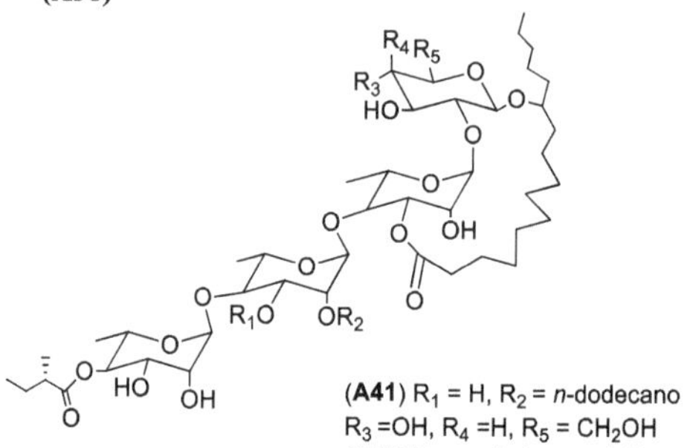

(A41) R$_1$ = H, R$_2$ = n-dodecano
R$_3$ =OH, R$_4$ =H, R$_5$ = CH$_2$OH
(A42) R$_1$ = n-dodecanoyl, R$_2$ =
R$_3$ =OH, R$_4$ =H, R$_5$ = CH$_2$OH
(A43) R$_1$ = n-dodecanoyl, R$_2$ =
R$_3$ = H, R$_4$ = OH, R$_5$ = CH$_3$
(A44) R$_1$ = H, R$_2$ = (2S)-
methylbutanoyl, R$_3$ = H, R$_4$ = O
R$_5$ = CH$_3$
(A45) R$_1$ = (2S)-methylbutanoy

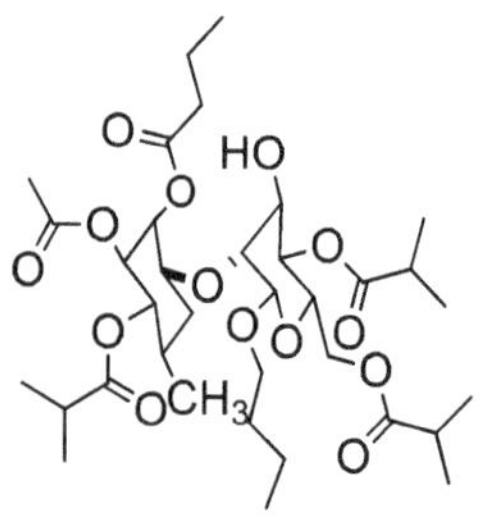

(A46)

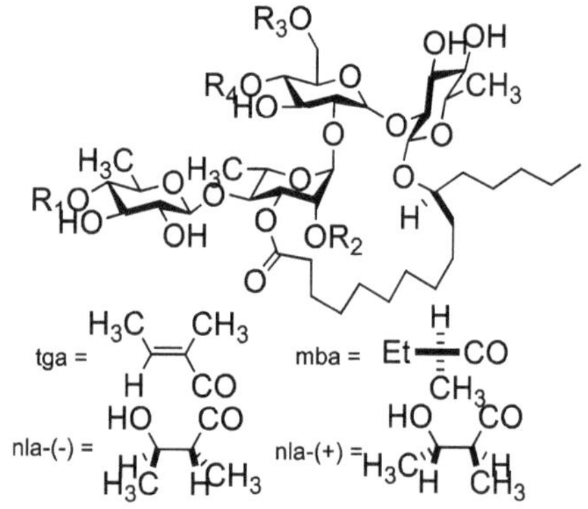

(A50)

tga =   mba =
nla-(-) =   nla-(+) =

(A47) R$_1$= mba, R$_2$= nla-(+), R$_3$= mba, R$_4$= H
(A48) R$_1$= tga, R$_2$= mba, R$_3$= H, R$_4$= nla-(-)
(A49) R$_1$= tga, R$_2$= nla-(+), R$_3$= mba, R$_4$= H

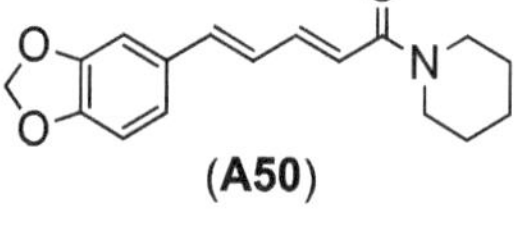

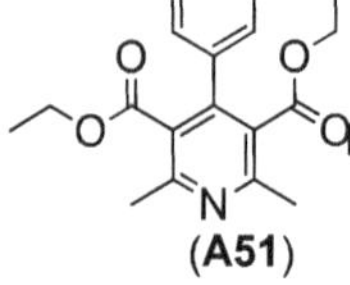

(A51) 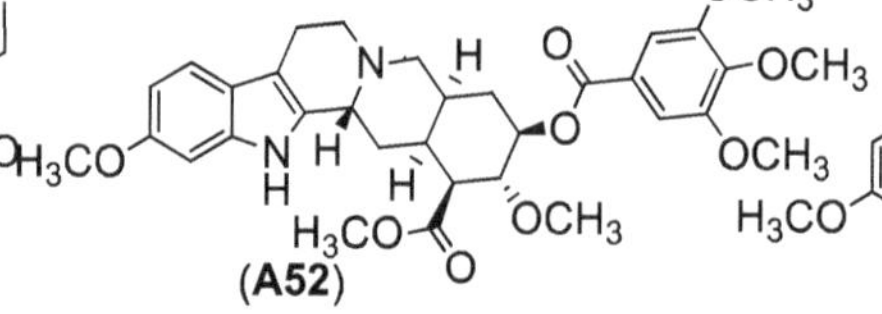 (A52) 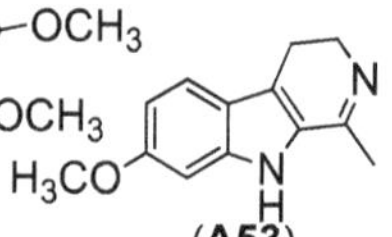 (A53)

Fig. 19.2 (continued)

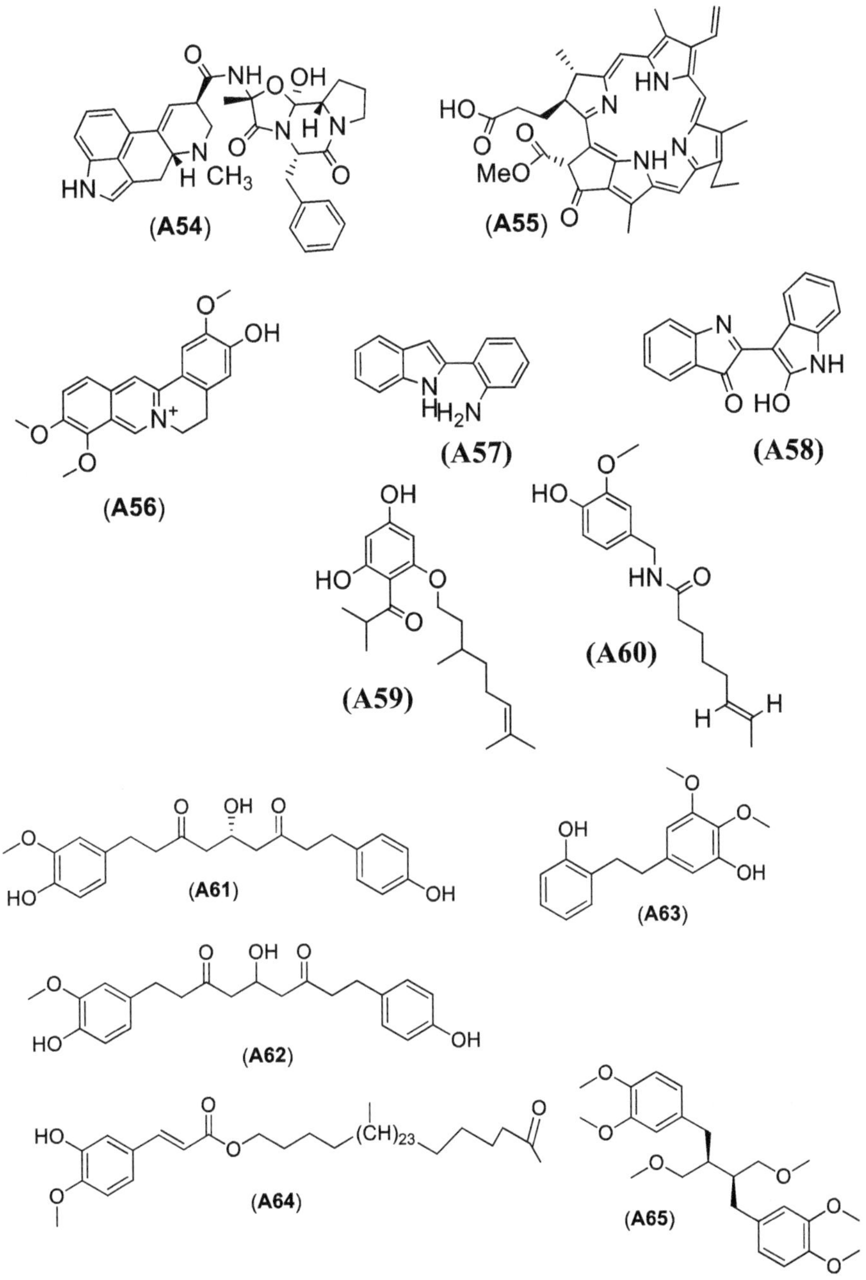

Fig. 19.2 (continued)

19.2.7 Polyphenols

19.2.7.1 2-Arylbenzofuran
SpinosanA (A1) from Dalea spinosa, at 48 µM, Reduced Berberine MIC by Eight-fold against Wild-Type S. aureus, whereas, (+)-Medicarpin (A2) at 56 µM Decreased Berberine MIC by Fourfold (Belofsky et al. 2006)

19.2.7.2 N-Caffeoylphenylkylamides
N-Trans-Feruloyl-4´-O-Methyldopamine (A3) Isolated from Mirabilis jalapa Exhibited Moderate EPI Activity against NorA-Overexpressed S. aureus 1199B Strain. It Showed an Eightfold Reduction in the Minimum Inhibitory Concentration of Norfloxacin at 100 µg/mL (Michalet et al. 2007)

19.2.7.3 Caffeoylquinic Acids
4´,5'-O-Caffeoylquinic Acid (A4) Isolated from Chloroform Extract of Artemisia absinthium Potentiated Activity of Berberine by Eight-Fold, EtBr by 4–Eight-Fold, and Fluoroquinolones, Ciprofloxacin, and Norfloxacin by 4–Eight-Fold in NorA-Overexpressed S. aureus Strain (Fiamegos et al. 2011)

Curcumin (A5) Was Reported to Exhibit Significant Inhibition of NorA EP in *S. aureus. There is an* Eightfold Reduction of Ciprofloxacin MIC at 25 µM by Curcumin. The Molecular Modeling Study of Curcumin with the Human Pgp and NorA Efflux Protein Revealed Favorable Binding Interactions (Joshi et al. 2014)

- *Chalcones.*

  A chalcone compound (**A6**) characterized from *Dalea versicolor* showed a fourfold increase in berberine activity against MDR S. aureus at 10 µg/mL (Holler et al. 2012b).

  3´,4'-Dihydroxy,3,4,5'-trimethoxy-chalcone (**A7**) isolated from the flowers of Arrabidaea brachypoda showed a significant decrease in MIC of norfloxacin by fourfold. A7 decreased MIC from 64 µg/mL to 16 µg/mL and increased EtBr accumulation in SA-1199B. A significant modulation in MIC was also observed for MepA gene by twofold in S. aureus K2068 (Rezende-Júnior et al. 2020).

- *Coumarins*:

  A 20-fold reduction of norfloxacin MIC in S. aureus-resistant strains (MRSA 16565, 9543, 5, and 7) was observed due to coumarins, viz., 4-{[(E)-5-(3,3-dimethyl-2-oxiranyl)-3-methyl-2-pentenyl]oxy}-7H-furo(3,2-g)chromen-7-one (**A8**) and 7-{[(E)-5-(3,3-dimethyl-2-oxiranyl)-3-methyl-2-pentenyl]oxy}-2H-2-chromenone (**A9**) isolated from grapefruit oil at 35.7 µg/mL and 30 µg/mL concentrations, respectively (Abulrob et al. 2004). The ciprofloxacin MIC reduced from 10–80 µg/mL to ≤2.5–5 µg/mL and of EtBr from 4–16 µg/mL to 0.5–2 µg/mL by galbanic acid (**A10**), a sesquiterpene coumarin isolated from the roots of Ferula szowitsiana, against several S. aureus-resistant clinical isolate strains, at 300 µg/mL (Bazzaz et al. 2010). A fourfold reduction of ciprofloxacin MIC by osthol (7-methoxy-8-prenylcoumarin) (**A11**) at 25 µM exhibited signifi-cant inhibition of the *S. aureus* NorA efflux pump (Joshi et al. 2014). Seven coumarins, viz., 5,7-dihydroxy-8-(2-methylbutanoyl)-6-[3,7-dimethylocta-2,6-dienyl]-4-phenyl-2H-chromen-2-one (**A12**), 5,7-dihydroxy-4-(1-hydroxypropyl)-8-(2-methylbutanoyl)-6-[3,7-dimeth ylocta-2,6-dienyl]-2H-chromen-2-one (**A13**), 5-hydroxy-8,8-dimethyl-6-(2-methylbutanoyl)-4-phenyl-2H-pyrano[2,3-h]chromen-2-one (**A14**), 5,7-dihydroxy-6-(2-methylbutanoyl)-8-(3-methyl but-2-enyl)-4-phenyl-2H-chromene-2-one (**A15**), 5,7-dihydroxy-8-(2-methylbutanoyl)-6-(3-methyl but-2-enyl)-4-phenyl-2H-chromene-2-one (**A16**), 5,7-dihydroxy-6-(2-methylbutanoyl)-4-phenyl-2H-chromene −2-one (**A17**), and 8,9-dihydro-5-hydroxy-8-(2-hydroxypropan-2-yl)-6-(2-methylbutanoyl)-4-phenyl furo[2,3-h]chromen-2-one (**A18**), isolated from Mesua ferrea flowering buds were studied for NorA EPI activity. Compounds **A12** and **A15–A18** exhibited modulation activity by displaying a ≥ two-fold reduction in MIC of EtBr against wild-type clinical strains of S. aureus 1199 and S. aureus 1199B, whereas compounds **A15–A18** showed a modulation effect by showing a ≥ 16-fold reduction in MIC of EtBr against MRSA 831. Compounds **A12** and **A15–A18** also reduced the MIC of norfloxacin by ≥eight-fold against S. aureus 1199B, and compounds **A16–A18** showed an ≥eight-fold reduction in MIC of norfloxacin against MRSA 831 at half of their MICs. Inhibition of EtBr efflux by NorA-overexpressed S. aureus 1199B and MRSA 831 confirmed the compounds **A16–A18** as potential NorA efflux pump inhibitors (EPI) (Roy et al. 2013).

- *Flavones and Flavonols.*
Chrysosplenol-D (**A19**) and chrysoplenetin (**A20**), methoxyflavonols reported from the herbaceous plant *Artemisia annua*, showed inhibition of S. aureus growth at a subinhibitory concentration (30 µg/mL) with MIC of 25 µg/mL and 6.25 µg/mL, respectively (Stermitz et al. 2002). Tiliroside (**A21**), purified and characterized from Herissantiatiubae aerial parts, did not exhibit antibacterial activity against S. aureus at 128 µg/mL (MIC, 256 µg/mL). The MIC of ciprofloxacin was decreased by 16- and eight-fold when tiliroside was introduced at 64 µg/mL (1/4 MIC) or 32 µg/mL (1/8 MIC) concentration, respectively (Falcão-Silva et al. 2009). Ethanol extract prepared from Persia lingue leaves was purified using a bioassay-guided process that resulted in the isolation and characterization of kaempferol-3-O-α-L-(2,4-bis-E-p-coumaroyl)-rhamnoside (**A22**) as NorA efflux protein inhibitor. This kaempferol glycoside showed the inhibition of EtBr efflux with an IC_{50} value of 2 µM. When studied for potentiation activity of ciprofloxacin, the kaempferol glycoside synergistically enhanced the antimicrobial effect of ciprofloxacin by eightfold at 1.56 µg/mL concentration, against a NorA-overexpressed S. aureus 1199B strain (Holler et al. 2012a).

Baicalein (**A23**) isolated from Scutellaria baicalensis revived the antibacterial effect of ciprofloxacin against the NorA efflux pump-overexpressed SA-1199B at 16 µg/ml in the time-kill and checkerboard dilution test. Baicalein showed a synergistic effect when it was combined with ciprofloxacin against ciprofloxacin-resistant clinical strains. The pharmacokinetic study revealed that baicalein exhibited a dose-dependent inhibition of MRSA pyruvate (Chan et al. 2011). Brachydin (BR-B) (**A24**) a dimeric flavonoid isolated from dichloromethane fraction of ethanolic extract of flowers of Arrabidaea brachypoda was evaluated for its EPI activity. BR-B decreased the MIC of norfloxacin by fourfold, i.e., 64 µg/mL to 16 µg/mL against SA1199-B strain. The addition of BR-B caused intracellular accumulation of EtBr by increasing fluorescence signal in EtBr accumulation assay in SA1199-B strain (de Sousa Andrade et al. 2020). Similarly, fisetinidol (A25) isolated from stems of Bauhinia pentandra showed a twofold reduction in MIC of norfloxacin by its synergistic effect (MIC/4) in S. aureus 1199-B strain. A twofold decrease in MIC of EtBr + fisetinidol was observed when compared with EtBr control (da Silva et al. 2020).

A dihydroflavonoid sophoraflavanone G (A26) isolated from traditional Chinese herb caused a 16-fold reduction in MIC of norfloxacin in NorA-overexpressed SA-1199B strain. An in vivo synergistic effect was also seen between oral SG (100 mg/kg) and norfloxacin against SA-1199B infection in female ICR mice (Sun et al. 2020).

- *Flavonolignans.*

5′-Methoxyhydnocarpin-D (5′-MHC-D), a flavonolignan (A27) reported from Berberis aetnensis leaves, exhibited EPI activity by reducing norfloxacin MIC to 0.25 µg/mL at 10 µg/mL concentration for wild-type S. aureus (Stermitz et al. 2000).

The EPI activity of silybin stereoisomeric mixture of flavonolignans, silybin a and b (28a and 28b), present in the extract of *Silybum marianum* was studied in MRSA. An RT-PCR study indicated that silybin exhibited reduced expression of quinolone-resistant efflux protein NorA in MRSA 41577 strain. MRSA 41577 strain when treated with silybin for 16 hours showed a 36% reduction in expression of NorA, thereby restoring the sensitivity of this strain towards quinolone antibiotics such as ciprofloxacin and norfloxacin. Thus, it revealed the EPI activity of silybin (Wang et al. 2018).

- *Isoflavones.*

Isoflavones, viz., genistein (A29), orobol (A30), and biochanin A (A31) reported from *Lupinus argenteus*, decreased norfloxacin MIC by 2–4-folds against S. aureus mutant strain, at 10 µg/mL [84]. Spinosan A (A32), an isoflavone compound characterized from *Dalea spinosa*, at 48 µM concentration showed an eightfold reduction of berberine MIC (89 µM) in wild-type S. aureus (Morel et al. 2003).

Boeravinone B (A33) isolated from Boerhaavia diffusa roots was evaluated in combination with ciprofloxacin in three strains of S. aureus, viz., NorA-overexpressed, wild-type, and knocked-out strains. Its mechanism as an EPI action was determined using EtBr accumulation and EtBr efflux assays. Boeravinone B decreased the ciprofloxacin MIC in S. aureus and MRSA strains, and in SA-1199B strain, the effect was more pronounced. The time-kill kinetics study showed that the combination of boeravinone B with ciprofloxacin at its subinhibitory concentration (0.25 × MIC) exhibited the same effect as that at its MIC level. Boeravinone B decreased the efflux of EtBr and hence can be a NorA inhibitor (Gibbons et al. 2004).

- *Tannins.*
  Catechin compounds, viz., epicatechin gallate (**A34**) and epigallocatechin gallate (**A35**), increased the norfloxacin MIC by fourfold in wild-type (SA 1199) and NorA-overexpressed S. aureus (SA1199B) strain, at 20 μg/mL concentration (Gibbons et al. 2004).

19.2.7.4 Terpenoids

Ferruginol (**A36**) characterized from *Chamaecyparis lawsoniana* exhibited NorA pump inhibitory activity in *S. aureus*-resistant strain. Ferruginol at a subinhibitory concentration (2 μg/mL) showed a twofold potentiation of norfloxacin against SA1199B strain (Smith et al. 2007b). A clerodane diterpene, 16α-hydroxycleroda-3,13(14)-Z-dien-15,16-olide (CD) (**A37**), isolated from the leaves of *Polyalthialongifolia* (Sonn.) in combination with fluoroquinolones decreased the MIC of fluoroquinolones by 16-fold (FICI 0.315–0.500). CD showed significant inhibition of EtBr efflux and post-antibiotic effect when studied using flow cytometry analysis. In clinical isolates of MRSA-ST2071, CD alone or in combination with antibiotics modulated the expression of several efflux pumps including NorA by twofold (Gupta et al. 2016). Oleanolic acid, (**A38**) and uvaol (**A39**) isolated from methanol extract of *Carpobrotus edulis* leaves, decreased the resistance of the *E. coli* AG100TET8 strain to tetracycline. There is a decrease in tetracycline MIC from 25 μg/mL to 6.25 μg/mL by these compounds. Significant reduction in the MIC of ciprofloxacin against *Salmonella* strains resistant to ciprofloxacin, *S. Enteritidis* 5408CIP, was achieved by uvaol. Uvaol also decreased the MIC of oxacillin in β-lactam-resistant strains of MRSA, MRSA COL$_{OXA}$ (Martins et al. 2011).

Among several cucurbitane skeleton-containing triterpenoids isolated from the aerial parts of *Momordica balsamina* L., balsaminagenin B (**A40**) showed a good EPI activity by inhibition of EtBr efflux in methicillin-resistant *S. aureus* pumps, viz., MRSA COL$_{OXA}$ at MIC of 50 μM and *E. faecalis* ATCC29212, respectively. A significant activity was also reported for Gram (+)ve *S. typhimurium* strains (Ramalhete et al. 2011).

19.2.7.5 Oligosaccharides

Five murucoidins (XII–XVI) (**A41–A45**) were isolated and characterized from *Ipomoea murucoides*. Murucoidin XIV at 5 μg/mL exhibited a fourfold increase in norfloxacin activity against *S. aureus* strains (Chérigo et al. 2009). A penta ester of neohesperidoside (**A46**) characterized from *Geranium caespitosum* decreased berberine MIC by 160 times at 10 μg/mL against *S. aureus* (NorA MDR strain) (Stermitz et al. 2003). Orizabin XIX (**A47**), orizabin IX (**A48**), and orizabin XV (**A49**) were purified and characterized from Mexican morning glory species. These oligosaccharides enhanced the effect of norfloxacin in SA-1199B strain. Orizabin XIX at 25 μg/mL potentiated the norfloxacin activity by fourfold (8 μg/mL from 32 μg/mL), whereas orizabin IX elevated norfloxacin activity by 16-fold at 1 μg/mL, and orizabin XV showed similar activity as that of orizabin IX in EtBr EPI assay (Pereda-Miranda et al. 2006).

19.2.7.6 Alkaloids

Piperine (**A50**), a piperidine alkaloid characterized from *Piper nigrum* fruits, displayed no growth of *S. aureus* mutant at 1 µg /mL concentration of ciprofloxacin when it was coadministered at the concentration of 50 µg /mL (Khan et al. 2006). A pyridine alkaloid, 2,6-dimethyl-4-phenylpyridine-3,5-dicarboxylic acid diethyl ester (**A51**), purified from *Jatropha elliptica*, decreased the ciprofloxacin MIC at 2 µg/mL against NorA overexpressed-*S. aureus* SA-1199B (Marquez et al. 2005). Reserpine (**A52**) enhanced the activity of tetracycline, by decreasing its MIC fourfold in two MRSA clinical isolates, viz., IS-58 and XU212. Reserpine elevated norfloxacin activity against *S. aureus* by fourfold (Gibbons and Udo 2000). Harmaline (**A53**) isolated from *Peganum harmala* reduced EtBr MIC by fourfold in *S. aureus* U949 (Mohtar et al. 2009). Ergotamine (**A54**), isolated from *Claviceps purpurea*, did not show antibacterial activity but exhibited a fourfold reduction in norfloxacin MIC at 20 µg/mL when co-administered with norfloxacin in *S. aureus*-resistant strain (Jachak et al. 2012). Pheophorbide A (**A55**) from *Berberis aetnensis*, when co-administrated at 0.5 µg/mL, enhanced the activity of ciprofloxacin by 16-fold (Musumeci et al. 2003). Jatrorrhizine (**A56**) a protoberberine alkaloid isolated from *Mahonia* species showed synergistic activity with norfloxacin against MRSA-NorA-overexpressed strain SA-1199B by an eightfold decrease in bacterial count. At mRNA level jatrorrhizine inhibited the expression of NorA when compared with vehicle ($p < 0.05$) (Yu et al. 2019). 2-(2-aminophenyl)indole (**A57**), a natural compound isolated from the microbial extract of soil isolate IMTB 2501, was evaluated for efflux pump inhibition potential on *S. aureus* SA-1199B, XU212, and RN4220-Msra. 2-(2-aminophenyl)indole reduced the MIC of ciprofloxacin, moxifloxacin, norfloxacin, and chloramphenicol in SA-1199B by 64-, 16-, 4-, and 4-folds. It also reduced the MIC of tetracycline in *S. aureus* XU212 and erythromycin in RN4220-Msra by 64- and 32-folds by inhibiting Tetk and Msra pumps. Postantibiotic effect of 3–3.2 h was also reported for ciprofloxacin, erythromycin, and tetracycline (Tambat et al. 2019). Indirubin (**A58**), a commonly used industrial dye isolated from the leaves of *Wrightia tinctoria* chloroform extract, decreased the MIC of ciprofloxacin by fourfold in checkerboard synergy assay on SA1199B. Indirubin showed a MIC value of 12.5 mg/L for *S. aureus* and a FICI value of 0.45 depicting synergism between indirubin and ciprofloxacin (Ponnusamy et al. 2010).

19.2.7.7 Miscellaneous NorA EPIs

Recently, a new phytoconstituent (**A59**) characterized from *Hypericum olympicum* L. cf. uniflorum caused a concentration-dependent increase in the accumulation of ^{14}C-labelled enoxacin in a radiometric accumulation assay in NorA-overexpressed *S. aureus* strain (Shiu et al. 2013). Capsaicin (8-methyl-N-vanillyl-6-nonenamide) (**A60**) significantly reduced the ciprofloxacin MIC in *S. aureus* SA-1199 and SA-1199B. It also showed PAE of ciprofloxacin by 1.1 h at its MIC concentration. When co-administered with capsaicin, the mutation prevention concentration (MPC) of ciprofloxacin was also decreased. Capsaicin as a NorA EPI was also confirmed by inhibition of EtBr efflux by NorA-overexpressed SA-1199B (Kalia et al. 2012). Two new diarylnonanoids (**A61**) and (**A62**), (*E*)-32-((3-(3-hydroxy-4- methoxyphenyl) isoferuloyl)oxy)dotriacontanoic acid (**A63**), and 5-(2-hydroxyphenethyl)-2,3-dimethoxyphenol) (**A64**) isolated from the rhizomes of *Dioscorea cotinifolia* were evaluated for their potential for antibiotic potentiation activity. Compounds (**A61**–**A64**) decreased the MIC of norfloxacin by 2–16-fold in *S. aureus* 1199B strain. Compound **A61** was found the most active showing a 512-fold reduction in MIC of tetracycline in TetK-XU212 strain with an MIC value <0.25 mg/L (Sibandze et al. 2020). The ethanolic extract of *Phyllanthus amarus* leaves (PLEE) and phyllanthin (PHY) were studied for antimicrobial activity against Gram-positive and Gram-negative bacteria and *Candida albicans* ATCC 10231 (yeast strain). PLEE showed activity against all tested microorganisms. PLEE and PHY (**A65**) decreased the MIC of norfloxacin in SA-1199B by fourfold (64 μg/mL to 16 μg/mL) and fivefold (64 μg/mL to 12.7 μg/mL), respectively (Ribeiro et al. 2019).

19.2.7.8 MsrA Efflux Pump Inhibitors of Natural Product Origin

To date, 15 natural products of plant origin have been reported to show MsrA efflux pump inhibition activity in *S. aureus* RN4220 strain in Fig. 19.3. Totarol (**B01**), a phenolic diterpene, showed an eightfold reduction in the MIC of erythromycin against *S. aureus* RN4220 strain that overexpresses macrolide-specific MsrA pump and was the first EPI reported against this pump (Smith et al. 2007a). Oleic acid (**B02**) and linoleic acid (**B03**) isolated from methanol extract of *Portulaca oleracea* leaves have been recently reported to exhibit EPI activity when combined with erythromycin against MsrA pump in *S. aureus* RN4220 strain. Linoleic acid at 16 μg/mL and oleic acid at 32 μg/mL reduced MIC of erythromycin by eightfold and fourfold, respectively (Fung et al. 2017). It was reported that caffeic acid (**B04**) caused a reversal of the resistance phenotype and it inhibited the MsrA pump belonging to the RN-4220 strain. Caffeic acid showed greater efficacy in the docking model, in agreement with the demonstrated experimental efficacy (Dos Santos et al. 2018). A diterpene, (4S, 9R, 14S)-4α-acetoxy-9β,14α-dihydroxydolasta-1 (15),7diene (**B05**) isolated from *C. cervicornis*, decreased the MIC of erythromycin by 16-fold in *S. aureus* 1199B strain overexpressing norA gene RN-4220 (de Figueiredo et al. 2019). Tannic acid (**B06**) at 1/8 of its MIC exhibited a significant reduction in MIC of erythromycin in *S. aureus* RN 4220 strain indicating EPI activity. It is proposed that this EPI activity of tannic acid may be due to its interaction with structures of membrane proteins involved in the efflux system of the RN4220 strain (Tintino et al. 2017). Furanocoumarins, viz., imperatorin (**B07**) and isopimpinellin (**B08**) isolated from Rutaceae family plants, showed a fourfold reduction in MIC of erythromycin in *S. aureus* RN4220 strain. It is reported that lipophilicity of putative EPIs of medicinal plant origin plays an important role in their EPI activity (Gibbons 2004). In the case of imperatorin and isopimpinellin, it is also reported that these furanocoumarins showed better EPI activity due to their lipophilicity (Madeiro et al. 2017). Menadione (vitamin K3) (**B09**), when assayed with erythromycin at 1/8 of its MIC, reduced the MIC of erythromycin significantly in *S. aureus* RN4220 strain. The combination of lipid-soluble vitamins like vitamin K3 with antibiotics is an attractive alternative to enhance the antibiotic activity of the

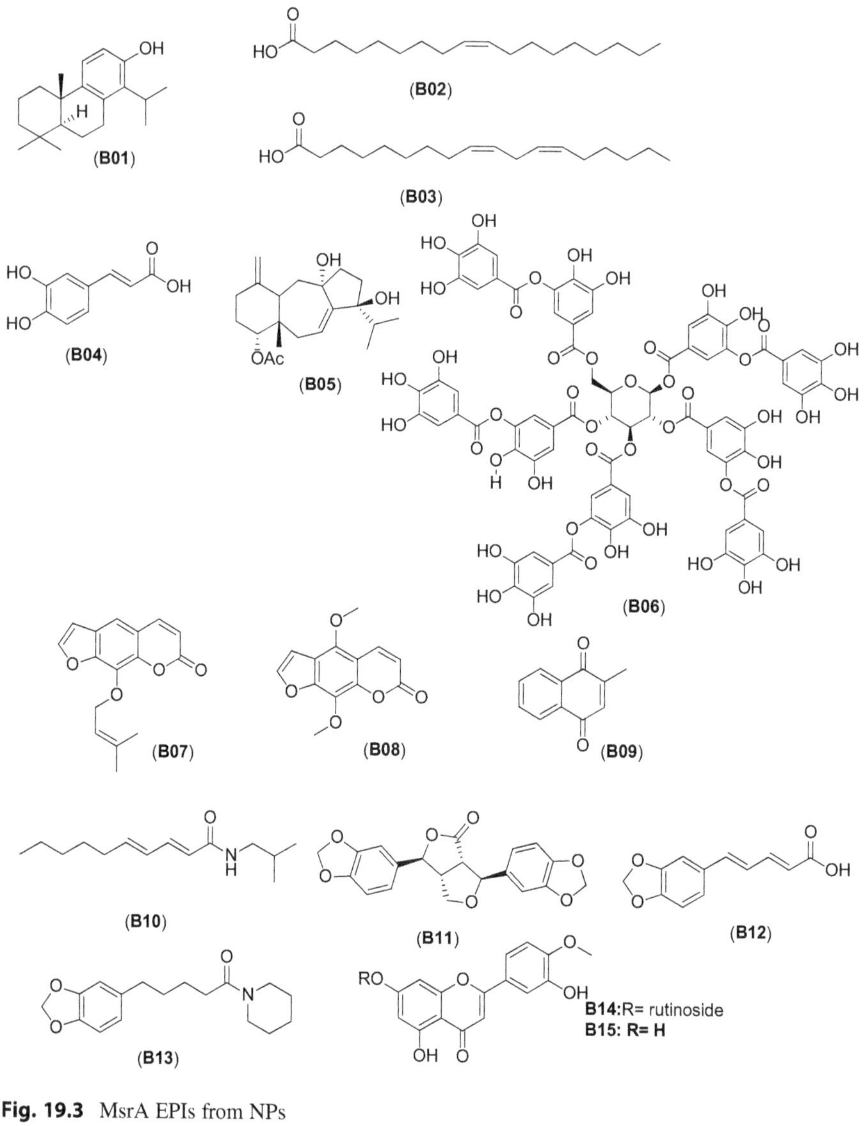

Fig. 19.3 MsrA EPIs from NPs

drug (erythromycin) in humans. Being lipophilic in nature, menadione can alter the
fluidity of bacterial membrane making it more susceptible to penetration by
antibiotics such as erythromycin (Tintino et al. 2018). A threefold reduction in
EtBr MIC was also reported when studied in SA-1199B NorA-overexpressed strain
(Tintino et al. 2020). Recently, we reported that the phytoconstituents isolated from
fruits of *Piper cubeba* exhibited MsrA efflux pump inhibitory activity. It was
observed that the MIC of erythromycin was decreased by 2–eight-fold when given
in combination with pellitorine (**B10**), sesamin (**B11**), piperic acid (**B12**), and

tetrahydropiperidine (**B13**). The MIC of erythromycin was decreased in MsrA pump-overexpressed *S. aureus* RN4220 strain. EtBr accumulation assay and real-time fluorometry-based efflux studies against *S. aureus* RN4220 confirmed the efflux pump inhibitory potential of these natural products (Ahirrao et al. 2020). Diosmetin (**B14**) and diosmin (**B15**) isolated from citrus fruits were evaluated against MRSA RN4220 and Pul5054 strains. Diosmetin showed synergistic activity against both the abovementioned strains, when given in combination with erythromycin. The FIC value, 0.28 (FIC < 0.5), indicated a synergistic effect (Chan et al. 2013).

19.2.7.9 Miscellaneous *S. aureus* and MRSA Efflux Pump Inhibitors of Natural Product Origin

The essential oil extracted from *Origanum vulgare* L. as well as its constituents, viz., carvacrol (**C01**) and thymol (**C02**), showed EPI activity. The essential oil showed a fourfold reduction of tetracycline MIC (64 µg/mL to 16 µg/mL) whereas carvacrol and thymol exhibited a twofold reduction of tetracycline MIC (64 µg/mL to 32 µg/mL) against *S. aureus* IS-58 strain overexpressing TetK efflux pump using agar dilution assay (Cirino et al. 2014). Essential oil and its major constituent, α-pinene (**C03**), extracted from *Croton grewioides* leaves were evaluated for their efflux pump inhibitory activity in SA-1199B (NorA-overexpressed strain) and IS-58 (TetK-overexpressed strain). A 64-fold (32 µg/mL to 0.5 µg/mL) and four-fold (64 µg/mL to 16 µg/mL) modulation in MIC of tetracycline and norfloxacin was seen for *C. grewioides* essential oil. α-Pinene lowered the MIC of tetracycline by 32-fold (32 µg/mL to 1.0 µg/mL) (de Medeiros et al. 2017). *Nigella sativa* essential oil and its constituents carvacrol (**C01**), thymoquinone (**C04**), and *p*-cymene (**C05**) were studied for their antibacterial effect and modulation of antibiotic resistance in methicillin-sensitive ATCC25923 and methicillin-resistant MRSA 272123 clinical isolate of *S. aureus*. All these constituents and essential oil displayed MIC values in mM range, indicating a weak antibacterial effect. Thymoquinone lowered the MIC of tetracycline and ciprofloxacin by 16-fold and eight-fold, respectively, in MRSA 272123 strain at half of their MIC value (62.5 µM). *N. sativa* essential oil exhibited a twofold reduction in MIC of ciprofloxacin at 3.25 µM (Mouwakeh et al. 2019).

Gambogic acid (**C06**) and neogambogic acid (**C07**) belonging to the xanthonoid class were evaluated against MRSA and MSSA. Both compounds showed potent antibacterial activity against 20 ATCC33591 strains. C06 and C07 exhibited a MIC of 0.5 µg/mL and 4 µg/mL in MSSA and MRSA strains, respectively, as compared to oxacillin that showed a MIC of 64 µg/mL in MRSA strains. Both the compounds inhibited biofilm formation by 87% at 8 µg/mL concentration (Hua et al. 2019).

Ailanthoidiol diacetate (**C08**) and ailanthoidiol di-2-ethyl butanoate (**C09**) isolated from the methanolic extract of roots of *Zanthoxylum capense* Thunb. showed EPI activity in EtBr accumulation assay in ATCC2593$_{EtBr}$ strain overexpressing NorA gene and MRSA strains, namely, SM39 and SM1. Ailanthoidiol diacetate showed a modulation factor of 8 (FIC = 0.25) for ATCC2593$_{EtBr}$ and *S. aureus* SM39 strains. Similarly, it showed a MF of 4 for *S. aureus* SM1 strain (FIC = 0.125). These natural products may be considered as

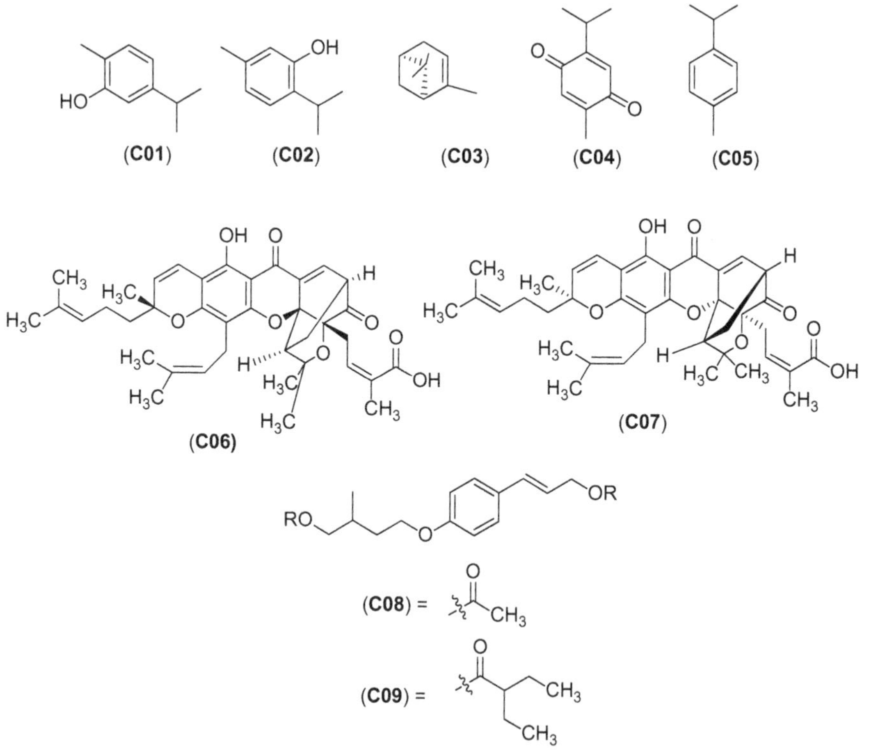

Fig. 19.4 Miscellaneous MRSA efflux pump inhibitors of natural product origin

lead molecules to tackle antibiotic resistance in MRSA and other *S. aureus* strains (Cabral et al. 2015) (Please, refer Fig. 19.4 for the chemical structures of miscellaneous MRSA EPIs).

19.3 Concluding Remarks

Antibiotic resistance is emerging at an alarming rate. Thus, there is an urgent unmet need to develop alternative therapies that either reduce the bacterial resistance to antibiotics or potentiate the activity of existing antibiotics. Efflux pumps are one of the major targets that confer resistance to clinically used antibiotics. Presently no efflux pump inhibitor combination with existing antibiotics is clinically approved to tackle antimicrobial resistance. Over the decades natural products have demonstrated multiple biological activities in biomedical research and served as an important source of lead molecules in drug discovery and development.

In this book chapter, we have described natural compounds with potential efflux pump inhibition activity. Among them, 2-(2-aminophenyl)indole (**A57**) isolated from *Streptomyces* sp. IMTB 2501 was found as the most potent NorA inhibitor, decreasing the MIC of ciprofloxacin, moxifloxacin, norfloxacin, and chloramphenicol in SA-1199B (NorA-overexpressed strain) by 64-, 16-, 4-, and 4-folds (FICI≤0.5), respectively. 2-(2-Aminophenyl)indole reduced the MIC of tetracycline in *S. aureus* XU212 and erythromycin in RN4220-Msra by 64- and 32-folds. 2-(2-Aminophenyl)indole can act as a lead molecule for the development of promising candidates that potentiate the activity of currently available antibiotics. (5R)-Hydroxy-1-(4-hydroxy-3-methoxyphenyl)-9-(4-hydroxyphenyl)nonane-3,7-dione (**A61**) isolated from dried rhizomes of *Dioscorea cotinifolia* was also found as a potent efflux pump inhibitor. It showed a 512-fold modulation in MIC of tetracycline against *S. aureus* in tetracycline-resistant XU 212 strain. Only a two-fold modulation was observed for norfloxacin in SA-1199B strain. Modification in the structure of the diarylnonanoids through synthesis could lead to enhancement in activity against other multidrug-resistant *S. aureus* strains. Essential oils and their active metabolites have been reported as good antibacterial agents. The essential oil of *Croton grewioides* (EOCg) and its major monoterpene, α-pinene (**C03**), were reported as a strong modulator of antibiotic resistance in SA-1199B and IS-58 strain of *S. aureus* overexpressing efflux proteins. EOCg reduced the MIC of norfloxacin by 64- and four-fold for tetracycline. α-Pinene significantly decreased the MIC of tetracycline by 32-fold when given in combination with tetracycline. Therefore, plant essential oils and their components may be considered as potential adjuvants of antibiotics in order to reverse or modulate bacterial resistance to antibiotics. This book chapter describes the natural products that showed activity against NorA, MsrA, and other efflux pumps responsible for extrusion of drugs in *S. aureus* and MRSA. However, no plant−/fungus−/marine-derived antibiotic is used clinically yet. Since isolation and identification of plant-derived drugs are tedious and time-consuming, in silico methodology and high-throughput screening can be utilized. A majority of bioactives discussed above have shown promising results as potent EPIs, determined using in vitro studies. The in vivo animal model studies and human clinical trials would be required to determine antibacterial action, efficacy, and toxicity studies to optimize a high therapeutic efficacy dosage of EPIs at an acceptable toxicity level.

References

Abreu AC, Coqueiro A, Sultan AR, Lemmens N, Kim HK, Verpoorte R, van Wamel WJ, Simões M, Choi YH (2017) Looking to nature for a new concept in antimicrobial treatments: isoflavonoids from Cytisus striatus as antibiotic adjuvants against MRSA. Sci Rep 7:1–16

Abulrob A-N, Suller MT, Gumbleton M, Simons C, Russell AD (2004) Identification and biological evaluation of grapefruit oil components as potential novel efflux pump modulators in methicillin-resistant Staphylococcus aureus bacterial strains. Phytochemistry 65:3021–3027

Ahirrao P, Tambat R, Chandal N, Mahey N, Kamboj A, Jain UK, Singh IP, Jachak SM, Nandanwar HS (2020) MsrA efflux pump inhibitory activity of Piper cubeba lf and its phytoconstituents against Staphylococcus aureus RN4220. Chem Biodivers 17:e2000144

Ahmed M, Borsch CM, Neyfakh A, Schuldiner S (1993) Mutants of the Bacillus subtilis multidrug transporter Bmr with altered sensitivity to the antihypertensive alkaloid reserpine. J Biol Chem 268:11086–11089

Akiba M, Lin J, Barton Y-W, Zhang Q (2006) Interaction of CmeABC and CmeDEF in conferring antimicrobial resistance and maintaining cell viability in campylobacter jejuni. J Antimicrob Chemother 57:52–60

Alekshun MN, Levy SB (2007) Molecular mechanisms of antibacterial multidrug resistance. Cell 128:1037–1050

Alibayov B, Baba-Moussa L, Sina H, Zdeňková K, Demnerová K (2014) Staphylococcus aureus mobile genetic elements. Mol Biol Rep 41:5005–5018

Bayot ML, Bragg BN (2020) Antimicrobial susceptibility testing. StatPearls, Treasure Island, FL

Bazzaz BSF, Memariani Z, Khashiarmanesh Z, Iranshahi M, Naderinasab M (2010) Effect of galbanic acid, a sesquiterpene coumarin from Ferula szowitsiana, as an inhibitor of efflux mechanism in resistant clinical isolates of Staphylococcus aureus. Braz J Microbiol 41:574–580

Belofsky G, Carreno R, Lewis K, Ball A, Casadei G, Tegos GP (2006) Metabolites of the "smoke tree", Dalea spinosa, potentiate antibiotic activity against multidrug-resistant Staphylococcus aureus. J Nat Prod 69:261–264

Belofsky G, Percivill D, Lewis K, Tegos GP, Ekart J (2004) Phenolic metabolites of dalea v ersicolor that enhance antibiotic activity against model pathogenic bacteria. J Nat Prod 67:481–484

Bohn C, Bouloc P (1998) The Escherichia coli cmlA gene encodes the multidrug efflux pump Cmr/MdfA and is responsible for isopropyl-β-d-thiogalactopyranoside exclusion and spectinomycin sensitivity. J Bacteriol 180:6072–6075

Braibant M, Guilloteau L, Zygmunt MS (2002) Functional characterization of Brucella melitensis NorMI, an efflux pump belonging to the multidrug and toxic compound extrusion family. Antimicrob Agents Chemother 46:3050–3053

Brown DG, Swanson JK, Allen C (2007) Two host-induced Ralstonia solanacearum genes, acrA and dinF, encode multidrug efflux pumps and contribute to bacterial wilt virulence. Appl Environ Microbiol 73:2777–2786

Cabral V, Luo X, Junqueira E, Costa SS, Mulhovo S, Duarte A, Couto I, Viveiros M, Ferreira M-JU (2015) Enhancing activity of antibiotics against Staphylococcus aureus: Zanthoxylum capense constituents and derivatives. Phytomedicine 22:469–476

Cao L, Srikumar R, Poole K (2004) MexAB-OprM hyperexpression in NalC-type multidrug-resistant Pseudomonas aeruginosa: identification and characterization of the nalC gene encoding a repressor of PA3720-PA3719. Mol Microbiol 53:1423–1436

Chan BC, Ip M, Gong H, Lui S, See RH, Jolivalt C, Fung K, Leung P, Reiner NE, Lau CB (2013) Synergistic effects of diosmetin with erythromycin against ABC transporter over-expressed methicillin-resistant Staphylococcus aureus (MRSA) RN4220/pUL5054 and inhibition of MRSA pyruvate kinase. Phytomedicine 20:611–614

Chan BC, Ip M, Lau CB, Lui S, Jolivalt C, Ganem-Elbaz C, Litaudon M, Reiner NE, Gong H, See RH (2011) Synergistic effects of baicalein with ciprofloxacin against NorA over-expressed methicillin-resistant Staphylococcus aureus (MRSA) and inhibition of MRSA pyruvate kinase. J Ethnopharmacol 137:767–773

Chen J, Kuroda T, Huda MN, Mizushima T, Tsuchiya T (2003) An RND-type multidrug efflux pump SdeXY from Serratia marcescens. J Antimicrob Chemother 52:176–179

Chérigo L, Pereda-Miranda R, Gibbons S (2009) Bacterial resistance modifying tetrasaccharide agents from Ipomoea murucoides. Phytochemistry 70:222–227

Choudhuri BS, Bhakta S, Barik R, Basu J, Kundu M, Chakrabarti P (2002) Overexpression and functional characterization of an ABC (ATP-binding cassette) transporter encoded by the genes drrA and drrB of mycobacterium tuberculosis. Biochem J 367:279–285

Cirino ICS, Menezes-Silva SMP, Silva HTD, de Souza EL, Siqueira-Júnior JP (2014) The essential oil from Origanum vulgare L. and its individual constituents carvacrol and thymol enhance the effect of tetracycline against Staphylococcus aureus. Chemotherapy 60:290–293

Costa SS, Sobkowiak B, Parreira R, Edgeworth JD, Viveiros M, Clark TG, Couto I (2019) Genetic diversity of norA, coding for a main efflux pump of Staphylococcus aureus. Front Genet 9:710

Costa SS, Viveiros M, Amaral L, Couto I (2013) Multidrug efflux pumps in Staphylococcus aureus: an update. Open Microbiol J 7:59

Craft KM, Nguyen JM, Berg LJ, Townsend SD (2019) Methicillin-resistant Staphylococcus aureus (MRSA): antibiotic-resistance and the biofilm phenotype. Med Chem Comm 10:1231–1241

da Silva HC, Leal ALAB, de Oliveira MM, Barreto HM, Coutinho HDM, dos Santos HS, Santiago GMP, de Freitas TS, Lima IKC, Teixeira AMR (2020) Structural characterization, antibacterial activity and NorA efflux pump inhibition of flavonoid fisetinidol. S Afr J Bot 132:140–145

Daigle DM, Cao L, Fraud S, Wilke MS, Pacey A, Klinoski R, Strynadka NC, Dean CR, Poole K (2007) Protein modulator of multidrug efflux gene expression in Pseudomonas aeruginosa. J Bacteriol 189:5441–5451

Danilchanka O, Mailaender C, Niederweis M (2008) Identification of a novel multidrug efflux pump of mycobacterium tuberculosis. Antimicrob Agents Chemother 52:2503–2511

de Figueiredo CS, de Menezes Silva SMP, Abreu LS, da Silva EF, da Silva MS, de Cavalcanti Miranda GE et al (2019) Dolastane diterpenes from Canistrocarpus cervicornis and their effects in modulation of drug resistance in Staphylococcus aureus. Nat Prod Res 33:3231–3239

de Kraker ME, Stewardson AJ, Harbarth S (2016) Will 10 million people die a year due to antimicrobial resistance by 2050? PLoS Med 13:e1002184

de Medeiros VM, do Nascimento YM, Souto AL, Madeiro SAL, de Oliveira Costa VC, Silva SMP et al (2017) Chemical composition and modulation of bacterial drug resistance of the essential oil from leaves of Croton grewioides. Microb Pathog 111:468–471

de Sousa Andrade LM, de Oliveira ABM, Leal ALAB, de Alcântara Oliveira FA, Portela AL, Neto JSL, de Siqueira-Júnior JP, Kaatz GW, da Rocha CQ, Barreto HM (2020) Antimicrobial activity and inhibition of the NorA efflux pump of Staphylococcus aureus by extract and isolated compounds from Arrabidaea brachypoda. Microb Pathog 140:103935

Dos Santos JF, Tintino SR, de Freitas TS, Campina FF, Irwin RA, Siqueira-Júnior JP, Coutinho HD, Cunha FA (2018) In vitro e in silico evaluation of the inhibition of Staphylococcus aureus efflux pumps by caffeic and gallic acid. Comp Immunol Microbiol Infect Dis 57:22–28

Edelsberg J, Weycker D, Barron R, Li X, Wu H, Oster G, Badre S, Langeberg WJ, Weber DJ (2014) Prevalence of antibiotic resistance in US hospitals. Diagn Microbiol Infect Dis 78:255–262

Falcão-Silva VS, Silva DA, Souza MFV, Siqueira-Junior JP (2009) Modulation of drug resistance in Staphylococcus aureus by a kaempferol glycoside from Herissantia tiubae (Malvaceae). Phytother Res 23:1367–1370

Fiamegos YC, Kastritis PL, Exarchou V, Han H, Bonvin AM, Vervoort J, Lewis K, Hamblin MR, Tegos GP (2011) Antimicrobial and efflux pump inhibitory activity of caffeoylquinic acids from Artemisia absinthium against gram-positive pathogenic bacteria. PLoS One 6:e18127

Fung K, Han Q, Ip M, Yang X, Lau C, Chan B (2017) Synergists from Portulaca oleracea with macrolides against methicillin-resistant Staphylococcus aureus and related mechanism. Hong Kong Med J 5(4):38–42

Gibbons S (2004) Anti-staphylococcal plant natural products. Nat Prod Rep 21:263–277

Gibbons S, Moser E, Kaatz GW (2004) Catechin gallates inhibit multidrug resistance (MDR) in Staphylococcus aureus. Planta Med 70:1240–1242

Gibbons S, Udo E (2000) The effect of reserpine, a modulator of multidrug efflux pumps, on the in vitro activity of tetracycline against clinical isolates of methicillin resistant Staphylococcus aureus (MRSA) possessing the tet (K) determinant. Phytother Res 14:139–140

Gupta VK, Tiwari N, Gupta P, Verma S, Pal A, Srivastava SK, Darokar MP (2016) A clerodane diterpene from Polyalthia longifolia as a modifying agent of the resistance of methicillin resistant Staphylococcus aureus. Phytomedicine 23:654–661

He G-X, Kuroda T, Mima T, Morita Y, Mizushima T, Tsuchiya T (2004) An H+-coupled multidrug efflux pump, PmpM, a member of the MATE family of transporters, from Pseudomonas aeruginosa. J Bacteriol 186:262–265

Higashi K, Ishigure H, Demizu R, Uemura T, Nishino K, Yamaguchi A, Kashiwagi K, Igarashi K (2008) Identification of a spermidine excretion protein complex (MdtJI) in Escherichia coli. J Bacteriol 190:872–878

Holler JG, Christensen SB, Slotved H-C, Rasmussen HB, Gúzman A, Olsen C-E, Petersen B, Mølgaard P (2012a) Novel inhibitory activity of the Staphylococcus aureus NorA efflux pump by a kaempferol rhamnoside isolated from Persea lingue Nees. J Antimicrob Chemother 67:1138–1144

Holler JG, Slotved H-C, Mølgaard P, Olsen CE, Christensen SB (2012b) Chalcone inhibitors of the NorA efflux pump in Staphylococcus aureus whole cells and enriched everted membrane vesicles. Bioorg Med Chem 20:4514–4521

Hua X, Jia Y, Yang Q, Zhang W, Dong Z, Liu S (2019) Transcriptional analysis of the effects of gambogic acid and neogambogic acid on methicillin-resistant staphylococcus aureus. Front Pharmacol 10:986

Huang J, O'Toole PW, Shen W, Amrine-Madsen H, Jiang X, Lobo N, Palmer LM, Voelker L, Fan F, Gwynn MN (2004) Novel chromosomally encoded multidrug efflux transporter MdeA in Staphylococcus aureus. Antimicrob Agents Chemother 48:909–917

Huda N, Lee E-W, Chen J, Morita Y, Kuroda T, Mizushima T, Tsuchiya T (2003) Molecular cloning and characterization of an ABC multidrug efflux pump, VcaM, in non-O1 Vibrio cholerae. Antimicrob Agents Chemother 47:2413–2417

Jachak SM, Roy SK, Gupta S, Ahirrao P, Gibbons S (2012) Small-molecule efflux pump inhibitors from natural products as a potential source of antimicrobial agents. In: Antimicrobial drug discovery: emerging strategies. CAB International, Wallingford, UK, pp 62–76

Jonas BM, Murray BE, Weinstock GM (2001) Characterization of emeA, anorA homolog and multidrug resistance efflux pump, inEnterococcus faecalis. Antimicrob Agents Chemother 45:3574–3579

Joshi P, Singh S, Wani A, Sharma S, Jain SK, Singh B, Gupta BD, Satti NK, Koul S, Khan IA (2014) Osthol and curcumin as inhibitors of human Pgp and multidrug efflux pumps of Staphylococcus aureus: reversing the resistance against frontline antibacterial drugs. Med Chem Comm 5:1540–1547

Kaatz GW, DeMarco CE, Seo SM (2006) MepR, a repressor of the Staphylococcus aureus MATE family multidrug efflux pump MepA, is a substrate-responsive regulatory protein. Antimicrob Agents Chemother 50:1276–1281

Kaatz GW, McAleese F, Seo SM (2005) Multidrug resistance in Staphylococcus aureus due to overexpression of a novel multidrug and toxin extrusion (MATE) transport protein. Antimicrob Agents Chemother 49:1857–1864

Kalia NP, Mahajan P, Mehra R, Nargotra A, Sharma JP, Koul S, Khan IA (2012) Capsaicin, a novel inhibitor of the NorA efflux pump, reduces the intracellular invasion of Staphylococcus aureus. J Antimicrob Chemother 67:2401–2408

Khan IA, Mirza ZM, Kumar A, Verma V, Qazi GN (2006) Piperine, a phytochemical potentiator of ciprofloxacin against Staphylococcus aureus. Antimicrob Agents Chemother 50:810–812

Kumar S, Mukherjee MM, Varela MF (2013) Modulation of bacterial multidrug resistance efflux pumps of the major facilitator superfamily. Int J Bacteriol 2013:204141

Li J, Liu D, Tian X, Koseki S, Chen S, Ye X, Ding T (2019) Novel antibacterial modalities against methicillin resistant Staphylococcus aureus derived from plants. Crit Rev Food Sci Nutr 59: S153–S161

Li J, Xie S, Ahmed S, Wang F, Gu Y, Zhang C, Chai X, Wu Y, Cai J, Cheng G (2017) Antimicrobial activity and resistance: influencing factors. Front Pharmacol 8:364

Li X-Z, Nikaido H, Poole K (1995) Role of mexA-mexB-oprM in antibiotic efflux in Pseudomonas aeruginosa. Antimicrob Agents Chemother 39:1948–1953

Li X-Z, Poole K, Nikaido H (2003) Contributions of MexAB-OprM and an EmrE homolog to intrinsic resistance of Pseudomonas aeruginosa to aminoglycosides and dyes. Antimicrob Agents Chemother 47:27–33

Lindsay JA, Holden MT (2004) Staphylococcus aureus: superbug, super genome? Trends Microbiol 12:378–385

Littlejohn TG, Paulsen IT, Gillespie MT, Tennent JM, Midgley M, Jones IG, Purewal AS, Skurray RA (1992) Substrate specificity and energetics of antiseptic and disinfectant resistance in Staphylococcus aureus. FEMS Microbiol Lett 95:259–265

Madeiro SA, Borges NH, Souto AL, de Figueiredo PT, Siqueira-Junior JP, Tavares JF (2017) Modulation of the antibiotic activity against multidrug resistant strains of coumarins isolated from Rutaceae species. Microb Pathog 104:151–154

Mahmood YH, Jamshidi S, Mark Sutton J, Rahman MK (2016) Current advances in developing inhibitors of bacterial multidrug efflux pumps. Curr Med Chem 23:1062–1081

Marquez B, Neuville L, Moreau NJ, Genet J-P, Dos Santos AF, De Andrade MCC, Sant'Ana AEG (2005) Multidrug resistance reversal agent from Jatropha elliptica. Phytochemistry 66:1804–1811

Martins A, Vasas A, Viveiros M, Molnár J, Hohmann J, Amaral L (2011) Antibacterial properties of compounds isolated from Carpobrotus edulis. Int J Antimicrob Agents 37:438–444

Matsuo T, Chen J, Minato Y, Ogawa W, Mizushima T, Kuroda T, Tsuchiya T (2008) SmdAB, a heterodimeric ABC-type multidrug efflux pump, in Serratia marcescens. J Bacteriol 190:648–654

McAleese F, Petersen P, Ruzin A, Dunman PM, Murphy E, Projan SJ, Bradford PA (2005) A novel MATE family efflux pump contributes to the reduced susceptibility of laboratory-derived Staphylococcus aureus mutants to tigecycline. Antimicrob Agents Chemother 49:1865–1871

Michalet S, Cartier G, David B, Mariotte A-M, Dijoux-Franca M-G, Kaatz GW, Stavri M, Gibbons S (2007) N-caffeoylphenalkylamide derivatives as bacterial efflux pump inhibitors. Bioorg Med Chem Lett 17:1755–1758

Minato Y, Shahcheraghi F, Ogawa W, Kuroda T, Tsuchiya T (2008) Functional gene cloning and characterization of the SsmE multidrug efflux pump from Serratia marcescens. Biol Pharm Bull 31:516–519

Mohtar M, Johari SA, Li AR, Isa MM, Mustafa S, Ali AM, Basri DF (2009) Inhibitory and resistance-modifying potential of plant-based alkaloids against methicillin-resistant Staphylococcus aureus (MRSA). Curr Microbiol 59:181–186

Morel C, Stermitz FR, Tegos G, Lewis K (2003) Isoflavones as potentiators of antibacterial activity. J Agric Food Chem 51:5677–5679

Mouwakeh A, Kincses A, Nové M, Mosolygó T, Mohácsi-Farkas C, Kiskó G, Spengler G (2019) Nigella sativa essential oil and its bioactive compounds as resistance modifiers against staphylococcus aureus. Phyther Res 33:1010–1018

Musumeci R, Speciale A, Costanzo R, Annino A, Ragusa S, Rapisarda A, Pappalardo M, Iauk L (2003) Berberis aetnensis C. Presl. Extracts: antimicrobial properties and interaction with ciprofloxacin. Int J Antimicrob Agents 22:48–53

Narui K, Noguchi N, Wakasugi K, Sasatsu M (2002) Cloning and characterization of a novel chromosomal drug efflux gene in Staphylococcus aureus. Biol Pharm Bull 25:1533–1536

Nishino K, Latifi T, Groisman EA (2006) Virulence and drug resistance roles of multidrug efflux systems of Salmonella enterica serovar typhimurium. Mol Microbiol 59:126–141

Odds FC (2003) Synergy, antagonism, and what the chequerboard puts between them. J Antimicrob Chemother 52:1–1

Opperman TJ, Nguyen ST (2015) Recent advances toward a molecular mechanism of efflux pump inhibition. Front Microbiol 6:421

Pagès J-M, Amaral L (2009) Mechanisms of drug efflux and strategies to combat them: challenging the efflux pump of gram-negative bacteria. Biochimica et Biophysica Acta (BBA)-proteins and. Proteomics 1794:826–833

Paixão L, Rodrigues L, Couto I, Martins M, Fernandes P, De Carvalho CC, Monteiro GA, Sansonetty F, Amaral L, Viveiros M (2009) Fluorometric determination of ethidium bromide efflux kinetics in Escherichia coli. J Biol Eng 3:1–13

Pereda-Miranda R, Kaatz GW, Gibbons S (2006) Polyacylated oligosaccharides from medicinal Mexican morning glory species as antibacterials and inhibitors of multidrug resistance in Staphylococcus aureus. J Nat Prod 69:406–409

Piddock LJ (2006) Clinically relevant chromosomally encoded multidrug resistance efflux pumps in bacteria. Clin Microbiol Rev 19:382–402

Ponnusamy K, Ramasamy M, Savarimuthu I, Paulraj MG (2010) Indirubin potentiates ciprofloxacin activity in the NorA efflux pump of Staphylococcus aureus. Scand J Infect Dis 42:500–505

Poole K, Gotoh N, Tsujimoto H, Zhao Q, Wada A, Yamasaki T, Neshat S, Ji Y, Li XZ, Nishino T (1996) Overexpression of the mexC–mexD–oprJ efflux operon in nfxB-type multidrug-resistant strains of Pseudomonas aeruginosa. Mol Microbiol 21:713–725

Pumbwe L, Randall LP, Woodward MJ, Piddock LJ (2005) Evidence for multiple-antibiotic resistance in campylobacter jejuni not mediated by CmeB or CmeF. Antimicrob Agents Chemother 49:1289–1293

Rahman MM, Matsuo T, Ogawa W, Koterasawa M, Kuroda T, Tsuchiya T (2007) Molecular cloning and characterization of all RND-type efflux transporters in Vibrio cholerae non-O1. Microbiol Immunol 51:1061–1070

Ramalhete C, Spengler G, Martins A, Martins M, Viveiros M, Mulhovo S, Ferreira M-JU, Amaral L (2011) Inhibition of efflux pumps in meticillin-resistant Staphylococcus aureus and enterococcus faecalis resistant strains by triterpenoids from Momordica balsamina. Int J Antimicrob Agents 37:70–74

Rezende-Júnior LM, Andrade LMDS, Leal ALAB, Mesquita ABDS, Santos ALP, Neto JDSL, Siqueira-Júnior JP, Nogueira CES, Kaatz GW, Coutinho HDM (2020) Chalcones isolated from Arrabidaea brachypoda flowers as inhibitors of NorA and MepA multidrug efflux pumps of Staphylococcus aureus. Antibiotics 9:351

Ribeiro AMB, de Sousa JN, Costa LM, de Alcântara Oliveira FA, Dos Santos RC, Nunes ASS, da Silva WO, Cordeiro PJM, Neto JSL, de Siqueira-Júnior JP (2019) Antimicrobial activity of Phyllanthus amarus Schumach. & Thonn and inhibition of the NorA efflux pump of Staphylococcus aureus by Phyllanthin. Microb Pathog 130:242–246

Rossi ED, Aínsa JA, Riccardi G (2006) Role of mycobacterial efflux transporters in drug resistance: an unresolved question. FEMS Microbiol Rev 30:36–52

Roy SK, Kumari N, Pahwa S, Agrahari UC, Bhutani KK, Jachak SM, Nandanwar H (2013) NorA efflux pump inhibitory activity of coumarins from Mesua ferrea. Fitoterapia 90:140–150

Schindler BD, Kaatz GW (2016) Multidrug efflux pumps of gram-positive bacteria. Drug Resist Updat 27:1–13

Sharma A, Gupta VK, Pathania R (2019) Efflux pump inhibitors for bacterial pathogens: from bench to bedside. Indian J Med Res 149:129

Shiu WK, Malkinson JP, Rahman MM, Curry J, Stapleton P, Gunaratnam M, Neidle S, Mushtaq S, Warner M, Livermore DM (2013) A new plant-derived antibacterial is an inhibitor of efflux pumps in Staphylococcus aureus. Int J Antimicrob Agents 42:513–518

Shriram V, Khare T, Bhagwat R, Shukla R, Kumar V (2018) Inhibiting bacterial drug efflux pumps via phyto-therapeutics to combat threatening antimicrobial resistance. Front Microbiol 9:2990

Sibandze GF, Stapleton P, Gibbons S (2020) Constituents of two Dioscorea species that potentiate antibiotic activity against MRSA. J Nat Prod 83:1696–1700

Smith EC, Kaatz GW, Seo SM, Wareham N, Williamson EM, Gibbons S (2007a) The phenolic diterpene totarol inhibits multidrug efflux pump activity in Staphylococcus aureus. Antimicrob Agents Chemother 51:4480–4483

Smith EC, Williamson EM, Wareham N, Kaatz GW, Gibbons S (2007b) Antibacterials and modulators of bacterial resistance from the immature cones of Chamaecyparis lawsoniana. Phytochemistry 68:210–217

Stavri M, Piddock LJ, Gibbons S (2007) Bacterial efflux pump inhibitors from natural sources. J Antimicrob Chemother 59:1247–1260

Stermitz FR, Cashman KK, Halligan KM, Morel C, Tegos GP, Lewis K (2003) Polyacylated neohesperidosides from Geranium caespitosum: bacterial multidrug resistance pump inhibitors. Bioorg Med Chem Lett 13:1915–1918

Stermitz FR, Scriven LN, Tegos G, Lewis K (2002) Two flavonols from Artemisa annua which potentiate the activity of berberine and norfloxacin against a resistant strain of Staphylococcus aureus. Planta Med 68:1140–1141

Stermitz FR, Tawara-Matsuda J, Lorenz P, Mueller P, Zenewicz L, Lewis K (2000) 5 '-Methoxyhydnocarpin-D and Pheophorbide a: Berberis species components that potentiate Berberine growth inhibition of resistant staphylococcus a ureus. J Nat Prod 63:1146–1149

Stoitsova SO, Braun Y, Ullrich MS, Weingart H (2008) Characterization of the RND-type multi-drug efflux pump MexAB-OprM of the plant pathogen pseudomonas syringae. Appl Environ Microbiol 74:3387–3393

Su X-Z, Chen J, Mizushima T, Kuroda T, Tsuchiya T (2005) AbeM, an H+-coupled Acinetobacter baumannii multidrug efflux pump belonging to the MATE family of transporters. Antimicrob Agents Chemother 49:4362–4364

Sun Z-L, Sun S-C, He J-M, Lan J-E, Gibbons S, Mu Q (2020) Synergism of sophoraflavanone G with norfloxacin against effluxing antibiotic-resistant Staphylococcus aureus. Int J Antimicrob Agents 56:106098

Tambat R, Jangra M, Mahey N, Chandal N, Kaur M, Chaudhary S, Verma DK, Thakur KG, Raje M, Jachak S (2019) Microbe-derived indole metabolite demonstrates potent multidrug efflux pump inhibition in Staphylococcus aureus. Front Microbiol 10:2153

Tintino SR, de Souza VC, Silva J, CDM O-T, Pereira PS, Leal-Balbino TC, Pereira-Neves A, Siqueira-Junior JP, da Costa JG, Rodrigues FF (2020) Effect of vitamin K3 inhibiting the function of NorA efflux pump and its gene expression on Staphylococcus aureus. Membranes 10:130

Tintino SR, Morais-Tintino CD, Campina FF, Costa MS, Menezes IR, de Matos YML, Calixto-Júnior JT, Pereira PS, Siqueira-Junior JP, Leal-Balbino TC (2017) Tannic acid affects the phenotype of Staphylococcus aureus resistant to tetracycline and erythromycin by inhibition of efflux pumps. Bioorg Chem 74:197–200

Tintino SR, Oliveira-Tintino CD, Campina FF, Limaverde PW, Pereira PS, Siqueira-Junior JP, Coutinho HD, Quintans-Júnior LJ, da Silva TG, Leal-Balbino TC (2018) Vitamin K enhances the effect of antibiotics inhibiting the efflux pumps of Staphylococcus aureus strains. Med Chem Res 27:261–267

Truong-Bolduc QC, Strahilevitz J, Hooper DC (2006) NorC, a new efflux pump regulated by MgrA of Staphylococcus aureus. Antimicrob Agents Chemother 50:1104–1107

Tsuji BT, Yang JC, Forrest A, Kelchlin PA, Smith PF (2008) In vitro pharmacodynamics of novel rifamycin ABI-0043 against Staphylococcus aureus. J Antimicrob Chemother 62:156–160

Turner RJ, Taylor DE, Weiner JH (1997) Expression of Escherichia coli TehA gives resistance to antiseptics and disinfectants similar to that conferred by multidrug resistance efflux pumps. Antimicrob Agents Chemother 41:440–444

Wang D, Xie K, Zou D, Meng M, Xie M (2018) Inhibitory effects of silybin on the efflux pump of methicillin-resistant Staphylococcus aureus. Mol Med Rep 18:827–833

Xu XJ, Su XZ, Morita Y, Kuroda T, Mizushima T, Tsuchiya T (2003) Molecular cloning and characterization of the HmrM multidrug efflux pump from Haemophilus influenzae Rd. Microbiol Immunol 47:937–943

Yamada Y, Shiota S, Mizushima T, Kuroda T, Tsuchiya T (2006) Functional gene cloning and characterization of MdeA, a multidrug efflux pump from Staphylococcus aureus. Biol Pharm Bull 29:801–804

Yerushalmi H, Lebendiker M, Schuldiner S (1995) EmrE, an Escherichia coli 12-kDa multidrug transporter, exchanges toxic cations and H+ and is soluble in organic solvents. J Biol Chem 270:6856–6863

Yu H, Wang Y, Wang X, Guo J, Wang H, Zhang H, Du F (2019) Jatrorrhizine suppresses the antimicrobial resistance of methicillin-resistant Staphylococcus aureus. Exp Ther Med 18:3715–3722

Zechini B, Versace I (2009) Inhibitors of multidrug resistant efflux systems in bacteria. Recent Pat Antiinfect Drug Discov 4:37–50

Zloh M, Kaatz GW, Gibbons S (2004) Inhibitors of multidrug resistance (MDR) have affinity for MDR substrates. Bioorg Med Chem Lett 14:881–885

CPSIA information can be obtained
at www.ICGtesting.com
Printed in the USA
LVHW050015050223
738687LV00005B/74